STATISTICS

Second Edition

Richard C. Weimer
Frostburg State University

 Wm. C. Brown Publishers

Book Team

Editor *Earl McPeek*
Developmental Editor *Janette S. Scotchmer*
Visuals Processor *Jodi Wagner*
Visuals/Design Freelance Specialist *Barbara J. Hodgson*

Wm. C. Brown Publishers
A Division of Wm. C. Brown Communications, Inc.

President *G. Franklin Lewis*
Vice President, Publisher *George Wm. Bergquist*
Vice President, Operations and Production *Beverly Kolz*
National Sales Manager *Virginia S. Moffat*
Group Sales Manager *Vincent R. Di Blasi*
Vice President, Editor in Chief *Edward G. Jaffe*
Marketing Manager *Elizabeth Robbins*
Advertising Manager *Amy Schmitz*
Managing Editor, Production *Colleen A. Yonda*
Manager of Visuals and Design *Faye M. Schilling*
Production Editorial Manager *Julie A. Kennedy*
Production Editorial Manager *Ann Fuerste*
Publishing Services Manager *Karen J. Slaght*

Wm. C. Brown Communications, Inc.

President and Chief Executive Officer *Mark C. Falb*
Chairman of the Board *Wm. C. Brown*

Cover photo credit: Image strip from film "Papillon" by Lillian Schwartz. Computer programs by John Chambers. Discussed in the Computer Artists Handbook—W. W. Norton. Copyright 1989 Computer Creations Corporation. All rights reserved. This image may not be reproduced without written authorization.

Cover and interior design: Deborah Schneck

Copyediting and Production Services: Patricia Steele

Illustrations rendered by: LM Graphics

Special acknowledgement is given to MINITAB, Inc. for their assistance. MINITAB is a registered trademark of MINITAB, Inc. For information about MINITAB Statistical Software contact:

MINITAB, Inc.
3081 Enterprise Drive
State College, PA 16801 USA

(814) 238–3280 Telephone
(814) 238–4383 FAX
881612 Telex

Library of Congress Catalog Card Number: 91–55590

ISBN 0–697–12146–1

Printed in the United States of America by Wm. C. Brown Publishers, 2460 Kerper Boulevard, Dubuque, IA 52001

10 9 8 7 6 5 4 3 2 1

This book is dedicated, with love, to my wife,
Marlene H. Weimer

Contents

Preface

Statistics was developed over a period of eight years from a set of lecture notes for classroom experiences of the introductory probability and statistics course taught to students majoring in business, the social sciences, the biological sciences, and the natural sciences. The book is intended as an introduction to probability and modern statistical techniques for students who have had at least one course in elementary algebra and desire a good understanding of the concepts and principles of basic probability and statistics. The logic and theory of applied statistics are developed through clear, logical exposition.

My primary purpose in writing this text was to produce an accurate, readable exposition of the subject to help students understand elementary statistical techniques. Therefore, the emphasis is on understanding and interpretation rather than on applying a collection of "cookbook results."

Approach

Students are motivated to study new information and each new idea by the use of clear exposition, practical applications, and examples. An applications approach is used throughout *Statistics* to communicate the main ideas of modern probability and statistics. Each topic is motivated by using a practical application and is followed by a generalization or rule. At least one application is then solved to further explain and justify the development. For example, most introductory texts present the mean, median, and mode as three measures of central tendency. No attempt is usually made to justify the need for studying these ideas and why all three measures need to be investigated. With *Statistics,* the student is introduced to central tendency by stating its purpose for study. This is then followed by a scenario using four different types of data, which serve to motivate the reason for studying four measures of central tendency. Each measure is then discussed and developed.

Throughout the text, mathematical rigor has been sacrificed in favor of thorough understanding; it is better to understand statistical concepts and principles than it is to memorize a list of formulas and terminology and not be able to apply them or to understand their implications. Language and notations are standard and precise.

About the Book

The book is organized into three units: Unit I, Descriptive Statistics; Unit II, Basic Probability; and Unit III, Inferential Statistics. The first two units serve as a springboard for inferential statistics, and probability is presented as a way to bridge the gap between descriptive and inferential statistics. A spiral approach is used throughout to present and organize material. For example, regression and correlation are introduced in chapter 4 as descriptive concepts and are used to characterize statistical populations. These ideas are again introduced in chapter 14 and expanded upon in the development of inferential statistics and model building. The concept of sum of squares is first presented in chapter 3, then is used throughout the text to simplify ideas and formulas involving variance. Sampling distributions are introduced in chapter 8, and then used throughout the development of inferential techniques. The importance of the role of sampling distributions in inference making is clearly demonstrated and emphasized as a central concept underlying the development of hypothesis testing and estimation procedures. The *t* distributions are introduced in chapter 8 and are used to construct confidence intervals for the mean in chapter 9. In this way the student feels comfortable learning statistics, since he/she has already seen many of the ideas before.

A more balanced coverage is given to descriptive statistics and probability than in most introductory texts. This coverage affords the instructor a great deal of flexibility in choosing course content. For example, the material on grouped data in section 3.3 may not be covered by most instructors, but some instructors will deem this material essential because their students will have to deal with published government reports. An adequate understanding of inferential statistics necessitates some level of understanding of introductory probability theory. The instructor is given a choice in level of coverage of probability. The rules for counting and probability are presented in separate sections of chapter 5 to minimize student confusion.

This book differs from most introductory statistics texts in its clear development of sampling distributions in chapter 8. In this chapter the student is first introduced to sampling, sampling error, sampling distributions, and the central limit theorem. The student is also introduced to the *t* distributions in this chapter. This has proven to be an important feature because students feel more comfortable with using these distributions to draw inferences in later chapters once they already understand them. In most treatments the *t* distributions and estimation of population means are dealt with simultaneously.

Statistics uses examples and applications to motivate and explain statistical concepts and principles. **Examples** are used to explain concepts and to show how statistical procedures are performed. **Applications** are used to explain solutions to statistical problems. Some authors use the terms "examples" and "applications" interchangeably; we choose to make the distinction. Students learn a concept by being exposed to examples and nonexamples of the concept. Then, applications are used to solve some type of problem that has been posed or to answer some question that has been asked. The applications are similar to the types of exercises that the student encounters in the **Further Applications** and **Going Beyond** exercises at the end of most sections. This distinction between examples and applications should enable students to learn statistics more easily. If the student doesn't understand a concept, then relevant examples can be reviewed; if the student has trouble solving an exercise, appropriate applications can be reviewed.

The chapters of the book, and the sections within chapters, are organized in a logical sequence. The first 11 chapters constitute the core content for a basic course. There are sections and topics within these 11 chapters that may be omitted without loss of continuity. No attempt, however, has been made to classify any material as optional; what one person deems essential, may be classified by another as optional, and vice versa.

New to This Edition

The basic philosophy and organization of the text have remained unchanged, but the content has been substantially expanded for the one and two-term course sequences in applied probability and statistics. This edition also possesses an increased emphasis on the use of the computer in all chapters. Some of the new content and features changes are described as follows.

▪ The beginning-of-chapter features include an overview of the chapter and an introductory Motivator.

▪ There are three levels of exercises, instead of two, and most sections are concluded by a large exercise set that is divided into three parts: Basic Skills, Further Applications, and Going Beyond. **Basic Skills** exercises offer additional drill and practice with computational types of problems that are essential to understanding. **Further Application** exercises are similar to the applications presented in the text. They represent practical applications from a wide variety of disciplines, and stress the development of the skills necessary to classify problems by distribution and also to translate problems from words into mathematical symbols and formulas for solution. **Going Beyond** exercises are more difficult in nature. They expand the ideas encountered in the text, ask the student to supply simple proofs of certain important facts, and often develop new ideas. Going beyond exercises are at a level appropriate to an honors section in introductory statistics. Altogether there are more than 2000 exercises in the text.

▪ Parallel exercises: There is usually an even-numbered exercise similar to an odd-numbered exercise for the basic skills and further applications exercises. The answers to all odd-numbered basic skills and further applications exercises are included in the **Answers to Odd-Numbered Exercises** section within the appendices. Answers to odd-numbered review exercises and to all chapter achievement test questions are also provided there.

▪ End-of-chapter features include **Computer Applications** and **Experiments with Real Data.** The computer applications typically involve large data sets, repetitive calculations, lengthy calculations, or some simulation. The experiments with real data are included in chapters 8–15 to provide experiences working with real data. A database consisting of selected health information involving 720 individuals is included in appendix C.

▪ Chapter 1 Introduction

> Section 1.5 The Role of the Computer in Statistics has been added to stress the role that computers play in statistics

▪ Chapter 3 Descriptive Statistics—Analyses of Univariate Data

> Section 3.2 Measures of Dispersion or Variability has been expanded to include a discussion of interquartile range. This concept is prerequisite to understanding box plots discussed in section 3.4.

> Measures of central tendency and dispersion for grouped data are now treated in a separate section (section 3.3, Central Tendency and Dispersion for Data Contained in Grouped Frequency Tables). This change will afford the instructor more flexability in designing his/her course. Instructors using computers in their courses may prefer not to cover this material.

> Section 3.4 Standard Scores and Outliers (old section 3.3) now includes content on box plots and outliers. This material has been added for those wanting to include more material on exploratory data analysis.

▪ Chapter 4 Analyses of Bivariate Data

> The material from Section 4.1 (Linear Equations and Their Graphs) of the first edition has been deleted. Section 4.1 now contains material on covariance. This development serves to better facilitate the student's understanding of correlation, presented in section 4.2.

▪ Chapter 5 Introduction to Elementary Probability

> The notation $\cup$ and $\cap$ is used instead of "or" and "and." This notation provides less confusion in the statements of results and allows the student to build on material learned from set theory in high school.

> Material on permutations and combinations have been added to section 5.3 to allow more sophisticated counting.

▪ Chapter 6 Discrete Distributions

> The chapter title has been changed from Binomial Distributions to Discrete Distributions.

To permit greater flexibility in teaching, three new sections (sections 6.4, 6.5, and 6.6) have been added.

▪ Chapter 7 Continuous Distributions

The chapter title has been changed from Normal Distributions to Continuous Distributions.

To permit greater flexibility in teaching, two new sections (sections 7.1 and 7.4) have been added.

▪ Chapter 11 Inferences Concerning Two Parameters

The material on small samples has been separated into two sections (sections 11.5 and 11.6) for greater clarity and less confusion on the part of the student.

▪ Chapter 13 Analysis of Variance

The chapter title has been changed from Single-Factor Analysis of Variance to reflect the chapter's expanded content.

To further afford the instructor greater flexibility in designing a course and to permit instructors using the computer in their courses to further develop ANOVA, two new sections (sections 13.4 and 13.5) have been added.

In section 13.3, Bonferroni's procedure is used instead of Sheffé's procedure. This change enables the student to build on the previous results about the t distributions and affords the student greater understanding.

▪ Chapter 14 Linear Regression Analysis

To permit instructors using the computer to further investigate the linear model, a new section, section 14.4 on multiple linear regression, has been added.

Section 14.3 from the first edition had been modified by moving material on Spearman's test to chapter 15. This change permits the instructor to deal with nonparametric techniques in a single chapter.

▪ Chapter 15 Nonparametric Tests for Large Samples

A new section 15.2 now includes a discussion of Wilcoxon's signed-ranks test. In most cases, this test is more powerful than the sign test presented in section 15.1.

A new section 15.4 discusses Friedman's test as the nonparametric analog to two-way ANOVA when interaction is not present.

The text is designed to be used in courses with or without computers and provides a concrete introduction to the use of computers, which may be optional. For those classes that desire coverage of computers, *Statistics* provides some of the following advantages:

▪ There is no need to discuss the computational formulas because these are handled most efficiently by the computer.

▪ Simulating values of random variables for empirical distributions can be used.

▪ Understanding of sampling distributions and the central limit theorem is facilitated.

▪ Understanding of sampling error and properties of probability is facilitated.

▪ The use of formulas in ANOVA and multiple linear regression can be deemphasized and the time saved can be used to interpret results.

▪ The student is able to deal with large quantities of real data.

Because of its large-scale prominence in colleges and universities throughout the country, MINITAB was chosen as the statistical software package to illustrate the use of the computer in applied probability and statistics. MINITAB computer displays are used in most sections of the book to illustrate the MINITAB commands and the corresponding output. The MINITAB computer displays serve at least the following five purposes:

1. Illustrate the ease of using MINITAB.
2. Illustrate the necessary commands that need to be supplied by the user to obtain desired statistical results.
3. Illustrate the format and notation used in computer output.
4. Illustrate the extent of the statistical computing power that is available with MINITAB (and other similar software packages).
5. Teach the use of MINITAB to perform certain standard statistical tasks.

Organization

The book is organized to allow the instructor a great deal of flexibility. Those who wish to pass quickly through chapters 1 and 2 need teach only the concept of the histogram in chapter 2. If you are not planning an in-depth discussion of probability, only sections 5.6 and 5.7 of chapter 5 need be taught as a basis for later sections (section 5.6 is background for chapter 12; section 5.7 is important for sections 6.3, 8.1, and 8.2). If you intend to omit binomial distributions and related concepts, sections 6.1, 6.2, 6.3, 7.3, 8.5, 9.3, part of 10.4, 11.3, and 12.1 may be omitted. After chapter 11 has been studied, the remaining chapters can be covered in any order.

Students often find it confusing to study normal distributions, the normal approximation to the binomial, and the central limit theorem in the same chapter. Great care has therefore been given to making these three topics sequential. In addition, the central limit theorem has been developed in a way that stresses its importance in the development of large-sample statistical methods for making inferences. The sampling distribution of the sample sum is also described as approximately normal and likened to binomial distributions being approximated by normal distributions.

Estimation for single parameters following a logical development is presented in chapter 9, and the hypothesis-testing procedures for one-sample tests are developed in chapter 10. The two procedures are presented side-by-side in chapter 11, showing students how to make comparisons between two parameters using two populations. Both procedures are shown to produce consistent results when nondirectional tests are used.

The concept of sampling error (introduced in chapter 8) serves as a unifying concept, explaining the natural variation in values of sample statistics when sampling is used. For example, in chapter 10, if there is no evidence to suggest that the null hypothesis is false, the difference between the value of test statistic and the hypothesized population value is attributed to sampling error.

Students are frequently asked to classify applications according to sampling distribution and necessary assumptions. Nonparametric tests in chapter 15 are presented as an alternative whenever assumptions are tenuous or known to be violated. Interpreting results and identifying the associated risks are stressed throughout chapters 8–15.

Learning Features

This text differs from other works in the field by presenting statistics as a way of dealing with uncertainty and variability, showing the true roles that variability, sampling error, and sampling distributions play in the development of inferential statistics. Students who intuitively understand the underlying principles and how they are related will be better equipped for the future than those who have been conditioned to simply substitute numbers into an appropriate formula. Other distinct features include the following:

1. The text motivates and explains the logic that underlies the principles of inferential statistics, encouraging readers to practice this logic themselves. Students are taught not only how to use statistical techniques, but also to understand why they work. The text material has been sequenced in a relevant, connected fashion.

2. Histograms, the building blocks of statistics, are used throughout to indicate basic properties of theoretical distributions.

3. Important relationships concerning hypothesis tests are pointed out whenever possible: for example, analysis of variance is presented as an extension of the two-sample t test; the relationship between t^2 and F is discussed; the chi-square test for testing more than two population proportions is presented and shown to be an extension of the two sample z test for proportions; and the relationship between z^2 and χ^2 is examined.

4. Proportions for dichotomous or binomial populations are first introduced by associating them with means for a population consisting of ones and zeros. The properties for the sampling distribution of the sample proportion are then found by using the properties for the sampling distribution of the sample mean.

5. Assumptions underlying the various hypothesis tests are emphasized. The F test for testing the assumption of homogeneity of variances is discussed before the two-sample t test, so it is possible to test this assumption before learning how to use the two-sample t test.

6. One-way ANOVA is introduced and promoted in relation to the concept of range. Bonferroni's procedure is provided for testing for pairwise differences and Hays' omega-square statistic is used to measure the strength of relationship following a significant F test. Two-way ANOVA is first introduced using a randomized block design. This is then followed by a discussion of interaction and factorial designs. There is a strong emphasis on using a computer approach to ANOVA for data in two-factor designs once the student understands the basic concepts involved.

7. A computer approach is taken to multiple linear regression. The approach is understandable and complete. With the aid of a computer, the linear model with more than one independent variable is discussed as an extension of the basic one-variable model and major results are demonstrated and explained by using the output from MINITAB. Once again, the emphasis is on understanding the use and implications of multiple linear regression and being able to interpret the computer output. The use of formulas to obtain results is not emphasized.

8. The pedagogy system will guide and assist students as they study. Each chapter begins with an introductory list of objectives and ends with a concise summary. In addition, each chapter has the following elements:

   ■ A **motivator** opens the chapter and motivates and emphasizes the need for discussing the chapter content. These motivators consist of brief summaries of actual research, real applications from the literature, or interesting problems. They serve to partially answer the question "Why should I study the material in this chapter?"

   ■ A **chapter overview** provides an overview for study of the chapter.

   ■ **Examples** and **applications** are used liberally to explain and justify new concepts and procedures. The applications provided are realistic, yet not overly technical in nature.

Because they are drawn from a broad variety of areas in the natural, biological, and social sciences, but do not include a lot of background information or technical language, these applications enable students from various discipline areas to successfully complete the exercises.

■ Exercises at the end of most sections provide the student a chance to practice what was presented in the section.

■ A list of important notation is found at the end of each chapter.

■ A list of important facts and formulas is found at the end of each chapter.

■ Review exercises, which do not follow any particular ordering of topics, are found at the end of each chapter.

■ Optional computer applications, found at the end of chapters 2–15, ask the student to perform some simulation or analyze a relatively large set of real data.

■ An achievement test is found at the end of chapters 2–15 so that the student can evalute his/her cumulative understanding of the material presented in the chapter.

Supplements

Statistics is accompanied by a full set of supplements for use by the instructor and the student. These supplements will support your elementary statistics course and enhance learning by the student.

For the Instructor

■ **Instructor's Resource Manual**—contains answers to problems, instructional hints, and transparency masters to support classroom instruction.

■ **wcb TestPak 3.0**—is a computerized testing service that contains a test bank and a quiz bank of problems organized by learning objectives, section references, and difficulty levels for easy test and quiz preparation. Also available is *Gradepak* to assist in recordkeeping of test results. The classroom management software of *TestPak 3.0* is available for IBM®, Apple®, and Macintosh®.

■ **Test Item File**—is a printout of all test and quiz problems available in *TestPak 3.0*. You may use the Wm. C. Brown call-in service to prepare your tests and have them mailed to you.

■ **Statistics Videos**—a set of 60 VHS and Beta videos containing discussions on all the topics found in *Statistics* is available. Contact your Wm. C. Brown sales representative for more information.

■ **Data Analysis Statistical Software**—several software packages are available. Contact your Wm. C. Brown sales representative for more information.

For the Student

■ **MINITAB Lab Manual**—contains lab exercises using MINITAB software for those courses that integrate MINITAB in them.

■ **Student Study Guide**—provides tutorial support as well as a data set diskette featuring the real data set found in appendix C of *Statistics*.

Acknowledgments

I appreciate the assistance of the many people who helped me prepare this book. I gratefully acknowledge the suggestions and constructive criticisms offered by hundreds of Frostburg State University students who studied statistics from the first edition and earlier drafts during the past seven years. Students Peter Tirrell, Stewart Crall, Donna Pope, and William Byers helped check and verify answers to the problems in the first edition. I also appreciate the help and advice given to me by my colleagues, Dr. Kil Lee, Dr. Lance Revennaugh, and Dr. Kurtis Lemmert. Numerous constructive criticisms and suggestions for the first edition were provided by the following reviewers:

James Baker, Jefferson Community College; James Baldwin, Nassau Community College; Pat Cerrito, University of South Florida; James Daly, California Polytechnic State University; Ken Eberhard, Chabot College; Antanas Gilvydis, Malcolm X College; Raymond Guzman, Pasadena City College; Gary Itzkowitz, Glassboro State College; Keith Nelson, Beloit College; Jim Ridenhour, Austin Peay State University; Larry Ringler, Texas A&M University; Ann Thomas, University of Northern Colorado; William Tomhave, Concordia College; and John Van Druff, Fort Steilacoom Community College.

I also wish to thank these reviewers of the second edition: Derek K. Chang, California State University, Los Angeles; William H. Beyer, The University of Akron; Nancy J. Carter, California State University, Chico; Kenneth R. Eberhard, Chabot College; Theodore S. Erickson, Wheeling Jesuit College; H. Joseph Heffelfinger, Anne Arundel Community College; Dr. Angela Hernandez, University of Montevallo; R. Bruce Lind, University of Puget Sound; Cameron Neal, Jr., Temple Junior College; Steve Patch, University of North Carolina at Asheville; Jim Ridenhour, Austin Peay State University; Joseph F. Stokes, Western Kentucky University; Deborah A. Vrooman, Coastal Carolina College of the University of South Carolina; June Miller White, St. Petersburg Junior College; and Ellen T. Wood, Stephen F. Austin State University.

I especially appreciate the expert advice offered by Professor Susan L. Reiland and Ms. Patricia Steele. Professor Reiland read the entire manuscript for the first edition before the final draft was prepared and served as copy editor for the first edition manuscript. Ms. Steele served as copy editor for the second edition and her professionalism, dedication, and meticulous care and treatment of details warrant a special thanks.

My thanks also go to the editorial and production staff of Wm. C. Brown Publishers, especially Jan Scotchmer for her developmental assistance and encouragement throughout the project.

Finally, I am also indebted to Dr. Larry Brant of the Gerontology Research Center and National Institute of Aging, Baltimore, Maryland for the use of a database consisting of 720 records of health-related data included in appendix C and used throughout the text. This data was collected as part of the Baltimore Longitudinal Study of Aging (BLSA) and represents only a fraction of the information that has been collected by the study.

Richard C. Weimer
January 1992

UNIT ONE

Descriptive Statistics

1 Introduction

CHAPTER OBJECTIVES

In this chapter we will investigate

▷ *The two meanings of the term statistics.*

▷ *The difference between a population and a sample.*

▷ *The difference between a statistic and a parameter.*

▷ *The difference between descriptive and inferential statistics.*

▷ *How induction and deduction relate to probability and statistics.*

▷ *Why it is important to study statistics.*

▷ *The role of the computer in statistics.*

|||| MOTIVATOR 1

*T*he most common reason why companies strike out with customers is poor service. Trainers and organization development professionals are being challenged to help their companies improve customer-service quality, customer satisfaction, and repeat purchase locality. According to an article by Stum and Church,[1] the research demands it. They cite the following facts, which were all arrived at using statistical techniques and which illustrate the many ways that statistical information is used by management in decision making.

■ According to a Forum study, the most common reason that customers switch to a competitor is poor service.

■ The American Management Association asserts that 60% of new sales should come from old customers, showing repurchase loyalty.

■ Consultant R. L. Desatnick points out that in the auto industry, for example, a loyal customer represents a lifetime revenue of $140,000.

■ The Consumer Affairs Office warns that seven out of ten people may stop doing business with a supplier based on the way they are treated during the first contact.

■ AT&T reports that the number of 800 numbers (often used by companies that want to provide customer information or assistance) grows 25% annually.

■ The Technical Assistance Research Project (TARP) states that a company will never hear from up to 90% of its unsatisfied customers (although those unhappy people will tell ten others about their negative experiences), but that when dissatisfied customers do complain, their loyalty significantly increases if their complaints are resolved to their satisfaction.

Chapter Overview

People view statistics differently. Statistics is commonly perceived as anything having to do with percentages, averages, charts, and graphs. For some, statistics is an area of study consisting of rules and methods for dealing with information. For others, statistics is a way of acting and thinking about worldly events, events that occur with irregularity and are governed by laws of uncertainty. This chapter introduces the basic ideas and language of statistics.

SECTION 1.1   *Why Study Statistics?*

There are at least four good reasons for studying statistics. By studying statistics, we are able to

1. Learn the rules and methods for dealing with statistical information.
2. Evaluate and assess the importance of published statistical findings.
3. Learn the aspects of statistical thinking as an important and essential component of a liberal arts education.
4. Better understand the empirical world around us.

Perhaps one of the most important reasons for studying statistics at this level is to enable us to critically appraise statistical information communicated by the media. For example, consider the following statements made in advertisements by the media (X is used instead of the brand name):

1. Brand X tires stop 35% faster.
   (Faster than what?)
2. Over a four-year period gasoline mileage for car X increased by 50%.
   (50% of what?)
3. Brand X soap is 99 44/100% pure.
   (Pure what? Soap?)
4. Ninety percent of all brand X autos sold in the last ten years are still on the road.
   (By this statement we are to wrongly assume that the cars were sold in approximately equal numbers each year. Most of the cars still on the road were bought during the preceding three or four years.)
5. Brand X contains twice as much pain reliever.
   (Does this mean that it is more effective than any other pain reliever for relieving pain?)
6. Four out of five dentists surveyed report that they prefer brand X toothpaste.
   (How many dentists were surveyed? How were they chosen?)
7. No aspirin works any harder at relieving pain than brand X.
   (This statement does not say that brand X is any better than any other brand; it does say that brand X works as hard as any other brand.)
8. Women who use brand X reported 80% relief during the first several hours.
   (Can relief be reported in percentage terms? What does this mean?)
9. Brand X pain reliever is recommended most by people who know the most.
   (Know the most about what? Everything?)

As consumers of statistical information and potential users of statistical techniques, we need to understand the basic ideas and tools of statistics. Most of us are influenced daily by some aspect of statistics in the information we get from radio, television, or newspapers and magazines. For example we may read or hear that

1. Studies suggest that about 50% of all drownings among teenagers and adults are associated with alcohol use.

2. One-parent families now account for 26% of all U.S. families with children under 18—up from only 13% in 1970.

3. Seven out of ten Americans do not have wills.

4. The prevalence of diabetes is nearly three times as high in overweight people as it is in nonoverweight people.

5. More than 3000 insurance companies pay more than $8.8 billion in claims each year.

6. There is a 50% chance that the victim will never race again.

7. Children who brush their teeth with brand X toothpaste have 35% fewer cavities.

8. The median net worth of newly retired Social Security beneficiaries in 1981–1982 was between $64,700 and $68,300 for couples, and between $17,000 and $30,000 for unmarried people.

9. In 1960 it was estimated that only 1% of high school seniors had tried marijuana, whereas in 1980 it was estimated that 60% had done so.

10. Studies suggest that feelings of helplessness are correlated with a marked decrease in several immune system cells that fight disease.

These examples indicate that statistical information is used for a variety of reasons. Among these reasons are to

- Inform the public (all of the preceding examples).
- Provide comparisons (examples 2, 4, 8, and 9).
- Explain actions that have taken place (examples 1, 4, and 10).
- Influence decisions that will take place (examples 1 and 7).
- Justify a claim or assertion (examples 1, 7, and 10).
- Predict future outcomes (example 6).
- Estimate unknown quantities (examples 1 and 9).
- Establish a relationship or association between two factors (examples 1, 4, and 10).

Since we are consumers of statistical information, we can use statistics to study and gain an understanding of many changing events that will contribute to our understanding of the world around us. Studying statistics will enable us to give a reasonable interpretation to each of the previous examples. For instance, the figure 35% in example 7 is open to interpretation because we do not know the basis of the comparison. It may be difficult, if not impossible, to find a toothpaste that produces 35% fewer cavities than any other toothpaste when tested under similar conditions on independent and similar groups of children. But it should be fairly simple to find one child who uses brand X toothpaste and who has 35% fewer cavities than some other child, perhaps one who does not brush his or her teeth at all. As daily consumers of statistical information, we should be knowledgeable in both the uses and misuses of statistics. We will learn in this course how each of the numbers in the examples of this section could have been derived.

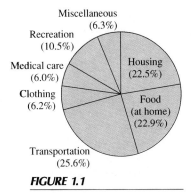

FIGURE 1.1

How $142,700 is spent to raise a child until age 17

As consumers of statistical information, we are frequently confronted with graphs displaying statistical information. For example, parents spend approximately $142,700 on a child by the time the child is 17 years of age. The circle graph in figure 1.1 indicates how this money is spent.[2] We can see at a glance that food, housing, and transportation costs are approximately equal and that recreation, medical care, and miscellaneous expenses share nearly equal amounts of the total budget.

As potential users of statistical methods and techniques, we need to be concerned with doing basic statistical research, describing the results of our scientific inquiry, making decisions based on this inquiry, and estimating unknown quantities. Examples 1.1, 1.2, and 1.3 further illustrate how statistics might be used in general, and examples 1.4, 1.5, and 1.6 demonstrate applications of statistics in answering current issues of interest.

EXAMPLE 1.1

A television commercial claims that one of a product brand is superior to all other brands. If the claim is based on scientific inquiry, the statement is used to educate the viewers. Suppose we doubt the claim. In an attempt to disprove it, we might gather relevant information concerning all brands of the product, analyze the results using proper statistical procedures, and make a decision regarding the advertised claim. Frequently, advertised claims are based on insufficient information or faulty analyses of the information.

EXAMPLE 1.2

Suppose we want to determine the best teacher at Excel College. How should we go about making this determination? We could ask Excel students who the best teacher is, analyze the results, and make our determination. Should we survey every student? How will the survey be conducted? How will the information be analyzed? How will the best teacher be determined? Answers to these, and other questions, are a focus of statistics.

EXAMPLE 1.3

A life-insurance company is considering offering reduced premiums to policyholders who engage in an ongoing exercise program. To aid the insurance company in making a decision, mortality information will be collected and analyzed. What type of exercise program will quality a policyholder for a reduced premium? How much should the reduction be? What risk factors would disqualify a policyholder in an ongoing exercise program from a premium reduction? A person with a strong background in statistics could assist the life-insurance company in evaluating the merits of the new program.

EXAMPLE 1.4

What role does diet play in coronary heart disease? The role of diet in coronary heart disease has been contested for nearly two generations. The heart-diet theory maintains that reducing the level of blood cholesterol through diet will reduce the risk of developing coronary heart disease. To partially test the relationship between blood cholesterol reduction and coronary heart disease, a study involving the drug cholestyramine, a cholesterol-lowering medication was undertaken. The study involved 3800 middle-aged men. All had blood-cholesterol levels of at least 265 milligrams (per deciliter of blood), which put them in the top 5% of adult Americans in blood cholesterol values, and all were found to be free of any signs of coronary heart disease on entering the study. Nineteen hundred men were randomly assigned to each of two groups: a treatment group and a control group. Participants in the treatment group received daily doses of cholestyramine and followed a cholesterol-reducing diet for an average of 7.4 years, while participants in the control group did not receive cholestyramine, but received a placebo that was indistinguishable from cholestyramine. The study concluded that the drug group had fewer heart attacks (155 versus 187 for the control group) and fewer deaths from heart attack (30 versus 38). The difference between the two groups was judged to be statistically significant (probability not due to chance factors alone). The findings support the belief that lowering blood cholesterol by using cholestyramine in middle-aged men with cholesterol levels above 265 milligrams would be effective in reducing coronary heart disease.[3]

EXAMPLE 1.5

Is new Coke that good? Early in 1985, the Coca-Cola Company announced that it was changing its secret formula for making Coke, a formula it had used since 1886. After the new Coke was marketed, the staff at *Consumer Reports* attempted to answer such questions as: What does the new Coke really taste like? Is it better than old Coke? How does it compare with Pepsi? The research staff conducted three separate blind taste tests involving 95 staff members and 532 plastic cups. The results of the study showed no difference in preferences between Pepsi and the new Coke. Both products were preferred over the old Coke by a 2-to-1 margin. All three formulations were found to consist of about 99% carbonated water and sugar, and all three were remarkably similar in sugar, each with between 6.14% and 6.22% fructose, and between 4.54% and 4.73% destrose (corn sugar).[4] (Miscellaneous traditions and other human factors can affect consumer preferences. Even though the experiment seems to indicate that new Coke is superior in taste to old Coke, Coca-Cola Classic is currently outselling new Coke in many regions of the United States.)

EXAMPLE 1.6

Is tobacco smoke harmful to nonsmokers? It has long been known that smoking is harmful to smokers. To determine if tobacco smoke is harmful to nonsmokers, a study involving lung-function tests was undertaken at the University of California at San Diego. Tests were conducted on 200 middle-aged nonsmokers whose environments were relatively free of tobacco smoke and on another 200 middle-aged nonsmokers who were routinely exposed to tobacco smoke at work for 20 years or more. Both groups were also compared with smokers who do not inhale, light smokers, moderate smokers, and heavy smokers. The researchers concluded that the two groups of nonsmokers did not differ significantly in lung-test results measuring forced vital capacity and initial expiratory rate. However they did report a statistically significant difference between the two groups in the amount of impairment in the small airways of the lungs. The nonsmokers who were passively exposed to smoke at work did not have scores that were judged to be indicative of lung disease, but their scores were similar to those for light smokers (one to ten cigarettes per day) and smokers who do not inhale. The study suggests that chronic exposure to tobacco smoke at work is harmful to the nonsmoker and significantly reduces small-airways function.[5]

EXERCISE SET 1.1

For each of the following statements, (a) state the conclusion you think a reader has after reading the statement, and (b) list what questions you think need to be asked about the statement to avoid a false conclusion.

1. Between the ages of 35 and 65, seven out of ten workers will suffer a disability lasting three months or longer.

2. At age 32, a disability lasting three months or longer is six times more likely than death.

3. Our auto service personnel have 100 years of repair service.

4. In more than 1,000,000 miles of testing, our automobiles had a repair record of less than 1%.

5. Ninety eight percent of the doctors prescribe the pain reliever found in brand X.

6. Foods containing oat bran help to reduce serum cholesterol levels in adults.

7. Low-fat food products help to reduce heart attacks.

8. Foods high in fiber help to reduce the chances of getting colon cancer.

SECTION 1.2   *Language of Statistics*

As with all sciences, statistics has its own language. We begin by examining the term **statistics.** There are two meanings for statistics.

> Statistics (singular) is the science of collecting, organizing, analyzing, and interpreting information. Statistics (plural) are numbers calculated from a set or collection of information.

As a science, statistics is concerned with describing the results of scientific inquiry, making decisions based on this inquiry, and estimating unknown quantities. The numerical characteristic that is used as an estimate serves as an example of a **statistic.**

EXAMPLE 1.7

Researchers estimate that IBM's entire family of personal computers controls about 40% of the microcomputer market in the United States. The number 40% is an example of a statistic.

A basic distinction in statistics concerns the difference between a population and a sample.

> A **population** is the totality of information or objects of concern to the statistician for a particular investigation.

> A **sample** is any subset of the population.

EXAMPLE 1.8

The collection of student grade-point averages (GPAs) at a local college could serve as a statistical population, and any subcollection (say the GPAs of the students in a Mathematics 101 class) could serve as a sample from this population.

EXAMPLE 1.9

For the Excel College example given in section 1.1, the population consists of the responses from the entire student body to the question "Who is the best teacher?" Since it would be extremely difficult and time consuming to poll each student, we might instead decide to poll a representative subset of the student body. This representative subset of the population represents a sample. The information from the sample can be used to **estimate** who is the best teacher at Excel College.

APPLICATION 1.1

A manufacturer of kerosene heaters wants to determine if its customers are satisfied with the performance of their heaters. Toward this goal, 5000 of its 200,000 customers are contacted and each is asked, "Are you satisfied with the performance of the kerosene heater you purchased?" Identify the population and sample for this situation.

Solution: The population is the hypothetical collection of responses from all 200,000 customers. We have not polled the entire population, but we hope to learn something about it from the sample. The sample is the collection of 5000 responses made by the customers polled. ■

EXAMPLE 1.10

A statistical population, however, need not exist. For example, if a researcher is interested in the possible selling prices of 1995 automobiles, the desired information does not yet exist. Even though the information is not available, the selling prices for cars over the past several years along with information concerning the current rate of inflation, could be used to predict the selling prices for 1995 automobiles.

A value used in statistics can be characterized as either a statistic or a **parameter,** depending on the extent of the information. We explore the following definitions in example 1.11 and further show their uses in applications 1.2 and 1.3.

> A statistic is any numerical characteristic of a sample.
>
> A parameter is any numerical characteristic of a population.

EXAMPLE 1.11

In a survey conducted in 1989 by the Food Marketing Institute on trends in supermarket shopping, a sample of shopper responses revealed that the average household's grocery bill was $74. The value $74 is an example of a statistic. The survey also revealed that for each additional minute spent in the store beyond the average 80 or 90 minutes per week, an additional $1.89 is spent. The figures 80, 90, and $1.89 are also examples of statistics. This survey was conducted to determine information about the population of all supermarket shoppers. The average grocery bill for all shoppers, the average time spent shopping per week for all shoppers, and the additional amount spent for each additional minute of shopping beyond the average shopping time per shopper are all examples of parameters that are unknown. If a survey of all supermarket shoppers revealed the average household makes 2.3 trips to the supermarket each week, and the responses of all supermarket shoppers comprise the population, then the value 2.3 is an example of a parameter.

A P P L I C A T I O N 1.2

In order to estimate the proportion of students at a certain college who smoke cigarettes, an administrator polled a sample of 200 students and determined the proportion of students from the sample who smoke cigarettes. Identify the parameter and the statistic.

Solution: The parameter is the proportion of all students at the administrator's college who smoke cigarettes, whereas the statistic is the proportion of all students in the sample of 200 who smoke cigarettes. ▪

A P P L I C A T I O N 1.3

A tip is the amount of money above the amount of the bill given for satisfactory service. Patrons in 1500 nightclubs were given a confidential questionnaire with their bills asking how much tip was being given. Calculations showed that the average tip was about 15% of the amount of the bill. Is 15% a parameter or a statistic? Explain.

Solution: If only the 1500 establishments are under study, then the tipping information from the 1500 establishments constitutes the population and 15% of the bill is a parameter. However, if the tipping data from the 1500 establishments form a sample from some larger population of tipping data, then 15% of the bill is a statistic. ▪

EXERCISE SET 1.2

1. In an attempt to reduce the number of highway accidents, the state of Maryland has implemented a campaign aimed at reducing the number of speeders and drivers under the influence of alcohol. A researcher is interested in determining the extent to which alcohol is a contributing factor in highway fatalities within the state of Maryland. The researcher secured accident information for the month of June from 5 of the 22 highway patrol offices in the state.
   a. What is the population of interest to the researcher?
   b. Describe the sample.
   c. How could the researcher use the sample information to estimate the extent to which alcohol is a contributing factor in highway fatalities within the state of Maryland?

2. The medical problem of acquired immune deficiency syndrome (AIDS) has created high levels of anxiety and concern among the public. Records reveal that 71% of all AIDS cases in the United States have been among gay or bisexual men and about 18% have been among intravenous drug users. Many people question the possibility of contracting AIDS through a blood transfusion. Although blood for transfusions is screened for AIDS, a medical researcher wants to study the medical records of 50 hospitals located in cities throughout the United States to determine the extent of AIDS cases verified to have been caused by blood transfusions.
   a. What is the population of interest to the researcher?
   b. Describe the sample.

3. A doctor recently made the claim that a tablespoon of cod liver oil daily can cure arthritis. A researcher is interested in testing the claim. Two groups, each containing 50 arthritis patients are used. Patients in only one group are to be administered a tablespoon of cod liver oil daily for one year, after which all subjects in both groups will be examined for the symptoms of arthritis.
   a. What are the two populations of interest?
   b. Describe the two samples.

4. From 1971 until early 1985, the National Highway Traffic Safety Administration (NHTSA) had attributed at least 207 deaths to Ford Motor Company vehicles unexpectedly backing into and over people. In addition, there had been 4597 reported injuries resulting from unexpected reverse movements of Ford vehicles. By June 1980, the NHTSA had received more than 23,000 reports about Ford cars failing to engage or hold in park. Instead of a government-ordered recall of affected vehicles, an agreement was negotiated with Ford in an attempt to prevent further injuries or deaths. Ford agreed to send out warning notices, with accompanying warning stickers to be displayed on the dashboards of the vehicles, to the owners of the affected vehicles—some 23 million cars and trucks. In mid-1981, the Center for Auto Safety checked 700 Ford vehicles in four cities to ascertain if warning stickers were being displayed on the dashboards. Only 7% of the cars observed actually displayed the warning sticker.[6]
   a. Identify the population of interest to the Center for Auto Safety for this example.
   b. Describe the sample.
   c. Would you say the sticker campaign was successful in reducing unexpected reverse movements in Ford vehicles? Explain.
   d. Identify a population of interest to the NHTSA.

5. A six-month study was conducted to determine if stress and mood are linked to the presence of certain immune system cells. Thirty-six people under different levels of stress participated in the study, which involved examining blood samples taken from them at regular intervals. The blood was analyzed for changes in the number of helper and T cells, which regulate immune functions. The results showed that increased levels of stress appear to be directly tied to the decreased number of T cells and outbreaks of herpes.[7]
   a. Identify the population of interest.
   b. Describe the sample.

6. A complete census of the student body of a university revealed that the number of students who were 50 years of age or older was 515. Is 515 a statistic or a parameter?

7. A telephone poll of 100 households in a community was conducted to determine citizen interest in paying higher taxes to improve the quality of public education. The poll revealed that 37% would pay higher taxes to support quality education. Is 37% a statistic or a parameter?

Descriptive and Inferential Statistics

The procedures and analyses encountered in statistics fall into two general categories, descriptive and inferential, depending on the purpose of the study.

> **Descriptive statistics** comprises those methods used to organize and describe information that has been collected.

These methods are used to analyze the information and to display information in graphical form to allow meaningful interpretations. The methods of descriptive statistics help us describe the world around us. We use descriptive statistics when we collect such information as the average yield of wheat per acre of a particular agricultural area, the number of people in various income categories, or the average number of points scored by a particular football team during first-quarter play. Through descriptive statistics we hope to learn how things are.

EXAMPLE 1.12

The following situations involve descriptive statistics.

1. A bowler wants to find his bowling average for the past 12 games.

2. A politician wants to know the exact percentage of votes cast for her in the last election.

3. Mary wants to describe the variation in her five test scores in first-quarter calculus.

4. Mr. Smith wants to determine the average weekly amount he spent on groceries during the past three months.

Inferential statistics, on the other hand, involves the theory of probability.

> **Inferential statistics** comprises those methods and techniques for making generalizations, predictions, or estimates about the population from the sample.

The ability to make generalizations about the population from the sample is an important aspect of statistics. We rarely have the complete information needed to form an absolute truth about some worldly event. Decisions and inferences are based on limited and incomplete information. The methods of inferential statistics, and the knowledge gained from their use, allow us to use limited available information to understand and deal with the uncertainties of our randomly changing world. We might, for example, predict the wheat yield for the coming year on the basis of growing trends over previous years. We could estimate the increase in average income over a five-year period, based on past knowledge of average income and other descriptive statistics. We could try to predict the total points scored for the season by a particular football team knowing the points scored in the first seven games. With inferential statistics, we state how things will be, *probably*—or sometimes only *maybe*. By using methods of probability, we will attempt to measure the degree of uncertainty associated with an inference.

EXAMPLE 1.13

The following situations (which parallel the descriptive situations given in example 1.12) involve inferential statistics.

1. A bowler wants to estimate his chance of winning an upcoming tournament based on his current season average and the averages of the competing bowlers.

2. Based on an opinion poll, a politician would like to estimate her chance for reelection in the upcoming election.

3. Based on the variation in test scores in her first-quarter calculus test scores, Mary would like to predict the variation in her second-quarter calculus test scores.

4. Based on last year's grocery bills, Mr. Smith would like to predict the average weekly amount he will spend on groceries for the upcoming year.

EXERCISE SET 1.3

1. Mr. Jackson, a candidate for mayor in a small town, wants to determine if he should campaign harder against his opponent. To determine this, he will poll 500 of the 1500 registered voters. If the results indicate he has 25% more votes than his opponent, he will not intensify his campaign efforts against the opponent.
   a. Identify the population.
   b. Identify the sample.
   c. Identify a statistic.
   d. Identify a parameter.
   e. What would Mr. Jackson do if he had 65% of the sample votes.

2. Give an example, not mentioned in the text, of each of the following concepts.
   a. population   c. statistic
   b. sample   d. parameter

3. A survey of grocery prices was conducted by an independent marketing agent at four of the ten grocery markets in a small city. The following are prices charged for a 5-pound bag of sugar: $1.25, $1.18, $1.20, and $1.30. The agent made the following four statements. Which were obtained using inferential statistics and which were obtained using descriptive statistics? Explain.
   a. The highest price charged in the town is $1.30.
   b. Two markets charge more than $1.20 for a 5-pound bag of sugar.
   c. One fourth of all markets charge more than $0.25 per pound for sugar.
   d. The prices at all markets for a 5-pound bag of sugar range from $1.18 to $1.30.

4. A commercial states "four out of five doctors recommend Preparation A." Do you think this conclusion is drawn from a sample or a population? Discuss.

5. The average cost for student textbooks last semester at a small college was determined to be $135.00, based on an enrollment of 1200 students. As a class project at the college, a statistics class polled 25 students to determine their average textbook cost last semester. It was determined to be $152.25.
   a. Identiy the population.
   b. Identify the sample.
   c. Identify two parameters.
   d. Identify two statistics.
   e. What might the statistics class conclude if the average book cost for the sample of 25 students is $400?

6. Classify the nature of each of the following statements as inferential or descriptive. Also, provide any assumptions you are making concerning a statement to support your answer.
   a. A family of five or more averages $109 a week on grocery purchases.
   b. Sixty-six percent of all grocery purchases are unplanned.
   c. In 1978 there was a total of 11,767 grocery items purchased.
   d. Each year 80% of new grocery products fail.
   e. The number of different grocery items stocked in a typical store in 1989 was 26,430.

SECTION 1.4   *Inferences and Deductions*

The study of statistics involves both induction and deduction (see fig. 1.2).

FIGURE 1.2

Induction versus deduction

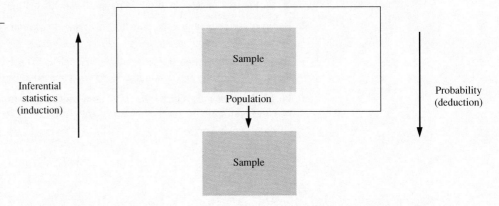

Inferential statistics (induction)   Sample   Population   Sample   Probability (deduction)

> **Induction** involves reasoning from specific instances to the general case.
>
> **Deduction** involves reasoning from the general case to specific instances.

When we make a generalization about a population parameter based on information derived from a sample, we are using induction.

EXAMPLE 1.14

If by tasting a number of grapes from a bowl of grapes, we arrive at the generalization that all the grapes in the bowl are sour, we have used inductive reasoning. The generalization that all the grapes in the bowl are sour is an example of an inference.

> An **inference** is a generalization arrived at by using induction.

In inferential statistics, inferences are typically made about a parameter of a population by employing only one specific sample (instead of many samples as one would expect by using induction). When this is done great care is taken to ensure that the sample is truly representative of the population.

When we ascribe properties of samples from a population, we are using deduction. The deductions will involve probability, the study of uncertainty. We will study probability in chapters 5–7 and inferential statistics in chapters 8–15.

EXAMPLE 1.15

(Deduction and probability.) Suppose 1000 automobiles were recently manufactured. Let's assume that 5% of them have a defective steering component and that a local dealer has a sample of ten of these cars. As an application of probability, we might attempt to determine the likelihood that at least two of these ten cars have defective steering components. Since the sample is a subset of the population, we would expect (deduce) that 5% represents the likelihood that a given car in the sample is defective. In chapter 5, we will learn how to determine the likelihood that at least two of ten cars have defective steering components.

EXAMPLE 1.16

(Inferential statistics.) Suppose 1000 cars were manufactured and the number of cars with defective steering systems is unknown. To estimate the percentage of cars in the population of 1000 cars with defective steering systems, we will inspect a sample of ten cars. If two of the ten cars are found to have defective steering systems, we might infer (using induction) that 20%, or 200, of the 1000 cars have defective steering systems. The proportion of defective steering systems in the sample is an example of a statistic; its value is 0.20. The percentage of cars in the population with defective steering systems is an example of a parameter. In chapter 9, we will learn how to use statistics to estimate unknown parameters.

The **reliability** of an inference is an important aspect of inferential statistics. An inference is *reliable* if it can be depended on with confident certainty; it cannot accurately describe a characteristic of a population if it is not reliable. In statistics, we are concerned only with reliable inferences. Probability theory must be used in order to determine the reliability of an inference. We shall investigate the reliability of inferences in chapters 9 and 10 and probability in chapters 5–7.

SECTION 1.5 *Role of the Computer in Statistics*

With the advent of microcomputers, the heavy computational work associated with large data sets or complicated analyses can be relegated to the computer. As the tedious data manipulations are done by the computer, the user can concentrate on carrying out data analyses. There are many user-friendly software programs available that permit students and statisticians to carry out tedious statistical calculations with little or no difficulty. Some of the more popular programs include MINITAB, SPSSx, SAS, BMDP, and SYSTAT. All of these programs allow the user to communicate with the computer system by supplying simple commands.

In this text we have chosen to use MINITAB to illustrate statistical applications involving the computer. MINITAB was originally developed at Pennsylvania State University as a tool for teaching statistics. Today it is widely used for both teaching and research all across the country. MINITAB can be installed on mainframe-, mini-, and microcomputers. It offers a great deal of computing power to both the student of statistics and the researcher using statistics.

MINITAB is an interactive, command-driven program. Once MINITAB is loaded into the computer system, the user communicates with the system by supplying commands that are executed immediately by the system. It is very easy to use. The system prompt, MTB >, displayed on the monitor informs the user that the system is ready to accept a command from an input device, such as a keyboard. Once the data is input to the program, the user supplies a command and depresses the return or enter key, and the system immediately responds with the desired value.

EXAMPLE 1.17

Computer display 1.1 contains the MINITAB commands and corresponding output used to determine the mean (average) of the numbers 34, 68, 39, 21, and 42. At the system prompt, MTB >, the user types SET C1 , followed by depressing the return or enter key (until the return or enter key is depressed, the computer does not know the command). This informs MINITAB to create a column labeled C1 in memory to hold the data. The computer system responds with the data prompt, DATA>. The user types the five numerals: 34, 68, 39, 21, and 42, followed by depressing the return or enter key to register the data with the system (spaces are used to separate the numbers). The computer responds with the data prompt again. Since there is no more data, the user types $\boxed{\text{END}}$ to inform the system that there is no more data. The system responds once again with the system prompt, MTB>, which informs the user that it is awaiting another command from the user. The user now types the command $\boxed{\text{MEAN C1}}$, followed by depressing the return or enter key, to request the average of the column of numbers labeled by C1. Immediately, the computer responds with MEAN = 40.800 and another system prompt, MTB >.

Notice that in this example the instructions $\boxed{\text{SET C1}}$ and $\boxed{\text{MEAN C1}}$ are the **commands** that are entered by the user. At the end of each command, the user must depress the key labeled *enter* or *return* in order to register the commands with the computer system. This means of communication with the computer system is why MINITAB is called a **command-driven system,** as opposed to a menu-driven system where a choice in a menu is entered to bring about a particular action of the computer.

Computer Display 1.1

```
MTB > SET C1
DATA> 34 68 39 21 42
DATA> END
MTB > MEAN C1
 MEAN = 40.800
MTB >
```

Computers are efficiently used when large amounts of data need to be processed, repetitive tasks need to be performed, or the results of data analyses need to be done quickly and accurately. The problems encountered in this book will involve relatively small data sets; but even with relatively small data sets, some of the calculations will be tedious when done with a calculator. It is desirable to understand the hand calculations done in the text and to use similar steps to solve several more exercises in the exercise sets. After this is done you should understand the uses and limitations of the procedures, as well as being able to understand and interpret the results they produce. If you have access to a computer, you can use a statistical software package (such as MINITAB) to carry out similar procedures in the future.

Some of the data sets in the applications are used to demonstrate the use of MINITAB. These applications (computer displays) are used to serve at least five purposes:

1. To illustrate the ease of using MINITAB.
2. To illustrate the commands that need to be supplied by the user to obtain desired statistical results.
3. To illustrate the format and notation used in computer output.
4. To illustrate the extent of the statistical computing power that is available with MINITAB (and other similar software packages).
5. To teach the use of MINITAB to perform certain statistical tasks.

CHAPTER REVIEW

■ *IMPORTANT TERMS* ■

The following chapter terms have been mixed in ordering to provide you better review practice. For each term, provide a definition in your own words. Then check your responses against the definitions given in the chapter.

command	estimate	command-driven system
deduction	descriptive statistics	induction
inference	inferential statistics	population
sample	statistic	statistics
MINITAB	reliability	parameter

CHAPTER OBJECTIVES

In this chapter we will investigate

▷ *Data.*

▷ *Two general types of data.*

▷ *Data classified according to the type of measurement scale used.*

▷ *How to organize and summarize data using tables.*

▷ *How to display data using various types of graphs.*

||||| MOTIVATOR 2 ➤

A survey involving two samples and containing two parts was conducted to measure customer-service perceptions.[8] One sample consisted of more than 1300 customers from a wide geographic range (including the United States, Canada, and Great Britain). The other sample contained nearly 900 customer-service people representing nine diverse organizations. The first part of both survey questionnaires was constructed around 17 dimensions of customer service. Respondents rated each dimension on two scales, importance and proficiency. The importance scales asked respondents, "How important do you feel this dimension is to effective customer service?" Possible responses ranged from 5 (extremely important) to 1 (not important). The proficiency scale asked respondents, "How well do you feel customer-service people use this dimension when they interact?" For this scale, the responses ranged from 5 (always done well) to 1 (never done well). The second part of both surveys was designed to measure the impact of customer service on customers' decisions to repeat business. One item asked "How much does good service affect your decision to do business again with that organization?" On the five-point scale, responses ranged from 5 (extremely great effect) to 1 (little or no effect). A second item asked respondents, "How often do you tell other people when you receive excellent or poor customer service?" Response categories were "never," "occasionally," and "frequently."

The results of both surveys indicated that customers' perceptions of service quality are different from the perceptions of service employees. Table 2.1 lists the mean ratings given by customers and customer-service personnel for 5 of the 17 dimensions of customer service; figures 2.1 and 2.2 show the impact of customer service on business. In this chapter we will learn about the various kinds of data and how to organize and present data by using tables like those used here as well as graphs.

TABLE 2.1

Customer and Customer-Service Personnel Ratings of Service

Dimension	Customer Sample		Customer-Service Personnel Sample	
	How important	How well done	How important	How well done
Communication	4.05	2.95	4.55	3.64
Customer sensitivity	3.92	2.67	4.38	3.56
Decisiveness	3.84	2.74	4.34	3.53
Job knowledge	4.10	2.96	4.54	3.56
Motivation to serve customers	3.97	2.73	4.27	3.32

FIGURE 2.1

Effects of good service

	Great Effect	Moderate Effect	Little or No Effect
Customer sample	97%	2%	1%
Customer-contact personnel sample	83%	13%	4%

FIGURE 2.2

Word-of-mouth discussions concerning service

	Excellent Service	Poor Service
Customer sample	38%	75%
Customer-contact personnel sample	57%	65%

Chapter Overview

The most basic aspect of statistics is the information involved. Without information to collect, organize, analyze, or interpret, there would be no reason to use or study statistics. The information that is used in statistics is referred to as **data.** In order for the information to be useful for decision making, it must be first organized and displayed properly. The type of data will indicate the methods to be used for analysis. We begin this chapter with a discussion of the various types of data.

SECTION 2.1

Data: The Building Blocks of Statistics

Data is the plural of datum, a piece of information. When used collectively, *data* is a synonym for *sample*. Data can be classified into two general categories, quantitative and qualitative.

Quantitative data refer to numerical information, such as how much or how many, and are measured on a numerical scale.

EXAMPLE 2.1

Weights in pounds, ages in years, prices in dollars, lengths in centimeters, and volumes in cubic inches are all examples of quantitative data.

> **Qualitative data** represent categorical or attributive information that can be classified by some criterion or quality.

EXAMPLE 2.2

Gender (male, female); color (red, green, blue); religion (Protestant, Catholic, Jewish); blood type (A, B, AB, O); favorite make of car (Ford, Chevrolet); and computer brand (IBM, Kaypro, Zenith, Compaq) are all examples of qualitative data.

Data consisting of numerals can be classified in quantitative or qualitative terms, depending on how it is used. If it is used as a label for identification purposes, it is qualitative; otherwise, it is quantitative (see example 2.3). Some measurements, however, can be made by using either quantitative or qualitative scales, as in example 2.4.

EXAMPLE 2.3

If a serial number on a radio is used to identify the number of radios manufactured up to that point, it is a quantitative measure. But if the serial number is used only for identification purposes, it is a qualitative bit of information.

EXAMPLE 2.4

If the height of an individual is measured in feet and inches, then the classification is quantitative, but if height is measured as short, medium, or tall, the classification is qualitative. In addition, height could be measured using quantitative data (feet and inches) but reported or classified using qualitative data (small, medium, or tall).

Quantitative data can be classified as discrete or continuous.

> Data resulting from a process that can be counted are **discrete data.**

> Data resulting from a measurement process where the characteristic being measured can assume any numerical value over an interval are **continuous data.**

EXAMPLE 2.5

The number of children in a family, the number of cars in the parking lot, the salary of an individual, and the number of people in a checkout line are all examples of discrete data. Your heart rate and your systolic blood pressure (as measured on a digital readout) are also examples of discrete data. The speed of a car (in miles per hour), however, could not result in discrete data; the speed of a car could be any rate from 0 miles per hour to the maximum speed of the car.

EXAMPLE 2.6

Continuous data cannot be counted. Weight (in kilograms), height (in inches), time (in minutes), distance (in feet) are all examples of continuous data. Barometric pressure and the length of time it takes you to get to school are also examples of continuous data. The number of people on the beach on a busy weekend is not continuous, since the number of people can be counted.

Every measurement process that results in continuous data is limited by the precision of the measurement instrument. For example, if a measuring device is precise to the nearest tenth of an inch and it is used to measure the height of an individual, then there are only a finite number of possible measurements that can result. A person's height is rounded to the nearest tenth of an inch. Such a measurement represents an approximation to the real measurement. The real measurements are theoretical and represent continuous data, while the approximate measurements represent discrete data, since there are only a finite number of ways to measure something with an instrument of given precision. In reality, all physical measurements are discrete. The restriction of limited precision applies only to the measuring instrument, not the data; the data are of a continuous nature that are rounded in value according to the precision of the measuring instrument.

Our main purpose in analyzing data is to make meaningful interpretations. As a general rule, the amount of information contained in the data depends on the type of data. The quantitative-versus-qualitative dichotomy and the discrete-versus-continuous dichotomy are not always adequate for the classification of data according to the amount of information it contains. Data can also be classified according to the type of measurement scale or procedure by which it is produced.

Consider the numeral 4 in the following four situations:

    a. John's football-jersey numeral is 4.

    b. John is in the 4th grade.

    c. John recorded the temperature as 4° Celsius.

    d. John grew a cucumber that measured 4 inches in length.

These four situations represent four different levels of information, resulting from the use of different measuring scales. The measurement in situation a, for example, is used only for identification or classification; it identifies John as football player number 4. The 4 in 4th grade, in situation b, is also a classification, but it denotes more information—4 is also a grade level, more advanced than the 3rd grade and less advanced than the 5th grade, although exactly how much more or less advanced is not something we can measure.

In situation c, we again see a comparison of levels—4° indicates that the temperature is higher than a temperature of 2° Celsius and lower than a temperature of 7° Celsius. What's more, a temperature of 4° is 1.5° higher than a temperature of 2.5° because the distance between 4° and 2.5° is 1.5°. A temperature of 4° Celsius, however, is not twice as warm as a temperature of 2° Celsius.

Finally, in situation d, the measurement 4 identifies the cucumber as a member of the class of all cucumbers measuring 4 inches in length. We also know that the cucumber is longer than a 3-inch cucumber, that it is 1 inch longer than a 3-inch cucumber, and that it is twice as long as a cucumber measuring 2 inches in length.

Situations a–d are representative of four different types of measurement scales that we will discuss in detail because the type of measurement scale determines the amount of information that is contained in any given datum.

Four Types of Measurement Scales Used in Statistics

1. Nominal
2. Ordinal
3. Interval
4. Ratio

Nominal Scale

Nominal scales exist for quantitative and qualitative data. A *nominal scale for numeric data* assigns numbers to categories to distinguish one from another, as in example 2.7. A *nominal scale for qualitative data,* as in example 2.8, is an unordered grouping of data into discrete categories, where each datum can go into only one group. Nominal scales are mainly used for identification or classification purposes.

EXAMPLE 2.7

Numeric data that are nominal include numerals on basketball jerseys, postal zip codes, telephone numbers, and football scores (6 points for a touchdown, 1 point for an extra-point kick, 2 points for an extra-point run or safety, and 3 points for a field goal).

EXAMPLE 2.8

Nominal data that are qualitative include gender, ethnic category, blood type, and religion.

Ordinal Scale

Data measured on a nominal scale that is ordered in some fashion are referred to as **ordinal data.** An ordinal scale places measurements into categories, each category indicating a different level of some attribute that is being measured.

EXAMPLE 2.9

Ordinal-data lists include

1. Letter grades: A, B, C, D, and F. These grades indicate categories of achievement, as well as levels of achievement.

2. Academic ranks: instructor, assistant professor, associate professor, and professor. A professor has higher academic rank than an instructor.

3. Residence numbers on a particular street: 421 North Street, 423 North Street, and the like. The residence at 423 North Street is located between the residences located at 421 North Street and 425 North Street.

4. Teacher ratings: poor, fair, good, and superior.

5. Grades of school: first, second, third, and so on.

The difference or distance between values measured on an ordinal scale cannot be determined. Even though we often code a letter grade of A as 4, B as 3, C as 2, D as 1, and F as 0, we should not say, for example, that an A is twice as good as a C or that the A-student knows twice as much as the C-student. All we can say is that an A grade is a better or higher grade than a C grade. An ordinal scale has no unit of distance.

Interval Scale

Data measured on an ordinal scale for which distances between values can be calculated are called **interval data.** The distance between two values is relevant. Interval data are necessarily quantitative. An interval scale does not always have a zero point, a point that indicates the absence of what we are measuring.

EXAMPLE 2.10

Interval data lists include

1. **IQ test scores:** An IQ score of 110 is five points higher than an IQ score of 105 (ordinal data). Not only can we say that an IQ score of 110 is higher than an IQ score of 105, we can also say that it is five points higher. But we cannot say that a person with an IQ score of 180 is twice as smart as a person with an IQ of 90. A given difference between two IQ scores does not always have the same meaning. For example, the differences (100 − 90) and (150 − 140) may have different interpretations even though they both equal 10. Although a person with an IQ of 140 is more intelligent (as measured by the intelligence test) than a person with an IQ of 100, we cannot say that a person having an IQ of 150 is as much more intelligent than a person with an IQ of 140 as a person with an IQ of 100 is more intelligent than a person with an IQ of 90.

2. **Celsius temperatures:** A temperature of 80° is 40° warmer than a temperature of 40°. But it is not correct to say that 80° is twice as warm as 40°. Note also that a temperature of 0° does not represent the absence of any heat. The zero point on the Celsius temperature scale was arbitrarily set at the icing point. This point indicates that some heat is present. (Theoretically, −273° C represents the absolute minimum temperature, the temperature for which the molecules of a substance approach a speed of 0.)

3. **Calendar dates:** Ronald Reagan was inaugurated as the 40th president of the United States in 1981, 192 years after George Washington (1789). We can specify the distance between these two ordered events—192 years—but year 0 (if it existed) would not represent the absence of time.

Ratio Scale

Data measured on an interval scale with a zero point meaning "none" are called **ratio data.** With data measured with a ratio scale, we can determine how many times greater one measurement is than another. Because the zero point of the Celsius temperature scale does not represent the complete absence of heat, the Celsius scale is not a ratio scale. On the other hand, the Kelvin temperature scale, where 0 K corresponds to −273° C, is an example of a ratio temperature scale.

EXAMPLE 2.11

Ratio scales include scales commonly used to measure units, such as feet, pounds, dollars, and centimeters. The results of counting objects are also ratio data. Ten apples are twice as many as five apples. With a ratio scale, a 200-pound person will always weigh twice as much as a 100-pound person, using any ratio scale (for example, a scale using ounces, grams, or kilograms).

APPLICATION 2.1

Suppose a group of teachers are surveyed concerning their religious affiliation and 15 indicated Protestant, 21 indicated Catholic, and 7 indicated Jewish. What type of data are involved?

Solution: The response from each teacher is either Protestant, Catholic, or Jewish, and these responses constitute nominal (categorical) or qualitative data. On the other hand, the numbers 15, 21, and 7 result from counting the response data; these results thus represent quantitative data. Numbers resulting from operations performed on the data, such as adding, should not be confused with the data collection. ▪

It is important to be able to classify data according to the measurement scale used. To draw an inference about a population of interest, the techniques used are dependent on the type of measurement scale. For example, if a sample of ordinal data is used, a statistical technique that uses ordinal data must be used. By classifying the data according to the type of measure scale used, a researcher can identify the best statistical technique to use to analyze the data.

Basic Skills

1. Classify the following data as quantitative or qualitative:
   a. heights (in inches) of five basketball players
   b. weights (in ounces) of 12 chickens
   c. ethnic classifications of 20 employees
   d. telephone numbers of friends

2. Classify the following data as quantitative or qualitative:
   a. GPAs of the members of the junior class
   b. letter grades of 15 students in Philosophy 209
   c. number of M & M's in a two-ounce package
   d. birthdays of family members

3. Classify the following data as discrete or continuous:
   a. the number of defects in each of 50 new cars
   b. SAT math scores of 30 high school seniors
   c. the distance (in yards) gained by a halfback each game during last season
   d. the weights (in pounds) lost by 20 people on a weight-reduction diet

4. Classify the following data as discrete or continuous:
   a. the number of runs scored each game for the Pirates during the 1900 season
   b. the wages (in dollars) earned last month by 50 college presidents
   c. the average temperatures (in degrees Fahrenheit) each day for the past 30 days
   d. the number of grains of sand for each of 100 beaches

5. The accompanying table contains the distribution of vehicles registered at Excel College.

Rank	Type of Vehicle	Number Registered
1	Car	150
2	Truck	25
3	Motorcycle	15
4	Bicycle	10

   a. Classify the data in each of the three columns as quantitative or qualitative.
   b. Classify the data in the third column as discrete or continuous.

c. Classify the data in each of the three columns as nominal, ordinal, interval, or ratio.

6. Memorial hospital maintains the following information on each patient:

   social security number
   date of last admission
   date of birth
   insurance company
   employer
   home address
   home telephone number

   a. Classify the information as quantitative or qualitative.
   b. Classify the data in each category as discrete, continuous, or neither.
   c. Classify the information as nominal, ordinal, interval, or ratio.

7. The accompanying figure shows a numerical scale to measure teaching effectiveness.

   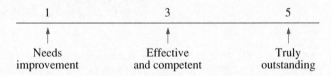

   a. Identify the type of measurement scale.
   b. Suppose this scale is used by 30 students to evaluate their statistics instructor. Would the results be easier to interpret than the results that would be obtained if the 30 students evaluated their instructor by submitting a written statement of free response? Explain.

8. College students are classified as freshman, sophomore, junior, or senior according to the number of credits earned. What type of measurement scale is this?

9. Give an example not given in this section of an ordinal scale for quantitative data.

10. Is all numerical information quantitative data? Explain.

11. Is all nonnumerical information qualitative data? Explain.

SECTION 2.2   *Organizing Data Using Tables*

The objective of data organization is to arrange a set of data into a useful form in order to reveal essential features and to simplify certain analyses. Data that are not organized in some fashion are referred to as **raw data.** One method of arranging raw data is to construct an ordered array; that is, arrange the data from low to high (or from high to low). If the number of data is large, the data array may be difficult to manage or comprehend. As a result, tables are often used as a general approach to organizing raw data. In this section, we will discuss various types of tables used for organizing data; in section 2.3, we will discuss graphical means for displaying raw data organized in tabular form. The type of data—nominal, ordinal, interval, or ratio—will determine the type of display.

The **frequency** of a measurement or category is the total number of times the measurement or category occurs in a collection of data. The use of frequencies is best suited for qualitative or discrete data. The symbol f is used to denote the frequency of a measurement. The following sample of data represent the number of free throws missed by a basketball team during the last seven games:

$$7 \quad 2 \quad 8 \quad 4 \quad 2 \quad 7 \quad 2$$

The number 7 occurs with a frequency of $f = 2$, 2 occurs with a frequency of $f = 3$, and 8 and 4 each occur with a frequency of $f = 1$.

There are two general types of tables for reporting data that employ frequencies: **ungrouped frequency tables** and **grouped frequency tables.** Both tables are referred to as **frequency tables.** We shall discuss ungrouped frequency tables first.

Ungrouped Frequency Tables

The free throw data cited above can be summarized as shown in table 2.2, where x denotes the measurements and f denotes the frequency of each measurement. Table 2.2 is an example of an ungrouped frequency table for discrete data.

TABLE 2.2

Frequency Table of Free-Throw Data

x	f
2	3
4	1
7	2
8	1

APPLICATION 2.2

Construct a frequency table for the following data representing the number of absences per class period during the 1988 fall term for students enrolled in Statistics 101.

$$
\begin{array}{cccccc}
9 & 8 & 7 & 8 & 4 & 3 \\
2 & 1 & 0 & 5 & 3 & 2 \\
1 & 1 & 7 & 3 & 2 & 8 \\
7 & 6 & 6 & 4 & 3 & 2 \\
2 & 0 & 9 & 4 & 6 & 9 \\
6 & 9 & 4 & 3 & 5 & 7 \\
3 & 2 & 1 & 4 & 4 & 2
\end{array}
$$

Solution: As an intermediate step, we use **tally marks** to aid in determining the frequency f for each observation, with x representing the number of absences.

Number of Absences (x)	Tally	Frequency (f)
0				2
1						4
2	⏦||	7
3	⏦|	6
4	⏦|	6
5				2
6						4
7						4
8					3
9						4
		42

Corresponding to each observation, we place a tally mark (|) in the tally column adjacent to its value. After all the tallies are placed, they are counted for each measurement x to determine the frequency. Note that the sum of the frequencies in a frequency table represents the size of the data collection. In this case, the sum of the frequencies (42) represents the 42 class periods for which absences were recorded. ■

APPLICATION 2.3

Five faculty members (Jones, Smith, Baker, Brown, and Thomas) at a small university were nominated for chair of the faculty senate. The accompanying data represent the results of the election. Construct a frequency table for the data.

Jones	Jones	Smith	Smith	Jones	Thomas
Smith	Baker	Baker	Jones	Thomas	Jones
Smith	Smith	Smith	Brown	Brown	Jones
Smith	Thomas	Smith	Brown	Brown	Brown

Solution: The frequency table is as follows:

Faculty Member	Frequency (f)
Baker	2
Brown	5
Jones	6
Smith	8
Thomas	3

■

Grouped Frequency Tables

Frequency tables, such as the one shown in Table 2.2, are correctly termed **ungrouped frequency tables** because each measurement has its own corresponding frequency. A **grouped frequency table,** on the other hand, shows frequencies according to groups or classes of measurements. Grouped frequency tables are typically used for summarizing large amounts of continuous data, containing relatively few repetitions. Such data summaries facilitate certain statistical calculations and graphic displays when the computer is not used. In order to use a grouped frequency table to summarize data, the data must be measured by at least an interval scale. For large amounts of data not measured by at least an interval scale, an ungrouped frequency table may be used.

Suppose Memorial Hospital wants to study whether its emergency room staffing is adequate. To start the study, the department manager tracks the number of people visiting the emergency room each day for a 12-day period, with the following results:

Day	1	2	3	4	5	6	7	8	9	10	11	12
No. of visitors	7	43	8	22	13	28	36	18	23	21	15	52

To simplify the data, the manager then constructs six groupings or classes: the first class will represent 1 to 10 patients; the second class, 11 to 20 patients; the third class, 21 to 30 patients; and so on. From this grouping, he or she then prepares a grouped frequency table (table 2.3) to show how often during the twelve days the number of patients fell into each group.

TABLE 2.3

Grouped Frequency Table for Emergency Room Data

Class	Frequency (f)
1–10	2
11–20	3
21–30	4
31–40	1
41–50	1
51–60	1

The classes in grouped frequency have what are called **class limits.** For the class 1–10, 1 is called the *lower class limit* and 10 is called the *upper class limit.* There are two measurement that fall between 1 and 10, inclusive; three measurements that fall between 11 and 20, inclusive; four measurements that fall between 21 and 30, inclusive; one measurement falls between 31 and 40, inclusive, and so on. The distance between any two consecutive upper limits or any two consecutive lower limits is called the **class width.** The width of each class in table 2.3 is 10. The distance between the upper limit of the first class and the upper limit of the second class is $20 - 10 = 10$. Each class in a frequency table has theoretical class limits called **class boundaries**; an upper theoretical class limit is called the *upper boundary,* and a lower theoretical class limit is called the *lower boundary.* The lower boundary for the first class is 0.5 and the upper boundary for the first class is 10.5. For this frequency table, the upper class boundary for each class is found by adding 0.5 to the upper class limit, and the lower class boundary for each class is found by subtracting 0.5 from the lower class limit.

Notice that when examining a grouped frequency table without the raw data (that is, the data before statistical processing), we do not know the individual measurements. For example, in table 2.3 we know that two measurements fall in the class 1–10, but we do not know what the two measurements are. This would not be the case for an ungrouped frequency table, where all the measurements are known.

> Every grouped frequency table should possess the following three characteristics:
>
> 1. Uniformity: Each class should have the same width.
> 2. Uniqueness: No two classes should overlap.
> 3. Completeness: Each piece of data should belong to some class.

Class boundaries and class widths for a grouped frequency table are determined by considering the **unit** or precision of measurement. For the classes in table 2.3, the precision of measurement is the nearest whole number (since we are counting people), so the unit of measurement is 1.

> The lower class boundary of an interval is located one-half unit below the lower class limit, and the upper class boundary of an interval is located one-half unit above the upper class limit.

For the first class in table 2.3, the lower class boundary is $[1 - 0.5(1)] = 0.5$ and the upper boundary is $[10 + 0.5(1)] = 10.5$. None of the raw data can fall at the boundaries of an interval. Thus, the measurements 0.5 and 10.5 could not fall in the first class, but that any measurement *between* 0.5 and 10.5 could. Of course, 0.5 and 10.5 are not possible measurements, so the class boundaries have mathematical meaning only.

The width w for any class in a grouped frequency table can be found by subtracting the lower class boundary from the upper class boundary.

Thus, for the first class in table 2.3, $w = 10.5 - 0.5 = 10$. Consider examples 2.12 and 2.13.

EXAMPLE 2.12

The following is a grouped frequency table for the number of seeds in 21 oranges.

Class	Frequency (f)
3–6	5
7–10	6
11–14	7
15–18	3

The precision of measurement for the classes is 1, since whole-number data are involved. For the class 7–10, if we add $(0.5)(1) = 0.5$ to the upper class limit 10, we get the upper class boundary 10.5. To get the lower class boundary, we subtract 0.5 from the lower class limit to get $7 - 0.5 = 6.5$ (refer to example 2.13). The width of the class 7–10 is then found by subtracting the lower class boundary from the upper class boundary, or $w = 10.5 - 6.5 = 4$.

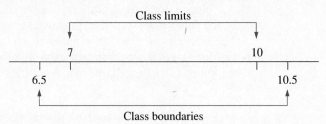

EXAMPLE 2.13

The following is a grouped frequency table for the weights (in pounds) of 18 new-born babies.

Class	Frequency (f)
3.0–4.4	1
4.5–5.9	1
6.0–7.4	7
7.5–8.9	8
9.0–10.4	1

The precision of measurement for the classes is 0.1 pound. For the class 7.5–8.9, subtracting one-half the unit from the lower class limit yields $7.5 - (0.5)(0.1) = 7.5 - 0.05 = 7.45$, the lower class boundary. The upper class boundary is found by adding one-half the unit to the upper class limit, yielding $8.9 + (0.5)(0.1) = 8.95$. Note that no weight can correspond to either boundary because the precision of measurement is to the nearest one-tenth of a pound.

In any grouped frequency table, the class width can be found simply by carrying out the following step:

To Determine the Class Width

Subtract two consecutive upper class limits or two consecutive lower class limits or two consecutive lower boundaries or two consecutive upper boundaries or subtract the lower boundary of a class from the upper boundary of the same class.

For the emergency room data originally given in table 2.3, we can compute the class width as indicated in table 2.4.

TABLE 2.4

Computation of Class Width for Table 2.3

Class	Frequency (f)
1–10	2
11–20	3
21–30	4
31–40	1
41–50	1
51–60	1

$$w = 20 - 10 = 10$$

$$w = 31 - 21 = 10$$

Notice, however, that the class width is *not* found by subtracting the lower class limit from the upper class limit.

Choosing Classes for Grouped Frequency Tables

In order to construct a grouped frequency table for a given collection of data, we need to answer three questions concerning the classes.

1. How many classes should be used?
2. What should be the width of each class?
3. At what value should the first class start?

Choosing the number of classes involves several considerations. If all the data are grouped into a small number of classes, characteristics of the original data are hidden and pertinent information may be lost. On the other hand, too many classes provide too much detail and defeat the purpose of grouping, which is to condense the data in a meaningful manner that is easy to interpret. In addition, too many classes may produce too many empty classes and render the grouping of data meaningless.

The number of classes, denoted by c, depends on the situation and the amount of data involved. Since there is no general agreement among statisticians regarding the number of classes to use and the choice is arbitrary, we will use between 5 and 15 classes, inclusive, in this text.

Number of Classes for a Grouped Frequency Table:

Between 5 and 15 classes (inclusively).

A useful suggestion for the number of classes is given by **Sturges' rule,** which states that the number of classes needed is approximately

> **Sturges' Rule**
>
> $$c = 3.3(\log n) + 1$$
>
> where n is the number of measurements and $\log n$ is the logarithm of n to the base 10

The value of c is usually rounded to the nearest whole number.

EXAMPLE 2.14

If $n = 25$ measurements are involved, Sturges' rule would suggest using six classes, since

$$c = 3.3(\log n) + 1$$
$$= 3.3(\log 25) + 1$$
$$= 3.3(1.3979) + 1 \approx 6$$

where $\approx$ means approximately equal to.

Some researchers believe that in most situations Sturges' rule provides a value for c that permits the construction of a grouped frequency table that presents a realistic picture of the raw data. Once we agree on the number of class intervals to use, the width of each class is found by using the **range** R, which is the difference between the largest measurement U and the smallest measurement L in the sample:

> **Range**
>
> $$R = U - L$$

Because c classes must cover the range, we divide the range by the number of classes to find the class width w:

> **Class Width**
>
> $$w = \frac{R}{c}$$

Since the smallest measurement should fall in the first class, the lower limit of the first class should be at or somewhat below the smallest measurement L. So that we can have general agreement on the classes in our grouped frequency tables, we will always start the first class with the smallest measurement. (This will be especially helpful when checking our answers.) In actual practice, the first class typically begins at a point that permits the classes to be expressed in convenient intervals, but there are occasions when exceptions to the guidelines are justified (see exercise 33 at the end of this section).

When the first class begins with the smallest measurement, the smallest value that w can assume depends on the unit of measurement. The smallest value for the class width w is determined by rounding the quotient R/c to the next highest integer value.

> The value of w is taken to be the least integer greater than R/c.

APPLICATION 2.4   Professor Smith gave a final examination consisting of 100 questions to his Introductory Accounting class. The following data represent the number of correct answers on each test. Construct a grouped frequency table with five classes to aid Professor Smith in analyzing the results.

17	15	78	21	10	32	7	65	18	87
4	22	34	42	9	9	82	79	98	4
44	64	62	77	2	81	45	37	83	44
77	13	41	16	17	13	82	37	5	54
7	67	88	41	61	22	92	16	67	85

Solution:

Step 1 We first determine the range R. Since the largest measurement is $U = 98$ and the smallest measure is $L = 2$, the range is

$$R = U - L$$
$$= 98 - 2 = 96$$

Step 2 We next determine w, the width of each class. Note that the number of classes is given as $c = 5$.

$$w = \frac{R}{c} = \frac{96}{5} = 19.2$$

Since the unit of precision for the examination scores is 1, we choose the least integer greater than 19.2 to be the value of our width. For our application, the least integer greater than 19.2 is 20; thus, $w = 20$.

Step 3 We start at $L = 2$ and construct the first class with a width of 20. Suppose the first class extends from 2 to x, where x represents the unknown upper class limit (refer to the accompanying diagram).

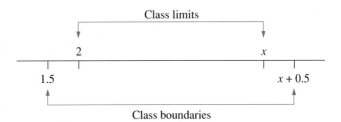

Since the unit of measurement is 1 and $0.5(1) = 0.5$, the upper class boundary can be represented as $x + 0.5$. The width of the first class is found by subtracting the lower class boundary from the upper class boundary. Thus,

$$w = (x + 0.5) - 1.5$$

Since $w = 20$, we have

$$20 = x - 1$$

Solving this equation for x we find that the upper class limit is $x = 21$. Thus, the first class becomes 2–21.

Step 4 To obtain each class following the first class, we add $w = 20$ to the lower and upper limits of each preceding class. Thus,

2–21
22–41   (Note: $41 = 21 + 20$)
42–61   (Note: $42 = 22 + 20$)
62–81
82–101   (Note: $101 = 81 + 20$)

Step 5 To determine the frequency of each class we use a tally column. If a piece of data falls in a class, we record a tally mark (|) in the tally column corresponding to that class. Table 2.5 contains our grouped frequency table for the 50 examination scores.

TABLE 2.5

Group Frequency Table of Final Examination Scores

Class	Tally	Frequency (f)
2–21	⌶⌶⌶⌶⌶⌶ III	18
22–41	⌶⌶⌶ III	8
42–61	⌶⌶⌶ I	6
62–81	⌶⌶⌶⌶⌶⌶	10
82–101	⌶⌶⌶ III	8
		50

■

APPLICATION 2.5

The accompanying data represent the number of customers visiting a store for a 22-day period. Use six classes and construct a grouped frequency table for the data.

28 42 52 50 29 31 34 45 48 38 28
33 33 49 32 37 41 43 46 49 34 49

Solution:
Step 1 The largest measurement is $U = 52$ and the smallest measurement is $L = 28$. We determine the range:

$$R = U - L$$
$$= 52 - 28 = 24$$

Step 2 We determine the width of each class:

$$w = \frac{R}{c} = \frac{24}{6} = 4$$

We should use a width of $w = 5$ in this case. If not, the six intervals may not contain all the data. With $w = 4$, there is no room above the largest data value or below the smallest data value. Since we have agreed to begin the first interval with the smallest data value, we risk the largest data value not belonging to the last interval. To see why this is the case, let's suppose our width is $w = 4$. The first class is 28–31, and the six classes are:

Class
28–31
32–35
36–39
40–43
44–47
48–51

Note that the value $U = 52$ does not belong to any class. To remedy this situation we should choose the least integer greater than 4. The least integer greater than 4 is 5. Thus, the class width should be $w = 5$.

Step 3 Table 2.6 is the completed grouped frequency table.

TABLE 2.6

Grouped Frequency Table for the Data in Application 2.5

Class	Tally	f
28–32	Ж╫	5
33–37	Ж╫	5
38–42	|||	3
43–47	|||	3
48–52	Ж╫		6
53–57		0

In this case the last class is empty. To remedy this situation we could start the first class at a smaller value, say 26. The frequency table would then become:

Class	f
26–30	3
31–35	6
36–40	2
41–45	4
46–50	6
51–55	1

■

Class Mark

The midpoint of each class is called the **class mark** and is denoted by X. When data are summarized into a grouped frequency table, some information is lost; we do not know the exact values of the measurements that lie in each class. The best we can do is to let each measurement within a given class be represented by the class mark of that class. Using class marks, instead of the raw data, makes the computations easier but results in a lack of accuracy. For a given class, the class mark is found using the formula

$$X = \frac{l_1 + l_2}{2}$$

where l_1 is the lower class limit and l_2 is the upper class limit of the class.

EXAMPLE 2.15

For the first class in table 2.5, the class mark is

$$X = \frac{l_1 + l_2}{2}$$
$$= \frac{2 + 21}{2} = 11.5$$

For the second class, the class mark is

$$X = \frac{l_1 + l_2}{2}$$
$$= \frac{22 + 41}{2} = 31.5$$

Notice that the class mark for class 2 can also be found by adding $w = 20$ to the class mark for class 1 ($11.5 + 20 = 31.5$). In general, each class mark for classes following the first class can be found by adding $w = 20$ to the preceding class mark. Thus, the remaining three class marks are found to be 51.5, 71.5, and 91.5. Table 2.7 displays a grouped frequency table containing the class marks.

Table 2.7

Class Marks for Grouped
Frequency Table of
Application 2.4 or Table 2.5

Class Number	Class	Tally	f	Class Mark, X
1	2–21	JHT JHT JHT III	18	11.5
2	22–41	JHT III	8	31.5
3	42–61	JHT I	6	51.5
4	62–81	JHT JHT	10	71.5
5	82–101	JHT III	8	91.5

APPLICATION 2.6

The following data set represents the amounts of cash (in dollars) spent in a particular weekend by 25 graduate students. Construct a grouped frequency table containing five classes.

39.78	28.30	28.31	17.95	44.47
46.65	31.47	33.45	29.17	48.39
$U \rightarrow$ 82.71	43.63	41.17	47.32	52.16
25.94	50.32	35.25	35.70	17.89 $\leftarrow L$
60.20	48.14	22.78	38.22	23.25

Solution:

Step 1 Compute the range R. Since $U = 82.71$ and $L = 17.89$, the range is

$$R = U - L$$
$$= 82.71 - 17.89 = 64.82$$

Step 2 Compute the class width w. Since $c = 5$, we have

$$\frac{R}{c} = \frac{64.82}{5} = 12.96$$

The least integer greater than 12.96 is 13. Thus, the class width is $w = 13$.

Step 3 We start at $L = 17.89$ and construct a class with width $w = 13$. The unit is 0.01 and $(0.5)(0.01) = 0.005$. Let x represent the upper boundary of the first class. Then the width is found by subtracting the lower class boundary from the upper class boundary.

$$w = x + 0.005 - 17.885$$
$$13 = x + 0.005 - 17.885$$
$$13 = x - 17.88$$
$$x = 30.88$$

Hence, the first class is 17.89–30.88.

Step 4 To obtain the remaining classes, we add 13 to the preceding class limits.

Class		
17.89–30.88	{	first class
13 13	{	add $w = 13$ to both limits of first class
30.89–43.88	{	second class
13 13	{	add $w = 13$ to both limits of second class
43.89–56.88	{	third class

The remaining two classes are found in a similar fashion. They are

$$56.89\text{--}69.88$$
$$69.89\text{--}82.88$$

Step 5 The frequencies are determined for the five classes using tallies, as shown in table 2.8.

Step 6 Class marks are found for each class using the midpoint formula given previously. The class mark for the first class is

$$X = \frac{l_1 + l_2}{2}$$
$$= \frac{17.89 + 30.88}{2} = 24.385$$

Each successive class mark is found by adding $w = 13$ to the preceding class mark. Table 2.8, the grouped frequency table for the data, also displays the class marks.

TABLE 2.8

Relative Frequency Table for Data of Application 2.6

Class Number	Class	Tally	f	X
1	17.89–30.88	JHT III	8	24.385
2	30.89–43.88	JHT III	8	37.385
3	43.89–56.88	JHT II	7	50.385
4	56.89–69.88	I	1	63.385
5	69.89–82.88	I	1	76.385
			25	

▪

Relative Frequency Tables

It is sometimes useful to express each value or class in a frequency table as a fraction or a percentage of the total number of measurements. The **relative frequency** for a measurement (or class) is found by dividing the frequency f of the measurement (or class) by the total number of measurements n. The table is then called a **relative frequency table** (see example 2.16). A relative frequency table has several advantages over a frequency table. One important advantage, as we see in example 2.17, is that we can make meaningful comparisons between similar sets of data having the same classes but different total frequencies.

EXAMPLE 2.16

The relative frequency for class 1 in table 2.8 is $n = 8/25 = 0.32$. Table 2.9 displays a relative frequency table for the data in application 2.6. Note that the sum of the relative frequency column is 1.00.

TABLE 2.9

Relative Frequency Table

Class	Relative Frequency
17.89–30.88	0.32
30.89–43.88	0.32
43.89–56.88	0.28
56.89–69.88	0.04
69.89–82.88	0.04
	1.00

EXAMPLE 2.17

Consider table 2.10, which shows the starting salaries of recent mathematics graduates at two state colleges, A and B.

TABLE 2.10

Frequency Tables for Starting Salaries at Two Colleges

College A		College B	
Salary class	f	Salary class	f
$10,000–12,999	0	$10,000–12,999	1
13,000–15,999	2	13,000–15,999	1
16,000–18,999	7	16,000–18,999	2
19,000–21,999	6	19,000–21,999	2
22,000–24,999	3	22,000–24,999	3
25,000–27,999	2	25,000–27,999	1

By examining the two parts of the table, we see that each college graduates three mathematics majors making starting salaries from $22,000 to $24,999. But comparing the relative frequencies, we see that college A has $3/20 = 15\%$ of its mathematics graduates earning from $22,000 to $24,999, while college B has $3/10 = 30\%$ of its mathematics graduates earning from $22,000 to $24,999 (see table 2.11).

TABLE 2.11

Relative Frequencies For Starting Salaries at Two Colleges

College A		College B	
Salary class	Rel f	Salary class	Rel f
$10,000–12,999	$0/20 = 0$	$10,000–12,999	$1/10 = 0.10$
13,000–15,999	$2/20 = 0.10$	13,000–15,999	$1/10 = 0.10$
16,000–18,999	$7/20 = 0.35$	16,000–18,999	$2/10 = 0.20$
19,000–21,999	$6/20 = 0.30$	19,000–21,999	$2/10 = 0.20$
22,000–24,999	$3/20 = 0.15$	22,000–24,999	$3/10 = 0.30$
25,000–27,999	$2/20 = 0.10$	25,000–27,999	$1/10 = 0.10$

In a relative frequency table, the sum of the relative frequency column (with no rounding error) is always one. This is not surprising since a relative frequency is equivalent to a percentage. To convert a relative frequency to a percentage, we multiply the relative frequency by 100%. We shall find relative frequency tables useful when studying probability in chapter 5 because the sum of the probabilities involved will always equal 1.

Cumulative Frequency Tables

The **cumulative frequency** for any measurement (or class) is the total of the frequency for that measurement (or class) and the frequencies of all measurements (or classes) of smaller value. There are many occasions where we are interested in the number of observations less than or equal to some value. Table 2.12 shows a cumulative frequency table for the data in application 2.4.

EXAMPLE 2.18

Occasions where cumulative frequency is of interest include the following:

1. A quality-control engineer might want to know how many days a production process produced at most 100 defective items.

2. A teacher might be interested in the number of students that received a score less than or equal to 70% on an examination.

3. A basketball coach might be interested in the number of games in which the opponents scored at most 60 points.

TABLE 2.12	Frequency Table		Cumulative Frequency Table	
	Class	*f*	Class	Cumulative frequency
Cumulative Frequency Table for Examination Score Data of Application 2.4	2–21	18	2–21	18
	22–24	8	22–24	26 = (18 + 8)
	42–61	6	42–61	32 = (26 + 6)
	62–81	10	62–81	42 = (32 + 10)
	82–101	8	82–101	50 = (42 + 8)
		50		

Cumulative Relative Frequency Tables

Cumulative frequency tables can also be constructed for tables containing relative frequencies or percentages. When this is done, the table is called a **cumulative relative frequency table.** A cumulative relative frequency table is displayed in table 2.13 for the data in application 2.4. It was obtained from table 2.12 by computing cumulative relative frequencies for the cumulative frequencies.

TABLE 2.13	Class	Cumulative Relative Frequency
Cumulative Relative Frequency Table for Data of Application 2.4	2–21	18/50 = 0.36
	22–41	26/50 = 0.52
	42–61	32/50 = 0.64
	62–81	42/50 = 0.84
	82–101	50/50 = 1.00

Cumulative relative frequencies have many uses. One use is the scoring of standardized scholastic tests such as the Scholastic Aptitude Test (SAT) and many college entrance exams. Test scores are usually given as **percentiles.** A *percentile score* tells what part of the tested population scored *less than the given score.*

EXAMPLE 2.19

If 590 is said to be the 90th percentile in the mathematics portion of the SAT, it means that 90% of the scores on the mathematics portion of the test were lower than 590.

EXAMPLE 2.20

Table 2.14 reports the heights (in inches) of 200 male freshmen at a college.

TABLE 2.14

Heights (in Inches) of Male Freshmen

Height	*f*	Relative Frequency	Cumulative Relative Frequency
59.5–62.5	2	0.01	0.01
62.5–65.5	12	0.06	0.07
65.5–68.5	24	0.12	0.19
68.5–71.5	46	0.23	0.42
71.5–74.5	62	0.31	0.73
74.5–77.5	36	0.18	0.91
77.5–80.5	16	0.08	0.99
80.5–83.5	2	0.01	1.00

This table, which employs class boundaries and cumulative relative frequencies, can be used to determine percentiles. The following results are apparent from observation of the cumulative relative frequency table:

1. A height of 74.5 inches is at the 73rd percentile.

2. The 50th percentile is between 71.5 inches and 74.5 inches.

3. The 19th percentile is 68.5 inches.

4. The 75th percentile is between 74.5 inches and 77.5 inches.

APPLICATION 2.7

Suppose the 70th percentile of weights of all adult males is 175 pounds and the 85th percentile is 195 pounds. What percentage of men have weights greater than 175 pounds and less than 195 pounds?

Solution: By definition, 70% of adult males weigh less than 175 pounds and 85% of adult males weigh less than 195 pounds. So, $0.85 - 0.70 = 0.15 = 15\%$ of adult males weigh strictly between 175 pounds and 195 pounds. ■

Bivariate Tables

If data result from measuring two different aspects of a member of a population, then the data are called **bivariate data.** Two variables are used to represent the two aspects of each member. A member might be an object, person, or source. Suppose we want to investigate the height and weight of all high school basketball players in Allegany County, Maryland. A member is a basketball player, and associated with each basketball player, we have two aspects, height (in inches) and weight (in pounds). If we let the variable x represent the height of a basketball player and the variable y represent the weight of a basketball player, then the ordered pair (x, y) can represent the height and weight, respectively, of a member (player). To date, we have only been measuring one aspect of each member of a population. So we only used one variable to represent the measurements. When only one variable is used to represent the data resulting from the members of a population, the data are called **univariate data.**

EXAMPLE 2.21

(Univariate data.) If the data are weights (in pounds) of a class of 30 statistics students, then a member is a student and the aspect of measurement is the weight of a student.

EXAMPLE 2.22

(Bivariate data.) Suppose we are interested in the average daily rainfall and the average daily temperature in Athens, Georgia, for the past ten years. The population consists of the past ten years. A member of the population is a year. In this case, a year is a source for two pieces of information. The two pieces of information are average daily rainfall and average daily temperature. Each year gives rise to two measurements.

A **bivariate frequency table** is an arrangement of data into two categories of classification. The data used to construct the bivariate frequency table usually result from frequency counts. Each category is identified with a symbol, called a variable. A variable is used to represent data within a category. The categories may be discrete numbers, number intervals, or qualitative values (such as gender, hair color, or religion).

EXAMPLE 2.23

Suppose data were collected from a sample of voters concerning their political philosophy and party affiliation. Each voter was asked to identify their political philosophy as liberal, conservative, or other, and their party affiliation as Democrat, Republican, or other. The two variables of classification are political philosophy and party affiliation. The political philosophy variable has three categories or levels of classification (liberal, conservative, other), while the party affiliation variable also has three categories or levels (Democrat, Republican, other). The data are tabulated in table 2.15.

TABLE 2.15

Bivariate Frequency Table

Party Affiliation	Party Philosophy			
	Liberal	Conservative	Other	Total
Democrat	78	65	37	180
Republican	84	79	7	170
Other	38	46	16	100
Total	200	190	60	450

Among other things, the following information can be easily read from the table:

1. There are 78 voters who indicated they were liberal Democrats.

2. There were 79 voters who indicated they were conservative Republicans.

3. There were 450 voters polled.

4. There were 170 Republicans polled.

5. There were 60 voters who classified their political philosophy as other.

APPLICATION 2.8

The statistics grades and gender for 32 college students are shown in the following table. Construct a frequency table for the bivariate data.

Student	Grade	Gender	Student	Grade	Gender
1	B	M	17	C	F
2	C	F	18	E	F
3	C	F	19	C	M
4	C	M	20	B	F
5	B	F	21	D	M
6	B	F	22	E	M
7	A	M	23	B	M
8	C	M	24	B	M
9	D	F	25	C	M
10	C	M	26	C	F
11	B	F	27	D	M
12	A	F	28	B	F
13	C	M	29	D	F
14	D	F	30	A	M
15	D	F	31	E	M
16	A	F	32	A	F

Solution:

Step 1 We use tally marks to determine the totals for each of the ten gender-grade combinations.

	Grade				
Gender	A	B	C	D	E
M	||	|||	|||||	||	||
F	|||	JH	||||	||||	|

Step 2 Next we find the totals for the two rows, five columns, and ten gender/grade combinations.

	Grade					
Gender	A	B	C	D	E	Total
M	2	3	6	2	2	15
F	3	5	4	4	1	17
Total	5	8	10	6	3	32

◼

APPLICATION 2.9

A study of college faculty was undertaken to consider their attitudes toward collective bargaining between administration and labor unions. The results are summarized in the following table.

Faculty Rank	Attitude Toward Collective Bargaining			
	Favor	Oppose	Unsure	Total
Professor	45	8	2	55
Associate Prof.	31	16	3	50
Assistant Prof.	42	19	4	65
Instructor	12	4	14	30
Total	130	47	23	200

Use the table to answer each of the following:

    a. What percentage of the faculty oppose collective bargaining?

    b. What percentage of the faculty are associate professors?

    c. What percentage of the professors favor collective bargaining?

    d. What percentage of instructors favor collective bargaining?

    e. What percent of those faculty that oppose collective bargaining are professors?

    f. What percent of the total faculty are represented by assistant professor or above who favor collective bargaining?

Solution:

    a. $45/200 = 23.5\%$.

    b. $50/20 = 25\%$.

    c. $45/55 = 81.8\%$.

d. $12/30 = 40\%$.

e. Forty-seven faculty oppose collective bargaining and, of these, 8 are professors. Therefore, the professors make up $8/47 = 17.02\%$ of those that oppose collective bargaining.

f. $(42 + 31 + 45)/200 = 59\%$. ▪

EXERCISE SET 2.2

Basic Skills

1. For each of the following sets of conditions, determine the class width w.
   a. $L = 17, U = 81, c = 8$
   b. $L = 14.5, U = 102.3, c = 7$
   c. $L = 23, U = 204, c = 11$
   d. $L = 23.65, U = 67.24, c = 10$
   e. $L = 13.6, U = 73.6, c = 12$

2. For each of the following sets of conditions, determine the class width w.
   a. $L = 27, U = 87, c = 7$
   b. $L = 24.3, U = 112.5, c = 9$
   c. $L = 39, U = 130, c = 13$
   d. $L = 13.64, U = 75.24, c = 10$
   e. $L = 15.2, U = 75.2, c = 12$

3. For each of the following sets of conditions, determine the upper limit for the first class interval.
   a. $L = 17, U = 81, c = 8$, unit of measurement $= 1$
   b. $L = 14.5, U = 102.3, c = 7$, unit of measurement $= 0.1$
   c. $L = 23, U = 204, c = 11$, unit of measurement $= 1$
   d. $L = 23.65, U = 67.24, c = 10$, unit of measurement $= 0.01$
   e. $L = 13.6, U = 73.6, c = 12$, unit of measurement $= 0.1$

4. For each of the following sets of conditions, determine the upper limit for the first class interval.
   a. $L = 27, U = 87, c = 7$, unit of measurement $= 1$
   b. $L = 24.3, U = 112.5, c = 9$, unit of measurement $= 0.1$
   c. $L = 39, U = 130, c = 13$, unit of measurement $= 1$
   d. $L = 13.64, U = 75.24, c = 10$, unit of measurement $= 0.01$

e. $L = 15.2, U = 75.2, c = 12$, unit of measurement $= 0.1$

5. Refer to exercise 3. For each set of conditions, determine the boundaries for the first class interval.

6. Refer to exercise 4. For each set of conditions, determine the boundaries for the first class interval.

7. Using the following grouped frequency table, construct
   a. a relative frequency table.
   b. a cumulative frequency table.
   c. a cumulative relative frequency table.

Class	f
1–4	14
5–8	18
9–12	12
13–16	16
17–20	20

8. Using the following grouped frequency table, construct
   a. a relative frequency table.
   b. a cumulative frequency table.
   c. a cumulative relative frequency table.

Class	f
10–15	13
16–21	10
22–27	9
28–33	17
24–39	22
40–45	6

9. Using the accompanying ungrouped frequency table, construct
   a. a relative frequency table.
   b. a cumulative frequency table.
   c. a cumulative relative frequency table.

x	f
4	1
7	3
8	6
9	4
10	2

10. Using the accompanying ungrouped frequency table, construct
    a. a relative frequency table.
    b. a cumulative frequency table.
    c. a cumulative relative frequency table.

x	f
12	8
15	10
20	7
22	13
35	10
40	2

11. For the accompanying table, identify the
    a. class marks.
    b. class boundaries.

Class	f
1–4	14
5–8	18
9–12	12
13–16	16
17–20	20

12. For the accompanying table, identify the
    a. class marks.
    b. class boundaries.

Class	f
10–15	13
16–21	10
22–27	9
28–33	17
34–39	22
40–45	6

13. Use Sturges' rule to determine the number of classes for a data collection having a size equal to
    a. 25.
    b. 50.
    c. 75.
    d. 100.
    e. 500.

14. Use Sturges' rule to determine the number of classes for a data collection having a size equal to
    a. 35.
    b. 80.
    c. 95.
    d. 200.
    e. 1000.

Further Applications

15. The following data represent the amounts (in dollars) spent by a sample of 25 students for snacks during a final examination period.

57	28	63	38	29	89	77	72	39
47	64	84	88	42	36	72	69	
68	41	52	39	72	45	52	84	

    By using six classes, construct a grouped frequency table.

16. The following observations represent the speeds in miles per hour (mph) of 30 cars recorded by police radar on a heavily traveled interstate highway:

57	63	70	53	61	60	67	79	64	62
66	73	71	78	84	53	48	80	54	60
67	65	62	55	52	69	73	72	66	58

    By using seven classes, construct a grouped frequency table.

17. Refer to the data given in exercise 15. Use Sturges' rule to construct a grouped relative frequency table.

18. Refer to the data given in exercise 16. Use Sturges' rule to construct a grouped relative frequency table.

19. The noon temperatures (in degrees Fahrenheit) recorded on July 1 for the pat 28 years in a small town are as follows:

66	83	77	90	78	84	83	80	77	79
75	88	72	66	83	85	94	88	79	79
72	78	76	84	81	73	80	90		

    Use this data and six classes to construct a grouped cumulative frequency table.

20. People who own their homes in the United States appear to change homes often. The accompanying data indicate how long (in years) owners in the 50 states stayed put in 1988:[9]

26.3	20.8	19.2	18.5	18.5	18.2	17.2	16.1
15.9	15.2	14.7	14.5	14.1	14.1	13.9	
13.7	13.5	13.3	13.3	13.2	12.3	12.3	
12.2	12.2	12.2	12.0	11.9	11.6	11.5	
11.5	11.4	11.4	11.4	11.1	11.0	10.9	
10.9	10.3	10.2	10.2	9.9	9.8	9.4	
9.1	8.8	8.5	8.5	8.4	8.0	7.9	

Use this data and seven classes to construct a grouped cumulative frequency table.

21. The accompanying data represent a sample of prices (in cents) charged for unleaded gasoline in a particular city during a certain month.

123.9	127.9	130.9	121.9	132.9	120.8	115.9
117.9	131.9	121.9	126.9	122.8	126.9	
137.9	115.9	115.9	121.9	126.9	119.9	
118.9	119.8	116.9	129.9	122.8	119.9	

Use this data and five classes to construct a grouped cumulative relative frequency table.

22. By using the data in exercise 20 and five classes, construct a grouped cumulative relative frequency table.

23. A sample of 30 students were asked how many books they purchased for classes last semester. Their responses are as follows:

5	6	5	5	4	5	4	5	3	6	4	4	4	6	2
9	5	4	3	3	8	11	7	8	7	4	10	4	3	6

a. Construct an ungrouped relative frequency table.
b. Construct an ungrouped cumulative relative frequency table.

24. The accompanying data represent the percentage of household income spent in 1988 on food in the largest metropolitan areas in the United States.[10]

14	13	29	22	14	12	15	12	15	16
16	16	16	17	17	12	12	11	14	12
12	9	19	17	16	11	13	15	14	12
14	15	13	13	11	13	14	13	13	12
12	14	15	11	14	11	12	13	12	11

a. Construct an ungrouped relative frequency table.
b. Construct an ungrouped cumulative relative frequency table.

25. Students in a small school were classified according to class standing and music preference. The results are recorded in the accompanying table.

Music Preference	Class Standing				
	Freshman	Sophomore	Junior	Senior	Total
Rock	16	11	7	6	40
Country	10	12	3	5	30
Classical	3	1	2	4	10
Jazz	23	11	2	4	40
Folk	3	0	6	1	10
Total	55	35	20	20	130

a. What percentage of the freshmen prefer classical music?
b. What percentage of the rock music fans are sophomores?
c. What percentage of the student body prefer country music?
d. What percentage of the student body are juniors?
e. What percentage of the student body are juniors or seniors?
f. What percentage of the student body prefer country or folk music?

26. A sample of voters were polled concerning their preference among three candidates running for mayor. The results are recorded by gender in the following table.

Gender	Candidate			
	A	B	C	Total
Male	15	16	4	35
Female	5	4	1	10
Total	20	20	5	45

a. What percentage of the voters prefer candidate B?
b. What percentage of the voters are male?
c. What percentage of the males prefer candidate C?
d. What percentage of the voters who prefer candidate A are female?
e. What percentage of the voters are female and prefer candidate B?
f. What percentage of the voters prefer candidates A or C?

27. The following data represent the monthly telephone bills (in dollars) of 25 residents of a small community:

19.80	36.05	28.50	21.48	21.15
25.12	23.47	27.81	26.66	20.35
30.22	25.49	20.80	23.83	25.35
23.48	25.81	26.83	20.77	19.98
35.87	22.02	21.07	30.96	33.38

a. What percentage of the group paid over $20.00?
b. What percentage of the group paid over $24.00 but less than $28.00?

28. Refer to the data given in exercise 20 to determine
a. what percentage of U.S. homeowners stayed in their homes over 10 years.
b. what percentage of U.S. homeowners stayed in their homes more than 12 years but less than 20 years.

29. Consider the following grouped frequency table.

Classes	f
4.5–9.4	2
9.5–14.4	3
14.5–19.4	4
19.5–24.4	1
24.5–29.4	8

a. Find w, the width of each class.
b. Find the five class marks.
c. Find the boundaries for the first class.
d. What percentage of the data are greater than 19.45?
e. What percentage of the data are less than 24.45?
f. What percentage of the data fall in the class 14.5–19.4?

Going Beyond

30. The following data represent the weekly amounts (in dollars) spent on food by 50 newlywed couples.

57.10	70.89	59.17	60.08	49.16
56.17	66.94	67.08	58.10	71.28
50.25	46.39	55.01	68.81	58.70
69.48	46.02	54.16	65.07	48.09
58.32	50.82	45.43	57.20	62.30
51.42	58.76	46.37	47.16	63.51
55.45	58.14	57.14	58.63	55.14
60.37	52.41	74.13	62.38	51.15
64.00	47.75	52.59	42.73	60.32
48.10	59.62	59.46	57.16	58.19

Construct a grouped frequency table having the smallest number of classes for which the class width is $w = 2.75$.

31. The miles-per-gallon calculations for 40 fuel refills of a new automobile are as follows:

26.6	28.7	29.2	26.4	29.3	25.8	28.7	29.0
30.0	28.1	28.3	27.8	27.6	31.9	26.6	28.4
30.3	30.4	29.2	29.3	26.5	28.7	28.8	28.3
28.8	27.1	28.9	31.2	30.2	29.2	30.3	32.0
28.4	30.3	29.5	28.4	27.4	30.8	29.5	31.5

a. Construct a grouped cumulative relative frequency table with eight classes.
b. Use this table to approximate the 50th percentile.

32. An efficiency study was conducted by a supermarket. The following data represent the lengths of time (in minutes) required to service 50 customers at checkout:

3.5	1.8	2.3	0.7	5.2	0.9	0.9	0.9	3.0	1.1
1.2	2.3	1.7	3.2	1.7	0.4	1.4	0.7	1.2	0.7
1.6	0.3	1.0	1.0	0.5	0.6	2.8	2.4	0.3	3.1
0.8	1.2	1.7	1.2	0.2	4.0	2.5	1.9	0.8	1.2
0.2	1.3	0.6	0.6	1.8	0.7	1.5	1.3	1.4	1.1

a. Construct a relative frequency table with ten classes having the smallest class width.
b. What percentage of the data fall in the first class?
c. What percentage of the data fall in the first class or the last class?

33. For the data in exercise 21, construct a grouped frequency table with seven classes having the smallest possible width and class marks that end in 0.9 cent.

34. The following data represent grade-point averages (based on a five-point system) last semester for a group of college juniors majoring in psychology.

1.3	1.4	2.0	3.7	2.4	3.6	4.0	3.8
1.9	1.1	1.7	2.3	4.2	4.3	3.4	3.0
4.0	4.1	3.6	2.1	2.0	1.5	2.7	2.6
3.6	1.2	3.5	3.3	4.0	3.8	2.6	1.9

Construct a grouped frequency table having eight classes with the smallest possible width that agrees with the measurement unit, and begin the first class with 1.0.

SECTION 2.3   ***Graphic Representation of Data***

A **graph** is a pictorial means for portraying and summarizing data. A pictorial presentation of the data often makes certain features of the data more apparent than a frequency table. One result of presenting data in graphical form is that new characteristics of the data are often discovered. Graphical displays of data have enjoyed increased use in the media; this is due, in part, to the popularity and use of computer graphics. Graphs are of many types. Some of the more common types are the circle graph, bar graph, line graph, stem-and-leaf diagram, histogram, and ogive. We will discuss each of these in some detail.

Bar and Circle Graphs

Two of the commonly used graphical types are **bar graphs** and **circle graphs.** Both types are generally used for categorical or nominal data. Circle graphs are used only to portray parts of a total; they are very popular for displaying budget information.

Both types of graphs will be illustrated using the data in table 2.16, which shows the recipients of charitable giving by U.S. citizens in 1983.[11] Figure 2.3 represents a vertical-bar graph for the data in table 2.16.

TABLE 2.16

Recipients of Charitable Giving

Recipients	Amount (in billions of dollars)
Religion	31.0
Arts and humanities	4.1
Social services	6.9
Education	9.0
Health	9.2
Other	4.7

FIGURE 2.3

Vertical-bar graph of recipients of 1983 charitable giving

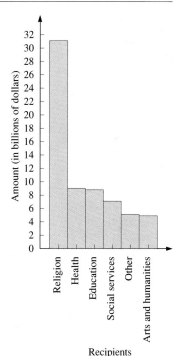

Table 2.17 organizes the computations necessary for constructing a circle graph for the charity-recipient data. Each entry in the percent column was obtained by dividing the amount by the total amount (64.9) and then multiplying by 100. Entries in the degrees column were obtained by multiplying the entries in the percent column by 360°, the number of degrees in a circle.

Table 2.17

Computations for Constructing Circle Graph for the Data of Table 2.16

Recipient	Amount	%	Degrees	
Religion	31.0	47.8	172.1	← (0.478 × 360)
Health	9.2	14.2	51.1	
Education	9.0	13.9	50.0	
Social Services	6.9	10.6	38.2	← $\left(\dfrac{6.9}{64.9}\right)(100)$
Arts & Humanities	4.1	6.3	22.7	
Other	4.7	7.2	25.9	
Totals	64.9	100.0	360.0	

The circle graph in figure 2.4 was constructed using a protractor, a device for measuring angles, and the information from table 2.17. We can see from a glance at the circle graph that religion got the biggest share—an amount approximately equal to the total amount for the remaining recipients. Arts and humanities got the smallest share.

FIGURE 2.4

Circle graph for recipients of 1983 charitable giving

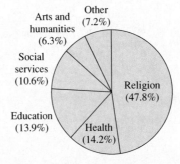

Many computer programs for data analysis allow the computer to draw circle graphs when given percentages or raw data.

Stem-and-Leaf Diagrams

The use of a grouped frequency table has one rather obvious disadvantage: the original data are lost in the grouping process. A **stem-and-leaf diagram** can be used to overcome this limitation. Stem-and-leaf diagrams offer a quick and novel way for displaying numerical data. If a numeral has two or more digits, then it can be split into a stem and a leaf. A *stem* is the leading digit (part) of the numeral, while the *leaf* constitutes the trailing digit(s). For example, the numeral 278 could be split two ways:

$$
\begin{array}{ccc}
2 & | & 78 \\
\uparrow & & \uparrow \\
\text{Stem} & & \text{Leaf}
\end{array}
\qquad
\begin{array}{ccc}
27 & | & 8 \\
\uparrow & & \uparrow \\
\text{Stem} & & \text{Leaf}
\end{array}
$$

A graphical display for data classified using stems and leaves is very simple to accomplish. Each piece of data serves as a leaf on some stem.

EXAMPLE 2.24

Let's construct a stem-and-leaf diagram for the following collection of 25 algebra test scores:

$$78 \quad 67 \quad 65 \quad 87 \quad 75 \quad 65 \quad 71 \quad 54 \quad 94$$
$$64 \quad 84 \quad 82 \quad 81 \quad 68 \quad 85 \quad 76 \quad 89$$
$$98 \quad 59 \quad 57 \quad 79 \quad 65 \quad 59 \quad 80 \quad 67$$

Since all the scores fall between 50 and 99, let's use the tens digit in each case to serve as the stem and the units digit to serve as the leaf.

Step 1 Place the stems in a vertical array using a vertical line segment, called the *trunk,* to separate the leaves from the stems.

```
5 |
6 |
7 |
8 |
9 |
```

Step 2 Place each leaf to the right of its stem. Since the first score is 78, we place the leaf, 8, on its stem, 7.

```
          5 |
          6 |
Stem → 7 | 8 ← Leaf
          8 |
          9 |
```

Continuing this process for each score, we obtain the following stem-and-leaf diagram:

```
5 | 9 7 4 9
6 | 4 5 7 5 7 8 5
7 | 8 6 1 9 5
8 | 5 4 2 9 7 1 0
9 | 8 4
```

By observing the above stem-and-leaf diagram we can conclude that

1. The largest score is 98.

2. The smallest score is 54.

3. The scores vary from 54 to 98.

4. Stem 9 has the fewest leaves.

5. Stems 6 and 8 contain the most leaves with seven leaves on each stem.

6. The total number of leaves represents the size of the data set.

Note that it doesn't make any difference in which order the leaves are placed on a stem. If the leaves on each stem are ordered from smallest to largest, the diagram is called an *ordered stem-and-leaf diagram*. An ordered stem-and-leaf diagram for the algebra-test scores is as follows:

```
5 | 4 7 9 9
6 | 4 5 5 5 7 7 8
7 | 1 5 6 8 9
8 | 0 1 2 4 5 7 9
9 | 4 8
```

An ordered stem-and-leaf diagram is useful for ordering data and computing position points. A **position point** is a number that has a certain percent of the data falling below it. For example, with some simple additions, one can quickly observe that the 40% position point for the algebra scores data set is 68; that is, $(0.40)(25) = 10$ scores fall below a score of 68. An ordered stem-and-leaf diagram represents a graphical display corresponding to a frequency table.

Sometimes it may be desirable to include fewer than ten leaf values on a single stem to spread out the data, particularly if a few stems contain a large number of leaves. When this is done, we have a visual display that corresponds to a grouped frequency table. When the number of classes is increased, certain salient characteristics of the data may become more apparent. If each stem in a stem-and-leaf diagram is split into two stems, called *substems,* each having the same number of leaf values, the resulting stem-and-leaf diagram is referred to as a **double stem diagram.** Applications 2.10 and 2.11 illustrate additional modifications of the construction procedures for stem-and-leaf diagrams.

EXAMPLE 2.25

Let's use the collection of 25 test scores and construct a double stem stem-and-leaf diagram with each stem having five possible leaf values. This is accomplished by splitting each stem into two substems, called a and b. Substem a will contain digits 0 through 4 as its leaves and substem b will contain digits 5 through 9 as its leaves. For example, substem 5a will contain leaves 0–4 and substem 5b will contain leaves 5–9. The resulting double-stem diagram is shown here.

5a	4
5b	7 9 9
6a	4
6b	5 5 5 7 7 8
7a	1
7b	5 6 8 9
8a	0 1 2 4
8b	5 7 9
9a	4
9b	8

APPLICATION 2.10

A national survey by utility regulators found that power costs vary widely across the United States. The power costs in the 25 most expensive cities (rated by average cost in cents per kilowatt hour) in 1984 are as follows:

16.5	14.3	14.3	13.9	13.8	11.2	11.1	11.1	10.8
13.1	12.8	12.1	12.0	11.8	10.8	10.8	10.7	
11.6	11.4	11.3	11.3	11.2	10.8	10.6	10.6	

Construct a stem-and-leaf diagram for the data.

Solution: We will ignore the decimal points in the data; each value in the final array can be converted back to its original value by multiplying by 0.1. Also, the numbers will be treated as three-digit numbers ranging from 106 to 165. By using double-digit stems, we get the following ordered diagram:

```
10 │ 6 6 7 8 8 8 8
11 │ 1 1 2 2 3 3 4 6 8
12 │ 0 1 8
13 │ 1 8 9
14 │ 3 3
15 │
16 │ 5
```

Leaf 5 on stem 16 represents 16.5 cents. We can easily determine that 20% of the average costs are above 13.1 cents. It would not be wise in this application to use double-digit leaves and one-digit stems, since all the leaves would be on one stem. What good is a stem-and-leaf diagram with only one stem? ∎

Computer display 2.1 illustrates the use of MINITAB to construct a stem-and-leaf diagram for the data in application 2.10.

Computer Display 2.1

```
MTB > SET C1
DATA> 16.5 14.5 14.3 13.9 13.8 13.1 12.8 12.1 12.0 11.8
DATA> 11.6 11.4 11.3 11.3 11.2 11.2 11.1 11.1 10.8 10.8
DATA> 10.8 10.8 10.7 10.6 10.6
DATA> END
MTB > STEM C1 ;
SUBC> INCREMENT = 1 .

STEM-AND-LEAF OF C1    N = 25
LEAF UNIT = 0.10


  7   10  6678888
 (9)  11  112233468
  9   12  018
  6   13  189
  3   14  33
  1   15
  1   16  1
```

Notice in computer display 2.1 that STEM is shorthand for stem-and-leaf diagram. Only the first four letters of a command need be entered by the user. Also notice that the STEM-AND-LEAF command has several subcommands. One of the available subcommands is INCREMENT, which is used to specify the stems. To use an available subcommand with a command, we must place a semicolon at the end of the command. MINITAB then responds with the subcommand prompt, SUBC>. The subcommand is entered and followed by a period. The period informs MINITAB that no further subcommands will follow and the main command and subcommand are to be executed. If the period does not follow the end of the subcommand, MINITAB will respond with another subcommand prompt, SUBC>.

Three columns are used by MINITAB in the output. The left-most column, called **depths,** tells how many leaves lie on that line or "beyond." For example, the 9 on the third line from the top means there are nine leaves on that line or below it. The 3 on the third line from the bottom means that there are three leaves on that line or below it. The line with the parentheses will contain the middle observation if the total number of observations is odd and the middle two observations if the number of observations is even. The parentheses enclose a count of the number of leaves on this line. (The second row down from the top contains nine values and the median. Seven values fall above the second row and nine values fall below the second row). The second column displays the stems, while the numbers to the right of the stem are the leaves.

Computer display 2.2 contains a double stem-and-leaf diagram for the data in application 2.10. Notice that the subcommand INCREMENT = 1 was not used in computer display 2.2 and a double stem-and-leaf diagram was constructed.

Computer Display 2.2

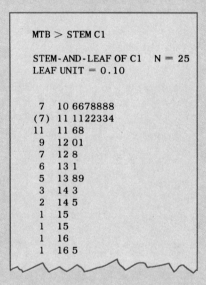

```
MTB > STEM C1

STEM-AND-LEAF OF C1    N = 25
LEAF UNIT = 0.10


   7    10 6678888
  (7)   11 1122334
  11    11 68
   9    12 01
   7    12 8
   6    13 1
   5    13 89
   3    14 3
   2    14 5
   1    15
   1    15
   1    16
   1    16 5
```

APPLICATION 2.11

The following data represent the one-year percentage changes in the prison populations at 25 federal and state prisons:[12]

0.6	12.9	10.8	11.7	0.4	−11.1	0.6	2.5	0.2	−4.4	−1.4	−3.2
−1.7	−1.2	7.0	−10.1	19.2	20.6	−0.5	9.8	2.1	16.3	8.8	20.8	4.1

Construct a stem-and-leaf diagram for the data.

Solution: Ignoring decimals, we observe that the data range from −111 to 208. Let's use stem values of −1, −0, +0, 1, and 2. To make all of the values three-digit numbers, we put a zero in front of any two-digit value. Thus, 4.1 is represented as 041. We need two stems for zero to indicate the signs of numbers. For example, the stem for the value 0.6 is +0, the stem for 7.0 is +0, and the stem for −1.7 is −0. The value 0.6 must be represented in the stem-and-leaf diagram as +006; 7.0 must be represented as +070; and −1.7 must be represented as −017. The stem-and-leaf diagram is shown here

−1	01 11
−0	05 12 14 17 32 44
+0	02 04 06 06 21 25 41 70 88 98
1	08 17 29 63 92
2	06 08

Remember that each value in the diagram must be converted by multiplying by 0.1 before interpretations are made.

The leaves in this application could also contain decimals. In this case we would not need to multiply the values in the diagram by 0.1 before making interpretations are made. The corresponding stem-and-leaf diagram is

−1	0.1 1.1
−0	0.5 1.2 1.4 1.7 3.2 4.4
+0	0.2 0.4 0.6 0.6 2.1 2.5 4.1 7.0 8.8 9.8
1	0.8 1.7 2.9 6.3 9.2
2	0.6 0.8 ▪

Stem-and-leaf diagrams are also useful in other adaptations as indicated by the following two situations.

1. Two similar distributions can be compared if they have common stems. In this case the leaves of one diagram might be placed to the right of the stems and the leaves of the other diagram might be placed to the left of the stems as shown here.

Leaves	Stem	Leaves
8 6	5	3 6 8
9 8 7 3	6	2 7
8 6	7	3 5 5 7 6
5 4 3 0	8	3 3 4 5

2. More than two distributions can be compared by arranging the diagrams in column form if they share common stems. The stems could be placed at the extreme right of such a diagram and the leaves attached as shown in the following diagram:

Stem	Distribution 1 Leaves	Distribution 2 Leaves	Distribution 3 Leaves
3	2 7	0	1 1
4	4 4 5	6 8	
5	7 9	3 4 7	5 6 8 9
6	3		3 4

Histograms

A **histogram** is a type of bar graph for a frequency distribution. Histograms can be constructed for grouped and ungrouped frequency distributions. We will first consider histograms for ungrouped frequency distributions.

Ungrouped Frequency Distributions

The general idea of constructing a histogram for ungrouped frequency data is to represent each frequency by a bar whose area is proportional to the frequency. Typically, the width of each bar is chosen to be 1 so that the area of the bar is equal to the frequency of the measurement.

A P P L I C A T I O N 2.12   The number of school-age children in each of a sample of 50 families is contained in the accompanying table. Construct a histogram for the data.

Number of School-Age Children	Frequency (f)
0	15
1	8
2	14
3	9
4	4

FIGURE 2.5

Histogram for frequency data contained in application 2.12

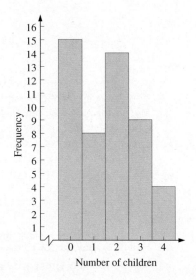

Solution: Our histogram will contain five bars. We will locate the number of school-age children along the horizontal axis and the frequency along the vertical axis. We will place the zero point (0) on the horizontal axis to the right of its usual position (the intersection of the two axes). This will enable us to center the bars over the values so that the vertical axis will not get in the way of the first bar. If we choose the width of each bar to be 1 and the height of each bar to be the frequency, then the area of each bar will be equal to the product of the frequency and 1. The sum of the areas of the five bars will be equal to the sum of the frequencies. The histogram is pictured in figure 2.5. Note that the horizontal axis is broken to draw attention to the fact that the horizontal scale does not start at zero. We break the axis to indicate that we are not deliberately trying to distort perspective. ■

Histograms for Grouped Frequency Data

Two steps are typically followed when constructing a histogram for data measured using an interval or ratio scale.

1. The data are organized into a grouped frequency table.
2. A bar graph is constructed using the class boundaries to locate the bar and the frequencies to indicate the heights of the bars.

APPLICATION 2.13

The following grouped frequency table represents the extent of unemployment (in percentages) for 27 eastern cities.[13]

Extent of Unemployment (in percentages)	Number of Cities
3.7–5.1	5
5.2–6.6	12
6.7–8.1	6
8.2–9.6	1
9.7–11.1	0
11.2–12.6	1
12.7–14.1	2
	27

Construct a histogram for the data.

Solution: The histogram is constructed by first locating the class boundaries along the horizontal axis and the frequencies on the vertical axis. For each class, a rectangular bar is constructed using the class boundaries to measure the width of the bar and the frequency to measure the height. Since all the classes in a grouped frequency table have the same width, the areas of the bars will be proportional to the heights of the bars; that is, to the frequency of the classes. To construct the histogram we follow these steps.

Step 1 We first calculate the class boundaries. Note that the unit of measurement is 0.1 (of a percent). Thus, $(0.5)(0.1) = 0.05$ is subtracted from the lower class limit of a class to find the lower class boundary and 0.05 is added to the upper class limit of a class to find the upper class boundary.

FIGURE 2.6

Histogram for unemployment data

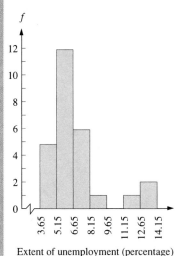

Extent of unemployment (percentage)

Class	Boundaries	f
3.7–5.1	3.65–5.15	5
5.2–6.6	5.15–6.65	12
6.7–8.1	6.65–8.15	6
8.2–9.6	8.15–9.65	1
9.7–11.1	9.65–11.15	0
11.2–12.6	11.15–12.65	1
12.7–14.1	12.65–14.15	2
		27

Step 2 A bar graph is now constructed using the class boundaries and frequencies. The boundaries are located along the horizontal axis and the frequencies are located along the vertical axis as shown in figure 2.6. ■

A histogram improves our ability to compare the frequencies of corresponding classes; a class frequency can be easily compared with those of its neighboring classes. We can immediately see that the second class in the histogram illustrated in figure 2.6 has the largest frequency and that the frequency of this class is twice the frequency of the third class. For the classes that measure unemployment between 8.15% and 11.15%, there is a rapid decline in the number of cities represented.

The shape of a histogram may change dramatically with a change in the number of intervals n or the width of the intervals w. For this reason, we should be careful about drawing conclusions about the shape of the sample distribution.

EXAMPLE 2.26

The three histograms shown in figure 2.7 represent a sample of 100 measurements for different values of n and w. The histogram in part (a) has $n = 5$ and $w = 9.95$; the histogram in part (b) has $n = 8$ and $w = 6.22$; and the histogram in part (c) has $n = 5$ and $w = 4.60$. Notice how the appearance changes as the number of intervals and class width vary.

FIGURE 2.7

Effect of changes in number of intervals and class width on appearance of histogram. The horizontal axis represents lengths in inches.

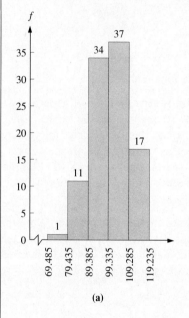

(a)

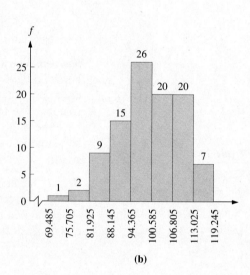

(b)

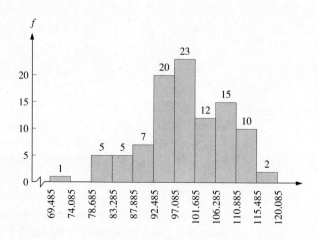

(c)

MINITAB can be used to construct a histogram from raw data. Computer display 2.3 contains a histogram constructed with MINITAB for the accompanying data representing the ages (in years) of a sample of 40 tourists traveling to Japan by United Airlines during a recent one-month period.

67	18	63	74	28	44	60	69	44	66
36	26	50	34	44	41	58	68	43	51
62	43	54	63	71	62	54	65	61	52
60	61	45	66	80	72	61	57	65	70

Computer Display 2.3

```
MTB > SET C1
DATA> 67 18 63 74 28 44 60 69 44 66
DATA> 36 26 50 34 44 41 58 68 43 51
DATA> 62 43 54 63 71 62 54 65 61 52
DATA> 60 61 45 66 80 72 61 57 65 70
DATA> END
MTB > HIST C1;
SUBC> INCREMENT = 8;
SUBC> START = 21.5.

HISTOGRAM OF C1  N = 40

MIDPOINT        COUNT
  21.50         1 *
  29.50         2 **
  37.50         3 ***
  45.50         6 ******
  53.50         6 ******
  61.50        12 ************
  69.50         8 ********
  77.50         2 **
```

Notice that the subcommand INCREMENT = 8 specifies that the width of each class interval is to be 8, and the subcommand START = 21.5 specifies that the midpoint of the first class is to be 21.5. Hence, the first class begins with 18 and the first class interval is 18–25. MINITAB does not display the intervals, only the midpoints of the intervals.

Relative Frequency Histograms

A **relative frequency histogram** can be constructed by changing the vertical scale of a frequency histogram. Instead of beginning with a group frequency table, we begin with a grouped relative frequency table. The height of the bars on a relative frequency histogram will indicate the *proportion* of the total represented by each class. The basic shape of a relative frequency histogram is similar to the shape of the corresponding frequency histogram.

EXAMPLE 2.27

For the data in application 2.13 we have the relative frequency table shown in table 2.18 and the corresponding relative frequency histogram depicted in figure 2.8.

Table 2.18

Relative Frequency Table for Unemployment Data

Class	Boundaries	f	Rel f
3.7–5.1	3.65–5.15	5	0.19
5.2–6.6	5.15–6.65	12	0.44
6.7–8.1	6.65–8.15	6	0.22
8.2–9.6	8.15–9.65	1	0.04
9.7–11.1	9.65–11.15	0	0.00
11.2–12.6	11.15–11.65	1	0.04
12.7–14.1	12.65–14.15	2	0.07
		27	1.00

FIGURE 2.8

Relative frequency for unemployment data of application 2.13

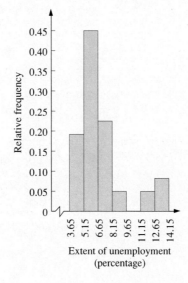

Line Graphs or Frequency Polygons

A **line graph** or **frequency polygon** is constructed using a grouped frequency table with class marks. The line graph offers a useful alternative to the histogram; the choice of which to use is usually a personal matter. A line graph creates the impression that the frequencies change more smoothly, whereas a histogram suggests that the frequencies change abruptly. A line graph or frequency polygon can be constructed for data displayed in a grouped frequency table by identifying each class mark and its corresponding frequency (X, f) with a point on the graph. These points are then joined by a sequence of line segments as seen in application 2.14.

APPLICATION 2.14

The following grouped frequency table reports the average annual salaries (to the nearest $100) of factory workers in 27 eastern cities.[14] Construct a frequency polygon for the data.

Average Salary	Number of Cities
$12,500–14,300	1
14,400–16,200	5
16,300–18,100	3
18,200–20,000	7
20,100–21,900	6
22,000–23,800	1
23,900–25,700	3
25,800–27,600	1

Solution:

Step 1 We first find the class marks, designated by X.

Average Salary	f	X
$12,500–14,300	1	13,400
14,400–16,200	5	15,300
16,300–18,100	3	17,200
18,200–20,000	7	19,100
20,100–21,900	6	21,000
22,000–23,800	1	22,900
23,900–25,700	3	24,800
25,800–27,600	1	26,700

Step 2 We now construct the line graph shown in figure 2.9. The class marks are located on the horizontal axis and the frequencies are located on the vertical axis. Note that the line graph is "tied down" at both ends by connecting the first and last points to the horizontal axis at points a distance of $w = 1900$ from the nearest class marks.

FIGURE 2.9

Average salaries of factory workers

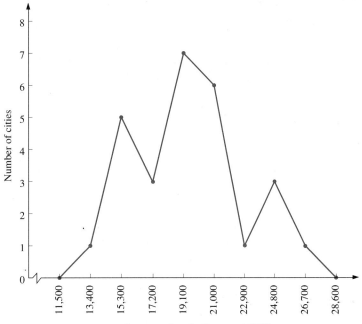

Number of cities

Average salary (to the nearest $100)

The following salient characteristics of the data are shown in the line graph of figure 2.9.

1. Most of the cities fall between the extreme ends of the scale. Only one city has factory workers earning an average annual salary of approximately $13,400, and only one city has factory workers earning an average annual salary of approximately $26,700.

2. The data appear to be centered at approximately $19,000. ■

Ogives

A line graph constructed from a cumulative frequency table or a cumulative relative frequency table is called an **ogive** (pronounced ō'jiv). Ogives offer a graphical means for interpolating or approximating the number or percentage of observations less than or equal to a specified value.

For an ogive, an upper class boundary and its corresponding cumulative frequency (or cumulative relative frequency) are used to locate a point. Consecutive points are then joined by line segments. Cumulative frequencies (or cumulative relative frequencies) are always located on the vertical axis. Application 2.15 illustrates the construction of an ogive.

APPLICATION 2.15   Construct an ogive using cumulative frequencies for the data in application 2.14.

Solution:

Step 1 We first find the cumulative frequencies.

Average Income	Upper Boundaries	f	Cumulative Frequency
$12,500–14,300	14,350	1	1
14,400–16,200	16,250	5	6
16,300–18,100	18,150	3	9
18,200–20,000	20,050	7	16
20,100–21,900	21,950	6	22
22,000–23,800	23,850	1	23
23,900–25,700	25,750	3	26
25,800–27,600	27,650	1	27

Step 2 We use the boundaries of the classes to locate the points on the horizontal axis and the frequencies for the vertical axis.

Step 3 We construct the ogive (figure 2.10). Notice that the cumulative frequency for the lower boundary of the first class is 0. We can determine at a glance the number of cities having average salaries for factory workres below any specified amount. ■

FIGURE 2.10

An ogive for annual incomes of factory workers for 27 eastern cities

Using Ogives to Determine Percentiles

A cumulative relative frequency ogive can be used to determine percentiles as described in application 2.16.

APPLICATION 2.16

For the data in application 2.15, construct a cumulative relative frequency ogive and use it to approximate the 50th percentile (P_{50}) and the 75th percentile (P_{75}). Recall that the 75th percentile is the measurement below which 75% of the measurements fall.

Solution:

Step 1 We first find the cumulative relative frequencies using the cumulative frequencies.

Average Income	Upper Boundary	Cum f	Rel Cum f
$12,500–14,300	14,350	1	0.037
14,400–16,200	16,250	6	0.222
16,300–18,100	18,150	9	0.333
18,200–20,000	20,050	16	0.593
20,100–21,900	21,950	22	0.815
22,000–23,800	23,850	23	0.852
23,900–25,700	25,750	26	0.963
25,800–27,600	27,650	27	1.000

Step 2 We use the boundaries of the classes to locate the points on the horizontal axis and use the relative cumulative frequencies for the vertical axis.

Step 3 We construct the ogive (figure 2.11). We see that P_{50}, the 50th percentile, is between $18,150 and $20,050—approximately $19,500—and that P_{75}, the 75th percentile, is a little less than $22,000. Hence, approximately 50% of the cities have factory workers earning an average salary less than $19,500, and 75% of the cities have factory workers earning an average salary less than $22,000. As a result, approximately 25% of the factory workers earn between $19,500 and $22,000. ■

FIGURE 2.11

Ogive of incomes of factory
workers

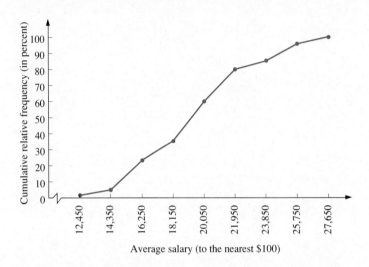

Histograms, Ogives, and
the Shapes of Populations

Histograms and ogives for sample data give the researcher an indication of the shape of the population from which the sample was selected. The histogram of a sample suggests the shape of the corresponding population frequency curve. A relative frequency histogram for a sample should have a shape similar to the relative frequency distribution for the population, and an ogive for a sample should have approximately the same shape as the ogive for the population. Since populations are often represented by relative frequency curves or cumulative relative frequency curves, it is important that we understand their sample counterparts.

EXAMPLE 2.28

Suppose an automatic bottle filler in a brewery is set to deliver 12 ounces of beer to each bottle. A sample of 50 bottles revealed the following contents in ounces:

12.335	12.111	12.166	11.900	11.889	12.057	11.848	
12.151	11.717	11.584	12.497	12.083	12.018	11.704	
12.187	12.082	12.491	11.929	11.743	12.035	12.335	
12.520	11.988	12.080	12.001	11.990	11.748	12.103	
12.185	12.100	11.846	12.240	12.339	11.611	11.856	
11.629	11.912	11.786	11.853	11.655	12.101	11.886	
12.410	11.956	12.108	11.923	11.853	11.919	12.130	12.408

The histogram in figure 2.12(a) illustrates the distribution of bottle contents for the sample of 50 bottles of beer. The histogram for the sample approximates a bell-shaped population (called a *normal distribution,* which we will examine in detail in chapter 7) illustrated in figure 2.12(b).

FIGURE 2.12

Sample histogram, population distribution, and ogives for beer-contents data

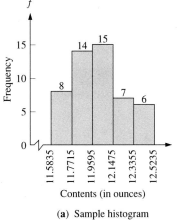

(a) Sample histogram

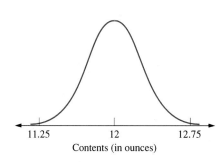

(b) Population relative frequency curve

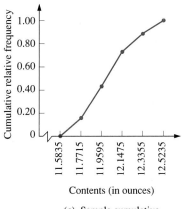

(c) Sample cumulative relative frequency curve

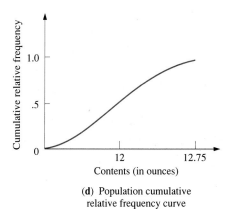

(d) Population cumulative relative frequency curve

The ogive for the sample illustrated in figure 2.12(c) approximates the S shape of the cumulative relative frequency distribution for the normal distribution illustrated in figure 2.12(d). A cumulative relative frequency distribution for a bell-shaped or normal population will always have an S-shaped appearance.

Frequently the data produced by a particular process or application will have a known form, such as a bell-shaped distribution. This information, as we shall later see, can then be used to evaluate sample data taken from the population.

EXERCISE SET 2.3

Basic Skills

1. Consider the following sample of grades:

   A C D B C C C D F F
   D F A D C B C D D B

   Construct a
   a. bar graph.
   b. circle graph.

2. Consider the following sample of move ratings:

   X R X PG PG X X R PG13 G
   G G PG13 R R R G R G PG

   Construct a
   a. bar graph.
   b. circle graph.

3. Construct a frequency histogram for the following data listed here. Use six bars.

   17 14 16   8 31 16 14   9 17 11
   25 24 28 10 48 24 12 13 43 24
   32 37 33 42 11 34 16 41 21 15

4. Construct a frequency histogram having 7 bars for the data in exercise 3.

5. Construct a frequency polygon for the data in exercise 3 using eight points (including the end points).

6. Construct a frequency polygon for the data in exercise 3 using nine points (including the end points).

7. Construct a stem-and-leaf diagram for the accompanying data.

   49 62 53 61 51 51 49 61 54 52
   48 49 62 60 45 62 49 53 51 45
   52 61 61 62 63 53 61 59 62 51
   50 50 65 54 67 66 59 72 75 62

8. Construct a stem-and-leaf diagram for the data in exercise 3.

9. For the accompanying histogram construct a relative frequency histogram.

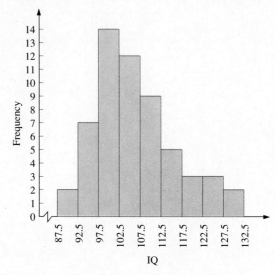

IQs of a random selection of graduating students at ABC High School

10. Construct an ogive corresponding to the following grouped frequency table:

Class	f
1–4	14
5–8	18
9–12	12
13–16	16
17–20	20

11. What kinds of graphs are appropriate for
    a. qualitative data?
    b. quantitative data?

12. What kinds of graphs are appropriate for
    a. nominal data?
    b. ordinal data?
    c. interval and ratio data?

Further Applications

13. The number of calories burned per hour by a 130-pound woman engaging in ten activities is as shown in the table:[15]

Activity	Calories Burned per Hour
Lying down, awake	72
Sitting at rest	95
Working at desk	128
Dressing, undressing	140
Walking level, 2.6 mph	190
Bicycling level, 5.5 mph	295
Tennis	384
Swimming, slow crawl	450
Hiking, no load	425
Jogging, 5.3 mi/hr	550

Construct a bar graph for the data.

14. The accompanying table lists the number of students from 10 foreign countries studying in U.S. universities during the 1988–1989 academic year.[16]

Country of Origin	Number of Students
China	29,040
Taiwan	28,760
Japan	24,000
India	23,350
Korea	20,610
Malaysia	16,170
Canada	16,030
Hong Kong	10,560
Iran	8,950
Indonesia	8,750

Construct a bar graph for the data.

15. Of 100 hospital patients, 30 had type O blood, 38 had type A blood, 22 had type B blood, and 10 had type AB blood. For this data, construct a
    a. bar graph.
    b. circle graph.

16. The accompanying data indicate the final grades in U.S. history for a sample of 50 students.

A	B	B	B	C	D	D	D	A	A
C	C	C	B	B	C	D	D	D	B
C	C	D	D	D	D	F	A	A	C
B	C	B	B	A	A	F	C	C	D
D	F	F	A	A	C	C	B	B	C

For this data, construct a
    a. bar graph.
    b. circle graph.

17. The accompanying table shows the ten most commonly performed plastic surgery operations.[17]

Operation	Number Performed
Hand surgery	160,000
Laceration repair	150,000
Tumor removal	100,000
Breast augmentation	75,000
Industrial injury	70,000
Eyelid surgery	57,000
Nose surgery	55,000
Burn repair	45,000
Reconstruction	45,000
Facelift	40,000

Construct a bar graph for the data.

18. Construct a bar graph for the following data representing base fares (in U.S. dollars) for the world's busiest subway systems:[18]

Subway System	Base Fare
Moscow	1¢
Tokyo	78¢
New York City	$1.15
Mexico City	11¢
Paris	87¢
Osaka, Japan	78¢
Leningrad	1¢
London	$1.17
Seoul	28¢
Hong Kong	32¢

19. Construct a stem-and-leaf diagram for the following data representing the test scores made by 20 students on an English test:

67 71 90 46 51 74 34 65 55 63
71 66 54 46 22 69 61 57 46 84

20. The accompanying data represent the Environmental Protection Agency (EPA) average miles per gallon estimates for 30 new cars.

22 31 20 27 21 29 27 35 47 29
27 23 51 41 30 34 27 35 27 27
31 38 25 27 44 35 34 32 21 19

Construct a stem-and-leaf diagram for the data.

21. Refer to exercise 20. Construct a relative frequency histogram using seven classes for the data.

22. Refer to exercise 19. Construct a relative frequency histogram using six classes for the data.

23. Construct a histogram for the following data representing the number of automobiles sold per week last year by an automobile salesperson.

Number of Cars Sold	Frequency (f)
0	4
1	17
2	12
3	13
4	2
5	3
6	0
7	1

24. Construct a relative frequency histogram for the data in exercise 23. What do you notice about the shapes of this relative frequency histogram and the histogram in exercise 23? In what respects do they differ?

Going Beyond

25. Convert the accompanying circle graph to a bar graph.

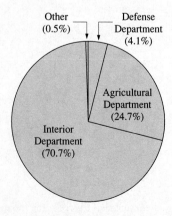

Land owned in 1975 by federal agencies (total of 761 million acres)

26. Convert the accompanying bar graph to a circle graph.

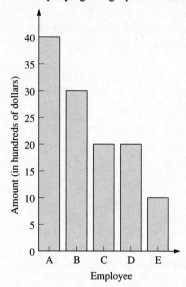

Last year's sales (in hundreds of dollars)

27. Refer to the histogram shown in exercise 9. Find the
    a. total number of IQ scores.
    b. number of IQ scores falling between 97.5 and 102.5.
    c. width of each bar.
    d. relative frequency for the IQ scores comprising the 110 IQ bar.
    e. percentage of IQ scores falling below 117.5.

28. Refer to the accompanying line graph. Find the
    a. total number of families involved.
    b. total number of children involved.
    c. percentage of families that have four children.
    d. percentage of families that have fewer than three children

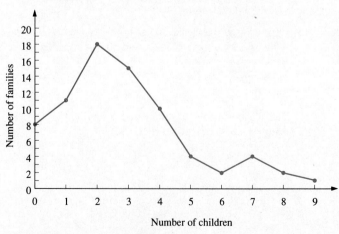

Number of children per family

29. The accompanying table expresses the time (in hours and minutes) bus drivers in five cities must work in order to purchase McDonald's Big Macs, small fries, and medium cokes for a family of four. Figures are given for years 1979 and 1984.[19]

City	1979	1984
Chicago	0:58	1:02
Tokyo	1:29	1:19
Paris	1:52	2:18
Dusseldorf	1:41	1:47
London	2:02	2:24

Construct a single bar graph comparing driver times for the five cities for both years.

30. For the data described by the line graph in exercise 28, construct a relative frequency ogive.

CHAPTER SUMMARY

In this chapter we learned that data can be classified as quantitative or qualitative. Quantitative data can be classified as discrete or continuous depending on whether it can be counted. Data can also be classified according to the type of measurement scale used. The four commonly used scales are nominal, ordinal, interval, and ratio. Tables and graphs are used to organize the data so that they can be more easily used and understood. We studied grouped and ungrouped frequency tables, including relative frequency and cumulative frequency tables. In addition, we saw that data distributions can also be represented graphically by using circle graphs, bar graphs, histograms, stem-and-leaf diagrams, frequency polygons or line graphs, and ogives.

CHAPTER REVIEW

▪ *IMPORTANT TERMS* ▪

The following chapter terms have been mixed in ordering to provide you better review practice. For each term, provide a definition in your own words. Then check your responses against the definitions given in the chapter.

relative frequency histogram	frequency	ordinal data
bar graph	frequency polygon	percentiles
bivariate frequency table	frequency tables	qualitative data
circle graph	bivariate data	quantitative data
double stem diagram	graph	position point
class boundaries	grouped frequency tables	ratio data
data	histogram	relative frequency
class mark	continuous data	relative frequency table
class limits	range	stem-and-leaf diagram
class width	depths	Sturges' rule
cumulative frequency	interval data	tally marks
raw data	line graph	unit
cumulative relative frequency table	nominal data	univariate data
ungrouped frequency tables	ogive	discrete data

▪ **IMPORTANT SYMBOLS** ▪

f, frequency

w, class width

c, number of classes

n, total number of measurements

R, range

U, largest measurement

L, smallest measurement

X, class mark

l_1, lower class limit

l_2, upper class limit

▪ **IMPORTANT FACTS AND FORMULAS** ▪

For a grouped frequency table, all classes have the same width.

For a grouped frequency table the number of classes should be between 5 and 15, inclusive.

Sturges' rule: The number of classes needed in a grouped frequency table is approximately equal to $c = 3.3(\log n) + 1$, where n is the number of measurements.

The class width is found by dividing the range by the number of classes and rounding the result to the least integer greater than R/c.

For a grouped frequency table, the first class always begins with the smallest measurement.

Class boundaries are used to construct histograms and ogives.

REVIEW EXERCISES

1. The heights of 50 female college students (in centimeters) are as follows:

157	155	171	150	163	150	172	161
154	174	163	148	152	163	149	158
176	164	157	153	169	161	160	164
155	162	151	167	167	167	170	158
163	175	169	169	158	150	156	157
174	162	150	151	165	170	156	170
153	154						

   a. Construct a grouped frequency table using ten classes.
   b. Construct a stem-and-leaf diagram.
   c. Construct an ogive using the result of part a.
   d. Construct a histogram using the result of part a.

2. Classify the following data as either quantitative or qualitative:
   a. the weights (in ounces) of 20 apples
   b. the colors of ten cars
   c. the length (in centimeters) of a 12-inch ruler
   d. the religious preferences of 15 people
   e. the letter grades for a class of students
   f. the percentage grades (class averages on a 100-point scale) for a class of students
   g. the gender classifications of 50 teachers
   h. the status (off/on) of 30 light switches
   i. the street addresses of 100 relatives
   j. the jersey numbers for the members of a baseball team
   k. the number π (pi)

3. For exercise 2 classify the data as nominal, ordinal, interval, or ratio.

4. For exercise 2 classify the quantitative data as either discrete or continuous.

5. Twenty people were asked to identify their religious preference. The results are

   C P P J J A J C P P
   C J J C P P A P C J

   where C denotes Catholic, P denotes Protestant, J denotes Jewish, and A denotes Atheist. Construct a
   a. frequency table.
   b. bar graph.
   c. circle graph.

6. Given their limits, find the widths of the following classes:
   a. 7–16
   b. 3.4–7.8
   c. 1.3–4.5
   d. 1.23–4.78
   e. 0.03–0.09

7. If a grouped frequency table is to contain eight classes, where the smallest measurement is 14 and the largest is 94, find the width of each class.

8. Consider the following bivariate frequency table:

	Pass	Fail	Total
Male	11	15	26
Female	14	10	24
Total	25	25	50

Find the
a. number of females that passed.
b. percentage of males that failed.
c. percentage who passed that are males.

9. An experiment was conducted to determine the effect of a certain drug on serum cholesterol levels (in mg/100 ml) in 30-year-old males. The following measurements were obtained:

245	185	230	225	265	210
245	165	195	170	205	225
160	240	285	175	260	225
235	120	145	185	195	210
190	220	140	215	195	

a. Construct a stem-and-leaf diagram.
b. Construct a grouped frequency table with ten classes.
c. Construct a relative frequency histogram using the table in part b.

10. The heights (to the nearest inch) of 33 students are as follows:

66	65	64	68	69	65	68	68	64	66	64
63	71	70	67	69	71	59	67	72	70	67
69	69	66	63	67	70	66	70	67	64	80

Construct a grouped frequency histogram having eight bars.

11. Construct a double stem diagram for the data in exercise 10.

12. The following table represents U.S. average life expectancies for years 1950 and 1983.

	1950		1983	
	Males	Females	Males	Females
Newborns	65.5	71.0	70.9	78.3
15-year-olds	68.6	73.5	72.2	79.3
25-year-olds	69.4	74.0	72.9	79.6
35-year-olds	70.2	74.5	73.7	80.0
45-year-olds	71.6	75.6	74.5	80.5
65-year-olds	77.7	80.0	79.5	83.8

a. Construct a bar graph for the average male life expectancies for 1983.
b. Construct a bar graph for the average female life expectancies for 1983.

13. The following data represent the weights in pounds for a sample of college students:

114	115	116	120	123	126	128	129	131
132	132	133	134	135	135	137	138	139
142	142	143	146	147	152	157	158	161
164	165	167	168	168	170	170	172	174
174	174	175	175	176	177	177	178	180
184	184	184	186	187	189	194	195	195
200	201	202	206	207	209			

a. Construct a stem-and-leaf diagram.
b. Construct a double stem diagram. Does this diagram reveal any characteristics of the data not revealed by the stem-and-leaf diagram in part a? Offer an explanation for the difference in shape.
c. By using the diagram in part a, construct a histogram for the raw data.

14. The accompanying frequency table contains the speeds (in miles per hour) of a sample of 60 cars traveling on 14th Avenue in New York as measured by a policeman with radar.

Class	f
28–33	1
34–39	3
40–45	6
46–51	28
52–57	14
58–63	8

Construct a relative frequency ogive for the data.

Computer Applications

1. The accompanying data represent the SAT math scores of a sample of 100 freshmen students at a large university.

411	606	425	444	507	300	548	387	432
527	508	294	578	469	640	444	261	436
442	508	520	423	556	546	363	569	457
554	624	515	527	450	509	506	374	316
566	415	576	298	401	589	474	571	455
615	439	404	447	676	333	496	559	430
660	494	449	421	690	682	349	485	505
648	475	309	531	499	503	400	550	522
553	555	473	372	505	460	550	653	560
327	458	490	557	337	513	579	403	489
454	470	495	552	600	651	519	698	568
408								

Use a computer program to construct
   a. a grouped frequency table having ten classes.
   b. a stem-and-leaf diagram for the data.
   c. the histogram that corresponds to the frequency table in part a.

2. The following data represent the I.Q. scores of a sample of 100 tenth-grade students at University High School:

132	103	94	78	108	105	98	114
89	112	95	82	86	124	118	120
87	120	107	95	104	100	81	91
94	99	89	93	86	98	122	78
117	115	91	90	97	79	97	104
71	95	86	87	97	107	78	149
124	87	80	71	92	80	106	121
123	117	114	90	109	90	72	86
75	94	94	83	100	86	105	116
95	83	93	109	116	128	94	69
134	111	116	94	135	88	88	102
130	99	94	73	93	98	110	80
120	92	99	80				

Use a computer program to construct
   a. a grouped frequency table having twelve classes.
   b. a stem-and-leaf diagram for the data.
   c. the histogram that corresponds to the frequency table in part a.

■ CHAPTER ACHIEVEMENT TEST ■

The following data indicate the weights in pounds that a group of women lost in the first two weeks of a daily exercise program:

1	2	12	3	15	5	12	11	3	4
3	5	0	7	17	6	17	13	2	5
5	7	1	11	3	9	9	8	18	8
10	9	4	12	1	8	8	7	11	9
15	11	8	4	5	11	3	14	12	10

Use this data set to answer exercises 1–4.

1. Construct a stem-and-leaf diagram.

2. Construct a grouped frequency table containing five classes.

3. Construct an ogive using relative frequencies and the table constructed in exercise 2.

4. Construct a frequency histogram having five bars using the table constructed in exercise 2.

5. a. Find the width of the class 10–20, where 10 and 20 are the class limits.
   b. If $L = 32.1$, $U = 89.7$, $c = 5$, and the unit of measurement is 0.1, determine the upper limit of the first class.

6. Classify each of the following data sets as nominal, ordinal, interval, or ratio:
   a. birthdays of family members
   b. zip codes of all college presidents living in the state of Maryland
   c. average Fahrenheit temperatures for the past 10 days
   d. academic ranks of twenty college teachers
   e. the amount of spending money that each of 50 freshman students possess
   f. favorite professional football team for each person in a group of 30

7. Refer to exercise 6. Classify each of the quantitative data sets as discrete or continuous.

8. Suppose 50 measurements have a smallest value of 17 and a largest value of 96. If a grouped frequency table is to be constructed having 11 classes with the first class limit starting at 17, find the smallest integer-valued class width.

9. Consider the following bivariate table representing the number of As and Bs for three sections of Speech 102 taught by Professor Smith:

Grade	Section		
	I	II	III
A	5	6	1
B	3	2	5

   a. Find the percentage of A grades that were earned in section I.
   b. Find the percentage of section II grades that are Bs.
   c. What percentage of the total number of grades are section II grades?
   d. What percentage of the total number of grades are A grades?
   e. What percentage of the total number of grades are the A grades made in section II?

3 Descriptive Statistics: Analyses of Univariate Data

CHAPTER OBJECTIVES

In this chapter we will investigate

▷ Four measures of central tendency.

▷ How to calculate the measures of central tendency for grouped and ungrouped data.

▷ How to find percentiles for grouped and ungrouped data.

▷ Advantages and disadvantages of using each measure of central tendency.

▷ The concept of skewness.

▷ The concept of the sum of squares.

▷ Four measures of dispersion.

▷ How to calculate the measures of dispersion for grouped and ungrouped data.

▷ Chebyshev's theorem.

▷ Standard scores.

▷ How to construct and use box plots.

MOTIVATOR 3

A large university has been suspected of discriminating against women in its hiring practices. To investigate the charges, the number of men and women who have applied for various advertised teaching positions during a three-year period are found by a Human Rights Commission, and the percentages of qualified males and females are calculated. The recruitment data are contained in the following table:

Area	Male No. applications	No. accepted	Percent accepted	Female No. applications	No. accepted	Percent accepted
Math	75	15	20	20	8	40
Chemistry	100	35	35	30	21	70
Physics	50	9	18	50	18	36
English	25	1	4	400	32	8
Total	250	50	20	500	79	16

Do the data support that there is discrimination against females? Note that for each area, the proportion of females hired is twice as high as the proportion of males hired, but the combined data for all categories indicate that the proportion of males hired is higher. After you finish studying this chapter, you should be able to answer the above question.

Chapter Overview

In chapter 2, we presented the methods for organizing data using tables and graphs. Those techniques represent visual means for discovering relationships, patterns, or trends within the data. In this chapter, we want to supplement the visual interpretations made possible by tables and graphs with numerical measures of characteristics enjoyed by most collections of quantitative data. These characteristics include center, spread, and position points for a set of data. We begin by discussing the concept of center for a data set.

SECTION 3.1   *Measures of Central Tendency and Location*

Measures of Central Tendency

The first characteristic of a set of data that we want to measure is the center or central tendency. The purpose of a **measure of central tendency** is to summarize a collection of data so that we can obtain a general overview; such a measure serves as a representative for the rest of the data.

A measure of central tendency of a set of data also provides a sense of the central value for a seemingly disorganized set of observations. Consider the following four examples:

1. Weights in pounds: 5, 6, 12, 15, and 20.
2. Grades for a test: 31, 74, 78, 79, 80, and 81.
3. Colors of cars: three white, four red, seven black, and one blue.
4. Faculty ranks: seven professors, three associate professors, two assistant professors, and ten instructors

In examples 1 and 2, the scale used is ratio; in example 3, nominal; and in example 4, ordinal. Which measurement(s) would you use to describe the central value or to represent the data set for each example? There are many measures of central tendency used to locate a center of a set of data. Four of the more common measures are the mean, the median, the mode, and the midrange.

The **mean** is the arithmetic average.

The **median** is the middle ordered score.

The **mode,** if it exists, is the most frequent score.

The **midrange** is the arithmetic average of the largest and smallest measurements.

To describe the center measurements for the four examples just given, we would use the mean for example 1, the median for example 2, and the mode for examples 3 and 4. Now let's examine each measure of central tendency in detail and learn the reasons for the choices in the four examples above.

Mean

The mean or arithmetic average of a set of numbers is found by adding the numbers and then dividing the sum by n, the number of measurements.

EXAMPLE 3.1

The following ten scores represent the number of points scored in ten basketball games by player A: 6 10 3 7 6 6 8 5 9 10. The mean is

$$\frac{6 + 10 + 3 + 7 + 6 + 6 + 8 + 5 + 9 + 10}{10} = \frac{70}{10} = 7$$

The value 7 represents, in some sense, the central or "middle" number of points scored in ten games by player A.

Means can be calculated for both samples and populations. They are computed the same way, but are denoted differently. The sample mean is denoted by $\bar{x}$ and the population mean is denoted by the Greek letter μ (pronounced mu). A formula for calculating the mean of a sample of numerical data is given by

Sample Mean
$\bar{x} = \dfrac{\Sigma x}{n}$

(3.1)

where $\bar{x}$ denotes the sample mean, x denotes a sample measurement, Σx denotes the sum of the sample measurements, and n is the size of the sample. A formula for finding the population mean is given by

Population Mean
$\mu = \dfrac{\Sigma x}{N}$

where μ is the mean of the population and N is the size of the population. In order to use this formula, each measurement in the population must be known (see application 3.1). However, the sample mean is not meaningful for all types of data, as we will see in example 3.2.

APPLICATION 3.1

The annual amounts (in billions of dollars) of U.S. agricultural exports from 1974 to 1983 are 21.9, 21.9, 23.0, 23.6, 29.4, 34.7, 41.2, 43.3, 39.1, and 33.7.[20] Determine the mean μ if the data constitute a population.

Solution: The sum of the measurements is $\Sigma x = 311.8$. As a result, the population mean is

$$\mu = \frac{\Sigma x}{N}$$

$$= \frac{311.8}{10} = 31.18$$

Thus, the average amount of agricultural exports over the ten-year period is $31.18 billion dollars. ▪

EXAMPLE 3.2

Suppose we have recorded the color of hair for each of ten college students. The phrase "average hair color" has no meaning. The data in this situation are qualitative and the mean can be computed only for quantitative data.

Sometimes many observations share common values as in ungrouped frequency distributions. Suppose we have the following sample of ages (in years) for ten college freshmen:

$$18 \quad 18 \quad 18 \quad 18 \quad 19 \quad 19 \quad 19 \quad 20 \quad 20 \quad 21$$

By applying the definition of the sample mean, formula (3.1), to the data on freshmen ages, we obtain

$$\bar{x} = \frac{\Sigma x}{n} = \frac{190}{10} = 19$$

In finding Σx, it might be simpler to add the four products $(4)(18)$, $(3)(19)$, $(2)(20)$, and $(1)(21)$. Each product can be written as fx, where f is the frequency of occurrence of an age x (see table 3.1). The sum of the f values equals n, and the sum of the fx values equals Σx.

TABLE 3.1

Raw Data and Frequency Table for Ages of Ten Freshmen

Raw Data		Frequency Table		
x	Tally	x	f	fx
18	IIII	18	4	72
19	III	19	3	57
20	II	20	2	40
21	I	21	1	21
			10	190

The sample mean is also equal to $\bar{x} = \Sigma fx / \Sigma f = 190/10 = 19$.

To find the mean for sample data displayed in a frequency table, we use the following formula:

Sample Mean for Data in a Frequency Table

$$\bar{x} = \frac{\Sigma fx}{\Sigma f} \tag{3.2}$$

Disadvantage to Using the Mean

The mean has one serious shortcoming—it is affected by extreme measurements on one end of a distribution. Because the mean depends on the value of every measurement, extreme values can lead to the mean misrepresenting the data.

EXAMPLE 3.3

Suppose a marathon runner has run in six of the largest marathon races in the country. He placed (in order of marathon run) in the following positions:

$$3 \quad 5 \quad 4 \quad 6 \quad 2 \quad 85$$

For the last race, in which he placed 85th, he went all out trying to win the race. He ran in first place for the first 22 miles, but developed extreme cramping and had to walk some of the last 4 miles. If the mean is used to describe the runner's ability, then the value 17.5 would be used. Since he finished in no more than sixth place in the first five races, it does not seem reasonable to use the mean to measure his overall running ability. Perhaps the median would provide a better measure since the mean is affected too much by the extreme value 85 in this example.

Median

For data measured on at least an interval scale, the median is the middle ordered score. For example, the median of the ordered test scores 9, 22, 37, 45, and 57 is 37.

To Determine the Median

1. Rank the data.
2. If the number of measurements is odd, then the median will be the measurement in the middle. If the number of measurements is even, the median is the mean of the two measurements occupying the middle positions.

The median for a population is denoted by μ and the median for a sample is denoted by $\tilde{x}$.

EXAMPLE 3.4

Suppose in their last seven games the Bobcats scored the following numbers of points:

$$6 \quad 10 \quad 3 \quad 21 \quad 0 \quad 35 \quad 14$$

The median number of points scored is found by first ranking the scores:

$$0 \quad 3 \quad 6 \quad 10 \quad 14 \quad 21 \quad 35$$

The median score is easily seen to be 10, since only one score occupies the middle position. If the Bobcats score 42 points their next game, then the eight scores would form the following sequence:

$$0 \quad 3 \quad 6 \quad 10 \quad 14 \quad 21 \quad 35 \quad 42$$

Since there is now an even number of scores, the values 10 and 14 occupy the middle positions and the median is found to be 12, the average of 10 and 14.

Computer display 3.1 illustrates the use of MINITAB to determine the mean and median for the Bobcat data (6, 10, 3, 21, 0, 35 and 14) of example 3.4.

Computer Display 3.1

```
MTB > SET C1
DATA> 6 10 3 21 0 35 14
DATA> END
MTB > MEAN C1
   MEAN = 12.714
MTB > MEDIAN C1
   MEDIAN = 10.000
```

The first three lines are used to store the data. At the system prompt MTB >, the user supplies the command $\boxed{\text{SET C1}}$ to create a column labeled C1 to hold the data. The system then responds with the data prompt, DATA> on the second line. The user then types in the data (the numbers are separated by using either a space or a comma). The END command on the third line is supplied by the user to indicate the end of the data set. On the fourth line, the command $\boxed{\text{MEAN C1}}$ supplied by the user, indicates that the value of the mean is requested. The system responds on the next line with the value of the mean (MEAN = 12.714). Similarly, the value of the median is requested on the next line by the user-supplied command $\boxed{\text{MEDIAN C1}}$. The system responds with MEDIAN = 10.000. Remember that at the end of each command, the user must depress the return or enter key in order to register the commands with the computer system.

EXAMPLE 3.5

Since the median is the middle value for a distribution, there may not be as many measurements below it as there are above it. For example, consider the following sample of five values:

$$6 \quad 6 \quad 6 \quad 7 \quad 8$$

The median value of 6 has no values below it and two values above it.

There are both advantages and disadvantages to using the median for interval data. One advantage is that the median is not affected by extreme scores on one end of the distribution. This was the reason for choosing the median to represent the "middle" measurement for the grade data illustrated in the example that began this section. A disadvantage to using the median is that it is not easily found for a large set of data, since the measurements must first be ranked (put in numerical order from smallest to largest or from largest to smallest).

For large sets of data that have been organized in a frequency table (where the values of x are ranked) or an ordered stem-and-leaf diagram, the median is found as follows:

> If n is odd, the median is the measurement with rank $(n + 1)/2$; and if n is even, the median is the average of the measurements with ranks $n/2$ and $n/2 + 1$.

Notice that $(n + 1)/2$ does not represent one of the measurements, rather it represents the number of values that must be counted to arrive at the median. For the five ranked values 4, 8, 12, 13, and 14, the measurement with rank $(5 + 1)/2 = 3$ is 12.

APPLICATION 3.2

Find the median for the sample data organized in table 3.2, a frequency table representing the number of absences for each class period during the 1988 spring term for an introductory philosophy class.

TABLE 3.2

Absence Data for
Application 3.2

Number of Absences	Frequency	Cum f
0	10	10
1	10	20
2	8	28
3	4	32
4	8	40

Solution: As a consequence of the given rule and the fact that there are 40 measurements involved, the median is the average of the 20th and 21st measurements. Notice that since there is an even number of measurements, two measurements occupy the middle position. We can count in either direction from the smallest measurement to the largest measurement or from the largest measurement to the smallest measurement to arrive at the median value. Since the 20th value counting from the smallest measurement is the 21st value counting from the largest measurement, we need only average the 20th and 21st values counting from the smallest value. Thus, the median for the data is $(1 + 2)/2 = 1.5$ absences. ▪

Mode

The mode, if it exists, is the most frequent measurement. It has two advantages: it is easily found for certain small samples and it is usually not influenced by extreme measurements on one end of an ordered set of data, as in example 3.7. When qualitative data are being analyzed, as in example 3.8, the mode is the only measure of central tendency that can be used. Finally, the mode can be used as a measure of central tendency for numerical data that is used in a qualitative sense (see example 3.9).

EXAMPLE 3.6

With the measurements

$$1 \quad 1 \quad 3 \quad 3 \quad 3 \quad 2 \quad 7 \quad 8$$

the mode is 3.

EXAMPLE 3.7

The mode is not affected by extreme measurements, as can be observed from the following two samples, A and B, each having a mode of 2.

$$A: 1, 2, 2, 2, 3, 78$$
$$B: 1, 2, 2, 2, 3, 8$$

The extreme measurement of 78 in sample A has no effect on the modal value.

EXAMPLE 3.8

Suppose that the blood types for a group of 12 student nurses are A, A, B, A, AB, O, O, B, O, A, B, and AB. The mode, or most frequent occurring blood type, is type A. For this data, it makes no sense to use either the mean or the median to locate a central observation. For this data the mode is the only measure of central tendency that makes sense.

EXAMPLE 3.9

The C & P Telephone Company provides local service only to points within a specific geographic area. Any call made from a point inside a specific geographic calling area to a point outside of the geographic calling area is assessed an additional toll charge by the phone company. The first three digits of a seven-digit phone number indicate the particular calling area, and the last four digits indicate the party called within the calling area. The listed calling areas can be called from a phone located within the calling area identified by phone numbers beginning with 689. Each of the following numbers represents the first three digits of a sample of calls made by a business from a phone located within the Frostburg calling area during a one-hour period:

264	Mt. Savage
324	Wellersburg
463	Lonaconing
689	Frostburg
697	Cumberland
722	Cumberland
724	Cumberland
729	Cumberland
759	Cumberland
777	Cumberland
895	Grantsville

264 324 463 689 697 722 689 895 324 324

Which observation should we use to represent the central value of this sample? Although the observations are numbers, they represent data measured on a nominal scale. The numbers are used in a qualitative sense; they only represent labels and no order is implied. As a result, the median makes no sense, since the data assume no particular order. For example, it makes no sense to ask what order relationship exists between 264 and 324 (or between Cumberland and Cumberland). It also makes no sense to average the ten observations, since the numbers aren't used in a quantitative sense and it makes no sense to add them. What interpretation would we give to $(264 + 324)$ in the context of this example? The only measure of central tendency that is appropriate for this application is the mode; the modal value is 324. This observation can serve to represent the central value of the ten observations.

A mode for data displayed in a frequency table is found by locating a largest frequency value, if all of the frequencies are not equal. The value of x that corresponds to the largest frequency value is then taken to be a mode. For application 3.2, the modes are easily seen to be 0 and 1.

Disadvantages of the Mode

The mode has several disadvantages as a measure of central tendency. One disadvantage is that there may be no mode for a given set of data; this situation arises when each measurement occurs with equal frequency. Another disadvantage is that the mode can exist, but not be unique, as in example 3.11.

EXAMPLE 3.10

The measurements

red   black   brown   blue

have no mode. The measurements

2   2   3   3   4   4   5   5

also have no mode.

EXAMPLE 3.11

With the measurements: red, red, red, black, blue, white, white, and white; both red and white are modes. In this case the collection of observations is called **bimodal**.

Midrange

The midrange of a set of data is the average of the largest and smallest measurements.

APPLICATION 3.3

The following are the number of twists that are required to break eight forged alloy bars: 32, 38, 45, 44, 27, 36, 40, and 38. Determine the midrange.

Solution: The midrange is the average of the largest and smallest measurements. The largest measurement is $U = 45$ and the smallest measurement is $L = 27$. The midrange is

$$\text{Midrange} = \frac{L + U}{2}$$
$$= \frac{27 + 45}{2} = 36 \quad \blacksquare$$

APPLICATION 3.4

Which measure of central tendency should be used to indicate the central salary of all wage earners in the United States?

Solution: The median is the preferred measure. Because of the extreme salaries on the high end of the salary scale, neither the mean nor the midrange should be used. Of course, the proper measure will depend on how the measure is to be used; to indicate our financial status on the international market, we would most likely use the mean value. One reason the modal value is not used is that there is no guarantee that a unique value will exist; there may be no value or a large number of values that occur with greatest frequency. ▪

Measures of Location

A **position point** for a distribution is that value for which a specified portion of the distribution falls at or below it. The median is an example of a position point, as are percentiles, deciles, and quartiles.

EXAMPLE 3.12

Fifty percent of the distribution is less than or equal to the median, and 50% of the distribution is greater than or equal to the median. Therefore the median is a position point.

Percentiles

> The *nth* **percentile**, denoted by P_n, is that value for which at least n% of the distribution falls at or below it and at least $(100 - n)$% falls at or above it.

A data set has 99 percentile points, which divide the data set into 100 parts; each part contains approximately 1% of the measurements. These percentile points are labeled $P_1, P_2, P_3, P_4, \ldots, P_{99}$.

EXAMPLE 3.13

Suppose we want to find the 25th percentile point (or 25th percentile) of the sample displayed in the following ordered stem-and-leaf diagram.

```
3 | 4  4  6  9
4 | 3  6  7  [8] [9]
5 | 0  1  1  5  7  7  8  9
6 | 0  0  4  4  7
7 | 1  5  8  8  8  9
8 | 4  6  8  8
```

The size of the sample is $n = 32$. The 25th percentile is that measurement that has at least 25% of the sample falling at or below it and at least 75% falling at or above it.

$$(25\%)(32) = \text{at least 8 values at or below it}$$
$$(75\%)(32) = \text{at least 24 values at or above it}$$

Counting eight leaves from the top of the trunk, we arrive at leaf 8 on stem 4. The value 48 has 8 values at or below it and 24 values above it. The value 49 also satisfies the conditions; 8 values are below it and 24 values are at or above it. The 25th percentile is the average of 48 and 49. Thus, $P_{25} = 48.5$.

EXAMPLE 3.14

Suppose we want to find the 30th percentile for the data in example 3.13. The 30th percentile is that measurement that has at least 30% of the sample falling at or below it and at least 70% of the sample falling at or above it.

$$(30\%)(32) = \text{at least 9.6 values at or below it}$$
$$(70\%)(32) = \text{at least 22.4 values at or above it}$$

Since counts must be whole numbers, the 20th percentile must have at least 10 values at or below it and at least 23 values at or above it. Note in both cases we choose the least integer greater than the product. By examining the stem-and-leaf diagram of example 3.18, we determine that 50 satisfies both these conditions. Thus, $P_{30} = 50$.

3	4 4 6 9
4	3 6 7 8 9
5	0 1 1 5 7 7 8 9
6	0 0 4 4 7
7	1 5 8 8 8 9
8	4 6 8 8

Quartiles and Deciles

> **Quartiles** are numbers that divide an ordered set of measurements extending from the smallest measurement to the largest measurement into four parts so that each part has approximately 25% of the measurements.

There are three quartile points, denoted by Q_1, Q_2, and Q_3. The *first quartile* Q_1 is the 25th percentile, the *second quartile* Q_2 is the 50th percentile or the median, and the *third quartile* Q_3 is the 75th percentile.

$$Q_1 = P_{25}$$
$$Q_2 = \tilde{x} = P_{50}$$
$$Q_3 = P_{75}$$

> **Deciles** are numbers that divide a set of measurements extending from the smallest measurement to the largest measurement into ten parts such that each part contains approximately 10% of the measurements.

There are nine deciles, denoted by D_1, D_2, D_3, . . . , and D_9; D_n is the nth decile. Each decile point corresponds to a percentile point. For example $D_4 = P_{40}$, $D_7 = P_{70}$, and so on.

APPLICATION 3.5

A sample of twelve workers was tested for gripping strength and the measurements (ordered from smallest to largest) were 80.6, 89.9, 101.4, 102.6, 115.0, 120.1, 123.4, 126.3, 131.8, 138.6, 151.6, and 160.5. Determine

    a. the first quartile.

    b. the second quartile.

    c. the third quartile.

    d. the second decile.

Solution:

    a. The first quartile is the 25th percentile.

        Q_1 will have at least $(0.25)(12) = 3$ values falling at or below it.

        Q_1 will also have at least $(0.75)(12) = 9$ values falling at or above it.

    At least three observations must be at or below Q_1 and at least nine observations must be at or above Q_1. The values 101.4 and 102.6 both meet these requirements. The first quartile Q_1 is therefore the average of 101.4 and 102.6. Hence,

$$Q_1 = \frac{101.4 + 102.6}{2} = 102$$

    b. The second quartile is the median. The median is the average of the sixth and seventh measurements. Thus, the second quartile is

$$Q_2 = \frac{120.1 + 123.4}{2} = 121.75$$

    c. The third quartile is the 75th percentile. From part a, we can determine that the number of observations at or above Q_3 is at least 3 and the number of observations at or below Q_3 is 9. Two values meet these requirements: 131.8 and 138.6. Thus Q_3 is the average of these values:

$$Q_3 = \frac{131.8 + 138.6}{2} = 135.2$$

    d. The second decile is the 20th percentile. At least $(0.2)(12) = 2.4$ values must be at or below D_2 and at least $(0.8)(12) = 9.6$ values must be at or above D_2. The value 2.4 must be rounded to 3 and the value 9.6 must be rounded to 10. Counts must be whole numbers; so we will always round up in order to satisfy the "least" criteria. So at least three values must be at or below D_2 and at least ten values must be at or above D_2. The measurement 101.4 satisfies these conditions. Thus, $D_2 = 101.4$. ■

Skewness

Same for both definitions

The shape of a histogram depends on the relative positions of the mean, median, and mode. In a **symmetrical histogram,** both sides of the histogram, determined by the mean, are identical (see example 3.15). When the sides of a histogram are not identical, we have what is called a **skewed histogram.** A histogram, or collection of data, for which there are more measurements below the mean than above the mean is said to be *skewed to the left,* as in example 3.16. On the other hand, as we see in example 3.17, a histogram, or collection of data, for which the measurements above the mean occur less frequently than the measurements below the mean is said to be *skewed to the right.*

EXAMPLE 3.15

As an illustration of a symmetrical distribution, consider the following data and corresponding frequency histogram:

x	f
1	1
2	3
3	5
4	7
5	5
6	3
7	1

From the table we can see that $\bar{x} = 4$, $\tilde{x} = 4$, and mode $= 4$. The corresponding frequency histogram is shown in figure 3.1. We can see that a symmetrical histogram has its mean equal to its median. In fact, for the histogram in figure 3.1, the mean, median, and mode are all identical. But this is not always the case for a symmetrical histogram, as indicated in figure 3.2. Here the mean and median are equal to 8, but the distribution is bimodal, with modes of 7 and 9.

FIGURE 3.1

Symmetrical frequency
histogram

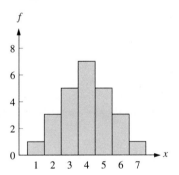

FIGURE 3.2

Symmetrical bimodal frequency
histogram

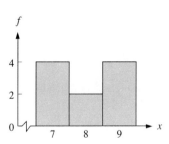

EXAMPLE 3.16

As an illustration of a distribution that is skewed to the left, consider the following data:

x	1	2	3	4	5	6	7	8	9
f	2	2	5	5	10	15	20	25	30

The mean is 6.94, the median is 7, and the mode is 9. For this distribution, there are 75 values above $\bar{x} = 6.94$ and 39 values below $\bar{x} = 6.94$. The histogram for this data set is shown in figure 3.3. Notice that the histogram has a long tail on the left side. As such, a histogram skewed to the left is sometimes referred to as being *negatively skewed*. A negatively skewed histogram will always have its median greater than its mean.

FIGURE 3.3

Distribution that is skewed to
the left

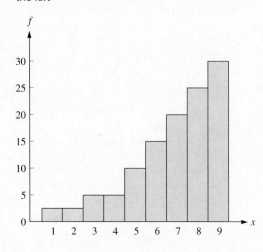

EXAMPLE 3.17

As an illustration of a distribution that is skewed to the right, consider the following data:

x	1	2	3	4	5	6	7	8	9
f	30	25	20	15	10	5	5	2	2

The mean is 3.06, the median is 3, and the mode is 1. For this distribution, there are 75 values below the mean $\bar{x} = 3.03$ and 39 values above the mean. The histogram for this set of data is shown in figure 3.4. Notice that the histogram has a long tail on the right end. Such a histogram is sometimes described as *positively skewed*. A positively skewed histogram will always have its mean greater than its median.

FIGURE 3.4

Distribution that is skewed to
the right

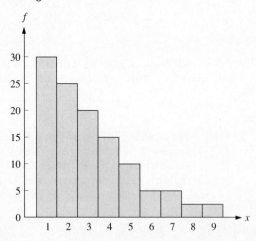

EXERCISE SET 3.1

Basic Skills

1. Calculate the mean, median, mode, and midrange for each of the following samples:
   a. 3, 9, 12, 7, 16, 20, 33, 3
   b. 5, 7, 22, 17, 5, 7, 20
   c. 8, 6, 0, 17, 12, 7, 5
   d. −4, 0, 13, 9, 4, 14, 20, 15

2. Calculate the mean, median, mode, and midrange for each of the following samples:
   a. 12, 7, 3, 20, 33, 2, 12
   b. 12, 15, 23, 7, 12, 40, 22, 16
   c. 5, 0, 7, 7, 13, 16, 9
   d. −5, 6, 13, 26, 0, 14, 25, 13

3. Calculate the mean for the sample where
   a. $\Sigma x = 37$ and $n = 12$.
   b. $\Sigma x = 20.6$ and $n = 56$.
   c. $\Sigma x = -12$ and $n = 33$.

4. Calculate the mean for the sample where
   a. $\Sigma x = 12.5$ and $n = 16$.
   b. $\Sigma x = 19$ and $n = 22$.
   c. $\Sigma x = -43.2$ and $n = 50$.

5. Calculate the mean, median, and mode for each of the following samples:
   a. 0, 0, 1, 1, 1, 0, 0, 0
   b. 3, 3, 3, 2, 2, 2, 4, 5, 3
   c. 0, 1, 1, 2, 2, 3, 3, 4, 4
   d. −1, 0, 0, 0, −1, 2, −2, 3

6. Calculate the mean, median, and mode for each of the following samples:
   a. 0, 1, 2, 3, 8, 50
   b. 0, 1, 2, 3, 8, 12, 50
   c. −12, −6, −5, 0, 13, 16, 0
   d. 0, 0, 0, 1, 1, 1, 1, 1, 0

7. Determine the skewness of the following samples:
   a. 12, 7, 16, 22, 17, 13, 16, 7, 10
   b. 14, 17, 2, 7, 13, 17, 22, 37, 0, 15
   c. 5, 10, 15, 25, 40, 65, 100
   d. 5, 10, 95, 90, 50

8. Determine the skewness of the following samples:
   a. 22, 13, 15, 2, 18, 34, 16
   b. 5, 17, 17, 17, 3, 100
   c. 0, 0, 0, 1, 1, 1, 1
   d. 2, 2, 3, 4, 4, 1, 5

9. An instructor accidentally erased the grade of one of his six students. The five remaining grades were 76, 85, 43, 89, and 65 and the mean of the six grades is 70. Find the grade that was erased.

10. The mean annual salary paid to four chief executive officers (CEOs) of large corporations is $125,000. Can one of them receive $600,000?

Further Applications

11. If the mean income for 20 workers is $40,000, what is the total income for the 20 workers?

12. If the mean height of a sample of 25 basketball players is 6.9 feet, what is the sum of the heights of the 25 players?

13. In an effort to cut down on his coffee consumption, an office worker recorded the following number of cups of coffee consumed for a 20-day period:

4	5	3	6	7	1	2	3	0	5
6	5	8	4	0	2	3	7	5	6

    Which measure of central tendency is most useful for this purpose? What is its numerical value?

14. Below is a collection of statistics test scores from 25 students on a 50-question test.

38	39	33	37	34	31	38	36	35	5

    Which measure of central tendency is most useful for describing the central value? What is its numerical value?

15. A bowler has been bowling regularly for the past five years. Her bowling scores for the past six games are: 201 187 162 234 208 198. For this sample, compute the values of the following statistics (if they exist):
    a. mean
    b. median
    c. mode
    d. midrange
    e. Q_1, Q_2, and Q_3
    f. D_4

16. In a survey conducted by a medical secretary to investigate waiting times (in minutes) of patients to see a doctor, a sample of patients for one day produced the following results:

35	25	35	50	25	55	30	50	35	35
5	5	60	35	30	30	25	55	30	20
60	25	25	40	80	20	20	5	5	10

a. Describe a typical waiting time by using the mean.
b. Describe a typical waiting time by using the median.
c. Which measure (mean or median) do you think is more representative of the data set? Explain.
d. Determine the three quartiles.
e. Determine the fourth decile.

17. The following table contains the annual salaries (in 100 dollars) for 25 laborers.

Annual Salary	Frequency
55	7
60	5
70	6
80	4
300	3
	25

a. What is the mode?
b. What is the mean?
c. What is the median?
d. What is the midrange?
e. Determine the skewness.
f. Which measure of central tendency would you use to determine the central value? Explain.
g. What is Q_1?
h. What is D_6?

18. A sample of 705 bus drivers were selected and the number of traffic accidents in which each was involved during a four-year period was recorded in the accompanying table.

Number of Accidents	Frequency
0	114
1	157
2	158
3	115
4	78
5	44
6	21
7	7
8	6
9	1
10	3
11	1

a. What is the mode?
b. What is the mean?
c. What is the median?
d. What is the midrange?
e. Determine the skewness.
f. Which measure of central tendency would you use to determine the central value? Explain.

g. What is Q_3?
h. What is D_4?

19. Which measure of central tendency would you use to select an accurate thermometer for purchase from a local hardware store? Explain.

Going Beyond

20. If 20 scores have a mean of 15 and 30 scores have a mean of 25, what is the mean of the total group of 50 scores?

21. Suppose 6 is the mean of a sample of four scores.
   a. If 5 is added to each of the scores, what is the mean of the new set of scores? (*Hint:* Try an example.)
   b. If each score is multiplied by 5, what is the mean of the new set of scores?

22. If the measurements (x) in a sample are transformed by use of the formula $y = ax + b$, determine a formula for the mean of the transformed measurements (y).

23. A teacher gave a standardized test to each of her three classes. From the data she determined the three medians and averaged them to determine the central point of her classes' ability. Can she be misled by doing this? Explain.

24. When averaging percentages, the geometric mean $\bar{x}_g$ is often used. The *geometric mean* is defined by

$$\bar{x}_g = \sqrt[n]{x_1 x_2 x_3 \ldots x_n}$$

where $x_1, x_2, \ldots, x_n$ are positive numbers. Find $\bar{x}_g$ for the following percentages: 95 125 140 100.

25. The *harmonic mean* $\bar{x}_h$, which often is used for averaging rates of travel for equal distances, is defined as the reciprocal of the average of the data reciprocals; that is,

$$\bar{x}_h = \frac{n}{\Sigma (1/x)}$$

where the n values of x are positive. Suppose one drives 20 miles at 30 miles per hour and 20 miles at 60 miles per hour. What is the average rate of speed for the 40-mile trip?

26. Would either $\bar{x}$ or $\bar{x}_g$ be appropriate for the data in exercise 24? Explain.

27. A race-car driver wants to average 60 miles per hour (mph) for two laps around a 1-mile track. For the first lap his time was 30 mph due to an electrical problem with the carburetion system. How fast must he travel for the second lap in order to accomplish his goal of 60 miles per hour for the two laps?

28. Suppose a commodity was priced $2 in 1988, $4 in 1989, and $2 in 1990. The percentage change from 1988 to 1990 is 200 and the percentage change from 1989 to 1990 is 50. Find the average percent change in price for the three-year period. Justify your answer.

29. The *root mean square* (rms) of a set of measurements is defined to be the square root of the average sum of squares of the measurements:

$$\text{rms} = \sqrt{\frac{\Sigma x^2}{n}}$$

It is useful in describing peak AC voltages and currents in electronics. Determine the root mean square of the following sample of voltages: 120 130 150 110 105.

30. If a constant C is added to each measurement in a data set, show that the mean of the new set of measurements is equal to the mean of the original set of measurements plus C.

31. If each measurement in a data set is multiplied by a constant C, show that the mean of the new set of measurements is equal to C times the mean of the original set of measurements.

32. Suppose we have a sample of n 1s and m 0s. Show that the mean of the sample is equal to the proportion of 1s in the sample.

33. Suppose a sample consists of all the even integers between 238 and 874, inclusive. Find the mean and the median of the sample.

34. Two professional baseball players have the career records shown in the accompanying table.

Player A				Player B			
Year	At bat	Hits	Ave	Year	At bat	Hits	Ave
1973	189	57	0.302	1973	85	27	0.318
1974	80	21	0.263	1974	144	42	0.292
1975	212	72	0.340	1975	53	19	0.358
1976	71	17	0.239	1976	207	52	0.251
1977	212	64	0.302	1977	55	19	0.345
1978	97	26	0.268	1978	263	74	0.281
1979	281	89	0.317	1979	107	35	0.327
1980	129	37	0.287	1980	175	52	0.297
1981	151	57	0.377	1981	75	29	0.387
1982	130	34	0.262	1982	163	45	0.276
Total	1552	474	0.305	Total	1327	394	0.297

If they are equal in other playing abilities and are negotiating next season's contract, which player should receive the higher salary, based on the better batting average? Explain.

35. With samples having extreme scores (called **outliers**), the mean is very sensitive to their presence, whereas the median is insensitive to their presence. Both measures can be suspect as a measure of central tendency in these cases. A *trimmed mean* offers a compromise. It is less affected by outliers and yet does not have the insensitivity of the median. A trimmed mean is found by ordering the measurements from smallest to largest, deleting a certain number of measurements from both ends of the ordered list, and averaging the remaining measurements. The percentage of values deleted from each end of the list is called the *trimming percentage*. For example, if $n = 20$ and the largest and smallest measurements are deleted, then the trimming percentage is $1/20 = 0.05 = 5\%$. Find the 10% trimmed mean for the sample data:

35	25	35	50	25	55	30	50	4	35
5	5	60	35	30	25	55	30	20	60
25	40	80	20	20	5	10	100	95	30

In this case, is the value of the trimmed mean a more accurate description of the center of the sample than the value of the mean? Explain.

36. Consider the following MINITAB printout:

```
MTB > SET C1
DATA> 9 14 12 17 11 20 13 18 22 12 15 16 5 7 9 19 8
DATA> END
MTB > DESCRIBE C1

         N    MEAN   MEDIAN   TRMEAN   STDEV   SEMEAN
C1      17   13.35    13.00    13.33    4.89     1.18
       MIN     MAX      Q1       Q3
C1    5.00   22.00    9.00    17.50
```

Determine the trimming percentage for the trimmed mean (TRMEAN).

37. Do the data in motivator 3 support that there is discrimination against females? Explain.

SECTION 3.2   *Measures of Dispersion or Variability*

Quite often the measures of central tendency alone do not adequately describe a characteristic being observed. For example, suppose Dave and Rick each shoot 25 arrows at a target. Their scores are as follows:

	Frequency	
Score	Dave	Rick
10	2	0
9	3	0
8	4	5
7	7	8
6	2	5
5	1	4
4	1	3
3	1	0
2	2	0
1	2	0

Dave and Rick have the same average score, $\bar{x} = 6.32$. But as figure 3.5 illustrates, Dave's performance with the bow is certainly different from Rick's performance; Dave's arrows are more spread out than Rick's arrows. We need a measure that is sensitive to this variability; the mean isn't.

FIGURE 3.5

Variability of scores for Rick and Dave

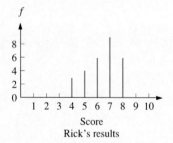

Rick's results

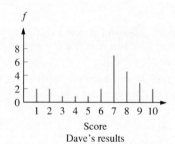

Dave's results

MINITAB has a built-in charting function that uses the DOTPLOT command. Computer display 3.2 contains two charts made with the DOTPLOT command: one for Rick's scores and one for Dave's scores. Each chart has different horizontal scales. These scales were assigned by MINITAB. It is difficult to compare these two dotplots and analyze the data spread because of the different interval sizes (1.8 for Rick's scores and 2 for Dave's scores). This difficulty is easily overcome by using another MINITAB subcommand (SAME), as shown in computer display 3.3. The dotplots use the same scale (2). It is now clear that Dave's scores are more variable.

Computer Display 3.2

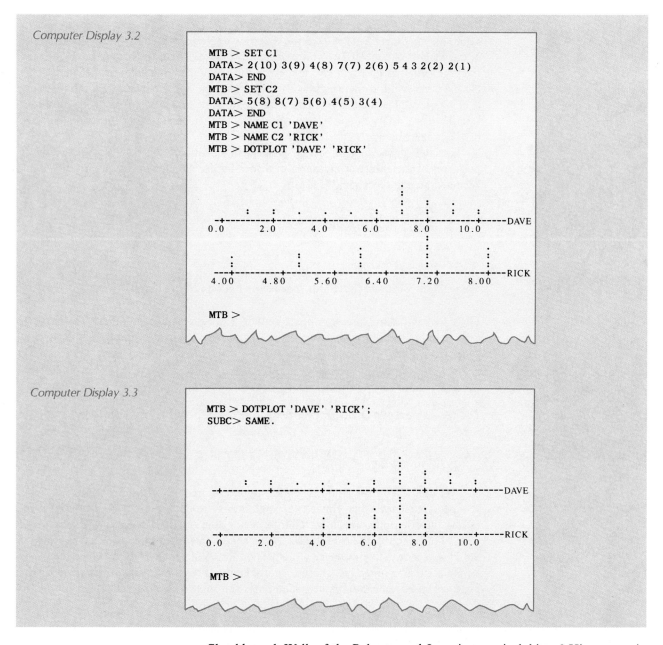

```
MTB > SET C1
DATA> 2(10) 3(9) 4(8) 7(7) 2(6) 5 4 3 2(2) 2(1)
DATA> END
MTB > SET C2
DATA> 5(8) 8(7) 5(6) 4(5) 3(4)
DATA> END
MTB > NAME C1 'DAVE'
MTB > NAME C2 'RICK'
MTB > DOTPLOT 'DAVE' 'RICK'
```

Computer Display 3.3

```
MTB > DOTPLOT 'DAVE' 'RICK';
SUBC> SAME.
```

Should coach Wells of the Bobcats send Jones in as a pinch hitter? His average is 0.310, but in some games he strikes out every time at bat and in other games he gets a hit every time at bat. Or should he put in Smith, who has a batting average of 0.290 and hits at least once in every game he is in? The answer appears to be obvious: put Smith in; his batting performance is less variable.

Any collection of measurements on the same unit will vary with the precision of the measuring instrument. For example, in a box of 24 2-ounce candy bars, not all the bars will weigh exactly 2 ounces. If they do, the scale is not sensitive or precise enough.

These same candy bars, weighed on a sensitive analytical scale, will not all weigh the same. The measurements will exhibit a certain degree of variability. A large variability in the weights of candy bars is undesirable. If the weights of the bars exceed 2 ounces, then the manufacturer will lose money on the production and sales of the candy bars. On the other hand, if the weights of the bars are less than 2 ounces, then the customer is being cheated, which causes consumer complaints and a potential loss of business. In either case, a large variability in the weights of the candy bars can not be tolerated by the management.

Variability is a very important concept in statistics. As a result, there are many measures of variability or **measures of dispersion** for a collection of quantitative data. Included among these measures are

- a. range
- b. interquartile range
- c. variance
- d. standard deviation

We will now examine these four measures of variability in detail.

Range

For a distribution of measurements (either a sample or a population), the **range** is defined to be the difference between the largest measurement U and the smallest measurement L, that is

$$R = U - L$$

EXAMPLE 3.18

The ages in years of a family group are: 30, 2, 1, 7, 4, 32, and 10. The range is

$$R = U - L$$
$$= 32 - 1 = 31$$

We used the range R in section 2.2 to determine the width of the intervals for a grouped frequency table. Since the range is easy to determine, it is often useful for estimating other measures of variability, such as the standard deviation, which are not as easy to compute (see exercise 52 at the end of this section). However, the range is not always a sensitive measure of dispersion for a collection of data, as we see in example 3.19. The range has another shortcoming: it can be drastically affected by the presence of extreme data values, sometimes referred to as *outliers*.

EXAMPLE 3.19

For the two sets of data pictured on number lines in figure 3.6, which is more dispersed, A or B? The answer is clearly set A, but notice that both A and B have the same range. This example illustrates that the range is not a sensitive measure of dispersion. For this reason, it is not an extremely useful measure of dispersion.

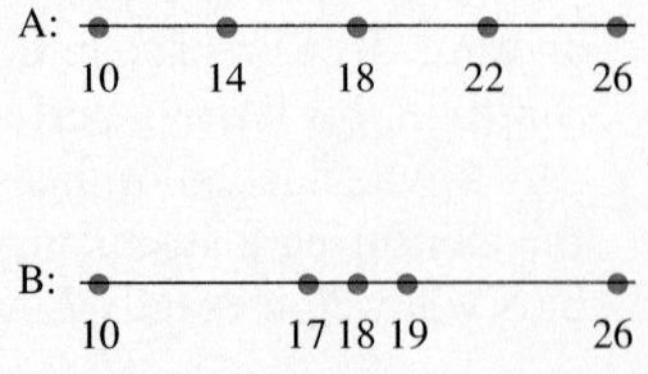

FIGURE 3.6

The range as a measure of dispersion

Interquartile Range

A measure of variation that is resistant to the presence of outliers is the **interquartile range,** denoted by IQR. It is defined as

Interquartile Range

$$IQR = Q_3 - Q_1$$

where Q_3 is the third quartile and Q_1 is the first quartile.

EXAMPLE 3.20

Consider the following ordered data set that represents the oxygen uptake values (mL/kg · min) of 21 male middle-aged runners while pedaling at 100 watts on a bicycle ergometer:[21]

12.81	14.95	15.83	15.97	19.90	18.27	18.34	19.82	19.94	20.62
20.88	20.93	20.98	20.99	21.15	22.16	22.24	23.16	23.56	35.78	36.73

The values 35.78 and 36.73 appear to be extreme values, or outliers, for this set of data. From the definition of the interquartile range, it is clear that these values will have no effect on the value of the interquartile range; these two values could be replaced by any two other values that could occupy the 20th and 21st positions in the ordering and also not affect the value of the interquartile range. Let's calculate the IQR by using the following three steps:

1. Calculate the first quartile. The first quartile is that value for which at least $(0.25)(21)$ = 5.25 measurements fall at or below it and at least $(0.75)(21) = 15.75$ measurements at or above it. Thus, 6 values fall at or below 18.27 and 16 values fall at or above 18.27. Thus, the first quartile is

$$Q_1 = 18.27$$

2. Calculate the third quartile. Counting six values from the right end of the ordering, we determine that the third quartile is

$$Q_3 = 22.16$$

3. Calculate the value of IQR. The value of the interquartile range is

$$IQR = Q_3 - Q_1$$
$$= 22.16 - 18.27 = 3.89$$

The interquartile range is not affected by the outliers 35.78 and 36.73, whereas the range is affected by 36.73.

We shall use the interquartile range in section 3.4 to construct box plots, data summaries that provide information about center, spread, symmetry versus skewness, and the presence of outliers.

The range and interquartile range are not sensitive measures of variation. The range is dependent only on the extreme measurements L and U, while the interquartile range does not take into account the measurements below Q_1 or above Q_3. The variance and standard deviation are both more sensitive measures of variation than the range or interquartile range, since they take into account all the measurements in a data set. But they share a common shortcoming in that they both are influenced by extreme scores. We shall examine these measures below. Both involve the concept of deviation score.

Deviation Score

In statistics, the quantity $(x - \bar{x})$ is called a **deviation score.**

$$\text{Deviation score} = x - \bar{x}$$

A positive deviation score for a measurement indicates that the measurement is above the mean, whereas a negative deviation score for a measurement indicates that the measurement is below the mean. A deviation score of 0 for a measurement indicates that the measurement is equal to the mean.

APPLICATION 3.6

Compute the deviation scores for the following data representing the number of defects found by an automobile inspector on an assembly line for the last five automobiles produced: 1 4 6 6 8.

Solution: The sample mean is easily found to be $\bar{x} = 5$. The deviation scores are presented in the following table:

x	$x - \bar{x}$
1	$1 - 5 = -4$
4	$4 - 5 = -1$
6	$6 - 5 = 1$
6	$6 - 5 = 1$
8	$8 - 5 = 3$

We can make the following observations:

a. The measurements 6 and 8 are above the mean and their deviation scores are positive.

b. The measurements 1 and 4 are below the mean and their deviation scores are negative.

c. The sum of the deviation scores is 0. ■

It can be easily shown that the sum of the deviation scores for any set of numbers is 0; that is,

$$\Sigma (x - \bar{x}) = 0, \text{ for any data set} \tag{3.3}$$

Equation (3.3) has an interesting physical interpretation.

EXAMPLE 3.21

The mean of a set of numbers can be geometrically described as the point on the number line that serves as the "center of gravity" for the numbers. If we imagine the number line supported by a point (fulcrum) located at the mean and 1-unit weights placed where the numbers in our sample are located on the line, then equation (3.3) implies that the weights below the mean will perfectly balance the weights above the mean. In other words, the mean serves as the center of gravity for the data. Consider the following diagram for the data on automobile defects:

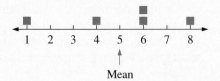

Mean

The number line with a fulcrum at 5 (the mean of the data set) would be perfectly balanced if unit weights were placed at the data values of 1, 4, 6, 6, and 8.

APPLICATION 3.7

The following data represent the annual amounts (in billions of dollars) that the United States spent for agricultural exports from foreign countries from 1974 to 1983, respectively: 10.2 9.3 11.0 13.4 14.8 16.7 17.4 16.8 15.4 16.2.[22] Find the deviation scores for each amount and verify that equation (3.3) holds for this data set.

Solution: The mean is found to be $\bar{x} = 14.12$. The deviation scores are contained in table 3.3.

TABLE 3.3

Data and Deviation Scores
for Application 3.7

Year	Amount	Deviation Score
1974	10.2	−3.92
1975	9.3	−4.82
1976	11.0	−3.12
1977	13.4	−0.72
1978	14.8	0.68
1979	16.7	2.58
1980	17.4	3.28
1981	16.8	2.68
1982	15.4	1.28
1983	16.2	2.08
		0

By adding the deviation scores, we have

$$\Sigma (x - \bar{x}) = 0. \quad ■$$

Sum of Squares

Deviation scores can be used to describe the dispersion of a given distribution of quantitative data. Recall that a deviation score represents the directed distance a measurement is from the mean of a set of data. As a result, we might think that the average for all the deviation scores should provide a measure of dispersion of all the measurements about their mean. But this is not the case, since by equation (3.3), the sum of all the deviation scores is 0. The positive deviation scores are canceled by the negative deviation scores when added. To avoid this problem caused by the negative deviation scores canceling the positive ones, we can first square each deviation score before adding. The resulting sum of squared deviation scores is called the **sum of squares** and is denoted by SS. As we will see later, SS is very useful in statistics for describing the dispersion of a collection of measurements about their mean.

We can compute a sum of squares for either a sample or a population. The formulas for both are as follows:

Sum of Squares Formulas
$SS = \Sigma (x - \bar{x})^2 \qquad SS = \Sigma (x - \mu)^2$
Sample $\qquad\qquad$ Population

(3.4)

The formulas differ, but the computational procedures are the same.

EXAMPLE 3.22

Let's find SS for the following sample of test scores made by five students on an American history test: 62, 80, 83, 72, and 73. We first find $\bar{x}$:

$$\bar{x} = \frac{62 + 80 + 83 + 72 + 73}{5} = 74$$

Then by using formula (3.4), we have

$$SS = \Sigma (x - \bar{x})^2$$
$$= (62 - 74)^2 + (80 - 74)^2 + (83 - 74)^2 + (72 - 74)^2 + (73 - 74)^2$$
$$= 144 + 36 + 81 + 4 + 1 = 266$$

In general, a sum of squares SS can be found as follows:

To Determine SS

a. Determine the mean.

b. Find the deviation score for each measurement.

c. Square each of the deviation scores.

d. Find the sum of these squares.

To simplify the computations involved in calculating SS, the following computational formulas will be useful:

Computational Formulas for SS

$$SS = \Sigma x^2 - \frac{(\Sigma x)^2}{n} \qquad SS = \Sigma x^2 - \frac{(\Sigma x)^2}{N} \qquad (3.5)$$

Sample   Population

where Σx^2 is the sum of the squares of the data, n is the size of the sample, and N is the size of the population. Both of the formulas given by (3.5) can be verified algebraically using the properties of summation found in appendix A.

EXAMPLE 3.23

Note that $\Sigma x^2 \neq (\Sigma x)^2$. This fact can be demonstrated by observing that

$$2^2 + 3^2 \neq (2 + 3)^2$$
$$13 \neq 25.$$

Example 3.24

Refer to example 3.22. Let's use formula (3.5) to compute SS for the American history test scores. We first organize the computations using the following two-column table:

x	x^2
62	3,844
80	6,400
83	6,889
72	5,184
73	5,329
370	27,646

By using formula (3.5) to calculate SS, we get

$$SS = \Sigma\, x^2 - \frac{(\Sigma\, x)^2}{n}$$

$$= 27{,}646 - \frac{370^2}{5} = 266$$

For purposes of computation, the formulas given by (3.5) are usually preferred to those given by (3.4). For one thing, the formulas of (3.5) are easier to use with a calculator, since there are fewer subtractions involved. Furthermore, the formulas in (3.5) do not require that the mean be found. If formulas (3.4) are used in situations where the mean does not terminate and rounding is involved, the calculations could lead to results lacking in precision. It's easy to find a situation where (3.5) is more accurate than (3.4).

EXAMPLE 3.25

Let's find SS for the values 0, 5, and 8. If the mean is rounded to the nearest tenth, then $\bar{x} = 13/3 \approx 4.3$. Then by using formula (3.4) we have

$$SS = \Sigma\, (x - \bar{x})^2$$
$$= (0 - 4.3)^2 + (5 - 4.3)^2 + (8 - 4.3)^2 = 32.670$$

By using formula (3.5), we have

$$SS = \Sigma\, x^2 - \frac{(\Sigma\, x)^2}{n}$$

$$= 89 - \frac{13^2}{3} = 32.667$$

To the nearest thousandth, the two answers differ by 0.003.

Variance

The **variance** of a population of measurements is defined as the average of the squared deviation scores and is denoted by σ^2 (read sigma squared). The symbol σ is the lower case Greek letter sigma. The variance of a population is given by formula (3.6).

Variance of a Population
$\sigma^2 = \dfrac{SS}{N}$

(3.6)

The variance of a sample is denoted by s^2 and is defined by the following formula:

Variance of a Sample
$s^2 = \dfrac{SS}{n - 1}$

(3.7)

In chapters 8–15, we will use the sample variance s^2 to estimate an unknown population variance σ^2. Note that

If we were to compute the sample variance s^2 by dividing SS by n, instead of $n - 1$, we would, on the average, underestimate σ^2.

For descriptive purposes only, some statisticians compute the sample variance by dividing SS by n. Of course, for large values of n, there is little difference between the values of SS/n and SS/$(n - 1)$. Used by itself as a descriptive measure of spread or dispersion, the variance is difficult to interpret, since the units of the variance are the squares of the units of measurement.

APPLICATION 3.8

Suppose the American history test scores given previously (62, 80, 83, 72, and 73) constitute a population. Find the population variance σ^2.

Solution: By using formula (3.6), we have

$$\sigma^2 = \frac{SS}{N}$$

$$= \frac{266}{5} = 53.2 \quad \blacksquare$$

APPLICATION 3.9

Table 3.4 shows the costs (in U.S. cents) per liter of high octane gasoline in 19 cities throughout the world. Determine the sample variance s^2.[23]

TABLE 3.4

Gasoline Costs in 19 World Cities

City	Cost Per Liter
Amsterdam	57
Brussels	53
Buenos Aires	38
Hong Kong	57
Johannesburg	48
London	56
Madrid	59
Manila	46
Mexico City	25
Montreal	47
Nairobi	57
New York	40
Oslo	65
Paris	58
Rio de Janeiro	42
Rome	76
Singapore	59
Sydney	43
Tokyo	79

Solution: We shall use formula (3.5) to compute SS. Toward this end, we first compute Σx and Σx^2. With the help of a calculator, we determine that $\Sigma x = 1005$ and

$\Sigma x^2 = 56,171$. Thus, the sum of squares is

$$SS = \Sigma x^2 - \frac{(\Sigma x)^2}{n}$$

$$= 56,171 - \frac{1005^2}{19} = 3011.7895$$

Now applying formula (3.6), we obtain

$$s^2 = \frac{SS}{n-1}$$

$$= \frac{3011.7895}{18} \approx 167.32$$

The sample variance of the 19 gasoline prices is 167.32 square cents. ▪

EXAMPLE 3.26

For the gasoline price-per-liter data in application 3.9, knowing that $s^2 = 167.32$ square cents has very little, if any, meaning by itself. We know that if the value of the variance is large, then the measurements are widely dispersed, while if the value of the variance is small, there is very little variability in the measurements.

EXAMPLE 3.27

If the variance is 0, all the measurements are equal. This is a consequence of the fact that SS is always greater than or equal to 0 and is equal to 0 only when each measurement is equal to the mean.

EXAMPLE 3.28

However, if we analyzed two samples of data, A and B, and found that $s^2_A = 10$ and $s^2_B = 5$, we would know that the measurements in sample A are more dispersed about their mean than the measurements in sample B are dispersed about their mean. For the most part and for descriptive purposes, the variance is used for comparisons as a relative measure of variation.

Standard Deviation

Another measure of dispersion, related to the variance, is the standard deviation. The **standard deviation** is defined as the positive square root of the variance. The population standard deviation is denoted by σ and the sample standard deviation is denoted by s. Hence, we have the following formulas:

Sample Standard Deviation

$$s = \sqrt{s^2} = \sqrt{\text{sample variance}}$$

Population Standard Deviation

$$\sigma = \sqrt{\sigma^2} = \sqrt{\text{population variance}}$$

EXAMPLE 3.29

For the data in application 3.8, the population standard deviation is $\sigma = \sqrt{53.2} = 7.29$, and for the data in application 3.9, the sample standard deviation is $s = \sqrt{167.32} = 12.94$ cents.

Computer display 3.4 illustrates the use of MINITAB for the data given in application 3.9. Note that MINITAB doesn't give the variance directly.

Computer Display 3.4

```
MTB > SET C1
DATA> 57 53 38 57 48 56 59 46 25 47 57 40 65 58 42 76 59 43 79
DATA> END
MTB > MEAN C1
  MEAN = 52.895
MTB > STDEV C1
  ST.DEV. = 12.935
MTB > LET K1 = STDEV(C1)**2
MTB > PRINT K1
K1 167.322
```

Notice that K1 is the value of the variance. K1 is called a **constant** in MINITAB. Constants are referred to by number, K1, K2, K3, ... Each constant can store one number. Constants can be created by using the LET command. The command LET K1 = STDEV(C1)**2 stores the variance in constant K1. Note also that the symbol "**" means exponentiation or raising a number to a power.

Why do we need both the variance and standard deviation as measures of dispersion? One answer to this question involves the unit of measurement. As we saw in application 3.9, if the set of data involves measurements in cents, then the unit of variance is square cents, and the unit of standard deviation is cents. Thus, an expression such as $x - \bar{x}$ would be meaningful but an expression such as $x - s^2$ would not, since in the first case the units match but in the second case they do not. In section 3.4 when we study standard scores, we will make use of the fact that a measurement from a distribution and the mean and standard deviation of the distribution all have the same unit of measurement. Applications 3.10 and 3.11 show the use of the sample variance and the sample standard deviation for making relative comparisons.

APPLICATION 3.10

The accompanying data represent the average miles-per-gallon per day for five days for cars A and B driven under similar conditions.

A	20	25	30	15	35
B	15	27	25	23	35

a. Find the mean and range of miles-per-gallon ratings for each car.

b. Which car seems to have obtained more consistent mileage if consistency is determined by examining the variances? Explain.

Solution:

a. For car A, we have

$$R_A = 35 - 15 = 20$$
$$\bar{x}_A = 25$$

For car B, we have

$$R_B = 35 - 15 = 20$$
$$\bar{x}_B = 25$$

Note that both cars have the same mean and the same range of miles-per-gallon ratings.

b. We calculate the variance for car A, s_A^2.

x	$x - \bar{x}$	$(x - \bar{x})^2$
20	-5	25
25	0	0
30	5	25
15	-10	100
35	10	100
		SS = 250

As a result of formula (3.7), we have

$$s_A^2 = \frac{\text{SS}}{n-1}$$

$$= \frac{250}{4} = 62.5$$

The variance for the gas mileages for car A is 62.5 square miles.
We next calculate the variance for car B, s_B^2.

x	$x - \bar{x}$	$(x - \bar{x})^2$
15	-10	100
27	2	4
25	0	0
23	-2	4
35	10	100
		SS = 208

As a result of formula (3.7) we have

$$s_B^2 = \frac{\text{SS}}{n-1}$$

$$= \frac{208}{4} = 52$$

The variance for the gas mileages for car B is 52 square miles. Since the variance for car B is smaller than the variance for car A, car B got more consistent gas mileages. Notice that if we had used the range, we would have concluded that both cars obtained equally consistent gas mileages. ▪

APPLICATION 3.11

The data in table 3.5 indicate the prices per pound (in U.S. dollars) for pork roast and cheddar cheese in 15 world capitals.[24]

TABLE 3.5

Pork and Cheese Prices in World Capitals

World Capital	Pork Roast (boneless)	Cheddar Cheese
Bern	$6.61	$4.00
Bonn	2.38	2.74
Brasilia	1.27	1.08
Buenos Aires	1.36	2.03
Canberra	2.06	2.60
London	1.56	1.81
Madrid	2.33	3.15
Mexico City	1.08	2.29
Ottawa	1.99	3.98
Paris	2.47	2.37
Pretoria	1.95	1.76
Rome	2.46	2.96
Stockholm	5.35	2.54
Tokyo	4.19	2.38
Washington	3.29	2.69

For which food, roast pork or cheddar cheese, are the world prices less variable (and more stable)?

Solution: With the use of a calculator we determine the following quantities:

$$\text{Pork data: } \Sigma\, x = 40.35,\ \Sigma\, x^2 = 143.01,\ n = 15$$
$$\text{Cheese data: } \Sigma\, x = 38.38,\ \Sigma\, x^2 = 106.67,\ n = 15$$

And as a consequence of formulas (3.5) and (3.7), we have:

Pork data

$$SS_p = \Sigma\, x^2 - \frac{(\Sigma\, x)^2}{n}$$
$$= 143.01 - \frac{(40.35)^2}{15} = 34.4685$$

The variance of the pork roast data is

$$s_p^2 = \frac{SS_p}{n-1}$$
$$= \frac{34.4685}{14} \approx 2.46$$

Cheese data

$$SS_c = 106.67 - \frac{(38.38)^2}{15} = 8.4684$$

And the variance of the cheese data is

$$s_c^2 = \frac{8.4684}{14} = 0.60$$

Thus, the world prices of cheddar cheese are more stable than the world prices of roast pork. ■

Estimate of s

It is interesting to note that for samples of at least size 20 having a bell-shaped distribution, we have the following estimate of the sample standard deviation:

> **Estimate of s**
>
> $$s \approx \frac{R}{4}$$
>
> (3.8)

where R denotes the range. This is a conservative estimate that can be used to check our computations for s, and it involves very little effort. The significance of dividing the range R by 4 will be discussed in chapter 7 when we examine the normal distributions.

APPLICATION 3.12

For the cheddar cheese data in application 3.11, estimate s by using formula (3.8) and check the estimate by computing the value for s.

Solution: The range for the cheddar cheese prices is

$$R = U - L$$
$$= 4.00 - 1.08 = 2.92$$

As a consequence of formula (3.8), we have

$$s_c \approx \frac{R}{4}$$
$$= \frac{2.92}{4} = 0.73$$

Since the standard deviation is the square root of the variance, we can use the result of application 3.11 to obtain

$$s_c^2 = 0.60$$
$$s_c = \sqrt{0.60} = 0.77$$

Since $R/4 = 0.73$ is in the same "ball park" as $s_c = 0.77$, we have little reason to suspect that an error has been committed. Don't be too concerned about a mistake unless one result is at least twice the size of the other. ▪

APPLICATION 3.13

Suppose that the largest measurement in a sample is 90 and the smallest measurement is 30. The standard deviation has been calculated to be 185. Does this answer seem reasonable? Explain.

Solution: No, the answer does not seem reasonable. The range is $90 - 30 = 60$, and using formula (3.8) we have

$$s \approx \frac{R}{4} = \frac{60}{4} = 15$$

Thus, we suspect an error has been made in calculating s to be 185. The calculations should be rechecked. ▪

Variance and Standard Deviation for Data in Frequency Tables

Frequently, we will have occasion to find the variance or standard deviation for data displayed in a frequency table. Both these measures can be calculated once SS is known. To find SS for data that have measurements with repetitions, we first determine the frequency for each measurement.

EXAMPLE 3.30

To find the sum of squares SS for the data 2, 2, 2, 2, and 7, representing the number of walks given up by a baseball pitcher in the last five games, we need only find the deviation scores for the measurements 2 and 7. The squared deviation score for 2 can then be multiplied by its frequency, $f = 4$, to get the sum of the squared deviations for the four values of 2. This sum is then added to the squared deviation score for 7 to arrive at SS. Since the mean of the five data points is 3, we have

$$SS = \Sigma (x - \bar{x})^2$$
$$= 4(2 - 3)^2 + 1(7 - 3)^2$$
$$= 4 + 16$$
$$= 20$$

Based on the ideas of example 3.30, we have the following formulas for finding the sum of squares when data are organized in a frequency table:

Sum of Squares for Data in a Frequency Table		
$SS = \Sigma f(x - \bar{x})^2$	$SS = \Sigma f(x - \mu)^2$	(3.9)
Sample	Population	

A P P L I C A T I O N 3.14

The following measurements represent the number of days it took express mail shipped from the west coast to reach its destination on the east coast for the past ten mailings: 2, 2, 2, 3, 3, 4, 4, 5, 5, and 10. Use formulas (3.9) to determine SS.

Solution: We first construct table 3.6, a frequency table to aid us with our calculations. The sample mean is easily found to be $x = 4$.

TABLE 3.6

Frequency Table for Application 3.14

x	f	$x - \bar{x}$	$(x - \bar{x})^2$	$f(x - \bar{x})^2$
2	3	-2	4	12
3	2	-1	1	2
4	2	0	0	0
5	2	1	1	2
10	1	6	36	36
				SS = 52

The value of SS is the sum of the entries in the last column, SS = 52.

For illustrative purposes, we shall also determine the value of SS by using formula (3.4) and table 3.7.

TABLE 3.7	x	$x - \bar{x}$	$(x - \bar{x})^2$

x	$x - \bar{x}$	$(x - \bar{x})^2$	
2	−2	4	
2	−2	4	$\}\, f = 3$ and $(3)(4) = 12$
2	−2	4	
3	−1	1	
			$\}\, f = 2$ and $(2)(1) = 2$
3	−1	1	
4	0	0	
			$\}\, f = 2$ and $(2)(0) = 0$
4	0	0	
5	1	1	
			$\}\, f = 2$ and $(2)(1) = 2$
5	1	1	
10	6	36	$\{\, f = 1$ and $(1)(36) = 36$
		SS = 52	

Calculation of SS Using Formula (3.4)

We see that SS = 52, as calculated using formula (3.9). Note that the first entry, 12, in the fifth column of table 3.6 corresponds to the sum of the first three entries of 4 listed in the last column of table 3.7 and so forth. ▪

The following computational formula can be used to find the sum of squares for data displayed in a frequency table.

Computational Formula for SS Using Frequencies

$$SS = \Sigma fx^2 - \frac{(\Sigma fx)^2}{\Sigma f} \qquad (3.10)$$

Formula (3.10) is often more convenient to use than the formulas expressed by (3.9). Notice that only one subtraction is involved with formula (3.10) and that the sample mean does not need to be calculated first.

APPLICATION 3.15 Find the sample variance for the following data representing the number of cigars smoked during a particular week by 15 cigar smokers:

x	10	15	17	20	22
f	1	3	5	2	4

Solution: The following table is used to organize the computations:

x	f	fx	x^2	fx^2
10	1	10	100	100
15	3	45	225	675
17	5	85	289	1445
20	2	40	400	800
22	4	88	484	1936
	15	268		4956

As a consequence of formula (3.10), we have

$$SS = \Sigma fx^2 - \frac{\Sigma (fx)^2}{\Sigma f}$$

$$= 4956 - \frac{268^2}{15} = 167.73$$

Hence, the sample variance is

$$s^2 = \frac{SS}{n-1}$$

$$= \frac{167.73}{14} = 11.981$$

Note also that the entries in the fx^2 column can be found either by (1) multiplying the corresponding entries in the x and fx columns or (2) squaring the entries in the x column and multiplying by the corresponding values of f. ■

Shortcoming of the Variance and Standard Deviation

The variance and standard deviation have a serious limitation: they can be seriously affected by the presence of outliers, since they both depend on the mean, which is affected by extreme measurements. When outliers are present in a data set and a measure resistant to outliers is desired, the interquartile range should be used.

Chebyshev's Theorem

The sample standard deviation s indicates the dispersion of the data about the sample mean. If the data values are clustered closely about the mean, then s is small; if the values are considerably spread about the mean, then s is large. But how shall we determine what values of s are large and what values are small? A theorem named after the Russian mathematician Pafnuty Lvovich Chebyshev (1821–1894) provides some useful insight as to how the magnitude of the standard deviation of any set of data relates to the concentration of the data about its mean. According to Chebyshev's theorem, the following statement is true for any set of quantitative data (both populations and samples):

Chebyshev's Theorem

The expression $1 - 1/k^2$ represents the minimum proportion of the data that will lie within k standard deviations of the mean provided that $k \geq 1$.

Note that the result of the calculation $1 - 1/k^2$ is a fraction. Multiplying this fraction by 100 yields the minimum percentage of the data that lie within k standard deviations of the mean. According to Chebyshev's theorem, for any set of measurements,

■ If $k = 1$, then $1 - 1/k^2 = 1 - 1/1^2 = 0$. Then at least 0% of the data will lie within 1 standard deviation of the mean (i.e., within $\bar{x} \pm s$). Thus, for $k = 1$, the interpretation offers no useful information concerning the spread of the data.

■ If $k = 3/2$, then $1 - 1/(3/2)^2 = 1 - 4/9 = 5/9 \approx 56\%$. Then at least 56% of the data will lie within 1.5 standard deviations of the mean (i.e., within $\bar{x} \pm 1.5s$).

■ If $k = 2$, then at least $1 - 1/2^2 = 3/4 = 75\%$. Then at least 75% of the data must fall within 2 standard deviations of the mean (i.e., within $\bar{x} \pm 2s$), as illustrated in figure 3.7.

FIGURE 3.7

Illustration of Chebyshev's theorem for $k = 2$

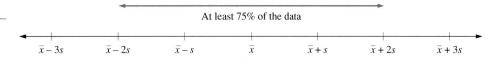

■ For $k = 3$, at least $(1 - 1/3^2)100\% = 89\%$ of the data in any sample must fall within 3 standard deviations of its mean (i.e., within $\bar{x} \pm 3s$), as shown in figure 3.8.

FIGURE 3.8

Illustration of Chebyshev's theorem for $k = 3$

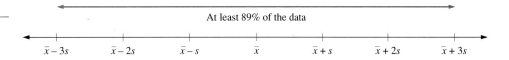

APPLICATION 3.16

The gasoline-cost data in application 3.9 is recalled here.

City	Cost Per Liter
Amsterdam	57
Brussels	53
Buenos Aires	38
Hong Kong	57
Johannesburg	48
London	56
Madrid	59
Manila	46
Mexico City	25
Montreal	47
Nairobi	57
New York	40
Oslo	65
Paris	58
Rio de Janeiro	42
Rome	76
Singapore	59
Sydney	43
Tokyo	79

a. Determine the interval specified by Chebyshev's theorem that will contain at least 75% of the data.

b. What percentage of the measurements actually falls within two standard deviations of the mean?

Solution:

a. By using a calculator we easily determine the mean to be $\bar{x} = 52.89$ cents. Earlier we determined the sample variance to be $s^2 = 167.32$. Thus, the standard deviation is $s = \sqrt{167.32} = 12.94$ cents. According to Chebyshev's theorem, at least $1 - 1/4 = 3/4 = 75\%$ of the data will lie within two standard deviations of the mean. For this data set,

$$\bar{x} - 2s = 52.89 - 2(12.94) = 27.01$$
$$\bar{x} + 2s = 52.89 + 2(12.94) = 78.77$$

Therefore, the interval (27.01, 78.77) will contain at least 75% of the data, as illustrated in the diagram.

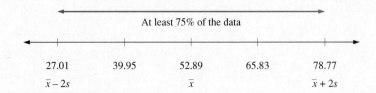

b. Seventeen of the 19 gasoline prices (89.47%) are found to fall between 27.01 and 78.77. This is consistent with our results in part a. Chebyshev's theorem specifies only a lower bound for the percentage of data that will lie within two standard deviations of the mean. As such it provides conservative estimates because little information is known about the shape of the sample. ▪

APPLICATION 3.17 Suppose the average attendance at a major league baseball park for home games during a particular season was 35,500, with a standard deviation of 4200. By using Chebychev's theorem, determine

    a. an interval that contains at least 80% of the attendances for the home games.

    b. at least what proportion of the home games has attendances between 25,000 and 46,000.

Solution:

    a. We set $1 - 1/k^2$ equal to 0.80 and solve for k.

$$1 - \frac{1}{k^2} = 0.80$$

$$\frac{1}{k^2} = 0.20$$

$$k^2 = \frac{1}{0.2} = 5$$

$$k = \sqrt{5} \approx 2.24$$

The interval is $\bar{x} \pm 2.24s = 35{,}500 \pm (2.24)(4200) = 35{,}500 \pm 9408$ or (26,092, 44,908). Thus Chebyshev's theorem guarantees that at least 80% of the attendances are between 26,092 and 44,908.

    b. Notice that Chebyshev's intervals are symmetric about their means. The width of an interval is

$$w = (\bar{x} + ks) - (\bar{x} - ks) = 2ks$$

We first determine the width of the interval (25,000, 46,000). The width is

$$w = 46{,}000 - 25{,}000 = 21{,}000$$

We set $2ks$ equal to 21,000 and solve the resulting equation for k:

$$2ks = 21{,}000$$
$$2k(4200) = 21{,}000$$
$$8400k = 21{,}000$$
$$k = \frac{21{,}000}{8400} = 2.5$$

Hence, at least $1 - 1/(2.5)^2 = 1 - 1/6.25 = 0.84 = 84\%$ of the home games has attendances between 25,000 and 46,000. ▪

It is sometimes convenient to think of Chebyshev's theorem in different terms. The following statement is equivalent to Chebyshev's theorem:

Alternate Form of Chebyshev's Theorem

At most $(1/k^2)100\%$ of the data in any data set will lie beyond k standard deviations of the mean.

For $k = 2$, we have the following diagram:

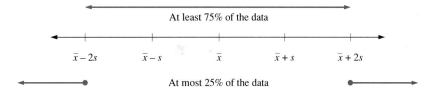

Chebyshev's theorem provides an explanation of how the standard deviation provides a measure of variation for a single sample or population. The validity of the theorem does not depend on the shape of the distribution. As such, it is a very useful and powerful result.

Summary of Notation Used   The following chart summarizes the notation frequently used in connection with samples and populations:

	Mean	Median	Variance	Standard Deviation	Size
Sample	$\bar{x}$	$\tilde{x}$	s^2	s	n
Population	μ	$\tilde{\mu}$	σ^2	σ	N

Note that $\bar{x}$, $\tilde{x}$, s^2, s, and n are examples of statistics, while μ, $\tilde{\mu}$, σ^2, σ, and N are examples of parameters. Recall from chapter 1 that statistics are values computed from a sample and parameters are values computed from a population. It is a popular convention in statistics to use Greek letters to denote most parameters. One exception to this rule is the notation for population size.

EXERCISE SET 3.2

Basic Skills

1. Find the range, variance, and standard deviation of the sample:

$$5 \quad 2 \quad 2 \quad 1 \quad 5 \quad 3 \quad 2 \quad 3 \quad 4$$

2. Find the range, variance, and standard deviation of the sample:

$$9 \quad 6 \quad 4 \quad 6 \quad 5 \quad 8 \quad 7 \quad 6 \quad 7 \quad 0$$

3. Determine the variance and standard deviation of the sample 1, 3, 11, 15, and 20.

4. Determine the variance and standard deviation of the sample 1, 2, 4, 10, 18, and 19.

5. Calculate $\bar{x}$, s^2, and s for the situation
   a. $\Sigma x^2 = 232$, $\Sigma x = 25$, and $n = 15$.
   b. $\Sigma x^2 = 515$, $\Sigma x = 101$, and $n = 20$.

6. Calculate the sample mean, sample variance, and sample standard deviation for the situation
   a. $\Sigma x^2 = 52$, $\Sigma x = 7$, and $n = 9$.
   b. $\Sigma x^2 = 25$, $\Sigma x = 12$, and $n = 13$.

7. The following values have been found for a sample:
   $\Sigma x^2 = 428$ $\Sigma x = 75$ $n = 10$
   Are they reasonable?

8. The following values have been found for a sample:
   $\Sigma x^2 = 48$ $\Sigma x = 7.5$ $n = 20$
   Are they reasonable?

9. Suppose we computed the variance of a sample of size 15 and got 10 by dividing SS by 15, instead of 14, and obtained a result of 10. Find the correct value for s^2.

10. If a calculator has a built-in program for calculating the variance, how could it easily be determined which variance (s^2 or σ^2) it is computing?

11. What is the sum of the deviation scores about the mean for any data set?

12. What is the average of any set of deviation scores?

13. Is the value of the standard deviation always smaller than the corresponding value of the variance?

14. Why does the expression $\bar{x} - s^2$ not make sense?

15. Is it possible for the range and standard deviation of a population to be equal? If so, give an example.

16. Is it possible for the range and variance to be equal? If so, give an example.

17. If the standard deviation for a data set is 0, what must be true concerning the data?

18. What can be said if the standard deviation of a sample is negative?

19. Suppose a sample has a mean of $\bar{x} = 25$ and a standard deviation of $s = 3.2$.
    a. Determine an interval that contains at least 90% of the measurements of the sample.
    b. At least what percentage of the sample are contained in the interval (17, 33)?

20. Suppose a sample has a mean of $\bar{x} = 540$ and a standard deviation of $s = 10.5$.
    a. Determine an interval that contains at least 92% of the measurements of the sample.
    b. At least what percentage of the sample are contained in the interval (524.25, 566.25)?

Further Applications

21. The following data set represents the final exam scores for a class of 30 philosophy students:

98	94	94	57	58	88	97	94	96	85
85	97	92	90	87	80	97	93	87	69
25	100	97	83	74	64	79	89	98	100

Find the percentage of scores that are actually within 2.1 standard deviations of the mean. Now use Chebyshev's theorem for $k = 3.6$. Are the results consistent with the theorem?

22. The following data represent the prices in cents for one pound of flour in 16 world capitals:

41	28	10	16	35	18	21	5
40	30	25	18	14	30	33	24

Find the percentage of prices that are within 1.5 standard deviation of the mean. Now use Chebyshev's theorem for $k = 1.5$. Are the results consistent with the theorem?

23. The average amount spent by customers at a grocery store is $8.34 and the standard deviation of the amount of sale is $8.33. Using Chebyshev's rule, what can be said about the proportion of customers that spend over $25?

24. The number of patients admitted to Memorial Hospital during a weekday has an average of 32 and a standard deviation of 4. On a particular day only 16 patients were admitted. Use Chebyshev's rule to decide if this is an unusual number of admittances for a weekday? Explain.

25. The following table gives a sample of lap times (in minutes) on a 2.5-mile track for two cars, A and B.

    A: 1.0   0.9   1.0   0.8   0.9   1.0   0.9   1.0
    B: 1.3   1.3   1.0   0.9   1.1   0.9   1.4   1.3

    a. Find the average lap times for cars A and B.
    b. Find the variance of the lap times for cars A and B.
    c. Which car had a lower average lap time?
    d. Which car performed more consistently if consistency is measured by the variance?
    e. Find the interquartile range for samples A and B.

26. The following table gives a sample of lap times (in minutes) on a 3-mile track for two cars, C and D.

    C: 1.1   0.8   1.1   0.9   1.0   1.0   0.9   1.1
    D: 1.2   1.4   1.3   0.9   1.1   0.8   1.5   1.4

    a. Find the average lap times for cars C and D.
    b. Find the variance of the lap times for cars C and D.
    c. Which car had a lower average lap time?
    d. Which car performed more consistently if consistency is measured by the variance?
    e. Find the interquartile range for samples C and D.

27. The accompanying table indicates the annual salaries (in dollars) for a sample of 25 laborers.

Annual Salary	Frequency
$5,500	7
6,000	5
7,000	6
8,000	4
30,000	3

    Find the
    a. range.
    b. mean.
    c. standard deviation.
    d. interquartile range.

28. The accompanying table shows the distribution for the number of defective transistors found in 215 lots produced by an electronics manufacturer.

Number of Defective Transistors	Number of Lots
0	25
1	78
2	54
3	33
4	16
5	7
6	2

    a. Find the range.
    b. Find the variance.
    c. Find the sample standard deviation.
    d. Find the interquartile range.

29. A large dairy continually monitors the level of butterfat content in its milk. In its 2% milk, the percentage of butterfat should not deviate much from this percentage. A standard deviation of 10% is acceptable. A sample of 20 cartons of milk was obtained and the percentage of butterfat in each carton was recorded. The results are recorded below.

1.85	2.25	2.01	1.90	1.97
1.80	2.05	2.23	1.65	1.86
2.02	2.09	2.04	2.07	2.14
1.93	2.08	2.17	1.91	1.93

    Calculate the mean and standard deviation for the sample of butterfat amounts. Is there any evidence that the butterfat content is too high? Explain.

Going Beyond

30. What effect does sample size have on the standard deviation and variance?

31. For the sample data illustrated by the accompanying line graph, find $\bar{x}$ and s for the number of children per family.

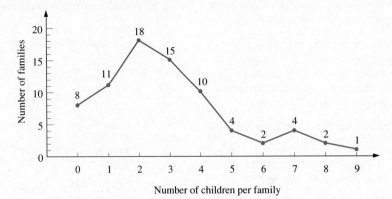

32. In some situations, data are dichotomous, consisting of only two distinct values. For example, dichotomous data result if responses are recorded as male-female, true-false, up-down, on-off, and so forth. In such cases, it is customary to use 0 to represent one value and 1 to represent the other. If 1, 0, 0, 0, 1, 1, 1, 1, 1, and 0 represent a population of values, find μ and σ for the population of 0s and 1s. If p represents the proportion of 1s, show that $\mu = p$ and $\sigma = \sqrt{p(1 - p)}$.

33. For any finite collection of data, determine the value of c that makes $\Sigma (x - c)^2$ as small as possible.

34. Consider the following three data sets:

    A: 20   30   40   50   60
    B: −20 −10   0   10   20
    C: −2   −1   0   1   2

    a. Find SS for each data set. Note that the values for set B were obtained by adding −40 to each measurement in set A and the values in set C were obtained by dividing each measurement in set B by 10.
    b. What relationship exists between SS_A and SS_B? Between SS_A and SS_C?
    c. What relationship exists between s_A^2 and s_B^2? Between s_A and s_C?
    d. What relationship exists between s_A and s_B? Between s_A and s_C?

35. If 3 is added to each measurement in a set of ten measurements having a standard deviation of 3, what is the standard deviation for the new data set?

36. The average grade on a statistics examination was 75 and the standard deviation was 10. After returning the examination to the students, the professor determined that one question was scored incorrectly and that each grade should be increased by 5 points. Find the mean, variance, and standard deviation for the corrected grades.

37. Consider the population of measurements X: 1.233, 1.236, 1.230, 1.236, 1.234, 1.237, 1.233, 1.235, 1.238, and 1.238. Suppose each measurement is transformed using $Y = 1000X - 1230$. Find the mean, variance, and standard deviation of the Y measurements. In addition, show that
    a. $\mu_y = 1000\mu_x - 1230$. As a result, $\mu_x = (0.001)(\mu_y + 1230)$.
    b. $\sigma_y^2 = (1000)^2\sigma_x^2$. As a result, $\sigma_y^2 = (0.0000001)\sigma_y^2$.
    c. $\sigma_y = (1000)\sigma_x$. As a result, $\sigma_x = (0.001)\sigma_y$.

38. If a constant C is added to each measurement in a data set, show that the variance of the new set of measurements is the same as the variance of the original set.

39. If each measurement in a data set is multiplied by a constant C, show that the sum of squares of the new set is equal to C^2 times the sum of squares of the original set.

40. If each measurement in a data set is multiplied by a constant C, is the standard deviation of the new set equal to C times the standard deviation of the original set?

41. Another measure of dispersion is the *mean absolute deviation* (MAD). It is defined by

$$\text{MAD} = \frac{\Sigma|x - \bar{x}|}{n}$$

Compute the value of MAD for the data in exercise 1.

42. The coefficient of variation provides a measure of variability that is independent of the measuring unit. As such, it can be used to compare the variability of two groups of data involving different units of measure. For example, it can be used to compare the standard deviation of the distribution of annual incomes (in dollars) and the standard deviation of the years of service for all the employees of a certain company. The *coefficient of variation* (CV) expresses the standard deviation as a percentage of the mean and is defined by CV $= (s/\bar{x})(100)$. Suppose a financial analyst for a stock brokerage firm wants to compare the variation in the price-earnings ratios for a group of common stocks with the variation in their net returns on investment. For the price-earnings ratios, the mean is 9.8 and the standard deviation is 2.4. The mean net return on investments is 20%, and the standard deviation of the net returns on investments is 4.3%. Use the coefficient of variation to compare the relative variation for the price-earnings ratios and the net returns on the investments.

43. Suppose the board of directors of a large corporation wants to compare the dispersion of incomes for its top executives with the dispersion of incomes for its unskilled employees. For a sample of executives, the mean salary is $400,000 and the standard deviation is $50,000, whereas for a sample of unskilled employees, the mean is $11,000 and the standard deviation is $1200. In which group is the relative dispersion greatest?

44. Can the coefficient of variation be used with data sets involving negative numbers? Explain.

45. The degree of skewness of a distribution is commonly measured by *Pearson's coefficient of skewness*, denoted by CS. For a sample, it is defined by

$$CS = \frac{3(\bar{x} - \tilde{x})}{s}$$

For a skewed distribution, the sign of CS will correspond to the direction of skewness. A distribution that is symmetric will have CS $= 0$. The following data represent the starting salaries (in thousands of dollars) of a sample of

1986 college graduates from a large midwestern university: 29.2 27.8 29.0 20.3 16.9 28.7 19.6 24.8 17.4 24.4 20.8 17.8 16.2 17.8. Calculate the coefficient of skewness for the salary data.

46. Find a value for the constant C that minimizes $\Sigma|x - C|$ for the following sample of measurements: 2 3 7 7 8.

47. Prove that $\Sigma(x - \bar{x})^2 = \Sigma x^2 - (\Sigma x)^2/n$.

48. If all the measurements in a population are within one standard deviation of the mean, characterize the population; that is, determine what kinds of numbers comprise the population.

49. Consider the sample of measurements: 1.2, 2, 3, 4, and 4.9. Create another sample of measurements having a
    a. mean three units higher.
    b. variance four times as large.
    c. mean three units higher and a variance four times as large.

50. For a population, can the standard deviation ever be larger than one half the range? Explain.

51. Show that for a sample of two measurements, $s = R/\sqrt{2}$.

52. If s is the standard deviation of a sample, it can be shown that

$$\frac{R}{2(n - 1)} \leq s \leq \left(\frac{R}{2}\right)\sqrt{\frac{n}{n - 1}}$$

where n is the sample size and R is the range. The following data represent the blood cholesterol levels for a sample of eight persons: 239 218 227 357 161 286 310 245.
    a. Find upper and lower bounds for s.
    b. Estimate s by using the midpoint of the interval determined by the above result.
    c. Calculate the value of s and compare the result with the estimated value found in part b.

S E C T I O N   3 . 3

Central Tendency and Dispersion for Data Contained in Grouped Frequency Tables

Measures of central tendency and dispersion for data that are displayed in grouped frequency tables can be calculated, but their values are not exact and only approximations. This is because the raw measurements are unknown and have been placed into class intervals. Before computers became commonplace, it required a great deal of computational effort to calculate measures of central tendency and dispersion for large data sets. In an attempt to deal with this problem and eliminate some of the computational drudgery, data were placed in grouped frequency tables and certain assumptions had to be made before carrying out the computations. The validity of these assumptions had a direct effect on the accuracy of the results.

Today high-speed computers make it possible to process huge lists of raw data very quickly and with highly accurate results, thus eliminating the computational advantages of frequency tables. You might then ask why we would want to compute approximate values of certain statistics from grouped frequency tables. There is a great deal of data summarized in grouped frequency tables supplied by others, such as the federal government, and the only way to calculate measures of central tendency is to use the grouped data.

Mean for Grouped Data

If we are finding the mean for data that have been displayed in a grouped frequency table, we use the class marks to represent the measurements for each class. Then formula (3.2) can be used to determine the **approximate sample mean** $\bar{x}_a$ since the original data are unknown and each observation is represented by its class mark.

APPLICATION 3.18

The following data represent the number of records sold each day for a 25-day period by a music shop located in a shopping mall:

$$\begin{array}{cccccccccccc}
60 & 36 & 61 & 56 & 19 & 35 & 51 & 42 & 21 & 28 & 33 & 67 & 30 \\
49 & 57 & 54 & 59 & 28 & 63 & 38 & 15 & 24 & 35 & 46 & 53
\end{array}$$

For convenience, the data have been displayed in the following grouped frequency table:

Number of Records Sold	Number of Days		Class mark
15–25	4	×	20
26–36	7	×	31
37–47	3	×	42
48–58	6	×	53
59–69	5	×	64

$\dfrac{T}{25}$

Find

a. $\bar{x}$, the average number of records sold per day.

b. $\bar{x}_a$, the approximate average number of records sold per day.

Solution:

a. With the help of a handheld calculator, we determine the sum of the 25 measurements to be $\Sigma x = 1060$. Hence, the sample mean is

$$\bar{x} = \frac{\Sigma x}{n}$$

$$= \frac{1060}{25} = 42.4$$

Thus, the average number of records sold per day is 42.40.

b. We first find the class marks, X. (Recall from chapter 2 that a class mark is the midpoint of a class interval.) Each class mark is then multiplied by its corresponding frequency as shown in table 3.8.

TABLE 3.8

Class Marks Multiplied
by Frequencies for
Application 3.18

Class	f	X	fX
15–25	4	20	80
26–36	7	31	217
37–47	3	42	126
48–58	6	53	318
59–69	5	64	320

By using formula (3.2), the approximate mean is

$$\bar{x}_a = \frac{\Sigma (fX)}{\Sigma f}$$

$$= \frac{1061}{25}$$

$$= 42.44$$

Note that $\bar{x}_a = 42.44$ is only an approximate value for the mean of the 25 original sample measurements; the approximation is considered good, compared with the exact value, $\bar{x} = 42.40$, obtained in part a. ■

Median for Grouped Data

There are two general methods for calculating the median of data that have been grouped into classes. The methods differ in the assumption regarding how the data are treated within the classes.

Method I Every value within the class occupies the class mark.

Method II The values in each class are distributed evenly throughout the class. This assumption will enable the median to have the following special property for a frequency histogram:

If a vertical line is drawn perpendicular to the horizontal axis of the histogram at the median value, then the area of the histogram that lies to the left of the vertical line is equal to the area of the histogram that lies to right of the vertical line.

Consider application 3.19 and note that the approximate values of the median produced by the two methods differ in value. Method II is the method that is typically used to approximate the median of data grouped in classes because of the equal distribution of area above and below the median in a histogram.

APPLICATION 3.19

Table 3.9 represents the speeds (in miles per hour) for a sample of 37 cars traveling through a 25 miles-per-hour school zone. Find the approximate median speed.

TABLE 3.9

Data for Application 3.19

Speed	Number of Cars	Cum f
1–5	3	3
6–10	2	5
11–15	5	10
16–20	10	20
21–25	7	27
26–30	10	37

Solution:

Method I The class marks (denoted by X) are contained in the following table. The class mark for the first class is $(1 + 5)/2 = 3$ and the other class marks are found by adding 5 (the class width) to the first class mark.

	Number of		
Speed	Cars	X	Cum f
1–5	3	3	3
6–10	2	8	5
11–15	5	13	10
16–20	10	18	20
21–25	7	23	27
26–30	10	28	37
	37		

From this point, we can determine the median according to the rule given in section 3.1. Since there are an odd number of measurements, the **approximate sample median** $\tilde{x}_a$ is the measurement (class mark) that occupies the 19th position in the table above. Thus, the approximate median is $\tilde{x}_a = 18$.

Method II Since $n = 37$, we want to locate the $n/2 = 37/2 = 18.5$th value. By observing the table, we notice that the 18.5th value (the approximate median) falls in the 16–20 class, because the first three classes contain a total of 10 values and the fourth class contains 10 values. Therefore, we must count $(18.5 - 10) = 8.5$ values into the 16–20 class under the assumption that the 10 values falling in this class are spread evenly throughout the class. In other words, we are seeking the measurement in the 16–20 class that is located a distance of 8.5/10 into the class. Since the width of each class is $w = 5$, to find the approximate median value $\tilde{x}_a$ we need only add $(8.5)/10$ of the width $w = 5$ to the lower boundary of the fourth class. Thus, the approximate value for the median is

$$\tilde{x}_a = 15.5 + \frac{8.5}{10}(5)$$

$$= 15.5 + 4.25 = 19.75 \qquad \blacksquare$$

EXAMPLE 3.31

A histogram for the data in application 3.19 is given in figure 3.9. We can easily verify that the sum of the areas of the rectangles below the value 19.75 is equal to the sum of the areas of the rectangles above 19.75.

FIGURE 3.9

Histogram for data of application 3.19

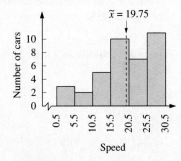

In general, if $\mathscr{L}$ is the lower boundary of the class within which the n falls, f is the frequency of the class containing the median, g is the numb that remain to be counted when we reach $\mathscr{L}$ (counting from the smallest w is the class width, then the approximate median for the data using method II is given by

$$\tilde{x}_a = \mathscr{L} + \left(\frac{g}{f}\right)(w)$$

For application 3.19, $\mathscr{L} = 15.5$, $g = 8.5$, $f = 10$, and $w = 5$. Substituting these values in the above expression yields

$$15.5 + \left(\frac{8.5}{10}\right)(5) = 19.75$$

the same value as we obtained in the application.

Mode for Grouped Data

A disadvantage to using the mode with a grouped frequency distribution of data is that the value of the mode often depends on the arbitrary grouping of the data. It is for this reason that a mode for a grouped frequency distribution is often referred to as a **crude mode** or **modal class.**

If the data are organized in a grouped frequency table, a crude mode or modal class, if it exists, can be easily identified; it corresponds to the class mark for a class having a largest frequency value. And for data displayed in a histogram, a mode is associated with a tallest bar.

EXAMPLE 3.32

For the histogram shown in figure 3.10, the crude modes are seen to be 20 and 40.

FIGURE 3.10

Histogram displaying two modes

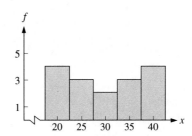

Midrange for Grouped Data

For data organized in a grouped frequency table the midrange is approximately the average of the lower class boundary of the first class and the upper class boundary of the last class.

EXAMPLE 3.33

The approximate midrange for the data in application 3.19 is

$$\frac{0.5 + 30.5}{2} = 15.5$$

Position Points for Data in a Grouped Frequency Table

Method II for finding the approximate median value for data in a grouped frequency table can also be used to find percentile points in a grouped frequency table.

APPLICATION 3.20 For the data displayed in the accompanying grouped frequency table, find P_{60}, the 60th percentile (or the 6th decile).

Speed	Number of Cars	Cum f
1–5	3	3
6–10	2	5
11–15	5	10
16–20	8	18
21–25	7	25
26–30	10	35

Solution: Starting with the first column in the table, we count 60% of the data, namely $(0.60)(35) = 21$ values. Thus, P_{60} must fall in the class containing the 21st measurement. This class is 21–25. The value P_{60} is located within the interval 21–25 at a distance of 2.14 from the left boundary point of the interval. [This distance of 2.14 was found by multiplying $(21 - 18)/7$ by 5, the width of the class.] Thus, the 60th percentile is

$$P_{60} = 20.5 + 2.14 = 22.64$$

Sixty percent of the data fall below a value of 22.64. ■

Variance and Standard Deviation

Class marks are typically used to represent the measurements occupying the classes within a grouped frequency table when the approximate variance or standard deviation is needed for the data. When this is done, the formulas for calculating the variance and standard deviation of an ungrouped frequency distribution from section 3.2 are used.

EXERCISE SET 3.3

Further Applications

1. The accompanying grouped frequency table shows the ages of a sample of 36 people attending an adult movie.

Class	f
8–13	2
14–19	7
20–25	13
26–31	5
32–37	9

a. Find the approximate mean age.
b. Approximate the median age by using methods I and II.
c. Find P_{40}, the 40th percentile, and P_{65}, the 65th percentile.
d. Find Q_3, the third quartile, and D_3, the third decile.
e. Determine the skewness of the frequency histogram.
f. Find the approximate variance.
g. Find the approximate standard deviation.

2. The grouped frequency table shown here gives the distribution of rainfall in a certain Maryland county for the month of June over the last 29 years.

Rainfall in Inches	Number of Years
2.0–2.5	3
2.6–3.1	5
3.2–3.7	6
3.8–4.3	8
4.4–4.9	7

a. Find the approximate mean amount of rainfall.
b. Approximate the median amount of rainfall by using method II.
c. Find P_{40}, the 40th percentile, and P_{75}, the 75th percentile.
d. Find Q_1, the first quartile, and D_4, the fourth decile.
e. Determine the skewness of the frequency histogram.
f. Find the approximate variance.
g. Find the approximate standard deviation.

3. The accompanying grouped frequency table indicates the ages of new-car purchasers at a large automobile dealership. Find
   a. the approximate mean age.
   b. the approximate median age by using methods I and II.
   c. the approximate sample variance.
   d. the approximate sample standard deviation.
   e. P_{40} and P_{69}.
   f. Q_1 and D_7.

Age Class	f
28–32	20
33–37	23
38–42	71
43–47	45
48–52	26

4. The accompanying data indicate the semimonthly amounts (in dollars) invested by a sample of 50 employees in a profit sharing plan:

Amount Invested	Number of Employees
30–34	5
35–39	11
40–44	14
45–49	8
50–54	5
55–59	7

   Find
   a. the approximate mean.
   b. the approximate median by using method II.
   c. the approximate variance.
   d. the approximate standard deviation.
   e. P_{60}, the 60th percentile, and P_{65}, the 65th percentile.
   f. Q_3, the third quartile, and D_8, the eighth decile

5. The following table contains automobile repair costs (in dollars) for an insurance company's minor claims category:

Repair Cost	Frequency
0–99	12
100–199	35
200–299	75
300–399	84
400–499	125

   Find the
   a. approximate mean.
   b. approximate median by using method II.
   c. approximate variance.
   d. approximate standard deviation.
   e. P_{20}, the 20th percentile, and P_{35}, the 35th percentile.
   f. Q_3, the third quartile, and D_9, the ninth decile

6. The accompanying table contains a grouped frequency distribution for the length of 50 long-distance telephone calls (rounded to the nearest minute) made by a certain agency. Compute the approximate variance and approximate standard deviation for the distribution.

Call Length	f
4–7	23
8–11	9
12–15	11
16–19	4
20–23	2
24–27	1

Going Beyond

7. For the data in exercise 1, what is the percentile corresponding to an age of 18 years? This percentage is commonly referred to as the *percentile rank* of 18.

SECTION 3.4   *Standard Scores and Outliers*

Standard Scores as Measures of Relative Standing

Suppose after taking a statistics examination you obtain your score. You are then interested in determining how your score compares to those of others that have also taken the examination. You might want to know if your score is above or below the mean and by how much. A **standard score** will provide you with the information on how well you did on the examination relative to the class; it will provide you with a measure of relative standing within the class.

Bob scores 700 on the mathematics portion of the SAT and Jim scores 24 on the College Placement Test (CPT) test of mathematical ability. The mean and the standard deviation of the SAT are 500 and 100, respectively, and the mean and standard deviation of the CPT are 18 and 6, respectively. If both tests are assumed to measure the same kind of ability, which person ranks higher? To answer this question, we need some method of comparison that will allow us to compare scores from different distributions. It is clear that the deviation of each score from its mean is not a correct basis for comparison in this case, since Jim's deviation score is

$$x - \bar{x} = 24 - 18 = 6$$

and Bob's deviation score is

$$x - \bar{x} = 700 - 500 = 200$$

Neither deviation score takes into account the spread of the scores.

By using standard scores, we shall see that Bob ranks higher than Jim in the ability measured by the test. A standard score takes into account the variability of measurements about their mean.

A measure that allows us to make comparisons from different distributions and takes into account the dispersion of the scores is the standard score. A *standard score* is defined as

$$\text{standard score} = \frac{\text{deviation score}}{\text{standard deviation}}$$

and is usually denoted by z. This relationship can be expressed as

Standard Scores

$$z = \frac{x - \mu}{\sigma} \qquad z = \frac{x - \bar{x}}{s} \qquad (3.11)$$

Population   Sample

depending on whether a population or sample is involved.

Since a standard score is defined as a deviation score divided by the standard deviation, it represents the number of standard deviations a score is from the mean.

A standard score is sometimes called a z score. Referring to the example above, Jim's standard score or z score is

$$z = \frac{x - \mu}{\sigma}$$

$$= \frac{24 - 18}{6} = 1$$

and Bob's standard score is

$$z = \frac{x - \mu}{\sigma}$$

$$= \frac{700 - 500}{100} = 2$$

Jim's score of 24 is one standard deviation above the mean for the CPT test, and Bob's score of 700 is two standard deviations above the mean for the SAT. Since both z scores are positive and Bob's z score is higher than Jim's, Bob ranks higher than Jim in the ability measured by the test.

APPLICATION 3.21   Suppose a set of scores has a mean of 10 and a standard deviation of 2.

    a. Fill in the missing entries in the following chart.

x	4	6	8	10	12	14	16
z							

    b. What does a z score of 0 indicate about a score?

    c. What does a positive z score indicate about a score?

    d. What does a negative z score indicate about a score?

    e. Other than indicating whether a score is above or below the mean, what additional information does a z score indicate?

Solution:

    a. By using formula (3.11), we obtain the following z scores:

x	4	6	8	10	12	14	16
z	-3	-2	-1	0	1	2	3

    b. A z score of 0 indicates the score is the mean.

    c. A positive z score indicates the score is above the mean.

    d. A negative z score indicates the score is below the mean.

    e. A z score also indicates the number of standard deviations a score is from the mean. ▪

APPLICATION 3.22   If a distribution of numbers resulting from measuring weights of small children has a mean of 20 lb and a standard deviation of 2 lb, what is the unit associated with any z score?

Solution: If x denotes the weight of a child in pounds, then x pounds minus 20 pounds is $(x - 20)$ pounds. Dividing $x - 20$ pounds by 2 pounds yields a quotient of $(x - 20)/2$. Thus, we observe that a z score has no unit of measure; it is a number only. ▪

APPLICATION 3.23   The data of application 3.11 concerning the prices of pork roast and cheddar cheese in 15 world capitals is repeated here.

World Capital	Pork Roast (boneless)	Cheddar Cheese
Bern	$6.61	$4.00
Bonn	2.38	2.74
Brasilia	1.27	1.08
Buenos Aires	1.36	2.03
Canberra	2.06	2.60
London	1.56	1.81
Madrid	2.33	3.15
Mexico City	1.08	2.29
Ottawa	1.99	3.98
Paris	2.47	2.37
Pretoria	1.95	1.76
Rome	2.46	2.96
Stockholm	5.35	2.54
Tokyo	4.19	2.38
Washington	3.29	2.69

Use z scores to determine which grocery item has the higher relative price in Washington relative to the prices in the other world capitals.

Solution: It can be shown that $\bar{x}_p = \$2.69$ and $\bar{x}_c = \$2.56$. We showed earlier that $s_c = \$0.77$ and it can be easily demonstrated that $s_p = \$1.57$. Since pork roast costs $3.29 per pound in Washington, the z score for pork roast z_p is

$$z_p = \frac{x - \bar{x}}{s}$$

$$= \frac{3.29 - 2.69}{1.57} = 0.38$$

Cheddar cheese costs $2.69 per pound in Washington. Its z score z_c is

$$z_c = \frac{x - \bar{x}}{s}$$

$$= \frac{2.69 - 2.56}{0.77} = 0.17$$

Thus, the price of roast pork is relatively higher in Washington than the price of cheddar cheese. ■

Suppose μ and σ are the mean and standard deviation, respectively, of a finite population. Each measurement x has a corresponding standard score z. The following important facts, which are explained in application 3.24, help to characterize the collection of all standard scores for a population:

> The population of all standard scores has a mean of 0 and a standard deviation of 1.

APPLICATION 3.24

a. Find μ and σ for the population consisting of the values 1, 2, and 3.

b. Find the three standard scores.

c. Show that the mean of the standard scores is 0 and the standard deviation is 1.

Solution:

a. The population mean is

$$\mu_x = \frac{1 + 2 + 3}{3} = 2$$

We use formula (3.6) to obtain the population variance:

$$\sigma_x^2 = \frac{SS}{N} = \Sigma \frac{(x - \mu)^2}{N}$$

$$= \frac{(1 - 2)^2 + (2 - 2)^2 + (3 - 2)^2}{3} = \frac{2}{3}$$

Thus, the standard deviation is

$$\sigma_x = \sqrt{\frac{2}{3}} = 0.816$$

b. We find the z scores using formula (3.11).
   For $x = 1$,

$$z = \frac{1 - 2}{0.816} = -1.225$$

For $x = 2$,

$$z = \frac{2 - 2}{0.816} = 0$$

For $x = 3$,

$$z = \frac{3 - 2}{0.816} = 1.225$$

c. The mean of the z scores is zero. To find SS for the z scores we organize our computations in the following table and then use formula (3.5).

z	z^2
-1.225	1.50
0	0
1.225	1.50
0	3

Formula (3.5) yields

$$SS = \Sigma z^2 - \frac{(\Sigma z)^2}{N}$$

$$= 3 - 0 = 3$$

By using formula (3.6), we have

$$\sigma_z^2 = \frac{SS}{N}$$

$$= \frac{3}{3} = 1$$

Hence, the standard deviation of the z scores is

$$\sigma_z = \sqrt{\text{variance}}$$

$$= \sqrt{1} = 1 \qquad \blacksquare$$

MINITAB can be used to show that the mean and standard deviation of the z scores in application 3.24 are 0 and 1, respectively. Computer display 3.5 contains the required commands and the corresponding output.

Computer Display 3.5

```
MTB > SET C1
DATA> 1 2 3
DATA> END
MTB > LET C2 = (C1 − MEAN(C1))/STDEV(C1)
MTB > LET K1 = MEAN(C2)
MTB > LET K2 = STDEV(C2)
MTB > PRINT C1 C2

ROW C1 C2

  1  1  −1
  2  2   0
  3  3   1

MTB > PRINT K1 K2
K1      0
K2      1.00000
MTB >
```

The column C2 in computer display 3.5 contains the z scores of the data in column C1. The constants K1 and K2 contain the mean and standard deviation, respectively, of the z scores in C2. The command PRINT C1 C2 displays the values in C1 and C2, and the command PRINT K1 K2 displays the mean and standard deviation, respectively, of the z scores. Notice that the results agree with those obtained in application 3.24.

Converting z Scores to x Scores

APPLICATION 3.25

For some applications, we want to convert z scores back to their original or **raw scores.** For example, if $\bar{x} = 10$ and $s = 2$, find the raw score x corresponding to a z score of $z = 16$.

Solution: We use the z score formula and solve for x.

$$z = \frac{x - \bar{x}}{s}$$

$$16 = \frac{x - 10}{2}$$

Multiplying both sides by 2, we have

$$32 = x - 10$$

Adding 10 to both sides, we get

$$x = 42. \quad ■$$

When the z score formula is solved for x, we obtain formula (3.12), which can be used for finding the raw score x given a standard score z (see application 3.26).

> **z Score to Raw Score**
>
> $$x = \mu + \sigma z \qquad (3.12)$$

APPLICATION 3.26 If a population has a mean of 70 and a standard deviation of 5, find the raw score corresponding to a z score of 1.5.

Solution: By using formula (3.12), we get

$$
\begin{aligned}
x &= \mu + \sigma z \\
  &= 70 + (5)(1.5) \\
  &= 70 + 7.5 = 77.5 \quad \blacksquare
\end{aligned}
$$

Box Plots

A **box plot** is a diagram that provides information about center, spread, and symmetry or skewness. The plot uses quartiles; thus, it is resistant to outliers. Sometimes box plots are referred to as **box-and-whisker diagrams.** The following steps are taken to construct a box plot:

Steps to Construct a Box Plot

1. Construct a number line and label the three quartiles on the line.
2. Draw a rectangular box above the line with end points located at the first and third quartiles (the height of the box is not important).
3. Draw a vertical line segment inside the box at the median.
4. Draw two horizontal lines (called whiskers), one extending from the median to the left extreme measurement and one extending from the median to the right extreme measurement.

EXAMPLE 3.34

Let's use the following data representing ozone readings in parts per million (ppm) taken at noon in a large city to construct a box plot:

9   14   12   17   11   20   13   18   22   12   15   16   5   7   9   19   8

We first arrange the data in numerical order from smallest to largest:

5   7   8   9   9   11   12   12   13   14   15   16   17   18   19   20   22

The median is $Q_2 = 13$, the lower quartile is $Q_1 = 9$, and the upper quartile is $Q_3 = 17$.

Step 1.

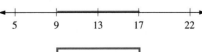

Step 2.

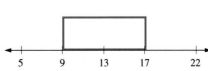

Step 3.

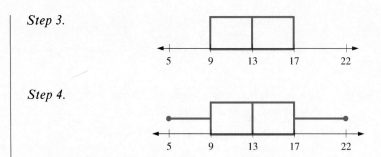

Step 4.

Since the median is somewhat to the right of middle and the longest whisker is to the right, the distribution is skewed to the right.

Computer Display 3.6 illustrates using MINITAB to construct a box plot for the ozone data.

Computer Display 3.6

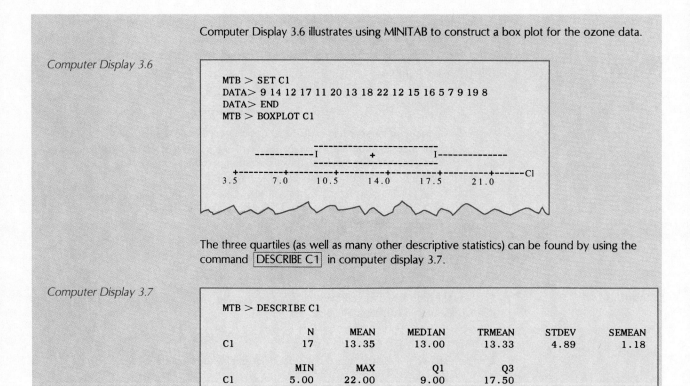

```
MTB > SET C1
DATA> 9 14 12 17 11 20 13 18 22 12 15 16 5 7 9 19 8
DATA> END
MTB > BOXPLOT C1


                  -----------------------------
      ------------I          +          I--------------
                  -----------------------------
      +---------+---------+---------+---------+---------+------C1
    3.5       7.0       10.5      14.0      17.5      21.0
```

The three quartiles (as well as many other descriptive statistics) can be found by using the command DESCRIBE C1 in computer display 3.7.

Computer Display 3.7

```
MTB > DESCRIBE C1

           N       MEAN     MEDIAN     TRMEAN      STDEV     SEMEAN
C1        17      13.35      13.00      13.33       4.89       1.18

         MIN        MAX         Q1         Q3
C1      5.00      22.00       9.00      17.50
```

EXAMPLE 3.35

Figure 3.11 indicates a box plot for a data set that is perfectly symmetrical. The median line is exactly in the middle of the box, and the two whiskers are of the same length. In actual practice we would not expect a sample of data to be perfectly symmetrical. The location of the median within the box is a good indicator of symmetry. The lengths of the whiskers are dependent on single data values, and thus are not as reliable as predictors of symmetry in the population as the location of the median within the box.

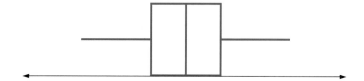

FIGURE 3.11

Symmetrical box plot

EXAMPLE 3.36

Figure 3.12 indicates two box plots for skewed data sets. The one on the left side is skewed to the left, and the one on the right is skewed to the right. Notice the location of the median in each case. If the distribution is skewed to the left, the median is to the right of the center of the box; and if the distribution is skewed to the right, then the median is to the left of the center of the box. Again, in actual practice with real data, the lengths of the whiskers are not a good indicator of skewness in the population, since they depend on single values. Notice that the width of the box is the interquartile range, and thus provides a measure of dispersion for the data. If one whisker were especially long, this would signal that the extreme measurement is a possible outlier.

FIGURE 3.12

Skewed box plots

EXAMPLE 3.37

Figure 3.13 summarizes the important features of a box plot.

FIGURE 3.13

Important features of a box plot

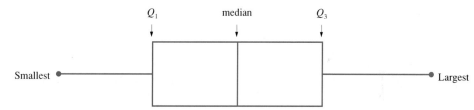

Detection of Outliers

An outlier, as we have said, is an extreme measurement in a data set. An outlier sometimes indicates that an error, such as a recording error, has been committed. It could also represent a very unusual measurement from the population. The investigation of outliers often reveals useful information, and it is quite possible that an outlier is the "jewel among the rocks" instead of the "rock among the jewels." Outliers can affect both the mean and standard deviation of the data set, thus distorting the center and variability. There is no general agreement among researchers as to what constitutes an outlier in a set of data. One of the following two rules of thumb are typically used to detect outliers in a data set.

A Measurement is an Outlier of a Sample if Either Rule Holds

Rule 1. The sample size is greater than 10, the frequency distribution is bell shaped, and the z score for the measurement is more than three standard deviations from the mean.

Rule 2. The measurement falls more than three IQRs below the lower quartile or more than three IQRs above the upper quartile.

APPLICATION 3.27 Consider the following ordered data set (originally presented in example 3.20) that represents the oxygen uptake values (mL/kg · min) of 21 male middle-aged runners while pedaling at 100 watts on a bicycle ergometer:

12.81 14.95 15.83 15.97 19.90 18.27 18.34 19.82 19.94 20.62 36.73
20.88 20.93 20.98 20.99 21.15 22.16 22.24 23.16 23.56 35.78

Determine if the measurement 35.78 is an outlier.

Solution: We shall use both rules of thumb.

  1. Rule 1—a stem-and-leaf diagram for the data is as follows:

12	81
13	
14	95
15	83 97
16	
17	90
18	27 34
19	82 94
20	62 88 93 98 99
21	15
22	16 24
23	16 56
.	
.	
.	
35	78
36	73

The measurements do not follow a bell shape and so the rule should not be used. But for the sake of illustration, let's calculate the z score for the measurement 35.78. The mean is $\bar{x} = 443.01/21 = 21.06$, and the standard deviation is $s = 5.75$. The z score for 35.78 is

$$z = \frac{35.78 - 21.06}{5.75} = 2.56$$

The measurement 35.78 is not an outlier, since it is only 2.56 standard deviations above the mean. Because of the extreme measurements the mean and standard deviation have been inflated and, as a consequence, the z score has been reduced.

  2. Rule 2—in example 3.20 we found that $Q_1 = 18.27$, $Q_3 = 22.16$, and IQR $= 3.89$. Since $22.16 + 3(3.89) = 33.83$ and $35.78 > 33.83$, we can conclude that 35.78 is an outlier. ▪

EXERCISE SET 3.4

Basic Skills

1. If $\mu = 47$ and $\sigma = 15$, fill in the missing values in the following table.

x	z
80	
	1.2
60	
	-2.37
47	
	3

2. If $\mu = 35$ and $\sigma = 16$, fill in the missing values in the following table.

x	z
50	
34	
	2.3
	-1.4
0	
	1

3. Consider the following population of data:

$$4 \quad 8 \quad 12 \quad 16 \quad 20$$

Find
   a. μ.
   b. σ.
   c. the z score for each of the raw scores.
   d. the mean and standard deviation for the z scores in part c.

4. Consider the following sample: 1, 2, 2, 6, 8, 11. Find
   a. $\bar{x}$.
   b. s.
   c. the z score for each measurement.
   d. the mean of the z scores.
   e. the standard deviation of the z scores.

5. Construct a box plot for the following data:

$$1.32 \quad 1.41 \quad 0.95 \quad 1.06 \quad 1.18$$
$$1.26 \quad 0.99 \quad 1.26 \quad 1.10$$

6. Construct a box plot for the following data:

$$65 \quad 74 \quad 77 \quad 83 \quad 89 \quad 92 \quad 96 \quad 95 \quad 103 \quad 109$$

Further Applications

7. Sue scores 625 on exam A in which $\mu = 600$ and $\sigma = 70$. Mary scores 525 on exam B in which $\mu = 500$ and $\sigma = 25$. If Sue and Mary both apply for a job and all other factors for both candidates are equal, who should be offered the job based on these exam scores. Use standard scores to defend your answer.

8. Dave and Rick are training for the Boston Marathon. Dave is training on a course in Cumberland, while Rick is training on a course in Frostburg. The mean time to complete the Cumberland course is 167.4 minutes, and the standard deviation is 25.9 minutes. The mean time to complete the Frostburg course is 143.1 minutes, and the standard deviation is 20.7 minutes. Dave says his course time on the Cumberland course is 91.5 minutes, and Rick says his course time on the Frostburg course is 86.2 minutes. Who do you think will do better in the Boston Marathon? Use standard scores to defend your answer.

9. The means and standard deviations of test scores for five classes are listed here. Suppose you obtain a score of 75 on a test. In which class would you have the highest relative standing?
   a. $\mu = 65, \sigma = 10$
   b. $\mu = 70, \sigma = 5$
   c. $\mu = 55, \sigma = 15$
   d. $\mu = 75, \sigma = 2$
   e. $\mu = 70, \sigma = 3$

10. The means and standard deviations of race times for four distance races are listed here. Suppose you obtain a time of 20 minutes in a race. In which race would you have the best relative standing?
   a. $\mu = 10, \sigma = 2$
   b. $\mu = 25, \sigma = 5$
   c. $\mu = 15, \sigma = 10$
   d. $\mu = 20, \sigma = 1$

11. Workers using machine A can produce daily quantities of product C with a mean of 75 and a standard deviation of 5, whereas workers using machine B can produce daily quantities of product C with a mean of 80 and a standard deviation of 8. Dick produced 83 units on machine A and John produced 92 units on machine B. Which worker produced the higher relative output? Why?

12. The mean annual salary of all male computer programmers in a large company is $35,000 and the standard deviation is $500. A female programmer earns an annual salary of $20,000 and believes she is being discriminated against. What do you think? Why?

13. The following data indicate the amounts (in cents) that states with a gasoline tax charge per gallon of gasoline:

9	9	13.5	7	6.5	11	9	11.7	11
11	12	9.8	5	13	8	11	9	9
8	9	10	13	13.7	8	13	8	
13	12	11	10.5	9	14	10	13	

   a. Construct a box plot for the data.
   b. Are there any outliers?

14. The Nielson Company collects data on the television viewing habits of Americans. The accompanying data indicates the weekly viewing times (in hours) of a sample of 20 college students:

16	36	22	27	38	51	30	25	10	5
29	21	26	31	11	25	33	25	15	16

   a. Construct a box plot for the data.
   b. Are there any outliers?

Going Beyond

15. The data in the accompanying stem-and-leaf diagram indicate the test scores made on a statistics examination.

```
2 | 7
3 | 0
4 | 8 8
5 | 5 5 7 9
6 | 0 0 3 5 5 6 7 9
7 | 1 1 4 6 7 8 8 8 9 9 9
8 | 2 3 3 5 5 7 9 9
9 | 0 4 4 7 9
```

   a. Construct a box plot for the data.
   b. Are there any outliers?

16. Can a score of 5 have a standard score of 3 if it is a member of a population having a mean of 7? Explain.

17. If a score of 13 is a member of a population with a mean of 7 and has a standard score of 3, find the variance of the population.

18. If a score of 10 is a member of a population with a variance of 9 and has a standard score of 5, find the mean of the population.

19. A population has a mean equal to 7 and a variance equal to 1. Find the value of the score that has a standard score equal to twice its value.

20. If only the z score constraint of Rule 1 is used to define an outlier, is the value 1,000,000 an outlier in the sample $\{0, 0, 0, 0, 1,000,000\}$? Explain.

21. If only the z score constraint of Rule 1 is used to define an outlier, is the value 1 an outlier of the sample $\{0, 0, 0, 1, 1, 1, 2, 2, 2, 5\}$?

22. Can you find a sample of size 4 whose largest value has a z score greater than $3/2$?

CHAPTER SUMMARY

In this chapter we introduced the concepts of central tendency, position points, and variability. We learned four measures of central tendency: mean, median, mode, and midrange. These measures provide central values for data sets. We learned that the relative positions of the mean, median, and mode in a distribution determine the symmetry or skewness of the distribution. Next, we studied four measures of dispersion or variability: range, variance, standard deviation, and interquartile range. These measures are used to describe the amount of spread in a data set. Chebyshev's theorem is important in understanding the concept of standard deviation. Finally, standardized scores and box plots were introduced. Standardized scores express the relative positions of measurements with respect to their mean. They are also useful for making relative comparisons of data from two different populations or samples. Box plots are useful for displaying center, variability, and skewness or symmetry all in one diagram. We also found them to be useful in helping to identify outliers in a set of data.

CHAPTER REVIEW

▪ *IMPORTANT TERMS* ▪

The following chapter terms have been mixed in ordering to provide you better review practice. For each term, provide a definition in your own words. Then check your responses against those given in the chapter.

constant	range	standard deviation
crude mode	box plot	sum of squares
box-and-whisker diagram	quartiles	symmetric histogram
bimodal	deciles	variance
deviation score	variability	z score
mean	approximate sample median	interquartile range
measures of central tendency	position point	outlier
measures of dispersion	raw score	percentiles
median	skewed histogram	approximate sample mean
midrange	standard score	modal class
mode		

▪ *IMPORTANT SYMBOLS* ▪

n, sample size	P_n, nth percentile	SS, sum of squares
$\bar{x}$, sample mean	Q_1, first quartile	σ^2, population variance
μ, population mean	Q_2, second quartile	s^2, sample variance
Σ, used to indicate addition	Q_3, third quartile	σ, population standard deviation
N, population size	D_n, nth decile	s, sample standard deviation
$\tilde{\mu}$, population median	R, range	$\bar{x}_a$, approximate sample mean
$\tilde{x}$, sample median	IQR, interquartile range	$\tilde{x}_a$, approximate sample median

▪ *IMPORTANT FACTS AND FORMULAS* ▪

Sample mean: $\bar{x} = \dfrac{\Sigma x}{n}$ (3.1)

Population mean: $\mu = \dfrac{\Sigma x}{N}$

Sample mean for data in a frequency table:

$\bar{x} = \dfrac{\Sigma fx}{\Sigma f}$ (3.2)

Interquartile range: $\text{IQR} = Q_3 - Q_1$

For any finite collection of data, the sum of the deviation scores is 0, or $\Sigma (x - \bar{x}) = 0$ (3.3)

Sum of squares for a population: $\text{SS} = \Sigma (x - \mu)^2$
(3.4)

Sum of squares for a sample: $\text{SS} = \Sigma (x - \bar{x})^2$ (3.4)

Computational formula for sample sum of squares:

$\text{SS} = \Sigma x^2 - \dfrac{(\Sigma x)^2}{n}$ (3.5)

Computational formula for population sum of squares:

$\text{SS} = \Sigma x^2 - \dfrac{(\Sigma x)^2}{N}$ (3.5)

Population variance: $\sigma^2 = \dfrac{\text{SS}}{N}$ (3.6)

Sample variance: $s^2 = \dfrac{\text{SS}}{n - 1}$ (3.7)

Sample standard deviation: $s = \sqrt{\text{sample variance}}$

Population standard deviation: $\sigma = \sqrt{\text{population variance}}$

Estimate of s: $s \approx \dfrac{R}{4}$ (3.8)

Sum of squares for grouped sample data:
$$SS = \Sigma f(x - \bar{x})^2 \quad (3.9)$$

Sum of squares for grouped population data:
$$SS = \Sigma f(x - \mu)^2 \quad (3.9)$$

Computational formula for sum of squares for data in a frequency table: $SS = \Sigma fx^2 - \dfrac{(\Sigma fx)^2}{\Sigma f}$ (3.10)

Chebyshev's theorem: At least $(1 - 1/k^2)100\%$ of any data set falls within k standard deviations of the mean, provided that k is a real number greater than or equal to 1.

z score or standard score of a measurement from a population: $z = \dfrac{x - \mu}{\sigma}$ (3.11)

z score or standard score of a measurement from a sample: $z = \dfrac{x - \bar{x}}{s}$ (3.11)

A population of z scores has a mean of 0 and a standard deviation of 1.

A measurement is an outlier of a sample if

  Rule 1. The sample size is greater than 10, the frequency distribution is bell shaped, and the z score is more than three standard deviations from the mean.

  Rule 2. The measurement falls more than three IQRs below the lower quartile or more than three IQRs above the upper quartile.

REVIEW EXERCISES

1. Calculate the mean, median, mode, midrange, range, variance, and standard deviation for each of the following populations:
   a. 3, 7, 4, 6, 8, 2
   b. 7, 8, 5, 2, 3
   c. 9, 6, 0, 1, 4
   d. 3, 3, 3

2. Calculate the mean, median, mode, range, variance, and standard deviation for each of the following samples:
   a. 4, 7, 2, 2
   b. 1, 8, 9, 4, 4
   c. 0, 0, 1, 1, 10
   d. 3, 3, 3
   e. 8, 14, 15, 16, 22

3. Calculate the z score for x in each of the following situations:
   a. $x = 22$, $\mu = 15$, $\sigma = 2$
   b. $x = -10$, $\mu = 5$, $\sigma = 8$
   c. $x = 0$, $\bar{x} = 12$, $s = 6$
   d. $x = 12.5$, $\bar{x} = 22$, $x = 0.4$
   e. $x = 17$, $\bar{x} = 15$, $s^2 = 4$

4. Calculate the mean, median, mode, variance, and standard deviation for the following frequency table of sample data:

x	f
0	1
1	3
2	2
3	4

5. Find the mean, median, mode, variance, and standard deviation for the sample data illustrated by the following line graph.

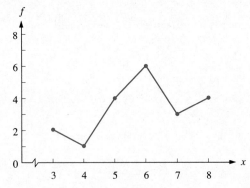

6. The following sample data were collected:

$$
\begin{array}{ccccc}
8 & 8 & 26 & 10 & 8 \\
8 & 8 & 18 & 8 & 14 \\
20 & 10 & 6 & 14 & 14
\end{array}
$$

a. Find $\bar{x}$ and s.

b. If a mistake was made in collecting the data and the original measurement of 26 should have been 20, would s increase or decrease? Explain.

c. If a mistake was made in collecting the data and the original measurement of 26 should have been 8, would s increase or decrease? Explain.

7. A calculus class has 30 members. The following test scores are from the students who sit in the first row: 87 83 89 71 95.

a. Is this collection of scores a sample or a population?

b. Calculate the mean and standard deviation for the data.

c. Find standard scores for the grades 71 and 95.

8. For each of the following data sets, specify an appropriate measure of central tendency and give its value. Justify your choice in each case.

a. Weight in pounds

3
2
4
13
4
4

b.

Rank	Number
Professor	25
Associate Professor	24
Assistant Professor	13
Instructor	10

c.

Party	Number
Democrat	200
Republican	300
Socialist	50
Independent	17

d.

Grade	Number
A	2
B	3
C	1

e.

Speed	Number
Fast	25
Slow	75

9. The following data represent the monthly charges (in U.S. dollars) for telephone service in 19 world cities: 7.28 8.54 15.28 5.51 3.17 6.34 3.80 4.59 5.12 9.98 7.04 10.00 11.96 5.48 2.30 5.85 9.39 8.73 7.66.[25]

a. Find $\bar{x}$.

b. Find s.

c. Find the z score for New York's monthly telephone service charge ($x = \$10.00$).

d. Construct a box plot for the data.

10. Fifty households were polled to determine the number of male inhabitants. The resulting data are listed here.

$$
\begin{array}{cccccccccc}
0 & 1 & 2 & 1 & 3 & 0 & 1 & 4 & 0 & 1 \\
1 & 1 & 1 & 1 & 1 & 0 & 1 & 3 & 2 & 3 \\
1 & 0 & 1 & 2 & 2 & 1 & 1 & 2 & 2 & 1 \\
0 & 0 & 0 & 0 & 0 & 0 & 0 & 1 & 1 & 1 \\
2 & 1 & 0 & 1 & 1 & 2 & 2 & 0 & 1 & 0
\end{array}
$$

a. Find $\bar{x}$.

b. Find s.

c. How many measurements fall within one standard deviation of the mean?

11. The following are EPA average miles-per-gallon ratings for 15 1989 compact and subcompact automobiles: 30, 31, 34, 31, 35, 41, 27, 35, 20, 47, 27, 29, 34, 38, and 32.

a. Find $\bar{x}$ and s.

b. Find the quartiles and the fourth decile.

c. Construct a box plot for the data.

12. The following data represent annual U.S. arms sales (in billions of dollars) to third-world nations from 1976 to 1983: 8.2, 9.8, 10.1, 9.2, 6.4, 6.8, 7.9, and 9.7. Find

a. $\bar{x}$.

b. s.

c. Q_3 and D_7.

Computer Applications

1. Listed here are the entrance examination scores of a sample of 100 entering freshmen at a university in the midwest.

```
432 257 502 506 425 479 387 394 282 423 606
417 596 395 517 512 501 620 142 556 671 633
340 489 646 394 440 323 367 554 544 347 576
320 505 356 428 797 353 532 294 555 512 433
454 563 299 355 455 452 412 436 562 602 561
630 375 338 244 283 452 412 326 564 350 664
279 284 221 432 446 284 492 348 401 267 372
617 285 195 309 637 314 415 546 577 282 370
353 457 394 485 276 377 170 690 583 273 393
258
```

Use a computer program to
a. find the sample mean.
b. find the sample standard deviation.
c. find the range.
d. find the quartile points.
e. find the 20th percentile.
f. find the fourth decile.
g. construct a histogram.
h. construct a stem-and-leaf diagram.
i. construct a boxplot.

2. The following data represent the weights (in hundredths of a pound) of a sample of newborn baby boys delivered during the past year at Memorial Hospital.

```
631 827 631 734 938 604 583 753 554 890 781
750 756 779 758 821 612 780 843 581 743 951
669 682 714 711 930 927 744 727 857 602 875
571 829 759 875 902 808 766 866 590 623 986
793 835 674 770 842 738 838 726 609 717 657
702 916 618 855 770 680 847 679 754 733 787
869 825 808 715 723 728 849 958 760 875 841
917 851 848 768 750 700 793 870 627 641 795
732 582 856 913 809 804 820 602 779 651 773
591
```

Use a computer program to
a. find the sample mean.
b. find the sample standard deviation.
c. find the range.
d. find the quartile points.
e. find the 60th percentile.
f. find the sixth decile.
g. construct a histogram.
h. construct a stem-and-leaf diagram.
i. construct a boxplot.

▪ CHAPTER ACHIEVEMENT TEST ▪

1. The final grades for a section of Math 209 are illustrated in the accompanying bar graph.

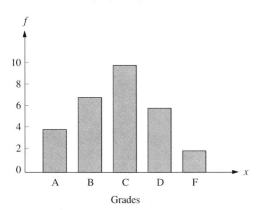

a. What measure of central tendency should be used to describe the central grade? Explain the reason for your answer.
b. Using your response for part a, find the center grade(s).
c. How many students are represented in the graph?
d. What percentage of students received a grade of C?
e. What percentage of students received a grade of C or better?

2. Consider the following sample: 3, 8, 7, 12, 10.
Find the
a. range.
b. mean.
c. median.
d. midrange.
e. variance.
f. standard deviation.
g. standard score for the measurement 10.
h. IRQ.

3. Consider the following frequency table for a population:

x	f
4	2
8	3
3	5

a. Find μ.
b. Find σ.

4. What can be said about x in relation to the remainder of the data set, if x
a. has a z score of 0?
b. has a standard score of 2?
c. has a z score of -1?

5. In which of the following situations is the raw score x largest relative to its data set?
a. $x = 37, \bar{x} = 20, s = 10$
b. $x = 500, \bar{x} = 200, s = 250$
c. $x = 3.0, \bar{x} = 1.0, s = 0.7$

6. If $\mu = 8$ and $\sigma^2 = 4$, find the raw score x corresponding to $z = -2$.

7. Refer to test question 2. Construct a box plot for the data.

8. Suppose a sample consists of five measurements 30, 80, 50, 40, and x. Determine the value of x so that the mean, median, and mode are all equal.

9. Do most people have more than the average number of feet? Explain your answer.

4 Descriptive Analyses of Bivariate Data

CHAPTER OBJECTIVES

In this chapter we will investigate

▷ What a scattergram is and how it is used.

▷ Covariance.

▷ Correlation.

▷ How to determine the correlation coefficient r.

▷ The least squares method for determining the prediction equation.

▷ How to determine the least squares equation, which estimates how two variables are related.

▷ How to use the regression equation for predictive purposes.

▷ How the correlation coefficient and the slope of the regression line are related.

▷ What the sum of squares for error is and how to calculate it.

MOTIVATOR 4

*T*he weather appears to have an effect on baseball offense. The accompanying table indicates a relationship between temperature and offense from 1987 through 1989.[26]

Temperature	Average	Runs Per Game	HRs Per Game
0°–59°	0.248	8.0	1.40
60°–69°	0.253	8.5	1.65
70°–79°	0.259	8.6	1.69
80°–89°	0.263	9.1	1.85
90°–up	0.263	9.1	1.83

As the temperature increases, the data suggest that the offense improves. An investigation of the relationship between temperature and offense involves regression and correlation, topics of study in this chapter.

Chapter Overview

Statistical analyses frequently involve quantitative data that are *bivariate* in nature; that is, for each unit in a sample, there corresponds a pair of measurements. The following are examples of **bivariate data:**

■ Salaries and ages of teachers in district A

■ Pulse rates and systolic blood pressures for Math 209 students

▪ Heights and weights for a group of Cub Scouts

▪ Daily rainfall and average daily temperatures for Frostburg for ten days

▪ 1985 spring and fall enrollments at 20 universities

This chapter will deal with graphs of bivariate data, measuring the strength of a linear relationship, and describing linear relationships between two variables. Throughout this chapter we will deal with only linear (**straight-line**) relationships.

SECTION 4.1   *Linear Dependency and Covariance*

Bivariate data can be viewed as a collection of ordered pairs (x, y), where the measurement x in the first data set is paired with the measurement y in the second data set. The value from the first data set is always written first in the pair. It is customary to call the x variable the **independent variable** and the y variable the **dependent variable.** The application will usually indicate which data set is to be associated with the independent variable. These ordered pairs can be plotted on a coordinate system. When this is done, the graph is called a **scattergram.**

EXAMPLE 4.1

Consider the accompanying collection of paired data representing the number of hours (x) studying for an exam and the grade received (y) on the exam by a sample of six students.

Student	A	B	C	D	E	F
x: hours	1	2	4	4	7	12
y: grade	71	71	74	80	80	86

A scattergram for the data is as follows.

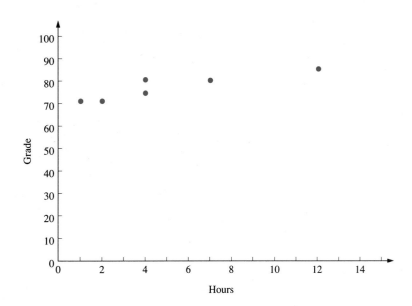

We are interested in determining whether there is any **linear dependency** between the two variables. That is, we want to determine if there is a tendency for the y variable to increase (or decrease) when the x variable increases. By examining the scattergram in example 4.1, it appears that there is a tendency for y to increase as x increases. In this case, we say there is some degree of linear dependency between x and y. If the tendency is for the variable y to increase as the variable x increases, the dependency is called *positive*. Whereas, if the tendency is for y to decrease as the variable x increases, the dependency is called *negative*. If there is no tendency for y to increase or decrease as the variable x increases, then there is no linear dependency.

If the paired data have a perfect linear relationship, the slope (steepness) of the straight line indicates the type of dependency relationship for the variables involved. If the line has a positive slope (is up hill from left to right), then the dependency is positive, while if the line has a negative slope (is down hill from left to right), the dependency is negative.

Consider the two lines in figure 4.1. Both lines go through the origin. The line labeled by l_1 has a positive slope, and the line labeled by l_2 has a negative slope. Notice that all the points (except the origin) on line l_1 fall in quadrants I and III; for a point in these quadrants both coordinates are positive or both coordinates are negative.

FIGURE 4.1

Straight lines through origin

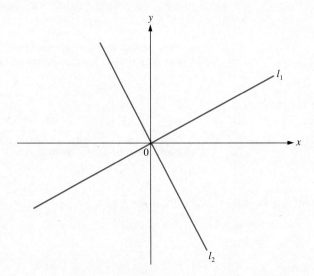

For every point contained on a line passing through the origin, a weight can be assigned. A weight is a number defined by the product of its coordinates. If the point is contained in quadrants I and III, the assigned weight is positive, whereas if the point is contained in quadrants II and IV, the weight is negative. In figure 4.1, each point on line l_2 (except the origin) has a negative weight, and each point on line l_1 (except the origin) has a positive weight. The origin is assigned a weight of 0.

To summarize:

1. Any line through the origin having positive slope has a product of coordinates that is non-negative.

2. Any line through the origin having negative slope has a product of coordinates that is non-positive.

Deviation scores for x and y can be used to create a formula for measuring the degree of linear dependency. Recall the following facts about deviation scores $x - \bar{x}$:

1. A measurement is below the mean if its deviation score is negative.
2. A measurement is above the mean if its deviation score is positive.
3. A measurement is equal to the mean if its deviation score is zero.

If we transform each measurement in a pair (x, y) to its corresponding pair of deviation scores $(x - \bar{x}, y - \bar{y})$, then the scattergram of the paired deviation scores takes on an interesting interpretation. The point $(\bar{x}, \bar{y})$ is called the **centroid** of the scattergram, and serves as a reference point. If we draw two lines through the centroid, one parallel to the x-axis and one parallel to the y-axis, then these two lines can serve as reference lines (or axes) for the deviation scores. We use $\bar{y}$ to label the axis parallel to the y-axis and $\bar{x}$ to label the axis parallel to the x-axis (see fig. 4.2 for the study data from example 4.1).

FIGURE 4.2

Scattergram of deviation scores of data in example 4.1

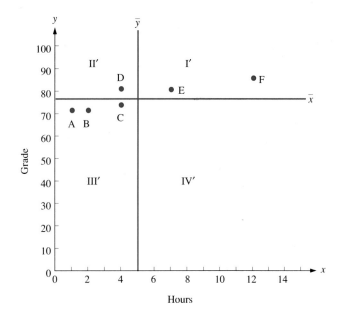

These new reference lines establish four quadrants: I′, II′, III′, and IV′. A pair of deviation scores will be plotted in quadrant I′ if its x deviation score is positive and its y deviation score is positive, in quadrant II′ if its x deviation score is negative and its y deviation score is positive, in quadrant III′ if its x deviation score is negative and its y deviation score is negative, and in quadrant IV′ if its x deviation score is positive and its y deviation score is negative. Perpendicular distances of the points measured from the $\bar{x}, \bar{y}$-axes represent the deviations from the centroid. Pairs of deviation scores $(x - \bar{x}, y - \bar{y})$ are plotted with respect to the $\bar{x}$- and $\bar{y}$-axes the same way that pairs (x, y) are plotted with respect to the x- and y-axes.

The product of the two deviation scores for a pair determines a weight. If all the points in a scattergram are contained in quadrants I′ and III′, then the weights are all positive; and if the points are contained in quadrants II′ and IV′, then the weights are all negative. The sum of the weights of all points in a scattergram provides an indication of strength of linear dependency. If the sum of the weights is positive, then the linear dependency is positive, whereas if the sum of the weights is negative, the dependency is negative. If the sum of the weights is zero, then there is no linear dependency between variables x and y. Table 4.1 contains information on the paired measurements used above. The mean of the x measurements is $\bar{x} = 5$ and the mean of the y measurements is $\bar{y} = 77$.

TABLE 4.1

Coordinates, Deviation Scores, Quadrants, and Weights for Data in Example 4.1

Student	x, y Coordinates	$x - \bar{x}$	$y - \bar{y}$	$\bar{x}, \bar{y}$ Coordinates	Quadrant	Weight
A	(1, 71)	-4	-6	$(-4, -6)$	III′	24
B	(2, 71)	-3	-6	$(-3, -6)$	III′	18
C	(4, 74)	-1	-3	$(-1, -3)$	III′	3
D	(4, 80)	-1	3	$(-1, 3)$	II′	-3
E	(7, 80)	2	3	$(2, 3)$	I′	6
F	(12, 86)	7	9	$(7, 9)$	I′	63
		0	0			111

Notice the following relationships from table 4.1:

1. All students except D are identified with pairs having positive weights that are plotted in quadrants I′ or III′.
2. Student D is identified with a pair having a negative weight and plotted in Quadrant II′.
3. $\Sigma (x - \bar{x}) = 0$ and $\Sigma (y - \bar{y}) = 0$.
4. The scattergram is dominated by points in quadrants I′ and III′ having positive weights. This is shown by the sum of the weights being 111.

Sample Covariance

Every weight assigned to a pair of deviation scores makes a contribution to the sum of all the weights. The sum of the weights of the deviation scores provides a total measure of dependency for the variables. It represents the combined tendency for the points to be in either (quadrant I′ or III′) or (quadrant II′ or IV′). If n represents the number of pairs and we divide the sum of products of the deviation scores by $n - 1$, we get, in some sense, an average measure of linear dependency, called the **sample covariance** and denoted by cov(x, y) or s_{zy}. Thus, the sample covariance is given by

Sample Covariance

$$\text{cov}(x, y) = s_{xy} = \frac{\Sigma (x - \bar{x})(y - \bar{y})}{n - 1} \tag{4.1}$$

EXAMPLE 4.2

The sample covariance for the test data of example 4.1 is

$$s_{xy} = \frac{\Sigma (x - \bar{x})(y - \bar{y})}{n - 1}$$

$$= \frac{111}{5} = 22.2$$

This result indicates a positive linear dependency between the amount of study time and the grade earned.

MINITAB can be used to determine the sample covariance for the above data. Computer display 4.1 contains the commands used and the corresponding output.

Computer Display 4.1

```
MTB > NAME C1 'HOURS' C2 'GRADE'
MTB > SET C1
DATA> 1 2 4 4 7 12
DATA> END
MTB > SET C2
DATA> 71 71 74 80 80 86
DATA> END
MTB > COVARIANCE C1 C2

          HOURS    GRADE
HOURS  16.0000
GRADE 22.2000   36.000
```

The entries 16.000 and 36.000 in the last two lines of the output represent the sample variance of hours and grades, respectively.

EXAMPLE 4.3

Figure 4.3 shows scattergrams that represent a negative linear dependency, a positive linear dependency, and a zero linear dependency. Notice that in each case the average product of deviation scores determines the type of dependency.

FIGURE 4.3

Scattergrams illustrating
positive, negative, and no
dependency

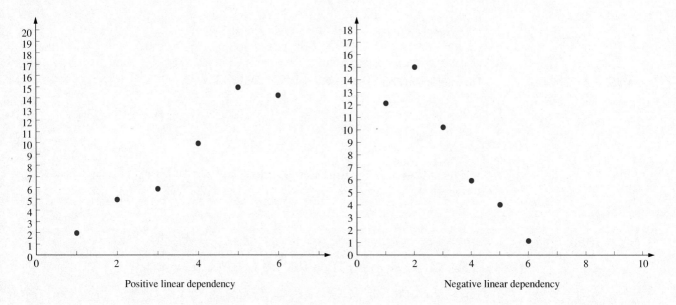

Positive linear dependency

Negative linear dependency

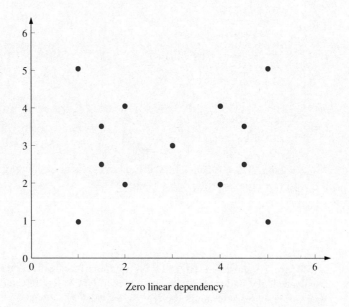

Zero linear dependency

The sample covariance is similar to sample variance in the sense that for the sample variance, the sum of squares SS is divided by $n - 1$, and for the sample covariance, the sum of products of the deviation scores is divided by $n - 1$.

$$s^2 = \frac{\Sigma (x - \bar{x})^2}{n - 1} \qquad\qquad s_{xy} = \frac{\Sigma (x - \bar{x})(y - \bar{y})}{n - 1}$$

↑   ↑

Sample variance   Sample covariance

The larger the value of the sample variance s^2, the more variation in the data. In general, the greater the magnitude of the sample covariance s_{xy}, the stronger the linear dependency. But both measures can be greatly affected by outliers; outliers have a general tendency to inflate the measures. The sample covariance has one serious drawback; two different scattergrams can have the same linear dependency yet have different sample covariances.

EXAMPLE 4.4

To see that two different scattergrams can have the same linear dependency but have different covariances, let's use the following two sets of paired data, which both have perfect linear dependency. The relationship used to generate the pairs is $y = x$.

Data Set A	
x	y
1	1
3	3

Data Set B	
x	y
2	2
6	6

The means for data set A are:
$$\bar{x} = 2 \qquad \bar{y} = 2$$

The means for data set B are:
$$\bar{x} = 4 \qquad \bar{y} = 4$$

The corresponding deviation scores and products of deviation scores are arranged in table 4.2.

TABLE 4.2

Sum of Products of Deviation Scores for Data Sets A and B

		Data Set A					Data Set B		
x	y	x − 2	y − 2	(x − 2)(y − 2)	x	y	x − 4	y − 4	(x − 4)(y − 4)
1	1	−1	−1	1	2	2	−2	−2	4
3	3	1	1	1	6	6	2	2	4
		0	0	2			0	0	8

The covariance for data set A is

$$s_{xy} = \frac{\Sigma (x - 2)(y - 2)}{n - 1} = \frac{2}{1} = 2$$

The covariance for data set B is

$$s_{xy} = \frac{\Sigma (x - 4)(y - 4)}{n - 1} = \frac{8}{1} = 8$$

Two points determine a line. The linear dependency in each case could not be stronger. It is perfect. But the covariance for data set B is four times the covariance for data set A. A result of this observation is that a change in the units of measurement of x or y will affect the value of the covariance. In section 4.2 we will use another measure of linear dependency that is not affected by the unit of measurement. The measure uses the sum of products of the deviation scores of x and y.

EXERCISE SET 4.1

Basic Skills

1. Consider the following set of bivariate data:

x	0	1	4	-5	2	6
y	-2	-1	2	-7	0	4

   a. Draw a scattergram.
   b. Using the scattergram, determine if the dependency is positive or negative.
   c. Compute the value of the sample covariance. What type of dependency relation does this indicate?

2. Consider the following set of bivariate data:

x	0	4	8	1	3	-1
y	2	6	-14	0	-4	4

   a. Draw a scattergram.
   b. Using the scattergram, determine if the dependency is positive or negative.
   c. Compute the value of $\text{cov}(x, y)$. What type of dependency relation does this indicate?

3. Determine the sample covariance for the accompanying paired data.

x	1	2	4	4	7	12
y	2	1	4	18	10	19

4. Determine the sample covariance for the accompanying paired data.

x	1	2	3	7	8	9
y	16	9	4	4	9	16

Further Applications

5. The grades of eight students in Math 101 (x) and English 101 (y) are as follows:

x	77	81	94	50	72	63	88	95
y	82	47	85	66	65	72	89	95

   a. Draw a scattergram.
   b. Using the scattergram, determine if the dependency is positive or negative.
   c. Compute the value of s_{xy}. What type of dependency relation does this indicate?

6. The following data represent SAT (math) score (x) and GPA (y) for a group of 10 students:

x	450	375	514	678	501	734	325	400	398	681
y	3.5	2.5	2.1	3.6	2.7	3.8	1.8	2.4	2.0	1.9

   a. Draw a scattergram.
   b. Using the scattergram, determine if the dependency is positive or negative.
   c. Compute the sum of the products of the deviation scores. What type of dependency relation does this indicate?

7. The following data represent the engine sizes in cubic inches and estimated miles per gallon for seven subcompact automobiles.

Car	Engine Size	Miles/ gallon
Chevette	98	31
Sentra	98	35
Colt	86	41
Isuzu I-Mark	111	27
Mercedes 190D	134	35
Firebird	173	20
VW Rabbit	97	47

   a. Draw a scattergram.
   b. Determine the sample covariance.

8. The following data represent price-earnings (PE) ratio and percent yield for seven stocks.

PE ratio	2.4	2.4	3.4	2.9	4.0	3.8	2.7
Percent yield	4.2	0.7	10	4.6	6.2	6.3	8.4

   a. Draw a scattergram.
   b. Determine the sample covariance.

9. The accompanying data indicate the number of alcoholic drinks consumed and the amount of blood alcohol concentrations for a sample of six subjects involved in an experiment and having similar body weights:

Number of drinks	2	3	4	5	6	7
Blood alcohol concentration	0.05	0.09	0.11	0.13	0.17	0.20

   a. Draw a scattergram for the data.
   b. Determine s_{xy}.
   c. What type of a dependency relationship exists between the number of drinks consumed and the level of blood alcohol?

10. Consider the following set of bivariate data:

x	1	2	3	4	5	6	7
y	12	7	4	3	4	7	12

   a. Draw a scattergram.
   b. Compute the value of the sample covariance.
   c. What type of dependency relationship exists between x and y? Discuss the results of parts a and b.

11. Consider the following set of bivariate data:

x	1	2	3	4	5
y	3	5	7	9	11

   a. Draw a scattergram.
   b. Calculate the value of the sample covariance.

12. Consider the following set of bivariate data:

x	1	2	3	4	5
y	-1	-3	-5	-7	-9

   a. Draw a scattergram.
   b. Calculate the value of the sample covariance.

Going Beyond

13. Show that
$$s_{xy} = \frac{n \Sigma xy - (\Sigma x)(\Sigma y)}{n(n-1)}.$$

SECTION 4.2   Correlation

One of the main objectives of statistics is to be able to estimate or predict the value of one variable from another variable. **Regression analysis** is a method used to study the relationship between two or more variables and to predict values for one of the variables. In many applications, a linear relationship exists between the variables that can be used for prediction purposes. **Correlation analysis** is a method used by statisticians to determine the strength of the linear relationship (or dependence) that exists between the variables. If the strength of the linear dependence is small, then it is usually not fruitful to use regression analysis to find the linear relationship to use for predictive purposes.

In section 4.1 we learned how to construct scattergrams. They represent a graphic means for determining if a linear relationship exists between two variables. If all the points fall exactly on a straight line, then we say the two variables have *perfect linear correlation*. If the points lie close to a straight line, the two variables are said to have a *strong degree of linear correlation*. If the straight line has a positive slope, we say the two variables have *positive linear correlation*, and if the line has a negative slope, we say the variables have *negative linear correlation*. And if the straight line has a slope of zero, we say there is *no linear correlation* between the two variables.

The first world record for the 1-mile run was 4:56 recorded in 1864. Since that time the 1-mile run has been lowered to 3:47.3, and the year 1945 was the last year that the 1-mile record was over 4 minutes. Table 4.3 shows the progress in the world-record times for the 1-mile run from 1945 to 1981. A scattergram for the record data is shown in figure 4.4.

Year	Country	Time
1945	Sweden	4:01.4
1954	United States	3:59.4
1954	Austria	3:58.0
1957	Great Britain	3:57.2
1958	Australia	3:54.5
1962	New Zealand	3:54.4
1964	New Zealand	3:54.1
1965	France	3:53.6
1966	United States	3.51.3
1967	United States	3:51.1
1975	Tanzania	3:50.0
1975	New Zealand	3:49.4
1981	Great Britain	3:47.3
1985	Great Britain	3:46.3

TABLE 4.3

World Records for the One-Mile Run from 1945 to 1985

A look at the scattergram in figure 4.4 suggests that a negative linear correlation exists for the data and that a linear approximation to the data would be reasonable. The correlation is negative since the points in the scattergram appear to lie closest to a straight line having a negative slope. Finding the linear approximation to the points in the scattergram involves regression analysis, which we will explore in section 4.3. Much speculation has been made by sports enthusiasts about the year of the 3:40 mile. Some field and track experts, using regression analysis, have predicted the year 2000.

Consider the SAT (math) scores and freshman grade point averages (GPAs) for each sophomore enrolled this semester at an eastern college. We might like to obtain answers to the following two questions:

1. Is there a linear relationship between SAT scores and GPAs?

2. If so, what is the relationship?

FIGURE 4.4

World-record times for the 1-mile run from 1945–1985

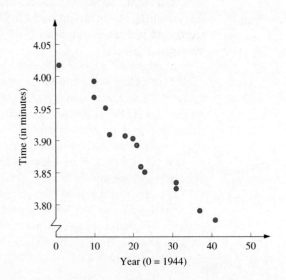

Answering the first question involves correlation and answering the second question involves regression. Consider the data in table 4.4, which contains SAT (math) scores and freshmen GPAs for a sample of ten sophomores enrolled at an eastern state college.

TABLE 4.4

SAT (Math) Scores and GPAs

Student Number	SAT Math Score	GPA
1	450	2.5
2	600	3.0
3	550	2.0
4	400	3.0
5	350	2.5
6	650	2.5
7	300	1.5
8	400	2.0
9	700	3.5
10	250	1.0

The data can also be displayed in a scattergram such as that shown in figure 4.5, where the SAT (math) scores are displayed on the horizontal axis and the GPAs are displayed on the vertical axis. Note that the correlation between GPAs and SATs appears to be positive, since the points seem to fall closest to a line having a positive slope.

In section 4.1, the sample covariance was use to determine the linear dependency between two variables. If the points in a scattergram scatter from lower left to upper right, then there is a positive dependency (see fig. 4.3), and if the points scatter from the upper left to the lower right, then there is negative dependency (see fig. 4.4). In other words, if the values of y tend to increase as the values of x increase, then we say the correlation is positive, whereas if the values of y tend to decrease as the values of x increase, we say the correlation is negative. If the points in a scattergram do not fall on a straight line, it is impossible to ascertain the magnitude of correlation r unless a formula is used that takes into account variations from a linear scattergram.

The linear correlation or strength of linear dependency for the scattergrams displayed in figures 4.4 and 4.5 can be measured by the sample covariance studied in section 4.1. But there are two problems with using the sample covariance to measure the strength of linear relationship (or dependency) between two variables. First, the covariance depends on the units of measurement. If we change the units for x and y,

FIGURE 4.5

Scattergram for SATs and GPAs

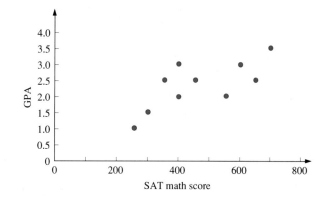

then the covariance changes. Second, there are no bounds on the values of the covariance. Fortunately we can remedy these two problems.

What we need to measure the strength of linear relationship is an index that possesses the following four properties:

1. It is free of the units of measure; its values do not depend on the units of measure for either variable.
2. Its value is equal to 1 if the points are on a straight line with positive slope.
3. Its value is equal to -1 if the points are on a straight line with negative slope.
4. Its value is 0 if there is no linear relationship between the variables.

Recall from section 3.4 that the mean and standard deviation of a collection of measurements have the same unit of measurement as the measurements in the collection. As a consequence, the deviation scores for a set of measurements have the same unit of measurement as the individual measurements. If we divide the deviation scores by the standard deviation, we have a set of measurements that have no unit of measure; they are unit free. Also recall from section 3.4 that these quotients are referred to as *z scores*. If a deviation score is positive (negative), its z score is also positive (negative).

If instead of using the sum of the products of the deviation scores for x and y in the numerator of the sample covariance formula to get an index of dependency, we use the sum of products of the z scores, $\Sigma z_x z_y$, where z_x represents the z score for x and z_y represents the z score for y, we would have an index that would satisfy all four of the properties listed above. This new index is called **Pearson's correlation coefficient,** and is denoted by r.

Pearson's Correlation Coefficient

$$r = \frac{\Sigma z_x z_y}{n - 1} \tag{4.2}$$

---

where n is the number of pairs involved in the sample. If the value of r equals 1 or -1, then there is perfect linear correlation between the variables or a perfect linear relationship, whereas if $r = 0$, there is no linear correlation or relationship. This means that as x tends to increase there is no definite tendency for the values of y to increase or decrease. Note that a value of $r = 0$ does not necessarily mean the lack of a relationship between x and y. A relationship that is nonlinear may exist [see fig. 4.7(c)].

EXAMPLE 4.5

The scattergrams in figure 4.6 exhibit perfect linear dependencies or relationships between x and y.

FIGURE 4.6

Scattergrams showing perfect linear relationships

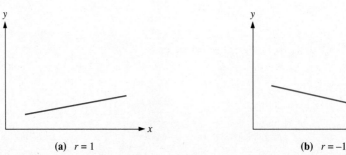

(a) $r = 1$ (b) $r = -1$

EXAMPLE 4.6

FIGURE 4.7

Scattergrams showing no
linear relationships

The scattergrams in figure 4.7 exhibit no linear relationship.

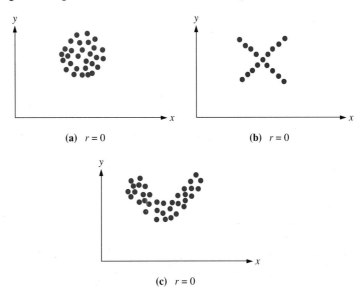

(a) $r = 0$ **(b)** $r = 0$

(c) $r = 0$

EXAMPLE 4.7

FIGURE 4.8

Scattergrams showing some
linear relationships

The scattergrams in figure 4.8 exhibit some linear relationship.

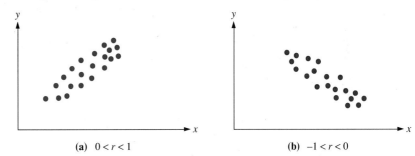

(a) $0 < r < 1$ **(b)** $-1 < r < 0$

APPLICATION 4.1

Compute the value of Pearson's correlation coefficient r for the following sample of
paired data from section 4.1 representing the number of hours (x) studying for an exam
and the grade received (y) on the exam by a sample of six students.

Student	A	B	C	D	E	F
x: hours	1	2	4	4	7	12
y: grade	71	71	74	80	80	86

Solution:

1. The mean and standard deviation for each group are as follows:

$$x: \bar{x} = 5 \text{ and } s = 4$$
$$y: \bar{y} = 77 \text{ and } s = 6$$

2. The z score for x is given by $z_x = (x - 5)/4$, and the z score for y is given by $z_y = (y - 77)/6$.

3. The computations are organized in the following table.

x	y	$x - 5$	$y - 77$	z_x	z_y	$z_x z_y$
1	71	-4	-6	-1	-1	1
2	71	-3	-6	-0.75	-1	0.75
4	74	-1	-3	-0.25	-0.5	0.125
4	80	-1	3	-0.25	0.5	-0.125
7	80	2	3	0.5	0.5	0.25
12	86	7	9	1.75	1.5	2.625
						4.625

4. By using formula (4.2), the value of Pearson's correlation coefficient is

$$r = \frac{\Sigma z_x z_y}{n - 1}$$

$$= \frac{4.625}{6 - 1} = \frac{4.625}{5} = 0.925 \quad \blacksquare$$

There is a computational formula for calculating the value of Pearson's correlation coefficient r that makes use of the sum of squares concept presented in section 3.2.

$$SS = \Sigma x^2 - \frac{(\Sigma x)^2}{n} \tag{4.3}$$

Since we are dealing with bivariate data (x, y), SS_x will denote the sum of squares for x, while SS_y will denote the sum of squares of y. Thus, we have

$$SS_y = \Sigma y^2 - \frac{(\Sigma y)^2}{n} \tag{4.4}$$

If we rewrite the right-hand side of formula (4.3) by replacing one occurrence of x with a y, we get the following expression:

$$\Sigma xx - \frac{(\Sigma x)(\Sigma x)}{n} \rightarrow \Sigma xy - \frac{(\Sigma x)(\Sigma y)}{n}$$

The resulting expression is called the **sum of cross products,** and is denoted by SS_{xy}.

> **Sum of Cross Products**
>
> $$SS_{xy} = \Sigma xy - \frac{(\Sigma x)(\Sigma y)}{n} \tag{4.5}$$

The values of SS_x, SS_y, and SS_{xy} are used to calculate the values of Pearson's correlation coefficient. The computational formula for calculating the value of Pearson's correlation coefficient r is

Computational Formula for r

$$r = \frac{SS_{xy}}{\sqrt{SS_x SS_y}}$$

(4.6)

A P P L I C A T I O N 4.2

Refer to application 4.1. Calculate the value of r by using formula (4.6) for the paired data.

Solution: We first calculate the values of SS_x, SS_y, and SS_{xy}. Table 4.5 organizes the computations needed to determine the proper sums used in the formulas.

TABLE 4.5

Computations for Calculating SS_x, SS_y, and SS_{xy}

x	y	x^2	y^2	xy
1	71	1	5,041	71
2	71	4	5,041	142
4	74	16	5,476	296
4	80	16	6,400	320
7	80	49	6,400	560
12	86	144	7,396	1032
30	462	230	35,754	2421

Determine SS_x.

$$SS_x = \Sigma x^2 - \frac{(\Sigma x)^2}{n}$$

$$= 230 - \frac{(30)^2}{6} = 80$$

Determine SS_y.

$$SS_y = \Sigma y^2 - \frac{(\Sigma y)^2}{n}$$

$$= 35,754 - \frac{(462)^2}{6} = 180$$

Determine SS_{xy}.

$$SS_{xy} = \Sigma xy - \frac{(\Sigma x)(\Sigma y)}{n}$$

$$= 2421 - \frac{(30)(462)}{6} = 111$$

Use formula (4.6) to determine r, Pearson's correlation coefficient.

$$r = \frac{SS_{xy}}{\sqrt{SS_x SS_y}}$$

$$= \frac{111}{\sqrt{(80)(180)}} = 0.925$$

which agrees with the result found in application 4.1. ■

Computer display 4.2 illustrates the use of MINITAB to obtain the correlation coefficient for the data of application 4.1.

Computer Display 4.2

```
MTB > READ C1 C2
DATA> 1 71
DATA> 2 71
DATA> 4 71
DATA> 4 80
DATA> 7 80
DATA> 12 86
DATA> END
   6 ROWS READ
MTB > NAME C1 'HOURS' NAME C2 'GRADE'
MTB > CORRELATION C1 C2

CORRELATION OF HOURS AND GRADE = 0.925
```

Coding to Simplify Computations of r

Frequently the values of x and y make it extremely cumbersome to calculate SS_x, SS_y, and SS_{xy}. To simplify the computations as much as possible, coding is often used. **Coding** involves using linear transformations with the data. The transformations are of the following type:

$$U = ax + b$$
$$V = cy + d$$

where $a \geq 0$ and $c \geq 0$. Then Pearson's correlation coefficient between U and V is identically equal to Pearson's correlation coefficient between x and y. Application 4.3 shows the process.

APPLICATION 4.3

Use coding to calculate the value of r for the bivariate data shown in the following table:

x	168	169	170	171
y	0.6	0.9	0.9	0.5

Solution: Let $U = x - 167$ ($a = 1$, $b = -167$), and let $V = 10y$ ($c = 10$, $d = 0$). To find the values for U, we substitute the values for x into the equation $U = x - 167$, and to find the values for V, we substitute the values for y into the equation $V = 10y$. Then the transformed data become

U	1	2	3	4
V	6	9	2	5

By using formula (4.6), the value of r for the bivariate data (U, V) is $r = -0.447$. Thus, Pearson's correlation coefficient for x and y is $r = -0.447$. ■

APPLICATION 4.4

For the 1-mile world record data in table 4.3, find the value of Pearson's correlation coefficient r.

Solution: The transformation $x = \text{year} - 1944$ will be used to code the data to simplify computations. The year 1945 will be coded as 1, the year 1946 as 2, and so on. In addition, the times will be expressed in minutes. In table 4.6, x represents the coded year and y represents the time in minutes.

	Year	Time	x	y	x^2	y^2	xy
TABLE 4.6	1945	4:01.4	1	4.023	1	16.1845	4.023
Coded Data for	1954	3:59.4	10	3.990	100	15.9201	39.900
Application 4.3	1954	3:58.0	10	3.967	100	15.7371	39.670
	1957	3:57.2	13	3.953	169	15.6262	51.389
	1958	3:54.5	14	3.908	196	15.2725	54.712
	1962	3:54.4	18	3.907	324	15.2646	70.326
	1964	3:54.1	20	3.902	400	15.2256	78.040
	1965	3:53.6	21	3.893	441	15.1554	81.753
	1966	3:51.3	22	3.855	484	14.8610	84.810
	1967	3:51.1	23	3.852	529	14.8379	88.596
	1975	3:50.0	31	3.833	961	14.6919	118.823
	1975	3:49.4	31	3.823	961	14.6153	118.512
	1981	3:47.3	37	3.788	1369	14.3489	140.156
	1985	3:46.3	41	3.772	1681	14.2280	154.652
			292	54.466	7716	211.9690	1125.363

The following steps are followed in calculating the value of r:

1. Calculate the value of SS_x.

$$SS_x = \Sigma x^2 - \frac{(\Sigma x)^2}{n}$$

$$= 7716 - \frac{(292)^2}{14} = 1625.7143$$

2. Calculate the value of SS_y.

$$SS_y = \Sigma y^2 - \frac{(\Sigma y)^2}{n}$$

$$= 211.9690 - \frac{(54.466)^2}{14} = 0.07291743$$

3. Calculate the value of SS_{xy}.

$$SS_{xy} = \Sigma xy - \frac{(\Sigma x)(\Sigma y)}{n}$$

$$= 1125.363 - \frac{(292)(54.466)}{14} = -10.6421429$$

4. Calculate the value of r.

$$r = \frac{SS_{xy}}{\sqrt{SS_x SS_y}}$$

$$= \frac{-10.6421429}{\sqrt{(1625.7143)(0.07291743)}} = -0.9774$$

The computed value of r indicates a strong linear relationship between year and time. Note that the negative value of r means that as time in years increases, the world record time for the 1-mile run decreases. ▪

Pearson's correlation coefficient r should be interpreted only as a mathematical measure of the strength of the linear relationship between two variables. Note:

> A high value of r should not be construed to mean necessarily that a cause and effect relationship exists between the variables, since both variables could have been influenced by other variables. Remember that a high value of r means that two variables tend simultaneously to vary in the same direction. The tendency is a mathematical phenomenon and does not necessarily imply a direct relationship between the variables.

EXAMPLE 4.8

If the number of religious meetings and the number of violent crimes were recorded each month for a group of cities with widely varying populations, the data would probably indicate a high positive correlation. But it would be ridiculous to conclude that the number of violent crimes and the number of religious meetings are directly related. A third variable—namely, population—is causing crime and religious meetings to vary in the same direction. The correlation between crime and religious meetings is an example of **spurious correlation,** correlation caused by a third variable. As a result, we should be cautious about concluding a causal relationship from an observed correlation, since the correlation may be spurious.

EXERCISE SET 4.2

Basic Skills

1. For the bivariate data shown in the accompanying table, find
   a. SS_x.
   b. SS_y.
   c. SS_{xy}.

x	1	5	2	4	8	9
y	3	7	2	6	7	4

2. For the bivariate data shown in the accompanying table, find
   a. SS_x.
   b. SS_y.
   c. SS_{xy}.

x	2	7	8	1	5	9	3
y	8	7	1	4	7	4	5

3. Would you expect positive correlation, negative correlation, or no correlation for each of the following bivariate data sets?
   a. shoe sizes and hat sizes
   b. average adult beer consumption for the 23 counties in Maryland last year and the number of births in the 23 counties in Maryland last year
   c. average teacher salaries and average SAT mathematics scores for the public school systems in Pennsylvania
   d. weights of cars and gas mileages

4. Would you expect positive correlation, negative correlation, or no correlation for each of the following bivariate data sets?
   a. weights and heights of 6-year-old children
   b. systolic blood pressures and heart rates for 30-year-old females
   c. average rainfall in inches and average peach-tree yield in bushels for Clark County, Georgia, over the past 10 years
   d. diameters and areas of circles

5. Consider the following paired data:

x	1	5	11	17
y	1	7	19	19

   By using formula (4.2) determine the value of Pearson's correlation coefficient.

6. Determine Pearson's correlation coefficient for the accompanying data by using formula (4.2).

x	1	1	4	10	10	16
y	1	2	2	10	13	14

7. Consider the following set of bivariate data:

x	0	1	4	−5	2	6
y	−2	−1	2	−7	0	4

   a. Draw a scattergram.
   b. By using the scattergram, determine if the correlation is positive or negative.
   c. Compute the value of r.

8. Consider the following set of bivariate data:

x	0	4	8	1	3	−1
y	2	6	−14	0	−4	4

   a. Draw a scattergram.
   b. Using the scattergram, determine if the correlation is positive or negative.
   c. Compute the value of r.

9. Consider the following set of bivariate data:

x	1	2	3	4	5	6	7
y	12	7	4	3	4	7	12

   a. Draw a scattergram.
   b. Compute the value of Pearson's correlation coefficient r.
   c. Discuss the results of parts a and b.

Further Applications

10. The grades of eight students taking both Math 101 (x) and English 101 (y) are as shown in the accompanying table.

x	77	81	94	50	72	63	88	95
y	82	47	85	66	65	72	89	95

   a. Draw a scattergram.
   b. Using the scattergram, determine if the correlation is positive or negative.
   c. Compute the value of r.

11. The data in the table represent SAT (math) score (x) and GPA (y) for a group of ten students.

x	450	375	514	678	501	734	325	400	398	681
y	3.5	2.5	2.1	3.6	2.7	3.8	1.8	2.4	2.0	1.9

   a. Draw a scattergram.
   b. Using the scattergram, determine if the correlation is positive or negative.
   c. Compute the value of r.

12. The accompanying data represent the engine sizes (in cubic inches) and estimated miles per gallon for seven subcompact automobiles.

Car	Engine Size	Miles/ Gallon
Chevette	98	31
Sentra	98	35
Colt	86	41
Isuzu I-Mark	111	27
Mercedes 190D	134	35
Firebird	173	20
VW Rabbit	97	47

Determine Pearson's correlation coefficient r.

13. The data in the accompanying table represent the 1984 price-earnings (PE) ratio and percentage yield for seven stocks. Calculate the correlation coefficient r.

PE ratio	2.4	2.4	3.4	2.9	4.0	3.8	2.7
Percent yield	4.2	0.7	10	4.6	6.2	6.3	8.4

14. The following data indicate the amounts in billions of dollars for U.S. agricultural imports and exports:[27]

Year	Import Costs	Export Prices
1974	21.9	10.2
1975	21.9	9.3
1976	23.0	11.0
1977	23.6	13.4
1978	29.4	14.8
1979	34.7	16.7
1980	41.2	17.4
1981	43.3	16.8
1982	39.1	15.4

Find Pearson's correlation coefficient for import costs and export prices. Verify your result by coding the import costs as $U = 10(\text{import cost}) - 219$ and the export prices as $V = 10(\text{export price}) - 93$ and determining Pearson's correlation coefficient for variables U and V.

15. The accompanying table reports the number of American schools (K–12) with video (VCRs) and computer hardware (PCs) has grown significantly during the past decade.[28]

Year	1982–1983	1983–1984	1984–1985	1985–1986	1986–1987	1987–1988	1988–1989
No. schools	83,648	82,952	81,971	81,461	81,408	80,999	82,089
No PCs	30,859	55,175	70,255	74,379	76,242	76,899	78,784
No. VCRs	25,663	36,545	56,151	64,744	70,037	73,495	80,776

   a. Calculate Pearson's correlation coefficient for the number of schools and the number of PCs.
   b. Calculate Pearson's correlation coefficient for the number of PCs and the number of VCRs.

16. Refer to exercise 15. Calculate the value of r for the number of schools and the number of VCRs. Verify your result by coding the number of schools as $U =$ (no. schools) $- 80,999$ and the number of VCRs as $V =$ (no. VCRs) $- 25,663$ and determining the value of r for U and V.

17. The following table gives systolic blood pressures (in mm Hg) for a group of hypertensive patients along with the dose level (in mg) of a certain hypertensive drug.

Dose	Pressure
1	275
2	235
3	193
4	128
5	106

Determine Pearson's correlation coefficient.

18. In an experiment to investigate the effect of increasing the dosage of a certain barbiturate (in $\mu M/kg$) on sleeping time (in hours), the following readings were made at each of three dose levels.

Dosage	Time	Dosage	Time
3	4	10	7
3	6	15	13
3	5	15	11
10	9	15	9
10	8		

Determine Pearson's correlation coefficient between dosage and sleeping time.

19. For the data in application 4.4, verify that the correlation coefficient between x and y is equal to Pearson's correlation coefficient between U and V.

Going Beyond

20. The income levels of blacks continue to lag far behind those of whites. The accompanying table lists the per capita income (in dollars) from 1979 to 1988 for whites and blacks.[29]

Year	White	Black
1979	12,342	7,241
1980	11,820	6,897
1981	11,686	6,675
1982	11,679	6,571
1983	12,026	6,836
1984	12,455	7,147
1985	12,832	7,520
1986	13,332	7,779
1987	13,687	7,961
1988	13,896	8,271

   a. Calculate the value of Pearson's r for the per capita incomes of blacks and whites.
   b. By coding the years (1979 $= 0$, 1980 $= 1$, etc.), determine the value of r for year and white income.
   c. By coding the years (1979 $= 0$, 1980 $= 1$, etc.), determine the value of r for year and black income.

21. For the data in exercise 5, show that $SS_{xy} = \Sigma (x - \bar{x})(y - \bar{y})$.

22. Show that $SS_{xy} = \Sigma (x - \bar{x})(y - \bar{y})$.

23. Show that $SS_{xy} = \Sigma (x - \bar{x})y$.

24. Show that $r = cov(x, y)/s_x s_y$, where s_x is the standard deviation of the x scores and s_y is the standard deviation of the y scores.

25. Let $U = ax + b$ and $V = cy + d$ with a and c greater than 0. Show that the Pearson's correlation coefficient between U and V is identically equal to Pearson's correlation coefficient between x and y.

26. Show that $cov(z_x, z_y) = \dfrac{\Sigma z_x z_y}{n - 1} = r$.

Regression and Prediction

In section 4.2, we learned how to determine the strength of linear relationship between two variables by using scattergrams and the correlation coefficient r. If the strength of the linear relationship found by using the correlation coefficient r is determined to be high, then it may be desirable to describe the linear relationship in terms of an equation. Determining the linear relationship involves the study of regression. As we shall see later, a regression equation can be used for predictive purposes.

A linear relationship between two variables x and y can be defined by the **linear equation** $y = b + mx$. The constant m represents the **slope** of the straight line and the constant b represents the **y-intercept.** The relationship between Fahrenheit and Centigrade temperatures is a linear relationship. Its equation is

$$F° = 32 + \frac{9}{5} C°$$

Given any centigrade temperature, say 25°, we can determine the Fahrenheit temperature:

$$F° = 32 + \frac{9}{5} C°$$

$$= 32 + \frac{9}{5}(25)$$

$$= 32 + 45 = 77°$$

Suppose we are interested in studying the relationship between SAT (math) scores and freshman GPAs. Further, suppose that after constructing a scattergram for the bivariate data from last year's freshman class and determining the correlation coefficient r, we decide to find the linear relationship, called the **regression equation.** By using the appropriate method (to be presented subsequently), suppose we determined the regression equation to be $\hat{y} = -1.33 + 0.007x$, where x represents the SAT score in mathematics and $\hat{y}$ (read "y hat") represents the predicted GPA at the end of the freshman year. This equation can be used for predictive purposes. Further, suppose Mary is currently in high school and has applied for admission to college. She took the SAT test and received a score of 480 in mathematics. By using the regression equation, $\hat{y} = -1.33 + 0.007x$, we can predict her success during her freshman year at college. To determine her predicted GPA, we substitute $x = 480$ into the regression equation $\hat{y} = -1.33 + 0.007x$ and solve for $\hat{y}$:

$$\hat{y} = -1.33 + 0.007x$$
$$= -1.33 + (0.007)(480) = 2.03$$

We would predict Mary's GPA at the end of her freshman year to be 2.03.

Suppose the college wanted to adopt an admissions policy whereby no student would be admitted who would not have a predicted GPA of at least 1.25 their freshman year. What would be the college's SAT "cutoff" score, the SAT score above which a student would have a predicted GPA of at least 1.25? This can be found by solving the following equation for x when $\hat{y} = 1.25$:

$$\hat{y} = -1.33 + 0.007x$$

By substituting 1.25 for $\hat{y}$, we have

$$1.25 = -1.33 + 0.007x$$

And solving for x, we get

$$2.58 = 0.007x$$
$$x = 369$$

Thus, the SAT cutoff score would be 369 and any student with an SAT score below 369 would be denied admission because of predicted poor success. Of course, the situation of predicting success is not so simple and many other variables come into play that must be taken into account.

In most practical applications involving bivariate data, the points in a scattergram do not all fall on a straight line. The task then is to identify a straight line that comes closest to all the points in the scattergram, where "close" will be judged by using the squares of vertical distances of the points from a straight line. This line is represented by the equation $\hat{y} = b + mx$ and is called the **line of best fit** or the **regression line.** The symbol m represents the slope of the line, and the symbol b represents the y-intercept. For this line, the sum of squares of the vertical distances is as small as possible. The procedure for determining the line of best fit is called the **least squares method.** Of all the straight lines that can be drawn on a scattergram, this method will identify the line that produces the smallest sum of squared deviations of the points in the scattergram from the line.

For illustrative purposes, suppose our scattergram has only four points. If we let e_1 represent the distance of the first point from some line represented by $\hat{y} = b + mx$, then the vertical distance corresponds to the error in using the line to predict y_1 using x_1 (see figure 4.9). Thus $e_i = y_i - \hat{y}_i$ expresses the error when predicting the ith value of y using the ith value of x.

FIGURE 4.9

Vertical distances of points from regression line

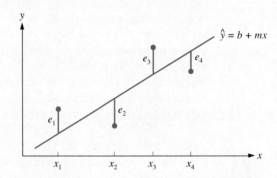

If $\hat{y} = b + mx$ is the line of best fit, then by principle of the method of least squares Σe_i^2 is a minimum. That is,

$$\Sigma e_i^2 = e_1^2 + e_2^2 + e_3^2 + e_4^2$$
$$= (y_1 - \hat{y}_1)^2 + (y_2 - \hat{y}_2)^2 + (y_3 - \hat{y}_3)^2 + (y_4 - \hat{y}_4)^2$$

is a minimum. In general, the sum Σe_i^2 is called the **sum of squares for error** and is denoted by SSE. Thus, we have

> **Sum of Squares for Error**
>
> $$SSE = \Sigma e_i^2 = \Sigma (y - \hat{y})^2$$
(4.7)

It should be pointed out that since $e_i = y_i - \hat{y}_i$, points above the line will have positive errors, points on the line will have 0 errors, and points below the line will have negative errors. If for all the points, the errors are added, a sum of 0 will always result. This is the reason that the errors are squared before they are added.

EXAMPLE 4.9

Consider the following data.

x	3	2	4	1
y	2	3	2	5

FIGURE 4.10

Scattergram showing line $\hat{y} = -2 + 2x$

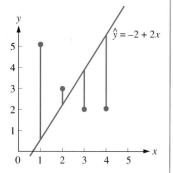

Let's compute SSE for some line, say $\hat{y} = -2 + 2x$, drawn on the scattergram. The line represented by the equation $\hat{y} = -2 + 2x$ is drawn on the scattergram shown in figure 4.10. The following table will be used to organize the computations:

x	y	$\hat{y}$	$y - \hat{y}$	$(y - \hat{y})^2$
3	2	4	−2	4
2	3	2	1	1
4	2	6	−4	16
1	5	0	5	25
				0

The first entry in the $\hat{y}$ column was found by substituting $x = 3$ into the equation $\hat{y} = -2 + 2x$ and solving for $\hat{y}$:

$$\hat{y} = -2 + 2x$$
$$= -2 + (2)(3) = 4$$

Note that the sum of the errors is

$$\Sigma (y - \hat{y}) = -2 + 1 - 4 + 5 = 0$$

and the sum of squares for error is

$$\text{SSE} = \Sigma (y - \hat{y})^2$$
$$= 4 + 1 + 16 + 25 = 46$$

Thus, for the line represented by $\hat{y} = -2 + 2x$, SSE = 46.

If no other line drawn on the scattergram in example 4.9 produces a value of SSE smaller than 46, then the line represented by the equation $\hat{y} = -2 + 2x$ is *the* regression line or line of best fit. Clearly, a trial-and-error method for selecting the best line according to the least squares criterion would not be productive. Fortunately, the determination of m and b in the equation $\hat{y} = b + mx$ to minimize SSE can be accomplished by using algebra or partial derivatives from calculus, and the details will be omitted. The least-squares formulas for finding m and b are as follows:

Constants for Regression

$$m = \frac{\text{SS}_{xy}}{\text{SS}_x} \tag{4.8}$$

$$b = \bar{y} - m\bar{x} \tag{4.9}$$

APPLICATION 4.5

For the data in example 4.9, find the regression equation and compute SSE.

Solution: The computations are organized using the following table.

x	y	x^2	y^2	xy
3	2	9	4	6
2	3	4	9	6
4	2	16	4	8
1	5	1	25	5
Sums 10	12	30	42	25

The sum of squares for x is

$$SS_x = \Sigma\, x^2 - \frac{(\Sigma\, x)^2}{n} = 30 - \frac{(10)^2}{4} = 5$$

The sum of cross products is

$$SS_{xy} = \Sigma\, xy - \frac{(\Sigma\, x)(\Sigma\, y)}{n} = 25 - \frac{(10)(12)}{4} = -5$$

By formula (4.8), the slope of the regression line is

$$m = \frac{SS_{xy}}{SS_x} = \frac{-5}{5} = -1$$

By formula (4.9), the y-intercept of the regression line is

$$b = \bar{y} - m\bar{x} = \frac{12}{4} - (-1)\frac{10}{4} = 3 + 2.5 = 5.5$$

FIGURE 4.11

Scattergram for application 4.5

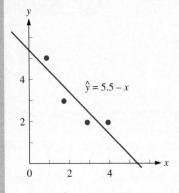

Thus, the regression equation is $\hat{y} = 5.5 - x$. Its graph appears on the scattergram in figure 4.11.

To find SSE, we organize our calculations in the following table:

x	y	$\hat{y}$	$y - \hat{y}$	$(y - \hat{y})^2$
3	2	2.5	−0.5	0.25
2	3	3.5	−0.5	0.25
4	2	1.5	0.5	0.25
1	5	4.5	0.5	0.25
			0	1

From the last column of the table, we have

$$SSE = \Sigma\, (y - \hat{y})^2$$
$$= 0.25 + 0.25 + 0.25 + 0.25$$
$$= 1 \qquad \blacksquare$$

For the data used in example 4.9, of all the lines that can be drawn on the scattergram, the regression line produces the smallest sum of squares for error, which is 1. Any other line will produce a sum of squares for error greater than 1. The line whose equation is $y = -2 + 2x$ produced a sum of squares for error of 46. It is interesting to observe that the point $(\bar{x}, \bar{y})$ is always on the regression line.

EXAMPLE 4.10

For the data in example 4.9, $\bar{x} = 2.5$ and $\bar{y} = 3$ and the point $(2.5, 3)$ satisfies the equation $\hat{y} = 5.5 - x$, since $3 = 5.5 - 2.5$. Thus, the point $(2.5, 3)$ is therefore on the line.

For a large set of bivariate data, it would be too time-consuming to follow the above method for finding SSE. Instead, the following computational formula can be used to find SSE.

Computational Formula for SSE

$$SSE = SS_y - m\, SS_{xy} \tag{4.10}$$

APPLICATION 4.6

For the data in application 4.5, find SSE by using formula (4.10).

Solution: By using the table in application 4.5, we first compute SSy:

$$SS_y = \Sigma\, y^2 - \frac{(\Sigma\, y)^2}{n}$$

$$= 42 - \frac{(12)^2}{4} = 6$$

By using formula (4.10), we obtain

$$SSE = SS_y - m\, SS_x$$
$$= 6 - (-1)(-5)$$
$$= 6 - 5 = 1$$

Hence, SSE $= 1$, as was computed in application 4.5. ▪

APPLICATION 4.7

For the 1-mile world record data in table 4.3, find the regression equation and use it to determine the year for which the predicted time for the mile run will be 3:40.

Solution: From application 4.4, we have

$$SS_x = 1625.7143$$
$$SS_{xy} = -10.6421429$$

Also, the mean of x is

$$\bar{x} = \frac{\Sigma\, x}{n} = \frac{292}{14} = 20.85714286$$

and the mean of y is

$$\bar{y} = \frac{\Sigma\, y}{n} = \frac{54.466}{14} = 3.890428571$$

By applying formula (4.8), we obtain the slope of the regression line:

$$m = \frac{SS_{xy}}{SS_x} = \frac{-10.6421429}{1625.7143} = -0.006546133$$

And by formula (4.9), the y-intercept of the regression line is

$$b = \bar{y} - m\bar{x} = 3.890428571 - (-0.006546133)(20.85714286) = 4.026962213$$

Thus, the regression equation is

$$\hat{y} = b + mx$$
$$= 4.026962213 - 0.006546133x$$

By letting $y = 3:40 = 3.667$ minutes and solving for x, we get $x = 55.0$. Since the transformation $x = $ year $- 1944$ was used to code the years, the predicted year in which the 3:40 mile will occur is

$$\text{year} = x + 1944$$
$$= 55 + 1944$$
$$= 1999$$

Thus, the predicted year for the 3:40 mile run is 1999. Note that the year 1999 was not arrived at by using prediction in the normal sense, where x is used to predict y. To do so, we would need to determine the regression equation using the least squares approach to predict y from x (see exercise 26 at the end of this section). ■

The regression equation in application 4.7 raises some interesting questions. What about the 3- or 2-minute mile? We find it hard to believe that any human will ever run a 2-minute mile. Yet the regression equation will produce a value of $y = 2$ if $x = 310$. By decoding the value of x, we find that the regression equation estimates that a 2-minute mile will be accomplished in the year 2254. We should always be cautious about making predictions far removed from the values of the variable x contained in the sample data. In our example, the coded values for x represent selected years from 1945 to 1985. Only values of x equal to (or close to) these values should be used for predictive purposes.

MINITAB can be used to determine the regression equation for the world record data. Computer display 4.3 shows the input and output information.

Computer Display 4.3

```
MTB > READ C1 C2
DATA> 1 4.023
DATA> 10 3.990
DATA> 10 3.967
DATA> 13 3.953
DATA> 14 3.908
DATA> 18 3.907
DATA> 20 3.902
DATA> 21 3.893
DATA> 22 3.855
DATA> 23 3.852
DATA> 31 3.833
DATA> 31 3.823
DATA> 37 3.788
DATA> 41 3.772
DATA> END
   14 ROWS READ
MTB > NAME C1 'CODEYEAR'
MTB > NAME C2 'CODETIME'
MTB > REGRESSION C2 1 C1
THE REGRESSION EQUATION IS CODETIME = 4.03 - 0.00655 CODEYEAR
```

The Relationship Between r and m

The correlation coefficient r and the regression slope m both involve the quantities SS_{xy} and SS_x. As a result, it is possible to solve for one in terms of the other. Using some elementary algebra, it can be shown that the following relationship holds.

> **Relationship between r and m**
>
> $$r = m\frac{s_x}{s_y}$$
> (4.11)

where s_x is the sample standard deviation of x and s_y is the sample standard deviation of y. Since s_x and s_y are greater than 0, the correlation coefficient r agrees in sign with the slope of the regression line. Thus formula (4.11) offers another explanation of why the correlation is positive if the points in a scattergram cluster from the lower left to the upper right and negative if the points cluster from upper left to lower right.

By solving formula (4.11) for m, we have

$$m = \frac{s_y}{s_x}r$$

If $r = 1$, then $m = \dfrac{s_y}{s_x}$, as illustrated in the diagram.

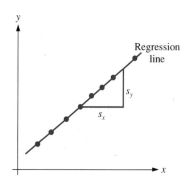

APPLICATION 4.8

A computer software package can be used to perform a regression analysis that provides values for b, m, and r. Consider the maximum heart rates and ages that were recorded for ten individuals on an intensive exercise program. The data are as follows:

Age	10	20	20	25	30	30	30	40	45	50
Heart rate	210	200	195	195	190	180	185	180	170	165

The printout shown here contains the appropriate regression analysis. Note that different notation is used: m represents the slope of the regression line, n represents the number of data pairs, B represents the y-intercept, R represents the correlation coefficient, and S.D. represents population standard deviation. The concept of *degrees of freedom* will be discussed later in the text. Computer software packages often provide more output than is needed at this stage and use different notations.

```
                CORRELATION AND LINEAR REGRESSION
    VARIABLE X: AGE              VARIABLE Y: HEART RATE
    MEAN OF X = 30               MEAN OF Y: 187
    S.D. OF X = 11.61895         S.D. OF Y: 13.07670
        NUMBER OF PAIRS (N) = 10
        CORRELATION COEFFICIENT (R) = −0.971
        DEGREES OF FREEDOM (DF) = 8
        SLOPE (M) OF REGRESSION LINE = −1.09259
        Y INTERCEPT (B) FOR THE LINE = 219.778
```

Computer printouts from commercial software packages often do not contain certain desired information. But many times the required information can be calculated from the information provided by the printout. For example, from this printout, we can determine the regression equation and the sum of squares for error, SSE. What peak heart rate would we predict for an age of 28?

Solution: The regression equation is $\hat{y} = 219.778 - 1.09259x$. For age 28, we would predict a peak heart rate of

$$\hat{y} = 219.778 - 1.09259(28) = 189.185$$

To find SSE, we shall use formula (4.10). We first need to find SS_x, SS_y, and SS_{xy}. Since the population variance is defined by $\sigma^2 = SS/N$, and the sample variance is defined by $s^2 = SS/(n - 1)$, to determine the value of SS given σ^2, we multiply σ^2 by N. That is, $s^2 = N/(n - 1)\sigma^2$ and $SS = N\sigma^2$. Hence,

$$SS_x = N\sigma_x^2 = 10(11.61895)^2 = 1350$$
$$SS_y = N\sigma_y^2 = 10(13.07670)^2 = 1710$$

Since the slope of the regression equation is defined by

$$m = \frac{SS_{xy}}{SS_x}$$

we can solve this equation for SS_{xy} to get

$$SS_{xy} = m\,SS_x$$
$$= (-1.09259)(1350) = -1474.9965$$

Thus, the sum of squares for error is

$$SSE = SS_y - m\,SS_x$$
$$= 1710 - (-1.09259)(-1474.9965) = 98.4336 \quad ■$$

EXERCISE SET 4.3

Basic Skills

1. For each of the following equations, find the slope and y-intercept of the line and sketch the graph:
   a. $y = 2x - 3$
   b. $y = x + 2$
   c. $2x + 3y = 6$
   d. $y = -2x$

2. For each of the following equations, find the slope and y-intercept of the line and sketch the graph:
   a. $y = -2x + 3$
   b. $y = 3x - 4$
   c. $4x - 3y = 12$
   d. $y = (2/3)x$

3. Find the regression equation and SSE for the data of exercise 1 of exercise set 4.2. The data are repeated here for convenience.

x	1	5	2	4	8	9
y	3	7	2	6	7	4

4. For exercise 2 of exercise set 4.2, find the regression equation and SSE. The data are repeated here for convenience.

x	2	7	8	1	5	9	3
y	8	7	1	4	7	4	5

5. For exercise 7 of exercise set 4.2, find the regression equation and SSE. The data are repeated here for convenience.

x	0	1	4	-5	2	6
y	-2	-1	2	-7	0	4

6. The data for exercise 8 in exercise set 4.2 are given in the table. Find the regression equation and SSE.

x	0	4	8	1	3	-1
y	2	6	-14	0	-4	4

7. The data for exercise 10 of exercise set 4.2 are reproduced here. Find the regression equation and SSE.

x	77	81	94	50	72	63	88	95
y	82	47	85	66	65	72	89	95

8. The following statistical values were obtained when nine pairs of bivariate data were analyzed:

   $$\bar{x} = 2.8, \qquad \bar{y} = 3.1, \qquad r = 0.049$$
   $$s_x = 1.476, \qquad s_y = 1.853, \qquad n = 10$$

   Find the regression equation $\hat{y} = b + mx$ and SSE.

9. The following statistical values were obtained when nine pairs of bivariate data were analyzed:

   $$\bar{x} = 7.27667, \qquad \bar{y} = 11.2722, \qquad r = 0.622$$
   $$s_x = 2.60702, \qquad s_y = 5.24589, \qquad n = 9$$

   Find the regression equation $\hat{y} = b + mx$ and SSE.

Further Applications

10. Consider the following bivariate data:

x	3	6	4	7
y	1	5	6	8

    Let $x' = x - \bar{x}$ and $y' = y - \bar{y}$ (recall that x' and y' are the deviation scores).
    a. Determine the regression equation $\hat{y}' = b + mx'$.
    b. Determine $\hat{y}'$ if $x' = 5$.

11. Consider the following bivariate data:

x	3	9	5	11
y	5	1	8	6

    Let $x' = x - \bar{x}$ and $y' = y - \bar{y}$ (recall that x' and y' are the deviation scores).
    a. Determine the regression equation $\hat{y}' = b + mx'$.
    b. Determine $\hat{y}'$ if $x' = 11$.

12. The weather appears to have an effect on baseball offense. The accompanying data indicate the relationship between temperature and offense from 1987 through 1989.[30] (This application was referred to in motivator 4.)

Temperature	Average	Runs Per Game	HRs Per Game
0°–59°	0.248	8.0	1.40
60°–69°	0.253	8.5	1.65
70°–79°	0.259	8.6	1.69
80°–89°	0.263	9.1	1.85
90°–up	0.263	9.1	1.83

    Use the class marks for temperatures 60°–89° to determine the regression equation that can be used to predict runs per game for given temperature.

13. Refer to exercise 12. Use the class marks for temperatures 60°–89° to determine the regression equation that can be used to predict home runs per game given temperature.

14. The accompanying table lists the 400 meter free style Olympic swimming times in seconds for women since 1924.

Year	Time	Year	Time
1924	362.20	1964	283.3
1928	342.8	1968	271.8
1932	328.5	1972	259.04
1936	326.4	1976	249.89
1948	317.8	1980	248.76
1952	312.1	1984	247.10
1956	294.6	1988	243.85
1960	290.6		

Find the regression equation and use it to predict the women's time for 1992. (*Hint:* Code the years using $x = $ year $- 1923$.)

15. The table lists the 400-meter free style Olympic swimming times in seconds for men since 1924.

Year	Time	Year	Time
1924	304.2	1964	252.2
1928	301.6	1968	249.0
1932	288.4	1972	240.27
1936	284.5	1976	231.93
1948	281.0	1980	231.31
1952	270.7	1984	231.23
1956	267.3	1988	226.95
1960	258.3		

Find the regression equation and use it to predict the men's time for 1992. (*Hint:* Code the years using $x = $ year $- 1923$.)

16. A study was conducted to test the effectiveness of a new drug for lowering heart rates in adult patients having heart disease. One thousand heart patients were involved in the study. The average reduction in heart rate (in beats per minute) for each of ten doses (in milligrams) of the drug is shown in the accompanying table.

Dose (x)	50	75	100	125	150	175	200	225	250	275
Average heart rate reduction	8	5	13	11	13	12	18	18	16	19

a. Determine the regression equation for predicting the average heart rate reduction given a fixed dosage of the drug.

b. Use the equation to predict the average heart rate reduction of a patient taking 300 mg of the drug.

Going Beyond

17. Using the regression equations found in exercises 14 and 15, determine the year in which the women's predicted time will equal the men's predicted time for the 400-meter free style race. Discuss your results.

18. For the data in exercise 10, find the z scores z_x for x and the z scores z_y for y. Then determine the regression equation $\hat{z}_y = b + mz_x$ and find r.

19. Refer to the accompanying computer printout.
    a. Find the equation for the regression line.
    b. Find SSE.

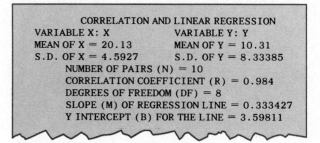

```
          CORRELATION AND LINEAR REGRESSION
VARIABLE X: X              VARIABLE Y: Y
MEAN OF X = 20.13          MEAN OF Y = 10.31
S.D. OF X = 4.5927         S.D. OF Y = 8.33385
     NUMBER OF PAIRS (N) = 10
     CORRELATION COEFFICIENT (R) = 0.984
     DEGREES OF FREEDOM (DF) = 8
     SLOPE (M) OF REGRESSION LINE = 0.333427
     Y INTERCEPT (B) FOR THE LINE = 3.59811
```

20. Prove formula 4.11.

21. Show that $SSE = SS_y - m^2 SS_x$.

22. A measure of how the points in a scattergram are scattered about the regression line is the **standard error of estimate** s_e. It is defined by $s_e = \sqrt{SSE/(n-2)}$. Find s_e for the data in exercise 10.

23. Show that $\Sigma (y - \hat{y}) = 0$, where $\hat{y} = b + mx$.

24. Show that $(\bar{x}, \bar{y})$ is a point on the regression line.

25. Show that the regression equation can be written as $\hat{y} = \bar{y} + m(x - \bar{x})$.

26. For the 1-mile world record data of application 4.4, find the regression equation to predict years given world record times. Use this equation to predict the year when the world record time for the 1-mile run will be 3:40. Which year—1999 or the year obtained here—would you predict for the world record time of 3:40?

CHAPTER SUMMARY

The concepts of linear equations, linear regression, and linear correlation were introduced in this chapter. To determine if a linear relationship exists between two variables, a scattergram is often used. We saw that the strength of a linear relationship can be measured by the correlation coefficient r and the sample covariance s_{xy}. The values of the correlation coefficient r can fall anywhere in the interval extending from -1 to 1, inclusive. If the points in the scattergram all fall on a straight line, the value of r is either -1 or 1, depending on whether the line has a positive or negative slope. A value of $r = 0$ indicates the lack of a linear relationship, and a value of r close to -1 or $+1$ does not necessarily imply any causation relationship. We also learned how to determine the regression equation by using the least-squares method. The sum of squares for error, SSE, is minimized when the least-squares method is used to determine the regression equation.

CHAPTER REVIEW

■ **IMPORTANT TERMS** ■

The following chapter terms have been mixed in ordering to provide you better review practice. For each term, provide a definition in your own words. Then check your responses against those given in the chapter.

standard error of estimate	regression equation	straight line
regression line	scattergram	sum of cross products
dependent variable	independent variable	sum of squares for error
slope	linear dependency	sample covariance
bivariate data	coding	centroid
Pearson's correlation coefficient	regression analysis	correlation analysis
least squares method	y-intercept	line of best fit
linear equation	spurous correlation	

■ **IMPORTANT SYMBOLS** ■

$\text{cov}(x, y)$, sample covariance	s_x, standard deviation of x	$\hat{y}$, predicted value of y
s_{xy}, sample covariance	s_y, standard deviation of y	e_i, prediction error
r, correlation coefficient	m, slope	SSE, sum of squares for error
SS_{xy}, sum of cross products	b, y-intercept	

■ **IMPORTANT FACTS AND FORMULAS** ■

Sample covariance: $\text{cov}(x, y) = s_{xy}$
$$= \frac{\Sigma (x - \bar{x})(y - \bar{y})}{n - 1} \quad (4.1)$$

Pearson's correlation coefficient:
$$r = \frac{\Sigma z_x z_y}{n - 1} \quad (4.2)$$

Sum of cross products:
$$SS_{xy} = \Sigma xy - \frac{(\Sigma x)(\Sigma y)}{n}$$

Computational formula for sample correlation coefficient:
$$r = \frac{SS_{xy}}{\sqrt{SS_x SS_y}} \quad (4.6)$$

Equation of a straight line: $y = b + mx$

Regression equation: $\hat{y} = b + mx$

Sum of squares for error: $SSE = \Sigma (y - \hat{y})^2$ (4.7)

Prediction error: $e_i = y_i - \hat{y}_i$

Slope of regression equation: $m = \dfrac{SS_{xy}}{SS_x}$ (4.8)

y-intercept of regression equation: $b = \bar{y} - m\bar{x}$ (4.9)

Computational formula for sum of squares for error:
$SSE = SS_y - m\,SS_{xy}$ (4.10)

Relationship between r and m: $r = m\dfrac{s_x}{s_y}$ (4.11)

REVIEW EXERCISES

1. In an attempt to determine the relationship between the amount spent on campaigning and the number of votes received during an election, the following data were collected:

Amount spent (in $1000): x	3	4	2	5	1
Votes received (in 1000s): y	14	12	5	20	4

a. Plot a scattergram.
b. Calculate the value of s_{xy}.
c. Calculate the value of r.
d. Find the regression equation and SSE.
e. Predict the number of votes received if $3500 is spent on the campaign.
f. For each additional $1000 spent, how many additional votes can be expected?
g. Draw a graph of the regression line on the scattergram.

2. In order to study the relationship between the number of times students cut classes and their final course grades, a Mathematics 209 instructor obtained the data shown here:

Number of class cuts	1	2	3	3	4	4	5	6	2	0
Grade	98	98	88	81	83	76	71	71	85	98

a. Plot a scattergram.
b. Calculate the value of s_{xy}.
c. Calculate the value of r.
d. Find the regression equation and SSE.
e. Predict the final course grade if a student has cut three classes.

f. For each additional cut, by how much will the predicted final grade be affected?
g. Draw a graph of the regression line on the scattergram.

3. The number of students offered admission (x) and the number of actual first-time enrollments (y) for the past seven years at a college are

x	3300	4100	5600	5200	5900	5500	5100
y	3000	3500	4200	4800	5000	5100	4700

a. Plot a scattergram.
b. Calculate the value of s_{xy}.
c. Calculate the value of r.
d. Find the regression equation and SSE.
e. Find $\hat{y}$ if $x = 5000$.
f. How many additional enrollments can be expected for every 1000 increase in offers made?
g. Draw a graph of the regression line on the scattergram.

4. A biometrician studied the effects of different doses (x) of a new drug on the pulse rate (y) of humans. The results for five individuals are as indicated in the table.

Dose (x)	2.5	3	3.5	4	4.5
Drop in rate (y)	8	11	9	16	19

a. Calculate the value of r and interpret the result.
b. Find the regression equation.
c. Find SSE.
d. Find $\hat{y}$ if $x = 3.75$.
e. For each unit increase in dose, what is the predicted drop in pulse rate?
f. Draw a graph of the regression line on a scattergram.

5. The following information was determined in a regression analysis:

$$\hat{y} = 25.1875x - 878.8583$$
$$s_y = 278.5247 \qquad s_x = 9.3956$$
$$\bar{x} = 51.5 \qquad \bar{y} = 418.3 \qquad n = 10.$$

a. Find the value of r.
b. Find SSE.
c. If $x = 45$, find $\hat{y}$.
d. For each unit increase in x, by how much will $\hat{y}$ change?

6. The accompanying data represent the annual arms sales (in billions of dollars) by the United States to third-world nations.

Year	1976	1977	1978	1979	1980	1981	1982	1983
Sales	8.2	9.8	10.1	9.2	6.4	6.8	7.9	9.7

a. Find the equation for predicting sales.
b. By using the regression equation found in part a, estimate the U.S. sales to third-world nations for the year 1984. (*Hint:* Code the years using $x = $ year $- 1975$.)

7. Nine goldfish were acclimated to a water temperature of $3°$ C and then subjected to gradually increasing water temperatures to determine if metabolic rate is related to temperature. Metabolic rate was determined by counting opercular beats per minute. The resulting data are listed in the accompanying table.

Temperature C°	Mean Number of Opercular Beats/Minute
5.0	33.0
7.5	44.8
10.0	54.0
12.5	52.5
15.0	70.2
17.5	99.8
20.0	110.5
22.5	117.0
25.0	129.1

a. Draw a scattergram.
b. Compute r.
c. Find the regression equation.
d. Determine SSE.
e. If the temperature were $0°$ C, how many opercular beats per minute would be expected?

The following data are for review exercises 8 and 9.

Distance to Nearest Tree (feet)	Ht (feet)	DBH (inches)
1	39	5.7
19.5	72	9.2
4.0	69	9.3
5.0	67	9.5
10.5	73	9.5
9.0	78	9.7
8.0	79	9.8
15.0	81	10.4
12.5	65	10.7
18.0	60	11.7

8. A forester wants to determine the correlation between total height (Ht) and diameter at breast height (DBH) of a sample of quaking aspen. For the tabled data determine
a. The value of r.
b. The regression equation.
c. SSE.

9. A forester wants to determine the correlation between size (measured as diameter at breast height) and distance to nearest tree for a sample of quaking aspen. Refer to the tabled data.
a. Draw a scattergram.
b. Compute the value of r.
c. Find the regression equation for predicting DBH from distance to the nearest tree.
d. Find SSE.

10. To determine if traffic flow, measured in number of vehicles per hour, and lead content contained in vegetation growing near highways are related, a study was undertaken at six different locations. The following data were obtained:

Number of vehicles	103	216	294	402	416	573
Lead content	4.6	7.4	26.1	37.2	24.8	38.7

Find the regression equation and use it to predict the lead content of vegetation experiencing a traffic flow of 300 vehicles per hour.

Computer Applications

1. The following table lists the selling price x (in thousands of dollars) and the area y (in hundreds of square feet) for a sample of 10 homes sold in Somerset County this past year.

x	98	118	94.6	92.5	68.5	77	62.5	89.1	120.5	33.6
y	16	39	24	17	17	13	18	15	20	13

Use a computer software program to
a. find the equation of the regression line for area in terms of selling price.
b. find the equation of the regression line for selling price in terms of area.
c. find the correlation coefficient.
d. find the sample covariance.
e. plot a scatter diagram.
f. estimate the selling price of a home having an area of 1500 square feet last year.
g. estimate the area of a home having a selling price of $100,000 last year.

2. The following table lists the selling price x (in thousands of dollars) and the yearly amount of real estate taxes y (in thousands of dollars) for a sample of 10 homes sold in Somerset County this past year.

x	98	118	94.6	92.5	68.5	77	62.5	89.1	120.5	33.6
y	3.2	2.9	2.4	2.7	2.1	2.6	1.6	2.7	4.2	0.73

Use a computer software program to
a. find the equation of the regression line for area in terms of amount of tax.
b. find the equation of the regression line for the amount of tax terms of selling price.
c. find the correlation coefficient.
d. find the sample covariance.
e. plot a scatter diagram.
f. estimate the selling price of a home having a tax of $2,500 last year.
g. estimate the real estate taxes of a home having a selling price of $100,000 last year.

3. According to a survey conducted by the Northwestern National Life Insurance Company, citizens of Utah, the state with the healthiest population, live two and a half years longer than citizens of Delaware, the state with the unhealthiest population. The survey ranked states from 1 to 50 (state 1 having the healthiest population and state 50 having the unhealthiest population) using six factors:

Life span—1 = highest life expectancy at birth

Disease—1 = the lower rate of major illness

Lifestyle—1 = the greatest participation in healthy habits and lowest rate of alcohol and cigarette consumption

Access—1 = the highest availability of doctors and highest portion of population and health of insurance

Lost time—1 = fewest missed school and work days and the fewest acute illnesses

Mortality—1 = the lowest total death rates, infant death rates, and premature death rates

The accompanying table lists the results of the survey.[31]

Rank	State	Life Span	Disease	Lifestyle	Access	Lost Time	Mortality
1.	Utah	3	1	1	38	10	1
2.	North Dakota	5	4	6	10	18	4
3.	Idaho	10	2	2	25	10	11
4.	Minnesota	2	5	18	3	31	6
5.	Hawaii	1	16	35	21	1	1
6.	Vermont	17	6	42	3	37	16
7.	Nebraska	6	7	3	17	37	22
8.	Colorado	9	9	29	20	7	5
9.	Wyoming	26	3	26	37	9	10
10.	Montana	25	7	18	16	13	18
11.	Washington	11	11	22	16	10	13
12.	Oregon	14	14	15	5	18	24
13.	New Mexico	22	12	17	43	2	8
14.	Wisconsin	7	12	31	17	31	14
15.	South Dakota	14	10	7	23	27	29
16.	Iowa	3	19	3	30	45	14
17.	Maine	19	17	25	12	45	23
18.	California	19	26	31	8	4	7
19.	Massachusetts	13	23	38	1	37	21
20.	Alaska	46	21	47	40	3	3
21.	Indiana	26	18	18	33	31	31
22.	Arizona	21	25	37	28	5	16
23.	Oklahoma	31	20	12	42	18	32
24.	New Hampshire*	14	15	49	11	37	9
25.	Kansas*	7	35	5	21	31	19
26.	Texas*	31	26	29	45	5	11
27.	Pennsylvania*	34	24	24	12	50	39
28.	Connecticut	11	36	38	1	27	20
29.	Kentucky	41	22	14	25	37	36
30.	New Jersey	22	28	38	19	18	28
31.	Missouri	26	31	13	33	45	32
32.	Ohio	35	30	26	25	37	30
33.	Virginia	36	29	38	31	18	27
34.	Arkansas	29	32	7	35	37	44
35.	West Virginia	43	34	9	28	45	47
36.	Illinois	37	37	31	24	13	36
37.	New York	29	42	36	7	13	38
38.	Louisiana*	50	33	28	47	13	40
39.	Tennessee*	39	41	10	35	37	40
40.	Rhode Island	18	43	45	9	45	26
41.	North Carolina	42	40	22	43	27	40
42.	Alabama	45	39	10	46	31	48
43.	Maryland	37	49	44	14	18	34
44.	Florida	22	46	48	15	31	45

*These states tied.

Rank	State	Life Span	Disease	Lifestyle	Access	Lost Time	Mortality
45.	Georgia	46	47	31	48	18	45
46.	South Carolina	49	45	18	49	13	49
47.	Nevada	44	38	50	32	8	24
48.	Michigan	31	50	42	40	18	35
49.	Mississippi	48	44	16	50	18	50
50.	Delaware	40	48	46	38	27	40

Calculate Pearson's correlation coefficient for the ranks of

  a. health of state and lifestyle.
  b. health of state and life span.
  c. lifestyle and lost time.
  d. access and lost time.
  e. disease and lifestyle.

4. Collect the following information from 25 drivers licensed to drive in the state where your college is located: birthday (coded from 1 to 366) and the last three digits of their drivers license number.

  a. Determine Pearson's correlation coefficient.
  b. Determine the regression equation for predicting the last three digits from the coded birthday.
  c. Discuss the significance of your results.

▪ CHAPTER ACHIEVEMENT TEST ▪

1. The number of push-ups $\hat{y}$ a normal, healthy child should be able to perform, based on age x, is given by $\hat{y} = 1.4x - 0.9$, where $4 \leq x \leq 17$.

  a. A ten-year-old child should be expected to do how many push-ups?
  b. As age increases, will the number of push-ups increase or decrease?
  c. For each additional one-year increase in age, a child should be expected to do how many more push-ups?
  d. Does the y-intercept have a meaningful interpretation in this case? Why or why not?

2. A study of the relationship between height (in inches) and weight (in pounds) of college males yielded data given here.

Height (x)	64	72	73	68	66	67
Weight (y)	165	158	173	125	125	139

Determine the following:

  a. SS_x
  b. SS_y
  c. SS_{xy}
  d. the slope, m
  e. they y-intercept, b
  f. the regression equation
  g. Pearson's correlation coefficient
  h. SSE

  i. $\hat{y}$ when $x = 65$
  j. the error e for the male in the sample whose height is $x = 72$ inches

3. There is a high positive correlation between teachers' salaries and annual consumption of beer in the United States.

  a. Does this mean that increased beer consumption has been caused by increasing teachers' salaries? Or, the more a teacher drinks, the more he or she will get paid? Explain.
  b. What additional factor(s) might cause teachers' salaries and beer consumption to increase together?

4. Since $\hat{y} = b + mx$ and $b = \bar{y} - m\bar{x}$, we have $\hat{y} = (\bar{y} - m\bar{x}) + mx$. Thus, $\hat{y} - \bar{y} = m(x - \bar{x})$. With the regression equation written in this form, show that the centroid $(\bar{x}, \bar{y})$ is on the regression line.

5. Find SSE for the following bivariate data (x, y):

x	1	3	5
y	2	4	0

6. Find the sample covariance for the accompanying paired data.

x	1	3	5	15	16
y	11	15	21	25	28

UNIT TWO

Basic Probability

5 *Introduction to Elementary Probability*

CHAPTER OBJECTIVES

In this chapter we will investigate

▷ *Experiments.*

▷ *Sample spaces.*

▷ *Events.*

▷ *What a compound event is and how to form compound events.*

▷ *Complementary events.*

▷ *How to use Venn diagrams to represent events.*

▷ *Mutually exclusive events.*

▷ *Probability.*

▷ *The methods for assigning probabilities.*

▷ *What mathematical odds are and how to calculate odds.*

▷ *The fundamental theorem of counting.*

▷ *How to use the fundamental theorem of counting to find probabilities.*

▷ *What permutations are and how to use them to count.*

▷ *What combinations are and how to use them to count.*

▷ *Some elementary rules of probability.*

▷ *Independent events.*

▷ *Random variables.*

▷ *How to find the mean and variance of a random variable.*

|||| MOTIVATOR 5 ▶

*T*he medical problem of acquired immune deficiency syndrome (AIDS) has created much activity within the research community to identify those people who have the virus. Did you know that if the probability of a person with the AIDS virus testing positive is 98/100, it does not necessarily mean that the probability of a person without the AIDS virus testing positive is 2/100?

The test most frequently used to screen donated blood for the presence of the AIDS virus is the ELISA test. It was developed to keep AIDS out of the blood supply. When the blood actually contains the AIDS virus, the ELISA test is positive with probability 98/100 and negative with probability 2/100. If the blood does not contain the antibodies for the virus, the test is positive with probability 7/100 and negative with probability 93/100. The higher probability for false positive results is viewed as acceptable because of the low probability of a false negative result.

If the ELISA test is applied to 10,000 blood samples of an at-risk population whose probability of having the disease is 5/1000, then we would expect to have 50 contaminated samples. Of these 50 contaminated samples, $0.98 \times 50 = 49$ will test positive. Out of the 10,000 samples, we would expect 9,950 of them to be uncontaminated, and of these, $0.07 \times 9950 = 696.5$ will test positive (false positive) using the ELISA test. The number of false negatives is $0.02 \times 50 = 1$. The higher number of false positives (696.5) in this case would be considered as acceptable because of the low number of false negatives (1) associated with the ELISA test.

The probability that a person has AIDS given that he/she has a positive blood test involves conditional probability, a topic that will be discussed in section 5.5. Ideally, a test is needed that has high predictive value when the test is positive or negative. That is, if the test is positive, then it is highly probable that the person has AIDS, or if the test is negative, it is highly unlikely that a person has AIDS. After studying the next several chapters, it should be reasonably clear why it is impossible to develop an AIDS test with a predictive value of 100%.[32]

Chapter Overview

The purpose of this chapter is to develop the basic ideas that will be needed for an adequate understanding of inferential statistics. *Inferential statistics* is a body of knowledge that treats the methods for characterizing a population using information calculated from samples drawn from that population. These methods always involve a certain degree of uncertainty. For example, we could address the following questions:

1. What is the *chance* that the AJAX Company will close down?
2. What is the *average* weight of month-old babies?
3. Is brand A *better* than brand B?

Every day we are faced with decision-making and probability statements. Statements involving the words *chance, likelihood, odds, likely, expected, possible, uncertain,* and *probably* are all addressing the same issue: uncertainty. Every day we make or hear statements similar to the following:

1. What is the *likelihood* we will have a test today?
2. The *odds* are 1 in 2 million that you will be struck by lightning.
3. The job will *likely* be completed on time.
4. The *chance* for rain today is 50%.
5. If a coin is tossed, there is a 50–50 *possibility* that a head will occur.

6. What is the *probability* that the new proposed method will lead to better results?

7. I am *confident* that I can pass this course.

We already have a good intuitive feeling for probability. Building on this basis, we will explore some of the not-so-obvious properties of probability theory to help us develop a better understanding of inferential statistics. Probability provides the foundation for developing the science of inferential statistics. Using probability theory, we can deduce the likelihood of certain samples occurring with specified properties. Such information will enable us to draw inferences about the population. We begin our discussion of probability with a discussion of experiments.

SECTION 5.1   *Experiments and Events*

Experiments

An **experiment** is any planned process that results in observations being made or data being collected. A very simple example of an experiment is tossing a six-sided die and observing the number showing face up when the die comes to rest. For this experiment, there could be six possible outcomes recorded, 1, 2, 3, 4, 5, or 6. On the other hand, the result of this experiment could be recorded as just an even or odd number. Thus it is important to carefully define the outcome to be recorded since the same basic situation could produce data that can be recorded from several different points of view. As another example of an experiment, let's say we test a selected sample of a finished product from an assembly line to determine whether or not the product is defective. This experiment has two outcomes: the product is defective or it is not defective. In both of these situations, we can repeat the experiments many times. For our purposes, we will mainly be interested in experiments that can be repeated or that can be conceived of as being repeatable. Experiments such as observing whether it will rain tomorrow or determining who will win the world series in baseball next year are not repeatable; we will not consider such experiments in this section. Frequently we will be interested in the outcomes resulting from repeating an experiment a given number of times.

All experiments have outcomes, and the outcomes of most experiments are uncertain and depend on chance. The outcomes from an experiment form a set called a sample space.

> A **sample space** for an experiment is a collection of all possible outcomes.

The simplest experiment involving uncertainty is one that has two outcomes and a single sample space, as we see in example 5.1. An experiment can, however, have more than one sample space; that is, more than one sample space can be used to describe the outcomes of an experiment. It is usually desirable to choose a sample space that provides the most information concerning the experiment. Consider examples 5.2 and 5.3 and application 5.1.

EXAMPLE 5.1

Observing the gender of the next baby born at Memorial Hospital is an experiment with two outcomes. A sample space for this experiment consists of the set $S = \{M, F\}$, where M represents a male, F represents a female, and the braces are used to indicate a collection or set.

EXAMPLE 5.2

If we observed the births of the next two babies born at Memorial Hospital, then a sample space for the experiment could be the set $S_1 = \{MM, MF, FM, FF\}$, where, for example, FM indicates that the first baby born was a female and the second baby born was a male. Another sample space for this experiment might consist of the number of possible male births, $S_2 = \{0, 1, 2\}$. Notice that these two sample spaces for the experiment provide different information. From the outcome of a single male in S_2, we can not determine if it was born first or second. For this experiment involving the same subjects, we have recorded the data in two different ways.

EXAMPLE 5.3

For the experiment of tossing a six-sided die, $S_1 = \{1, 2, 3, 4, 5, 6\}$ is a sample space for the experiment. The set $S_2 = \{\text{even number, odd number}\}$ is also a sample space. Sample space S_1 contains more information than sample space S_2. Knowing that the outcome of a roll of a six-sided die is even does not make it possible to determine whether it was a 2, 4, or 6.

APPLICATION 5.1

List a sample space for each of the following experiments:

   a. Toss a dime and a quarter, in that order, and observe how they fall.
   b. Toss a penny, a nickel, and a dime, in that order, and observe how they fall.
   c. Select a female college student at random and ask her how old she is in years.
   d. Toss a coin first, followed by a six-sided die, and observe how they fall.
   e. Toss a coin until a head occurs.

Solution:
   a. $S = \{HT, TH, TT, HH\}$. The outcome HT means the dime lands heads-up and the quarter lands tails-up.
   b. $S = \{HHH, HHT, HTH, HTT, THH, TTH, THT, TTT\}$. The outcome HTH means that the penny shows a head, the nickel shows a tail, and the dime shows a head.
   c. $S = \{10, 11, 12, 13, \ldots, 98, 99, 100\}$.
   d. $S = \{H1, H2, H3, H4, H5, H6, T1, T2, T3, T4, T5, T6\}$.
   e. $S = \{H, TH, TTH, TTTH, TTTTH, \text{etc.}\}$. The outcome TTTH means a head was obtained on the fourth toss. ▪

Events

For a given experiment we may be interested in determining the likelihood of a collection of outcomes occurring instead of the likelihood of a single outcome happening. For example, when three coins are tossed once, we might be interested in the outcomes that indicate that at least two heads are obtained. This collection of outcomes, $\{HHT,$ $HTH, THH, HHH\}$, is called an event.

> An **event** is any subcollection (or **subset**) of a sample space S.

EXAMPLE 5.4

Suppose the experiment is tossing a penny followed by tossing a dime. A sample space for this experiment could be $S = \{HH, HT, TH, TT\}$. Some possible events are

$$E_1 = \{HH\} \qquad E_4 = \{HH, TT\}$$
$$E_2 = \{HT\} \qquad E_5 = \{HT, TH\}$$
$$E_3 = \{TH\} \qquad E_6 = \{TT\}$$

There are 16 possible events. The **empty set** $\emptyset$ and the sample space S are also events. The event E_6 can be described as getting a tail on the penny and a tail on the dime. The event E_4 can be described as getting two heads or two tails.

> A **simple event** is an event that contains only one outcome.

EXAMPLE 5.5

The event $E_1 = \{HH\}$ for example 5.4 is a simple event, whereas the event E_4 is not a simple event.

Remember that an event is always a collection of outcomes from the collection of all outcomes identified as the sample space. **Venn diagrams** can be used for graphically portraying sample spaces and relationships between events. A rectangle commonly denotes the sample space and events are represented by circles drawn inside the rectangle as indicated in figure 5.1.

FIGURE 5.1

Venn diagram representing sample space and event

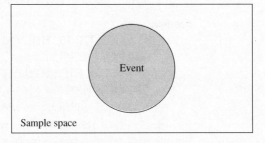

Sample space

APPLICATION 5.2

Suppose an experiment consists of examining three fuses. Each fuse can be defective (D) or nondefective (N). Eight possible outcomes are {NDD, NDN, NND, NDD, DND, DNN, DDN, DDD}. List the outcomes that comprise each of the following events:

a. E_1 = first fuse is defective.

b. E_2 = first and last fuses are defective.

c. E_3 = fuses are all good.

d. E_4 = at least one fuse is defective.

e. E_5 = at most one fuse is defective.

Solution:

a. $E_1 = \{DNN, DND, DDN, DDD\}$.

b. $E_2 = \{DDD, DND\}$.

  c. $E_3 = \{NNN\}$.

  d. Note that *at least one* means one or more. $E_4 = \{NDD, NDN, NND, DNN, DND, DDN, DDD\}$.

  e. Note that *at most one* means one or less. $E_5 = \{NDN, NND, NNN, DNN\}$. ▪

APPLICATION 5.3

Consider the experiment of tossing a red six-sided die and a black six-sided die and observing how they fall. A sample space of 36 possible outcomes is as follows, where the first entry is the outcome on the red die and the second entry is the outcome on the black die.

$$
\begin{array}{cccccc}
\{(1, 1) & (1, 2) & (1, 3) & (1, 4) & (1, 5) & (1, 6) \\
(2, 1) & (2, 2) & (2, 3) & (2, 4) & (2, 5) & (2, 6) \\
(3, 1) & (3, 2) & (3, 3) & (3, 4) & (3, 5) & (3, 6) \\
(4, 1) & (4, 2) & (4, 3) & (4, 4) & (4, 5) & (4, 6) \\
(5, 1) & (5, 2) & (5, 3) & (5, 4) & (5, 5) & (5, 6) \\
(6, 1) & (6, 2) & (6, 3) & (6, 4) & (6, 5) & (6, 6)\}
\end{array}
$$

The sample space S can be represented by the following diagram:

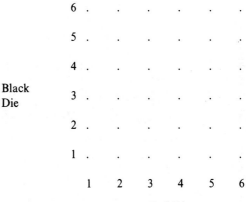

Give a description for the following events:

  a. $\{(1, 1), (2, 1), (3, 1), (4, 1), (5, 1), (6, 1)\}$

  b. $\{(1, 1), (2, 2), (3, 3), (4, 4), (5, 5), (6, 6)\}$

  c. $\{(3, 4), (4, 3), (5, 2), (2, 5), (6, 1), (1, 6)\}$

  d. $\{(5, 6), (6, 5)\}$

  e. $\{(1, 1)\}$

  f. $\{(4, 4)\}$

Solution:

  a. The black die shows 1.

  b. The two dice match.

  c. The sum of the dice equals 7.

d. The sum of the dice equals 11.

e. Both dice show a 1 (a pair of 1s is called "snake eyes").

f. Both dice show a 4 (a pair of 4s is called "boxcars"). ■

APPLICATION 5.4 For the dice-tossing experiment in application 5.3 list the outcomes for the following events:

a. The sum is even.

b. The sum is divisible by 5.

c. The sum is a prime number. (A *prime number* is a number greater than 1 that is divisible only by 1 and itself).

d. The number on the black die is two units larger than the number on the red die.

e. The sum is odd.

f. The sum is not exactly divisible by 5.

Solution:

a. The following pairs have a sum that is even:

$$\{(1, 1) \quad (1, 3) \quad (1, 5) \quad (2, 2) \quad (2, 4) \quad (2, 6)$$
$$(3, 1) \quad (3, 3) \quad (3, 5) \quad (4, 2) \quad (4, 4) \quad (4, 6)$$
$$(5, 1) \quad (5, 3) \quad (5, 5) \quad (6, 2) \quad (6, 4) \quad (6, 6)\}$$

b. The following pairs have a sum that is divisible by 5:

$$\{(1, 4), \quad (4, 1), \quad (3, 2), \quad (2, 3), \quad (5, 5), \quad (6, 4), \quad (4, 6)\}$$

c. The following pairs have a sum that is a prime number:

$$\{(1, 2), \quad (2, 1), \quad (1, 4), \quad (4, 1), \quad (1, 6), \quad (6, 1), \quad (2, 5), \quad (5, 2),$$
$$(3, 4), \quad (4, 3), \quad (5, 6), \quad (6, 5), \quad (2, 3), \quad (3, 2)\}$$

d. The following pairs have a number on the black die that is 2 greater than the number on the red die:

$$\{(1, 3), \quad (2, 4), \quad (3, 5), \quad (4, 6)\}$$

e. The following pairs have a sum that is odd:

$$\{(1, 2) \quad (1, 4) \quad (1, 6) \quad (2, 1) \quad (2, 3) \quad (2, 5)$$
$$(3, 2) \quad (3, 4) \quad (3, 6) \quad (4, 1) \quad (4, 3) \quad (4, 5)$$
$$(5, 2) \quad (5, 4) \quad (5, 6) \quad (6, 1) \quad (6, 3) \quad (6, 5)\}$$

f. The following pairs have a sum that is not divisible by 5:

$$\{(1, 1) \quad (1, 2) \quad (1, 3) \quad (1, 5) \quad (1, 6) \quad (2, 1)$$
$$(2, 2) \quad (2, 4) \quad (2, 5) \quad (2, 6) \quad (3, 1) \quad (3, 3)$$
$$(3, 4) \quad (3, 5) \quad (3, 6) \quad (4, 2) \quad (4, 3) \quad (4, 4)$$
$$(4, 5) \quad (5, 1) \quad (5, 2) \quad (5, 3) \quad (5, 4) \quad (5, 6)$$
$$(6, 1) \quad (6, 2) \quad (6, 3) \quad (6, 5) \quad (6, 6)\} ■$$

Event (Not E)

> If E is an event contained in a sample space S, then the *event not E*, denoted by $\overline{E}$, is the event containing all the outcomes in S that are not contained in E.

In the Venn diagram shown in figure 5.2, $\overline{E}$ is the shaded area inside of S and outside of E.

FIGURE 5.2

Venn diagram of the event "not E"

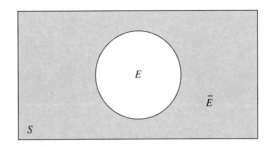

EXAMPLE 5.6

Consider the experiment of tossing a six-sided die. If E is the event of getting a 4 or a 6, then the event $\overline{E}$ contains the outcomes 1, 2, 3, and 5. That is, if $E = \{4, 6\}$, then $\overline{E} = \{1, 2, 3, 5\}$.

Suppose that E is an event for some experiment. Since an event E either does or does not occur, a sample space for the experiment is $S = \{E, \overline{E}\}$. As a result, any experiment has a sample space with just two outcomes, E and $\overline{E}$. Consider application 5.5.

APPLICATION 5.5

For each of the following experiments, list a sample space with only two outcomes.

    a. Toss a coin five times and observe the number of heads.

    b. Toss one die, followed by another.

    c. Spin the spinner and observe where it lands.

Solution: For each experiment we list two of the possible sample spaces containing two outcomes. There are many other possible sample spaces, which you are encouraged to explore.

    a. $S_1 = \{$obtain 5 heads, do not obtain 5 heads$\}$
       $S_2 = \{$obtain 2 heads, do not obtain 2 heads$\}$
    b. $S_1 = \{$sum is even, sum is odd$\}$
       $S_2 = \{$get two 3s, do not get two 3s$\}$
    c. $S_1 = \{$lands on 1, does not land on 1$\}$
       $S_2 = \{$lands on 2, does not land on 2$\}$ ■

Compound Events

Since events are sets, the union and intersection operators from set theory can be used with events to form **compound events.** If E and F are events, then the events $(E \cup F)$ and $(E \cap F)$ are examples of compound events.

$(E \cup F)$ is the event that E will occur or F will occur or they both occur.

$(E \cap F)$ is the event that both E and F will occur at the same time.

Venn diagrams can be used to illustrate compound events. In the Venn diagrams in figure 5.3, the compound events $(A \cup B)$ and $(A \cap B)$ are represented by the shaded regions.

FIGURE 5.3

Venn diagrams of compound events

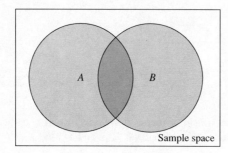

The event $A \cup B$

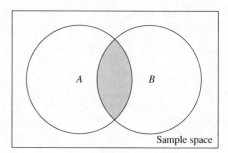

The event $A \cap B$

APPLICATION 5.6

Consider the dice-tossing experiment in application 5.3. Let E be the event that the number on the red die is 4 and let F be the event that the sum of the numbers showing is 7. Find the outcomes making up the following events:

a. E
b. F
c. $E \cup F$
d. $E \cap F$

Solution: See application 5.3 for the listing of S.

a. $E = \{(4, 1), (4, 2), (4, 3), (4, 4), (4, 5), (4, 6)\}$.
b. $F = \{(1, 6), (6, 1), (5, 2), (2, 5), (3, 4), (4, 3)\}$.
c. $E \cup F = \{(4, 1), (4, 2), (4, 3), (4, 4), (4, 5), (4, 6), (1, 6), (6, 1), (5, 2), (2, 5), (3, 4)\}$. Note that $(4, 3)$ is not included twice in the set.
d. $E \cap F = \{(4, 3)\}$. ■

APPLICATION 5.7

Suppose Harry goes to a quick-shop service station to buy a quart of transmission fluid and a quart of oil for his car. If there are three brands of transmission fluid available (X, Y, Z) and six brands of oil available (A, B, C, D, E, F), there are 18 different purchases that Harry can make. Suppose further that the three brands of transmission fluid are priced equally and that the six brands of oil are priced equally. Since Harry is not knowledgeable about oil for his car, he decides on his purchases by guessing. Let

E be the event that Harry purchases brand X or brand Z transmission fluid and let F be the event that he purchases brand A oil or brand C oil. List each of the following:

a. A sample space for the experiment of choosing a quart of transmission fluid and a quart of oil
b. The event $\overline{E}$
c. The event $(E \cup F)$
d. The event $(E \cap F)$

Solution: Figure 5.4 illustrates a sample space containing the events $\overline{E}$, $(E \cup F)$, and $(E \cap F)$. Note that S has 18 outcomes, event E has 12 outcomes, and event F has 6 outcomes.

FIGURE 5.4

Sample space and events for application 5.7

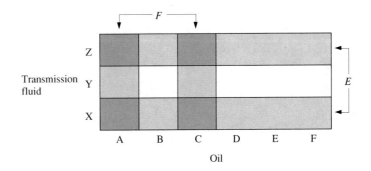

a. $S = \{$XA, XB, XC, XD, XE, XF, YA, YB, YC, YD, YE, YF, ZA, ZB, ZC, ZD, ZE, ZF$\}$.
b. $\overline{E} = \{$YA, YB, YC, YD, YE, YF$\}$.
c. $(E \cup F) = \{$XA, XB, XC, XD, XE, XF, YA, YC, ZA, ZB, ZC, ZD, ZE, ZF$\}$.
d. $(E \cap F) = \{$XA, ZA, XC, ZC$\}$. ▪

Mutually Exclusive Events

If E and F are events that have no outcomes in common, then E and F are called **mutually exclusive events.** That is,

---

Events E and F are mutually exclusive events if $(E \cap F) = \emptyset$.

Mutually exclusive events can be illustrated by a Venn diagram as in figure 5.5.

FIGURE 5.5

Mutually exclusive events

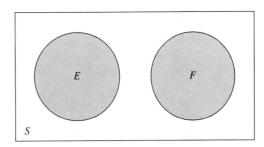

APPLICATION 5.8   Suppose an experiment consists of tossing a penny followed by tossing a quarter. Are the following pairs of events mutually exclusive?

    a. E = two heads, F = two tails
    b. E = {HT, TT} F = {HT, TH}
    c. $E \neq \emptyset$, F = {TT}
    d. $E, \overline{E}$

Solution:

    a. Yes, they have no outcomes in common.

    b. No, they have HT in common.

    c. Yes, they have no outcomes in common.

    d. Yes, they have no outcomes in common. ▪

APPLICATION 5.9   Suppose a town has three automobile dealerships, Ford, GM, and Chrysler. The GM dealer sells Pontiacs, Oldsmobiles, and Cadillacs; the Ford dealer sells Fords and Mercurys; and the Chrysler dealer sells Dodges, Plymouths, and Chryslers. The experiment consists of observing the order in which the next two cars are sold in town. If A is the event that the two cars are Ford products, B is the event that the two cars are GM products, and C is the event that the next two cars sold are a Pontiac and a Dodge, find the set of outcomes making up each of the following events:

    a. A
    b. B
    c. C
    d. $A \cap B$
    e. $A \cup C$

Are A and B mutually exclusive events?

Solution: Suppose F indicates that the next car sold is a Ford, M that the next car sold is a Mercury, O that the next car sold is an Oldsmobile, P that the next car sold is a Pontiac, C that the next car sold is a Cadillac, and D that the next car sold is a Dodge.

    a. A = {FM, MF}.
    b. B = {PO, OP, PC, CP, CO, OC}.
    c. C = {PD, DP}.
    d. $A \cap B = \emptyset$.
    e. $A \cup C$ = {FM, MF, PD, DP}.

Events A and B are mutually exclusive events, since $(A \cap B) = \emptyset$. ▪

EXERCISE SET 5.1

Basic Skills

1. List a sample space for each of the following experiments.
   a. Toss a nickel and a penny, in that order, and record the way they fall.
   b. Toss a six-sided die and a dime, in that order, and record the way they fall.
   c. Select a male college student at random and ask him if he owns a car.

2. List a sample space for each of the following experiments.
   a. Choose two coins from 1 dime, 1 quarter, and 1 nickel, and record them.
   b. Choose two adults and ask them if they are married.
   c. Toss a six-sided die until a 1 shows.

3. Describe at least two different sample spaces for the experiment of choosing a group of two students from a class of five students and record the results.

4. Describe at least two different sample spaces for the experiment of selecting one card from a set of ten cards, numbered 1 through 10, and record the results.

5. Let C be the event that tomorrow's weather is hot, and D the event that it rains tomorrow. Describe the following compound events:
   a. $C \cup D$ b. $C \cap D$ c. $\overline{C} \cap D$ d. $\overline{C} \cup \overline{D}$

6. Let C be the event that tomorrow's weather is hot, and D the event that it rains tomorrow. Describe the following compound events:
   a. $\overline{C} \cup D$ b. $\overline{C} \cap \overline{D}$ c. $\overline{C \cup D}$ d. $\overline{C \cap D}$

7. For the sample space of 36 outcomes when two dice are tosses (see application 5.3), find the number of outcomes for which
   a. both dice are even.
   b. exactly one die is even.
   c. at most one die is even.

8. For the sample space of 36 outcomes when two dice are tossed (see application 5.3), find the number of outcomes for which
   a. neither die is even.
   b. the sum is even.
   c. the quotient of the number on the red die divided by the number on the black die is a whole number.

9. A penny, a nickel, and a dime are tossed, in that order.
   a. List a sample space of eight outcomes.
   b. List the outcomes for the event for which the dime shows heads.
   c. Count the number of outcomes for which the dime shows heads.
   d. List the outcomes for the event for which either the nickel or the dime shows heads.
   e. Count the number of outcomes for which either the nickel or the dime shows heads.
   f. Count the number of outcomes for which either the nickel or dime shows heads, but not both of them.
   g. Count the number of outcomes for which the penny and the nickel agree (two coins agree if they both show heads or they both show tails).
   h. Count the number of outcomes for which the penny agrees with the nickel, but not with the dime.

10. A dime is tossed four times.
    a. List a sample space of 16 outcomes.
    b. Count the number of outcomes for which exactly three heads occur.
    c. Count the number of outcomes for which the first four tosses result in a head.
    d. Count the number of outcomes for which two heads and two tails occur.
    e. Count the number of outcomes for which all tails or all heads occur.
    f. Count the number of outcomes for which three heads and one tail occur.

11. Refer to the two-dice tossing experiment of application 5.3, and consider the sample space of 36 outcomes. Illustrate three different pairs of mutually exclusive events.

12. Refer to the two-dice tossing experiment of application 5.3, and consider the sample space of 36 outcomes. Illustrate three different pairs of events that are not mutually exclusive.

Use the accompanying Venn diagram containing events A, B, and C for exercises 13 and 14. Let $S = \{1, 2, 3, 4, 5, 6, 7, 8\}$, $A = \{1, 2, 3, 5\}$, $B = \{1, 2, 4, 6\}$, and $C = \{1, 3, 4, 7\}$.

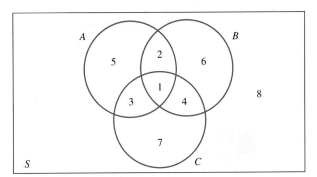

13. List the numbers that make up the following events:
    a. $A \cap B$
    b. $A \cap B \cap C$
    c. $A \cup B$
    d. $(\overline{A} \cup B) \cap C$
    e. $\overline{A \cap B}$
    f. $\overline{A}$

14. List the numbers that make up the following events:
    a. $\overline{A \cup B \cup C}$
    b. $(A \cap B) \cup (\overline{C} \cap A)$
    c. $\overline{A} \cup C$
    d. $\overline{A \cup B}$
    e. $(A \cap B) \cup C$
    f. $A \cup (B \cap C)$

15. Which of the following pairs of events are mutually exclusive events?
    a. E = Mrs. Smith gives birth to twins.
       F = A mother gives birth to a girl.
    b. E = Henry fails the last statistics test.
       F = Henry passes the course.
    c. E = Obtain a head and a tail from two tosses of a fair coin.
       F = Obtain two heads from two tosses of a fair coin.

16. Which of the following pairs of events are mutually exclusive events?
    a. E = Joe goes to the movies.
       F = Joe eats a candy bar.
    b. E = Mr. Doe files the 1040 EZ income-tax form.
       F = Mr. Doe files the 1040 long income-tax form.
    c. E = Our team loses the last baseball game.
       F = Our team loses the baseball tournament.

Further Applications

17. An experiment consists of asking three shoppers at random if they buy Brand A peanut butter. Let Y denote yes and N denote no. For example, YYN denotes the simple event that the first two people polled buy brand A and the third does not. List a sample space for this experiment.

18. Refer to the experiment in exercise 17, and let E be the event that at least two people say yes and let F be the event that the first person polled says no. List the simple events making up the following events:
    a. E
    b. F
    c. $\overline{E}$
    d. $E \cup F$
    e. $E \cap F$
    f. $E \cap \overline{F}$

Going Beyond

19. If a sample space contains n outcomes, how many possible events are there?

20. If a coin is tossed n times, what is the sample space if all the heads and tails are recorded?

21. Use Venn diagrams to illustrate each of the compound events in exercise 5.

22. Use Venn diagrams to illustrate each of the compound events in exercise 6.

SECTION 5.2   *Concept of Probability*

Hardly a day goes by that some aspect of probability or chance is not encountered. The weather forecaster calls for an 80% chance of rain. The San Francisco Giants, according to a sports announcer, have a 50–50 chance of winning the World Series. The odds are good that you will pass Calculus I if you study hard.

The probability of an event is a number between 0 and 1, inclusive, that is assigned to the event. If E is an event, then $P(E)$ denotes the probability of E; it is the likelihood of the occurrence of event E. If the probability is 0, then event E is certain not to occur, and if the probability is 1, the event is certain to occur. The closer $P(E)$ is to 1, the more likely it is to occur. The closer $P(E)$ is to 0, the more unlikely it is to occur, as shown in figure 5.6.

FIGURE 5.6

Interpretation of *P(E)*

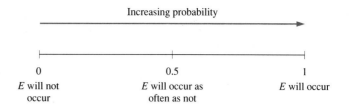

Probability Satisfies the Following Properties:

1. $P(E_i) \geq 0$
2. $P(E_i) \leq 1$
3. $\Sigma\, P(E_i) = 1$
   where $\Sigma\, P(E_i)$ is the sum of the probabilities for all the outcomes (simple events) in a sample space.

The probability of an event A is defined as the sum of the probabilities for the outcomes contained in A. Application 5.10 shows the assignment of probabilities to events once the probabilities of the outcomes in the sample space are known.

APPLICATION 5.10   Suppose a six-sided die is tossed once and the probability of any side landing face-up is $1/6$. If E is the event of getting an even number and F is the event of getting an odd number, find

    a. $P(E)$.

    b. $P(F)$.

    c. $P(E \cup F)$.

    d. $P(E \cap F)$.

Solution: The sample space is $S = \{1, 2, 3, 4, 5, 6\}$, the event E is $\{2, 4, 6\}$, and the event F is $\{1, 3, 5\}$. We thus have

    a. $P(E) = P(2) + P(4) + P(6) = 1/6 + 1/6 + 1/6 = 3/6 = 1/2$.

    b. $P(F) = P(1) + P(3) + P(5) = 1/6 + 1/6 + 1/6 = 3/6 = 1/2$.

    c. $P(E \cup F) = P(S) = 1$.

    d. $P(E \cap F) = 0$, since $E \cap F = \emptyset$ and $P(\emptyset) = 0$. ■

Assigning Probabilities to Events

The three properties satisfied by probabilities do not tell us how to assign probabilities to the outcomes in a sample space. All they do is to rule out certain assignments that are not consistent with our intuitive notions. There are two general methods for assigning probabilities to events: the objective method and the subjective method. The **objective method** involves assigning probabilities to events based on counting or repeated experimentation. The **subjective method,** on the other hand, involves assigning probabilities to events based on intuition or personal belief. When using the subjective method to assign probabilities to events, two knowledgeable people may not agree on the assignments.

Objective Method

EXAMPLE 5.7

Suppose a coin is tossed. The two outcomes are heads (H) and tails (T). What number $P(H)$ should we assign to H and what number $P(T)$ should we assign to T? Suppose we assign 0.7 to H and 0.3 to T. Is this a valid assignment of probabilities? The answer is yes, based on the three given properties, since (1) both numbers are greater than 0; (2) both numbers, 0.7 and 0.3, are less than 1; and (3) the sum of 0.7 and 0.3 is 1. But these assignments go against our intuition if we believe the coin is fair. The correct assignments, most of us would agree, are 0.5 for H and 0.5 for T. For if we tossed the coin a large number of times N and found the frequency f for the occurrence of a head, we would expect the **relative frequency** f/N for the occurrence of a head to be close to 0.5. Table 5.1 contains the number of heads obtained when a fair coin was tossed N times, as well as the relative frequency for the number of heads obtained in each case. Since H and T are the only two outcomes of the experiment (standing on edge is not allowed), then $P(T)$ must be 0.5 so that $0.5 + 0.5 = 1$.

TABLE 5.1

Frequency and Relative
Frequency of Heads in
Repeated Tossing of a Coin

Number of Coins, N	Number of Heads, f	Relative Frequency, f/N
1	0	0.000
2	1	0.500
3	1	0.333
4	1	0.250
5	3	0.600
6	3	0.500
7	4	0.571
8	3	0.375
9	6	0.667
10	4	0.400
.	.	.
.	.	.
.	.	.
100	51	0.510
1000	447	0.447
10000	5047	0.505

The limiting value of the relative frequency f/N for obtaining a head when a fair coin is tossed N times will approach 0.5 as N becomes large, as illustrated in figure 5.7.

Since 0.7 is not equal to the limiting value of the relative frequency for a head occurring (which is 0.5), we would not recommend assigning 0.7 as the probability for the outcome H. With the value 0.5 assigned to H, what probability $P(T)$ should be assigned to T? As a consequence of the third probability property, we know that

$$P(H + P(T) = 1$$

or

$$0.5 + P(T) = 1$$

Hence, $P(T) = 0.5$ must be the probability assignment for getting a tail.

FIGURE 5.7

Limiting values of *f/N* as *N* gets large

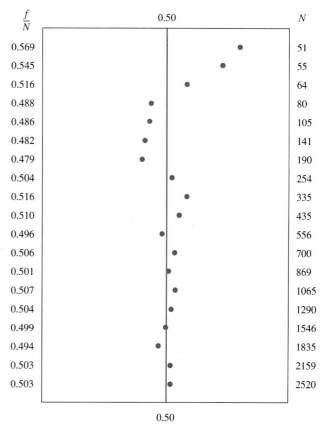

$\frac{f}{N}$	0.50	*N*
0.569		51
0.545		55
0.516		64
0.488		80
0.486		105
0.482		141
0.479		190
0.504		254
0.516		335
0.510		435
0.496		556
0.506		700
0.501		869
0.507		1065
0.504		1290
0.499		1546
0.494		1835
0.503		2159
0.503		2520

0.50

N = number of times a coin is tossed

If an experiment is repeatable, we can assign probabilities to the outcomes in accordance with their limiting relative frequencies in a fashion similar to what we did for the coin above. The only problem with doing this is that the limiting values of relative frequencies are not always known. To use this method, we need to have much repetitive data available, and then only approximations to the relative frequency limits can be found.

Empirical Probability

When probability is based on experience and the exact limiting values of relative frequencies are unknown, approximations to these limiting values must be used. When this is done, the objective method of assigning probabilities is said to be empirical. According to the method of **empirical probability**, if E is an event, $P(E)$ is approximately equal to f/N, where f is the number of favorable outcomes and N is the number of repetitions of the experiment. In this case, we have

$$P(E) \approx \frac{f}{N}$$

EXAMPLE 5.8

Suppose we consider the experiment of tossing a thumbtack. It can land one of two ways.

How can we determine the probability that the thumbtack will land point up? We could ask someone. But who would know the answer? Probably nobody. We could toss the thumbtack ten times and record the number of times it landed point-up. This relative frequency could serve as an estimate of the desired probability. But to get a better estimate, we could toss the thumbtack 100 or 1000 times, or even more, and record the frequency of landing point up. We would thereby obtain a good approximation for the probability of the thumbtack landing point up.

EXAMPLE 5.9

As our experience changes, so does the relative frequency. For example, if we tossed a coin six times and obtained three heads, we would estimate the chance of getting a head to be $3/6 = 0.5$. If the coin were tossed once more and it showed a head, then the estimated chance for getting a head would be $4/7 = 0.5714$; or if it fell tails, then the estimated chance for getting a head would be $3/7 = 0.4286$. This change in relative frequency reflects our changing knowledge. As was pointed out above, over the long run, the relative frequency should change very little, and the limiting value that the relative frequency is approaching is called the **probability.**

If you have access to a computer program (such as MINITAB) you can imitate the tossing of a coin as often as you like. Such an imitation is referred to as a **computer simulation.** The first two lines in computer display 5.1 use MINITAB to direct the computer to simulate 150 fair coin tosses by a random process and store the results in a column labeled C1. The results of the simulated tosses are then printed, followed by a tally of the results. Results of the first 15 tosses are shown in the first row of 1s and 0s. Each 1 represents a head and each 0 represents a tail. If we used only the first 15 tosses of the coin to estimate the probability of getting a head on a single toss of a fair coin, the estimate of P(head) would be $7/15 \approx 0.467$. Reading from the tally given at the bottom of the display, we see that the simulation produced 69 heads and 81 tails; thus based on 150 tosses, P(head) $\approx 69/150 = 0.46$. Another simulation would most likely produce different results.

Computer Display 5.1

```
MTB > RANDOM 150 C1 ;
SUBC> BERNOULLI P = 0.5 .
MTB > PRINT C1

C1
  1 0 0 1 0 0 1 0 1 0 1 0 1 1 0
  0 1 0 0 1 0 0 1 0 1 1 0 1 0 0
  1 1 1 0 0 0 1 0 1 0 1 0 0 1 0
  1 1 0 1 0 1 0 0 0 1 1 1 1 1 1
  0 0 0 0 0 0 0 1 0 0 1 1 1 0 0
  1 0 1 1 1 0 1 0 1 0 0 0 1 0 1
  1 1 1 1 0 0 0 0 0 0 0 0 0 1 1
  1 1 0 0 0 0 1 0 1 1 0 0 0 0 1
  1 0 1 0 1 0 1 1 0 1 0 0 0 0 0
  0 1 0 1 1 1 0 1 1 1 1 0 1 0 0

MTB > TALLY C1 ;
SUBC>ALL .

    C1    COUNT   CUMCNT    PERCENT   CUMPCT
     0       81       81      54.00    54.00
     1       69      150      46.00   100.00
   N = 150

MTB >
```

APPLICATION 5.11

An insurance company wants to estimate the probability of a police car being involved in an accident in a certain city during a one-month period. Last month, 7 out of 20 police cars were involved in accidents.

    a. What would you estimate the desired probability to be?

    b. What would you estimate to be the chance of a police car not being involved in an accident?

Solution:

    a. $7/20 = 0.35$.

    b. $0.65 = 1 - 0.35$. ■

APPLICATION 5.12

The SAT math scores for students at a large university are displayed in the following grouped frequency table:

SAT	f
200–299	3,600
300–399	11,900
400–499	12,000
500–599	5,500
600–699	1,500
700–799	500

If a student is selected at random, what is the probability that the student's SAT math score

    a. Exceeds 399?

    b. Is at most 599?

    c. Is between 600 and 699, inclusive?

    d. Is not between 400 and 499, inclusive?

    e. Is less than or equal to 699?

Solution: We first form a grouped relative frequency table (table 5.2). Three decimal accuracy was used in the computations. Recall that the relative frequency for a class is found by dividing the frequency f of the class by the total number of measurements N.

TABLE 5.2

Grouped Relative Frequency Table for Application 5.12

SAT	f	Rel F
200–299	3,600	0.103
300–399	11,900	0.340
400–499	12,000	0.343
500–599	5,500	0.157
600–699	1,500	0.043
700–799	500	0.014
	35,000	

    a. $P(\text{SAT} > 399) = 0.343 + 0.157 + 0.043 + 0.014 = 0.557$.

    b. $P(\text{SAT} \leq 599) = 0.103 + 0.340 + 0.343 + 0.157 = 0.943$.

c. $P(600 \leq SAT \leq 699) = 0.043$.

d. $P(SAT < 400 \text{ or } SAT > 499) = 1 - 0.343 = 0.657$ (since the sum of the relative frequencies is 1).

e. $P(SAT \leq 699) = 1 - 0.014 = 0.986$ (since the sum of the relative frequencies is 1). ▪

APPLICATION 5.13

In a small city each person was classified according to religion and political party affiliation. The results are summarized in the following table:

Religion	Political Party		
	Democrat	Republican	Independent
Protestant	10,000	8,000	2,000
Jewish	5,500	6,000	500
Catholic	8,500	9,500	1,500

If a person is chosen at random from this city, what is the probability that the person is a

a. Republican?

b. Catholic?

c. Protestant and a Republican?

d. Catholic and an Independent?

Solution: We first find the total for each row and column as indicated in table 5.3.

Table 5.3

Row and Column Totals for Application 5.13

Religion	Political Party			Total
	Democrat	Republican	Independent	
Protestant	10,000	8,000	2,000	20,000
Jewish	5,500	6,000	500	12,000
Catholic	8,500	9,500	1,500	19,500
Total	24,000	23,500	4,000	51,500

a. There are 23,500 Republicans out of a total of 51,500 persons. Therefore, $P(R) = 23,500/51,500 = 0.456$.

b. There are 19,500 Catholics out of a total of 51,500 persons. Therefore, $P(C) = 19,500/51,500 = 0.379$.

c. There are 8,000 people out of a total of 51,500 that are Protestant and Republican. Therefore, $P(P \text{ and } R) = 8,000/51,500 = 0.155$.

d. There are 1,500 people out of a total of 51,500 that are Catholic and Independent. Therefore $P(C \text{ and } I) = 1,500/51,500 = 0.029$. ▪

Classical Probability

If an experiment results in outcomes that all have the same probability of occurring, then the outcomes are called **equally likely outcomes.** If an experiment has n outcomes that we believe to be equally likely, we can assign each outcome in the sample space S a probability value of $1/n$. This is a consequence of probability property 3, which states that the sum of the probabilities for a sample space must be 1. Then if E is an event containing *f* outcomes from a **sample space** containing *n* equally likely outcomes, then the probability of *E* occurring is simply *f/n*. **Thus,** we have the following basic fact:

If *S* is a sample space of **equally likely outcomes** and *E* is an event, then

$$P(E) = \frac{f}{n} \tag{5.1}$$

where *f* is the number of outcomes in *E* and *n* is the number of outcomes in *S*. It is common to call the outcomes contained in *E*, **favorable outcomes.** The objective method for assigning probabilities using sample spaces of equally likely outcomes is referred to as the **classical probability method.**

EXAMPLE 5.10

Suppose a penny and a dime are both tossed once. They can fall in four ways. A sample space for this experiment is $S = \{HH, TH, HT, TT\}$. The letter written first in a pair stands for the outcome on the penny and the letter written second stands for the outcome on the dime. The possible outcomes may be visualized as follows:

$$
\begin{array}{ccc}
 & \text{2nd Coin} & \\
 & \text{H} \quad\quad \text{T} & \\
 \text{H} & \text{HH} \quad \text{HT} & \\
\text{1st Coin} & & \\
 \text{T} & \text{TH} \quad \text{TT} & \\
 & \text{Possible Outcomes} &
\end{array}
$$

The four outcomes are equally likely to occur if both coins are fair and they are tossed so that the outcome on one is not influenced by and does not influence the outcome of the other. As a result, we assign a probability value of 1/4 to each of the four outcomes. If *E* is the event $\{TH, HT\}$, then we have the following two methods for assigning a probability value to *E*:

1. Add the probabilities for the outcomes contained in *E*. Thus, $P(E) = 1/4 + 1/4 = 0.5$.

2. Use formula (5.1). The number of favorable outcomes is $f = 2$, the number of outcomes in *E*, and the number of equally likely outcomes in *S* is $n = 4$. Hence,

$$P(E) = \frac{f}{n} = \frac{2}{4} = \frac{1}{2}$$

APPLICATION 5.14

Consider the previous experiment of tossing a penny and a dime once. When two coins are tossed, we could record that 0, 1, or 2 heads appear. Thus the sample space is $S = \{0H, 1H, 2H\}$. Since there are three outcomes in *S*, we may be tempted to conclude that $P(1H) = 1/3$. But by the above argument $P(1H) = P(\{HT, TH\}) = 1/2$. Resolve this apparent contradiction.

Solution: The three outcomes in *S* are not equally likely to occur. The outcome 1H is twice as likely to occur as 0H or 2H. Thus, formula (5.1) is not valid for computing $P(1H)$, and we have $P(1H) = 1/2$. ▪

APPLICATION 5.15   For the two dice tossing experiment described in application 5.3, find the probability that

    a. Both show an even number.

    b. A sum of 7 shows.

    c. A sum of 7 or 11 shows.

    d. Both show a prime number (a prime number is a number greater than 1 that has no divisors other than 1 and itself)

    e. The red die shows a 2.

    f. A sum of 13 shows.

Solution: A sample space is as follows:

$$
\begin{array}{llllll}
\{(1, 1) & (1, 2) & (1, 3) & (1, 4) & (1, 5) & (1, 6) \\
(2, 1) & (2, 2) & (2, 3) & (2, 4) & (2, 5) & (2, 6) \\
(3, 1) & (3, 2) & (3, 3) & (3, 4) & (3, 5) & (3, 6) \\
(4, 1) & (4, 2) & (4, 3) & (4, 4) & (4, 5) & (4, 6) \\
(5, 1) & (5, 2) & (5, 3) & (5, 4) & (5, 5) & (5, 6) \\
(6, 1) & (6, 2) & (6, 3) & (6, 4) & (6, 5) & (6, 6)\}
\end{array}
$$

There are 36 equally likely outcomes in S.

    a. Let E_1 represent the event that both dice show an even number. A favorable outcome is one in which both dice show an even number: $(2, 2)$, $(2, 4)$, $(2, 6)$, $(4, 2)$, $(4, 4)$, $(4, 6)$, $(6, 2)$, $(6, 4)$, and $(6, 6)$. There are nine favorable outcomes. Hence,

$$
P(E_1) = \frac{f}{n} = \frac{9}{36} = \frac{1}{4}
$$

    b. Let E_2 represent the event that a sum of 7 shows. The set of favorable outcomes is $E_2 = \{(3, 4), (4, 3), (5, 2), (2, 5), (6, 1), (1, 6)\}$. Therefore,

$$
P(E_2) = \frac{f}{n} = \frac{6}{36} = \frac{1}{6}
$$

    c. Let E_3 represent the event that a sum of 7 or 11 shows. The set of favorable outcomes is $E_3 = \{(1, 6), (2, 5), (3, 4), (4, 3), (5, 2), (6, 1), (5, 6),$ and $(6, 5)\}$. Hence,

$$
P(E_3) = \frac{f}{n} = \frac{8}{36} = \frac{2}{9}
$$

    d. Let E_4 represent the event that both die show a prime number. Since $E_4 = \{(2, 2), (2, 3), (2, 5), (3, 2), (3, 3), (3, 5), (5, 2), (5, 3), (5, 5)\}$,

$$
P(E_4) = \frac{f}{n} = \frac{9}{36} = \frac{1}{4}
$$

    e. Let E_5 represent the event that the first die shows a 2. The set of favorable outcomes is $E_5 = \{(2, 1), (2, 2), (2, 3), (2, 4), (2, 5), (2, 6)\}$. Thus,

$$
P(E_5) = \frac{f}{n} = \frac{6}{36} = \frac{1}{6}
$$

    f. Let E_6 represent the event that a sum of 13 shows. Since a sum of 13 is impossible, $E_6 = \emptyset$. Thus,

$$
P(E_6) = 0 \quad ■
$$

The dice experiment of application 5.15 can be simulated by using MINITAB. Two dice are tossed 1000 times; the result of the first die is stored in C1 and the result of the second die is stored in C2. The sum of each pair is stored in C3, and the results are shown in computer display 5.2 along with the commands.

Computer Display 5.2

```
MTB > RANDOM 1000 C1 C2;
SUBC> INTEGER 1 6.
MTB > RSUM C1 C2 C3
MTB > TALLY C3;
SUBC> PERCENT.

C3 PERCENT
 2   2.40
 3   5.70
 4  10.00
 5  11.70
 6  13.60
 7  16.90
 8  14.60
 9   8.70
10   8.70
11   5.40
12   2.30

MTB >
```

Notice that the probability of a sum of 7 showing is approximately 0.169 and that the probability of a sum of 7 or 11 showing is approximately $0.169 + 0.054 = 0.223$. These two results compare favorably with the results found in application 5.15.

APPLICATION 5.16

What is the probability of correctly guessing all the answers to a true/false test containing four questions?

Solution: Let T denote true and F denote false. Since only one of these outcomes represents a favorable outcome, $f = 1$. A sample space of equally likely outcomes contains the following 16 outcomes:

TTTT	TFFT	FFTF	FTFT
TTFT	TFFF	FFFT	FTTF
TTTF	TFTT	FFTT	FTFF
TTFF	TFTF	FFFF	FTTT

The event TTTT corresponds to getting all questions correct and it has one outcome among the 16 outcomes listed. By using formula (5.1), we have

$$P(\text{TTTT}) = \frac{f}{n} = \frac{1}{16} \quad ■$$

Subjective Method

In many situations we have limited or no information concerning the outcomes of an experiment. We would have no frequency data available if the situation had never occurred before. A doctor treating a patient with a rare disease must make a prognosis based on his experience with the patient and the patient's overall medical record. A candidate running for public office for the first time will estimate his chance for election based on his intuition and hearsay. When probabilities are assigned to events based on intuition and personal beliefs, the assignment method is called *subjective*. Of course, when using the subjective method for assigning probabilities to events, the assignments will depend on the individual making the assignments and may vary widely from individual to individual.

Probability Histograms

For data that have been displayed in a grouped frequency table, a special type of relative frequency histogram can be constructed and used to estimate the probability that a measurement falls in a particular class. If instead of letting the ith bar in a relative frequency histogram have a height of f_i/N, as was done in section 2.3, we let the height h of the ith bar be

$$h = \frac{f_i}{wN}$$

where f_i is the frequency associated with the ith class, w is the class width, and N is the total number of measurements, then the area bar approximates the probability that a measurement will fall in that particular class. In addition, the total area of the histogram is then equal to 1, the sum of the probabilities. For this reason, such a histogram is referred to as a **probability histogram.** Note that the area of a relative frequency histogram, as constructed in section 2.3, has an area equal to w, the class width.

APPLICATION 5.17

For the accompanying data from application 5.12, construct a probability histogram.

SAT	f
200–299	3,600
300–399	11,900
400–499	12,000
500–599	5,500
600–699	1,500
700–799	500

Solution: Table 5.4 contains the class boundaries, the heights of the bars, and the areas of the bars.

TABLE 5.4

Values For Probability
Histogram for Application 5.17

Boundaries	h	Area of Bar
199.5–299.5	0.00103	0.103
299.5–399.5	0.00340	0.340
399.5–499.5	0.00343	0.343
499.5–599.5	0.00157	0.157
599.5–699.5	0.00043	0.043
699.5–799.5	0.00014	0.014
		1.000

The heights h were found by dividing the relative frequencies by $w = 100$, size $h = f_i/wN$. For example, the relative frequency of the first class in table 5.4 is $f_1/N = 3,600/35,000 = 0.103$. If we divide 0.103 by $w = 100$, we get

FIGURE 5.8

Probability histogram for application 5.17

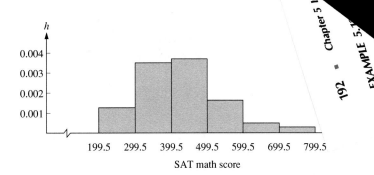

SAT math score

$(0.103)/(100) = 0.00103$. The graph of the probability histogram is shown in figure 5.8. The sum of the areas of the two tallest bars represents the probability $P(299.5 \leq x \leq 499.5) = 0.683$. ∎

Mathematical Odds

In gambling situations, such as horse races and sporting contests, the chance of an event occurring is often arrived at by using subjective probability. And these probabilities are often stated in terms of **odds.** If E is an event, then the *odds in favor of E,* written as $\text{Odds}(E)$, is defined by

> **Odds in Favor of E**
>
> $$\text{Odds}(E) = \frac{P(E)}{P(\overline{E})}$$

The *odds against E,* written as $\text{Odds}(\overline{E})$, is defined by

> **Odds Against E**
>
> $$\text{Odds}(\overline{E}) = \frac{P(\overline{E})}{P(E)} = \frac{1}{\text{Odds}(E)}$$

EXAMPLE 5.11

If E is the event that a favorite horse wins the race and $P(E) = 1/3$, then the odds in favor of winning, $\text{Odds}(E)$, is

$$\text{Odds}(E) = \frac{P(E)}{P(\overline{E})} = \frac{1/3}{2/3} = \frac{1}{2}$$

This result is sometimes written as 1:2.

Odds can be converted to probabilities. As a general rule we have

> If Odds $(E) = a{:}b$, then
>
> $$P(E) = \frac{a}{a + b}$$

If the odds in favor of the Pittsburgh Pirates winning the baseball game is 2:3, then the probability that the Pirates will win, $P(E)$, is

$$P(E) = \frac{a}{a + b}$$

$$= \frac{2}{2 + 3}$$

$$= \frac{2}{5} = 0.4$$

APPLICATION 5.18

If the odds are 3:2 against a favorite horse winning a race, what is the probability of

a. the horse losing the race?

b. the horse winning the race?

Solution: Let E be the event the horse will win the race. Then $\overline{E}$ is the event that the horse will lose the race.

a. $P(\overline{E}) = 3/(2 + 3) = 3/5$.

b. Since $P(E) + P(\overline{E}) = 1$, $P(E) = 1 - P(\overline{E})$ or

$$P(E) = \frac{1 - 3}{5} = \frac{2}{5}$$

We can also get the same result by using the odds rule:

$$P(E) = \frac{2}{2 + 3} = \frac{2}{5} \quad \blacksquare$$

APPLICATION 5.19

During the Vietnam war in 1967 approximately 15 million men were eligible for the military draft during any month. If 20,000 men were drafted each month, what were the odds against John Jones being drafted in April 1967?

Solution: Let E be the event that John was drafted. Then

$$P(E) = \frac{20,000}{15,000,000} = \frac{1}{750} \quad \text{and} \quad P(\overline{E}) = \frac{749}{750}$$

Hence, the odds against John being drafted were

$$\text{Odds}(\overline{E}) = \frac{P(\overline{E})}{P(E)} = \frac{749/750}{1/750} = 749:1 \quad \blacksquare$$

EXERCISE SET 5.2

Basic Skills

1. Match each of the following probabilities with one of the statements that follow:

   0   0.01   0.3   0.99   1

   a. The event is impossible. It can never occur.
   b. The event is certain.
   c. The event is very unlikely, but it will occur once in a while, in a long sequence of trials.
   d. The event will occur more often than not.

2. Which of the following numbers cannot be the probability of some event?
   a. 0.74   d. 0.5   g. 2/3
   b. −1   e. 0   h. 0.999 . . .
   c. 1.02   f. 1   i. 0.67

3. For each of the following situations, express the situation by using a probability value in decimal form.
   a. There is a 30–70 chance for getting funding for our project.
   b. The odds against striking oil are 100 to 1.

4. For each of the following situations, express the situation by using a probability value in decimal form.
   a. There is a 75% chance that surgery will be needed to correct the problem.
   b. The odds that the Giants will win the game tomorrow are 5 to 3.

Further Applications

5. A meteorologist forecasts rain with a probability of 70%.
   a. What are the odds in favor of rain?
   b. What are the odds against rain?

6. A meteorologist forecasts the odds of raining are 7 to 13.
   a. What is the probability that it will rain?
   b. What is the probability that it will not rain?

7. A bag contains three red marbles, two white marbles, and five blue marbles. One marble is selected at random. What is the probability that the marble is
   a. red?
   b. white?
   c. blue?

8. A box has three blue marbles, four yellow marbles, and two green marbles. One marble is selected at random. What is the probability that the marble is
   a. blue?
   b. yellow?
   c. green?

9. Refer to the experiment of exercise 7. List two sample spaces, one containing equally likely outcomes, and one not containing equally likely outcomes. In each case, list the probabilities associated with the outcomes.

10. Refer to the experiment of exercise 8. List two sample spaces for the experiment, one containing equally likely outcomes, and one not containing equally likely outcomes. In each case, list the probabilities associated with the outcomes.

11. The odds against being dealt three of a kind in a five-card poker hand are about 49:1. What is the probability of being dealt three of a kind?

12. The odds that a sixth-grader will continue and graduate from high school are three to one. What is the probability that a sixth-grader will not graduate from high school?

13. An American roulette wheel contains compartments numbered 1 through 36 plus 0 and 00. Of the 38 compartments, 0 and 00 are colored green, 18 of the others are red, and 18 are black. A ball is spun in the direction opposite to the wheel's motion, and bets are made on the number where the ball comes to rest. Suppose the wheel is fair.
   a. What is the probability of a black outcome?
   b. What are the odds against a red outcome?
   c. What is the probability of a red outcome?
   d. What are the odds in favor of a red outcome?
   e. What is the probability of an odd number occurring?

14. Refer to exercise 13.
   a. What is the probability of a red or black outcome?
   b. What is the probability of a 0?
   c. What are the odds in favor of a green outcome?
   d. What are the odds against a green outcome?
   e. What is the probability of an even number occurring?

15. To estimate the number of books required per course for a small college the student council took a sample of 100 courses and obtained the following results:

Number of Books Required (x)	Number of Courses (f)
0	5
1	48
2	21
3	12
4	9
5	3
6	2

Use empirical probabilities to find
a. the probability value for each x.
b. $\Sigma P(x)$.
c. $P(x$ is at least 5).
d. $P(x$ is 2 or 4).

16. Each person in a sample of forty was asked to name his or her favorite soft drink. The responses are shown below.

Soft Drink	Number Preferring
Pepsi-Cola	14
Coca-Cola	12
Sprite	8
Seven-Up	3
Dr. Pepper	2
Nehi Orange	1

By using empirical probabilities, what is the probability that a person's favorite soft drink is
a. Pepsi-Cola?
b. Coca-Cola?
c. Dr. Pepper?
d. Mr. Pibb?
e. Pepsi-Cola or Coca-Cola?

17. Each teacher at a college was classified according to gender and academic rank. The results are displayed in the following table:

		Rank		
Gender	Instr.	Asst. prof.	Assoc. prof.	Prof.
Male	300	400	700	300
Female	350	450	300	200

If a teacher is selected at random from this college, what is the probability that the teacher is a
a. male?
b. female
c. professor?
d. male professor?
e. female assistant professor?

18. A survey of employees at a large business yielded the following breakdown according to marital status and gender:

	Marital Status			
Gender	Married	Single	Divorced	Widowed
Male	25%	11%	10%	3%
Female	30%	8%	7%	6%

Suppose an employee is selected at random. What is the probability that the employee is
a. married?
b. female and divorced?
c. widowed?
d. male?
e. married or single?

19. The following data represent the average annual salaries (in thousands of dollars) for workers in the 50 states covered by unemployment insurance:[33]

28.7	18.0	17.1	16.0	15.2
19.7	17.9	16.9	15.9	15.0
19.7	17.8	16.8	15.9	14.8
19.3	17.3	16.8	15.5	14.7
17.8	17.3	16.7	15.5	14.6
18.8	17.2	16.5	15.5	14.6
18.7	17.2	16.5	15.5	14.3
18.1	17.2	16.4	15.4	14.1
18.1	17.1	16.1	15.2	13.9
18.1	17.1	16.1	15.2	13.2

Construct a probability histogram having eight bars.

Going Beyond

20. Finger prints are classified into eight generic types and possess many characteristics. One of these is the ridge count, which is useful in criminal investigation work. The accompanying grouped frequency table displays the ridge counts for 800 males.

Ridge Count	f
0–19	10
20–39	12
40–59	24
60–79	40
80–99	73
100–119	100
120–139	90
140–159	112
160–179	124
180–199	95
200–219	67
220–239	36
240–259	10
260–279	4
280–299	3
	800

a. Suppose fingerprints are found at the scene of a crime and it is determined that the ridge count is at least 220. If a male suspect has a ridge count of 241, what would you conclude? Why?

b. What is the probability of a male having a ridge count of at least 200?

c. What is the probability that a male has a ridge count of at least 120 and no more than 199?

21. Roll a pair of dice 100 times and record the sum of the numbers showing for each roll. What is the relative frequency of a sum of 8? (By mathematical logic, in the long run this relative frequency will approach the probability of a sum of 8, which is about 0.14, if the dice are fair.)

22. Two dice are tossed and the large number showing face up is recorded. If {1, 2, 3, 4, 5, 6} is a sample space for the experiment, find the probability for each of the six outcomes. Are the outcomes equally likely? Why or why not?

23. If a deck of 52 playing cards is well shuffled, what would you estimate to be the probability that the top three cards contain a king or a queen, or that a king and a queen are next to each other somewhere among the remaining 49 cards? Test your guess by simulating the experiment 20 times.

24. Can an experiment have two different sample spaces of equally likely outcomes? Explain.

25. To decide whether to bunt in a baseball game, a manager studies the accompanying data obtained from several hundred baseball games.[34]

Base(s) Occupied	Number of Outs	Proportion of Cases—No Runs Scored in Inning	Average No. of Runs in Inning	Number of Cases Observed
First	0	0.604	0.813	1728
Second	1	0.610	0.671	657
Second	0	0.381	1.194	294

a. If a player is on first base with no outs, in how many cases did runs score?

b. If a player is on first base with no outs, find the probability that at least one run scored.

c. If a player is on first base with no outs and a sacrifice bunt succeeds in the normal way of advancing the player on first base, is this a better situation than we had (in the sense of average number of runs scored)?

d. If a player is on second base with no outs, find the probability that at least one run scored in the inning.

e. If a player is on second base with one out, in how many cases did no runs score in the inning?

26. When two people seat themselves at a table with a square top and with one chair at each side, it is not unusual for them to sit in adjacent, rather than opposite seats. If two people randomly choose seats at such a table, determine the probability that they occupy adjacent seats at the table.

SECTION 5.3　　　*Counting*

If an experiment contains a large number of outcomes, it may be difficult to count the number of outcomes in an event. For such experiments (or sophisticated counting problems), we need to employ special counting techniques. Examples 5.13 and 5.14 illustrate sophisticated counting problems. They both require systematic and careful approaches in order to be able to properly list the outcomes.

EXAMPLE 5.13　　Suppose three six-sided dice are tossed and we are asked to determine the number of ways they can fall. A sample space, as we shall see later, may contain 216 equally likely outcomes.

EXAMPLE 5.14　　Consider the experiment of going to your new-car dealer to buy a new car. Once there, you find you can select from four models, each with 15 power options, with a choice of five exterior colors and eight interior colors. How many different choices (outcomes) are possible for you to make? The answer is 2400. In this case, the choices may not all be equally likely unless you choose randomly.

The special counting techniques that we shall study, which can be used to solve difficult counting problems similar to the problems in examples 5.13 and 5.14, are the fundamental theorem of counting, counting permutations, and counting combinations. We shall study the fundamental theorem of counting first.

Fundamental Theorem of Counting

The *fundamental theorem of counting* (*FTC*) is stated as follows:

> If an event can occur in *m* ways, and if after it has occurred, a second event can follow it and occur in any one of *n* ways, then the two events can occur together, in the order stated, in (*m*)(*n*) different ways. (This rule can be extended to any number of events.)

Since the three dice in example 5.13 are in no way related when tossed, and since each can fall in six different ways, the total number of ways they can fall, one following another, is (6)(6)(6) = 216 by the FTC.

For example 5.14, there are four events that can occur, one following another. They are as follows:

$$E_1 = \text{Choose one of four models.}$$
$$E_2 = \text{Choose one of 15 power options.}$$
$$E_3 = \text{Choose one of five exterior colors.}$$
$$E_4 = \text{Choose one of eight interior colors.}$$

The first event can occur in 4 ways, the second event in 15 ways, the third event in 5 ways, and the fourth event in 8 ways. Therefore, as a consequence of the FTC, the four events, one following another, can occur in (4)(15)(5)(8) = 2400 different ways. Note that this is an oversimplified example since in reality, some options are linked by the manufacturers and car dealers. Let's look at a simpler application.

APPLICATION 5.20

Suppose there are three roads from town A to town B and two roads from town B to town C. How many different ways can a person travel from A to C by way of B?

Solution: Consider the diagram, called a **tree diagram.**

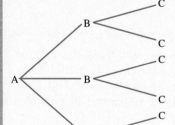

There are two decisions to be made:

  a. At A, which road to B?
  b. At B, which road to C?

The first decision can be made in three ways, and after this decision is made, the second decision can be made in two ways. According to the FTC, the total number of ways to go from A to C is (3)(2) = 6. This can also be seen by counting the branches on the diagram going from A to C.　■

APPLICATION 5.21

There are five different books a teacher wants to arrange on his desk from left to right. How many different arrangements are possible?

Solution: This problem can be viewed two different ways, as an arrangement problem or as a selection problem.

An Arrangement Problem

The five events are

$E_1 =$ Arrange the first book in one of five spaces.

$E_2 =$ After the first book is arranged, arrange the second book in one of the four remaining spaces.

$E_3 =$ After the first two books are arranged, arrange the third book in one of the three remaining spaces.

$E_4 =$ After the first three books are arranged, arrange the fourth book in any of the two remaining spaces.

$E_5 =$ Arrange the last book in only one way.

E_1 can be done in five ways, E_2 in four ways, E_3 in three ways, E_4 in two ways, and E_5 in one way. Thus, by the fundamental theorem of counting, the total number of arrangements is $(5)(4)(3)(2)(1) = 120$.

A Selection Problem

The five events are

$E_1 =$ Select a book for the first space.

    This can be done in five ways.

$E_2 =$ After the first space is filled, selecting the second book from the remaining four books.

    This can be done in four ways.

$E_3 =$ After the first two spaces are filled, selecting the third book from the three books.

    This can be done in three ways.

$E_4 =$ After the first three spaces are filled, selecting the fourth book from the remaining two books.

    This can be done in two ways.

$E_5 =$ Place the fifth book in the fifth space.

    This can be done in one way.

By the FTC, the total number of selections $(5)(4)(3)(2)(1) = 120$. ▪

APPLICATION 5.22 In how many ways can three items be chosen from a group of seven items and arranged on a shelf from left to right?

Solution: There are three events.

$E_1 =$ Choose an item from seven items for the first space.

    This can be done in seven ways.

$E_2 =$ Choose an item from the remaining six items for the second space.

    This can be done in six ways.

$E_3 =$ Choose an item from the remaining five items for the third space.

    This can be done in five ways.

By using the FTC, we see that the total number of ways is $(7)(6)(5) = 210$. ▪

By examining the previous applications, we note that there are two questions to be asked when solving any counting problem:

1. How many events are there?
2. In how many ways can each occur?

After determining the answers to questions 1 and 2, we use the FTC to solve the problem.

APPLICATION 5.23   There are five candidates for president of a club and two candidates for vice president. Use a tree diagram to determine the number of ways the two offices can be filled.

Solution: There are two events of interest: choosing a president and choosing a vice president. The president can be chosen in five ways and the vice president can be chosen in two ways. As a consequence of the FTC, the two offices can be filled in $(5)(2) = 10$ different ways. A tree diagram can be constructed to illustrate the 10 possibilities. Let A_1, A_2, A_3, A_4, and A_5 represent the candidates for president, and let B_1 and B_2 represent the candidates for vice president. Then we have the tree diagram shown in figure 5.9.

FIGURE 5.9

Tree diagram for application 5.23

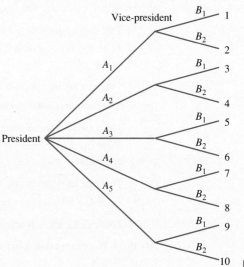

Permutations

An ordered arrangement of n objects is called a **permutation.** There are six permutations of the three letters A, B, and C. They are

$$ABC \quad ACB \quad BAC \quad BCA \quad CAB \quad CBA$$

We can use the FTC to determine the number of permutations of n objects, denoted by P_n^n. To do this, we imagine the n objects being placed into n boxes strung out in a row. The first box can be filled in n ways, the second box can be filled in $(n - 1)$ ways, the third box can be filled in $(n - 2)$ ways, and so on to the last box. It can be filled in only one way, since there is only one object remaining. Therefore, the number of permutations of n objects is equal to

$$P_n^n = n(n - 1)(n - 2)(n - 3) \cdots (2)(1) \tag{5.2}$$

The expression on the right side of equation (5.2) is called **n factorial** and is
by $n!$. By definition, $0! = 1$. Thus, $3! = (3)(2)(1) = 3$ and $5! = (5)(4)(.$
$= 24$.

> **Number of Permutations of n Objects Taken n at a Time**
>
> $$P_n^n = n!$$
> (5.3)

EXAMPLE 5.15

The number of permutations of five objects is

$$P_n^n = n!$$
$$P_5^5 = 5!$$
$$= (5)(4)(3)(2)(1) = 120$$

EXAMPLE 5.16

Suppose we want to determine the number of permutations of five objects taken three at a time, written as P_3^5. We can think of this problem as placing five distinct objects into three boxes. The first box can be filled in five ways. After this is done, the second box can be filled in four ways. And after this is done, the third box can be filled in three ways. As a consequence of the FTC, the number of permutations of five distinct objects taken three at a time is $(5)(4)(3) = 60$ ways. Thus, $P_3^5 = 60$.

In order to develop a formula for computing the value of P_r^n, let's make the following four observations concerning P_3^5:

1. Of the five objects, three of them are being arranged or permuted, and $(5 - 3) = 2$ of them are not being arranged.

2. The number of permutations of those objects that are being arranged is $(5)(4)(3) = 60$, and the number of permutations of those objects that are not being arranged is $2! = 2$. For each of the 60 permutations of the objects that we rearrange, we have two permutations of the objects that aren't being rearranged.

3. By the FTC, the total number of permutations of the five distinct objects is $(5)(4)(3)(2)(1) = 5!$. Hence, by observation 2 we have $5! = 60 \cdot 2 = 120$ or $P_3^5 \cdot (5 - 3)! = 5!$. By dividing both sides of this later equation by $(5 - 3)!$, we have the following relationship: $P_3^5 = 5!/(5 - 3)!$.

The observations in example 5.16 suggest the following general formula for calculating P_r^n, the number of permutations of n distinct objects taken r at a time $(n \geq r)$:

> **Number of Permutations of n Objects Taken r at a Time**
>
> $$P_r^n = \frac{n!}{(n - r)!}$$
> (5.4)

Formula (5.4) is referred to as the **permutation rule.**

APPLICATION 5.24   Suppose that ten students are available for three different jobs on campus. In how many ways can the jobs be filled?

Solution: We need to determine how many ways there are of assigning the three jobs to the ten students, or the number of permutations of ten objects taken three at a time. By the permutation rule, formula (5.4), the number of permutations of ten objects taken three at a time is

$$P_3^{10} = \frac{10!}{(10-3)!}$$
$$= \frac{10!}{7!} = \frac{(10)(9)(8)(7!)}{7!}$$
$$= (10)(9)(8) = 720 \quad ■$$

Combinations

When dealing with permutations of objects, the order of selection or arrangement of the objects is important. There are occasions when we want to consider collections of objects where the order of the objects is not important. When this is done, the selection is called a **combination.** In other words, a selection of r objects from a collection of n distinct objects without regard to the order in which the r objects are selected is called a **combination,** and the number of combinations of n objects taken r at a time is denoted by $\binom{n}{r}$. The symbol $\binom{n}{r}$ is also called a *binomial coefficient.* For example, suppose an experiment involves the selection of a committee of three women from a collection of eight women. Here we are not concerned with the order of selection. Each selection is a combination. The number of combinations (or selections) of eight women taken three at a time is denoted by $\binom{8}{3}$.

The number of combinations of n objects taken r at a time is related to the number of permutations of n objects taken r at a time. Each combination of r objects can be rearranged in $r!$ different ways. By applying the FTC, the total number of permutations of n distinct objects taken r at a time is equal to the product of $r!$ and the number of combinations of n distinct objects taken r at a time. That is,

$$P_r^n = r!\binom{n}{r}$$

Hence, the number of combinations of n distinct objects taken r at a time is given by

$$\binom{n}{r} = \frac{P_r^n}{r!} \qquad (5.5)$$

EXAMPLE 5.17

The number of ways of choosing three women from a collection of eight women is

$$\binom{n}{r} = \frac{P_r^n}{r!}$$
$$\binom{8}{3} = \frac{P_3^8}{3!} = \frac{(8)(7)(6)}{6} = 56$$

By using formula (5.4), we can express formula (5.5) as

> **The Number of Combinations of *n* Objects Taken *r* at a Time**
>
> $$\binom{n}{r} = \frac{n!}{r!(n-r)!}$$ (5.6)

APPLICATION 5.25 Evaluate the following binomial coefficients:

a. $\binom{6}{4}$

b. $\binom{6}{2}$

c. $\binom{5}{3}$

d. $\binom{5}{2}$

Solution: By using formula (5.6), we have

a. $\binom{n}{r} = \dfrac{n!}{r!(n-r)!}$

$\binom{6}{4} = \dfrac{6!}{4!(6-4)!}$

$= \dfrac{(6)(5)(4!)}{4!\,2!}$

$= \dfrac{(6)(5)}{2} = 15$

b. $\binom{6}{2} = \dfrac{6!}{2!\,4!} = 15$

c. $\binom{5}{3} = \dfrac{5!}{3!\,2!}$

$= \dfrac{(5)(4)(3!)}{3!\,2!}$

$= \dfrac{(5)(4)}{2} = 10$

d. $\binom{5}{2} = \dfrac{5!}{2!3!} = 10,$

by part c. ■

It can be shown in general that the number of combinations of *n* objects taken *r* at a time is the same as the number of combinations of *n* objects taken *n − r* at a time. That is,

$$\binom{n}{r} = \binom{n}{n-r}$$

By using this fact, we can often simplify computations involving binomial coefficients.

EXAMPLE 5.18

a. $\binom{10}{8} = \binom{10}{2} = 10!/(2!8!) = (10)(9)(8!)/(2!8!) = 45$. To get 45, we multiply the largest two factors of 10!, 10 and 9, and divide this product by $2! = (2)(1) = 2$. Note that both numerator and denominator of $\binom{10}{2} = (10)(9)/(2)(1)$ have the same number of factors, namely 2.

b. $\binom{10}{7} = \binom{10}{3} = 10!/(3!7!) = (10)(9)(8)(7!)/(3!7!) = 120$. To get 120, we multiply the largest three factors of 10!, 10, 9, and 8, and divide this product by $3! = (3)(2)(1) = 6$. Note again that both the numerator and denominator of $\binom{10}{3} = (10)(9)(8)/(3)(2)(1)$ have three factors.

APPLICATION 5.26

A study is to be conducted to determine the attitudes of faculty at a university concerning abortion. If a sample of four faculty is selected from a total of 45, how many different samples could be collected?

Solution: Since the order within a sample is unimportant, we are interested in determining the number of combinations of 45 faculty taken four at a time. By formula (5.6), we have

$$\binom{45}{4} = \frac{45!}{(45-4)!4!}$$

$$= \frac{45!}{41!4!}$$

$$= \frac{(45)(44)(43)(42)(41!)}{41!4!}$$

$$= \frac{(45)(44)(43)(42)}{4!} = \frac{3{,}575{,}880}{24} = 148{,}995 \quad ■$$

Pascal's Triangle

Binomial coefficients can also be obtained from a triangular array of numbers, called **Pascal's triangle.** The following array is a partial triangle containing seven rows:

	x Successes						
n trials	0	1	2	3	4	5	6
0	1						
1	1	1					
2	1	2	1				
3	1	3	3	1			
4	1	4	6	4	1		
5	1	5	10	10	5	1	
6	1	6	15	20	15	6	1

It is easy to construct the triangle. The first column ($x = 0$) has all 1s. In order to obtain a number in any row (not in the first column), we just add the number directly above the one we want and the one above and to the left.

EXAMPLE 5.19

To determine $\binom{6}{2}$, we go to the row corresponding to $n = 6 - 1 = 5$ and add 5 and 10. Thus, $\binom{6}{2} = 15$. To find $\binom{6}{5}$ we add 5 and 1. The process of adding numbers in the previous row may be continued to construct as many rows as desired.

n Trials	0	1	2	3	4	5	6
0	1						
1	1	1					
2	1	2	1				
3	1	3	3	1			
4	1	4	6	4	1		
5	1	5	10	10	5	1	
6	1	6	15	20	15	6	1

The entries in the row corresponding to $n = 7$ are

$$1 \quad 7 \quad 21 \quad 35 \quad 35 \quad 21 \quad 7 \quad 1$$

EXERCISE SET 5.3

Basic Skills

1. Evaluate the following expressions:

   a. P_2^4 b. $\binom{5}{2}$ c. P_3^6

   d. P_5^5 e. $\binom{7}{3}$ f. $\binom{9}{4}$

2. Evaluate the following expressions:

   a. P_1^5 b. $\binom{6}{4}$ c. P_2^7

   d. P_6^8 e. $\binom{9}{2}$ f. $\binom{10}{3}$

Further Applications

3. By assuming all combinations are possible, how many different homes can be built if a builder offers a choice of five basic plans, three roof styles, and two exterior finishes?

4. A business school gives courses in typing, shorthand, transcription, business English, technical writing, and accounting. In how many ways can a student choose three courses to take during three class periods?

5. In how many ways can seven of ten monkeys be arranged in a row for a genetics experiment?

6. In an experiment of social interaction, six people will sit in six seats in a row. In how many ways can this be done?

7. In assembling some electronic equipment, six wires enter a box that has six terminals. In how many ways can the wires be connected to the terminals, one wire to each terminal?

8. In how many ways can a judge award first, second, and third places in a contest with 15 entries?

9. A clothing store stocks socks made of either cotton or wool, each in five colors and seven sizes. How many items must be stocked in order to have available a complete assortment?

10. A man tries to choose the winner of each of eight football games. Excluding ties, how many different predictions are possible? Including ties?

11. Answer exercise 6, if the six people sit around a circular table and the relative positions of the people make a difference and not the seats they sit in.

12. If you are taking a true-false test with ten questions, in how many different ways can you fill in your answer sheet if you know none of the answers and guess at each one?

13. In how many ways can nine men be assigned to nine different jobs?

14. In how many ways can five jobs be assigned to nine men?

15. If five cards are to be selected in sequence, without re-placement, from a standard deck of 52 playing cards, how many different selections are possible?

16. Refer to exercise 15. If the cards are selected with replacement, how many different selections are possible?

17. A special computer keyboard can be purchased from six different suppliers. In how many ways can four suppliers be chosen from the six?

Going Beyond

18. The Upper Crust Pizza Shop advertises ten different pizza toppings. How many different pizzas can be ordered?

19. Refer to exercise 4. How many ways can a schedule of three courses be selected?

S E C T I O N   5 . 4

Finding Probabilities Using the Fundamental Theorem of Counting

The fundamental theorem of counting can be used to solve some of the more difficult probability problems. Applications 5.27–5.30 show the use of the FTC to solve probability problems.

A P P L I C A T I O N 5.27

If four coins are tossed, what is the probability of getting four heads?

Solution: We first determine the number of ways the four coins can land when tossed, one following another. Let E_1, E_2, E_3, and E_4 represent tossing the four coins, respectively. Each can land in exactly two ways. By the FTC, the four coins can land in $2^4 = 16$ different ways. Of these, there is only one favorable outcome, HHHH. Hence,

$$P(HHHH) = \frac{1}{16} \quad ■$$

A P P L I C A T I O N 5.28

If four coins are tossed, what is the probability that the first two coins are heads?

Solution: The total number of outcomes is 16, from application 5.27. A favorable outcome is an outcome in which the first two coins are heads. The first coin must land heads up, the second coin must land heads up, and the third and fourth coins can land either heads up or tails up. By the fundamental theorem of counting, the total number of favorable outcomes is $(1)(1)(2)(2) = 4$. Therefore, the probability that the first two coins are heads when four coins are tossed is $4/16 = 1/4$. ■

A P P L I C A T I O N 5.29

From a set of 100 cards numbered 1 to 100, one is selected at random. What is the probability that the number on it

a. Is exactly divisible by 5?

b. Ends in a 1 or a 2?

Solution:

a. Let E represent the event that the number on a card is divisible by 5. A number is divisible by 5 if it ends in a 0 or a 5. The number of one-digit numbers divisible by 5 is 1. The number of two-digit numbers divisible by 5 is $(9)(2) = 18$. The number of three-digit numbers divisible by 5 is 1. Therefore, the total number of cards with numbers divisible by 5 is $f = 1 + 18 + 1 = 20$. Since there are 100 cards, we have

$$P(E) = \frac{f}{n} = \frac{20}{100} = \frac{1}{5}$$

b. Let F represent the event that the number on the card ends in a 1 or 2. A number ending in a 1 or 2 can be a one-digit number or a two-digit number. There are $f_1 = 2$ one-digit numbered cards ending in a 1 or 2. For the two-digit numbered cards, the units digit can be one of two values and the tens digit can be one of nine values. By the FTC, the number of two-digit numbered cards ending in a 1 or 2 is $f_2 = (9)(2) = 18$. Hence,

$$P(F) = \frac{f_1 + f_2}{n} = \frac{2 + 18}{100} = \frac{20}{100} = \frac{1}{5} \quad ▪$$

APPLICATION 5.30

There are ten different books on a bookshelf; seven are mathematics books and three are science books. If the books are randomly placed on the bookshelf, what is the probability that the science books are all together?

Solution: Let E be the event that the science books are arranged together. By the FTC, the ten books can be arranged in $(10)(9)(8)(7)(6)(5)(4)(3)(2)(1) = 3,628,800$ different ways. To find the number of ways of arranging the 10 books so that the science books are together, we identify the following events:

$E_1 =$ Arranging the mathematics books

$E_2 =$ Arranging the science books

$E_3 =$ Inserting the set of science books among the mathematics books

The number of ways that E_1 can occur is $f_1 = (7)(6)(5)(4)(3)(2)(1) = 5040$. The number of ways that E_2 can occur is $f_2 = (3)(2)(1) = 6$. And the number of ways that E_3 can occur is $f_3 = 8$ (putting the science books before the mathematics books, after the first mathematics book, after the second mathematics book, after the third mathematics book, and so forth). By the FTC, the number of ways of arranging the ten books so that the science books are together is

$$f = (f_1)(f_2)(f_3)$$
$$= (5040)(6)(8) = 241,920$$

Thus, the probability that the science books are arranged together is

$$P(E) = \frac{f}{n}$$

$$= \frac{241,920}{3,628,800} \approx 0.06667 \quad ▪$$

EXERCISE SET 5.4

Further Applications

1. The tickets in a box are numbered 1 to 20 inclusive, and two tickets are drawn, one following the other without replacing the first ticket drawn. What is the probability that
   a. both are even?
   b. the first is even and the other odd?
   c. both are even or both are odd?

2. Refer to exercise 1. Do parts a–c if the first ticket is replaced before the second ticket is drawn.

3. Four nuts and four bolts are mixed together. If two parts are chosen at random, what is the probability that
   a. both are nuts?
   b. the first is a nut and the second is a bolt?
   c. one will be a nut and one a bolt?
   d. either both will be nuts or both will be bolts?

4. The letters of the word *Chance* are written on cards, one letter to a card. The cards are shuffled and land face up one after another. What is the probability that they spell *Chance*? (Note that the first letter is a capital.)

5. What is the probability that a student will spell the word, MATH, by randomly arranging the letters M, A, T, and H?

6. What is the probability that a four-letter *word* chosen from the letters in the word *figure* starts with a consonant?

7. If ten pool balls, each having a unique number from 1 to 10, are lined up along the side of a pool table, what is the probability that the balls numbered 5 and 6 occur together?

8. What is the probability that a three-digit number formed from the digits 1, 2, 3, 4, and 5 is even? What is the probability that it is odd? (Assume in both cases that digits can be repeated.)

9. Solve exercise 8 under the assumption that the digits cannot be repeated.

Going Beyond

10. If four men and four women are to be arranged in a lineup, what is the probability that a random arrangement of the eight individuals has
    a. the men and women alternating?
    b. the men all next to each other?

11. If four men and four women are to be seated at a round table, what is the probability that a random arrangement of the eight individuals has
    a. the men and women alternating?
    b. the men all next to each other?

12. From a set of cards numbered from 1 to 10,000, one is selected at random. What is the probability that the number on it
    a. is exactly divisible by 5?
    b. ends in an even number?

13. In an eight-cylinder engine the even numbered cylinders are on the left side and the odd numbered cylinders are on the right side. A good firing order is an arrangement in which the two sides of the engine alternate when fired, starting with cylinder 1. (For example, 1, 4, 5, 8, 3, 2, 7, and 6 is a good firing order.) If the engine is wired by a novice who knows nothing about what he is doing, determine the probability that a good firing order was chosen.

SECTION 5.5   Some Probability Rules

Probability of E or F, $P(E \cup F)$

If E and F are two events, we would like to develop a formula for $P(E \cup F)$. Let's examine figure 5.10, a Venn diagram for $(E \cup F)$. Let the area of circle E represent $P(E)$, and let the area of circle F represent $P(F)$. The event $(E \cup F)$ is illustrated by the entire shaded region within the two circles and $P(E \cup F)$ is the area of this region. Also, $P(E \cap F)$ is the area common to both regions E and F and represents the area of the common region. The region of $E \cap F$ is counted twice, once in the counting of

FIGURE 5.10

Venn diagram for $(E \cup F)$

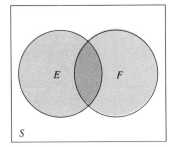

E and once in the counting of F, hence we need to subtract the extra counting of E and F. Thus,

$$P(E \cup F) = (\text{Area } E) + (\text{Area } F) - (\text{Area common to both } E \text{ and } F)$$
$$= P(E) + P(F) - P(E \cap F)$$

Hence, the probability of $(E \cup F)$ can be found by using the following rule:

> **Sum Rule**
>
> $$P(E \cup F) = P(E) + P(F) - P(E \cap F) \qquad (5.7)$$

Rule (5.7) is referred to as the **sum rule,** and application 5.31 illustrates its use.

APPLICATION 5.31

If a single card is drawn from an ordinary deck of playing cards, find the probability that it will be red or a face card (jack, queen, or king).

Solution: Let E represent drawing a red card and F represent drawing a face card. Then $P(E) = 26/52$, $P(F) = 12/52$, and $P(E \cap F) = 6/52$. Thus, using the sum rule, we have

$$P(E \cup F) = P(E) + P(F) - P(E \cap F)$$
$$= \frac{26}{52} + \frac{12}{52} - \frac{6}{52} = \frac{32}{52} = \frac{8}{13} \qquad \blacksquare$$

If E and F are mutually exclusive events, then $(E \cap F) = \emptyset$, the empty set. In this case $P(E \cap F) = 0$ and the sum rule (5.7) becomes $P(E \cup F) = P(E) + P(F)$. Thus, we have the following rule:

> **If E and F are mutually exclusive events, then**
>
> $$P(E \cup F) = P(E) + P(F) \qquad (5.8)$$

Probability of not E, $P(\overline{E})$

Since E and $\overline{E}$ are mutually exclusive events, we can apply rule (5.8) to obtain

$$P(E \cup \overline{E}) = P(E) + P(\overline{E})$$

In addition, since $P(E \cup \overline{E}) = P(S) = 1$, we have $P(E) + P(\overline{E}) = 1$ or $P(\overline{E}) = 1 - P(E)$. Hence, we have the following rule:

> **Probability of $\overline{E}$**
>
> $$P(\overline{E}) = 1 - P(E) \qquad (5.9)$$

APPLICATION 5.32

The probability that Bob will finish his term paper is $3/7$. Find the probability that he will not finish his term paper.

Solution: Let E be the event that Bob will finish his term paper. Then $\overline{E}$ is the event that he will not finish. Since $P(E) = 3/7$, by using rule (5.9) we have

$$P(\overline{E}) = 1 - P(E)$$
$$= 1 - \frac{3}{7} = \frac{4}{7} \qquad \blacksquare$$

Conditional Probability

We would now like to determine a rule for computing $P(E \cap F)$, the probability of $(E \cap F)$. In order to do so, we need the notion of conditional probability. The probability of event E occurring when we know that event F has already occurred is called **conditional probability** and is written as $P(E|F)$. $P(E|F)$ means the probability that the event E will occur, given the condition that the event F has occurred, or simply the probability of E given F. Applications 5.33–5.35 show conditional probability by using selection-type problems. For these applications, **selection with replacement** means that the first object is returned before the second object is drawn. Similarly, **selection without replacement** means that the first object is not returned before the second object is drawn. Note, in particular, that applications 5.33 and 5.34 show that the type of sampling has an effect on the results.

APPLICATION 5.33

Two balls are drawn without replacement from a bag containing three white and two black balls. Find the probability that

    a. The second ball is black given that the first ball is white.

    b. The second ball is black given that the first ball is black.

Solution:

    a. If the first ball selected is white, there are four balls left in the bag, of which two are black. Therefore,

$$P(B|W) = \frac{2}{4} = \frac{1}{2}$$

See the following diagram:

$$4 \to \begin{cases} \cancel{W} \\ W \\ W \\ B \\ B \end{cases} \Big\} \leftarrow 2$$

    b. If the first ball selected is black, there are four balls left in the bag, of which one is black. Therefore,

$$P(B|B) = \frac{1}{4}$$

See the following diagram:

$$4 \to \begin{cases} W \\ W \\ W \\ B \} \leftarrow 1 \\ \cancel{B} \end{cases} \quad ■$$

APPLICATION 5.34

Suppose the balls in application 5.33 are drawn with replacement. Two balls are drawn with replacement from a bag containing three white and two black balls. Find the probability that

    a. The second ball is black given that the first ball is white.

    b. The second ball is black given that the first ball is black.

Solution:

a. Since the white ball is returned to the bag before the second ball is drawn, it has no effect on the selection of the next ball. Since there are five balls in the bag and two of them are black, the probability of selecting a black ball is 2/5.

b. Since the black ball is returned to the bag before the next ball is drawn, it has no effect on the selection of the next ball. As in part a, the probability of selecting a black ball is 2/5. ▪

APPLICATION 5.35 A study was undertaken at a certain college to determine what relationship, if any, exists between mathematics ability and interest in mathematics. The ability and interest for 150 students were determined, with the results in the following table:

	Interest		
Ability	Low	Average	High
Low	40	8	12
Average	15	17	18
High	5	10	25

If one of the participants in the study is chosen at random, what is the probability

a. Of selecting a person who has low interest in mathematics?

b. Of selecting a person with average ability?

c. That the person has high ability in mathematics given that the person selected has high interest in mathematics?

d. That the person has high interest in mathematics given that the person selected has average ability in mathematics.

Solution: We first find the row and column totals, as shown in table 5.5.

TABLE 5.5

Row and Column Totals for Application 5.35

	Interest			
Ability	Low	Average	High	Total
Low	40	8	12	60
Average	15	17	18	50
High	5	10	25	40
Total	60	35	55	150

a. Since there are 60 participants with low interest out of a total of 150, the probability is 60/150 = 2/5.

b. Since there are 50 participants with average ability out of a total of 150, the probability is 50/150 = 1/3.

c. Of the 55 participants with high interest, 25 have high ability. Therefore, the probability is 25/55 = 5/11.

d. Since 50 participants have average ability and of these, 18 have high interest, the probability is 18/50 = 9/25. ▪

Note that parts c and d of application 5.35 involve conditional probability. For part c, we know the person chosen has high interest in mathematics. There are 55 college students who have high interest in mathematics; of these, 25 have high ability. Let HA denote high ability, HI denote high interest, and AA denote average ability. We therefore have

$$P(\text{HA}|\text{HI}) = \frac{25}{55} = \frac{5}{11}$$

This probability can be written as

$$P(\text{HA}|\text{HI}) = \frac{P(\text{HA} \cap \text{HI})}{P(\text{HI})}$$

$$= \frac{25/150}{55/150}$$

$$= \frac{25}{55} = \frac{5}{11}$$

Also, with part d of application 5.35, we find

$$P(\text{HI}|\text{AA}) = \frac{P(\text{HI} \cap \text{AA})}{P(\text{AA})}$$

$$= \frac{18/150}{50/150}$$

$$= \frac{18}{50} = \frac{9}{25}$$

These results suggest the following formula for computing $P(E|F)$.

Conditional Probability Formula

$$P(E|F) = \frac{P(E \cap F)}{P(F)} \tag{5.10}$$

provided $P(F) \neq 0$

Notice that condition F in $E|F$ has the effect of reducing the size of the sample space. In application 5.35, part c, the size of the original sample space is 150, and after the condition HI is stipulated, the size of the reduced sample space (only those with high interest) is 55.

Conditional probability $P(E|F)$ can be illustrated using a Venn diagram in which probabilities are interpreted as areas (see figure 5.11). If we are given the condition F, we are restricted to the outcomes in F. Hence, the elements in F comprise the reduced sample space. Then we are interested in the outcomes of E that are also contained in F. These are just the outcomes in E and F. The probability of E given F is then obtained by dividing the area of ($E \cap F$) by the area of F. Thus,

$$P(E|F) = \frac{\text{Area } (E \cap F)}{\text{Area } F}$$

$$= \frac{P(E \cap F)}{P(F)} \tag{5.11}$$

FIGURE 5.11

Venn diagram illustrating conditional probability

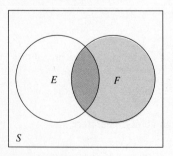

Probability of E and F,
$P(E \cap F)$

The probability of $(E \cap F)$ is given by the formula

Product Rule

$$P(E \cap F) = P(F)\, P(E|F) \qquad (5.12)$$

Formula (5.12) was obtained by multiplying both sides of formula (5.10) by $P(F)$. Rule (5.12) is often referred to as the **product rule** and is illustrated in application 5.36. Note that $(E \cap F)$ is logically equivalent to $(F \cap E)$. As a consequence, we have

$$P(E \cap F) = P(F)\, P(E|F)$$
$$= P(E)\, P(F|E)$$

A crucial point to remember when using the product rule to calculate $P(E \cap F)$ is the location of the symbol for the event occupying the "$*$" position in the following expression:

$$P(E \cap F) = P(*)\, P(\ |*) \qquad (5.13)$$

For example, if $*$ is replaced by E, then expression (5.13) would read

$$P(E \cap F) = P(E)\, P(F|E)$$

If $*$ is replaced by F, then expression (5.13) would read as

$$P(E \cap F) = P(F)\, P(E|F)$$

APPLICATION 5.36

If two balls are drawn without replacement from a bag containing three red, two black, and one white ball, what is the probability of getting two red balls?

Solution: Let R_1 denote the event of getting a red ball on the first draw and R_2 denote the event of getting a red ball on the second draw. By using rule (5.12), we have

$$P(R_1 \cap R_2) = P(R_1)P(R_2|R_1)$$
$$= \frac{3}{6}\,\frac{2}{5} = \frac{1}{5} \quad ■$$

APPLICATION 5.37

Find the probability of drawing two red balls in application 5.36 if the balls are drawn with replacement.

Solution: By using rule (5.12), we have

$$P(R_1 \cap R_2) = P(R_1)\, P(R_2|R_1)$$
$$= \frac{3}{6}\,\frac{3}{6} = \frac{1}{4}$$

Note that in this case we have

$$P(R_2|R_1) = P(R_2) \quad ■$$

APPLICATION 5.38　Refer to application 5.35.

    a. Find the probability of selecting a person with high ability and high interest in mathematics.

    b. Find the probability of selecting a person with low interest in mathematics and high ability in mathematics.

Solution:

    a. By using the product rule, we have

$$P(HA \cap HI) = P(HA)\, P(HI|HA)$$

$$= \frac{40}{150}\frac{25}{40}$$

$$= \frac{25}{150} = \frac{1}{6}$$

    b. By the product rule, we have

$$P(LI \cap HA) = P(HA)\, P(LI|HA)$$

$$= \left(\frac{40}{150}\right)\left(\frac{5}{40}\right)$$

$$= \frac{5}{150} = \frac{1}{30} \quad ■$$

The following two observations should be kept in mind when finding probabilities associated with compound events:

1. To find $P(E \cup F)$, we associate $\cup$ with the sum rule.
2. To find $P(E \cap F)$, we associate $\cap$ with the product rule.

EXERCISE SET 5.5

Basic Skills

1. Let $P(E) = 0.4$, $P(F|E) = 0.3$, and $P(E|F) = 0.4$.
    a. Find the following:
        i. $P(E \cap F)$
        ii. $P(F)$
        iii. $P(E \cup F)$
    b. Are E and F mutually exclusive events?

2. Let $P(E) = 0.5$, $P(F) = 0.4$, and $P(E \cup F) = 0.8$.
    a. Find the following:
        i. $P(E \cap F)$
        ii. $P(F|E)$
        iii. $P(E|F)$
    b. Are E and F mutually exclusive events?

3. Suppose $S = \{1, 2, 3, 4, 5, 6\}$ is a sample space for an experiment. The accompanying table lists the probabilities associated with each element of the sample space.

x	1	2	3	4	5	6
$P(x)$	0.1	0.3	0.1	0.2	0.1	0.2

Let $E = \{1, 2, 5\}$ and $F = \{1, 6\}$. Find the following probabilities:

    a. $P(E)$　　　　b. $P(F)$　　　　c. $P(E \cup F)$
    d. $P(E \cap F)$　　e. $P(E|F)$　　　f. $P(\overline{E})$
    g. $P(\overline{F} \cap E)$　　h. $P(\overline{E} \cap \overline{F})$　　i. $P(F|\overline{E})$

4. Suppose $S = \{1, 2, 3, 4, 5\}$ is a sample space for an experiment. The accompanying table lists the probabilities associated with each element of the sample space.

x	1	2	3	4	5
$P(x)$	0.2	0.2	0.1	0.3	0.2

Let $E = \{1, 2\}$ and $F = \{2, 3\}$. Find the following probabilities:

a. $P(E)$ b. $P(F)$ c. $P(E \cup F)$

d. $P(E \cap F)$ e. $P(E|F)$ f. $P(\overline{E})$

g. $P(\overline{F} \cap E)$ h. $P(\overline{E} \cap \overline{F})$ i. $P(F|\overline{E})$

Further Applications

5. If a fair six-sided die is tossed, find the probability of rolling
   a. a 2, given that the number rolled was odd.
   b. a 4, given that the number rolled was even.
   c. an even number, given that the number rolled was a 6.

6. If a ball is drawn from a box containing three red, two white, and five black balls, find the probability that
   a. the ball is red.
   b. the ball is not red.
   c. the ball is red or black.
   d. the ball is not red and not white.

7. Two balls are drawn without replacement from a bag containing two white balls and two black balls. Find the probability that
   a. the second ball is white given that the first ball is white.
   b. the second ball is white given that the first ball is black.
   c. both balls are white.
   d. the first ball is white and the second ball is black.
   e. one ball is white and one ball is black.

8. If two cards are drawn without replacement from a 52-card deck, find the probability that
   a. the second card is a heart, given that the first is a heart.
   b. they are both hearts.
   c. the second is black, given that the first is a spade.
   d. the second is a face card, given that the first is a jack.
   e. the first card is a jack and the second card is a face card.

9. Do exercise 7 if the first ball is replaced before the second ball is drawn.

10. Do exercise 8 if the first card is replaced before the second card is drawn.

11. Two marbles are drawn without replacement from a jar with four black and six white marbles. Find the probability

that
   a. both are white.
   b. both are black.
   c. the second is white, given that the first is black.
   d. they are different colors.

12. Do exercise 11 if the first marble is replaced before the second marble is drawn.

13. Suppose a hospital survey indicates that 35% of patients admitted have high blood pressure, 53% have heart trouble, and 22% have both high blood pressure and heart trouble. What is the probability that a patient selected at random has
   a. either high blood pressure or heart trouble or both?
   b. high blood pressure given that he has heart trouble?
   c. heart trouble given that he has high blood pressure?
   d. neither heart trouble nor high blood pressure?

14. The probability that a person swims is 0.45, and the probability that a person hunts is 0.58. If the probability that a person swims given that the person hunts is 0.21, find the probability that a person
   a. hunts and swims.
   b. hunts given that the person swims.
   c. hunts and doesn't swim.
   d. hunts or swims.

15. The accompanying table displays relative frequencies for red-green color blindness for males and females, where M represents males, F represents females, C represents color blind, and $\overline{C}$ represents not color blind.

	M	F
C	0.042	0.007
$\overline{C}$	0.485	0.466

If a person is chosen at random, use the table to find the following probabilities:

a. $P(M)$ e. $P(F|\overline{C})$

b. $P(M|C)$ f. $P(M \cap C)$

c. $P(F|C)$ g. $P(C|M)$

d. $P(C)$

16. A bank has observed that most customers at the tellers' windows either cash a check or make a deposit. The table indicates the transactions for teller A in one day.

	Checks Cashed	No. Checks Cashed
Deposits	50	20
No. deposit	30	10

Let C represent cashing a check and D represent making a deposit. Suppose a customer is chosen at random from teller A's customers. Express each of the following in words and find its value.

a. $P(C)$
b. $P(D)$
c. $P(C \cap D)$
d. $P(C \cup D)$
e. $P(C|D)$
f. $P(\overline{C}|D)$
g. $P(\overline{D}|D)$

17. In an attitudinal survey on strict gun-control legislation administered to 800 U.S. adults, the following results were obtained:

| | Stand | |
	In favor	Against
Shot a gun	75	200
Never shot a gun	425	100

If one of the 800 adults is chosen at random, use relative frequencies to approximate probabilities and determine each of the following:

a. P(in favor)
b. P(shot a gun and against)
c. P(against | never shot a gun)
d. P(shot a gun and in favor)
e. P(shot a gun)

Going Beyond

18. Let $P(E) = 0.2$ and $P(F) = 0.3$. Answer each of the following questions. Where appropriate, give an example.

a. Can $P(E \cup F) = 0.5$?
b. Can $P(E \cup F) = 0.7$?
c. Can $P(E \cup F) = 0.4$?
d. Can $P(E \cap F) = 0.2$?
e. Can $P(E \cap F) = 0.3$?
f. Can $P(E \cap F) = 0.1$?
g. Can $P(E \cap F) = 0.4$?

19. During a class lecture a history teacher stated that the probability of Israel and Syria both sending representatives to a peace conference is 0.8. Later during the same lecture, he stated that the probability of Syria sending a representative is 0.5. Do you believe these two statements? Explain.

20. If $P(E) = 0.2$, $P(F) = 0.3$, and $P(F|E) = 0.5$, rank the following events according to increasing probability: E, F, $\overline{E}$, $(E \cap F)$, $(E \cup F)$, $(E \cup \overline{E})$, and $(\overline{F} \cap F)$.

21. Give convincing arguments for the following two facts, and illustrate each using an example.

a. If $P(E) < P(F)$, then $P(E \cup F)$ is at least as large as $P(E)$.
b. If $P(E) < P(F)$, then $P(E \cap F)$ is at most as large as $P(E)$.

22. Determine a formula for $P(E \cap \overline{F})$ not involving conditional probability. (*Hint:* draw a Venn diagram.)

23. Refer to Motivator 5. Determine the probability that a person

a. tested for AIDS has AIDS.
b. has AIDS and has a positive test for AIDS.
c. has AIDS and has a negative test for AIDS.
d. has AIDS given a positive test result.
e. has AIDS given a negative test result.

24. Refer to Motivator 5. Suppose the probability of AIDS for an at-risk population is 0.003. Out of 25,000 blood samples from this population,

a. how many false positive results would you expect?
b. how many false negative results would you expect?
c. how many persons with AIDS would you expect?
d. how many persons with AIDS given a positive test result would you expect?

SECTION 5.6   *Independent Events*

If E and F are events such that the occurrence of F in no way influences the occurrence of E, then E and F are called **independent events**. Stated differently, E and F are independent events if the probability of E occurring given that event F has occurred is identically equal to the probability of event E occurring.

> **E and F are independent events if**
> $$P(E|F) = P(E)$$ (5.14)

If two events are not independent, we say they are **dependent events.**
As a consequence of rule (5.12), we also have

$$P(E \cap F) = P(F)P(E|F)$$

If E and F are independent events, then

$$P(F)P(E|F) = P(F)P(E)$$

Hence, we have the following **multiplication rule** for independent events.

Multiplication Rule
$$P(E \cap F) = P(E)P(F) \qquad\qquad (5.15)$$

Note that, as a general rule, sampling with replacement assures that two events will be independent, whereas sampling without replacement produces two events that will be dependent events.

APPLICATION 5.39

Which of the following pairs of events are independent?

    a. E = getting a head on a toss of a penny.
       F = getting a head on a toss of a dime.

    b. E = Mary's first child being a boy.
       F = Mary's second child being a girl.

    c. E = it will rain in Frostburg today.
       F = John fails his Math 101 exam today.

Solution:

    a. Since the events are clearly unrelated, they are independent.

    b. The two events are independent.

    c. These two events most likely are unrelated; hence they are independent events. ▪

APPLICATION 5.40

Two balls are drawn from a bag containing three white balls and two black balls. Let W_1 be the event of drawing a white ball on the first draw and let B_2 be the event of drawing a black ball on the second draw.

    a. If the balls are drawn with replacement, determine if W_1 and B_2 are independent events.

    b. If the balls are drawn without replacement, determine if W_1 and B_2 are independent events.

Solution: We shall use expression (5.14) to determine if the events are independent.

    a. Since the first ball is returned to the bag after the first draw and there are two black balls in the bag of five balls, $P(B_1|W_1) = 2/5 = P(B_2)$. Hence, the events are independent.

    b. The probability of selecting a black ball on the second draw given that a white ball was selected on the first draw is $P(B_2|W_1) = 2/4 = 1/2$, since there are four balls in the bag and two of them are black balls. There are two possibilities for selecting a black ball on the second draw: the first ball was white or the first ball was black.

Therefore,

$$P(B_2) = P[(W_1 \cap B_2) \cup (B_1 \cap B_2)]$$

Since the events $(W_1 \cap B_2)$ and $(B_1 \cap B_2)$ are mutually exclusive, we have

$$P[(W_1 \cap B_2) \cup (B_1 \cap B_2)] = P(W_1 \cap B_2) + P(B_1 \cap B_2)$$
$$= P(B_2|W_1)P(W_1) + P(B_2|B_1)P(B_1)$$
$$= \left(\frac{2}{4}\right)\left(\frac{3}{5}\right) + \left(\frac{1}{4}\right)\left(\frac{2}{5}\right) = \frac{8}{20} = \frac{2}{5}$$

The probability of selecting a black ball on the second draw given that a white ball was selected on the first draw is $P(B_2|W_1) = 2/4 = 1/2$. Since $2/5 \neq 1/2$, the two events are not independent. ■

APPLICATION 5.41   If two coins are tossed, find $P(HT)$.

Solution: The probability $P(HT)$ is the probability of getting a head on the first coin and a tail on the second coin. Since getting a head on one coin and getting a tail on another are independent events, by using rule (5.15) we have,

$$P(HT) = P(H)\,P(T)$$
$$= \frac{1}{2}\,\frac{1}{2} = \frac{1}{4} \quad ■$$

APPLICATION 5.42   Refer to the problem illustrated in application 5.35. If a student is selected and E is the event that the student has low ability in mathematics and F is the event that he has high interest in mathematics, are E and F independent events?

Solution: No. The probability of E given F is

$$P(E|F) = \frac{12}{55}$$

and

$$P(E) = \frac{60}{150} = \frac{2}{5}$$

As a result of rule (5.14), E and F are not independent events, since

$$P(E|F) \neq P(E)$$

That is,

$$\frac{12}{55} \neq \frac{2}{5} \quad ■$$

Care must be taken not to confuse the concepts of mutually exclusive events and independent events. There is no general relationship between the two types of events. Events can be mutually exclusive and not independent or they can be independent and not mutually exclusive. Or events can be dependent and not mutually exclusive (see exercises 13–16).

EXERCISE SET 5.6

Basic Skills

1. $P(E) = 0.5$, $P(F) = 0.6$, and $P(E \cap F) = 0.1$.
   a. Are E and F independent events? Why?
   b. Are E and F mutually exclusive events? Why?

2. $P(E) = 0.3$, $P(F) = 0.2$, and $P(E \cap F) = 0.06$.
   a. Are E and F independent events? Why?
   b. Are E and F mutually exclusive events? Why?

3. If events E and F are mutually exclusive events with $P(E) = 0.2$ and $P(F) = 0.3$, find
   a. $P(E \cup F)$.
   b. $P(E|F)$.

4. If events E and F are mutually exclusive events with $P(E) = 0.3$ and $P(F) = 0.4$, find
   a. $P(E \cup F)$.
   b. $P(E|F)$.

5. If E and F are events such that $P(E) = 0.4$, $P(F) = 0.3$, and $P(E|F) = 0.4$, are E and F independent events? Explain.

6. If E and F are events such that $P(E) = 0.3$, $P(F) = 0.2$, and $P(E|F) = 0.3$, are E and F independent events? Explain.

Further Applications

7. If two cards are drawn with replacement from a 52-card deck, find the probability that
   a. the second card is a heart, given that the first is a heart.
   b. both cards are hearts.
   c. the second is black, given that the first is a spade.
   d. the second is a face card, given that the first is a jack.
   e. the first card is a club or the second card is an ace.

8. Two marbles are drawn with replacement from a jar with four black and six white marbles. Find the probability that
   a. both are white.
   b. both are black.
   c. the second is white, given that the first is black.
   d. the first is black and the second is white.
   e. one is black and the other is white.

9. The accompanying table displays relative frequencies for red-green color blindness, where C represents the event a person is color blind, $\overline{C}$ represents the event that a person is not color blind, M represents the event a person is male, and $\overline{M}$ represents the event that a person is female.

	M	$\overline{M}$
C	0.042	0.007
$\overline{C}$	0.485	0.466

Are events C and M independent events?

10. If three coins are tossed once, find
    a. $P(HHH)$.
    b. $P(THT)$ (note: THT means getting a tail on the first coin, a head on the second coin, and a tail on the third coin).
    c. P(exactly one head).
    d. P(at least one head).
    e. P(at most two heads).

11. Freshmen at a small college are classified according to GPA in high school and SAT score in mathematics. The results are summarized in the accompanying table.

SAT Score	High School GPA		
	Low	Average	High
Low	50	30	50
Average	20	30	20
High	30	40	30

A student is selected at random. If E is the event that the student has an average high school GPA and F is the event that the student has a low SAT mathematics score, are E and F independent events? Explain.

12. Workers at a small industrial plant are classified according to religion and gender. The results are displayed in the table. If E is the event a person is male and F is the event a person is Jewish, are E and F independent events?

Religion	Gender	
	Male	Female
Protestant	30	20
Catholic	45	30
Jewish	45	30
Other	7	8

Going Beyond

13. Suppose a coin is tossed once. Let $E = \emptyset$ and $F = \{H\}$.
    a. Are E and F independent events? Explain.
    b. Are E and F mutually exclusive events? Explain.

14. Two coins are tossed. Let E represent the event of getting a head on the first coin and F represent getting two heads or two tails. Show that
    a. E and F are independent events.
    b. E and F are not mutually exclusive events.

15. Suppose a coin is tossed twice. Let $E = \{HH\}$ and $F = \{TT\}$.
    a. Are E and F mutually exclusive events? Explain.
    b. Are E and F independent events? Explain.

16. Suppose two six-sided dice are tossed once, let $E = \{(1, 2), (2, 1)\}$ and $F = \{(1, 2), (3, 4)\}$.
    a. Are E and F independent events? Explain.
    b. Are E and F mutually exclusive events? Explain.

17. If E and F are mutually exclusive events, what is the value of $P(E|F)$?

18. Suppose E and F are events with $P(E) > 0$ and $P(F) > 0$. If E and F are independent events, can E and F be mutually exclusive events? Explain.

19. A given manufacturing process produces 5% defective parts. Twenty percent of all parts are produced by machine A. There is a 10% probability that a part is defective given that it was produced by machine A.
    a. What is the probability that a tested part is defective and was produced by machine A?
    b. What is the probability that a part came from machine A given that it is defective?

20. Suppose the sample space is equal to the event $(E_1 \cup E_2 \cup E_3)$ and E_1, E_2, and E_3 have no common outcomes (the three events are called *mutually exclusive* and *exhaustive*). Show that for any other event B,

    $$P(B) = \Sigma \, [P(B|E_i)P(E_i)]$$

    This result is called the *law of total probability*. (*Hint:* See application 5.42.)

21. In a parking lot at a large airport, 60 percent of the cars are manufactured in the United States and 10 percent of these are sport cars; 20 percent of the cars are manufactured in Japan and 10 percent of these are sport cars; and finally, 20 percent of the cars are manufactured in Europe and 20 percent of these are sport cars. If a car is selected at random from the parking lot, what is the probability that it is a sport car?

22. Let E_1, E_2, and E_3 be mutually exclusive and exhaustive events, and suppose A is an event so that $P(A) > 0$. Then for any event E_k ($k = 1$, 2, or 3), show that

    $$P(E_k|A) = \frac{P(A|E_k) \, P(E_k)}{\Sigma \, [P(A|E_i) \, P(E_i)]}$$

    This result is known as *Bayes' theorem*. (*Hint:* Use the law of total probability of exercise 20.)

23. Three boxes contain red, white, and blue balls. Box 1 contains seven red balls, one white ball, and five blue balls; box 2 contains three red balls, four white balls, and two blue balls; and box 3 contains one red ball, six white balls, and two blue balls. A box is picked at random and a ball drawn from it is random. Given that the ball is red, what is the probability that it came from box 3?

24. Suppose you are on a game show and are given a choice of three doors. Behind one is a car; behind the other two are goats. You pick door number 1, and the game show host, who knows what's behind all three doors, opens door number 3, which has a goat. He then asks if you want to pick door number 2.
    a. Should you switch?
    b. What is the probability of winning the car if you do switch?
    c. What is the probability of winning the car if you don't switch?

SECTION 5.7   *Random Variables*

It is convenient in later work to relate the outcomes in an experiment to real numbers. When outcomes of an experiment can be associated with real numbers, they are easier to analyze. Unfortunately, not all experiments result in outcomes that are real numbers.

Suppose you work in the quality control department for a manufacturer of microcomputers. You have been assigned the task of inspecting four microcomputers chosen at random from the last batch produced and classifying them as defective (D) or non-defective (N). This experiment results in the following sample space of qualitative data:

DNDN	DDND	NDND	NNDN
DNND	DDDN	NDDN	NNND
DNDD	DDNN	NDDD	NNDD
DNNN	DDDD	NDNN	NNNN

These outcomes are not real numbers, but if each outcome is associated with the number of defective microcomputers, we can associate a unique real number with each outcome. For example, the outcome DNDN can be assigned the number 2, the outcome DNND can be assigned the number 2, the outcome NDNN can be assigned the number 1, and the outcome NNNN can be assigned the number 0. Such a pairing of the outcomes in a sample space of an experiment with unique real numbers is called a **random variable.** The random variable in the quality control example is the number of defective microcomputers in a batch of four. It has the five possible values 0, 1, 2, 3, and 4.

Random Variable

A *random variable* is a rule (or function) that assigns unique real numbers to each outcome in a sample space of an experiment. The following are other examples of random variables:

EXAMPLE 5.20

Consider the following experiments and corresponding random variables.

Experiment	Random Variable
1. Toss five coins.	number of heads obtained
2. Observe customers at a bank during one hour.	number of customers
3. Buy 12 computers.	number of defective computers
4. Weigh a person.	weight in pounds
5. Hit a golf ball.	distance it travels
6. Audit ten tax returns.	number of returns containing errors
7. Audit ten tax returns.	number of mistakes on line 33
8. Observe an employee work for an eight-hour period.	time spent by the employee in unproductive work
9. Poll ten people concerning candidate A's chances for election.	proportion in favor of A
10. Treat 50 people with a headache, using pill A.	the number cured

The values of a random variable are typically equated with the random variable. This usually causes no problems and is a matter of convenience. For experiment 1 above, if we let X denote the number of heads obtained when five coins are tossed, the possible values of the random variable are 0, 1, 2, 3, 4, and 5; we denote this by $X = \{0, 1, 2, 3, 4, 5\}$. Strictly speaking, this is incorrect, since X is the rule (function) and $\{0, 1, 2, 3, 4, 5\}$ represents its range values. Throughout this section, capital letters, such as X, Y, and Z will denote random variables and small letters such as x, y, and z will denote their values (range values).

If the values of a random variable can be counted, it is called a **discrete random variable;** if the values of the random variable can not be counted, it is called a **continuous random variable.** Experiments 4, 5, and 8 in the previous list represent continuous random variables, whereas the remaining random variables are discrete. The number of units sold, the number of defective computers, and any other random variable concerned with counting is discrete. Any random variable dealing with measurement, such as weight, time, temperature, and distance, is continuous. Experiment 8 above involves a continuous random variable since the number of hours of unproductive work may be any value in the time interval 0–8.

Probability Distributions

Probabilities can be associated with the values of a discrete random variable. If X is a discrete random variable and we are interested in determining the probability of the outcomes associated with k, a value of the random variable X, we write $P(X = k)$. When there is no chance for confusion we write $P(X = k)$ as $P(k)$.

EXAMPLE 5.21

Suppose the random variable X represents the number of heads obtained when two coins are tossed. The possible values for X are 0, 1, 2. Then,

$$P(X = 0) = P(0) = P(TT) = \frac{1}{4}$$

$$P(X = 1) = P(1) = P(HT, TH) = \frac{1}{2}$$

$$P(X = 2) = P(2) = P(HH) = \frac{1}{4}$$

A *probability distribution table,* or **probability table,** for a discrete random variable is a two-column table, with one column representing the values of the random variable and the other representing the associated probabilities. For example 5.21 a probability distribution table is shown in table 5.6.

TABLE 15.6

Probability Distribution
Example 5.21

x	$P(x)$
0	1/4
1	1/2
2	1/4
	1

Notice that the probability column must always sum to 1.

Probability Functions

Sometimes it is convenient to express the relationship between the values of a random variable and its associated probabilities in terms of a rule. Such a rule is called a **probability function** or probability distribution function. Consider application 5.43.

APPLICATION 5.43

Suppose five balls of the same size are numbered 1 to 5 and placed in a bag. An experiment involves selecting a ball at random. Let the random variable X denote the value of the ball selected.

a. Construct a probability table for X.

b. Find the probability function for X (call it g).

Solution: The five balls are equally likely to be selected. The following is the probability table for the discrete random variable X.

a.

x	$P(x)$
1	1/5
2	1/5
3	1/5
4	1/5
5	1/5

b. The function that relates the values of x with its associated probabilities is g, defined by $g(x) = 1/5$; $x = 1, 2, 3, 4, 5$. For any value of x not contained in the set $\{1, 2, 3, 4, 5\}$, $g(x)$ is defined to be zero. ▪

Probability Graphs

Graphs can be constructed for probability functions. For a discrete random variable, a graph of the probability function can be constructed using vertical line segments. The values of the random variable are displayed on the horizontal axis and the probabilities are displayed on the vertical axis. At each value of the random variable, a vertical line segment is constructed with a height equal to the probability of the random variable (see example 5.22). Note that the sum of the lengths of the vertical line segments must equal 1. For continuous random variables, on the other hand, we use areas instead of vertical line segments to represent probabilities, as in example 5.23.

EXAMPLE 5.22

Suppose the following is a probability table for a discrete random variable X.

x	$P(x)$
1	0.2
2	0.3
3	0.4
4	0.1

The graph of the corresponding probability function is shown in figure 5.12. Note that the sum of the lengths of the vertical line segments is 1.

FIGURE 5.12

Graph of probability function for discrete random variable

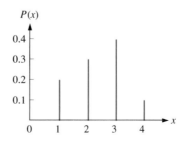

EXAMPLE 5.23

Suppose the values of the random variable X are all real numbers between 0 and 1 and the probability function for X is f, defined by

$$f(x) = \begin{cases} 2x & \text{if } 0 \le x \le 1 \\ 0 & \text{elsewhere} \end{cases}$$

A graph for the probability function f is shown in figure 5.13. The area of the region bounded by the line $f(x) = 2x$, the x-axis, and the lines $x = 0$ and $x = 1$ is 1. The region is a right triangle whose area is found as follows:

$$A = \left(\frac{1}{2}\right)(\text{base})(\text{height})$$

$$= \left(\frac{1}{2}\right)(1)(2) = 1$$

FIGURE 5.13

Graph of probability function for continuous random variable

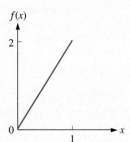

Since, for continuous random variables, probabilities are associated with areas, such probabilities can be found only for intervals, such as $a < x < b$. The probability of any particular value x for a continuous random variable must necessarily equal 0, since a vertical line segment drawn at the value x and extending upward to the probability function must have an area of 0. As a result, if X is a continuous random variable, $P(a < X < b) = P(a \le X \le b)$. A thorough study of continuous random variables involves calculus and is beyond the scope of this text. So, after example 5.24, for the remainder of this section, we will deal only with discrete random variables and their probabilities.

EXAMPLE 5.24

For the continuous random variable of figure 5.13, suppose we want to find the probability that the random value X is between 0.25 and 0.5; that is, $P(0.25 \le X \le 0.5)$. This probability is equal to the area under the line whose equation is $f(x) = 2x$ bounded by the x-axis and the lines $x = 0.25$ and $x = 0.5$, as illustrated in figure 5.14. The figure formed is a trapezoid. Recall that

FIGURE 5.14

$P(0.25 \le X \le 0.5)$

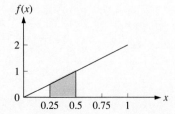

the area of a trapezoid is given by $A = h(b_1 + b_2)/2$, where h is the altitude and b_1 and b_2 are the bases. This formula yields the area (or probability) for the shaded region.

$$A = \frac{(0.25)(0.5 + 1)}{2}$$

$$= \frac{(0.25)(1.5)}{2} = 0.1875$$

APPLICATION 5.44   Consider the following probability function for a discrete random variable X.

$$f(x) = \frac{x}{10} \quad \text{for } x = 1, 2\ 3, 4$$

a. What are the values of the random variable X?
b. What probability is associated with each value of X?
c. Construct a probability table for X.
d. Construct a probability graph for X.

Solution:

a. The values of X are 1, 2, 3, and 4.
b. The probability associated with $x = 1$ is

$$P(X = 1) = f(1) = \frac{1}{10} = 0.1$$

Similarly, the probabilities for $x = 2$, 3, and 4 are found to be 0.2, 0.3, and 0.4, respectively.

c. A probability table is as follows:

x	$P(x) = f(x)$
1	0.1
2	0.2
3	0.3
4	0.4
	1

d.

Mean of a Discrete Random Variable

Recall that one formula for calculating the value of population mean μ is

$$\mu = \frac{\Sigma\,(fx)}{N}$$

where f is the frequency of a distinct measurement x and N is the size of the population.

This formula can be rewritten as

$$\mu = \Sigma \left[x \frac{f}{N} \right]$$

Since the relative frequency, f/N, represents the probability of x occurring, $P(x)$, the population mean can be written as $\mu = \Sigma [x\, P(x)]$. As a consequence of these observations, if X is a random variable, we define the **mean** of X as follows:

> **Mean of a Random Variable X**
>
> $$\mu_X = \Sigma [xP(x)] \tag{5.16}$$

where μ_X represents the mean of the random variable X.

EXAMPLE 5.25

A large industrial plant ran a campaign encouraging car pooling among its employees. The data in table 5.7 were recorded for all plant employees to monitor the effects of the campaign.

TABLE 5.7

Car-Pooling Data

Number of Occupants Per Car, x	f	xf	Relative Frequency, f/N
1	425	425	0.450
2	235	470	0.249
3	205	615	0.217
4	52	208	0.055
5	22	110	0.023
6	6	36	0.006
	945	1864	1.000

The population mean is $\mu = \dfrac{\Sigma (fx)}{N} = \dfrac{1864}{945} = 1.97$.

Now let's choose a car at random carrying employees to work and count the number of occupants per car. The number of occupants per car represents a random variable. If we let the random variable X represent the number of occupants per car, the values of the random variable are 1, 2, 3, 4, 5, and 6 with respective probabilities 0.45, 0.249, 0.217, 0.055, 0.023, and 0.006. The mean of this random variable is then

$$
\begin{aligned}
\mu_X = \Sigma (xP(x)) &= (1)P(1) + (2)P(2) + (3)P(3) + (4)P(4) + (5)P(5) + (6)P(6) \\
&= (1)(0.450) + (2)(0.249) + (3)(0.217) \\
&\quad + (4)(0.055) + (5)(0.023) + (6)(0.006) \\
&= 1.97
\end{aligned}
$$

Note that this agrees with the value calculated above.

The mean of a random variable X is also called the **expected value** of X and is denoted by $E(X)$. As a result, we have the following results for a discrete random variable:

> **Mean of a Discrete Random Variable X**
>
> $$\mu_X = E(X) = \Sigma (xP(x))$$

Analogous formulas also exist for the case of continuous random variables.

APPLICATION 5.45

A large hotel has found that the number of air-conditioning units that must be replaced each summer has the following probability table.

Number Replaced	Probability
0	0.35
1	0.30
2	0.20
3	0.10
4	0.05

Find the expected number of air conditioners that must be replaced each summer and interpret your answer.

Solution: Let the random variable X denote the number of air conditioners replaced. We will compute the product of x and $P(x)$ for each value of X. The results are organized in table 5.8. From the table we can see that $E(X) = \Sigma\,(xP(x)) = 1.20$. Hence, if the number of air-conditioning units that must be replaced is recorded for a large number of years, the average number of replacements would be 1.20 units per year.

TABLE 5.8

Calculations of $E(X)$ for Application 5.45

x	$P(x)$	$xP(x)$
0	0.35	0
1	0.30	0.30
2	0.20	0.40
3	0.10	0.30
4	0.05	0.20
		1.20

▪

Variance of a Discrete Random Variable

Recall from chapter 3 that the population variance can be defined as

$$\sigma^2 = \frac{SS}{N} = \frac{\Sigma f(x - \mu)^2}{N}$$

where f is the frequency associated with the measurement x and N is the size of the population. This can be rewritten as

$$\sigma^2 = \Sigma\left[(x - \mu)^2 \frac{f}{N}\right]$$

Again, if we let the relative frequency f/N represent the probability of x occurring $P(x)$ then we have the following formula for the **variance of a random variable** X.

> **Variance of a Random Variable X**
>
> $$\sigma_X^2 = \Sigma\,[(x - \mu)^2 P(x)] \tag{5.17}$$

where σ_X^2 is the variance of the random variable X and μ is the mean of X.

EXAMPLE 5.26

For the car-pooling data presented in table 5.7, the population mean is $\mu = 1.97$ and the population variance is

$$\sigma^2 = \frac{\Sigma\,[(x - \mu)^2 f]}{N}$$

$$= [(1 - 1.97)^2(425) + (2 - 1.97)^2(235) + (3 - 1.97)^2(205) + (4 - 1.97)^2(52)$$
$$+ (5 - 1.97)^2(22) + (6 - 1.97)^2(6)]/945$$
$$= 1.197$$

These computations can be rewritten as

$$\sigma^2 = (1 - 1.97)^2\frac{425}{945} + (2 - 1.97)^2\frac{235}{945} + (3 - 1.97)^2\frac{205}{945} + (4 - 1.97)^2\frac{52}{945}$$

$$+ (5 - 1.97)^2\frac{2}{945} + (6 - 1.97)^2\frac{6}{945}$$

$$= 1.197$$
$$= (1 - 1.97)^2 P(1) + (2 - 1.97)^2 P(2) + (3 - 1.97)^2 P(3)$$
$$+ (4 - 1.97)^2 P(4) + (5 - 1.97)^2 P(5) + (6 - 1.97)^2 P(6)$$

But this is just an instance of formula (5.17).

Standard Deviation of a Discrete Random Variable

The **standard deviation of a random variable** is defined to be the positive square root of the variance and is denoted by σ.

Standard Deviation of a Random Variable X
$\sigma_X = \sqrt{\text{variance}}$

APPLICATION 5.46

For the random variable X in application 5.45, find the variance σ_X^2 and standard deviation σ_X.

Solution: In application 5.45, μ was found to be 1.2. We organize our computations for σ_X^2 in the following table.

x	$P(x)$	$x - \mu$	$(x - \mu)^2$	$(x - \mu)^2 P(x)$
0	0.35	−1.2	1.44	0.504
1	0.30	−0.2	0.04	0.012
2	0.20	0.8	0.64	0.128
3	0.10	1.8	3.24	0.324
4	0.05	2.8	7.84	0.392
				1.36

Hence $\sigma_X^2 = 1.36$ and $\sigma_X = \sqrt{1.36} = 1.17$. ■

To find the variance σ_X^2 of a random variable X the following formula is computationally much easier to use than (5.17) and is mathematically equivalent.

Computational Formula for σ_X^2	
$\sigma_X^2 = \Sigma\,[x^2 P(x)] - \mu_X^2$	(5.18)

EXAMPLE 5.27

Let's use formula (5.18) to find σ_X^2 for the random variable illustrated in application 5.45. The calculations are organized in the following table. We have,

x	x^2	$P(x)$	$x^2P(x)$
0	0	0.35	0
1	1	0.30	0.30
2	4	0.20	0.80
3	9	0.10	0.90
4	16	0.05	0.80
			2.80

The mean μ_X was previously found to be $\mu_X = 1.2$. Thus, the variance of the random variable X is

$$\sigma_X^2 = \Sigma\,[x^2\,P(x)] - \mu_X^2$$

$$= 2.80 - (1.2)^2 = 1.36$$

Note that this agrees with the result obtained in application 5.46.

EXERCISE SET 5.7

Basic Skills

1. Suppose an experiment consists of tossing a coin four times and the random variable X denotes the number of heads obtained.
   a. Construct a probability table for X.
   b. Is X a discrete or continuous variable?
   c. Construct a probability graph for X.

2. Suppose an experiment consists of tossing two six-sided dice and that the random variable Y denotes the sum of the two dice.
   a. Construct a probability table for Y.
   b. Is Y a discrete or continuous variable?
   c. Construct a probability graph for Y.

3. Given the probability function f defined by

   $$f(x)\ \frac{5-x}{10} \quad \text{for } x = 1, 2, 3, 4$$

   for a random variable X, find the mean and standard deviation of X.

4. Given the probability function g denoted by $g(y) = 0.2$ for $y = 2, 3, 4, 5, 6$ for a random variable Y, find the mean and standard deviation of Y.

5. Let X represent the number of heads obtained when three coins are tossed. Find μ_X and σ_X.

6. Given the probability function h defined by

   $$h(x) = \frac{x}{10} \quad \text{for } x = 1, 2, 3, 4$$

   for a random variable X, find μ_X and σ_X.

7. Suppose X is a discrete random variable with probability function f defined by

   $$f(x) = \frac{x}{55} \quad \text{for } x = 1, 2, 3, \ldots, 10$$

   a. Show that $\Sigma\,P(x) = 1$.
   b. Find $P(X = 4)$.
   c. Find $P(X \le 3)$.
   d. Find $P(X \le 9)$.
   e. Find $P(2 \le X \le 4)$.
   f. Find $P(2 < X < 4)$.
   g. Find μ_X.
   h. Find σ_X.

8. Suppose X is a discrete random variable with probability function f defined by

   $$f(x) = \frac{x}{15} \quad \text{for } x = 1, 2, 3, 4, 5$$

   a. Show that $\Sigma\,P(x) = 1$.
   b. Find $P(X = 4)$.
   c. Find $P(X \ge 3)$.
   d. Find $P(X \le 5)$.
   e. Find $P(2 \le X \le 4)$.
   f. Find $P(2 < X < 4)$.
   g. Find μ_X.
   h. Find σ_X.

Further Applications

9. The probability table for the number of telephone calls X received by Mr. Jones in a day is as follows:

x	$P(x)$
0	0.40
1	0.23
2	0.17
3	0.09
4	0.11

Find
a. $P(X = 1)$.
b. $P(1 < X \le 3)$.
c. $P(X \ge 1)$.
d. $E(X)$.
e. σ_X^2.

10. The number X of fish dinners sold in one hour at a local restaurant is described by the following probability table:

x	0	1	2	3	4	5	6
$P(x)$	0.14	0.16	0.30	0.14	0.13	0.08	0.05

Construct a probability graph and find
a. $P(X \le 1)$.
b. $P(X > 1)$.
c. $P(2 \le X \le 4)$.
d. $P(X \le 5)$.
e. μ_X.
f. σ_X.

11. Based on past history, the accompanying frequency table reports X, the number of automobiles sold per day for a car dealer.

X	Number of Days
0	44
1	87
2	128
3	234
4	297
5	155
6	30
7	25

a. Construct a probability table for the random variable X.
b. Find the mean of X.
c. Find the standard deviation of X.

d. Find $P(X \le 5)$.
e. Find the expected number of cars sold. Interpret your result.

12. A raffle offers a first prize of $1000, two second prizes of $500, and 20 prizes of $20 each. If 10,000 tickets are sold at $0.50 each, find
a. the expected net profit (winnings) if one ticket is bought.
b. the variance of the net profit (winnings) if one ticket is bought.

13. A builder is considering a job that promises a profit of $25,000 with a probability of 0.8 or a loss (due to poor weather, strikes, and so forth) of $10,000 with a probability of 0.2.
a. What is the builder's expected profit?
b. What is the standard deviation of the profit?

14. To promote its products the Pepsi-Cola Company advertised a Nintendo holiday game. Instant-win prizes were offered to lucky customers who received special messages under bottle caps of selected marked products. The prize information is contained in the accompanying table.

Instant Prize	Odds	Estimated Retail Value
2L of Pepsi product	1:25.5	$1.49
$ Mario Money coupon	1:1,400	$5.00
Nintendo Game Pak	1:23,000	$30.00
Broderbund U-Force	1:70,000	$70.00
Nintendo Game Boy	1:140,000	$90.00
Nintendo Action Set	1:140,000	$100.00

Calculate the expected retail value of a prize received by a customer who purchased one bottle of a Pepsi product covered by the game.

15. Find the expected number of male children in a family with five children. Assume that male and female births are equally likely to occur.

Going Beyond

16. You have to be at school in 20 minutes and there are two routes that you can take to get there. The mean times to reach school are 12 and 16 minutes, respectively. Is the 12-minute route the better route to take? Explain.

17. Mr. Baker wants to insure his house for $80,000. The insurance company estimates a total loss may occur with a probability of 0.004, a 50% loss with a probability of 0.02, and a 25% loss with a probability of 0.08. If the insurance

company will pay no benefits for any other partial loss, what premium will Mr. Baker be required to pay each year if the insurance company wants to make an average profit of $500 per year on all policies of this type?

18. If two dice are tossed and the sum is recorded, let the random variable X denote this sum. Find μ and σ for X.

19. If X is a continuous random variable, with $P(X \leq 3)$ $= 0.45$, and $P(X \geq 4) = 0.40$, find
   a. $P(X < 3)$.
   b. $P(4 < X)$.
   c. $P(3 \leq X \leq 4)$.

20. If two cards are drawn with replacement from a standard deck of 52 cards, find the expected number of diamond cards that can result.

21. Prove that $\sigma^2 = \Sigma\,[x^2 P(x)] - \mu^2$.

22. The following problem is known as the *St. Petersburg Paradox*. Peter agrees to toss a coin until a head shows face up. If a head shows on the first toss, he agrees to pay you $1. If not, he agrees to give you $2 if he gets a head on the second toss, $4 if a head shows on the third toss, $8 if a head shows on the fourth toss, and so on. The number of dollars he will give you doubles with each additional throw. In order to break even, how much should Peter charge the game? Explain.

CHAPTER SUMMARY

The concepts and principles of elementary probability were introduced in this chapter. The fundamental theorem of counting, permutations, and combinations were introduced as aids in calculating more difficult probabilities. We saw that random variables can be used to provide numerical descriptions of experimental outcomes. We learned how to associate probability distributions with random variables and how to calculate means, variances, and standard deviations for discrete random variables.

CHAPTER REVIEW

▪ IMPORTANT TERMS ▪

The following chapter terms have been mixed in ordering to provide you better review practice. For each term, provide a definition in your own words. Then check your responses against those given in the chapter.

classical probability method	computer simulation	tree diagram
compound events	objective method	variance of a random variable
conditional probability	dependent events	permutation
continuous random variable	probability function	combination
discrete random variable	product rule	favorable outcomes
empirical probability	mutually exclusive events	independent events
empty set	odds	mean of a random variable
equally likely outcomes	probability	multiplication rule
event	probability table	permutation rule
expected value of a random variable	random variable	subjective method
experiment	relative frequency	standard deviation of a random
Venn diagram	sample space	variable
selection with replacement	simple events	probability histogram
selection without replacement	subset	Pascal's triangle
n factorial	sum rule	

■ *IMPORTANT SYMBOLS* ■

S, sample space

$\overline{E}$, the event not E

$(E \cup F)$, event E or F

$(E \cap F)$, event E and F

$P(E)$, the probability of event E

$a{:}b$, the odds are a to b

P_n^n, the number of permutations of n objects taken n at a time

$n!$, n factorial

P_r^n, the number of permutations of n objects taken r at a time

$\begin{pmatrix} n \\ r \end{pmatrix}$, binomial coefficient, the number of combinations of n objects taken r at a time

$P(E|F)$, conditional probability

μ_X, mean of a random variable X

σ_X^2, variance of a random variable X

σ_X, standard deviation of a random variable X

$E(X)$, expected value of X

■ *IMPORTANT FACTS AND FORMULAS* ■

If S is a sample space of equally likely outcomes and E is an event, then $P(E) = f/n$ where f is the number of outcomes contained in E and n is the number of outcomes contained in S.   (5.1)

Odds in favor of E: $\mathrm{Odds}(E) = \dfrac{P(E)}{P(\overline{E})}$.

Odds against E: $\mathrm{Odds}(\overline{E}) = \dfrac{P(\overline{E})}{P(E)} = \dfrac{\cdot}{\mathrm{Odds}(E)}$.

If $\mathrm{Odds}(E) = a{:}b$, then $P(E) = \dfrac{a}{a+b}$.

Fundamental theorem of counting. If one event can be performed in m different ways, and if after it has been done, another event can be performed in n different ways, then the two events can be performed, one following the other, in a total of $m \cdot n$ different ways.

n factorial: $n! = n(n-1)(n-2)(n-3) \cdots (2)(1)$.

The number of permutations of n objects taken n at a time: $P_n^n = n!$.   (5.3)

The number of permutations of n objects taken r at a time: $P_r^n = \dfrac{n!}{(n-r)!}$.   (5.4)

Binomial coefficient: $\begin{pmatrix} n \\ r \end{pmatrix} = \dfrac{P_r^n}{r!}$.   (5.5)

The number of combinations of n objects taken r at a time: $\begin{pmatrix} n \\ r \end{pmatrix} = \dfrac{n!}{r!(n-r)!}$.   (5.6)

Sum rule: $P(E \cup F) = P(E) + P(F) - P(E \cap F)$.   (5.7)

If E and F are mutually exclusive events, $P(E \cup F) = P(E) + P(F)$.   (5.8)

Probability of not E: $P(\overline{E}) = 1 - P(E)$.   (5.9)

Conditional probability rule: $P(E|F) = P(E \cap F)/P(F)$, provided $P(F) \neq 0$.   (5.10)

Product Rule: $P(E \cap F) = P(F)\, P(E|F)$.   (5.12)

E and F are independent events if $P(E|F) = P(E)$.   (5.14)

Multiplication rule for independent events: $P(E \cap F) = P(E)P(F)$.   (5.15)

Mean of random variable X: $\mu_X = E(X) = \Sigma\, (xP(x))$.   (5.16)

Variance of random variable X: $\sigma_X^2 = \Sigma\, [(x - \mu)^2 P(x)]$.   (5.17)

Standard deviation of random variable X: $\sigma_X = \sqrt{\text{variance}}$.

Computational formula for variance of random variable X: $\sigma_X^2 = \Sigma\, [x^2 P(x)] - \mu_X^2$.   (5.18)

REVIEW EXERCISES

1. The records of a particular hospital indicate that 18% of all its patients are admitted for surgery, 30% are admitted for obstetrics, and 5% are admitted for both surgery and obstetrics.
   a. What is the probability that a randomly selected patient will be admitted for surgery, obstetrics, or both?
   b. What is the probability that a randomly selected patient is not admitted for surgery?
   c. What is the probability that a randomly selected patient will be a surgery patient and will not receive obstetrical treatment?

2. Suppose E and F are two events with $P(E) = 1/2$ and $P(F) = 1/3$.
   a. If E and F are mutually exclusive events, find $P(E \cup F)$.
   b. If E and F are independent events, find $P(E|F)$, $P(F|E)$, $P(E \cap F)$, and $P(E \cap F)$.
   c. Can $P(E \cup F) = 11/12$? Explain.
   d. Can $P(E \cup F) = 1/6$? Explain.
   e. Can $P(E \cap F) = 0.6$? Explain.

3. Seventy-five students were polled and asked to name their favorite drink. Their responses are as follows:

beer	13	wine	2
soft drink	40	water	7
iced tea	4	hot tea	1
coffee	7	whiskey	1

   Suppose one of the surveyed students is chosen at random.
   a. Find the probability that the student indicated beer or wine.
   b. Find the probability that the student did not indicate coffee.

4. Employees at a particular plant were classified according to gender and political party affiliation. The results follow:

	Political Affiliation		
Gender	Democrat	Republican	Independent
Male	40	50	5
Female	18	8	4

   If an employee is chosen at random, find the probability that the employee is a
   a. male.
   b. Republican.
   c. female Democrat.
   d. Republican given that she is a female.
   e. male given that he is a Republican.

5. If a six-sided die is tossed, let the random variable X denote the number showing face up.
   a. Construct a probability table for X.
   b. Construct a probability graph for X.
   c. Find $E(X)$ and interpret your results.
   d. Find σ_X.

6. A person tosses three coins. If he gets three heads or three tails, he receives $10. If he does not receive three heads or three tails, he pays out $10. What are his expected winnings?

7. Mr. Jones can make $5000 with probability 0.4 or lose $2000 with probability 0.6 if he invests in Ajax Company stock. What is his expected gain? What is the standard deviation of his gain?

8. If three dice are tossed, let the random variable X denote the number of 2s obtained.
   a. Construct a probability table for X.
   b. Construct a probability graph for X.
   c. Find $E(X)$.
   d. Find σ_X^2.

9. A coin is biased (not fair) in such a way that a head is twice as likely to occur as a tail when the coin is tossed. If the coin is tossed twice, let the random variable X denote the number of heads obtained.
   a. Construct a probability table for X.
   b. Find μ_X and σ_X.
   c. Suppose the random variable Y denotes the number of tails obtained. Find μ_Y and σ_Y.

10. The Ace Bank stock is currently selling for $10 a share. An investor plans to buy shares and to hold the stock for one year. Let X denote the price of the stock (in dollars) after one year. The probability table for X is shown here.

x	$P(x)$
10	0.35
11	0.25
12	0.20
13	0.15
14	0.05

    a. Find the expected price of the stock after one year.
    b. What is the expected gain per share of stock over the one-year period?
    c. What percentage return on the investment is reflected by the expected gain per share?

d. Find the variance in the price of the stock over the one-year period.

e. Find the variance in the gain per share of the stock over the one-year period.

f. Construct a graph for the probability function.

Computer Applications

1. Use a computer program to simulate the birthday problem: estimate the probability that among 25 randomly selected people, at least two share the same birthday. Use commands similar to

```
RANDOM 25 C1;
INTEGERS 1 365.
SORT C1 C2
PRINT C2
```

Repeat this procedure nineteen more times, and estimate the probability that two have the same birthday. (*Hint:* With MINITAB use the STORE command to create a command file that can be executed 20 times.)

2. Use a computer simulation to do exercise 21 in exercise set 5.2: Roll a pair of dice 100 times and record the sum of the number showing for each roll. Determine the relative frequency of a sum of 8. (If the dice are fair, the probability of rolling a sum of 8 is approximately 0.14.)

3. Use a computer simulation to do exercise 23 in exercise set 5.2: If a deck of 52 playing cards is well shuffled, what do you estimate the probability is that the top three cards contain either a king or a queen, or that a king and a queen are next to each other somewhere among the remaining 49 cards? Test your guess by simulating the experiment 20 or more times.

▪ CHAPTER ACHIEVEMENT TEST ▪

1. A restaurant offers six sandwiches, ten types of drinks, four varieties of soups, and three desserts on its menu. How many different lunches can be ordered if one type of sandwich, one drink, one soup, and one dessert are desired?

2. A study is made of religious affiliation and political party. The following results were tabulated:

Political Party	Religion		
	Protestant	Catholic	Jewish
Democrat	10	15	25
Republican	20	30	40
Independent	5	15	5

A person is randomly selected from the group survey where D, R, and I denote Democrat, Republican, and Independent, respectively, and P, C, and J denote Protestant, Catholic, and Jewish, respectively.

a. Find $P(R)$.

b. Find $P(J)$.

c. Find $P(R \cap J)$.

d. Find $P(R \cup J)$.

e. Find $P(R|J)$.

f. Find $P(C|D)$.

g. Are events J and R independent? Explain.

h. Are J and R mutually exclusive events? Why?

3. Two males and three females apply for two teaching positions. Since all the applicants are equally qualified, two are randomly selected to fill the positions.

a. List a sample space for this experiment.

b. How many ways can two males be selected?

c. How many ways can two females be selected?

d. Find the probability that two females are hired.

e. Find the probability that one female and one male are hired.

4. The following table reports the annual deaths of male cancer patients over age 65 from the five major types of cancer:

Location	Number of Deaths
Colon	8,000
Lung	12,500
Pancreas	3,000
Prostate	10,800
Stomach	3,200

If a deceased male cancer patient is selected, what is the probability that he

a. died from lung cancer?

b. died from one of the two leading causes?

c. died from cancer of the colon or from cancer of the pancreas?

d. did not die from cancer of the stomach?

5. You pay $0.50 to spin a roulette wheel with 38 equally-spaced positions. If the wheel stops on the position you choose, you win $5.00; otherwise, you lose. If you play one game, find your expected winnings.

6. If two cards are drawn with replacement from a standard deck of playing cards, let the random variable X denote the number of hearts. Find μ_X and σ_X.

6

Discrete Distributions

CHAPTER OBJECTIVES

In this chapter we will investigate

▷ *Binomial experiments.*

▷ *The language associated with binomial experiments.*

▷ *How to calculate binomial coefficients.*

▷ *How to calculate binomial probabilities.*

▷ *How to use the binomial tables for calculating binomial probabilities.*

▷ *How to calculate the mean of a binomial random variable.*

▷ *How to calculate the variance and standard deviation of a binomial random variable.*

▷ *How to construct probability distribution graphs for binomial random variables.*

▷ *Multinomial experiments.*

▷ *How to calculate multinomial coefficients.*

▷ *How to calculate multinomial probabilities.*

▷ *Hypergeometric experiments.*

▷ *How to calculate hypergeometric probabilities.*

▷ *Poisson experiments.*

▷ *How to calculate Poisson probabilities.*

|||| **MOTIVATOR 6**

*S*ome airlines follow the practice of overbooking, selling more tickets than they have seats. The reason for this is that a certain percentage of those who make reservations don't show up at flight time. Suppose that the planes for a commuter airline can accommodate 35 passengers and it has been determined that on the average 5% of those who make reservations don't show up for the flight. If the airline sells 37 tickets for a particular flight, what is the probability that all 37 passengers show up? What is the probability that all those who show up for the flight can be accommodated? After studying section 6.2 you should be able to answer both these questions.

Chapter Overview

Binomial distributions form an important class of discrete distributions in statistics. They are used to describe a wide variety of processes in many ways. Binomial distributions result from repeated binomial experiments. We shall examine the properties and some of the applications of binomial experiments in this chapter. Related to binomial experiments are the multinomial and hypergeometric experiments. These experiments relax certain properties of the binomial, thus making them applicable to a wider variety of situations. Finally, we explore Poisson experiments as another variation of binomial experiments. These four types of experiments give rise to discrete random variables having discrete probability distributions. We shall learn to calculate probabilities associated with each type of random variable.

SECTION 6.1   *Binomial Distributions*

Consider the experiment of tossing a fair coin three times and observing the number of heads that result. Each toss of the coin is referred to as a **trial.** This experiment possesses the following characteristics:

1. It consists of three identical trials of tossing one coin.
2. Each trial results in exactly one of two outcomes, heads or tails.
3. The trials are independent; the outcome of one trial does not affect the outcome of any other trial.
4. The probability of getting a head remains constant from trial to trial, that probability being 0.5.

There are eight different equally likely possibilities that can result from tossing a coin three times. They are listed as follows:

HHT   THT   HTH   TTH   HHH   THH   HTT   TTT

If the discrete random variable X denotes the number of heads obtained, then the following probability distribution table can be used to organize the results.

x	$P(x)$
0	1/8
1	3/8
2	3/8
3	1/8

FIGURE 6.1

Probability distribution for X = number of heads in three tosses of a fair coin

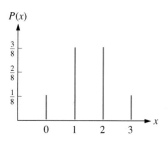

This probability distribution has the graph shown in figure 6.1. Notice that the graph is symmetrical about $x = 1.5$ heads. The experiment just described is an example of a class of experiments called binomial experiments.

In general, a **binomial experiment** is an experiment that possesses the following four properties:

Properties of a Binomial Experiment

1. The experiment consists of n identical trials.
2. Each trial results in exactly one of two outcomes, called success or failure. A success is denoted by S and a failure is denoted by F.
3. The n trials are independent.
4. The probability of success p remains constant from trial to trial.

The probability distribution for the number of successes is called a **binomial distribution.**

EXAMPLE 6.1

Each of the following situations can be modeled using a binomial experiment.

▪ Administering a cold medication having a success rate of 0.90 to ten people with colds and observing the number of people getting relief

▪ Tossing a thumbtack eight times and observing the number of times it lands point up

▪ Guessing on a ten-question true-false test and observing the number of correct answers

▪ Tossing a die six times and observing the numbers of 3s that result

▪ Observing a baseball player with a 0.400 batting average and counting the number of base hits he gets in the next three times at bat

The following symbols are used to describe binomial experiments:

$$S = \text{success}$$
$$F = \text{failure}$$
$$p = P(\text{success})$$
$$1 - p = P(\text{failure})$$
$$n = \text{number of trials}$$
$$x = \text{number of successes}$$

Note that for a binomial experiment with n trials, x can be any one of the $(n + 1)$ values 0, 1, 2, 3, . . . , n. Consider applications 6.1 and 6.2. Although success is the term used to describe the outcome of interest for a binomial experiment, it does not necessarily correspond to a "good" event, as application 6.3 shows.

APPLICATION 6.1

A cross-fertilization of related species of white-flowered and blue-flowered plants produces offspring of which 20% are white-flowered plants. Six blue-flowered plants were paired and crossed with six white-flowered plants, and it was found that there were two white-flowered plants among the six offspring. Is this a binomial experiment? If so, identify a trial, a success, and the values for *p, n,* and *x*.

Solution: This problem is an application of Mendel's theory of inherited characteristics. A trial consists of crossing a blue-flowered plant with a white-flowered plant. A *success* is obtaining a white-flowered offspring and $p = P(S) = 0.20$. According to Mendelian theory, the trials are independent. As a result, the experiment is binomial with $n = 6$ and $x = 2$. ■

APPLICATION 6.2

Mary tossed a six-sided die ten times to determine the number of 1s resulting. She obtained three 1s. Is this a binomial experiment? If so, what constitutes a trial, a success, and a failure? What are the values of n, p, and x?

Solution: This is a binomial experiment in which a trial consists of tossing a die. A success is getting a 1, and a failure is getting an outcome other than 1. The probability of success is $p = 1/6$, whereas the probability of failure is 5/6. There are $n = 10$ trials, and x, the number of successes, is three. If instead of recording the number of 1s that resulted, Mary had recorded the number of times each value appears face up on the die, the experiment would not have been binomial, since each trial would have resulted in one of six outcomes. ■

APPLICATION 6.3

A certain type of medication will not cause a skin reaction in 90% of the people who use it. We are interested in the number of people out of the next five who use it and have a skin reaction. Identify a trial, a success, the values for p and n, and the possible values of x.

Solution: A trial is treating a person with the medication, and a *success* is having a skin reaction. Hence, $p = 1 - 0.90 = 0.10$, $n = 5$, and x can be any one of the six values 0, 1, 2, 3, 4, or 5. ■

The assumption of independence of trials for a binomial experiment implies that the probability of success p remains constant from trial to trial. However, the converse is not true. The probability of success can remain constant from trial to trial, but the trials need not be independent. In many situations the independence assumption is not satisfied, particularly in situations where sampling is done without replacement. Caution must be taken because the binomial is not always appropriate.

EXAMPLE 6.2

Suppose a town has five licensed restaurants, of which two currently have at least one serious health-code violation. There are two inspectors, each of whom will inspect one restaurant during the coming week. The names of the restaurants are written on different slips of paper and thoroughly mixed. Each inspector randomly draws one of the slips, without replacement of the first before the second is drawn. A trial is inspecting a restaurant, and the number of trials is 2. A success S is interpreted as observing no health-code violations. Let S_1 denote a success on the first trial, S_2 denote a success on the second trial, and F_1 denote a failure on the first trial. Then the probability of getting a success on the first trial is

$$P(S_1) = \frac{3}{5} = 0.6$$

There are two possibilities for getting a success on the second trial:

$$S_2 = (S_1 \cap S_2) \text{ or } (F_1 \cap S_2)$$

By using the sum rule, we have

$$P(S_2) = P(S_1 \cap S_2) + P(F_1 \cap S_2)$$

And by using the product rule, we obtain

$$P(S_2) = P(S_1 \cap S_2) + P(F_1 \cap S_2)$$

$$= \frac{3}{5} \cdot \frac{2}{4} + \frac{2}{5} \cdot \frac{3}{4}$$

$$= \frac{3}{10} + \frac{3}{10} = 0.6$$

However, the trials are not independent, since

$$P(S_2|S_1) = \frac{P(S_1 \cap S_2)}{P(S_1)}$$

$$= \frac{3/10}{6/10} = 0.5$$

and

$$P(S_2|S_1) \neq P(S_2)$$

Binomial Coefficients

Suppose it has been determined over a long period of time that 80% of the people making plane reservations follow through with their flight plans; 20% of the potential customers do not show up at departure time. If four people make plane reservations and we are interested in the number of people who show up at flight time, we have an example of a binomial experiment with $n = 4$ and $p = 0.80$. The number of people who show up at flight time is a random variable with values $x = 0, 1, 2, 3$, and 4. Let's label the outcome "shows up" as a success (S) and the outcome "no show" as a failure (F). If x denotes the number of successes, the following 16 possibilities can result:

$x = 0$	$x = 1$	$x = 2$	$x = 3$	$x = 4$
FFFF	SFFF	SSFF	SSSF	SSSS
	FSFF	SFSF	SSFS	
	FFSF	SFFS	SFSS	
	FFFS	FSSF	FSSS	
		FSFS		
		FFSS		

By summarizing we see that the number of possibilities that can result for each value of x is as follows:

Value of x (number of Ss)	0	1	2	3	4
Number of outcomes	1	4	6	4	1

For a given value of x, the number of possible outcomes containing x successes is given by the **binomial coefficient** $\binom{n}{x}$. Recall from section 5.3 that $\binom{n}{x}$ represents the number of combinations of n outcomes containing x successes. For our illustration we have

x	0	1	2	3	4
$\binom{4}{x}$	1	4	6	4	1

Note that this agrees with our summary above.

In section 5.3, the formula for the binomial coefficient $\binom{n}{x}$ was given by

$$\binom{n}{x} = \frac{n!}{x!(n-x)!} \tag{6.1}$$

where n is the number of trials, x is the number of successes, and $(n - x)$ is the number of failures. Note that the number of successes plus the number of failures equals the number of trials.

APPLICATION 6.4

Trees in a particular forest are infested with a certain parasite. If 15 trees are selected at random for study, how many outcomes can result in

    a. 3 infested trees?

    b. No infested trees?

    c. 15 infested trees?

    d. At most 2 infested trees?

Solution: By using Pascal's triangle or formula (6.1), we have

a. $\dbinom{15}{3} = \dfrac{(15)(14)(13)(12!)}{(3)(2)(1)(12!)} = 455.$

b. $\dbinom{15}{0} = 1.$

c. $\dbinom{15}{15} = 1.$

d. $\dbinom{15}{0} + \dbinom{15}{1} + \dbinom{15}{2} = 1 + 15 + 105 = 121.$ ∎

EXERCISE SET 6.1

Basic Skills

1. Interpret the following binomial coefficients in terms of a binomial experiment. Do not calculate the coefficients.

    a. $\dbinom{8}{4}$ b. $\dbinom{6}{0}$

2. Interpret the following binomial coefficients in terms of a binomial experiment. Do not calculate the coefficients.

    a. $\dbinom{10}{1}$ b. $\dbinom{7}{7}$

For exercises 3–8, determine if the experiment is binomial. If the experiment is binomial, indicate the values for n, p, and x, and identify a trial and a success.

3. Forty percent of all campers at a certain summer camp contract poison ivy. Eight students attend and we are interested in the number who contract poison ivy.

4. A survey of the residents in a certain town indicated that 30% of the residents favor building a community center and 70% are opposed. Ten residents are randomly surveyed and asked if they favor the proposed new community center.

5. A six-sided die is tossed five times and the sum of the five faces showing is to be determined.

6. Out of the next ten babies born at Memorial Hospital, the number of males is to be determined. (Assume that male and female births are equally likely.)

7. At a certain college, 40% of entering freshmen eventually graduate. Of 30 freshmen who enroll next semester, the number who eventually graduate is to be determined.

8. Four candidates are running for governor. A survey is conducted to determine voter support for the four candidates.

9. Evaluate each of the following:

 a. 5!   b. $\binom{18}{7}$ c. 0!

 d. $\binom{12}{5}$ e. $\binom{4}{0}$

10. Evaluate each of the following:

 a. $\binom{5}{5}$ b. $\binom{25}{23}$ c. $\binom{11}{4}$

 d. $\binom{4}{2}(0.2)^2(0.8)^2$ e. $\binom{12}{7}$

Going Beyond

11. Show that $\binom{n}{x} = \binom{n}{n-x}$. How does this fact relate to Pascal's triangle? Explain.

12. For a given positive integer n, find the sum of all the binomial coefficients.

13. Show that $\binom{n+1}{x} = \binom{n}{x} + \binom{n}{x-1}$. How does this fact relate to Pascal's triangle? Explain.

SECTION 6.2   *Calculating Binomial Probabilities*

In this section, we will be concerned with calculating the probabilities associated with the outcomes in a binomial experiment.

Binomial Probability Formula

Let's return to the plane-reservation illustration in section 6.1, a binomial experiment with $n = 4$ and $p = 0.80$. Recall the following 16 possibilities:

$x = 0$	$x = 1$	$x = 2$	$x = 3$	$x = 4$
FFFF	SFFF	SSFF	SSSF	SSSS
	FSFF	SFSF	SSFS	
	FFSF	SFFS	SFSS	
	FFFS	FSSF	FSSS	
		FSFS		
		FFSS		

These can be summarized as follows:

Value of x	0	1	2	3	4
Number of outcomes	1	4	6	4	1

Note that each outcome associated with a given value of x has the same probability, since the trials are independent. For example, if $x = 2$,

$$P(SSFF) = P(S)P(S)P(F)P(F) = (0.80)^2(0.20)^2$$

and

$$P(SFSF) = P(S)P(F)P(S)P(F) = (0.80)^2(0.20)^2$$

The other four possibilities having $x = 2$ successes also have probabilities equal to $(0.80)^2(0.20)^2$. The probabilities associated with the five possible values of x are summarized in table 6.1.

TABLE 6.1

Probability of Individual Outcomes for Binomial Experiment with $n = 4$ and $p = 0.80$

x	Probability of Each Outcome
0	$(0.80)^0(0.20)^4$
1	$(0.80)^1(0.20)^3$
2	$(0.80)^2(0.20)^2$
3	$(0.80)^3(0.20)^1$
4	$(0.80)^4(0.20)^0$

Four observations concerning the entries in the table are

1. The exponent of $p = P(S)$ is equal to the number of successes x.
2. The exponent of $(1 - p) = P(F)$ is equal to the number of failures $(n - x)$.
3. The two bases, p and $(1 - p)$, must sum to 1.
4. The sum of the exponents, x and $(n - x)$, must be n.

Since for each value of x there are $\binom{4}{x}$ possible outcomes containing x successes, we have the following probability table, shown in table 6.2, for the number of successes x.

TABLE 6.2

Probabilities for Binomial Experiment with $n = 4$ and $p = 0.80$

Number of Successes, x	Number of Outcomes Having x Successes	Probability of Each Outcome	Probability of x Successes, $P(x)$
0	1	$(0.80)^0(0.20)^4$	$(1)(0.80)^0(0.20)^4 = 0.002$
1	4	$(0.80)^1(0.20)^3$	$(4)(0.80)^1(0.20)^3 = 0.026$
2	6	$(0.80)^2(0.20)^2$	$(6)(0.80)^2(0.20)^2 = 0.154$
3	4	$(0.80)^3(0.20)^1$	$(4)(0.80)^3(0.20)^1 = 0.410$
4	1	$(0.80)^4(0.20)^0$	$(1)(0.80)^4(0.20)^0 = 0.410$

Recall from algebra that $p^0 = 1$ for all nonzero values of p. Observe from the table that the probabilities fit the following pattern:

$$\binom{n}{x}p^x(1 - p)^{n - x}$$

A general formula for calculating $P(x)$, the probability of obtaining x successes in a binomial experiment having n trials with $P(S) = p$, is referred to as the **binomial probability formula.**

Binomial Probability Formula

$$P(x) = \binom{n}{x}p^x(1 - p)^{n - x} \tag{6.2}$$

EXAMPLE 6.3

Let's find the probability of $x = 2$ successes when $n = 5$ and $p = 0.3$. By using formula (6.2) we have

$$P(x) = \binom{n}{x}p^x(1 - p)^{n - x}$$

$$P(2) = \binom{5}{2}(0.3)^2(1 - 0.3)^{5 - 2}$$

$$= \binom{5}{2}(0.3)^2(0.7)^3$$

$$= (10)(0.09)(0.343) = 0.3087 \approx 0.309$$

Note that we will round all binomial probabilities to three decimal places in this chapter.

We can also use MINITAB to determine binomial probabilities. Computer display 6.1 shows the use of the PDF (probability density function) command to obtain $P(2)$ for a binomial experiment for which $n = 5$ and $p = 0.3$.

Computer Display 6.1

```
MTB > PDF 2;
SUBC> BINOMIAL WITH N = 5, P = 0.3.
    K        P(X = K)
  2.00        0.3087
MTB >
```

The following procedure will prove to be useful for determining probabilities associated with a binomial experiment:

1. Determine what constitutes a trial and a success.
2. Determine the probability of success, p.
3. Determine the number of trials, n.
4. Determine the value(s) of x, the number of successes.
5. Use the binomial probability formula to find $P(x)$:

$$P(x) = \binom{n}{x} p^x (1 - p)^{n - x}$$

APPLICATION 6.5 A recent survey showed that 60% of college students smoke. What is the probability that of five students surveyed, three of them smoke?

Solution: By using the above five-step procedure, we have
Step 1 A trial consists of determining whether a college student smokes. A success is finding a student who smokes.

Step 2 $p = 0.6$
Step 3 $n = 5$
Step 4 $x = 3$
Step 5

$$P(x) = \binom{n}{x} p^x (1 - p)^{n - x}$$

$$P(3) = \binom{5}{3} (0.6)^3 (1 - 0.6)^{5 - 3}$$

$$= \binom{5}{3} (0.6)^3 (0.4)^2$$

$$= (10)(0.216)(0.16) = 0.3456 \approx 0.346 \quad ■$$

APPLICATION 6.6 If $P(12) = \dfrac{18!}{12!6!}(0.7)^{12}(0.3)^6$ represents a binomial probability, find

a. The number of successes.
b. $P(S)$.
c. The number of failures.
d. $P(F)$.

Solution:

    a. $P(12)$ indicates that $x = 12$.

    b. $P(S) = p = 0.7$, the base that has the exponent of 12.

    c. $n - x = 6$, the exponent of the base, 0.3.

    d. $P(F) = 1 - 0.7 = 0.3$, the other base. ▪

Binomial Probability Tables

Table 1 of appendix B gives values for $P(x)$ for selected values of p and values of n through 25. To use the table, locate the proper section for the value of n. Then under the column labeled by p and across the row labeled by x, find the value of $P(x)$. Consider application 6.7.

A P P L I C A T I O N 6.7

If a baseball player with a batting average of 0.600 comes to bat five times in a game, what is the probability he will get three hits?

Solution: The following values are identified: $n = 5$, $p = 0.6$, and $x = 3$. By using the binomial probability tables, we find

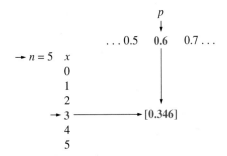

Thus, $P(3) = 0.346$.

By using the binomial probability formula, we get

$$P(3) = \binom{5}{3}(0.6)^3(0.4)^2$$
$$= (10)(0.6)^3(0.4)^2$$
$$= 0.3456 \approx 0.346$$

which agrees with the answer obtained from the binomial probability tables. ▪

Note that for given values of n and p, the sum of the probabilities for all values of x in the table may not equal 1.000 exactly. For example, for $n = 4$ and $p = 0.8$, the sum of the table values is $0.002 + 0.026 + 0.154 + 0.410 + 0.410 = 1.002$. This is because the entries in the table have been rounded to three decimal places. An entry listed in the table as 0.000 means that the entry is 0 rounded to three decimal places.

MINITAB can be used to generate binomial probability tables. Computer display 6.2 contains the MINITAB commands and output for the binomial probabilities listed in table 6.2.

Computer Display 6.2

```
MTB > PDF;
SUBC> BINOMIAL N = 4 P = 0.8.

   BINOMIAL WITH N = 4 P = 0.800000
     K        P(X = K)
     0          0.0016
     1          0.0256
     2          0.1536
     3          0.4096
     4          0.4096
MTB >
```

It is usually much easier to use a statistical software package (such as MINITAB) to calculate binomial probabilities than the binomial probability formula. The binomial table in appendix B was generated by the computer. We can use it to look up most of the binomial probabilities used in this book. The computer comes in handier than the binomial probability formula when the probability of success isn't one of the values listed in the table or the number of trials exceeds 25. For example, if we want to find $P(5)$ for a binomial experiment where $p = 0.178$ and $n = 10$, we couldn't use the binomial table in appendix B, since there is no column heading for $p = 0.178$. In this case we must either use a computer package or the binomial probability formula. Computer display 6.3 shows the results of using MINITAB to generate the table. The value of $P(5)$ is found to be 0.0169. I think you would agree that using MINITAB in this case is easier than using the binomial probability formula.

Computer Display 6.3

```
MTB > PDF;
SUBC> BINOMIAL N = 10, P = 0.178.

   BINOMIAL WITH N = 10 P = 0.178000
     K        P(X = K)
     0          0.1408
     1          0.3050
     2          0.2972
     3          0.1716
     4          0.0650
     5          0.0169
     6          0.0030
     7          0.0004
     8          0.0000
MTB >
```

EXERCISE SET 6.2

Further Applications

For each of the following problems, you should verify that the problem involves a binomial experiment and identify the values of *p, n,* and *x* before solving the problem. Your answers to the probability problems may differ from those given in the text, depending on whether you use the binomial probability formula or the binomial probability tables. The differences are due to rounding errors in the tables.

1. It was found that 40% of the campers at a certain summer camp contracted poison ivy. If eight students attend camp this summer, find the probability that
   a. all will contract poison ivy.
   b. two will contract poison ivy.
   c. at most three will contract poison ivy.
   d. at least seven will contract poison ivy.

2. A survey of the residents in a certain town showed that 30% favored brand X toothpaste. If ten of the residents are randomly surveyed at a grocery store, what is the probability that
   a. no one favors brand X?
   b. four favor brand X?
   c. at least eight favor brand X?
   d. at most two favor brand X?
   e. all favor brand X?

3. Dick has been observed to make 65% of his free-throw shots during basketball games. What is the probability that Dick will make
   a. three of the next six shots?
   b. five of the next ten shots?
   c. all of the next four shots?

4. In a certain city, 40% of the registered voters are Democrats. If nine voters are randomly selected, find the probability that
   a. two of them are Democrats.
   b. at least one of them is a Democrat.
   c. at least eight are Democrats.
   d. at most three are Democrats.

5. Assume that males and females are equally likely to be born (technically this is not correct, since approximately 100 female births occur for every 106 male births). Find the probability that of six babies born,
   a. there are three boys.
   b. there are at most three boys.
   c. there are four girls.
   d. none are boys.

6. At a certain college, 35% of the entering freshmen graduate. Of the next five freshmen who enroll, what is the probability that
   a. they all graduate?
   b. four graduate?
   c. three do not graduate?
   d. at least four graduate?
   e. none graduate?

7. A certain type of seed has a germination rate of 83%. If 12 seeds are planted, find the probability that
   a. they all germinate.
   b. 10 germinate.
   c. 11 germinate.
   d. at most 2 germinate.
   e. at least 10 germinate.

8. If five dice are tossed, what is the probability that
   a. three of them show a one?
   b. all of them show a one?
   c. at least three of them show a one?
   d. four of them don't show a one?

9. A test consists of ten multiple-choice questions with five possible answers. If a person guesses at each question, what is the probability that
   a. all questions are answered correctly?
   b. at most three questions are answered correctly?
   c. five questions are answered correctly?
   d. seven questions are answered incorrectly?

10. If 20 fair coins are tossed, find the probability that
    a. 11 heads are obtained.
    b. 15 heads are obtained.
    c. at least 16 heads are obtained.
    d. either no heads or 20 heads are obtained.

Going Beyond

11. The calculation of binomial probabilities can be obtained by recursion. Let $b(x; n, p)$ represent the probability of obtaining x successes with n trials where p is the probability of success on a single trial.
    a. Show that $b(x + 1; n, p) = p(n - x)/[(x + 1)(1 - p)], b(x; n, p)$. That is, if we know the probability of obtaining x successes $b(x; n, p)$, we can obtain the probability of obtaining $x + 1$ successes, $b(x + 1; n, p)$, by multiplying $b(x; n, p)$ by $p(n - x)/[(x + 1)(1 - p)]$.
    b. Use the recursion formula in part a to calculate the binomial probabilities for $n = 5$ and $p = 0.3$.

12. Suppose we permit the number of trials to vary in a binomial experiment; that is, suppose we repeat a trial until a success occurs. The random variable X, instead of the number of successes, becomes the number of repetitions required up to and including the first success. The probability formula for X is

$$P(X = k) = (1 - p)^{k-1}p$$

where $k = 1, 2, 3, \ldots$. The random variable X is called a *geometric random variable*. Show that the sum of the probabilities equals one.

13. When a fair coin is tossed, what is the probability that the first head will occur on the fourth toss?

14. An ordinary six-sided die is tossed until a 6 occurs for the first time. What is the probability that a 6 occurs on the
    a. third toss?
    b. fifth toss?

15. Solve the problems posed in motivator 6.

SECTION 6.3   *Finding Parameters for Binomial Distributions*

Recall that the probability distribution for the number of successes x resulting from a binomial experiment is referred to as a *binomial distribution*. In this section we want to find the mean and variance of a binomial distribution and examine the shape of its probability graph.

A binomial probability distribution can be displayed as a table or as a collection of $(n + 1)$ ordered pairs $(x, P(x))$ whose first component is the number of successes and whose second component is the probability associated with the given number of successes. This display can then be used to construct a graph for the distribution.

EXAMPLE 6.4

Consider the binomial experiment of tossing a fair coin five times. Table 6.3 is a probability distribution table for the number of heads obtained.

TABLE 6.3

Probability Distribution Table for Number of Heads Obtained in Five Tosses of a Fair Coin

x	$\binom{5}{x}$	$P(x)$
0	1	1/32
1	5	5/32
2	10	10/32
3	10	10/32
4	5	5/32
5	1	1/32
		1

We note that since $p = 0.5$, the probability of obtaining any particular arrangement of the five coins is $(1/2)(1/2)(1/2)(1/2)(1/2) = 1/32$. Thus, $P(x) = \binom{5}{x}\left(\dfrac{1}{32}\right)$ for $x = 0, 1, 2, 3, 4,$ and 5. This information can be used to construct the probability distribution graph for the number of successes (see figure 6.2). Note that the probability distribution is symmetrical about $x = 2.5$, which appears to be the "center" of the distribution. When $p = 0.5$, the binomial distribution will always be symmetrical.

FIGURE 6.2

Probability distribution graph
for number of heads in five
tosses of a fair coin

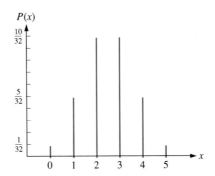

A binomial distribution accurately describes, to a reasonable degree, the repetitions of binomial experiments involving repeated trials, each with two outcomes. A binomial distribution describes all possible outcomes and corresponding probabilities for a binomial experiment. It is very similar to a relative frequency distribution that describes what has occurred. But it differs from a relative frequency distribution since it projects into the future and describes—using probability—what theoretically should occur.

If a binomial experiment is repeated a fixed number of times, the resulting distribution of relative frequencies for each of the $(n + 1)$ possible values for the number of successes x is called an **empirical binomial distribution.** As the number of repetitions for the binomial experiment increases, the empirical distributions approach the theoretical binomial distribution. The two distributions are illustrated in figure 6.3 for the binomial experiment consisting of tossing a fair coin six times and recording the number of heads obtained. For the empirical distribution, the experiment was repeated 100 times. Note that the graph of the empirical distribution resembles the graph of the theoretical distribution.

FIGURE 6.3

Comparison of empirical and
theoretical binomial
distributions

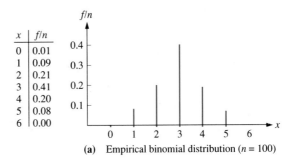

x	f/n
0	0.01
1	0.09
2	0.21
3	0.41
4	0.20
5	0.08
6	0.00

(a) Empirical binomial distribution ($n = 100$)

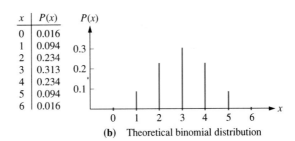

x	$P(x)$
0	0.016
1	0.094
2	0.234
3	0.313
4	0.234
5	0.094
6	0.016

(b) Theoretical binomial distribution

Many people confuse binomial experiments with binomial distributions; there is a difference. A binomial experiment involving n trials results in exactly one of the $(n + 1)$ possible outcomes for the associated binomial random variable. A binomial distribution, on the other hand, describes the probabilities associated with the $(n + 1)$ values (outcomes) for the random variable X denoting the number of successes that can result.

We now want to find the average of a binomial distribution; that is, we want to find the mean for a binomial random variable. For a binomial distribution, the mean μ is the expected value $E(x)$ for the number of successes x. Recall from section 5.7 that

$$E(X) = \Sigma \, [xP(x)] \tag{6.3}$$

EXAMPLE 6.5

For the experiment of tossing a coin five times, the mean of the random variable X denoting the number of heads is

$$\mu = \Sigma \, [xP(x)]$$
$$= 0P(0) + 1P(1) + 2P(2) + 3P(3) + 4P(4) + 5P(5)$$
$$= 0 + 1\left(\frac{5}{32}\right) + 2\left(\frac{10}{32}\right) + 3\left(\frac{10}{32}\right) + 4\left(\frac{5}{32}\right) + 5\left(\frac{1}{32}\right)$$
$$= \frac{80}{32} = 2.5$$

How should we interpret this answer of 2.5 heads? Surely, it does not mean that if we toss a fair coin five times, we will obtain 2.5 heads. Instead, if a large number of people each tossed a coin five times and we recorded the number of heads obtained for each person, the average of the number of heads obtained would be close to 2.5.

Formula (6.3) for calculating the **mean μ of** a **binomial distribution** could require considerable computation. Consider that it is intuitively reasonable that a machine that produces defective parts 1% of the time when batches of 500 are produced would, over a long period of time, produce an average number of defective parts per batch equal to $(0.01)(500) = 50$. Similarly, it is reasonable that in a large number of repeated experiments of tossing a fair coin 300 times, we can expect to obtain $(0.5)(300) = 150$ heads. In both of these situations, the expected value (or average) was obtained by multiplying the number of trials n by the probability of success p. As a consequence, to find the mean for a binomial distribution, we have the following formula, which is mathematically equivalent to formula (6.3) and is simpler to use.

Mean of a Binomial Distribution

$$\mu = np \tag{6.4}$$

EXAMPLE 6.6

For example 6.5, the mean is

$$\mu = np$$
$$= (5)(0.5) = 2.5$$

Note that this answer agrees with the answer found by using formula (6.3).

APPLICATION 6.8

The probability that a patient recovers from lung surgery is 0.95. If 25 people have this surgery, find the mean number of recoveries and interpret the result.

Solution: By using formula (6.4), we have

$$\mu = np$$
$$= (25)(0.95) = 23.75$$

We interpret this as follows: if a large number of hospitals each performed lung surgery on 25 patients and recorded the number who recovered, the average number of recoveries for all the hospitals involved would be close to 23.75. ▪

Variance of a Binomial Distribution

Recall from section 5.7 that the variance of a random variable X is given by

$$\sigma_X^2 = \Sigma\,[(x - \mu)^2 P(x)] \qquad (6.5)$$

Application 6.9 shows how we find the variance for a binomial distribution.

APPLICATION 6.9

Dave and Rick are gambling by rolling a die. If the die shows a 2, 3, or 4 Dave wins; otherwise, Rick wins. If the die is tossed three times, find the mean and variance for the distribution of number of times Dave wins.

Solution: The experiment is binomial with $n = 3$ and $p = 1/2$. By using formula (6.4), we have

$$\mu = np$$
$$= (3)\left(\frac{1}{2}\right) = 1.5$$

To find the variance, we need the following probability distribution table for X, the number of times Dave wins:

x	$P(x)$
0	1/8
1	3/8
2	3/8
3	1/8

We use the following table to facilitate the computations involved in calculating σ^2:

x	$P(x)$	$x - \mu$	$(x - \mu)^2$	$(x - \mu)^2 P(x)$
0	1/8	−1.5	2.25	0.28125
1	3/8	−0.5	0.25	0.09375
2	3/8	0.5	0.25	0.09375
3	1/8	1.5	2.25	0.28125
				0.75

Hence, $\sigma_X^2 = 0.75$. ■

For a binomial distribution, formula (6.5) is mathematically equivalent to the following formula, which is simpler to use:

> **Variance of a Binomial Distribution**
>
> $$\sigma^2 = np(1 - p) \tag{6.6}$$

This is the preferred formula for computing the **variance σ^2 of a binomial distribution,** because it involves fewer computations.

EXAMPLE 6.7

By using formula (6.6) to calculate the variance σ^2 for the binomial data in application 6.9, we get

$$\sigma^2 = np(1 - p)$$
$$= (3)(0.5)(0.5) = 0.75$$

which agrees with the result obtained by using formula (6.5).

The **standard deviation for a binomial distribution** is the square root of its variance.

> **Standard Deviation of a Binomial Random Variable**
>
> $$\sigma = \sqrt{np(1 - p)} \tag{6.7}$$

Let's consider its use in applications 6.10 and 6.11.

APPLICATION 6.10

Refer to application 6.9. Suppose Dave wins if the die shows a 2 or 4. Find the mean and standard deviation for the distribution of number of times Dave wins.

Solution: For this binomial experiment, $p = 1/3$. By using formula (6.4), we get

$$\mu = np$$
$$= (3)\left(\frac{1}{3}\right) = 1$$

So Dave expects to win once out of three trials. To obtain the variance, we apply formula (6.6) to obtain

$$\sigma^2 = np(1 - p)$$

$$= (3)\left(\frac{1}{3}\right)\left(\frac{2}{3}\right) = \frac{2}{3}$$

Hence, by using formula (6.7) we find the standard deviation σ to be

$$\sigma = \sqrt{\frac{2}{3}} \approx 0.82 \quad ▪$$

APPLICATION 6.11

A student takes a multiple-choice test with 50 questions, each with five possible choices. If the student guesses at each question, find the mean and standard deviation of the distribution of the number of questions answered correctly. Also, find the mean and standard deviation for the distribution of the number of questions missed by the student.

Solution: For this binomial experiment, $n = 50$ and $p = 0.20$. According to formula (6.4), the mean is

$$\mu = np$$
$$= (50)(0.2) = 10$$

The student could expect to get 10 correct out of the 50 simply by guessing. To determine the standard deviation, we use formula (6.7):

$$\sigma = \sqrt{np(1 - p)}$$
$$= \sqrt{(50)(0.2)(0.8)} \approx 2.83$$

The number of questions missed forms a binomial distribution with $n = 50$ and $p = 0.80$. Thus, the mean is

$$\mu = np$$
$$= (50)(0.80) = 40$$

and the standard deviation is

$$\sigma = \sqrt{(50)(0.80)(0.20)} \approx 2.83$$

Note that the standard deviations are equal and the sum of the two means equals n. ▪

Shapes of Binomial-Distribution Graphs

The values of n and p determine the shape of a binomial distribution. Figures 6.4 (a–g) indicate how the graphs of binomial distributions change for various values of p when $n = 6$. Seven different values for p are used. Notice that in figure 6.4(g) the graph is symmetrical about its mean $\mu = 3$. In general, if $p = 1/2$, then we have

$$\mu = np$$
$$= \frac{n}{2}$$

FIGURE 6.4

Binomial distribution graphs for $n = 6$ and different values of p

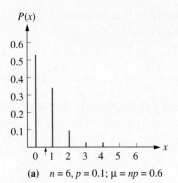

(a) $n = 6, p = 0.1; \mu = np = 0.6$

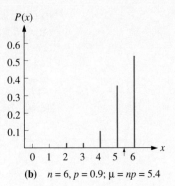

(b) $n = 6, p = 0.9; \mu = np = 5.4$

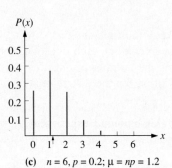

(c) $n = 6, p = 0.2; \mu = np = 1.2$

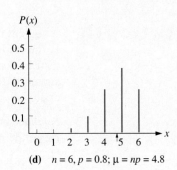

(d) $n = 6, p = 0.8; \mu = np = 4.8$

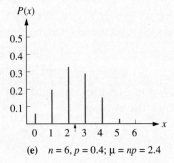

(e) $n = 6, p = 0.4; \mu = np = 2.4$

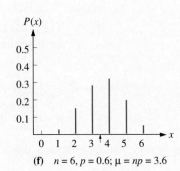

(f) $n = 6, p = 0.6; \mu = np = 3.6$

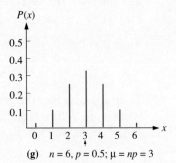

(g) $n = 6, p = 0.5; \mu = np = 3$

and the graph of the binomial distribution is symmetrical about its mean $\mu = n/2$. And if $p \neq 0.5$, notice that a binomial graph is not symmetrical about its mean. In figures 6.4(a), (c), (e), and (g) notice that as the value of p approaches the value 0.5 from the left side, the mean μ (denoted by the arrow) approaches the value $n/2 = 3$ from the left side. By examining the graphs in figures 6.4(b), (d), (f), and (g), we also see that as the values of p approaches 0.5 from the right, the values of μ also approach the value $n/2 = 3$ from the right side.

The graphs in figures 6.5(a–f) illustrate the effect on a binomial distribution of varying the values of n for the fixed value of $p = 0.1$. Probability graphs are given for $n = 2, 5, 10, 20, 25,$ and 30. Notice that as the sample size n increases, the mean (indicated by the arrow) increases and the distributions appear to become more symmetrical about their means.

FIGURE 6.5

Binomial distribution graphs for $p = 0.1$ and different values of n

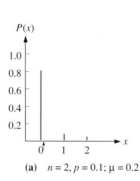

(a) $n = 2, p = 0.1; \mu = 0.2$

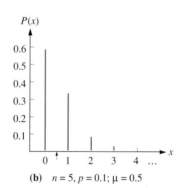

(b) $n = 5, p = 0.1; \mu = 0.5$

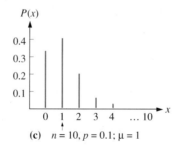

(c) $n = 10, p = 0.1; \mu = 1$

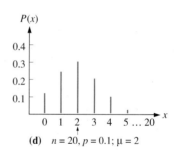

(d) $n = 20, p = 0.1; \mu = 2$

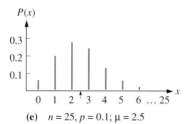

(e) $n = 25, p = 0.1; \mu = 2.5$

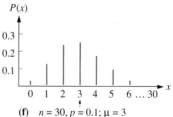

(f) $n = 30, p = 0.1; \mu = 3$

APPLICATION 6.12

Construct a graph for the binomial distribution having $n = 10$ and $p = 0.6$ and find μ.

Solution: The following probabilities were found using the binomial probability table (table 2 of appendix B).

x	0	1	2	3	4	5	6	7	8	9	10
$P(x)$	0.000	0.002	0.011	0.042	0.111	0.201	0.251	0.215	0.121	0.040	0.006

The corresponding graph is shown in figure 6.6. The mean μ is

$$\mu = np$$
$$= (10)(0.6)$$
$$= 6$$

FIGURE 6.6

Binomial distribution with $n = 10$ and $p = 0.6$

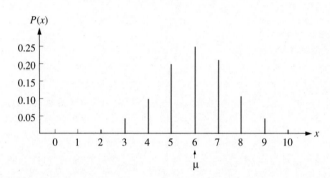

Although the graph is not symmetric about $\mu = 6$, the graph reveals that the distribution is not badly skewed. This is not surprising, since the value $p = 0.6$ is close to the value $p = 0.5$. ■

EXERCISE SET 6.3

Basic Skills

1. Assume that the conditions for a binomial experiment hold in the following situations. For each, find μ, σ^2, and σ.
   a. $n = 10, p = 0.6$
   b. $n = 20, p = 0.85$
   c. $n = 40, p = 0.3$
   d. $n = 5, p = 0.2$

2. Assume that the conditions for a binomial experiment hold in the following situations. For each, find μ, σ^2, and σ.
   a. $n = 15, p = 0.9$
   b. $n = 6, p = 0.45$
   c. $n = 500, p = 0.1$
   d. $n = 4, p = 0.60$

Further Applications

3. If 2% of all radios manufactured by the AJAX Company are defective, find the mean and standard deviation of the number of defective radios in groups of 50.

4. Find the mean and variance of the number of girls in families with six children. Assume that for every 106 males born, there are 100 females born.

5. What are the mean and standard deviation of the number of heads that will occur when 1000 fair coins are tossed?

6. The probability that Joe shoots a bulls-eye with his rifle is 68%. If he shoots in rounds of ten shots, find the mean and standard deviation of the number of perfect hits per round.

7. Over a long period of time it has been determined that 90% of all students enrolled in Math 209 pass the course. If groups of 30 students take the course, determine the mean and standard deviation of the number of students per course who pass.

8. Construct graphs similar to those of figure 6.5 and 6.6 for the following probability distributions. Assume that the binomial conditions hold. Locate μ on each graph.
   a. $n = 5, p = 0.2$
   b. $n = 6, p = 0.45$
   c. $n = 4, p = 0.60$

9. Construct a graph of the probability distribution for each of the following binomial distributions.
   a. $n = 20, p = 0.5$ d. $n = 20, p = 0.3$
   b. $n = 20, p = 0.4$ e. $n = 20, p = 0.8$
   c. $n = 20, p = 0.6$ f. $n = 20, p = 0.2$

Going Beyond

10. Suppose a binomial distribution results from n trials, where the probability of success is p. Then consider the binomial distribution with n trials in which the probability of success is $1 - p$.

a. How do the means of the two binomial distributions compare?
b. How do their standard deviations compare?

11. For a binomial distribution with a fixed value of n, are there values of p for which $\sigma^2 = 0$? Explain.

12. For a binomial distribution with a fixed value of n, for what value of p is σ^2 largest? Explain.

13. If $b(x; n, p)$ denotes the probability of x successes for a binomial experiment with n trials and $P(S) = p$, show that

$$b(x; n, 1 - p) = b(n - x; n, p)$$

SECTION 6.4   Multinomial Distributions

In sections 6.1–6.3 we examined binomial experiments and their corresponding probability distributions. These experiments were characterized by the following basic properties:

1. The experiment consists of n identical trials.
2. Each trial results in exactly one of two outcomes, called *success* and *failure*.
3. The n trials are independent.
4. The probability of success remains constant from trial to trial.

If we modify property 2 to allow for more than two outcomes to each trial, the experiment is called a **multinomial experiment.** Since a multinomial experiment contains more than two outcomes, we do not use the terms success and failure to describe the outcomes. The probability for each outcome remains constant from trial to trial. An experiment that has each trial resulting in exactly one of three outcomes is called a **trinomial experiment.**

Trinomial Experiments

EXAMPLE 6.8

Consider this example of a trinomial experiment. A health clinic specializes in using a particular method for treating arthritic patients. Past records show that one half of all patients benefit from the treatment, one third of them are not affected by the treatment, and one sixth of them have adverse side effects. An experiment is to be performed by treating two new patients suffering from arthritis. A trial is treating an arthritis patient. Since there are two patients, there are $n = 2$ trials. The trials are independent, since one patient's treatment should not have an influence on another patient's treatment. Each trial results in one of three outcomes:

E_1: the patient receives benefit

E_2: the patient receives no benefit

E_3: the patient has an adverse reaction

For treating a single patient, the resulting probabilities are:

$$p_1 = P(E_1) = \frac{1}{2}$$

$$p_2 = P(E_2) = \frac{1}{3}$$

$$p_3 = P(E_3) = \frac{1}{6}$$

Note that since $p_1 + p_2 + p_3 = 1/2 + 1/3 + 1/6 = 1$, the probability $P(E_3)$ can be written as $p_3 = (1 - p_1 - p_2)$. These three probabilities remain constant from trial to trial.

For $n = 2$ trials, the experiment can result in one of nine outcomes. These are listed in table 6.4.

TABLE 6.4

Outcomes from the Trinomial Experiment with $n = 2$

$(E_1\ E_1)$	(E_2, E_1)	(E_3, E_1)
(E_1, E_2)	(E_2, E_2)	(E_3, E_2)
(E_1, E_3)	(E_2, E_3)	(E_3, E_3)

For example, the outcome (E_1, E_3) indicates that the first patient receives benefit from the treatment and the second patient has an adverse reaction to the treatment. Probabilities can be computed for each of the nine outcomes.

To derive the general formula for the probability distribution of a trinomial experiment, we proceed as we did for the binomial distribution discussed in section 6.2. Recall the binomial probability formula:

$$P(x) = \binom{n}{x} p^x (1 - p)^{n-x}$$

The right-hand side of the binomial probability formula consists of two factors: the binomial coefficient $\binom{n}{x}$ and $p^x(1 - p)^{n-x}$. The binomial coefficient counts the total number of ways that one can get x successes and $(n - x)$ failures in n trials. The second factor, $p^x(1 - p)^{n-x}$, represents the probability of getting any one of the $\binom{n}{x}$ outcomes.

A probability for a trinomial experiment will also consist of two factors. The first factor is the trinomial coefficient. If each of n trials can result in three possible outcomes, E_1, E_2, and E_3, then the **trinomial coefficient** $\binom{n}{x_1\ x_2\ x_3}$ provides the number of ways in n trials that one can get x_1 outcomes of type E_1, x_2 outcomes of type E_2, and x_3 outcomes of type E_3. The following formula, called the **trinomial coefficient formula,** provides the value of the trinomial coefficient:

Trinomial Coefficient Formula

$$\binom{n}{x_1\ x_2\ x_3} = \frac{n!}{x_1! x_2! x_3!} \tag{6.8}$$

EXAMPLE 6.9

Refer to example 6.8. The number of ways that two patients can be divided into three outcome categories in such a way that the first category has $x_1 = 0$ patients, the second category has $x_2 = 1$ patient, and the third category has $x_3 = 1$ patient can be represented as the trinomial coefficient, $\begin{pmatrix} 2 \\ 0 \ 1 \ 1 \end{pmatrix}$. Its value can be determined by using formula (6.8):

$$\begin{pmatrix} 2 \\ 0 \ 1 \ 1 \end{pmatrix} = \frac{2!}{0! \ 1! \ 1!} = 2$$

This result can be verified by examining table 6.4. The two outcomes are (E_2, E_3) and (E_3, E_2).

APPLICATION 6.13

Evaluate the following trinomial coefficients:

a. $\begin{pmatrix} 4 \\ 1 \ 2 \ 1 \end{pmatrix}$

b. $\begin{pmatrix} 6 \\ 1 \ 2 \ 3 \end{pmatrix}$

Solution: We shall use the trinomial coefficient formula (6.8).

a. $\begin{pmatrix} 4 \\ 1 \ 2 \ 1 \end{pmatrix} = \frac{4!}{1! 2! 1!} = \frac{24}{2} = 12$

b. $\begin{pmatrix} 6 \\ 1 \ 2 \ 3 \end{pmatrix} = \frac{6!}{1! 2! 3!} = \frac{(6)(5)(4)(3)(2)(1)}{(2)(6)} = 60$ ▪

Note that, in general, the trinomial coefficient provides the number of ways that n objects can be arranged into three categories for which there are x_1 objects in the first category, x_2 objects in the second category, and x_3 objects in the third category $(x_1 + x_2 + x_3 = n)$.

APPLICATION 6.14

How many ways can five different balls be put into three boxes so that the first box has two balls, the second box has one ball, and the third box has two balls?

Solution: The trinomial coefficient $\begin{pmatrix} 5 \\ 2 \ 1 \ 2 \end{pmatrix}$ provides the answer. By using formula (6.8), we have

$$\begin{pmatrix} 5 \\ 2 \ 1 \ 2 \end{pmatrix} = \frac{5!}{2! \ 1! \ 2!} = 30 \quad ▪$$

For a trinomial experiment, since the n trials are independent, any specified order yielding x_1 outcomes of type E_1, x_2 outcomes of type E_2, and x_3 outcomes of type E_3 will occur with probability $p_1^{x_1} \cdot p_2^{x_2} \cdot p_3^{x_3}$. The total number of orderings of the outcomes yielding this probability is given by the trinomial coefficient $\begin{pmatrix} n \\ x_1 \ x_2 \ x_3 \end{pmatrix}$. Hence the product of these two factors provides the trinomial probability. The following formula, called the **trinomial probability formula,** can be used to determine probabilities associated with a trinomial experiment.

Trinomial Probability Formula

$$P(x_1, x_2, x_3) = \binom{n}{x_1\ x_2\ x_3} p_1^{x_1} \cdot p_2^{x_2} \cdot p_3^{x_3} \qquad (6.9)$$

where $x_1 + x_2 + x_3 = n$ and $p_1 + p_2 + p_3 = 1$.

EXAMPLE 6.10

Refer to example 6.8. Suppose the random variable X_1 represents the number of patients receiving benefit, the random variable X_2 represents the number of patients that receive no benefit, and the random variable X_3 represents the number of patients having adverse reactions. Note that the number of patients having adverse reactions can be written as $(2 - X_1 - X_2)$. The three random variables can assume any of the three values, 0, 1, and 2 that satisfy the inequalities $0 \le 2 - X_1 - X_2 \le 2$. By using the trinomial probability formula, we can obtain the probability distribution for the experiment. The probabilities are listed in table 6.5. Notice that the sum of the probabilities equals 1.

TABLE 6.5

Trinomial Probability Distribution for $n = 2$, $p_1 = 1/2$, $p_2 = 1/3$, and $p_3 = 1/6$

(x_1, x_2, x_3)	$P(x_1, x_2, x_3)$
$(0, 0, 2)$	$\dfrac{2!}{0!\ 0!\ 2!}\left(\dfrac{1}{2}\right)^0\left(\dfrac{1}{3}\right)^0\left(\dfrac{1}{6}\right)^2 = \dfrac{1}{36}$
$(0, 1, 1)$	$\dfrac{2!}{0!\ 1!\ 1!}\left(\dfrac{1}{2}\right)^0\left(\dfrac{1}{3}\right)^1\left(\dfrac{1}{6}\right)^1 = \dfrac{1}{9}$
$(0, 2, 0)$	$\dfrac{2!}{0!\ 2!\ 0!}\left(\dfrac{1}{2}\right)^0\left(\dfrac{1}{3}\right)^2\left(\dfrac{1}{6}\right)^0 = \dfrac{1}{9}$
$(1, 0, 1)$	$\dfrac{2!}{1!\ 0!\ 1!}\left(\dfrac{1}{2}\right)^1\left(\dfrac{1}{3}\right)^0\left(\dfrac{1}{6}\right)^1 = \dfrac{1}{6}$
$(1, 1, 0)$	$\dfrac{2!}{1!\ 1!\ 0!}\left(\dfrac{1}{2}\right)^1\left(\dfrac{1}{3}\right)^1\left(\dfrac{1}{6}\right)^0 = \dfrac{1}{3}$
$(2, 0, 0)$	$\dfrac{2!}{2!\ 0!\ 0!}\left(\dfrac{1}{2}\right)^2\left(\dfrac{1}{3}\right)^0\left(\dfrac{1}{6}\right)^0 = \dfrac{1}{4}$
	$\overline{1.0}$

APPLICATION 6.15

Refer to example 6.8. If each of five new patients receive the treatment, what is the probability that exactly two of them will benefit from the treatment and one of them will have adverse side effects?

Solution: Note that if two benefit from the treatment and one has adverse side effects, then there must be two patients that receive no relief. We therefore want to determine the probability $P(2, 2, 1)$. Note that $x_1 = 2$, $x_2 = 2$, and $x_3 = 1$. By using formula (6.9), we have

$$P(x_1, x_2, x_3) = \binom{n}{x_1\ x_2\ x_3} p_1^{x_1} \cdot p_2^{x_2} \cdot p_3^{x_3}$$

$$P(2, 2, 1) = \binom{5}{2\ 2\ 1}\left(\dfrac{1}{2}\right)^2\left(\dfrac{1}{3}\right)^2\left(\dfrac{1}{6}\right)^1$$

$$= \dfrac{5!}{2!\ 2!\ 1!}\left(\dfrac{1}{4}\right)\left(\dfrac{1}{9}\right)\left(\dfrac{1}{6}\right)$$

$$= (30)\left(\dfrac{1}{216}\right) = \dfrac{5}{36} \approx 0.139 \quad ■$$

Multinomial Experiments

If a trial in a multinomial experiment has more than three outcomes, formulas (6.8) and (6.9) are easily generalized to formulas for the **multinomial coefficient** and the **multinomial probability formula.**

Multinomial Coefficient

$$\binom{n}{x_1 \; x_2 \; x_3 \; \cdots \; x_k} = \frac{n!}{x_1! x_2! x_3! \cdots x_k!} \tag{6.10}$$

where $x_1 + x_2 + x_3 + \cdots + x_k = n$.

Multinomial Probability Formula

$$P(x_1, x_2, \ldots, x_k) = \binom{n}{x_1 \; x_2 \; x_3 \; \cdots \; x_k} p_1^{x_1} \cdot p_2^{x_2} \cdot p_3^{x_3} \cdots p_k^{x_k} \tag{6.11}$$

where $x_1 + x_2 + \cdots + x_k = n$ and $p_1 + p_2 + \cdots + p_k = 1$.

APPLICATION 6.16

The members of an educational organization are classified according to one of four political classifications: registered Republican, registered Democrat, registered Independent, and unregistered. The percentages of registered Republicans, registered Democrats, registered Independents, and unregistered educators are: 30%, 40%, 10%, and 20%, respectively. What is the probability that in a random sample of ten educators, there are five registered Republicans, four registered Democrats, and one registered Independent?

Solution: This is a multinomial experiment with four outcome categories. For $n = 10$, let the random variable X_1 represent the number of registered Republicans, the random variable X_2 represent the number of registered Democrats, the random variable X_3 represent the number of registered Independents, and the random variable X_4 represent the number of unregistered educators. Also, let $p_1 = 0.3$, $p_2 = 0.4$, $p_3 = 0.1$, and $p_4 = 0.2$. We need to determine the probability $P(X_1 = 5, X_2 = 4, X_3 = 1, X_4 = 0)$.

$$P(5, 4, 1, 0) = \binom{10}{5 \; 4 \; 1 \; 0} (0.3)^5 (0.4)^4 (0.1)^1 (0.2)^0$$

$$= \frac{10!}{5! \; 4! \; 1! \; 0!} (0.00243)(0.0256)(0.1)(1) = 0.0078 \quad \blacksquare$$

EXERCISE SET 6.4

Basic Skills

1. Evaluate the following multinomial coefficients:

   a. $\binom{7}{3 \; 2 \; 0 \; 2}$

   b. $\binom{8}{1 \; 2 \; 3 \; 1 \; 1}$

2. Evaluate the following multinomial coefficients:

   a. $\binom{9}{1 \; 0 \; 0 \; 8}$

   b. $\binom{10}{0 \; 0 \; 5 \; 3 \; 2}$

3. Is $\begin{pmatrix} 9 \\ 1\ 2\ 0\ 7 \end{pmatrix}$ a valid multinomial coefficient? Why?

4. Is $\begin{pmatrix} 10 \\ 1\ 2\ 3\ 2 \end{pmatrix}$ a valid multinomial coefficient? Why?

5. Suppose X_1 and X_2 are trinomial random variables with $p_1 = 0.3$ and $p_2 = 0.2$. If $n = 6$ find the following probabilities:
   a. $P(X_1 = 3 \text{ and } X_2 = 2)$
   b. $P(X_1 = 1 \text{ and } X_2 = 1)$
   c. $P(X_1 = 0 \text{ and } X_2 = 0)$
   d. $P(X_1 = 2 \text{ and } X_2 = 4)$

6. Suppose X_1 and X_2 are trinomial random variables with $p_1 = 0.4$ and $p_2 = 0.3$. If $n = 7$ find the following probabilities:
   a. $P(X_1 = 2 \text{ and } X_2 = 2)$
   b. $P(X_1 = 3 \text{ and } X_2 = 1)$
   c. $P(X_1 = 0 \text{ and } X_2 = 0)$
   d. $P(X_1 = 3 \text{ and } X_2 = 4)$

Further Applications

7. A fair six-sided die is tossed five times. Find the probability of getting
   a. three 1s, one 2, and one 6.
   b. three 4s and two 3s.
   c. all 1s.

8. A fair six-sided die is tossed seven times. Find the probability of getting
   a. two 1s, two 2s, one 3, and two 4s.
   b. one 4, one 2, one 3, one 4, one 5, and two 6s.
   c. two 1s, one 2, one 3, one 4, and two 6s.

9. A box contains a large number of marbles. Fifty percent of these marbles are white, 30% are red, and 20% are black. Suppose that ten marbles are chosen at random from the box. What is the probability that
   a. there are five red marbles, two white marbles, and three black marbles?
   b. there are three white and four black marbles?
   c. there are five white marbles and five red marbles?

10. Repair calls for refrigerators fall in the following four categories: freon leakage, compressor failure, blown fuse, and miscellaneous. From past records, the probabilities associated with these four categories are 0.4, 0.2, 0.3, and 0.1, respectively. In the next six service calls, what is the probability that
    a. half will involve freon leakage and half will involve blown fuses?

    b. two will involve compressor failure, one will involve a blown fuse, and two will involve miscellaneous problems?
    c. four will involve compressor failure and two will involve blown fuses?

11. The members of an educational organization are classified according to one of three political classifications: registered Republican, registered Democrat, and unregistered. The percentages of registered Republicans, registered Democrats, and unregistered educators are 50%, 40%, and 10%, respectively. What is the probability that in a random sample of five educators, there are
    a. five registered Republicans?
    b. three registered Republicans and one registered Democrat?
    c. three registered Republicans and two registered Democrats?

12. A health clinic specializes in using a particular method for treating arthritic patients. Past records show that one half of all patients benefit from the treatment, one third of them are not affected by the treatment, and one sixth of them have adverse side effects. Among the next seven patients, what is the probability that
    a. three will benefit from treatment and four will not be affected?
    b. all will have relief?
    c. four will have relief and one will not be affected?

13. Suppose a red die and a green die are rolled 18 times.
    a. Find the probability that the red die is higher seven times, and they are tied four times.
    b. Find the probability that they are tied exactly four times.

14. A committee of size 12 is to be chosen from a large university (so that we may assume sampling with replacement). Suppose the university has 30% black students, 40% white students, and 30% Hispanic students.
    a. Find the probability that the committee has three Hispanics and four whites.
    b. What is the probability that there are an equal number of Hispanics, whites, and Blacks?

15. If a pair of fair dice is tossed eight times, what is the probability of obtaining a total of 7 three times, a matching pair once, and any other combination four times.

16. A college plays ten football games during a season. In how many ways can the team end the season with four wins, five losses, and one tie?

17. How many ways can eleven people go on a trip using three cars that will hold 2, 4, and 5 passengers, respectively?

18. If four balanced six-sided dice are tossed, what is the probability that all faces are even?

19. If four balanced six-sided dice are tossed, what is the probability that two fall with even faces showing and two fall with odd faces showing?

Going Beyond

20. Show that the binomial coefficient $\binom{n}{x_1}$ can be written as $\binom{n}{x_1\ x_2}$, where $x_1 + x_2 = n$.

SECTION 6.5 *Hypergeometric Distributions*

In this section we want to consider experiments that obey three of the four properties of a binomial experiment; the independence property among the trials will be relaxed. That is, the individual trials will be considered dependent. The resulting experiment is called a **hypergeometric experiment.** Hypergeometric experiments are widely used when sampling is done without replacement. Consider the following example of a hypergeometric experiment.

EXAMPLE 6.11

Suppose an urn contains two red chips and three black chips. An experiment consists of drawing three chips at random, without replacement, from the urn. Of interest to the experimenter is the number of red chips selected.

If we let the random variable X denote the number of red chips selected, then the values of the random variable are $\{0, 1, 2\}$. Let's determine the probability distribution for the random variable X. The number of ways of selecting a sample of three chips from a collection of five chips is just the number of combinations of five objects taken three at a time, $\binom{5}{3} = 10$. Thus, a sample space consists of ten equally likely outcomes.

a. Let's determine the number of samples of three that contain no red chips and three black chips. There is only one way to draw no red chips from two red chips, $\binom{2}{0} = 1$, and there is only one way to draw three black chips from a sample of three black chips, $\binom{3}{3} = 1$.

Therefore, the probability of selecting a sample of size three with no red chips is

$$P(X = 0) = \frac{\binom{2}{0}\binom{3}{3}}{\binom{5}{3}} = \frac{1}{10}$$

b. The number of samples of size three that contain one red chip is found by multiplying the number of ways of drawing one red chip from two red chips and the number of ways of drawing two black chips from three black chips.

The number of ways of drawing one red chip from two red chips is $\binom{2}{1} = 2$, and the number of ways of drawing two black chips from three black chips is $\binom{3}{2} = 3$. Thus, there are $(2)(3) = 6$ samples that have exactly one red chip.

Thus, the probability of drawing a sample of size three with one red chip is

$$P(X = 1) = \frac{\binom{2}{1}\binom{3}{2}}{\binom{5}{3}} = \frac{6}{10} = \frac{3}{5}$$

c. The number of ways of drawing two red chips is equal to the number of ways of drawing two red chips from two red chips multiplied by the number of ways of drawing one black chip from three black chips.

The number of ways of drawing two red chips is $\binom{2}{2} = 1$. And the number of ways of drawing one black chip is $\binom{3}{1} = 3$. Therefore, the number of ways of drawing a sample of three containing exactly two red chips is $1 \cdot 3 = 3$.

The probability of drawing a sample of three chips, without replacement, containing two red chips is

$$P(X = 2) = \frac{\binom{2}{2}\binom{3}{1}}{\binom{5}{3}} = \frac{3}{10}$$

Note that $P(X = 0) + P(X = 1) + P(X = 2) = \frac{1}{10} + \frac{6}{10} + \frac{3}{10} = \frac{10}{10} = 1$.

Table 6.6 shows the probability distribution for the hypergeometric experiment. Notice the pattern with the numerical entries.

TABLE 6.6

Probability Distribution for the Hypergeometric Distribution Involving Two Red Chips and Three Black Chips

x	$P(X = x)$
0	$\dfrac{\binom{2}{0}\binom{3}{3}}{\binom{5}{3}} = \dfrac{1}{10}$
1	$\dfrac{\binom{2}{1}\binom{3}{2}}{\binom{5}{3}} = \dfrac{6}{10}$
2	$\dfrac{\binom{2}{2}\binom{3}{1}}{\binom{5}{3}} = \dfrac{3}{10}$

To generalize the method we used in example 6.11, suppose that we are to choose n objects from a collection of n_1 objects of one kind (called *successes*) and n_2 objects of another kind (called *failures*). The selection is without replacement and we are interested in the number of successes chosen. The total number of ways of choosing n objects from $(n_1 + n_2)$ objects is the binomial coefficient, $\binom{n_1 + n_2}{n}$. The number of ways of choosing x successes and $(n - x)$ failures is the product $\binom{n_1}{x}\binom{n_2}{n - x}$. Thus, the probability of drawing x successes in n trials is given by the called the **hypergeometric probability formula.**

Hypergeometric Probability Formula

$$P(x) = \frac{\binom{n_1}{x}\binom{n_2}{n-x}}{\binom{n_1+n_2}{n}} \quad \text{for } x = 0, 1, 2, 3, \ldots, \text{or } n \qquad (6.12)$$

where $n \leq n_1 + n_2$. Note that formula (6.12) follows the pattern established in table 6.6.

APPLICATION 6.17 Electric fans are shipped in lots of ten. Before a lot is accepted, an inspector chooses three of these fans and inspects them. If none of the tested fans are defective, the lot is accepted. If one or more fan is found to be defective, the entire shipment is inspected. Suppose there are, in fact, two defective fans in the lot. What is the probability that 100% inspection is required?

Solution: Let the random variable X be the number of defective fans in a lot. We use formula (6.12) to determine $P(X = 0)$. We note that $x = 0$, $n_1 = 2$, $n_2 = 8$, and $n = 10$.

 $n = 3$

$$P(x) = \frac{\binom{n_1}{x}\binom{n_2}{n-x}}{\binom{n_1+n_2}{n}}$$

$$P(0) = \frac{\binom{2}{0}\binom{8}{3}}{\binom{10}{3}}$$

$$= \frac{56}{120} = 0.467$$

One hundred percent inspection will be required if $x \geq 1$. Since $P(X \geq 1) = 1 - P(X = 0)$, the probability that 100% inspection is required is $1 - 0.467 = 0.533$. ■

EXERCISE SET 6.5

Basic Skills

1. Suppose an urn contains five red chips and seven black chips. An experiment consists of drawing three chips at random, without replacement, from the urn. Let the random variable X denote the number of red chips selected. Construct the probability distribution for X.

2. Suppose an urn contains eight red chips and seven black chips. An experiment consists of drawing five chips at random, without replacement, from the urn. Let the random variable X denote the number of red chips selected. Construct the probability distribution for X.

3. A committee of size four is to be selected at random from four mathematicians and six computer scientists. Find the probability distribution for the number of mathematicians on the committee.

4. A committee of size four is to be selected at random from eight mathematicians and seven computer scientists. Find the probability distribution for the number of mathematicians on the committee.

Applications

5. A carton of 24 calculators contains 4 that are defective. If 4 calculators are randomly selected from this carton, what is the probability that
   a. exactly 3 are defective?
   b. at most 1 is defective?
   c. all 4 are defective?

6. From a 12-member basketball team, 5 members are selected by lot to represent the team at a rules convention.
   a. What is the probability that the 5 selected include the two best players on the team?
   b. What is the probability that the 5 selected include the five starters?
   c. What is the probability that the 5 selected include none of the five starters?

7. A quality-control inspector examines a random sample of 5 batteries from each carton of 24 batteries coming off the assembly line. If, in fact, a carton of batteries contains 4 defective batteries, find the probability that a random sample of 5 batteries will contain
   a. none of the defective batteries.
   b. exactly 2 defective batteries.
   c. at least 1 defective battery.

8. An IRS auditor randomly selects 4 tax returns from 15 returns. If 5 of the 15 returns contain illegal deductions, what is the probability that the auditor will catch
   a. exactly 1 illegal return?
   b. 4 illegal returns?
   c. exactly 2 illegal returns?

9. A gardner plants five bulbs selected at random from a box containing six daffodil bulbs and five tulip bulbs. What is the probability that he planted
   a. three daffodil bulbs and two tulip bulbs?
   b. five tulip bulbs?

10. Suppose a storeroom has 20 tires of which 3 are defective. If 5 tires are chosen at random from the storeroom, what is the probability that
    a. at least 2 of the 5 tires are defective?
    b. 1 defective tire is selected?

11. A student board consists of two freshmen, five sophomores, eight juniors, and ten seniors. If four members of the board are selected at random to attend a national convention, what is the probability that each class is represented at the national convention?

Going Beyond

12. The *mean of a hypergeometric random variable* is $\mu = n \, n_1/(n_1 + n_2)$, where n is the sample size, n_1 represents total successes, and n_2 represents total failures. Verify this for the random variable given in exercise 1.

13. Refer to exercise 1. Use formula (5.18) of section 5.7 to determine the variance of X.

14. The *variance of a hypergeometric random variable* is given by

$$\sigma_X^2 = \frac{n \, n_1 \, n_2 (n_1 + n_2 - n)}{(n_1 + n_2)^2 (n_1 + n_2 - 1)}$$

Use this formula to determine the variance of the random variable given in exercise 1, and check the answer against the answer obtained in exercise 13.

15. The binomial distribution can be used to approximate the **hypergeometric distribution** if $20n \le n_1 + n_2$. A sample of five students is to be selected from a class of 1000 students. If 25% of the class are engineering students, what is the probability that two of the five students are engineering students if the sample was obtained
    a. without replacement?
    b. with replacement?
    c. Compare your answers to parts a and b.

16. Suppose in a city of 10,000 voters, 6000 of the voters watch a particular television program. If a random sample of five voters is selected from the city, what is the probability that three of the five are watching the show?

SECTION 6.6   Poisson Distributions

The computation of binomial probabilities can be tedious, especially if the number of trials is large. When the number of trials is large ($n \ge 100$) and $\mu = np$ is small ($\mu < 10$), binomial probabilities can be approximated by a special form of the Poisson probability function. The **Poisson probability distribution** is defined by the **Poisson probability formula:**

Poisson Probability Formula

$$P(x) = \frac{\lambda^x e^{-\lambda}}{x!} \tag{6.13}$$

---

where the parameter $\lambda > 0$, $e \approx 2.71828$, and $x = 0, 1, 2, 3, \ldots$. The mean of a binomial random variable is $\mu = np$. If we substitute np for λ, which is the **mean of a Poisson random variable,** in formula (6.13), then we get the following formula, which can be used to approximate binomial probabilities.

$$P(x) = \frac{(np)^x e^{-np}}{x!} \tag{6.14}$$

APPLICATION 6.18

If the probability is 0.0001 that any one commercial airplane will be highjacked while in flight, what is the probability that 10 of the next 30,000 airplanes will be highjacked?

Solution: If we consider an airflight as a trial and a highjacking as a success, then we want to determine the binomial probability $P(10)$, where $n = 30,000$ and $p = 0.0001$. By using the binomial probability formula (6.2), we have

$$P(x) = \binom{n}{x} p^x (1-p)^{n-x}$$

$$P(10) = \binom{30,000}{10}(0.0001)^{10}(0.9999)^{29,990}$$

This binomial probability is extremely difficult to compute without the use of a computer. By using MINITAB, we determine that $P(10) = 0.0008$ (see computer display 6.4 for a list of the commands used and the corresponding output).

Computer Display 6.4

```
MTB > PDF 10;
SUBC> BINOMIAL N=30000, P=0.0001.
     K        P(X = K)
  10.00        0.0008
```

Since $n = 30,000$ and $\mu = np = (30,000)(0.0001) = 3$, we can use the Poisson probability distribution with $\lambda = 3$ to approximate $P(10)$. With the aid of a calculator, we find that

$$P(x) = \frac{(np)^x e^{-np}}{x!}$$

$$P(10) = \frac{(3)^{10} e^{-3}}{10!} = 0.00081$$

which, to four decimal places, agrees with the result found above using MINITAB.

To evaluate $P(10) = (3)^{10} e^{-3}/10!$ directly, we can use table 2 in appendix B or MINITAB. If we use table 2 in appendix B, we locate the column labeled by $\lambda = 3$

across the top of the table and the row labeled $X = 10$ along the left-hand side. The probability is where the row and column intersect. This value is 0.0008. If we use MINITAB, computer display 6.5 contains the MINITAB commands used to obtain $P(10)$ and the corresponding output.

Computer Display 6.5

```
MTB > PDF 10;
SUBC> POISSON 3.
        K       P(X = K)
    10.00         0.0008
    MTB >
```

The following MINITAB printouts show how closely the binomial probability distribution with $n = 100$ and $p = 0.05$ is approximated by the Poisson probability distribution with $\lambda = np = 5$.

```
MTB > PDF;
SUBC> BINOMIAL 100 0.05.

BINOMIAL WITH N = 100 P = 0.050000
   K        P(X = K)
   0         0.0059
   1         0.0312
   2         0.0812
   3         0.1396
   4         0.1781
   5         0.1800
   6         0.1500
   7         0.1060
   8         0.0649
   9         0.0349
  10         0.0167
  11         0.0072
  12         0.0028
  13         0.0010
  14         0.0003
  15         0.0001
  16         0.0000
```

```
MTB > PDF;
SUBC> POISSON 5.

POISSON WITH MEAN = 5.000
   K        P(X = K)
   0         0.0067
   1         0.0337
   2         0.0842
   3         0.1404
   4         0.1755
   5         0.1755
   6         0.1462
   7         0.1044
   8         0.0653
   9         0.0363
  10         0.0181
  11         0.0082
  12         0.0034
  13         0.0013
  14         0.0005
  15         0.0002
  16         0.0000
```

The binomial probability formula is used for a binomial experiment to determine the probability of obtaining a specific number of successes in a fixed number of trials if we know the probability of a single success. The Poisson probability formula (6.13), on the other hand, is used to determine the probability of a number of successes that take place per unit of time or space, rather than for a specific number of trials. In addition to approximating binomial probabilities when $n \geq 100$ and $\mu = np < 10$, the Poisson probability formula can also be used to determine probabilities associated with experiments involving random phenomena, such as the

▪ Number of incoming telephone calls (0, 1, 2, and so on) to a switchboard during a specific period of time.

- Number of misprints on a typical page of print by a book publisher.
- Number of Big Macs sold by McDonald's during a lunch period.
- Number of patients arriving at the emergency room of a particular hospital during a given day.
- Number of fish caught in a large lake during a given day.
- Number of flaws per bolt of a certain type of cloth.
- Number of accidents per hour on a given stretch of highway.
- Number of defective components found in a newly manufactured automobile.
- Number of chocolate chips found in a chocolate chip cookie made by a certain bakery.
- Number of customers arriving at checkout during a given hour.
- Number of claims processed on a given day by an insurance company.

The number of chocolate chips observed in a cookie selected at random is an example of a space-related phenomena. Another example of a space-related phenomena is the number of errors on a randomly selected printed page.

An experiment that involves a space-related process satisfying the following three conditions is called a **Poisson experiment:**

1. The average number of occurrences (μ) is constant for each unit of time or space.
2. The probability of more than one occurrence in a given unit of time or space is very small.
3. The number of occurrences in disjoint intervals of time or space are independent of each other.

If the random variable X represents the number of occurrences resulting from a Poisson experiment, then the variable is called a **Poisson random variable.** The Poisson probability formula is used to compute probabilities associated with a Poisson random variable.

APPLICATION 6.19

Let the random variable X represent the number of chocolate chips in a randomly selected chocolate chip cookie baked by a particular bakery. Explain why it is reasonable to assume that X is a Poisson random variable.

Solution: It is reasonable to assume that a cookie can be subdivided into n regions of equal area so that each region contains at most one chocolate chip. A trial is inspecting a region to determine if it contains a chocolate chip. Observing a region with a chocolate chip is considered to be a success. If p is the probability that a region contains a chocolate chip, it is also reasonable to assume that p is constant from trial to trial and that the trials are independent. We may assume that the size of the cookie is large relative to the size of a chocolate chip, and we may also assume that there are few chocolate chips per cookie. That is, we may assume that n is large and p is small. Under the foregoing assumptions, we have a binomial experiment. Therefore, we may approximate the binomial probabilities by using the Poisson probability formula. Hence, X may be assumed to be a Poisson random variable. ■

It can be shown that a Poisson random variable has its mean equal to its variance. That is,

Properties of a Poisson Random Variable X with Parameter $\lambda > 0$
1. $\mu_X = \lambda$ 2. $\sigma_X^2 = \lambda$

A P P L I C A T I O N 6.20

A bank receives an average of three bad checks per day. What is the probability that on a given day it will receive either four or five bad checks?

Solution: We note that $\lambda = \mu = 3$, and use formula (6.11). The solution is $P(4) + P(5)$, where

$$P(x) = \frac{\lambda^x e^{-\lambda}}{x!}$$

With the aid of a calculator or table 2 in appendix B, we find that

$$P(4) = \frac{3^4 e^{-3}}{4!} = 0.168$$

and

$$P(5) = \frac{3^5 e^{-3}}{5!} = 0.101$$

Hence, $P(4) + P(5) = 0.168 + 0.101 = 0.269.$ ▪

A P P L I C A T I O N 6.21

Calls at a switchboard follow a Poisson probability distribution and occur at an average of six calls per hour. What is the probability that there would be two or more calls in an hour?

Solution: Since the sum of the Poisson probabilities is 1, we have

$$P(X = 0) + P(X = 1) + P(X \geq 2) = 1$$

Therefore, the probability of two or more calls is equal to

$$P(X \geq 2) = 1 - [P(0) + P(1)]$$

We can use formula (6.11) and a calculator or table 2 in appendix B to calculate values of $P(0)$ and $P(1)$.

$$P(0) = \frac{6^0 e^{-6}}{0!} = 0.0025$$

$$P(1) = \frac{6^1 e^{-6}}{1!} = 0.0149$$

Hence, the probability of two or more calls per hour is

$$1 - (0.0025 + 0.0149) = 1 - 0.0174 = 0.9826 \quad ▪$$

EXERCISE SET 6.6

Basic Skills

1. For each case, compute Poisson probabilities associated with the values shown.
   a. $x = 4, \lambda = 3$
   b. $x = 7, \lambda = 0.6$
   c. $x = 10, \lambda = 10$
   d. $x = 5, \lambda = 1$

2. For each case, compute Poisson probabilities associated with the values shown.
   a. $x = 5, \lambda = 4$
   b. $x = 2, \lambda = 9$
   c. $x = 1, \lambda = 1$
   d. $x = 0, \lambda = 1.5$

3. Suppose X is a binomial random variable with $n = 500$ and $p = 0.01$. Determine a Poisson approximation for
   a. $P(X = 4)$.
   b. $P(X < 1)$.
   c. $P(X = 10)$.
   d. $P(X > 1)$.

4. Suppose X is a binomial random variable with $n = 1000$ and $p = 0.002$. Determine a Poisson approximation for
   a. $P(X = 1)$.
   b. $P(X < 2)$.
   c. $P(X = 3)$.
   d. $P(X > 1)$.

Further Applications

5. One hundred and fifty first-year law students take the bar examination in a certain state. The probability that a first-year law student will pass the exam is 0.05. Approximate the probability that
   a. eight law students pass the exam.
   b. at least three law students pass the exam.
   c. at most two students pass the exam.
   d. at least two students pass the exam.

6. Two thousand medical students take the Medical College Admissions Test (MCAT) exam. The probability that a medical student will get an average score of 16 on the MCAT is 0.001. Approximate the probability that
   a. two medical students average 16.
   b. at least two medical students average 16.
   c. at most two medical students average 16.
   d. no medical students average 16.

7. Police records show that there are an average of three accidents per week on Route 40. By assuming that the accidents follow a Poisson distribution, determine the prob-

ability that for a week selected at random there are
   a. four accidents.
   b. four or five accidents.
   c. at most three accidents.
   d. at least three accidents.

8. The number of errors per page for a local newspaper is distributed as a Poisson random variable with parameter $\lambda = 2.5$. Determine the probability that a selected at random page has
   a. five errors.
   b. two or four errors.
   c. at least two errors.
   d. at most two errors.

9. Suppose that the number of letters lost in the mail on a given day is a Poisson random variable with parameter $\lambda = 4$. What is the probability that on a given day
   a. exactly three letters will be lost in the mail?
   b. four or five letters will be lost in the mail?
   c. at least one letter will be lost in the mail?
   d. at most two letters will be lost in the mail?

10. The number of cars passing over a certain bridge is a random variable with parameter $\lambda = 10$. For a particular day, determine the probability that
    a. five cars pass over the bridge.
    b. no cars pass over the bridge.
    c. at least five cars pass over the bridge.
    d. at most two cars pass over the bridge.

Going Beyond

11. The number of telephone calls arriving at a local switchboard during any time interval of length t (measured in minutes) is a Poisson random variable with parameter $\lambda = 5$.
    a. Determine the probability of 12 calls arriving at the switchboard during a given two minute period.
    b. Determine the probability that calls will arrive at the rate of at least one per minute during a given period of two minutes.

12. If X is a Poisson random variable with parameter λ, and if $P(X = 0) = 0.2$, find $P(X > 2)$.

13. If X is a Poisson random variable with parameter λ, find the value of a such that $P(X = a)$ is largest.

14. If X is a Poisson random variable such that $3P(X = 2) = 2P(X = 1)$ find
    a. $P(X = 0)$.
    b. $P(X = 3)$.

CHAPTER SUMMARY

In this chapter we studied binomial experiments and their properties. We learned how to associate the outcomes of a binomial experiment with a discrete binomial random variable. We learned how to calculate probabilities associated with a binomial random variable and how to find the mean, variance, and standard deviation of the binomial probability distributions. We also investigated multinomial, hypergeometric, and Poisson experiments as variations of binomial experiments. With each, we used the probability function to compute probabilities.

CHAPTER REVIEW

■ IMPORTANT TERMS ■

The following chapter terms have been mixed in ordering to provide you better review practice. For each of the following terms, provide a definition in your own words. Then check your responses against those definitions given in the chapter.

binomial coefficient
binomial distribution
binomial experiment
binomial probability formula
mean of a binomial distribution
multinomial coefficient
multinomial experiment
multinomial probability formula
hypergeometric experiment

hypergeometric distribution
empirical binomial distribution
standard deviation of a binomial
  distribution
trial
Poisson probability formula
variance of a binomial distribution
Poisson random variable

mean of a Poisson random variable
hypergeometric probability formula
Poisson experiment
Poisson probability distribution
trinomial experiments
trinomial coefficient
trinomial probability formula
trinomial coefficient formula

■ IMPORTANT SYMBOLS ■

n, the number of trials

S, success

F, failure

p, the probability of success

$1 - p$, the probability of failure

x, the number of successes

$E(x)$, the mean number of successes

$\binom{n}{x}$, binomial coefficient

$\binom{n}{x_1 \, x_2 \, x_3}$, trinomial coefficient

$\binom{n}{x_1 \, x_2 \, x_3 \, \cdots \, x_k}$, multinomial coefficient

n_1, number of successes for a hypergeometric experiment

n_2, number of failures for a hypergeometric experiment

λ, mean of a Poisson random variable

■ IMPORTANT FACTS AND FORMULAS ■

Binomial coefficient: $\binom{n}{x} = \dfrac{n!}{x!(n-x)!}$. (6.1)

$\binom{n}{x} = \binom{n}{n-x}$.

Binomial probability formula: $P(x) = \binom{n}{x} p^x (1-p)^{n-x}$. (6.2)

Mean of a binomial distribution: $\mu = np$. (6.4)

Variance of a binomial distribution: $\sigma^2 = np(1-p)$. (6.6)

Standard deviation of a binomial distribution: $\sigma = \sqrt{np(1-p)}$. (6.7)

Trinomial coefficient formula: $\begin{pmatrix} n \\ x_1 \, x_2 \, x_3 \end{pmatrix} = \dfrac{n!}{x_1! \, x_2! \, x_3!}$.

(6.8)

Trinomial probability formula: $P(x_1, x_2, x_3)$

$= \begin{pmatrix} n \\ x_1 \, x_2 \, x_3 \end{pmatrix} p_1^{x_1} \, p_2^{x_2} \, p_3^{x_3}$, where $n = x_1 + x_2 + x_3$ and $p_1 + p_2 + p_3 = 1$. (6.9)

Multinomial coefficient: $\begin{pmatrix} n \\ x_1 \, x_2 \, x_3 \, \cdots \, x_k \end{pmatrix}$

$= \dfrac{n!}{x_1! \, x_2! \, x_3! \cdots x_k!}$, where $n = x_1 + x_2 + \cdots + x_k$.

(6.10)

Multinomial probability formula: $P(x_1, x_2, \ldots, x_k)$

$= \begin{pmatrix} n \\ x_1 \, x_2 \, x_3 \, \cdots \, x_k \end{pmatrix} p_1^{x_1} \, p_2^{x_2} \, p_3^{x_3} \cdots p_k^{x_k}$ where $n = x_1 + x_2 + \cdots + x_k$ and $p_1 + p_2 + \cdots + p_k = 1$. (6.11)

Hypergeometric probability formula:

$P(x) = \dfrac{\begin{pmatrix} n_1 \\ x \end{pmatrix} \begin{pmatrix} n_2 \\ n - x \end{pmatrix}}{\begin{pmatrix} n_1 + n_2 \\ n \end{pmatrix}}$, where $n \le n_1 + n_2$. (6.12)

Poisson probability formula: $P(x) = \dfrac{\lambda^x e^{-\lambda}}{x!}$, $\lambda > 0$. (6.13)

REVIEW EXERCISES

1. Thirty-five percent of all automobiles sold in the United States are foreign-made. Five new automobiles are randomly selected.
   a. Construct a binomial probability distribution table for the number of automobiles that are foreign made.
   b. Construct a vertical line graph for the binomial distribution.
   c. Find the mean number of automobiles that are foreign-made.
   d. Find the standard deviation of the number of automobiles that are foreign-made.
   e. Find the probability that at most four are foreign-made.

2. The probability is 0.5 that a new-born baby will be a boy. If eight babies are born at a local hospital, find
   a. the probability that three are boys.
   b. the probability that none are boys.
   c. the probability that two are girls.
   d. the expected number of boys.

3. If a baseball player with a batting average of 0.250 comes to bat ten times in a series, what is the probability that he will get
   a. three hits?
   b. at least one hit?
   c. at most two hits?

4. If it is assumed that a golfer will hit a drive into a sand trap 17% of the time, what is the probability that the player will hit the ball into a sand trap exactly four out of the first nine holes?

5. Refer to exercise 4. Find the expected number of times that the ball will be hit into a sand trap.

6. A particular type of birth-control pill is 90% effective. If 500 people used the pill, how many unplanned births would you expect?

7. It is estimated that 85% of all household plants are over-watered. For ten household plants, find the mean and standard deviation of the number of plants that are overwatered and interpret your results.

8. If there are five red balls and 15 black balls in a bag and three are selected one at a time with replacement, what is the probability that all three are red?

9. If an unprepared student takes a multiple-choice test, each question having five choices, and guesses at each of the 25 questions, what is the probability that she receives a score of 28%?

10. On the average, a store has had gross sales of over $5000 a day on seven days in ten over the past several months. If we assume this trend to continue, what is the probability that the store will have gross sales over $5000 at least four times in the next seven days?

11. If X is a Poisson random variable with $\lambda = 4$, find
   a. $P(X = 1)$.
   b. $P(X = 3)$.
   c. $P(X \le 3)$.
   d. $P(X \ge 4)$.

12. Telephone calls enter a switchboard on the average of one every three minutes. If the number of calls follow a Poisson process, what is the probability of two or more calls arriving in a three-minute period?

13. If a manufacturer of integrated circuits knows that 2% of its integrated circuits are defective, approximate the probability that a batch of 150 integrated circuits contains at most one defective circuit.

14. In 1927 Babe Ruth had the following statistics:

G	AB	H	2B	3B	HR	R	RBI	BB	SO	SB	BA
151	540	192	29	8	60	158	164	138	89	7	0.356

Suppose he had four official bats in a certain game. Estimate the probability of getting two home runs and two outs, not necessarily in that order.

15. The game of Scrabble involves 44 vowels and 54 consonants. If you draw seven letters at random, what is the probability that you have
   a. all vowels?
   b. 4 vowels?
   c. 3 consonants?

Computer Applications

1. Use a computer program to find the probability that among 200 babies born, the total number of girls is 95 or fewer. Assume that the probability of a girl being born is 0.485.

2. The accompanying data represent the results of performing a binomial experiment. If 1 represents a success and 0 represents a failure, use a computer program to
   a. estimate the probability of a success.
   b. determine the mean and standard deviation of the distribution.

```
1 0 0 1 0 0 0 0 0 1 0 1 1 1 0
0 0 0 1 0 0 0 0 0 0 0 1 0 1 0
1 0 0 0 0 1 1 1 0 1 0 0 1 1 0
0 0 0 0 0 0 0 0 1 0 0 1 0 0 0
1 0 1 1 0 1 1 1 0 1 0 0 1 1 0
1 1 1 1 1 1 1 1 0 1 0 0 1 0 1
1 1 1 0 0 0 1 0 0 0
```

3. a. Use a computer program to simulate 100 binomial experiments having parameters $n = 4$ and $p = 0.4$. Determine the mean and standard deviation of the 100 results and use a histogram to summarize the results.
   b. Sort the results for part a and construct a relative frequency table for the results.
   c. Determine a probability table for the binomial distribution having parameters $n = 4$ and $p = 0.4$. Compare this table with the table constructed in part b.

4. Simulate tossing a fair six-sided die 100 times.
   a. Find the mean and standard deviation of the 100 results.
   b. Sort the 100 results and construct a relative frequency table for the 100 results.

1. It is estimated that 5% of type A transistors manufactured by the AJAX Company are defective. If a lot of 20 transistors is examined, find the probability that
   a. none are defective.
   b. at most 1 is defective.
   c. 19 are defective.
   d. at least 5 are defective.

2. A multiple-choice test contains ten questions, each with five choices. If an unprepared student guesses at each question, find the probability that she/he gets
   a. five correct.
   b. at most two correct.
   c. at most nine correct.
   d. gets two or three correct.

3. For exercise 2, find and interpret your answers for
   a. the expected number of correct answers.
   b. the variance of the number of correct answers.

4. Each face of a four-sided pyramid is numbered 1 to 4. Suppose this pyramid is tossed five times and each side is equally likely to land face down.
   a. Construct a probability table for the number of 1s obtained.
   b. Construct a graph depicting the probabilities for the number of 1s obtained.
   c. Find μ and σ for the number of 1s obtained.

5. The number X of births at Memorial Hospital on any particular day follows a Poisson distribution with mean equal to two births per day. What is the probability that the number of births on a particular day is three?

6. Suppose a large bin consists of three types of washers: small, medium, and large. Twenty percent of the washers are small, 30% are medium, and 50% are large. If six washers are randomly drawn, what is the probability that two are small, two are medium, and two are large?

7. Telephone calls enter a switchboard on the average of two every three minutes. If the number of calls follow a Poisson process, what is the probability of two calls arriving in a three-minute period?

8. Suppose an urn contains three white balls and four black balls. If two balls are drawn from the urn without replacement what is the probability of getting one white ball and one black ball?

7

Continuous Distributions

CHAPTER OBJECTIVES

In this chapter we will investigate

▷ Uniform distributions.

▷ How to calculate probabilities associated with a uniform distribution.

▷ Normal distributions.

▷ Standard normal distributions.

▷ How to associate areas under the standard normal curve with probabilities.

▷ How to calculate probabilities associated with the standard normal distribution.

▷ How to calculate probabilities associated with any normal distribution.

▷ Practical applications for normal distributions.

▷ How to use the normal distributions to approximate binomial distributions.

▷ Exponential distributions.

▷ How to calculate probabilities associated with exponential distributions.

MOTIVATOR 7

*T*he normal distributions are of great importance in statistics. They were first discovered by Abraham de Moivre (1667–1754), a French mathematician living in London, in the course of solving problems for wealthy gamblers. Later on they were further studied by Pierre Laplace and Karl Gauss. Gauss (1777–1855), a German mathematician, was the first to explore the properties of the normal distributions, and they came to be known as the Gaussian law of error, a reference to his use of them to describe the distribution of errors of estimation in the method of least squares, a method he devised in 1795 at the age of 18. Besides their usefulness in solving gambling problems, they are found also to apply to errors of observation in astronomy, and in fact to all kinds of physical measurements.

The term "normal distribution" arose from the observation that the errors in repeated measurements of the same physical quantity tended to exhibit considerable regularity in their frequency distributions. It was commonplace to expect errors in repeated measurements of a quantity, such as height, weight, or intelligence, to possess a bell-shaped or normal distribution. The normal distributions are also called Gaussian distributions, especially in the natural science areas. In France, they are also referred to as Laplacian distributions.

Chapter Overview

An experiment involving measurements that can assume any value in a continuum can be identified with a continuous random variable. The values of the random variable consist of the real-number measurements falling in the continuum. The infinite set of measurements that comprise the possible outcomes of an experiment involving some form of measurement is called a **continuous distribution.** Recall from section 5.7 that probabilities associated with a continuous distribution are identified as areas under a curve. The total area under the curve above the *x*-axis must equal 1.

As an example of a continuous distribution, consider the time it takes an individual to react to a given stimulus and suppose the maximum range in reaction times is 2 seconds. If reaction times were measured to the nearest millisecond, there would be 2000 possible outcomes; there are only 2000 outcomes because of the limitations of the measuring device. Theoretically, the reaction times could be any of the infinite number of values from the shortest reaction time to the longest reaction time. The theoretical reaction times comprise a continuous distribution.

In this chapter we will study three very important types of continuous probability distributions: uniform, normal, and exponential.

SECTION 7.1   *Uniform Distributions*

In section 5.2 we constructed probability histograms for continuous data by selecting class intervals and constructing a rectangle for each class interval. The area of a rectangle corresponded to the proportion of measurements falling within the interval. And the sum of the areas of all the rectangles was 1. If we had a large number of measurements and permitted a large number of class intervals, then each class interval would have a narrow width. The area of each narrow rectangle would correspond to the probability that a measurement fell in the narrow interval. Thus, the more measurements we have, the closer the histogram can be made to look like a continuous curve (see figure 2.12 in chapter 2). This suggests that if we have a smooth curve for which the area under the curve is 1, we can identify it with a continuous probability distribution. With a smooth curve for the probability function for a random variable, the random variable must necessarily have an infinite number of values. The probability of any single value must be zero; otherwise, the sum of an infinite number of nonzero values would exceed 1. With a continuous random variable we will consider probability only in terms of the likelihood of a particular value of the random variable falling within a certain interval. Probability functions of continuous random variables are referred to as **probability density functions,** since nonzero probabilities are identified with areas under their graphs and to distinguish them from probability functions of discrete random variables.

The simplest continuous random variable is the **uniform random variable,** for which we can identify a **uniform probability distribution.** The probability function for a uniform random variable has a graph that is a horizontal line segment situated above the *x*-axis. The area under the line segment above the *x*-axis is equal to 1.

Let *a* and *b* be any two real numbers with $a < b$.

Uniform Probability Density Function

$$f(x) = \frac{1}{b-a} \quad \text{for } a < x < b \tag{7.1}$$

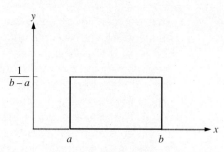

If x is not between a and b, then, we define $f(x) = 0$. Notice that the area of the rectangle formed by the x-axis, the horizontal line segment, and the vertical lines $x = a$ and $x = b$ is 1.

$$\text{Area} = (\text{length})(\text{height}) = (b - a)\,\frac{1}{(b - a)} = 1$$

The inclusion or exclusion of an end point of an interval does not affect the probability that a continuous random variable falls within an interval because the probability of a particular end point is zero. Thus, if X is a continuous random variable

$$P(a < X < b) = P(a \le X < b) = P(a < X \le b) = P(a \le X \le b)$$

EXAMPLE 7.1

Suppose that a wheel of a streetcar has a radius of r inches and travels along steel streetcar tracks. Also suppose that a reference point of 0 is selected somewhere on the circumference of the wheel and that the random variable X denotes the distance of a point on the circumference from the reference point. When the brakes are applied to the car, some point on the wheel makes contact with the rail and slides momentarily along the rail. This point can be labeled by the distance x it is from the fixed reference point on the wheel. For repeated braking action of the wheel, the value of x represents a random variable that is uniformly distributed over the interval from 0 to $2\pi r$ (the circumference of the wheel). If some points along the wheel made contact more often than others, then flat areas along the wheel would result and the wheel would show uneven wear at these spots and would be out of round. The probability function for the random variable X is

$$f(x) = \frac{1}{2\pi r}$$

for x between 0 and $2\pi r$.

APPLICATION 7.1

An electronic machine makes 3/8-inch bolts that are to have a length of 3 inches. If, in fact, the lengths of the 3/8-inch bolts are uniformly distributed over the interval ranging from 2.5 inches to 3.5 inches, what is the probability that a bolt selected at random from a finished batch of 3/8-inch bolts will have a length that is

    a. Between 2.75 and 3.25 inches?

    b. Greater than 3.25 inches?

    c. Exactly 3 inches long?

Solution: Let the random variable X denote the length of 3/8-inch bolt. Also, $a = 2.5$ and $b = 3.5$. By using formula (7.1) the probability density function f is defined by

$$f(x) = \frac{1}{b - a} = \frac{1}{3.5 - 2.5} = 1$$

The graph of the probability function for X is shown in figure 7.1.

FIGURE 7.1

Graph of probability function for application 7.1

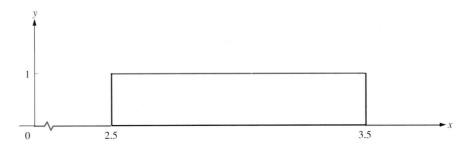

a. To determine the probability that the value of the random variable falls somewhere between 2.75 and 3.25, we find the area of the rectangle determined by the graph of the x-axis and the lines $x = 2.75$ and $x = 3.25$. This shaded area in figure 7.2

FIGURE 7.2

Graph of probability that random variable is between 2.75 and 3.25

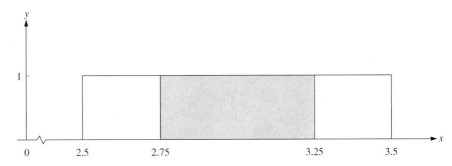

represents the probability $P(2.75 < X < 3.25)$. The length of the shaded rectangle is $3.25 - 2.75 = 0.5$ and the height is 1; therefore, the area is

$$\text{Area} = (\text{length})(\text{height}) = (0.5)(1) = 0.5$$

Hence, the probability that a bolt selected at random from a finished batch of 3/8-inch bolts will have a length between 2.75 inches and 3.25 inches is 0.5.

b. The probability that a bolt will have a length greater than 3.25 inches is the area of the rectangle formed by the graph of the x-axis and the lines $x = 3.25$ and $x = 3.5$. The shaded area in figure 7.3 represents the probability $P(X > 3.25)$ $= P(3.25 < X < 3.5)$. The area of this rectangle is

$$\text{Area} = (\text{length})(\text{width}) = (3.5 - 3.25)(1) = 0.25$$

FIGURE 7.3

Graph of probability that random variable is greater than 3.25

Thus the probability that a bolt will have a length greater than 3.25 inches is 0.25.

c. The probability that a random bolt has an exact length of 3 inches is 0. ■

APPLICATION 7.2   A telephone answering service has been designed so that the minimum time a caller must wait for a response from the service is 20 seconds and the maximum time is 50 seconds. If the response times are uniformly distributed, find the probability that a random caller will have a response time

    a. Falling between 25 seconds and 45 seconds.

    b. Less than 30 seconds or greater than 40 seconds.

Solution: Note that $a = 20$ and $b = 50$. The probability density function is

$$f(x) = \frac{1}{b-a} = \frac{1}{50-20} = \frac{1}{30}$$

where x is between 20 and 50. The graph of the probability density function is given in figure 7.4.

FIGURE 7.4

Graph of probability density function for application 7.2

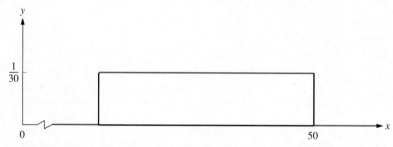

    a. We need to find the area under the curve $f(x) = 1/30$, between $x = 25$ and $x = 45$ (see the shaded area in figure 7.5). The area of the shaded rectangle is $(45 - 25)$ $(1/30) = 15/30 = 0.667$. Therefore the probability that a random caller will have to wait between 25 seconds and 45 seconds for a response is 0.667.

FIGURE 7.5

Graph of probability that X is between 25 and 45

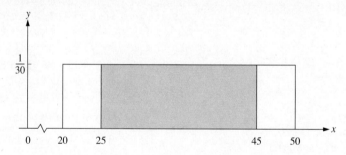

    b. The probability, as indicated by the shaded regions in figure 7.6, is

FIGURE 7.6

Graph of probability that X is less than 30 or greater than 40

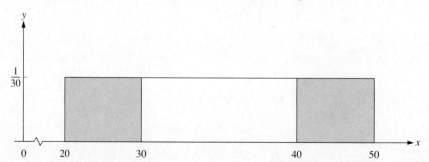

$$P[(X < 30) \text{ or } (P(X > 40)] = P(20 < X < 30) + P(40 < X < 50)$$

$$= \left(\frac{1}{30}\right)(30 - 20) + \left(\frac{1}{30}\right)(50 - 40)$$

$$= \frac{1}{3} + \frac{1}{3} = \frac{2}{3}$$

Hence, the probability that a random caller will have to wait less than 30 seconds or more than 40 seconds for a response is 2/3. ▪

EXERCISE SET 7.1

Basic Skills

1. If the random variable X is uniformly distributed for values greater than 20 but less than 80, find
   a. $P(X = 25)$.
   b. $P(30 < X < 50)$.
   c. $P(X < 70)$.
   d. $P(X > 45)$.
   e. $P(10 < X < 20)$.

2. Suppose the random variable X is uniformly distributed over the interval extending from 25 to 75. Find
   a. $P(30 < X < 70)$.
   b. $P(X < 37)$.
   c. $P(X = 50)$.
   d. $P(X > 70)$.
   e. $P(80 < X < 90)$.

3. Suppose the random variable Y is uniformly distributed over the interval extending from -20 to 50. Find
   a. $P(-5 < Y < 5)$.
   b. $P(Y > -10)$.
   c. $P(Y < 40)$.
   d. $P(Y = 0)$.

4. Suppose the random variable Y is uniformly distributed over the interval extending from -30 to 30. Find
   a. $P(-10 < Y < 5)$.
   b. $P(Y > 20)$.
   c. $P(Y < -5)$.
   d. $P(Y = -30)$.

Further Applications

5. Suppose that an automatic coffee dispenser never dispenses less than 6 ounces of coffee or more than 10 ounces of coffee. Any amount of coffee between 6 ounces and 10 ounces is equally likely to occur. If an amount is dispensed, determine the probability that the amount
   a. is less than 7 ounces.
   b. is more than 6 ounces.
   c. is between 7 and 9 ounces.

6. Suppose that the annual rainfall in a certain region is uniformly distributed with values ranging from 10 to 14 inches. In a randomly selected year, determine the probability that the annual rainfall will be
   a. less than 10 inches.
   b. greater than 11 inches.
   c. between 11 and 14 inches.

7. Suppose that the annual average winter temperatures (in degrees Fahrenheit) for a northern region are uniformly distributed with temperatures ranging from $-20°$ to $20°$.

If a random winter season is selected, determine the probability that the average temperature will be
   a. less than 10°.
   b. between $-5°$ and 15°.
   c. less than 10° or greater than 5°.
   d. between $-10°$ and 10°.

8. Suppose a teacher never arrives for a class on time. She is never more than 5 minutes early or more than 5 minutes late, and within this range of time she is as likely to arrive at one time as at any other time. Find the probability that the teacher will be
   a. at least 3 minutes late.
   b. at most 4 minutes early.
   c. right on time.
   d. no more than 2 minutes early or late.

Going Beyond

9. If X is uniformly distributed over the interval extending from a to b, then the mean or expected value of X is given by $\mu_X = E(X) = (a + b)/2$ and the variance of X is given by $\sigma_X^2 = (b - a)^2/12$. Find mean and variance for the random variable given in exercise 5.

10. Find the mean and standard deviation for the random variable given in exercise 6.

11. If X is a random variable that is uniformly distributed over the interval extending from a to b, determine $E(X^2)$.

12. Suppose that a continuous random variable X has a probability density function g given by $g(x) = (1/2)x$ for values of x ranging from 0 to 2. Find the probability that X is
   a. less than 1.
   b. between 1 and 2.
   c. greater than 0.5.
   d. less than 0.5 or greater than 1.5.

13. Given the function f defined by $f(x) = (5/4)x$ for values of x from 3 to b. Find the value of b so that f is a probability density function.

SECTION 7.2   *Normal Distributions*

One of the most important classes of continuous distributions is the **normal distribution.** Since its discovery more than 350 years ago, it has developed into an indispensable tool for every branch of science, industry, and commerce. Many real and natural occurrences have frequency distributions that are very close to the normal distribution in shape.

EXAMPLE 7.2

The frequency distribution of the nitrogen content of leaves on a tree tends to be normal. Physical measurements are often normally distributed. Systolic and diastolic blood pressures, heart rates, blood cholesterol levels, heights of adult men, and weights of two-year-old girls are all examples of distributions of data that tend to follow the normal distribution. In addition, test scores for many standardized tests, such as the SAT and the American College Test (ACT), possess distributions that are bell shaped.

FIGURE 7.7

Graph of normal distribution

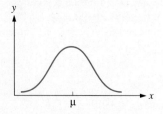

A normal distribution is mound shaped or takes on the appearance of a bell, as illustrated in figure 7.7. The equation of the bell-shaped curve is given by

$$y = \frac{1}{\sqrt{2\pi}\sigma}e^{-(x-\mu)^2/2\sigma^2}$$

where μ represents the mean of the population, σ is the standard deviation of the population, $e \approx 2.718$, and x is any real number.

The two parameters μ and σ completely specify the position and shape, respectively, for a normal distribution. A small value of σ means that the normal curve is a narrow, peaked bell, whereas a large value of σ means that the normal curve is a wide, flat bell (see figure 7.8). Figure 7.9 illustrates three normal distributions having a mean of 20 but different standard deviations. And figure 7.10 illustrates three normal distributions with different means but the same standard deviation. In many practical applications, we have collections of measurements whose graphs approximate a bell-shaped curve.

FIGURE 7.8

Effect of μ and σ on position and shape of graphs of normal distribution

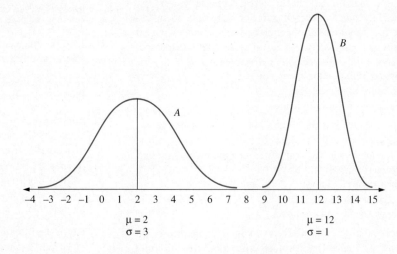

FIGURE 7.9

Graphs of normal distributions with same mean but different standard deviations

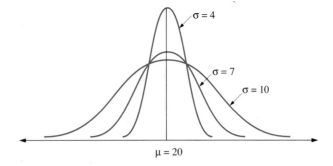

FIGURE 7.10

Graphs of normal distributions with different means but same standard deviation

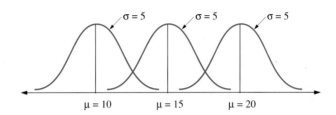

EXAMPLE 7.3

Consider the following collection of 100 systolic blood pressure measurements of a group of individuals aged 20–24 years.

98	141	127	112	107	111	116	104	87	130
131	124	126	129	65	126	139	136	137	102
128	152	115	124	96	114	116	119	108	113
126	116	107	134	82	141	85	126	131	141
128	85	145	150	97	125	109	119	159	114
117	131	122	104	145	119	132	138	109	102
115	109	143	128	136	118	125	124	109	133
131	127	100	110	109	112	125	166	93	92
113	118	116	136	96	131	141	153	85	119
92	112	140	121	108	92	98	159	91	125

The following stem-and-leaf diagram for this data set suggests that the blood-pressure measurements are normally distributed.

```
 6 | 5
 7 |
 8 | 2 5 5 5 7
 9 | 1 2 2 2 3 6 6 7 8 8
10 | 0 2 2 4 4 7 7 8 8 9 9 9 9 9
11 | 0 1 2 2 2 3 3 4 4 5 5 6 6 6 6 7 8 8 9 9 9 9
12 | 1 2 4 4 4 5 5 5 5 6 6 6 6 7 7 8 8 8 9
13 | 0 1 1 1 1 1 2 3 4 6 6 6 7 8 9
14 | 0 1 1 1 1 3 5 5
15 | 0 2 3 9 9
16 | 6
```

A grouped frequency table for the data is as follows:

Class Limits	f
65–75	1
76–86	4
87–97	9
98–108	11
109–119	27
120–130	20
131–141	19
142–152	5
153–163	3
164–174	1

A frequency histogram for the data is shown in figure 7.11. Although the frequency histogram is not symmetrical, it is approximately bell shaped and suggests that the blood-pressure measurements may be normally distributed.

FIGURE 7.11

Histogram for blood-pressure measurements

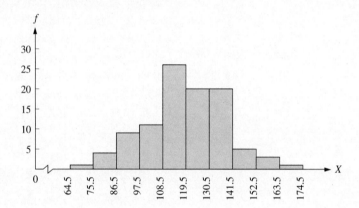

Properties of Normal Distributions

Some of the more important properties of normal distributions are listed here.

Properties of Normal Distributions

1. A normal distribution is mound or bell shaped.
2. The area under a normal curve and above the horizontal axis always equals 1.
3. The mean is located at the center of the distribution and a curve is symmetrical about the line perpendicular to the horizontal axis at the value of the mean.
4. The mean, median, and mode are all equal.
5. A curve for a normal distribution extends indefinitely to the left and to the right of the mean and approaches the horizontal axis.
6. A curve for a normal distribution never touches the horizontal axis.
7. The shape and position of a normal distribution depend on the parameters μ and σ; and, as a result, there are an infinite number of normal distributions.

APPLICATION 7.3   Explain whether each of the following curves could represent a normal distribution.

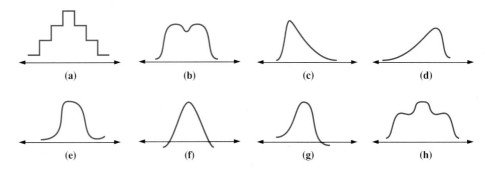

Solution:

    a. No, the curve is not mound shaped.

    b. No, the curve is not mound shaped.

    c. No, the curve is not symmetrical.

    d. No, the curve is not symmetrical.

    e. No, the curve does not get close to the horizontal axis.

    f. No, the curve crosses the horizontal axis.

    g. No, the curve crosses the horizontal axis.

    h. No, the curve is not mound shaped. ▪

APPLICATION 7.4   Find the area of the shaped regions under the normal curves shown here.

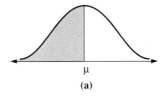 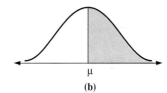

Solution: In both cases the shaded area is 0.5, since the total area under a normal curve is 1 and a curve is symmetrical about its central measurement μ. ▪

Empirical Rule   The **empirical rule** applies to any normal distribution. Figure 7.12 illustrates the empirical rule.

Empirical Rule

    a. Approximately 68% of the measurements fall within one standard deviation of the mean, that is, within $(\mu \pm \sigma)$.

    b. Approximately 95% of the measurements fall within two standard deviations of the mean, that is, within $(\mu \pm 2\sigma)$.

    c. Approximately 99.7% of the measurements fall within three standard deviations of the mean, that is, within $(\mu \pm 3\sigma)$.

FIGURE 7.12

Graphical illustration of empirical rule

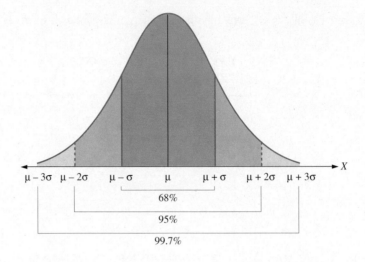

APPLICATION 7.5   Suppose $\sigma = 2$ and $\mu = 30$ for the normal distribution pictured here.

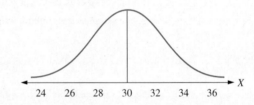

Find the percentage of the measurements that fall

    a. between 30 and 32.

    b. between 32 and 34.

    c. above 32.

Solution:

    a. Since the curve is symmetrical about $\mu = 30$ and approximately 68% of the scores fall between 28 and 32, approximately (1/2)(68%) = 34% of the measurements fall between 30 and 32.

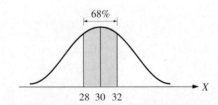

b. This problem requires us to find the shaded areas illustrated in the following drawings:

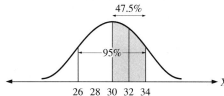

Approximately 95% of the scores fall between 26 and 34. Since the curve is symmetrical about $\mu = 30$, $(1/2)(0.95) = 0.475 = 47.5\%$ of the scores fall between 30 and 34. Hence, the percentage illustrated in the first drawing is 47.5%. As a result of part a, the percentage illustrated in the second drawing is 34%. Therefore, we subtract 34% from 47.5% to get 13.5%, the percentage of measurements that fall between 32 and 34.

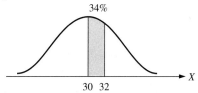

c. By part a, 34% of the scores fall between 30 and 32. Since 50% fall above 30, it follows that $0.50 - 0.34 = 0.16 = 16\%$ of the scores fall above 32.

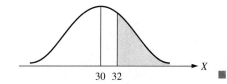

Recall from section 3.4 that a z score for a population is defined by

$$z = \frac{x - \mu}{\sigma}$$

and indicates the number of standard deviations a score is from the mean. A measurement equal to $(\mu + \sigma)$ is one standard deviation above the mean and has a z score equal to 1:

$$z = \frac{x - \mu}{\sigma}$$
$$= \frac{(\mu + \sigma) - \mu}{\sigma}$$
$$= \frac{\sigma}{\sigma} = 1$$

By the same reasoning, the measurement $(\mu + 2\sigma)$ is two standard deviations above the mean and has a z score equal to 2:

$$z = \frac{x - \mu}{\sigma}$$

$$= \frac{(\mu + 2\sigma) - \mu}{\sigma} = 2$$

And the measurement $(\mu + 3\sigma)$ is three standard deviations above the mean and has a z score of 3.

$$z = \frac{x - \mu}{\sigma}$$

$$= \frac{(\mu + 3\sigma) - \mu}{\sigma} = 3$$

As a consequence of these results, the empirical rule can be restated as follows:

Restatement of the Empirical Rule

a. In any normal distribution, approximately 68% of the z scores fall between -1 and $+1$.

b. In any normal distribution, approximately 95% of the z scores fall between -2 and $+2$.

c. In any normal distribution, approximately 99.7% of the z scores fall between -3 and $+3$.

Approximating σ and s

In section 3.2 it was stated that $R/4$ provides a good approximation for s, where R is the range and s is the sample standard deviation. The range of a normal distribution is infinite, and approximately 99.7% of a normal distribution falles within $\mu \pm 3\sigma$. If we restrict the distribution to $\mu \pm 3\sigma$, then the range of the restricted distribution is approximately equal to 6σ. Thus, for any finite bell-shaped distribution, the range R is approximately equal to 6σ, as illustrated in figure 7.13. Thus, we have

$$\sigma \approx \frac{R}{6}$$

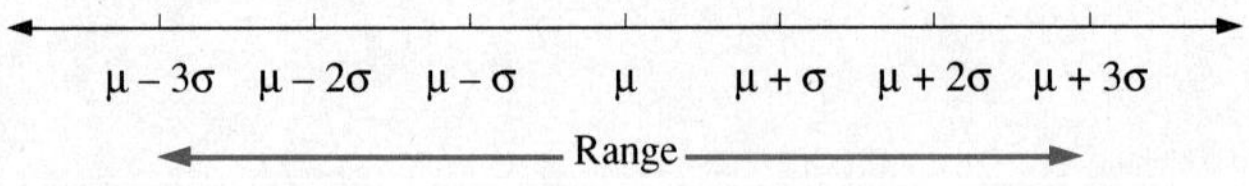

FIGURE 7.13

Range $\approx 6\sigma$ for a bell-shaped distribution

Hence, $R/6$ provides an approximation for σ and also for s. In addition, since 95% of the measurements in a normal distribution fall within $\mu \pm 2\sigma$, $R/4$ also provides an approximation for s based on the same line of reasoning. Since $R/4$ always provides a larger—and thus more conservative—approximation for s than $R/6$, the estimate $R/4$ is typically used. In most cases when a distribution is symmetrical, $R/4$ provides a reasonable approximation for s. But for badly **skewed distributions**—distributions in which measurements occur with greater frequency either above or below the mean— $R/4$ usually provides a poor approximation for s.

Probability and Area

Probabilities related to a normal distribution are identified with areas. For example, the probability that a measurement x from a normal distribution is between 20 and 40 can be identified with the shaded area in the following figure:

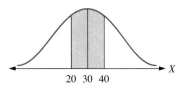

Similarly, the shaded area under the following normal curve can be interpreted as the probability that a measurement x is between 6 and 8 and is written as $P(6 < X < 8)$.

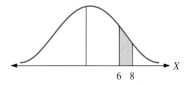

APPLICATION 7.6 Write a corresponding probability statement for each of the following areas under a normal curve:

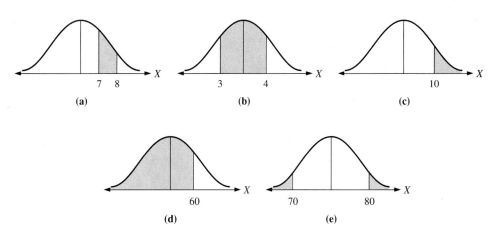

Solution: The following are the probability statements and their corresponding interpretations:

Using Inequalities	Interpretation
a. $P(7 < X < 8)$	the probability that x is between 7 and 8
b. $P(3 < X < 4)$	the probability that x is between 3 and 4
c. $P(X > 10)$	the probability that x is greater than 10
d. $P(X < 60)$	the probability that x is less than 60
e. $P(X < 70 \text{ or } X > 80)$	the probability that x is less than 70 or greater than 80 ■

Standard Normal Distribution

Recall from section 3.4 that a z score has a mean of zero and a standard deviation of one. If X is a normal random variable with mean μ and standard deviation σ, then the variable $Z = (X - \mu)/\sigma$ is a standardized variable; its units are standard deviations. The standard normal variable Z has a mean of 0 and a standard deviation of 1. The probability distribution of Z is called the **standard normal distribution,** and the symbol Z is used to denote the standard normal distribution.

Finding Probabilities Using the Standard Normal Table

We want to be able to determine the probability of the random variable Z having a value in a specified interval. The empirical rule can be used to find certain probabilities or areas associated with the standard normal distribution. However, most practical applications will not involve intervals of exactly one, two, or three standard deviations from the mean. For example, the empirical rule cannot be use to find $P(0 < Z < 1.23)$. In such a case, the probability must be found using a **standard normal table** (z table; see the front endpaper). This table gives the areas under the standard normal curve, above the horizontal axis, to the right of 0, and below a specified z value, as illustrated in figure 7.14. To find an area (or probability) in the z table, we locate the units and tenths digits of z down the left-hand column and the hundredths digit across the top; the number where the row and column intersect is the desired area.

FIGURE 7.14

Area under standard normal curve given in the z table

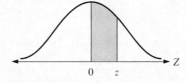

EXAMPLE 7.4

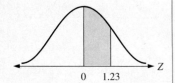

Consider the shaded area in the figure. To find the area between 0 and $z = 1.23$ under a standard normal curve, we locate the 1.2 row at the left of the z table and the 0.03 column along the top; the entry in the table body at the intersection of the 1.2 row and 0.03 column is 0.3907, as shown here:

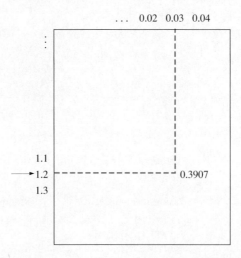

Thus, 0.3907 is the area of the shaded region, which is also the probability that z is between 0 and 1.23.

APPLICATION 7.7   Find the following probabilities associated with the standard normal distribution, using the *z* table:

  a. $P(Z$ is between -1.34 and $1.68)$
  b. $P(Z$ is between 1.34 and $2.38)$
  c. $P(Z$ is less than $-1.25)$

Solution:

  a. An important first step to finding any normal probability value is to draw a diagram illustrating the area corresponding to the probability to be found.

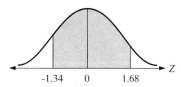

The steps are as follows:

  1. Find the area from $z = 0$ to $z = 1.68$.
  2. Find the area from $z = -1.34$ to $z = 0$.
  3. Add these areas.

     The area from $z = 0$ to $z = 1.68$ is 0.4535, and, because of symmetry, the area from $z = -1.34$ to $z = 0$ is the same as the area from $z = 0$ to $z = 1.34$, which is 0.4099. Adding these two areas we obtain 0.8634. Thus,

$$P(-1.34 < Z < 1.68) = 0.4535 + 0.4099 = 0.8634$$

  b.

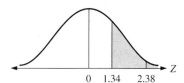

The steps are as follows:

  1. Find the area from $z = 0$ to $z = 2.38$.
  2. Find the area from $z = 0$ to $z = 1.34$.
  3. Subtract the second area from the first.

     Thus, $P(1.34 < Z < 2.38) = 0.4913 - 0.4099 = 0.0814$.

  c.

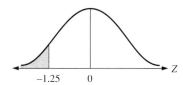

The steps are as follows:

  1. Find the area between $z = -1.25$ and $z = 0$.
  2. Subtract this area from 0.5.

     Thus, $P(Z < -1.25) = 0.5 - 0.3944 = 0.1056$. ▪

EXAMPLE 7.5

Occasionally we need to calculate probabilities associated with z values that fall in the extreme regions of the standard normal distribution. For example, we might need to determine $P(Z < -5)$ or $P(Z > 6)$. In both these cases the probabilities are small, but not equal to 0, since the curve extends indefinitely far in both directions. For very small probabilities (probabilities written as 0.0000 to four decimal places), we shall use the symbol 0^+. In addition, we shall indicate cumulative probabilities associated with large z values, such as $P(Z < 7)$, by using the symbol 1^-.

Verifying the Empirical Rule

Probabilities listed in the standard normal table (z table; see front endpaper), can be used to verify the empirical rule. Note that:

$$P(-1 < Z < 1) = 2P(0 < Z < 1) = 2(0.3413) = 0.6826 \approx 0.68$$
$$P(-2 < Z < 2) = 2P(0 < Z < 2) = 2(0.4772) = 0.9544 \approx 0.95$$
$$P(-3 < Z < 3) = 2P(0 < Z < 3) = 2(0.4987) = 0.9974 \approx 0.997$$

**Finding z Scores
Given Areas**

Frequently, we must find the z value, given an area under the standard normal curve. This is the reverse process of finding probabilities or areas corresponding to z values. Consider applications 7.8–7.11. If the area or probability we seek in the standard normal table is equally close to two values in the table, we choose the larger z value, as illustrated in application 7.9.

APPLICATION 7.8

Find a standard normal value z_0 so that $P(Z$ is between $-z_0$ and $z_0) = 0.95$.

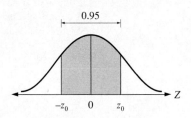

Solution: The shaded area in the accompanying figure is 0.95. Since the z table lists areas under the normal curve from $z = 0$ to $z = z_0$, we divide 0.95 by 2 to get an area of 0.475 between $z = 0$ and $z = z_0$. We now locate the area 0.475. in the body of the z table, as shown.

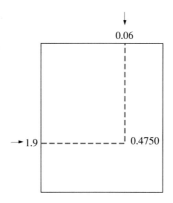

Thus, the value of z_0 is 1.96 and $P(-1.96 < Z < 1.96) = 0.95$ ▪

APPLICATION 7.9

Find a standard normal value z_0 so that $P(Z > z_0) = 0.05$.

Solution: $P(0 < Z < z_0) = 0.45$, since $P(0 < Z < z_0) + P(Z > z_0) = 0.5$. From the z table, we find

Area	z Value
0.4495	1.64
0.4500	
0.4505	1.65

Since 0.45 is midway between 0.4495 and 0.4505 and 1.65 is the larger z value, we choose $z_0 = 1.65$. Hence, $P(Z > 1.65) \approx 0.05$. ▪

APPLICATION 7.10

Find a standard normal value z_0 so that $P(Z < z_0) = 0.75$.

Solution: Since $P(Z < 0) = 0.5$, $P(0 < Z < z_0) = 0.25$. Checking the z table, we find $z_0 = 0.67$, since 0.25 is closer to 0.2486 than to 0.2517.

	Area	z Value
	0.2486	0.67
0.0014		
	0.2500	
0.0017		
	0.2517	0.68

Thus, $P(Z < 0.67) \approx 0.75$. ▪

APPLICATION 7.11   Find a standard normal value z_0 so that $P(-1.14 < Z < z_0) = 0.25$.

Solution: By symmetry, $P(-1.14 < Z < 0) = P(0 < Z < 1.14) = 0.3729$. Hence, z_0 must be less than 0, and we have the following drawing:

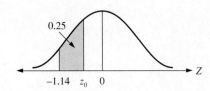

Thus, $P(0 < Z < -z_0) = 0.3729 - 0.25 = 0.1229$. Reading the z table, we see that the area closest to 0.1229 is 0.1217.

Area	z Value
0.1217	0.31
0.1229	
0.1255	0.32

Hence, $-z_0 = 0.31$ and $z_0 = -0.31$. ▪

EXERCISE SET 7.2

Basic Skills

1. Suppose a given normal distribution has a mean of 20 and a standard deviation of 4. By using the empirical rule, approximate the following probabilities:
   a. $P(X > 24)$ c. $P(12 < X < 24)$
   b. $P(16 < X < 24)$ d. $P(X > 28)$

2. Suppose a given normal distribution has a mean of 20 and a standard deviation of 4. By using the empirical rule, approximate the following probabilities:
   a. $P(8 < X < 12)$ c. $P(X < 12)$
   b. $P(24 < X < 32)$ d. $P(20 < X < 28)$

3. Find the following probabilities using the z table on the front endpaper:
   a. $P(Z$ is greater than 0)
   b. $P(Z$ is between -2.5 and 1.5)
   c. $P(Z$ is between -2.7 and -1.3)
   d. $P(Z$ is between 2 and 3)

4. Find the following probabilities using the z table on the front endpaper:
   a. $P(Z$ is greater than 2.8)
   b. $P(Z$ is greater than 3.7)
   c. $P(Z$ is between -4 and 4)
   d. $P(Z$ is between 2.1 and 2.9)

5. Find the following probabilities using the z table on the front endpaper:
   a. $P(-1.23 < Z < 0)$
   b. $P(0 < Z < 2.34)$
   c. $P(-1.78 < Z < 2.38)$
   d. $P(-1.12 < Z < -1.01)$

6. Find the following probabilities using the z table on the front endpaper:
   a. $P(1.23 < Z < 2.75)$ c. $P(Z > 2.97)$
   b. $P(Z < 2.34)$ d. $P(Z > 4.38)$

7. Find the area under the standard normal curve
   a. between $z = 1.8$ and $z = 2.4$.
   b. above $z = -1.78$.
   c. below $z = 2.5$.

8. Find the area under the standard normal curve
   a. below $z = -2.17$.
   b. above $z = 1.98$.
   c. between $z = -1.28$ and $z = 2.35$.

9. For each case, find the standard normal value z_0 so that
   a. $P(Z$ is between $-z_0$ and $z_0) = 0.90$.
   b. $P(Z$ is greater than $z_0) = 0.025$.

c. $P(Z > z_0) = 0.10$.

d. $P(-z_0 < Z < z_0) = 0.99$.

10. For each case, find the standard normal value z_0 so that

    a. $P(Z > z_0) = 0.85$.

    b. $P(Z$ is between $-z_0$ and 2$) = 0.94$.

    c. $P(Z < z_0$ or $z > 1.1) = 0.90$.

    d. $P(Z < z_0$ and $z > 1.3) = 0.24$

Going Beyond

11. A handy approximation for areas under the standard normal curve from 0 to z ($z > 0$) is given by the following rule:

z Value	Approximate Area
$0 \leq z \leq 2.2$	$\dfrac{z(4.4 - z)}{10}$
$2.2 < z < 2.6$	0.49
$z \geq 2.6$	0.50

The maximum absolute error for the approximation is 0.0052.[35] Use this approximation to find the probabilities in exercise 3 and compare the results.

12. If u_1 and u_2 are two random values from the interval (0, 1), then the following formulas can be used to generate two standard normal random variables:

$$z_1 = \sqrt{-\ln(u_2)} \, \cos(2\pi u_1)$$
$$z_2 = \sqrt{-\ln(u_2)} \, \sin(2\pi u_1)$$

where the angle is expressed in radians. The values of u_1 and u_2 can be obtained by using a random number table and selecting three digits at a time. Use these formulas to simulate two observations from the standard normal distribution.

13.   a. Write a computer program to simulate 25 observations from the standard normal distribution.

    b. Calculate the mean and standard deviation for the 25 measurements in part a, and construct a histogram to verify that the sample is from the standard normal distribution

---

SECTION 7.3   *Applications of Normal Distributions*

The standard normal table cannot be used directly to calculate probabilities associated with a normal distribution other than the standard normal distribution.

EXAMPLE 7.6

Suppose we are interested in finding the following area for the normal distribution with a mean of 50 and a standard deviation of 5.

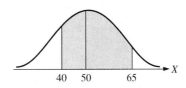

We cannot directly use the standard normal table (z table; see front endpaper), since the random variable of interest is not the standard normal random variable. In order to determine probabilities associated with a nonstandard normal distribution, we need a rule that will enable us to transform any normal distribution to the standard normal distribution, as indicated in figure 7.15. Then we could determine probabilities by using the z table.

FIGURE 7.15

Transformation to standard
normal distribution

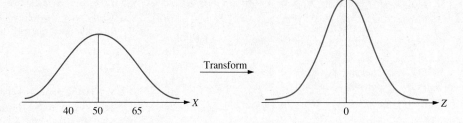

The transformation rule that we use is the standard-score formula:

$$z = \frac{x - \mu}{\sigma}$$

It is used to transform any normal x score to a standard normal z score. Thus, for the above illustration, we have

$$z = \frac{40 - 50}{5} = -2$$

and

$$z = \frac{65 - 50}{5} = 3$$

Hence, we have the following diagrams where the two shaded areas are equal.

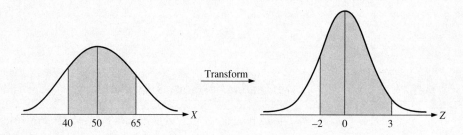

That is,

$$P(40 < X < 65) = P(-2 < Z < 3)$$

The probability that z is between -2 and 3 is

$$P(-2 < Z < 3) = 0.4772 + 0.4987 = 0.9759.$$

Thus, $P(40 < X < 65) = 0.9759$.

It is always a good idea to draw normal-curve illustrations when attempting to calculate probabilities associated with a normal distribution. The illustrations need not be drawn accurately, since their primary use is to identify the appropriate regions to calculate areas. The normal-curve illustrations used throughout the text may not be technically accurate; they are intended for conceptual use only.

The z score formula allows us to think of the standard normal probability table as a list of probabilities associated with intervals beginning with the mean and extending to z standard deviations to the right of the mean for any normal curve. Consider applications 7.12 and 7.13.

APPLICATION 7.12

A company manufactures light bulbs with a mean life of 500 hours and a standard deviation of 100 hours. If it is assumed that the useful lifetimes of the light bulbs are normally distributed (that is, the lifetimes form a normal distribution), find the number of bulbs out of 10,000 that can be expected to last between 650 and 780 hours.

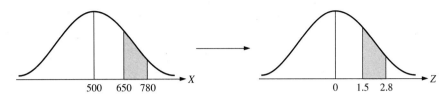

Solution: We first compute the z scores for $x = 650$ and $x = 780$. The z score for $x = 650$ is

$$\frac{650 - 500}{100} = 1.5$$

The z score for $x = 780$ is

$$\frac{780 - 500}{100} = 2.8$$

Thus,

$$P(650 < X < 780) = P(1.5 < Z < 2.8)$$
$$= 0.4974 - 0.4332 = 0.0642$$

Therefore, we can expect $(10,000)(0.0642) = 642$ light bulbs to last between 650 and 780 hours. ■

In practical applications of statistics we deal primarily with finite distributions of raw-data measurements. These distributions can be only approximations to normal distributions, since normal distributions are continuous and contain infinite numbers of measurements. It is common practice in statistical literature to refer to such distributions as "normal" even when the context makes it clear that such distributions can be only approximately normal. Such is the case in the statement of the problems in applications 7.12 and 7.13.

APPLICATION 7.13

If systolic blood pressure measurements X for the 20–24 year age group are normally distributed with a mean of 120 and a standard deviation of 20, find

  a. $P(X > 135)$.
  b. $P(X < 146)$.
  c. $P(105 < X < 110)$.

Solution:

  a. The z score for $x = 135$ is

$$z = \frac{x - \mu}{\sigma}$$

$$= \frac{135 - 120}{20} = 0.75$$

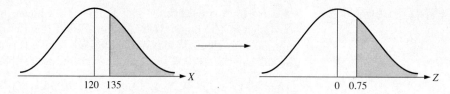

By using the z table (see the front endpaper), we find

$$P(X > 135) = P(Z > 0.75)$$
$$= 0.5 - P(0 < Z < 0.75)$$
$$= 0.5 - 0.2734 = 0.2266$$

b. The z score for $x = 146$ is

$$z = \frac{x - \mu}{\sigma}$$
$$= \frac{146 - 120}{20} = 1.3$$

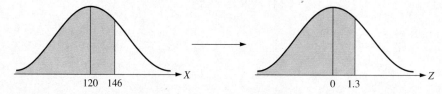

From the z table, we find

$$P(X < 146) = P(Z < 1.3)$$
$$= 0.5 + P(0 < Z < 1.3)$$
$$= 0.5 + 0.4032 = 0.9032$$

c. The z score for $x = 105$ is

$$\frac{105 - 120}{20} = -0.75$$

and the z score for $x = 110$ is

$$\frac{110 - 120}{20} = -0.5$$

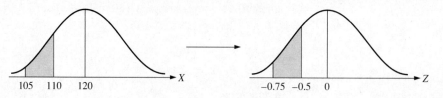

By using the z table, we find

$$P(105 < X < 110) = P(-0.75 < Z < -0.5)$$
$$= P(0 < Z < 0.75) - P(0 < Z < 0.5)$$
$$= 0.2734 - 0.1915 = 0.0819 \quad ■$$

Percentiles, Quartiles, and Deciles Associated with Normal Distributions

Recall that position points were discussed in chapter 3. Three commonly used measures of position are: percentiles, quartiles, and deciles. **Percentiles** are numbers that divide an interval of measurements ranging from the smallest measurement to the largest measurement into 100 parts so that each part contains 1% of the measurements. There are 99 percentile points, denoted by $P_1, P_2, \ldots, P_{99}$. In other words, we label the nth percentile P_n. The 75th percentile, P_{75}, is that measurement below which 75% of the measurements fall.

Quartiles are numbers that divide an interval of measurements extending from the smallest measurement to the largest measurement into four parts so that each part contains 25% of the measurements. There are three quartile points; these are denoted by Q_1, Q_2, and Q_3, and we label the nth quartile Q_n. The second quartile is the 50th percentile or the median. The first quartile Q_1 is that number so that 25% of the measurements fall below Q_1. The first quartile Q_1 is also equal to P_{25}, the 25th percentile.

Deciles are numbers that divide an interval of measurements extending from the smallest measurement to the largest measurement into 10 parts so that each part contains 10% of the measurements. We term D_n, the nth decile, and denote the nine decile points by $D_1, D_2, D_3, \ldots, D_9$. Applications 7.14 and 7.15 show how these position points are found for normal distributions.

A P P L I C A T I O N 7.14

Math scores on the SAT are normally distributed with a mean of 500 and a standard deviation of 100. Find P_{80}, the 80th percentile.

Solution: P_{80} is that number (or score) below which 80% of the scores fall. Then the area under the normal curve between $x = 500$ and P_{80} is 0.30 (see fig. 7.16). If we locate the z score in the z table that corresponds to an area of 0.30, we can then use the z score formula to solve for P_{80}, since

$$z = \frac{P_{80} - 500}{100}$$

FIGURE 7.16

Determination of P_{80} for SAT math scores

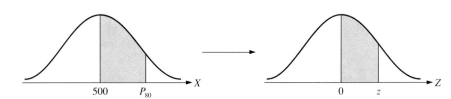

The approximate value of z that produces an area of 0.30 is found in the z table to be 0.84. Hence, we have

$$0.84 = \frac{P_{80} - 500}{100}$$

Multiplying both sides of this equation by 100 and then adding 500 to both sides, we get

$$P_{80} = 584 \quad ▪$$

APPLICATION 7.15

For the SAT math scores with $\mu = 500$ and $\sigma = 100$, find

a. D_3, the 3rd decile.

b. Q_1, the first quartile.

Solution:

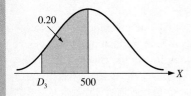

a. We find the z-score for D_3:

$$z = \frac{D_3 - 500}{100}$$

By using the z table (see the front endpaper), we find $P(0 < Z < 0.52)$ ≈ 0.20. Thus, the z score for the third decile is

$$-0.52 = \frac{D_3 - 500}{100}$$

Multiplying both sides by 100 and then adding 500 to both sides, we get

$$D_3 = 448$$

b. We find the z score for Q_1 as follows:

$$z = \frac{Q_1 - 500}{100}$$

From the z table, we find

$$P(0 < Z < 0.67) = 0.25$$

Thus,

$$-0.67 = \frac{Q_1 - 500}{100}$$

Solving for Q_1, we have

$$Q_1 = 433 \quad ■$$

A given measurement x has a **percentile rank** of n if x equals (or is approximately equal to) the nth percentile, that is, if $x = P_n$. Application 7.16 illustrates finding the percentile rank for a given measurement.

APPLICATION 7.16

John took the SAT math test and received a score of 572. Find John's percentile rank.

Solution: Consider figure 7.17. To find the percentile rank of John's score, we find the shaded area in the figure and round it to the nearest percentage. We will find this area by first finding the z score for $x = 572$:

$$z = \frac{x - \mu}{\sigma}$$

$$= \frac{572 - 500}{100} = 0.72$$

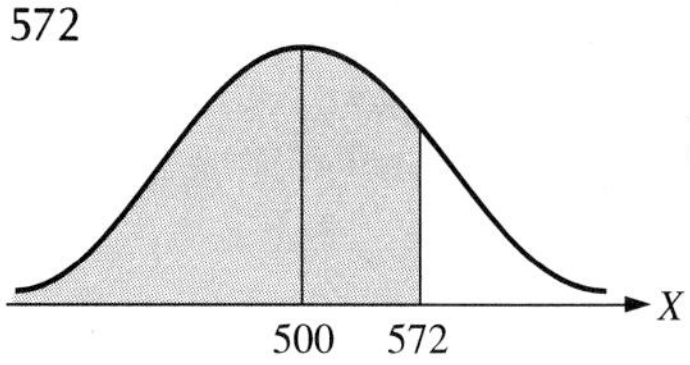

FIGURE 7.17

Determination of percentile rank for SAT math score of 572

The shaded area in figure 7.17 is

$$P(X < 572) = P(Z < 0.72)$$
$$= 0.5 + P(0 < Z < 0.72)$$
$$= 0.5 + 0.2642 = 0.7642$$

Hence, $P_{76} \approx 572$ and the percentile rank of 572 is 76. ■

MINITAB can be used to determine the percentile rank of John's score. The commands and output are contained in the computer display 7.1.

Computer Display 7.1

```
MTB > CDF 572;
SUBC> NORMAL 500, 100.
  572.0000 0.7642
MTB >
```

The CDF, cumulative distribution function, command and its associated NORMAL subcommand determine the probability $P(X \le 572)$ for the normal random variable X having $\mu = 500$ and $\sigma = 100$.

Checking the Plausibility that a Sample Is from a Normal Distribution

To determine whether a normal distribution can serve as a reasonable model for the population that produced the sample, graphical procedures are helpful in detecting serious departures from normality. Histograms can be checked for lack of symmetry, and the empirical rule can be used to check the proportions of the data that fall within the intervals determined by $\bar{x} \pm s$, $\bar{x} \pm 2s$, and $\bar{x} \pm 3s$.

A **z score plot** of the data and corresponding percentile points of the standard normal distribution can be an effective procedure to check the plausibility of a normal model when the sample size is at least 20. If the sample size is n, n standard normal values can be found that divide the standard normal distribution into $(n + 1)$ intervals, each interval having the same probability. The measurements in the sample are ordered from smallest to largest and paired with the n percentile points. The pairs are then plotted to obtain a z score plot. For purposes of discussion, let's suppose the sample is of size 4. We would first find the four percentile points, P_{20}, P_{40}, P_{60}, and P_{80}, that divide the standard normal distribution into five intervals, each having a probability of 0.20. The smallest sample value is paired with P_{20}, the next largest sample value is paired with P_{40}, the next with P_{60}, and the largest sample value with P_{80}. These four pairs are then plotted in a graph. A straight-line pattern supports the plausibility that the sample was drawn from a normal population. A departure from normality is indicated by a graph displaying a curved appearance. Pearson's correlation coefficient can also be used to detect a straight-line pattern; values of $|r|$ close to 1 indicate a strong linear relationship. Application 7.17 illustrates the ideas only; in practical applications the sample size should be at least 20.

APPLICATION 7.17

Construct a z score plot of the following sample of data to determine if the sample could have been produced by a normal population.

$$y: \quad 48 \quad 55 \quad 57 \quad 60$$

Solution: The four percentile points that divide the distribution into five equally likely intervals are found with the aid of the z table on the front endpaper (see the accompanying figure): $P(Z \leq -0.84) = 0.2$, $P(Z \leq -0.25) = 0.4$, $P(Z \leq 0.25) = 0.6$, and $P(Z \leq 0.84) = 0.8$.

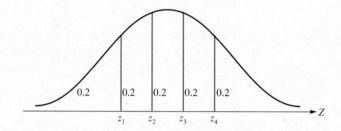

These z scores are then paired with the four ordered observations:

Sample Observations, y	Percentile Points, z Scores
48	$-0.84 \ (= z_1)$
55	$-0.25 \ (= z_2)$
57	$0.25 \ (= z_3)$
60	$0.84 \ (= z_4)$

The z score plot is shown in figure 7.18.

FIGURE 7.18

z score plot for application 7.17

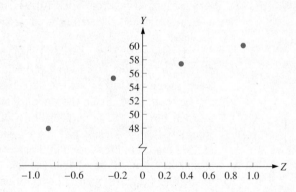

Since the four data points appear to follow a linear trend, we have no reason to believe that the sample wasn't produced by a normal population. In addition, Pearson's correlation coefficient for the four data pairs is $r = 0.965$, supporting the strong linear trend for the z score plot. If the distribution (X) that produced the sample has mean μ and standard deviation σ, each percentile point z can be converted to an x score by using the transformation $x = \mu + \sigma z$. We would expect each sample value to be close to its corresponding x value if the population is normal. That is, a plot of the sample values versus the normal x scores should produce a straight-line pattern $(y = \mu + \sigma z)$, where the slope of the line is σ and the vertical axis intercept is μ.

Sample Value (y)	x Value
48	$\mu + \sigma z_1$
55	$\mu + \sigma z_2$
57	$\mu + \sigma z_3$
60	$\mu + \sigma z_4$

EXERCISE SET 7.3

Basic Skills

1. If a random variable X is normally distributed with a mean of 70 and a standard deviation of 5, find the probability that X is
   a. between 60 and 80.
   b. less than 65.
   c. less than 81.
   d. greater than 67.5.
   e. less than 67 or greater than 77.

2. If a random variable X is normally distributed with a mean of 50 and a standard deviation of 8, find the probability that X is
   a. greater than 60.
   b. less than 65.
   c. between 40 and 60.
   d. less than 38.
   e. less than 39 or greater than 65.

Further Applications

3. A machine produces bolts whose diameters are normally distributed with an average diameter of 0.25 inch and a standard deviation of 0.02 inch. What is the probability that a bolt will be produced with a diameter
   a. greater than 0.3 inch?
   b. between 0.2 and 0.3 inch?
   c. less than 0.19 inch?
   d. greater than 0.1 inch?

4. A certain brand of light bulbs has a mean life of 750 hours and a standard deviation of 75 hours. If the lifetimes of the bulbs are normally distributed, what is the probability that a light bulb will last
   a. between 750 hours and 850 hours?
   b. between 785 hours and 800 hours?
   c. more than 700 hours?
   d. less than 800 hours?

5. For the SAT math scores, which are normally distributed with a mean of 500 and a standard deviation of 100, find
   a. P_{30}. b. P_{50}. c. P_{70}. d. P_{90}.

6. For the SAT math scores, which are normally distributed with a mean of 500 and a standard deviation of 100, find
   a. Q_1. b. D_7. c. D_3. d. Q_3.

7. Refer to exercise 3. If 100,000 bolts were manufactured, find the expected number of bolts with a diameter
   a. greater than 0.3 inch.
   b. between 0.2 and 0.3 inch.
   c. less than 0.19 inch.
   d. greater than 0.1 inch.

8. Refer to exercise 4. If 100,000 light bulbs were manufactured, find the expected number of light bulbs with a lifetime
   a. between 750 hours and 850 hours?
   b. between 785 hours and 800 hours?
   c. greater than 700 hours?
   d. less than 800 hours?

9. The Apex Taxi Company has found that taxi fares are normally distributed with $\mu = \$4.30$ and $\sigma = \$1.25$. If a driver takes a taxi call at random, what is the probability the fare will be less than $3.05?

10. The mean GPA at a certain university is 2.2 and the standard deviation is 0.5. At a special honors convocation, students with the top 3% of the GPAs are to be given special recognition. If the GPAs are normally distributed, what is the minimum GPA a student needs in order to receive recognition?

11. The heights of adult males are normally distributed with a mean of 70 inches and a standard deviation of 2.6 inches. How high should a doorway be constructed so that 90% of the men can pass through it without having to bend?

12. A coffee dispenser is set to fill cups with an average of 8 ounces of coffee. If the amounts dispensed are normally distributed with a standard deviation of 0.4 ounce, what percentage of 9-ounce cups will overflow?

13. The number of hours a college student studies per week is normally distributed with a mean of 25 hours and a standard deviation of 10 hours.
    a. What percentage of students will study less than 30 hours?
    b. What percentage will study less than 20 hours?
    c. What percentage will study more than 30 hours?
    d. Out of a class of 100 students, approximately how many will study between 12 and 38 hours per week?

14. If X is normally distributed with $\mu = 30$ and $\sigma = 5$, find the value x_0 so that
    a. $P(X > x_0) = 0.05$.
    b. $P(X < x_0) = 0.005$.

Going Beyond

15. The useful life of a particular brand of cathode-ray tube (CRT) used in color televisions is normally distributed with mean life of 7.7 years and a standard deviation of 1.9 years.
    a. What is the probability that a CRT will last more than 8 years?
    b. If the CRTs are guaranteed for 2 years, what percentage of CRTs will have to be replaced?
    c. If the manufacturer will replace only 2% of the CRTs, what guarantee period (to the nearest month) should be stated on the warranty card?

16. A manufacturer of electric motors wants to determine the length of time to guarantee its motors so no more than 4% of them will have to be replaced. If the lives of the motors are normally distributed with a mean life of 10 years and a standard deviation of 0.75 year, find the guarantee period to the nearest month.

17. If the random variable X is normally distributed with mean μ and variance σ^2, find the percentile rank of
    a. $\mu + \sigma$.　　c. $\mu + 2\sigma$.
    b. $\mu - \sigma$.　　d. $\mu - 2\sigma$.

18. Construct a z score plot to determine if a normal distribution could serve as a model for the following sample. That is, determine if the following sample could have been drawn from a normal population.

36	58	42	37	24
52	49	56	34	32
44	53	51	47	37
51	30	31	47	50
43	31	53	44	27

19. If u_1 and u_2 are two random values from the interval $(0, 1)$, then the following formulas can be used to generate two standard normal random variables:

$$z_1 = \sqrt{-\ln(u_2)}\,\cos(2\pi u_1)$$
$$z_2 = \sqrt{-\ln(u_2)}\,\sin(2\pi u_1)$$

where the angle is expressed in radians. The values of u_1 and u_2 can be obtained by using a computer's random number generator or a random number table and selecting three digits at a time. Use these formulas to simulate two observations from the normal distribution having a mean of 25 and a standard deviation of 5 by using the relations $x_1 = \mu + \sigma z_1$ and $x_2 = \mu + \sigma z_2$.

20. a. Write a computer program to simulate 25 observations from the normal distribution having a mean of 30 and a standard deviation of 5.
    b. Calculate the mean and standard deviation for the 25 measurements in part a and construct a histogram to verify that the sample is from the normal distribution having a mean of 30 and a standard deviation of 5.

SECTION 7.4　　*Using Normal Distributions to Approximate Binomial Distributions*

There are many binomial experiments for which the calculation of binomial probabilities involves a number of long and tedious calculations.

EXAMPLE 7.7

Consider the following binomial experiment.

Over a long period of time, it has been determined that 70% of the lawyers who take the bar examination pass the examination. Of 500 lawyers who take the examination next, find the probability that at least 370 of them pass.

In order to solve this problem, we need to find $P(370) + P(372) + \cdots + P(499) + P(500)$. This would involve using the binomial probability formula 131 times. Let's attempt to calculate $P(370)$ first.

By using formula (6.2), we have

$$P(x) = \binom{n}{x} p^x (1 - p)^{n-x}$$

$$P(370) = \binom{500}{370} (0.7)^{370} (0.3)^{130}$$

We now try to compute the binomial coefficient $\binom{500}{370}$.

$$\binom{500}{370} = \frac{(500)!}{(370)! \, (130)!}$$

A hand-held calculator does not help us with this calculation. And do not be surprised if the college's computer also fails you. Later, in application 7.18, we will use a normal distribution to help us approximate the probability that at least 370 lawyers will pass the bar examination.

Bar Graphs for Binomial Distributions

Recall that we use vertical line segment graphs in chapter 6 to represent binomial probability distributions. Another type of probability graph, called a *bar graph* or *histogram,* employs vertical bars instead of vertical line segments. At each value of x on the horizontal axis, two points are marked at $x - 0.5$ and $x + 0.5$. The distance between these points, which is 1, becomes the width of one side of a rectangular bar and the length of the other side becomes $P(x)$. The area of the xth bar is then $P(x)$, as indicated in figure 7.19. Note that the area under a probability bar graph is 1, since $\Sigma P(x) = 1$. Although areas are generally associated with probabilities of continuous distributions, we should be careful not to classify a binomial distribution as continuous. The use of a bar graph for a binomial distribution is for convenience only; it is especially convenient when approximating binomial probabilities.

FIGURE 7.19

Probability bar graph

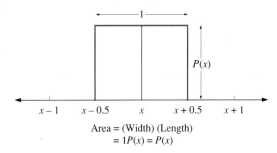

Area = (Width) (Length)
= 1P(x) = P(x)

For a binomial distribution that is symmetrical (or approximately symmetrical) about its mean, the bar graph closely resembles a normal distribution. As a result, a normal distribution can be used to approximate a binomial distribution (see fig. 7.20).

Recall that in any normal distribution with mean μ and standard deviation σ, 99.7% of the distribution (or, for practical purposes, almost all of the distribution) lies between $\mu - 3\sigma$ and $\mu + 3\sigma$. Suppose a binomial distribution has mean $\mu = np$ and standard deviation $\sigma = \sqrt{np(1 - p)}$. If it is to be approximated by a normal distribution, the normal distribution should have a mean equal to np and a standard deviation equal to $\sqrt{np(1 - p)}$. Since the values for the binomial distribution are greater

FIGURE 7.20

Approximating a binomial
distribution with a normal
distribution

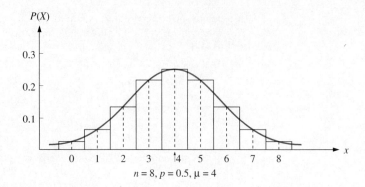

$n = 8, p = 0.5, \mu = 4$

than or equal to 0 and less than or equal to n, we must have $0 \le \mu - 3\sigma$ and $\mu + 3\sigma$ $\le n$, where μ and σ represent the mean and standard deviation, respectively, for the binomial. Thus, the following two inequalities must be satisfied:

1. $0 \le np - 3\sqrt{np(1 - p)}$
2. $np + 3\sqrt{np(1 - p)} \le n$

If $p \le 0.5$, it can be shown that the inequalities are equivalent to $np \ge 4.5$ and if $p \ge 0.5$, they are equivalent to $n(1 - p) \ge 4.5$. Thus, we state the following rule:

> A binomial distribution can be approximated by a normal distribution if $np \ge 5$ and $n(1 - p) \ge 5$.

or the smaller # of $\left(n\frac{p}{(1-p)}, n\frac{(1-p)}{p}\right)$ is > 9

Note that for large values of n, both inequalities should hold, especially for small values of p or $1 - p$.

EXAMPLE 7.8

If $p = 0.01$ and $n = 600$, then this binomial distribution can be approximated by a normal distribution, since $np = (600)(0.01) = 6$ and $n(1 - p) = (600)(0.99) = 594$.

To use a normal distribution to approximate a binomial distribution, we follow these steps.

1. Check to see if $np \ge 5$ and $n(1 - p) \ge 5$ hold.
2. If so, calculate μ and σ for the binomial distribution.
3. The normal distribution with mean μ and standard deviation σ is used (the parameters were found in step 2) as the approximating distribution.
4. Construct a bar graph for the binomial distribution and determine the limits for finding areas under the normal curve.
5. Find z scores for these limits and use the standard normal to find the corresponding area(s). This number is the approximation to the binomial probability sought.

APPLICATION 7.18

Find $P(x \geq 370)$ for the binomial experiment discussed in example 7.7, at the beginning of this section, where $n = 500$ and $p = 0.70$.

Solution: Since $np = (500)(0.7) = 350 \geq 5$ and $n(1 - p) = (500)(0.3) = 150 \geq 5$, a normal approximation is possible. By using formulas (6.4) and (6.7), we have

$$\mu = np = 350$$
$$\sigma = \sqrt{np(1 - p)}$$
$$= \sqrt{(350)(0.3)} = 10.25$$

Thus, the normal distribution with $\mu = 350$ and σ 10.25 will be used for approximation purposes.

A bar graph for the binomial is shown here along with the normal distribution with $\mu = 350$ and $\sigma = 10.25$.

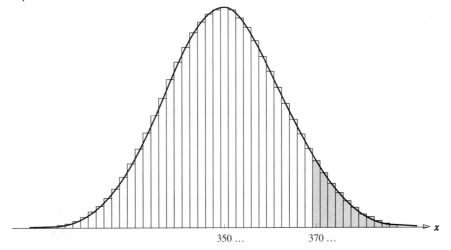

We want to find the sum of the areas of the bars that are shaded, and as an approximation to the sum of these areas, we will find the area under the normal curve to the right of 369.5. (To see this note that the bar labeled 370 has a width of 1 and extends from 369.5 to 370.5, a distance of 1, as indicated in the following figure.)

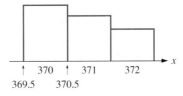

Hence, we find the shaded area under the following normal curve:

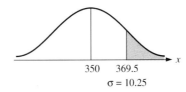

To find the shaded area, we find the z score for 369.5:

$$z = \frac{x - \mu}{\sigma}$$

$$= \frac{369.5 - 350}{10.25} = 1.90$$

From the standard normal table, we find

$$P(Z \geq 1.90) = 0.5 - 0.4713$$
$$= 0.0287$$

Thus,

$$P(X \geq 370) \approx 0.0287 \quad ■$$

MINITAB can be used to determine $P(X < 370)$ for the bar examination problem. Then by subtraction we can determine $P(X \geq 370)$. Computer display 7.2 illustrates the MINITAB commands used and the resulting output.

Computer Display 7.2

```
MTB > CDF 369.5;
SUBC> NORMAL 350 10.25.
  369.5000 0.9714
MTB >
```

The third line in the display indicates that $P(X < 370) = 0.9714$. By subtraction we find that

$$P(X \geq 370) = 1 - 0.9714 = 0.0286$$

The difference of 0.0001 between the answer we found in application 7.18 and the answer computed with the help of MINITAB is due to the greater numerical accuracy obtained with MINITAB.

APPLICATION 7.19
If it is known that 60% of cattle inoculated with a serum are protected from a certain disease, find the probability that 8 or 9 of 15 cows inoculated will not contract the disease by using

    a. the binomial probability formula.

    b. a normal distribution to approximate the binomial.

    c. the binomial probability tables.

Then compare the results.

Solution: We have $n = 15$, $p = 0.6$, and $x = 8, 9$.

    a. By using the binomial probability formula (6.2), we have

$$P(8) = \binom{15}{8}(0.6)^8(0.4)^7$$

$$= (6435)(0.6)^8(0.4)^7 = 0.177$$

and

$$P(9) = \binom{15}{9}(0.6)^9(0.4)^6$$
$$= (5005)(0.6)^9(0.4)^6 = 0.207$$

Hence, the probability that 8 or 9 of the 15 inoculated cows will not contract the disease is

$$P(8) + P(9) = 0.177 + 0.207 = 0.384$$

b. We first note that $np = (15)(0.6) = 9$ and $n(1 - p) = (15)(0.4) = 6$, both of which are greater than 5. By using formula (6.4) we have

$$\mu = np = 9$$

and by using formula (6.7), we have

$$\sigma = \sqrt{np(1 - p)}$$
$$= \sqrt{(9)(0.4)} = 1.897$$

Next, we draw part of the binomial distribution to determine the limits for the area under the normal curve with $\mu = 9$ and $\sigma = 1.897$.

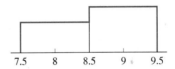

From the figure we see that we want the area under the normal curve, between 7.5 and 9.5.

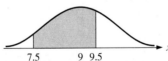

Next, we find the z scores for 7.5 and 9.5. The z score for $x = 7.5$ is

$$z = \frac{7.5 - 9}{1.897} = -0.79$$

and the z score for $x = 9.5$ is

$$z = \frac{9.5 - 9}{1.897} = 0.26$$

From the standard normal table, we find that the area under the standard normal curve above the horizontal axis between $z = -0.79$ and $z = 0.26$ is

$$P(-0.79 < Z < 0.26) = 0.2852 + 0.1026 = 0.3878$$

This value compared with 0.3840 represents less than 1% error; the approximation is good, even for an n as small as 15.

c. By using the binomial probability tables, we find $P(8) = 0.177$ and $P(9) = 0.207$. Thus,

$$P(8 \text{ or } 9) = P(8) + P(9)$$
$$= 0.177 + 0.207$$
$$= 0.384$$

which agrees exactly with the result in part a. ▪

EXERCISE SET 7.4

Basic Skills

1. For a binomial experiment with $p = 0.5$ and $n = 10$, find $P(4 \le X \le 6)$ by using
   a. the binomial probability formula.
   b. a normal approximation.
   c. the binomial probability tables.
   Compare your results.

2. For a binomial experiment with $p = 0.6$ and $n = 15$, find $P(8 \le X \le 10)$ by using
   a. the binomial probability formula.
   b. a normal approximation.
   c. the binomial probability tables.
   Compare your results.

Further Applications

3. The probability that an adverse reaction will result from a flue shot is 0.02. If 1000 individuals are randomly selected to receive the flu shot, find the approximate probability that
   a. at most 20 people will have an adverse reaction.
   b. at least 24 people will have an adverse reaction.
   c. between 21 and 29 (inclusive) will have an adverse reaction.
   d. exactly 210 will have an adverse reaction.

4. Suppose the probability that a certain thumbtack will land point up is 0.7. If 1000 thumbtacks are tossed, find the approximate probability that
   a. 715 or more will land point up.
   b. between 685 and 720 (inclusive) will land point up.
   c. exactly 675 will land point up.
   d. at most 730 will land point up.

5. If a fair coin is tossed 1000 times, find the approximate probability that
   a. at most 530 coins will land heads up.
   b. between 485 and 520 (inclusive) tosses will land heads up.
   c. exactly 525 tosses will land heads up.
   d. the number of tosses that land heads up is greater than 490.

6. In a certain city, 60% of the families own their own homes. If 500 families are selected, at random from the city, what is the approximate probability that
   a. at least 315 families own their homes?
   b. between 290 and 310 families (inclusive) own their homes?
   c. exactly 305 families own their homes?
   d. at most 320 families own their homes?

7. In a quality control program for a production process, if a sample of 100 items produced yields less than 10 defective items, the production is considered satisfactory; otherwise it is considered unsatisfactory. What is the approximate probability that in a production of 100 items the production process is considered
   a. satisfactory when, in fact, the process produces 15% defective items?
   b. unsatisfactory when, in fact, the process produces 15% defective items?

8. To determine whether a new drug is effective, it will be given to 500 patients. If 440 or more achieve positive results, the drug will be classified as effective. Otherwise, the drug will be classified as ineffective. Find the approximate probability that the drug will be classified ineffective if, in fact, the drug is 90% effective.

9. The probability that a certain brand of thumbtack will land point up when dropped is 0.60. If 500 tacks are dropped, what is the approximate probability that 325 or less land point up?

10. If 1000 coins are tossed the most likely outcome is 500 heads. Use the normal approximation to find the probability of getting 500 heads.

Going Beyond

11. The equation for the *standard normal curve* is

$$f(z) = \frac{1}{\sqrt{2\pi}} e^{-z^2/2}$$

The result in exercise 10 could be approximated by multiplying $f(0)$ by $1/\sigma$, since the width of the 500th bar in x units is 1 and in z units is $1/\sigma$. Perform this calculation and compare your answer to the answer obtained in exercise 10.

12. Prove that if $0 \le np - 3\sqrt{np(1-p)}$ and $np + 3\sqrt{np(1-p)} \le n$, then $np \ge 4.5$ or $n(1-p) \ge 4.5$.

SECTION 7.5   *Exponential Distributions*

A distribution that is closely related to the Poisson distribution, discussed in section 6.6, is the *exponential distribution*. The Poisson distribution deals with the number of events occuring in a time interval, whereas the exponential distribution can be used to model the times between occurrences of successive events. The distribution of elapsed times between occurrences of two successive events very often follows an exponential distribution. Examples are times of successive

■ Customers arriving at a service facility.

■ Calls coming into a switchboard.

■ Breakdowns in a certain system.

■ Earthquakes in a certain region.

■ Failures of certain types of electronics components.

A random variable X is said to have an exponential distribution if the probability density function (pdf) has the following form:

FIGURE 7.21

Graph of exponential density functions

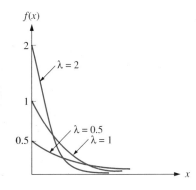

Exponential Probability Density Function

$$f(x) = \lambda e^{-\lambda x}$$
where $x > 0$ and $\lambda > 0$

An **exponential random variable** has the following properties:

1. Its mean is equal to $1/\lambda$.
2. Its variance is equal to $1/\lambda^2$.

Note that an exponentially distributed random variable has its mean equal to its standard deviation. Since the mean and standard deviation of an exponential random variable depend only on the constant λ (called a **parameter**), the shape of the graph of its pdf depends on the parameter λ. Figure 7.21 displays graphs of three **exponential density functions** corresponding to $\lambda = 0.5$, $\lambda = 1$, and $\lambda = 2$. Notice that each curve crosses the y-axis at λ.

Since an exponential random variable is continuous, to find probabilities associated with it we must be able to determine areas under the graph of its pdf above the x-axis. It can be shown by calculus that the area under the curve, above the x-axis, and between the lines $x = 0$ and $x = c$ $(c > 0)$ can be found by evaluating the expression $1 - e^{-\lambda c}$. The function F defined by $F(x) = 1 - e^{-\lambda x}$ $(x > 0)$ is called the **cumulative distribution function** (cdf) for the exponential random variable X having parameter λ.

Cumulative Distribution Function for Exponential Random Variable

$$F(x) = 1 - e^{-\lambda x} \text{ where } x > 0 \text{ and } \lambda > 0. \tag{7.2}$$

EXAMPLE 7.9

Suppose X is an exponentially distributed random variable with parameter $\lambda = 0.5$. Let's calculate the probability that the value of the random variable will be less than 3, that is, let's determine $P(X < 3) = F(3)$. To determine the value $F(3)$ we can use formula (7.2).

$$F(3) = 1 - e^{-(0.5)(3)} = 1 - e^{-1.5}$$

At this point we can either use table 3 in appendix B or a calculator having the capability to calculate the value of the exponential, e^{-6}. Table 3 provides the values of e^{-x} for values of x from 0 to 10 in increments of 0.05. To use table 3 to approximate the value of $e^{-1.5}$, we locate the value of $x = 1.5$ in the first column of the table and, adjacent to it, in the second column of the table is the value 0.2231. Thus, the approximate value of $F(3)$ is

$$F(3) = 1 - e^{-1.5} \approx 1 - (0.2231) = 0.7769$$

APPLICATION 7.20

If a random variable has the exponential distribution with $\lambda = 1.5$, find $P(1.8 \leq X \leq 2.4)$.

Solution: Note that $P(1.8 \leq X < 2.4) = P(X \leq 2.4) - P(X < 1.8) = F(2.4) - F(1.8)$, where F is the cumulative distribution function.

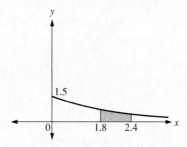

We determine the cumulative probabilities $F(2.4)$ and $F(1.8)$ by using formula (7.2):

$$F(2.4) = 1 - e^{-(1.5)(2.4)} = 1 - e^{-3.6}$$

and

$$F(1.8) = 1 - e^{-(1.5)(1.8)} = 1 - e^{-2.7}$$

We can use table 3 in appendix B or a calculator having the capability to evaluate $e^{-3.6}$ and $e^{-2.7}$. By using either, we determine the approximations

$$e^{-3.6} \approx 0.0273$$

and

$$e^{-2.7} \approx 0.0672$$

Hence,

$$F(2.4) - F(1.8) = (1 - e^{-3.6}) - (1 - e^{-2.7})$$
$$= (1 - 0.0273) - (1 - 0.0672)$$
$$= 0.9727 - 0.9328 \approx 0.0399$$

Hence, $P(1.8 \leq X \leq 2.4)$ is approximately equal to 0.04. ■

APPLICATION 7.21

Suppose a computer user is working at a computer terminal connected to a host computer in a networked system. The time (in seconds) it takes the host computer to respond to a user's inquiry has an exponential distribution with an expected response time of 2 seconds. Determine the probability that the system's response to a user's request is at most 5 seconds.

Solution: Let the random variable X represent the system's response time. Since we are given that $E(X) = 2$ seconds and $E(X) = 1/\lambda$, we know that $\lambda = 1/2$. By using the cumulative probability function (7.2), we have

$$P(X \leq 5) = F(5)$$
$$= 1 - e^{-(0.5)(5)}$$
$$= 1 - e^{-2.5}$$
$$= 1 - 0.0821 \quad \text{(from table 3, appendix B)}$$
$$= 0.9179$$

Thus the probability that a user will wait at most 5 seconds for the system to respond is 0.9179. ▪

MINITAB can be used to determine $F(5) = P(X < 5)$ in application 7.21. The CDF command and the EXPONENTIAL subcommand are used. Computer display 7.3 contains the commands used and corresponding output. Note that the value of the mean is supplied with the subcommand and not the value of the parameter.

Computer Display 7.3

```
MTB > CDF 5;
SUBC> EXPONENTIAL 2.
  5.0000   0.9179
```

APPLICATION 7.22

The time (in minutes) between telephone calls arriving at a switchboard is exponentially distributed with parameter $\lambda = 1.8$.

a. Find the probability that there will be a wait of at least 1 minute between the first and the second telephone calls arriving at this switch board on a given day.

b. Find the probability that there will be a wait of at least 1 minute each for the first three telephone calls arriving at this switch board on a given day.

Solution: Let X represent the time in minutes between successive calls. The cumulative distribution function is $F(x) = 1 - e^{-1.8x}$.

a. The probability that there will be a wait of at least a minute between the first and second calls is

$$P(X \geq 1) = 1 - F(1)$$
$$= 1 - (1 - e^{-(1.8)(1)})$$
$$= e^{-1.8}$$
$$= 0.1653 \quad \text{(from table 3, appendix B)}$$

b. Since the calls arriving at the switchboard are independent, the probability that there will be a wait of at least 1 minute each for the first three telephone calls arriving at this switch board on a given day is $(0.1653)^3 = 0.0045$. ▪

The *reliability of a component (or system) at time t, R(t),* is defined by $R(t) = P(T > t)$, where T is the life length of the component (or system). If the length of time between successive breakdowns of a component (or system) is distributed as an exponential random variable with parameter λ, then the **reliability function** is defined as

Reliability Function

$$R(t) = e^{-\lambda t}$$

APPLICATION 7.23

Suppose the length of life (in hours) of an electronic device is exponentially distributed with $\lambda = 0.002$. If the device is 95% reliable, determine the number of hours the device will operate before a breakdown.

Solution: Since $R(t) = e^{-\lambda t}$ and $\lambda = 0.002$, the reliability function for the device is $R(t) = e^{-0.002t}$. Hence, we must have $0.95 = e^{-0.002t}$. We can use logarithms to solve this equation for t.

$$-0.002t = \ln(0.95)$$

$$t = \frac{\ln(0.95)}{-0.002} = 25.65 \text{ hours}$$

Therefore, if each of 100 such devices is operating for 25.65 hours, approximately 95 of them will not fail during that period. ■

EXERCISE SET 7.5

Basic Skills

1. Use a calculator to determine the value of $e^{-c\lambda}$ for each of the following pairs:
   a. $\lambda = 2, c = 3$ b. $\lambda = 1.5, c = 2.3$
   c. $\lambda = 0.75, c = 4.1$

2. Use a calculator to determine the value of $e^{-c\lambda}$ for each of the following pairs:
   a. $\lambda = 1.2, c = 2.3$ b. $\lambda = 1.8, c = 3.3$
   c. $\lambda = 1.65, c = 10.4$

3. Suppose that X is exponentially distributed with parameter $\lambda = 6.4$. Determine the mean and variance of X.

4. Suppose that X is exponentially distributed with parameter $\lambda = 0.02$. Determine the mean and variance of X.

5. Suppose that X is exponentially distributed with parameter $\lambda = 0.12$. Find the following probabilities:
   a. $P(X > 8.3)$ b. $P(X < 9.3)$
   c. $P(8.2 < X < 11.2)$

6. Suppose that X is exponentially distributed with parameter $\lambda = 0.32$. Find the following probabilities:
   a. $P(X > 2.3)$ b. $P(X < 3)$
   c. $P(2.8 < X < 4.1)$

Further Applications

7. Suppose the length of life (in hours) of an electronic device is exponentially distributed with $\lambda = 0.02$. If the device is 95% reliable, determine the number of hours that the device will operate before a breakdown.

8. Suppose the length of life (in hours) of an electronic device is exponentially distributed with $\lambda = 0.012$. If the device is 89% reliable, determine the number of hours that the device will operate before a breakdown.

9. Suppose that the life length of an electronic device is exponentially distributed with a mean of 500 hours. It is known that the reliability of the device is 0.88. How many hours must the device be operated to achieve a reliability of 0.95?

10. Suppose that the life length of an electronic device is exponentially distributed with a mean of 450 hours. It is known that the reliability of the device is 0.80. How many hours of operation must the device be operated to achieve a reliability of 0.90?

11. Let X represent the time (in minutes) between successive arrivals at the emergency room of a hospital and suppose X is exponentially distributed with parameter $\lambda = 0.10$.

Find

a. the expected time between two successive arrivals.
b. the standard deviation of two successive arrivals.
c. $P(X \leq 10.5)$.
d. $P(X > 9.6)$.

12. Suppose the operating time (in years) of a special battery is exponentially distributed with an expected life of two years. Find the probability that a battery of this type will have an operating time
    a. longer than three years.
    b. between one and three years.
    c. less than two years.

13. Suppose the time (in minutes) between successive arrivals at a drive-up window at a bank is exponentially distributed with parameter $\lambda = 0.5$. Determine the probability that at least one out of the next four arrivals will have to wait longer than 3 minutes.

14. Suppose the time (in minutes) between successive arrivals at the emergency room of a city hospital is exponentially distributed with parameter $\lambda = 0.4$. Determine the probability that at least two out of the next five arrivals will have to wait less than 2 minutes.

15. Let X be an exponential random variable with parameter $\lambda = 0.125$. Find the
    a. 60th percentile.
    b. third quartile.
    c. median.
    d. fourth decile.

16. Let X be an exponential random variable with parameter $\lambda = 0.08$. Find the
    a. 40th percentile.
    b. first quartile.
    c. median.
    d. seventh decile.

17. The length of time that it takes for one individual to be served at a cafeteria is exponentially distributed with a mean of 3 minutes. What is the probability that a person is served in less than 2 minutes on at least three of the next six days?

18. Suppose a system contains an integrated circuit whose lifetime (in years) is distributed exponentially with an expected value of seven years. If six of these integrated circuits are installed in different systems, what is the probability that at least two are still functioning at the end of ten years?

19. A manufacturer of electric fans wants to determine the length of time to guarantee its fans so that no more than 5% of them will have to be replaced. If the life lengths of the fans is exponentially distributed with a mean of seven years, find the guarantee period to the nearest month.

20. A manufacturer of automobile batteries wants to determine the warranty period for one type of battery it manufactures. If the life lengths of this type of battery is exponentially distributed with a mean life of two years, find the warranty period, to the nearest month, if it expects to replace no more than 6% of this type of battery.

CHAPTER SUMMARY

In this chapter we studied uniform, normal, and exponential probability distributions. We learned how to associate areas under a curve with probabilities. We learned how to find probabilities associated with the uniform distributions and the standard normal distribution. We also learned how to use the standard normal distribution to compute probabilities associated with any normal distribution. We learned how to use the normal distribution to approximate binomial distributions. These approximations can be used whenever np and $n(1 - p)$ are greater than or equal to 5. Finally, we learned how to find probabilities associated with exponential distributions. The distribution of elapsed times between occurrences of two successive events often follow an exponential distribution.

CHAPTER REVIEW

■ **IMPORTANT TERMS** ■

The following chapter terms have been mixed in ordering to provide you better review practice. For each of the following terms, provide a definition in your own words. Then check your responses against the definitions given in the chapter.

uniform random variable	probability density functions	skewed distribution
continuous distribution	exponential random variable	standard normal distribution
percentile rank	cumulative distribution function	standard normal table
deciles	exponential density function	uniform probability distribution
empirical rule	quartiles	parameter
normal distribution	z score plot	reliability function
percentiles		

■ **IMPORTANT SYMBOLS** ■

Z, the standard normal random variable

0^+, a number slightly larger than 0

1^-, a number slightly smaller than 1

P_n, the nth percentile

Q_n, the nth quartile

D_n, the nth decile

λ, exponential parameter

■ **IMPORTANT FACTS AND FORMULAS** ■

Uniform probability density function: $f(x) = \dfrac{1}{b - a}$ for values of x between a and b where $a < b$.   (7.1)

The area under a normal curve is one.

Empirical Rule: In any normal distribution, approximately

68% of the measurements fall within $\mu \pm \sigma$,
95% of the measurements fall within $\mu \pm 2\sigma$, and
99% of the measurements fall within $\mu \pm 3\sigma$.

Transformation rule: $z = \dfrac{x - \mu}{\sigma}$.

A normal distribution with $\mu = np$ and $\sigma = \sqrt{np(1 - p)}$ can be used to approximate a binomial distribution with parameters n and p if both $np \geq 5$ and $n(1 - p) \geq 5$.

The probability density function for a random variable that is exponentially distributed with parameter λ: $f(x) = \lambda e^{-\lambda x}$ for $\lambda > 0$ and $x > 0$.

Mean of an exponential random variable with parameter λ: $\mu = \dfrac{1}{\lambda}$.

Standard deviation of an exponential random variable with parameter λ: $\sigma = \dfrac{1}{\lambda}$.

Cumulative distribution function for an exponential random variable with parameter λ: $F(x) = 1 - e^{-\lambda x}$ for $\lambda > 0$ and $x > 0$.   (7.2)

Reliability function for an exponential random variable with parameter λ: $R(t) = e^{-\lambda t}$.

REVIEW EXERCISES

1. Suppose X is a uniform random variable with values in the range extending from 10 to 90. Determine the following probabilities:
   a. $P(20 < X < 50)$
   b. $P(10 < X < 70)$
   c. $P(X < 40)$
   d. $P(X > 60)$
   e. $P[(X < 50) \text{ or } (X > 79)]$

2. Find the following probabilities:
   a. $P(-1.2 < Z < 2.2)$ f. $P(Z > 1.58)$
   b. $P(Z > 1.24)$ g. $P(Z < -2.14 \text{ or } Z > 2.18)$
   c. $P(Z > -1.78)$ h. $P(1.14 < Z < 2.76)$
   d. $P(Z < -2.14)$ i. $P(-2.11 < Z < -1.17)$
   e. $P(Z < 1.58)$

3. For each of the following, find a value of z_0 so that
   a. $P(-z_0 < Z < z_0) = 0.70$.
   b. $P(Z > z_0) = 0.60$.
   c. $P(Z > z_0) = 0.40$.
   d. $P(Z < z_0) = 0.80$.
   e. $P(Z < z_0) = 0.30$.

4. Let X be normally distributed with $\mu = 40$ and $\sigma = 5$. Find the following probabilities:
   a. $P(X > 47)$
   b. $P(X > 32)$
   c. $P(X < 36)$
   d. $P(X < 51)$
   e. $P(36 < X < 48)$
   f. $P(42 < X < 61)$

5. Let X be normally distributed with $\mu = 50$ and $\sigma = 10$. For each of the following, find the value of x_0 so that
   a. $P(X > x_0) = 0.70$.
   b. $P(X > x_0) = 0.30$.
   c. $P(X < x_0) = 0.30$.
   d. $P(X < x_0) = 0.60$.

6. If the test scores of 400 students are normally distributed with a mean of 100 and a standard deviation of 10, approximately how many students scored between 90 and 110?

7. A statistics professor announces that 10% of the grades he gives are A's. The results of the final examination indicate that the mean score is 73 and the standard deviation is 6. What minimum score must a student obtain to receive an A? (Assume that the grades are normally distributed.)

8. The daily output of an assembly line is normally distributed with a mean of 165 units and a standard deviation of 5 units. Find the probability that the number of units produced per day is
   a. less than 162 units.
   b. greater than 173 units.
   c. between 152 units and 174 units.
   d. between 171 units and 177 units.

9. Refer to exercise 8. Find the number of units x so that on 75% of the days the production output is at least x units.

10. Suppose the typical speeds on Maryland highways are normally distributed with a mean of 59 mph and a standard deviation of 4 mph. If the state police are instructed to ticket the fastest 10% of the motorists, what is the fastest you could travel on Maryland highways and not receive a ticket?

11. A certain type of battery has an operating lifetime which is exponentially distributed with a mean of two years. Find the probability that a battery will have an operating lifetime
    a. more than three years.
    b. between one and three years.

12. A student with an IQ of 135 claims her IQ score is in the top 3% of her class. Is her claim true if the scores of her classmates are normally distributed with a mean of 120 and a standard deviation of 7?

13. A particular type of birth-control pill is 90% effective. If 50 people use the pill, what is the approximate probability that the pill will result in one birth? Assuming that male and female births are equally likely, what is the probability that this birth is a male?

14. It is estimated that 85% of all household plants are overwatered. In a group of 500 plants, what is the approximate probability that
    a. 433 are overwatered?
    b. between 420 and 435 (inclusive) are overwatered?
    c. at least 430 are overwatered?
    d. at most 440 are overwatered?

15. If it is estimated that 80% of all students who take the Math 209 final exam pass, what is the approximate probability that at least half of a group of 30 students who take the final exam pass?

16. Let X be an exponentially distributed random variable having an expected value of 5. Find
    a. $P(X > 5.5)$.
    b. $P(X < 4)$.
    c. $P(3.5 < X < 5.5)$.
    d. $P(X = 5)$.

Computer Applications

1. SAT math scores have a mean of 500 and a standard deviation of 100. Use a computer program to find the probability of getting a score less than 550 (*Hint:* Use commands similar to

        CDF 500;
        NORMAL MU = 500 SIGMA = 100.)

2. Generate a random sample of 200 scores from a normal population having a mean of 500 and a standard deviation of 100. Display the sorted measurements and determine the proportion of measurements less than 550. Compare this result with the answer obtained in computer application 1. (*Hint:* Use commands similar to

        RANDOM 200 C1;
        NORMAL MU = 500 SIGMA = 100.
        SORT C1 C2
        PRINT C2.)

3. Repeat computer application 2 for 500 scores instead of 200. (*Hint:* If MINITAB is used, the CODE command can be used to count the number of measurements less than 550.)

4. Determine if the following sample of data could have come from a normal population. What would you estimate the mean and standard deviation of the population to be?

68	80	68	89	82	61	67	80	76	74	63
74	90	74	78	73	64	71	93	60	74	87
74	83	73	88	90	80	87	81	68	100	79
70	91	83	68	79	74	79	65	84	71	74
62	69	95	76	97	62	71	71	80	85	67
78	57	66	76	75	84	84	62	71	82	82
67	66	86	75	60	60	85	76	80	95	79
79	74	71	79	91	74	62	75	80	75	77
70	71	72	74	73	61	64	67	91	70	79
77										

5. If X is a normally distributed random variable with $\mu = 30$ and $\sigma = 5$, determine a value x_0 so that $P(-x_0 < X < x_0) = 0.95$.

◾ CHAPTER ACHIEVEMENT TEST ◾

1. Suppose X is uniformly distributed between the values 7 and 13. Find the following probabilities:
   a. $P(X > 8)$
   b. $P(X < 12)$
   c. $P(8 < X < 12)$

2. Find the following probabilities:
   a. $P(Z < 0.51)$
   b. $P(-1.2 < Z < 2.3)$
   c. $P(-2.4 < Z < -0.78)$
   d. $P(Z > -1.14)$
   e. $P(Z < -0.24)$

3. For each of the following, find the value of z_0 so that
   a. $P(Z > z_0) = 0.30$.
   b. $P(Z > z_0) = 0.75$.
   c. $P(Z < z_0) = 0.40$.

4. Suppose X is normally distributed with $\mu = 20$ and $\sigma = 3$. Find
   a. $P(X > 27)$.
   b. $P(X > 16)$.
   c. $P(13 < X < 15)$.
   d. $P(25 < X < 28)$.

5. For test question 4, find the value of z_0 for each of the following:
   a. $P(X > z_0) = 0.90$
   b. $P(X > z_0) = 0.40$
   c. $P(X < z_0) = 0.78$

6. If X is normally distributed with $\mu = 40$ and $\sigma = 6$, find
   a. Q_3.
   b. P_{40}.

7. The useful lifetimes of a certain brand of light bulb are normally distributed with $\mu = 250$ hours and $\sigma = 50$ hours. If the manufacturer guarantees that its lights will last at least 125 hours, what percentage of the bulbs does it expect to replace under this guarantee?

8. A manufacturer of computer chips claims that only 1% of its chips are defective. If we assume the claim to be true, what is the probability that no more than 25 chips are defective in the next batch of 2000 chips?

9. If X is exponentially distributed with $\lambda = 3$, determine
   a. $P(X < 4)$.
   b. $P(X > 2)$.
   c. $P(2 < X < 5)$.
   d. the median.

UNIT THREE

Inferential Statistics

8

Sampling Theory

CHAPTER OBJECTIVES

In this chapter we will investigate

▷ *Bias.*

▷ *Sampling error.*

▷ *How to use the random number table.*

▷ *Various methods for drawing a random sample.*

▷ *Sampling distributions.*

▷ *The sampling distribution of the mean.*

▷ *What the mean, variance, and standard deviation are for the sampling distribution of the mean.*

▷ *How to find probabilities associated with the sampling distribution of the mean when the population is normal.*

▷ *t distributions.*

▷ *The central limit theorem.*

▷ *Applications of the central limit theorem.*

▷ *The sampling distribution of the sample sum.*

▷ *The mean, variance, and standard deviation of the sampling distribution of the sample sum.*

▷ *The sampling distribution of the sample proportion.*

▷ *The mean, variance, and standard deviation of the sampling distribution of the sample proportion.*

▷ *Applications using the sampling distribution of the sample proportion.*

|||| MOTIVATOR 8 ➤

Consider the experiment of tossing a fair, six-sided die ten times and determining the sum of the ten results. For example, suppose ten tosses result in the following sequence: 2, 3, 3, 6, 5, 1, 4, 4, 1, and 5. The sum of the results is 34. Further, suppose the experiment is repeated 1000 times and a histogram is constructed for the 1000 sums. Would you expect the histogram to have any particular shape? What would you expect for the mean of the 1000 sums? The answers to these two questions are quite surprising. Let's use MINITAB to simulate the experiment 1000 times, plot a histogram of the results, and calculate certain descriptive statistics. Computer display 8.1 contains the commands used and the corresponding output.

Computer Display 8.1

```
MTB > RANDOM 1000 C1-C10;
SUBC > INTEGER 1 6.
MTB > RSUM C1-C10 C11
MTB > HISTOGRAM C11

HISTOGRAM OF C11 N = 1000
EACH * REPRESENTS 10 OBS.

MIDPOINT      COUNT
      16          1 *
      20          2 *
      24         28 ***
      28        124 *************
      32        221 **********************
      36        284 ****************************
      40        208 ********************
      44        102 **********
      48         28 ***
      52          2 *


MTB > DESCRIBE C11

           N     MEAN  MEDIAN  TRMEAN  STDEV  SEMEAN
C11     1000   35.232  35.000  35.230  5.397   0.171

          MIN      MAX      Q1      Q3
C11    16.000   50.000  31.000  39.000
```

Notice that the histogram is bell shaped and approximates a normal distribution. The mean is very close to 35, the product of 10 and the mean of the six possible outcomes that result when the die is tossed. That is, $35.232 \approx (10)[(1 + 2 + 3 + 4 + 5 + 6)/6] = (10)(3.5) = 35$. If it was possible to repeat the experiment indefinitely and record the results, the mean of the distribution of sums would be exactly equal to 35. These two phenomena did not occur by accident; they are related to a remarkable result, called the *central limit theorem,* which we will discuss in section 8.4. We shall examine this result and other related results in this chapter.

Chapter Overview

A major concern of inferential statistics is to estimate unknown population characteristics (parameters) by examining information gathered from a subcollection (sample) of the population. The focus of concern is on the sample. A sample should be representative of the population if it is to be used to study the population. We will follow certain selection procedures to ensure that our samples accurately reflect the sampled population. Only when representative samples are used can probabilistic statements be made about the population being sampled. For example, suppose we are interested in the attitudes of college students toward studying. The relevant population in this case is the collection of responses from all college students regarding studying. A sample is a subcollection from the population that we will use to gauge the attitudes of all college students. If we use the responses of only social fraternity members, then we will most likely obtain biased or atypical results. Or, if we use only the responses from members of honorary fraternities, we would probably obtain results biased in favor of studying. By using a biased sample, we would obtain a distorted view of the attitudes of the entire student body.

SECTION 8.1   *Types of Errors and Random Samples*

When it is desirable to study the characteristics of large populations, samples are used for many reasons. A complete enumeration of the population, called a **census,** may not be economically possible; or, there may not be enough time to examine the whole population. In some situations a census may not be possible; for example, a census of the sea life population in the Atlantic Ocean is not possible.

EXAMPLE 8.1

The following show the uses of sampling in different fields:

1. *Politics*—Samples of voters' opinions are used by candidates to gauge public opinion and support in elections.

2. *Education*—Samples of students' test scores are used to determine the effectiveness of a teaching technique or program.

3. *Industry*—Samples of products coming off an assembly line are used for quality control purposes.

4. *Medicine*—Samples of blood-sugar measurements for diabetic patients are used to test the efficacy of a new technique or drug.

5. *Agriculture*—Samples of corn yields per plot are used to test the effects of a new fertilizer on crop yields.

6. *Government*—Samples of voters' opinions are used to determine public opinions on matters related to national welfare and security.

Two general types of errors may occur when sample values (or statistics) are used to estimate population values (or parameters): sampling error and nonsampling error. **Sampling error** refers to the natural variation inherent among samples taken from the same population. This type of error will exist when the sample is not a perfect copy of the population. Even if great care is taken to ensure that two samples of the same size are representative of the same population, we would not expect both samples to be identical in every detail. For example, we would not expect the two sample means to be identical. Sampling error is an important concept that will help us better understand the nature of inferential statistics. We will further examine the concept of sampling error later in this chapter.

Errors that arise with sampling but that cannot be classified as sampling errors are called **nonsampling errors.** Sampling bias is one type of nonsampling error. **Sampling bias** refers to a systematic tendency inherent in a sampling method that gives estimates of a parameter that are, on the average, either smaller (negative bias) or larger (positive bias) than the actual parameter. Errors arising from data collection fall into this category (see examples 8.2 and 8.3). Errors that result from data accumulation or processing are also classified as nonsampling errors (see example 8.4).

EXAMPLE 8.2

If we wanted to obtain information concerning attitudes toward abortion and we obtained a sample consisting predominantly of men, then we would encounter sampling bias.

EXAMPLE 8.3

A classic example of sampling bias is the survey of the attitudes of several million people taken by *Literary Digest,* a popular journal of the period, to forecast the 1936 presidential winner. Republican Alfred Landon was running against Democrat Franklin Roosevelt. Names of people to be included in the survey were obtained by the *Digest* from telephone directories and magazine subscription lists. A majority of those surveyed indicated their preference for Landon, and the journal predicted he would win by a landslide. As we know, Landon lost. Many people who were most likely to vote for Roosevelt did not have telephones or subscribe to magazines. The sample thus contained strong bias in favor of Landon.

EXAMPLE 8.4

When gathering data nonsampling errors could result when instruments used in making measurements are out of adjustment or not properly calibrated. Processing errors could result if data are misplaced or lost in storage, or answers obtained from people during a survey are not truthful. The latter case might arise with questions dealing with age, to which some people will lie out of vanity.

Sampling bias can be removed (or minimized) by using randomization. **Randomization** refers to any process for selecting a sample from the population that involves impartial or unbiased selection. A sample chosen by randomization procedures is called a **random sample.** The most popular types of sampling techniques involving randomization are simple random sampling, stratified sampling, cluster sampling, and systematic sampling. Each of these will now be explained.

Random Samples

If a random sample is chosen in such a way that all samples of the same size are equally likely to be chosen, then the sample is called a **simple random sample.**

EXAMPLE 8.5

Suppose we want to select a random sample of 5 students from a statistics class of 20 students. The binomial coefficient $\binom{20}{5}$ gives the total number of ways of selecting an unordered sample of 5 students from a class of 20 students.

$$\binom{20}{5} = \frac{20!}{5!15!}$$
$$= \frac{(20)(19)(18)(17)(16)}{(5)(4)(3)(2)(1)} = 15{,}504$$

If we listed the 15,504 possibilities on separate pieces of paper (a tremendous task), placed them into a container, and thoroughly mixed, then we could choose a random sample of 5 by selecting one slip with five names. A simpler procedure to select a random sample would be to write each of the 20 names on separate pieces of paper, place them into a container, thoroughly mix them, and then draw five slips at once.

Another method of obtaining a random sample of 5 students from a class of 20 students is based on a table of random numbers. Table 8.1 is a brief table of random numbers. It was created by using a computer with a built-in random number generator. Numbers are grouped into sets of five digits for ease of reading. For example, the boxed digit 5 in the table is in the third row and tenth column.

TABLE 8.1

Random Number Table

50242	94818	05825	87975	77496
11021	62231	88043	88062	71692
59323	5222⑤	32913	02586	31934
00576	29013	84384	00445	98538
53228	13826	36564	74019	32067
⟦26⟧406	32693	68126	99353	23898
64122	44928	96161	39435	31767
90911	26356	91554	43233	39124
13835	62884	31326	28818	05167
36471	28587	30432	74161	07649
27201	65409	40210	72802	20198
95370	73018	94933	15544	26640
60880	56768	47370	38451	63468
15130	08631	55031	48008	04123
01239	72330	94403	71123	37141
02928	17216	79729	27076	58220
10220	70101	10902	80993	19104
88707	83518	68517	40570	40291
61461	91259	97508	56729	71972
51245	79283	10366	38290	30673

A **random number table** can be constructed using a computer. However, without using a computer we can construct a random number table in the following manner. If the ten digits from 0 to 9 are written on slips of paper, placed into a container, and thoroughly mixed, then the first slip chosen determines the first number in the table. The slip of paper is returned to the container and a second slip is drawn after the slips are again mixed thoroughly. The second slip determines the second number in the random number table. This process is followed until a table with the desired number of digits is obtained.

EXAMPLE 8.6

We will illustrate the use of the random number table by choosing a random sample of 5 students from the following list of 20 students enrolled in Statistics 209:

01	Mike Able	11	Sarah Kemp
02	Mary Baker	12	James Lum
03	Joe Cable	13	Robert Moon
04	Ed Doe	14	Karen Nie
05	Jake Eldon	15	Beth Oboe
06	Sue Fum	16	Dave Poe
07	Pete Gum	17	Rick Quest
08	Harry Hoe	18	Bart Rat
09	Ora Ida	19	Stella Star
10	Helen Jewel	20	Maud Tuck

The names of the 20 students are listed in some order (we have used alphabetical order for convenience) and each student is assigned an identification number. In our example, we need only *two-digit* identification numbers. A starting place in table 8.1 is selected at random, perhaps by closing your eyes and pointing to some number in the table with the tip of your pencil. Random numbers in groups of *two* are then read either horizontally or vertically starting at the random starting place. If a two-digit number greater than 20 or a number that has been previously obtained is chosen, it is omitted and the next pair is selected. Suppose the digit 2 in the sixth row and first column is selected as the starting point. If two-digit numbers are read horizontally starting at this point, then the following names constitute a random sample of 5 students:

    12 James Lum
    17 Rick Quest
    13 Robert Moon
    07 Pete Gum
    01 Mike Able

EXAMPLE 8.7

The procedure used in example 8.6 for selecting a random sample of size 5 from a class of 20 students could be made more efficient in terms of the time required to identify the 5 students and, as a result, require less effort. Instead of assigning each of the 20 students a single two-digit number, we could assign each of them 5 two-digit numbers as follows;

Student	Two-Digit Number Assignments				
Able	00	01	02	03	04
Baker	05	06	07	08	09
Cable	10	11	12	13	14
Doe	15	16	17	18	19
Eldon	20	21	22	23	24
Fum	25	26	27	28	29
.					
.					
.					
Star	90	91	92	93	94
Tuck	95	96	97	98	99

If we use this procedure to identify the students for our random sample and start with digit 2 in the sixth row and first column of the random number table, we will select the following five students:

Random Digits	Student
26	Sue Fum
40	Ora Ida
63	Robert Moon
93	Stella Star
68	Karen Nie

Note that we had to draw six two-digit numerals to choose our sample since the random number 26 was chosen twice.

APPLICATION 8.1

It is impossible to determine by inspection whether a sample is random or not. For example, suppose we want to select four months of the year to study certain biological phenomena and that March, June, September, and December are chosen. Do these four months represent a random sample of months?

Solution: It is impossible to tell from the information given whether the sample is random. To determine whether a sample is random, we must know what selection process was used. These months might have been chosen because they are spread through the year: if so, the sample is not random. However, if these months were selected with the aid of a random number table or other randomization procedures, then they do represent a random sample. ■

There are many situations in which simple random sampling is impractical, impossible, or undesirable. Although it would be desirable to use simple random samples for national opinion polls on product information or presidential elections, such would be very costly and time consuming.

Stratified sampling involves separating the population into nonoverlapping groups, called *strata,* and then selecting a simple random sample from each stratum. The information from the collection of simple random samples would then constitute the overall sample.

EXAMPLE 8.8

Suppose we want to sample the opinions of faculty at a large university on an important issue. It might be difficult to sample all university faculty, so let's suppose we select a random sample from each college (or academic department). The strata would be the colleges (or academic departments) within the university.

Stratified sampling is commonly used for national opinion polls because opinions tend to vary more among different localities than within localities. For this application, the strata might be counties, states, or regions. The criteria for forming the strata should ensure that the observations within each stratum are as much alike as possible. The observations within a given stratum should have less variation than the observations among strata.

Cluster sampling involves selecting a simple random sample of heterogeneous subunits, called *clusters,* from the population. Each element in the population belongs to exactly one cluster, and the elements within each cluster are usually heterogeneous or unalike.

EXAMPLE 8.9

Suppose a cable television service company is considering opening a branch office in a certain large city. The company plans to conduct a survey to determine the percentage of households that would use its service. Since it is not practical to survey every household, the company decides to choose a city block at random and then to survey every household in the block. The city block is a cluster.

With cluster sampling, clusters are formed to represent, as closely as possible, the entire population. A simple random sample of the clusters is then used to study the population. Studies of social institutions, such as churches, hospitals, schools, and prisons, generally involve cluster sampling. In such studies, a single cluster, or several clusters, are selected randomly for study. The entire population can be effectively studied by studying its miniature copies or clusters. If a cluster is too large to be studied effectively, elements within the cluster(s) can be randomly selected.

Systematic sampling is a sampling technique that involves an initial random selection of observations followed by a selection of observations obtained by using some rule or system.

EXAMPLE 8.10

To obtain a sample of telephone subscribers within a large city, a random sample of page numbers within the phone directory might first be obtained. Then by selecting the 20th name on each page, we would obtain a systematic sample. Or we could select one name from the first page of the directory and then select every 100th name thereafter. An alternate approach would involve first selecting a number at random from those numbered 1 to 100; assume the number 40 is selected. Then select names from the directory that correspond to the numbers 40, 140, 240, 340, and so forth.

Sampling Error

Any measurement involves some error. If the sample mean is used to measure (estimate) the population mean μ, then the sample mean, as a measurement, involves some error. As an example, suppose a random sample of size 25 is drawn from a population with a mean of $\mu = 15$. If the mean of the sample is $\bar{x} = 12$, then the observed difference $\bar{x} - \mu = -3$ is referred to as *sampling error*. A sample mean $\bar{x}$ can be thought of as the sum of two quantities, the population mean μ and sampling error. If e denotes the sampling error, then we have

$$\bar{x} = \mu + e$$

EXAMPLE 8.11

Suppose random samples of size 2 are selected from a population consisting of three values, 2, 4, and 6. In order to simulate a "large" population so that sampling can be carried out a large number of times, we will assume that sampling is done with replacement, that is, the sampled number is replaced before the next number is drawn. In addition, we will select ordered samples. In an ordered sample, the order in which an observation is selected is important. Thus, the ordered sample {2, 4} is different from the ordered sample {4, 2}. In the ordered sample {4, 2}, 4 was drawn first, and then 2. Table 8.2 contains a list of all possible ordered samples of size 2 that can be selected with replacement from the population of values 2, 4, and 6. In addition, the table contains sample means and corresponding sampling errors. Note that the population mean is equal to $\mu = (2 + 4 + 6)/3 = 4$.

Table 8.2

Ordered Samples of Size 2 from the Population of Values 2, 4, 6

Ordered Samples	$\bar{x}$	Sampling Error $e = \bar{x} - \mu$
{2, 2}	2	$2 - 4 = -2$
{2, 4}	3	$3 - 4 = -1$
{2, 6}	4	$4 - 4 = 0$
{4, 2}	3	$3 - 4 = -1$
{4, 4}	4	$4 - 4 = 0$
{4, 6}	5	$5 - 4 = 1$
{6, 2}	4	$4 - 4 = 0$
{6, 4}	5	$5 - 4 = 1$
{6, 6}	6	$6 - 4 = 2$

The following is a frequency graph for the sample means.

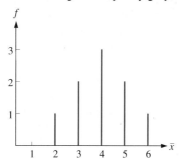

Note the following interesting relationships contained in table 8.2:

1. The mean of the collection of all sample means is 4, the mean of the population from which the samples were drawn. If $\mu_{\bar{x}}$ denotes the mean of all the sample means, then we have

$$\mu_{\bar{x}} = \frac{3 + 4 + 3 + 4 + 5 + 5 + 2 + 4 + 6}{9} = \frac{36}{9} = 4$$

2. The sum of the sampling errors is 0.

$$e_1 + e_2 + e_3 + \cdots + e_9$$
$$= (-2) + (-1) + 0 + (-1) + 0 + 1 + 0 + 1 + 2 = 0$$

Thus, if $\bar{x}$ is used to measure (estimate) the population mean μ, the average of all the sampling errors is 0.

When a statistic, such as $\bar{x}$, is used to measure or estimate a parameter and the average of all the sampling errors is 0, the statistic is said to be **unbiased.** In addition, we say a statistic has **zero bias** when used to measure or estimate a parameter if the average of all the sampling errors is 0. Thus, in example 8.11, $\bar{x}$ has zero bias and is an unbiased statistic. Strictly speaking, when the sample mean is used for estimation purposes, it is referred to as an *estimator* and a value of the sample mean is referred to as an **estimate.** Since the value of the sample mean varies from sample to sample, it is an example of a random variable. All estimators are random variables. We will discuss estimators in more detail in chapter 9.

When two different random samples of size 25 are drawn from the same population and the mean of each sample is calculated, we would not expect both sample means to be identical. Rather, each mean would select a different sampling error. If $\bar{x}_1$ denotes the mean of the first sample, $\bar{x}_2$ denotes the mean of the second sample, e_1 denotes the sampling error associated with $\bar{x}_1$, and e_2 denotes the sampling error associated with $\bar{x}_2$, then we have

$$\bar{x}_1 = \mu + e_1$$
$$\bar{x}_2 = \mu + e_2$$
$$\bar{x}_1 - \bar{x}_2 = e_1 - e_2$$

The difference between sample means reflects only the difference between the sampling errors. If repeated pairs of random samples were obtained and the differences in sample means were obtained for each pair, we would expect to obtain an average difference in sample means (sampling errors) of 0.

EXERCISE SET 8.1

Basic Skills

1. By using the first digit in the fourth row of table 8.1 as the starting point and moving horizontally to the right, select a random sample of size 10 from the list of Statistics 209 students given in example 8.6.

2. By using the first digit in the fifth row of table 8.1 as the starting point and moving horizontally to the right, select a random sample of size 15 from the list of Statistics 209 students given in example 8.6.

3. Simulate tossing a coin ten times by using table 8.1. Start with the first digit in the fifth row of table 8.1 and move horizontally to the right.

4. Simulate tossing a six-sided die 15 times by using table 8.1. Start with the first digit in the third row of table 8.1 and move horizontally to the right.

5. A new-car dealer wants to select a random sample of opinions concerning a new model from 15 of the 97 customers on the mailing list who have purchased the new model during the past year. Explain how the sample could be selected with the help of a random number table.

6. Describe a procedure involving the random number table to choose a sample of 50 from a population of 100,000 households.

7. Consider the first ten rows in table 8.1 and record the frequency that each digit occurs. How many times would you expect each digit to occur?

8. Consider the first ten columns in table 8.1 and record the frequency that each digit occurs. How many times would you expect each digit to occur?

9. For exercise 7, do you think the variation between the observed frequency and expected frequency for each digit indicates a variation due to sampling error? Complete the accompanying table and find the average of the sampling errors.

Digit	Frequency	Expected Frequency	Sampling Error
0			
1			
2			
3			
4			
5			
6			
7			
8			
9			

10. For exercise 8, do you think the variation between the observed frequency and expected frequency for each digit indicates a variation due to sampling error? Complete the accompanying table and find the average of the sampling errors.

Digit	Frequency	Expected Frequency	Sampling Error
0			
1			
2			
3			
4			
5			
6			
7			
8			
9			

11. Start at the first digit in the sixth row and move horizontally to the right in table 8.1 to select a random sample of 12 tosses of a six-sided die. Construct a table similar to the one in exercise 9 and find the average of the sampling errors.

12. Start at the first digit in the fourth row and move horizontally to the right in table 8.1 to select a random sample of ten tosses of a six-sided die. Construct a table similar to the one in exercise 9 and find the average of the sampling errors.

13. A random sample of 5 teachers is to be selected from a population of 200 teachers to participate in special workshop.
   a. Label the teachers from 001 to 200. Which teachers will be selected for the workshop if table 8.1 is used and the starting point is the first digit in the fourth row and the second column and the digits are read horizontally moving to the right.
   b. A more efficient selection process involves labeling the teachers as in part a, and assigning the numbers 001, 201, 401, 601, 801 to the first teacher, numbers 002, 202, 402, 602, 802 to the second teacher, numbers 003, 203, 403, 603, 803 to the third teacher, . . . , and numbers 200, 400, 600, 800, 000 to the last teacher. By using this scheme and starting at the same point as in part a, select a random sample of five teachers.

14. A random sample of 5 teachers is to be selected from a population of 150 teachers to participate in special workshop.
   a. Label the teachers from 001 to 150. Which teachers will be selected for the workshop if table 8.1 is used and the starting point is the first digit in the fifth row and second column and the digits are read horizontally moving to the right?
   b. By using a more efficient scheme and starting at the same point as in part a, select a random sample of five teachers. Describe your scheme.

15. Consider all possible ordered samples of size 2 drawn with replacement from the population of values 0, 2, 4, 6. Construct a table similar to table 8.2 and determine if $\bar{x}$ is an unbiased estimator of μ.

16. Consider all possible ordered samples of size 3 drawn with replacement from the population of values 0, 3, 6, 9. Construct a table similar to table 8.2 and determine if $\bar{x}$ is an unbiased estimator of μ.

17. Do the following procedures produce random samples? Explain why or why not?
   a. To obtain a random sample of students in a class, select all students in the class who wear glasses.

b. To obtain a random sample of coeds, select every third person entering the gym through the front door.

c. To obtain a random sample of college students attending a large university, select students by using a random number table and the last four digits of their Social Security numbers.

18. Do the following procedures produce random samples? Explain why or why not.

a. To select a random sample of college athletes, select every second person coming out of the gym through the front door.

b. To select a random sample of people who have phones, select every other person on a randomly selected page of the phone book.

c. To select a random sample of size 5 from a class of 25 students, write each name on a slip of paper, place the slips in a hat, stir the slips, and choose one at a time, without replacement, until five names are drawn.

19. Would the unordered sample {2, 4, 6, 8, 10} constitute a random sample from the population of whole numbers from 1 to 10, inclusive? Explain.

Going Beyond

20. Starting at the first digit in the sixth row of table 8.1 and reading horizontally, select two random samples of ten distinct two-digit numbers, one sample at a time, and compute the means for both samples. If a sample mean is used to estimate the population mean μ, what is the difference between the sampling errors for these two sample means? If this experiment is repeated a large number of times, what kind of bias would you expect for the differences between sample means? Explain.

21. Indicate an appropriate sampling technique (simple random sampling, stratified sampling, cluster sampling, or systematic sampling) for obtaining the following samples. For each, give reasons for your choice.

a. A sample of students in a college to study student opinions on nuclear power as an alternative power source

b. A sample of students to study student attitudes on campus parking

c. A sample of corn plants to study yields and quality

d. A sample of rabbits in western Maryland to study weights of rabbits

e. A sample of fish in Piney Dam to study species of fish

f. A sample of school teachers in Allegany County to study teachers' opinions on the issue of merit pay

g. A sample of school teachers in Maryland to study teachers' opinions on the issue of merit pay

h. A sample of car owners to study their preferences for engine oil

i. A sample of retired persons to study use of leisure time by retired persons

22. Suppose two upper-level mathematics classes, A and B, take the same ten-point quiz and the results are as follows:

Class A	3	9	6	5	4	8	4	7	6	5
Class B	6	5	4	2	3	7	7	3	5	6

If in fact both classes know the material equally well, estimate the probability that $\bar{x}_A - \bar{x}_B \geq 0.9$. (*Hint:* Write each of the 20 scores on a 3 × 5 card, shuffle the 20 cards well, and deal two hands of 10 cards. Then calculate the mean of each hand, A and B, and determine if $\bar{x}_A - \bar{x}_B \geq 0.9$. After dealing 20 pairs of 10-card hands, determine the proportion of 10-card hands resulting in $\bar{x}_A - \bar{x}_B \geq 0.9$. This number estimates $P(\bar{X}_A - \bar{X}_B \geq 0.9)$ if in fact both classes did equally well.)

23. The device shown in the accompanying figure is called a *hextat*. A ball is dropped into the device. At each peg it hits, it goes either left or right. After hitting five pegs it lands in one of the slots *A* through *F*. Simulate dropping 16 balls into the hextat as follows: use the random number table starting with the digit 1 in the second row and first column and moving horizontally to the right to choose 80 digits, which determine the final positions of the 16 balls. Let an even digit represent left and an odd digit represent right. The first 5 digits will determine the position of the first ball; the next 5 digits the position of the second ball,

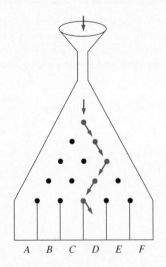

$A \quad B \quad C \quad D \quad E \quad F$

and so on. For example, the first 5 random digits drawn are 1, 1, 0, 2, and 1. This means that the first ball moves right, right, left, left, and right as shown in the figure. (If a large number of balls are used and the final position of each ball measures its horizontal distance from the starting point, then the horizontal distances have a mean of 0 and form a normal distribution. The hextat is used to depict, among other things, Brownian movement of plant spores or small particles suspended in a liquid.)

SECTION 8.2   *Sampling Distributions*

Random samples drawn from a population are by their basic nature unpredictable. We would not expect two random samples of the same size taken from the same population to possess the same sample mean or be exactly alike. Any statistic (such as the sample mean) computed from the measurements in a random sample can be expected to vary in its value from random sample to random sample. Accordingly, we shall want to study the distribution of all possible values of a statistic. Such distributions will be very important in the study of inferential statistics, since inferences about populations will be made using sample statistics. By studying the distributions associated with sample statistics, we will be able to judge the reliability of a sample statistic as an instrument for making an inference about an unknown population parameter.

Since the values of a statistic, such as $\bar{x}$, vary from random sample to random sample, the statistic can be considered a *random variable* with a corresponding frequency distribution. Frequency distributions for sample statistics are referred to as **sampling distributions.** For convenience, we shall denote the population by X and the sampling distribution of the mean by $\bar{X}$. In general, the **sampling distribution of a statistic** is the distribution of all possible values of the statistic computed from samples of the same size. Consider application 8.12.

EXAMPLE 8.12

Suppose random samples of size 20 are selected from a large population. For each sample, the sample mean $\bar{x}$ is computed. The collection of all these sample means is called the **sampling distribution of the sample mean.** This can be illustrated as in figure 8.1.

FIGURE 8.1

Sampling distribution of sample mean

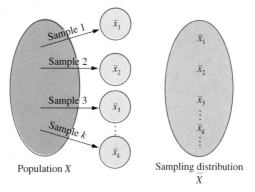

Population X Sampling distribution $\bar{X}$

EXAMPLE 8.13

Suppose samples of size 20 are randomly selected from a large population and the sample standard deviation of each sample is computed. The collection of all these sample standard deviations is called the **sampling distribution of the sample standard deviation.** This can be illustrated as in figure 8.2.

FIGURE 8.2

Sampling distribution of sample standard deviation

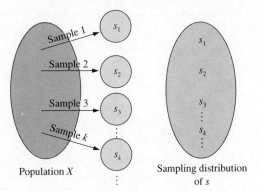

Population X Sampling distribution of s

APPLICATION 8.2

Suppose ordered samples of size 2 are selected, with replacement, from the population of values 0, 2, 4, 6. Find

  a. μ, the population mean.

  b. σ, the population standard deviation.

  c. $\mu_{\bar{x}}$, the mean of the sampling distribution of the mean.

  d. $\sigma_{\bar{x}}$, the standard deviation of the sampling distribution of the mean.

In addition, draw frequency graphs for the population distribution of X and the sampling distribution $\bar{X}$.

Solution:

  a. The population mean is

$$\mu = \frac{0 + 2 + 4 + 6}{4} = 3$$

  b. The population standard deviation is

$$\sigma = \sqrt{\frac{(0 - 3)^2 + (2 - 3)^2 + (4 - 3)^2 + (6 - 3)^2}{4}}$$

$$= \sqrt{\frac{9 + 1 + 1 + 9}{4}} = \sqrt{5}$$

  c. We next list the elements of the sampling distribution of the mean and the corresponding frequency distribution.

Sample	$\bar{x}$
{0, 0}	0
{0, 2}	1
{0, 4}	2
{0, 6}	3
{2, 0}	1
{2, 2}	2
{2, 4}	3
{2, 6}	4
{4, 0}	2
{4, 2}	3
{4, 4}	4
{4, 6}	5
{6, 0}	3
{6, 2}	4
{6, 4}	5
{6, 6}	6

Frequency Distribution of $\bar{x}$

$\bar{x}$	f
0	1
1	2
2	3
3	4
4	3
5	2
6	1
	16

The mean of the sampling distribution of the sample mean is

$$\mu_{\bar{x}} = \frac{\Sigma\,(f\bar{x})}{\Sigma\,f}$$
$$= \frac{(0)(1) + (1)(2) + (2)(3) + (3)(4) + (4)(3) + (5)(2) + (6)(1)}{16} = \frac{48}{3} = 3$$

FIGURE 8.3

Population distribution and sampling distribution of sample mean for application 8.2

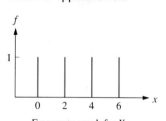

Frequency graph for X

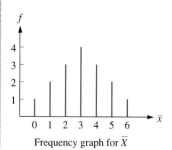

Frequency graph for $\bar{X}$

d. To find the standard deviation of the distribution of the sample mean, we organize the calculations in the following table:

$\bar{x}$	f	$\bar{x} - \mu_{\bar{x}}$	$(\bar{x} - \mu_{\bar{x}})^2$	$(\bar{x} - \mu_{\bar{x}})^2 f$
0	1	$0 - 3 = -3$	9	9
1	2	$1 - 3 = -2$	4	8
2	3	$2 - 3 = -1$	1	3
3	4	$3 - 3 = 0$	0	0
4	3	$4 - 3 = 1$	1	3
5	2	$5 - 3 = 2$	4	8
6	1	$6 - 3 = 3$	9	9
				40

The standard deviation of the sampling distribution of the sample mean is

$$\sigma_{\bar{x}} = \frac{\sigma}{\sqrt{n}}$$
$$= \sqrt{\frac{40}{16}}$$
$$= \sqrt{2.5} \approx 1.58$$

The frequency graphs for the population and the sampling distribution of the sample mean are shown in figure 8.3. ▪

As with any random variable, the distribution of the sample mean has a mean (or expected value), a variance, and a standard deviation. It can be shown that the sampling distribution of the sample mean has a mean equal to the population mean. That is,

$$\mu_{\bar{x}} = E(\bar{x}) = \mu$$

Notice that for application 8.2, $\mu_{\bar{x}} = 3 = \mu$.

APPLICATION 8.3

For the sampling distribution of the mean in application 8.2, find the

a. Sampling error for each mean.

b. Mean of the sampling errors.

c. Standard deviation of the sampling errors.

Solution:

a. The samples, sample means, and sampling errors are shown in the following table:

Sample	Sample Mean, $\bar{x}$	Sampling Error, $e = \bar{x} - \mu$
{0, 0}	0	$0 - 3 = -3$
{0, 2}	1	$1 - 3 = -2$
{0, 4}	2	$2 - 3 = -1$
{0, 6}	3	$3 - 3 = 0$
{2, 0}	1	$1 - 3 = -2$
{4, 0}	2	$2 - 3 = -1$
{6, 0}	3	$3 - 3 = 0$
{2, 4}	3	$3 - 3 = 0$
{4, 2}	3	$3 - 3 = 0$
{2, 6}	4	$4 - 3 = 1$
{6, 2}	4	$4 - 3 = 1$
{4, 6}	5	$5 - 3 = 2$
{6, 4}	5	$5 - 3 = 2$
{2, 2}	2	$2 - 3 = -1$
{4, 4}	4	$4 - 3 = 1$
{6, 6}	6	$6 - 3 = 3$

b. The mean of the sampling errors μ_e is

$$\mu_e = \frac{(-3) + (-2) + (-1) + 0 + \cdots + 1 + 31}{16} = 0$$

c. The standard deviation of the sampling errors is simply found after condensing the information obtained thus far into the following frequency table:

e	f	$e - \mu_e$	$(e - \mu_e)^2 f$
-3	1	$3 - 0 = 3$	9
-2	2	$2 - 0 = 2$	8
-1	3	$1 - 0 = 1$	3
0	4	$0 - 0 = 0$	0
1	3	$-1 - 0 = -1$	3
2	2	$-2 - 0 = -2$	8
3	1	$-3 - 0 = -3$	9
	$N = 16$		40

The standard deviation of the distribution of sampling errors σ_e is thus

$$\sigma_e = \sqrt{\frac{\Sigma[(e - \mu_e)^2 f]}{N}}$$

$$= \sqrt{\frac{40}{16}} = 1.58 \quad ▪$$

The standard deviation of the sampling distribution of a statistic is referred to as the **standard error of the statistic.** For application 8.2, the standard error of the mean, denoted by $\sigma_{\bar{x}}$, is 1.58. Notice that for application 8.3, the standard deviation of the 16 sampling errors σ_e is also equal to 1.58. Thus, it is reasonable to call $\sigma_{\bar{x}}$ the **standard error of the mean.** It can be shown that if samples of size n are selected with replacement from a population, then the standard error of the mean is equal to the standard deviation of the distribution of sampling errors. Thus, in general, we have

$$\sigma_e = \sigma_{\bar{x}}$$

where $e = \bar{x} - \mu$.

EXAMPLE 8.14

Suppose a fair coin is tossed ten times and the number of heads obtained X is recorded. If this experiment is repeated a large number of times, the number of heads obtained each time will vary from 0 to 10. Thus, X, the number of heads obtained when ten fair coins are tossed, is a random variable with a sampling distribution. The sampling distribution is binomial with a mean equal to

$$\mu_X = np$$
$$= (10)(0.5) = 5$$

and a standard deviation equal to

$$\sigma_X = \sqrt{np(1 - p)}$$
$$= \sqrt{(10)(0.5)(0.5)} \approx 1.58.$$

EXAMPLE 8.15

Suppose samples of size 1 are drawn from a normal population with mean $\mu = 10$ and standard deviation $\sigma = 2$, and for each sample, a z score $Z = (X - \mu)/\sigma$ is recorded. Since the values for Z will vary with the sample, Z can be considered a random variable with a sampling distribution. The sampling distribution of Z is the standard normal distribution. As was seen in chapter 7, the mean of Z is

$$\mu_Z = 0$$

and the standard deviation of Z is

$$\sigma_Z = 1$$

A P P L I C A T I O N 8.4

Suppose ordered samples of size 2 are selected with replacement from the population of values 0, 2, 4, 6, and 8. For each sample, the range R is recorded. For the sampling distribution of R

a. Construct a frequency table for R and a frequency graph for R.

b. Find the mean of the sampling distribution of R.

c. Find the standard error of R.

Solution:

a. We first list the ordered samples along with their corresponding ranges in the following table:

Sample	Range R	Sample	Range R
$\{0, 0\}$	0	$\{4, 6\}$	2
$\{0, 2\}$	2	$\{4, 8\}$	4
$\{0, 4\}$	4	$\{6, 0\}$	6
$\{0, 6\}$	6	$\{6, 2\}$	4
$\{0, 8\}$	8	$\{6, 4\}$	2
$\{2, 0\}$	2	$\{6, 6\}$	0
$\{2, 2\}$	0	$\{6, 8\}$	2
$\{2, 4\}$	2	$\{8, 0\}$	8
$\{2, 6\}$	4	$\{8, 2\}$	6
$\{2, 8\}$	6	$\{8, 4\}$	4
$\{4, 0\}$	4	$\{8, 6\}$	2
$\{4, 2\}$	2	$\{8, 8\}$	0
$\{4, 4\}$	0		

The frequency graph and frequency table are shown here.

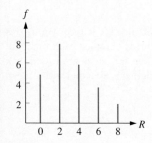

R	f
0	5
2	8
4	6
6	4
8	2

b. The mean of the sampling distribution of the range is

$$\mu_R = \frac{\Sigma\,(fR)}{\Sigma f}$$

$$= \frac{(5)(0) + (8)(2) + (6)(4) + (4)(6) + (2)(8)}{25}$$

$$= \frac{80}{25} = 3.2$$

c. The standard deviation of the sampling distribution of R, or the standard error of R, is found next.

$$\sigma_R = \sqrt{\frac{\Sigma\,[(R - \mu_R)^2 f]}{\Sigma f}}$$

$$= \sqrt{\frac{(0 - 3.2)^2(5) + (2 - 3.2)^2(8) + (4 - 3.2)^2(6) + (6 - 3.2)^2(4) + (8 - 3.2)^2(2)}{25}}$$

$$= 2.4$$

Hence, the standard error of R is $\sigma_R = 2.4$. ■

Note that the sampling distribution of a statistic must necessarily involve samples of the same size. If the samples involved are of different sizes, then it makes no sense to talk about the shape of the sampling distribution.

Sampling Distribution of the Mean

The sampling distribution of the sample mean has parameters that are related to the parameters of the population sampled. As we previously pointed out, the mean of the sampling distribution of the sample mean is equal to the mean of the population from which the samples were selected. This can be expressed by the following formula:

> **Mean of the Sampling Distribution of the Mean**
>
> $$\mu_{\bar{x}} = \mu \qquad (8.1)$$

This fact was illustrated in application 8.2.

EXAMPLE 8.16

Suppose a large hardware store carries three different brands of paint, priced at $8, $10, and $14 per gallon. Based on past experience, the probability distribution table for the price paid by a customer selected at random is as follows:

x	8	10	14
$P(x)$	0.5	0.3	0.2

The price a randomly selected customer pays for paint is a discrete random variable with values 8, 10, and 14. Suppose two randomly chosen customers make independent paint purchases. Let X_1 denote the price paid by the first customer and let X_2 denote the price paid by the second customer. Table 8.3 lists the possible purchases with their corresponding probabilities and average purchase price $\bar{x}$.

TABLE 8.3

Probabilities of Average Purchase Price for Paint Example

x_1	x_2	$P(x_1, x_2)$	$\bar{x}$
8	10	$(0.5)(0.3) = 0.15$	9
8	8	$(0.5)(0.5) = 0.25$	8
8	14	$(0.5)(0.2) = 0.10$	11
10	8	$(0.3)(0.5) = 0.15$	9
10	10	$(0.3)(0.3) = 0.09$	10
10	14	$(0.3)(0.2) = 0.06$	12
14	8	$(0.2)(0.5) = 0.10$	11
14	10	$(0.2)(0.3) = 0.06$	12
14	14	$(0.2)(0.2) = 0.04$	14

Note that if E is the event that the first customer purchases paint priced at $8, and F is the event that the second customer purchases paint priced at $10, then since the paint purchases are independent, we have $P(E \cap F) = P(E)P(F)$. In addition, since $P(E) = P(X_1 = 8)$ and $P(F) = P(X_2 = 10)$, to calculate the value of $P(8, 10) = P(X_1 = 8$ and $X_2 = 10)$, we need only multiply $P(8)$ by $P(10)$. Thus, $P(8, 10) = P(8)P(10) = (0.5)(0.3) = 0.15$. The remaining probabilities in table 8.3 were found in a similar fashion.

A probability table for the sampling distribution of $\bar{x}$ is shown below.

$\bar{x}$	8	9	10	11	12	14
$P(\bar{x})$	0.25	0.30	0.9	0.20	0.12	0.04

The probability $P(9) = P(\bar{X} = 9)$ was found by adding 0.15 and 0.15, obtained from table 8.3 to get 0.30. The other probabilities were found similarly. A probability histogram for the sampling distribution of mean purchases is shown in figure 8.4.

FIGURE 8.4

Probability histogram for sampling distribution of $\bar{x}$ in paint example

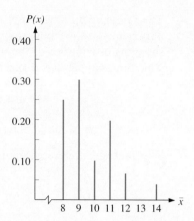

The mean of the sampling distribution of the sample mean is

$$\mu = \Sigma\,[\bar{x}P(\bar{x})]$$
$$= (8)(0.025) + (9)(0.30) + (10)(0.09) + (11)(0.20) + (12)(0.12) + (14)(0.04) = 9.8$$

This result agrees with the value for $\mu_{\bar{x}}$, since

$$\mu_{\bar{x}} = \Sigma\,[\bar{x}P(\bar{x})]$$
$$= (8)(0.5) + (10)(0.3) + (14)(0.2) = 9.8$$

The standard error of the mean is found by using formula (5.18):

$$\sigma_{\bar{x}} = \sqrt{\Sigma\,[\bar{x}^2P(\bar{x})] - \mu^2}$$
$$= \sqrt{98.62 - 9.8^2}$$
$$= \sqrt{2.58} \approx 1.61$$

Method of Sampling

The standard deviation of the sampling distribution of the mean $\sigma_{\bar{x}}$ is influenced by the method of sampling. If sampling is done with replacement from a small population, then each measurement in the sample is independent of any other value, and the sampling can be done indefinitely, as would be the case for a very large population. In large populations where sampling is done without replacement, the statistical dependence of any one value on another is so slight that it is usually ignored. But if sampling is done without replacement from a small population, the sample values are not statistically independent, and this fact must be taken into consideration when computing $\sigma_{\bar{x}}$.

From a practical point of view, what really matters when sampling is done without replacement and $\sigma_{\bar{x}}$ is calculated is the size of the samples relative to the size of the population. There is general agreement among researchers that when the size of the sample is no more than 5% of the size of the population, the population can be considered large relative to the size of the sample, and statistical independence of the sample values is usually assumed as an approximation to reality (see example 8.17). However, when we sample without replacement from small populations, the statistical dependence between values must be taken into consideration, as in example 8.18.

EXAMPLE 8.17

Suppose the sample {5, 68} were drawn without replacement from the population of positive integers less than or equal to 1000. Then, $P(68) = 0.001$ and $P(68\,|\,5) = 1/999 \approx 0.001001$. For most practical purposes, P(choosing 68 given that 5 has been selected) can be considered to be equal to $P(68)$ and statistical independence of the sample values can be assumed.

EXAMPLE 8.18

Suppose a sample of size 2 is drawn without replacement from a population having the three values 2, 6, and 8. The probability of choosing 2 first is 1/3, but the probability for the next value chosen cannot be 1/3. It must be 1/2. Here P(choosing 3 given that 2 has been chosen) = 1/2, instead of 1/3. Hence, the sampled values cannot be considered independent of one another.

Sampling from Large Populations

The following formula can be used to calculate $\sigma_{\bar{x}}$ when sampling is done from a large population (or a small population with replacement).

Standard Error of the Mean for Large Samples

$$\sigma_{\bar{x}} = \frac{\sigma}{\sqrt{n}} \tag{8.2}$$

where σ is the standard deviation of the population sampled and n is the sample size.

EXAMPLE 8.19

To find the standard error of the mean $\sigma_{\bar{x}}$ for the paint illustration of example 8.16, we first find the standard deviation of the population σ by using formula 5.18:

$$\sigma = \sqrt{\Sigma [x^2 P(x)] - \mu^2}$$
$$= \sqrt{101.2 - (9.8)^2} = \sqrt{5.16} = 2.27$$

Then, by using formula (8.2), we have

$$\sigma_{\bar{x}} = \frac{\sigma}{\sqrt{n}}$$
$$= \frac{2.27}{\sqrt{2}} = 1.61$$

Note that this value agrees with the result found earlier.

APPLICATION 8.5

For application 8.2, use formula (8.2) to find $\sigma_{\bar{x}}$, the standard error of the mean (the standard deviation of the sampling distribution of the sample mean).

Solution: The standard error of the mean is

$$\sigma_{\bar{x}} = \frac{\sigma}{\sqrt{n}}$$
$$= \frac{\sqrt{5}}{\sqrt{2}} = \sqrt{2.5} \approx 1.58$$

Note that this result agrees with the answer found in application 8.2 by finding the standard deviation of the 16 measurements in the sampling distribution of the mean. ■

The standard error of the mean is a very important concept in statistics. Let's examine how the size of the samples used to form the sampling distribution of the mean relates to the standard error of the mean. Suppose the standard deviation of a large sampled population is $\sigma = 12$ and that samples of size $n = 4$ are selected. The standard error of the mean is

$$\sigma_{\bar{x}} = \frac{\sigma}{\sqrt{n}}$$
$$= \frac{12}{\sqrt{4}} = 6$$

If instead of using samples of size $n = 4$, we increased the sample size to 16, the standard error of the mean would become

$$\sigma_{\bar{x}} = \frac{\sigma}{\sqrt{n}}$$

$$= \frac{12}{\sqrt{16}} = 3$$

We see from this demonstration that as the sample size n increases, the standard error of the mean $\sigma_{\bar{x}}$ decreases. This means that as the values of n increase, the sample means cluster more closely about their mean μ. We shall rely heavily on this fact in chapter 9 when we use a sample mean to estimate an unknown population mean. If the size of the random sample is large, we shall feel confident that our sample mean is as close to the population mean as any other sample mean in the same sampling distribution of the mean, since the sample means cluster closely about μ. Of course, the larger the sample, the closer our sample mean should be to the population mean.

Sampling from Small Populations

When sampling is done from a small population without replacement, the following formula can be used to find $\sigma_{\bar{x}}$:

> **Standard Error of the Mean When Sampling Without Replacement**
>
> $$\sigma_{\bar{x}} = \frac{\sigma}{\sqrt{n}} \sqrt{\frac{N - n}{N - 1}}$$
>
> (8.3)

where σ is the standard deviation of the sampled population, n is the sample size, and N is the size of the population.

As a rule of thumb, if sampling is done without replacement and the size of the population is at least 20 times as large as the sample size (i.e., if $N \geq 20n$), then either formula (8.2) or (8.3) can be used. Otherwise, formula (8.3) is used. Formula (8.3) is equivalent to formula (8.2) when N is infinitely large because the values of $\sqrt{(N - n)/(N - 1)}$ approach 1 as the values of N get large. The factor $\sqrt{(N - n)/(N - 1)}$ is sometimes referred to as a **correction factor** for the population being finite.

APPLICATION 8.6

Suppose the lengths of service (in years) of three college mathematics teachers is as shown in the table.

Math Teacher	Length of Service
Smith	6
Jones	4
Doe	2

Further, suppose we select random samples of size 2 without replacement and compute the mean length of service for each. The collection of sample means will constitute the sampling distribution of the mean. Find the standard error of the mean $\sigma_{\bar{x}}$

    a. Without using formula (8.3).

    b. Using formula (8.3).

Solution: All possible samples of size 2 are listed in the table with their sample means.

Sample Names	Length of Service	Sample Mean $\bar{x}$
Jones, Doe	{4, 2}	3
Doe, Jones	{2, 4}	3
Smith, Doe	{6, 2}	4
Doe, Smith	{6, 2}	4
Jones, Smith	{4, 6}	5
Smith, Jones	{6, 4}	5

The population mean is

$$\mu = \frac{2 + 4 + 6}{3} = 4$$

a. The following table organizes the computations for finding the standard error of the sample means:

Sample	$\bar{x}$	$\bar{x} - \mu_{\bar{x}}$	$(\bar{x} - \mu_{\bar{x}})^2$
{4, 2}	3	3 − 4	1
{2, 4}	3	3 − 4	1
{6, 2}	4	4 − 4	0
{6, 2}	4	4 − 4	0
{4, 6}	5	5 − 4	1
{6, 4}	5	5 − 4	1

The standard error of the mean is

$$\sigma_{\bar{x}} = \sqrt{\frac{\Sigma (\bar{x} - \mu_{\bar{x}})^2}{N}}$$

$$= \sqrt{\frac{4}{6}} \approx 0.82$$

b. We first find the population standard deviation. The calculations are organized in the following table:

x	$x - \mu$	$(x - \mu)^2$
2	2 − 4 = −2	4
4	4 − 4 = 0	0
6	6 − 4 = 2	4
		8

The standard deviation of the population is

$$\sigma = \sqrt{\frac{\Sigma (x - \mu)^2}{N}}$$

$$= \sqrt{\frac{8}{3}} \approx 1.63$$

By using formula (8.3) with $n = 2$, $N = 3$, and $\sigma = 1.63$, we have

$$\sigma_{\bar{x}} = \frac{\sigma}{\sqrt{n}} \sqrt{\frac{N - n}{N - 1}}$$

$$= \frac{1.63}{\sqrt{2}} \sqrt{\frac{3 - 2}{3 - 1}} \approx 0.82$$

Note that this agrees with the result found in part a. ▪

APPLICATION 8.7

If the breaking point of 4000 fuses has a mean of 5 amperes and a standard deviation of 1.5 amperes, and samples of 150 fuses are selected from this population, find $\mu_{\bar{x}}$ and $\sigma_{\bar{x}}$.

Solution: From formula (8.1) we have $\mu_{\bar{x}} = 5$ amperes. Since $4000 > 20(150)$, we can use either formula (8.2) or formula (8.3) to find $\sigma_{\bar{x}}$. By using formula (8.2), we have

$$\sigma_{\bar{x}} = \frac{\sigma}{\sqrt{n}}$$

$$= \frac{1.5}{\sqrt{150}} \approx 0.122$$

By formula (8.3), we have

$$\sigma_{\bar{x}} = \frac{\sigma}{\sqrt{n}} \sqrt{\frac{N-n}{N-1}}$$

$$= \frac{1.5}{\sqrt{150}} \sqrt{\frac{4000-150}{4000-1}} \approx 0.120$$

Note that the answers agree to two decimal places. When the condition $N > 20n$ is satisfied, using formula (8.3) instead of formula (8.2) is usually not worth the additional effort. ■

The flowchart in figure 8.5 summarizes the decisions to be made when the value of the standard error of the mean is calculated.

FIGURE 8.5

Flowchart for calculating standard error of the mean

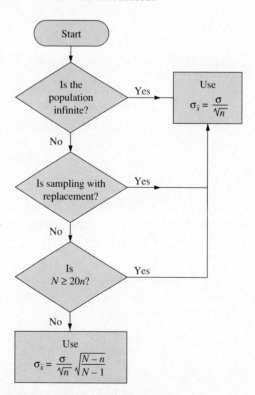

EXAMPLE 8.20

A six-sided die can be used to simulate sampling from the population of values 1, 2, 3, 4, 5, and 6. By rolling the die five times, we can generate a sample of size 5. And by repeatedly generating samples and recording the sample means, we can generate an empirical sampling distribution. The following 32 samples were generated by using a die:

Sample	$\bar{x}$	Sample	$\bar{x}$	Sample	$\bar{x}$	Sample	$\bar{x}$
1, 2, 3, 5, 4	3	5, 1, 1, 3, 2	2.4	5, 2, 4, 3, 5	3.8	5, 3, 1, 6, 5	4
3, 3, 6, 5, 2	3.8	2, 2, 4, 4, 1	2.6	2, 6, 3, 6, 2	3.8	4, 5, 3, 3, 4	3.8
2, 1, 3, 5, 4	3	5, 3, 5, 4, 1	3.6	1, 4, 2, 2, 1	2	2, 2, 5, 6, 1	3.2
3, 3, 4, 6, 3	3.8	2, 4, 5, 6, 1	3.6	1, 6, 6, 5, 1	3.8	1, 5, 5, 3, 8	4.4
3, 5, 4, 6, 2	4	4, 4, 4, 2, 5	3.8	4, 2, 2, 6, 4	3.6	3, 4, 6, 4, 6	4.6
1, 2, 5, 2, 5	3	2, 3, 6, 1, 2	2.8	5, 3, 3, 1, 5	3.4	1, 3, 1, 1, 1	1.4
6, 5, 6, 6, 5	5.6	4, 6, 3, 4, 5	4.4	1, 4, 4, 3, 1	2.6	6, 2, 6, 5, 1	4
1, 6, 3, 3, 3	3.2	2, 5, 3, 1, 2	2.6	4, 5, 5, 1, 4	3.8	3, 1, 1, 2, 3	2

A frequency table for the 32 sample means is as shown.

The empirical distribution can be thought of as a sample from the sampling distribution of the mean. As a sample, it has a mean of 3.42 and a standard deviation of 0.85. The standard deviation of the population of values 1, 2, 3, 4, 5, and 6, is $\sigma = 1.71$. As a result, the standard error of the mean is

$\bar{x}$	f	$\bar{x}$	f
1.4	1	3.4	1
2	2	3.6	3
2.4	1	3.8	8
2.6	3	4.0	3
2.8	1	4.4	2
3	3	4.6	1
3.2	2	5.6	1

$$\sigma_{\bar{x}} = \frac{\sigma}{\sqrt{n}}$$

$$= \frac{1.71}{\sqrt{5}} = 0.76$$

Note that the mean and standard deviation for the empirical distribution can be used to estimate $\mu_{\bar{x}}$ and $\sigma_{\bar{x}}$, respectively.

Empirical Distribution of $\bar{x}$	**Sampling Distribution of $\bar{x}$**
$\bar{x} = 3.42$	$\mu_{\bar{x}} = 3.5$
$s_{\bar{x}} = 0.85$	$\sigma_{\bar{x}} = 0.76$

MINITAB can be used to simulate sampling from the population of values 1, 2, 3, 4, 5, and 6. Computer display 8.2 contains the commands for generating 32 samples of five values, along with the output. Each of the five columns contains 32 integers from 1 to 6, inclusive. The first entries in each of the five columns make up the first sample, the second entries in each column make up the second sample, and so on. The RMEAN command is used to determine the mean of each sample (row), and these 32 means are stored in column C6.

Computer Display 8.2

```
MTB > RANDOM 32 C1-C5;
SUBC > INTEGER 1 6.
MTB > RMEAN C1-C5 C6
MTB > DESCRIBE C6


         N      MEAN     MEDIAN    TRMEAN     STDEV    SEMEAN
C6      32    3.4375    3.4000    3.4357    0.5014    0.0886


         MIN       MAX        Q1        Q3
C6    2.2000    4.6000    3.0500    3.8000
```

Notice that the mean of the empirical distribution is 3.4375 and the standard deviation is 0.5014. The display also contains the smallest mean, the largest mean, the three quartiles for the 32 sample means, and the trimmed mean, TRMEAN. The trimmed mean was discussed in exercise 35 of exercise set 3.1.

EXERCISE SET 8.2

Basic Skills

1. Characterize or identify the elements in the following sampling distributions:
   a. the sampling distribution of the median
   b. the sampling distribution of the variance

2. Characterize or identify the elements in the following sampling distributions:
   a. sampling distribution of the range
   b. the sampling distribution of the midrange

3. For each of the following situations, where sampling is without replacement, specify what formula(s) can be used for determining $\sigma_{\bar{x}}$:
   a. $N = 5000, n = 500$
   b. $N = 3500, n = 100$
   c. $N = 1000, n = 40$

4. For each of the following situations, where sampling is without replacement, specify what formula(s) can be used for determining $\sigma_{\bar{x}}$:
   a. $N = 500, n = 50$
   b. $N = 350, n = 10$
   c. $N = 1000, n = 50$

5. What do the symbols $\mu_{\bar{x}}$ and μ denote?

6. What do the symbols $\sigma_{\bar{x}}$ and σ denote?

7. A certain population has a standard deviation of 15. Random samples of size n are taken with replacement and the means of the samples are computed. What happens to the standard error of the mean as n is increased from 400 to 900? What happens to $\sigma_{\bar{x}}$ as n increases?

8. A population of size 10,000 has a standard deviation of 25. Random samples of size n are taken without replacement and the means of the samples are computed. What happens to the standard error of the mean as n gets larger and larger.

9. If $\sigma = 25$, find the standard error of the mean if sampling is without replacement and
   a. $N = 750, n = 15$.
   b. $N = 800, n = 45$.

10. If $\sigma = 20$, find $\sigma_{\bar{x}}$ if sampling is without replacement and
    a. $N = 1000, n = 100$.
    b. $N = 750, n = 15$.

11. If ordered samples of size 2 are selected with replacement from the population whose values are 0, 2, 4, and 8, find $\mu_{\bar{x}}$ and $\sigma_{\bar{x}}$ without using formulas (8.2) or (8.3). (*Hint:* there are 16 ordered samples.)

12. If the sampling in exercise 11 is done without replacement, find $\mu_{\bar{x}}$ and $\sigma_{\bar{x}}$ without using formulas (8.2) or (8.3).

13. Repeat exercise 11, but use formula (8.2) or (8.3) to find $\sigma_{\bar{x}}$.

14. Repeat exercise 12, but use formula (8.2) or (8.3) to find $\sigma_{\bar{x}}$.

15. Under what conditions will the sampling distribution of the mode exist?

Further Applications

16. Costs for a tune-up at a large service station are $30 for a four-cylinder car, $36 for a six-cylinder car, and $42 for an eight-cylinder car. From past records, it is known that 10% of its tune-ups are done on four-cylinder cars, 40% on six-cylinder cars, and 50% on eight-cylinder cars. Suppose two cars that need a tune-up are randomly chosen.
    a. Construct a probability distribution table for the average service charge.
    b. Find $\mu_{\bar{x}}$.

17. Costs for an oil change and lubrication at a large service station are $12 for a four-cylinder car, $15 for a six-cylinder car, and $18 for an eight-cylinder car. From past records, it is known that 40% of oil changes and lubrication jobs are done on four-cylinder cars, 20% on six-cylinder cars, and 40% on eight-cylinder cars. Suppose two cars that need an oil change and lubricated are randomly chosen.
    a. Construct a probability distribution for the average service charge.
    b. Find $\mu_{\bar{x}}$.
    c. Find $\sigma_{\bar{x}}$.

Going Beyond

18. Ordered samples of size 2 are drawn with replacement from the population of values 0, 2, 4, and 6 and the range is computed for each sample.
    a. Construct a frequency table for values of the range.
    b. Find the mean of the sampling distribution of the range.
    c. Find the standard error of the range.
    d. Draw a graph for the frequency distribution of the range.

19. Use a die to simulate sampling from the population of values 1, 2, 3, 4, 5, and 6. Let the size of each sample be 10, select 25 samples and find the mean for each sample. Then find the mean and standard deviation for the empirical sampling distribution of the mean. Compare these values with the theoretical values. Why do they differ?

20. Show that if $20n \leq N$, then
    $$0.97 \leq \sqrt{(N - n)/(N - 1)} \leq 1.$$

21. a. Suppose a finite population consists of the values 0, 3, and 9. Find the parameters μ and σ^2 for the population.
    b. Suppose ordered samples of size 2 are selected with replacement from the population. Construct the sampling distribution of $\bar{x}$ by listing the nine samples and finding the mean for each.
    c. For the sampling distribution of $\bar{x}$, find $\mu_{\bar{x}}$ and $\sigma_{\bar{x}}$.
    d. Show that $\bar{x}$ is an unbiased estimator of μ.
    e. For each sample in part b, find s^2.
    f. For the sampling distribution of s^2, show that s^2 is an unbiased estimator for σ^2.
    g. Show that s is a biased estimator of σ.
    h. Find the standard error of the sample variance.

22. Refer to exercise 16.
    a. Construct a probability distribution table for the total service charge for the two cars.
    b. Find the mean of the sampling distribution of the total service charge.
    c. Find the standard error of the total service charge.

23. Prove that $\sigma_e = \sigma_{\bar{x}}$ where $e = \bar{x} - \mu$.

SECTION 8.3   *Sampling from Normal Populations*

Many sampling distributions in statistics are based on the assumption that sampling is done from a normal population. It is reasonable to expect that if the sampled population is normal, then the sampling distribution of the sample mean is also normal. This is the substance of the following important result:

> **Sampling from a Normal Population**
>
> If samples of size n are drawn from a normal population, then the sampling distribution of the mean is normal.

In section 8.2 we learned that if sampling is done from a large population with mean μ and standard deviation σ, then the mean of the sampling distribution of the mean is equal to

$$\mu_{\bar{x}} = \mu$$

and the standard deviation of the distribution of sample means (standard error of the mean) is given by

$$\sigma_{\bar{x}} = \frac{\sigma}{\sqrt{n}}$$

where n is the size of the samples.

EXAMPLE 8.21

Suppose the heights of American males are normally distributed with a mean of 70 inches and a standard deviation of 2.6 inches. If samples of size 20 are selected from the population of heights, the sampling distribution of the mean is normally distributed with a mean equal to $\mu = 70$ inches and a standard deviation equal to $\sigma_{\bar{x}} = \sigma/\sqrt{n} = 2.6/\sqrt{20} = 0.58$ inches. These two distributions are illustrated in figure 8.6. Note that the sampling distribution of the mean has a smaller standard deviation than the population distribution.

FIGURE 8.6

Population distribution and sampling distribution of mean for heights of American males

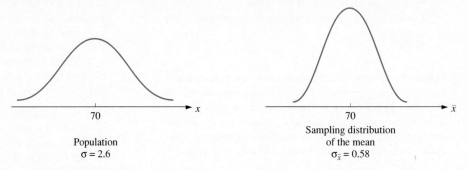

Recall from section 7.2 that the z score formula

$$z = \frac{x - \mu}{\sigma}$$

can be used to transform any normal random variable X with mean μ and standard deviation σ into the standard normal variable Z with mean 0 and standard deviation 1. As a result, the sampling distribution of z is the standard normal distribution.

If a sample of size n is taken from a normal population and the sample mean $\bar{x}$ is calculated, then the z score for $\bar{x}$ is given by

$$z = \frac{\bar{x} - \mu_{\bar{x}}}{\sigma_{\bar{x}}}$$

Since $\mu_{\bar{x}} = \mu$ and $\sigma_{\bar{x}} = \sigma/\sqrt{n}$, we have

$$z = \frac{\bar{x} - \mu}{\sigma/\sqrt{n}}$$

Hence, the sampling distribution of the z statistic $(\bar{x} - \mu)/(\sigma/\sqrt{n})$ is the standard normal distribution. Note that the z statistic can also be written as

$$z = \sqrt{n}\,\frac{\bar{x} - \mu}{\sigma}$$

Consider applications 8.8 and 8.9.

APPLICATION 8.8

Suppose a random sample of size 9 is selected from a normal population with mean $\mu = 25$ and standard deviation $\sigma = 6$. What is the probability that the sample mean $\bar{x}$ is greater than 28?

Solution: The sampling distribution of the mean is normal with $\mu_{\bar{x}} = \mu = 25$ and $\sigma_{\bar{x}} = \sigma/\sqrt{n} = 6/\sqrt{9} = 2$. To determine the probability $P(\bar{X} > 28)$, we need to find the area under the normal curve, above the horizontal axis, and to the right of

$\bar{x} = 28$, as indicated in figure 8.7. The z value for $\bar{x} = 28$ is

$$z = \frac{\bar{x} - \mu}{\dfrac{\sigma}{\sqrt{n}}}$$

$$= \frac{28 - 25}{2} = 1.5$$

Thus, we have

$$P(\bar{X} > 28) = P(Z > 1.5)$$
$$= 0.5 - 0.4332 = 0.0668$$

FIGURE 8.7

$P(\bar{X} > 28)$ for application 8.8

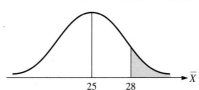

Hence, the probability that a random sample of size 9 has a mean greater than 28 is $P(\bar{X} > 28) = 0.0668$. ▪

APPLICATION 8.9

The times required for workers to complete a certain task are normally distributed with a mean of 30 minutes and a standard deviation of 9 minutes. If a random sample of 25 workers is drawn from the work force, find the probability that the mean completion time for the sample of workers is between 28 and 33 minutes.

Solution: The population distribution of task completion times and the sampling distribution of the mean are shown in figure 8.8. Since $\mu = 30$,

$$\mu_{\bar{x}} = 30$$

FIGURE 8.8

Population distribution and sampling distribution for application 8.9

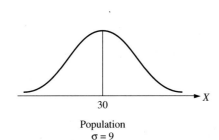

Population
$\sigma = 9$

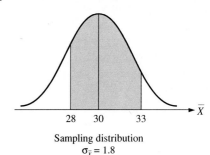

Sampling distribution
$\sigma_{\bar{x}} = 1.8$

Also, since $\sigma = 9$, the standard error of the mean is

$$\sigma_{\bar{x}} = \frac{\sigma}{\sqrt{n}}$$

$$= \frac{9}{\sqrt{25}} = 1.8$$

As a result, the z value for $\bar{x} = 28$ is

$$z = \frac{28 - 30}{1.8} = -1.11$$

and, the z value for $\overline{x} = 33$ is

$$z = \frac{33 - 30}{1.8} = 1.67$$

Thus,

$$P(28 < \overline{X} < 33) = P(-1.11 < Z < 1.67)$$
$$= 0.3665 + 0.4525 = 0.819 \quad \blacksquare$$

It is important to recognize when to use the sampling distribution of the mean. Consider the following two problems:

a. If the weights of newborn children are normally distributed with a mean of 115 ounces and a standard deviation of 12 ounces, find the probability that a randomly selected newborn child weighs more than 125 ounces.

b. If the weights of newborn children are normally distributed with a mean of 115 ounces and a standard deviation of 12 ounces, find the probability that the average weight of a random sample of 16 newborn children exceeds 125 ounces.

Which problem involves the sampling distribution of the mean? If you chose problem b you are correct. Problem b involves locating a sample mean based on a sample of size 16 and problem a involves locating a single measurement. In problem b we are interested in the location of $\overline{x}$ within the sampling distribution of the mean. Both problems involve finding normal probabilities. The solution to problem a is as follows (see figure 8.9). The z value for $\overline{x} = 125$ is

$$z = \frac{125 - 115}{12} = 0.83$$

FIGURE 8.9

Normal distribution and transformed normal distribution for problem a

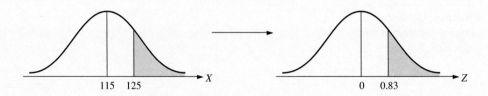

The probability that a newborn child weighs more than 125 ounces in problem b is

$$P(\overline{X} > 125) = P(Z > 0.83)$$
$$= 0.5 - 0.2967 = 0.2033$$

The solution to problem b refers to figure 8.10. The z value for a mean of 125 ounces is

$$z = \frac{(\overline{x} - \mu_{\overline{x}})}{\sigma_{\overline{x}}}$$

$$= \frac{125 - 115}{3} = 3.33$$

FIGURE 8.10

Population distribution and sampling distribution for problem b

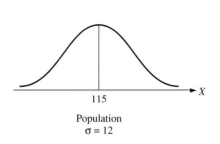

Population
$\sigma = 12$

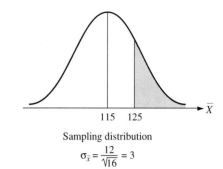

Sampling distribution
$\sigma_{\bar{x}} = \dfrac{12}{\sqrt{16}} = 3$

The probability that a sample mean is greater than 125 ounces is

$$P(\bar{X} > 125) = P(Z > 3.33)$$
$$= 0.5 - 0.4996 = 0.0004$$

Thus, it is rather unlikely that the mean weight of a sample of 16 newborn children will exceed 125 ounces.

t Distributions

When sampling is from a normal population, the sampling distribution of the mean is normal. However, if the standard deviation of the population is unknown, we cannot standardize the sample mean to a standard normal. In most practical situations, the population standard deviation σ is unknown, and the sample standard deviation s is used to estimate σ. Consequently, the following statistic does not have the standard normal sampling distribution:

$$\frac{\bar{x} - \mu}{s/\sqrt{n}}$$

This statistic is denoted by t and is called the **t statistic.** Thus, the t statistic is given by the formula

---

t Statistic

$$t = \frac{\bar{x} - \mu}{s/\sqrt{n}}$$

In 1908, W. Gosset, a staff member of an Irish brewery, published a research paper concerning the equation for the probability distribution of t. Since employees of the brewery were not permitted to publish research findings, Gosset published his results under the pen name Student. Since that time, the sampling distribution of the t statistic has been referred to as the *Student t distribution,* or simply the **t distribution.** The actual equation of the t distribution is very complicated and will be omitted here. Instead, the t table (see the back endpaper) contains a collection of t values and their associated probabilities.

The sampling distribution of t is similar to the standard normal. Both are bell shaped, both have means equal to 0, and both are symmetrical about their means. The sampling distribution of t is more variable than the standard normal distribution. For the statistic z, $\bar{x}$ is the only quantity that varies from sample to sample, whereas for t, both $\bar{x}$ and s vary from sample to sample. The exact shape of a t distribution is completely specified by a single value (parameter) referred to as the number of **degrees of freedom** (df). The sample size n is related to df by the following relationship:

$$df = n - 1$$

Actually, df is the divisor in the formula for computing the sample variance s^2:

$$s^2 = \frac{SS}{n-1} = \frac{SS}{df}$$

The variance of a t distribution depends on the sample size and is always greater than 1, the variance of the standard normal distribution. When df > 2, the variance of a t distribution is given by

$$\sigma_t^2 = \frac{df}{df - 2}$$

In summary, a sample of size n is drawn from a normal population with mean μ and unknown standard deviation and the sample mean $\bar{x}$ is computed. The t value for $\bar{x}$ is given by

$$t = \frac{\bar{x} - \mu}{s/\sqrt{n}}$$

The sampling distributions of t have the following properties:

1. They have a mean $\mu = 0$.
2. They are symmetric about $\mu = 0$.
3. They are more variable than the standard normal distributions.
4. They are bell shaped.
5. Their exact shape depends on df $= n - 1$.
6. Their variances depend on df and $\sigma^2 = df/(df - 2)$, provided df > 2.
7. As n gets large, the sampling distributions of t approach the standard normal distribution Z.

Since the sampling distributions for t are more variable than the standard normal distribution, the t distributions have thicker tail areas than the standard normal distribution, as figure 8.11 illustrates.

FIGURE 8.11

t distributions compared to standard normal distribution

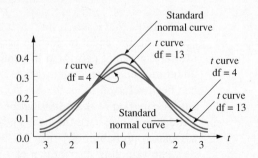

APPLICATION 8.10   Suppose a random sample of size 11 is selected from a normal population with a mean of 14. If the sample mean is 18 and the sample standard deviation is 14.3, calculate the value of the t statistic.

Solution: The value of t is

$$t = \frac{\bar{x} - \mu}{s/\sqrt{n}}$$

$$= \frac{18 - 14}{14.3/\sqrt{11}} \approx 0.93 \quad ■$$

To locate a t value in t table (see back endpaper) for one of the five given probabilities (values of α) listed at the top of the table, we locate the row labeled by df at the left of the table and the column identified by the probability at the top of the table; the t value is determined by where the row and column intersect.

EXAMPLE 8.22

To locate the t value for a probability (or area) of $\alpha = 0.025$ and a sample size of 13, we locate the row labeled df $= n - 1 = 13 - 1 = 12$ and the column labeled $\alpha = 0.025$; the t value is listed at the intersection of this row and column as shown here:

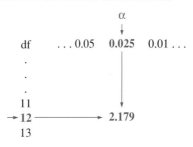

Thus, $t = 2.179$.

APPLICATION 8.11

Suppose a random sample of size 16 is selected from a normal population with $\mu = 20$. If the standard deviation of the sample is $s = 4$, find $P(\bar{X} > 21.753)$.

Solution: We find the t value for 21.753. Since $t = (\bar{x} - \mu)/(s/\sqrt{n})$,

$$t = \frac{21.753 - 20}{4/\sqrt{16}} = 1.753$$

Thus,

$$P(\bar{X} > 21.753) = P(t > 1.753)$$

Since df $= n - 1 = 16 - 1 = 15$, we read across the row labeled 15 in the t table (see back endpaper) until we locate 1.753 under the column heading 0.05. See table 8.4 for a portion of the t table.

TABLE 8.4

Portion of the t Table
(See the Back Endpaper)

df	0.01	0.05	0.025	0.01	0.005
.			α		
.					
.					
12	1.356	1.782	2.179	2.681	3.055
13	1.350	1.771	2.160	2.650	3.012
14	1.345	1.761	2.145	2.624	2.977
→ 15	1.341	1.753	2.131	2.602	2.947
16	1.337	1.746	2.120	2.583	2.921

Thus, $P(\bar{X} > 21.753) = 0.05$. ■

Unlike the standard normal table (z table; see the front endpaper), the t table contains only five probabilities (or areas). This may appear at first to be a big disadvantage, but for our applications, these five probabilities will be sufficient. Our future applications involving a t distribution will require t values corresponding to df $= n - 1$ for one of the five given probability values in the table.

EXERCISE SET 8.3

Basic Skills

1. If a sample of size 37 is taken from a normal population that has a mean of 50 and a standard deviation of 30, find
   a. $P(\overline{X} > 60)$.
   b. $P(45 < \overline{X} < 58)$.
   c. $P(\overline{X} < 47)$.
   d. $P(\overline{X} > 45)$.
   e. $P(\overline{X} < 62)$.

2. A sample of size 100 is taken from a normal population that has a mean of 25 and a standard deviation of 15. Find the probability that the sample mean is
   a. between 26 and 27.
   b. greater than 28.
   c. less than 23.5.
   d. between 23.4 and 24.5.
   e. less than 23.5 or greater than 26.5.

3. If $n = 6$, find $P(t > 4.032)$.

4. If $n = 21$, find $P(t < 2.845)$.

5. If $n = 16$, find $P(-2.131 < t < 2.131)$.

6. If $n = 15$, find $P(t > 1.953)$.

7. If $n = 22$, find $P(t < -2.080)$.

8. If $n = 13$, find $P(1.782 < t < 3.055)$.

9. If $P(t < -1.796) = 0.05$, find n.

10. If $P(-1.96 < t < 1.96) = 0.95$, find n.

11. If $\overline{x} = 15$, $\mu = 20$, $s = 7$, and df $= 8$, calculate the value of the t statistic.

12. If $\overline{x} = 20$, $\mu = 13$, $s = 5$, and df $= 15$, calculate the value of the t statistic.

13. Given a random sample of size 12 from a normal distribution, find a value for t_0 such that
    a. $P(1.363 < t < t_0) = 0.09$.
    b. $P(t_0 < t < 3.106) = 0.045$.

14. If a random sample of size 80 is obtained from a normal distribution with $\mu = 32$ and $\sigma = 10$, find a value for t_0

so that
   a. $P(\overline{X} > t_0) = 0.30$.   c. $P(\overline{X} < t_0) = 0.60$.
   b. $P(\overline{X} < t_0) = 0.40$.   d. $P(33 < \overline{X} < t_0) = 0.05$.

15. A sample of size 26 is taken from a normal population that has a mean of 30. If the sample standard deviation is 10, find the probability that the sample mean is less than 34.04.

Further Applications

16. If it is assumed that SAT math scores are normally distributed with a mean of 500 and a standard deviation of 100, find the probability that the sample mean obtained from a sample of 49 scores is
    a. greater than 520.
    b. between 475 and 520.
    c. less than 525.
    d. less than 470.

17. Suppose the hourly wages of all workers in a certain auto plant are normally distributed with a mean of $12.50 and a standard deviation of $0.95. If a random sample of 100 workers is selected from this plant, find the probability that the sample mean hourly wage is
    a. less than $12.60.
    b. between $12.45 and $12.65.
    c. more than $12.30.

18. The average life of a certain type of electric heater is 10 years and the standard deviation is 1.5 years. If the lives of the electric heaters are known to be normally distributed, find
    a. the probability that the mean life of a random sample of 16 heaters is less than 10.5 years.
    b. the value of $\overline{x}$ so that 20% of the means computed from random samples of size 25 fall below $\overline{x}$.
    c. P_{60}, the 60th percentile of the sample distribution of the mean with $n = 9$.

Going Beyond

19. Is one likely to obtain the random sample 27, 25, 26, 24, 23, and 22 from a normal population that has a mean of 22? Explain.

SECTION 8.4   *Sampling from Nonnormal Populations*

Recall from section 8.3 that if the sampled population is normal, the sampling distribution of the mean is normal regardless of the sample size. What is the sampling distribution of the mean if the sampled population isn't normal? In actual practice we sample without replacement from populations that are usually large, but whose exact shape and parameters are usually unknown. As a result, one might suspect that the sampling distribution of the mean would not follow any basic shape. But if the sample size n is large ($n \geq 30$), the sampling distribution of the mean is bell shaped. This remarkable fact is the substance of the well-known **central limit theorem,** which is stated as follows:

The Central Limit Theorem

The sampling distribution of the mean is approximately normal for nonnormal sampled populations. The larger the sample size, the closer the sampling distribution is to being normal.

For most purposes, the normal approximation is considered good provided $n \geq 30$. That the shape of the sampling distribution of the mean is approximately normal, even in cases where the original population is bimodal (see figure 8.12), is indeed remarkable.

FIGURE 8.12

Bimodal population and sampling distribution of mean

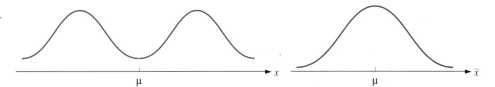

The following three situations illustrate the central limit theorem for three nonnormal sampled populations. The first population has a histogram, which is rectangular-shaped; the second population has a histogram, which is U shaped; and the third population has a histogram, which is J shaped. For each of these nonnormal sampled populations, three sampling distributions of the mean are illustrated, one for $n = 2$, one for $n = 5$, and one for $n = 25$. For each population, MINITAB was used to generate an empirical sampling distribution of the mean involving 500 samples for each of the three sample sizes. Notice the shape the sampling distribution of the mean in each situation as the sample size n increases. For $n = 2$, the sampling distributions of the mean take on appearances different from the sampled populations. For $n = 5$, the sampling distributions of the mean take on bell-shaped appearances. And for $n = 25$ the sampling distributions of the mean appear to be approximately normal. For each of the three situations notice also that the variance of the sampling distribution of the mean decreases as n increases.

For convenience, the histograms are drawn horizontally using asterisks to form the bars. The mean and standard error of the theoretical sampling distribution of the mean and the empirical sampling distribution of the mean are given for each sample size for purposes of comparison.

Situation A

The histogram for the population has a rectangular shape, as shown here.

```
EACH * REPRESENTS 2 OBS.

MIDPOINT     COUNT
   0          100  *************************************************
   1          100  *************************************************
   2          100  *************************************************
   3          100  *************************************************
   4          100  *************************************************
   5          100  *************************************************
   6          100  *************************************************
   7          100  *************************************************
   8          100  *************************************************
   9          100  *************************************************
```

The empirical sampling distribution of the mean based on 500 samples of size $n = 2$ is as follows:

```
EACH * REPRESENTS 2 OBS.

MIDPOINT     COUNT
   0            5  ***
   1           20  **********
   2           44  **********************
   3           60  ******************************
   4           94  ***********************************************
   5           87  ********************************************
   6           75  *************************************
   7           58  *****************************
   8           39  ********************
   9           18  *********
```

Empirical Distribution	Theoretical Distribution
Mean = 4.5970	Mean = 4.5
Standard error = 2.050	Standard error = 2.032

The empirical distribution of 500 sample means for a sample size of $n = 5$ is as follows:

```
EACH * REPRESENTS 2 OBS.

MIDPOINT     COUNT
   1.0          2  *
   1.5          6  ***
   2.0         18  *********
   2.5         22  ***********
   3.0         47  ************************
   3.5         52  **************************
   4.0         83  ******************************************
   4.5         63  ********************************
   5.0         92  **********************************************
   5.5         37  ******************
   6.0         40  ********************
   6.5         19  **********
   7.0         13  *******
   7.5          3  **
   8.0          3  **
```

Empirical Distribution	Theoretical Distribution
Mean = 4.4084	Mean = 4.5
Standard error = 1.2777	Standard error = 1.2852

The empirical distribution for 500 sample means for a sample size of $n = 25$ is as follows:

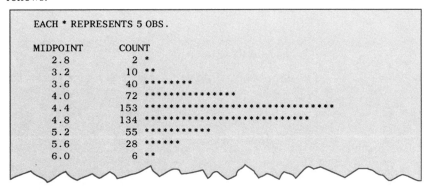

```
EACH * REPRESENTS 5 OBS.

MIDPOINT        COUNT
   2.8             2  *
   3.2            10  **
   3.6            40  ********
   4.0            72  **************
   4.4           153  ******************************
   4.8           134  **************************
   5.2            55  ***********
   5.6            28  ******
   6.0             6  **
```

Empirical Distribution	Theoretical Distribution
Mean = 4.5114	Mean = 4.5
Standard error = 0.5458	Standard error = 0.5747

Situation B

The histogram for the population has a **U** shape, as shown here:

```
EACH * REPRESENTS 5 OBS.

MIDPOINT        COUNT
    0            180  ****************************************
    1             90  ******************
    2             80  ****************
    3             30  ******
    4             20  ****
    5             20  ****
    6             30  ******
    7             80  ****************
    8             90  ******************
    9            180  ****************************************
```

The empirical distribution of sample means for 500 samples of size $n = 2$ is as follows:

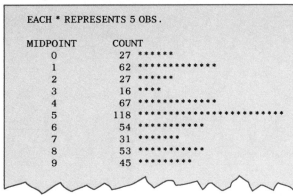

```
EACH * REPRESENTS 5 OBS.

MIDPOINT        COUNT
    0             27  ******
    1             62  *************
    2             27  ******
    3             16  ****
    4             67  *************
    5            118  ************************
    6             54  ***********
    7             31  *******
    8             53  ***********
    9             45  *********
```

Empirical Distribution	Theoretical Distribution
Mean = 4.534	Mean = 4.5
Standard error = 2.603	Standard error = 2.580

The empirical distribution of 500 sample means for a sample size of $n = 5$ is given here:

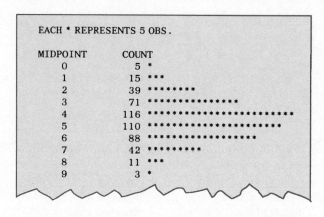

```
EACH * REPRESENTS 5 OBS.

MIDPOINT          COUNT
    0                 5 *
    1                15 ***
    2                39 ********
    3                71 **************
    4               116 ***********************
    5               110 **********************
    6                88 ******************
    7                42 *********
    8                11 ***
    9                 3 *
```

Empirical Distribution	Theoretical Distribution
Mean = 4.5084	Mean = 4.5
Standard error = 1.6561	Standard error = 1.6319

The empirical distribution for 500 sample means for a sample size of $n = 25$ is as follows:

```
EACH * REPRESENTS 5 OBS.

MIDPOINT          COUNT
   2.5                3 *
   3.0               24 *****
   3.5               48 *********
   4.0              109 **********************
   4.5              127 **************************
   5.0              106 *********************
   5.5               57 ***********
   6.0               23 *****
   6.5                1 *
   7.0                2 *
```

Empirical Distribution	Theoretical Distribution
Mean = 4.5175	Mean = 4.5
Standard error = 0.7522	Standard error = 0.7298

Situation C

The histogram for the population is J shaped:

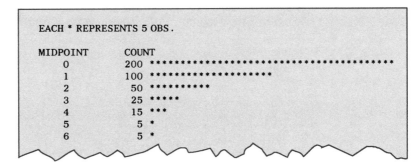

```
EACH * REPRESENTS 5 OBS.

MIDPOINT     COUNT
    0         200    ******************************************
    1         100    ********************
    2          50    *********
    3          25    *****
    4          15    ***
    5           5    *
    6           5    *
```

The empirical distribution for 500 sample means based on a sample size of $n = 2$ is as follows:

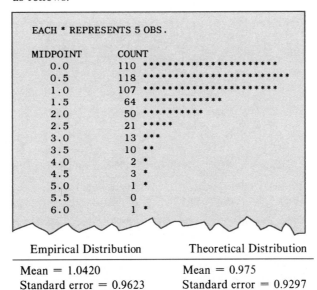

```
EACH * REPRESENTS 5 OBS.

MIDPOINT     COUNT
   0.0        110    **********************
   0.5        118    ***********************
   1.0        107    *********************
   1.5         64    *************
   2.0         50    *********
   2.5         21    *****
   3.0         13    ***
   3.5         10    **
   4.0          2    *
   4.5          3    *
   5.0          1    *
   5.5          0
   6.0          1    *
```

Empirical Distribution	Theoretical Distribution
Mean = 1.0420	Mean = 0.975
Standard error = 0.9623	Standard error = 0.9297

The empirical distribution of 500 sample means based on a sample size of $n = 5$ is as follows:

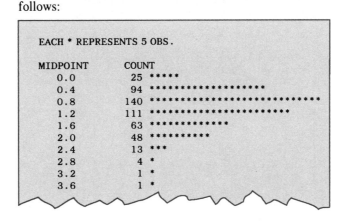

```
EACH * REPRESENTS 5 OBS.

MIDPOINT     COUNT
   0.0         25    *****
   0.4         94    *******************
   0.8        140    ****************************
   1.2        111    **********************
   1.6         63    *************
   2.0         48    *********
   2.4         13    ***
   2.8          4    *
   3.2          1    *
   3.6          1    *
```

Empirical Distribution	Theoretical Distribution
Mean = 0.9656	Mean = 0.9750
Standard error = 0.5981	Standard error = 0.588

The empirical distribution of 500 sample means based on a sample size of $n = 25$ is as follows:

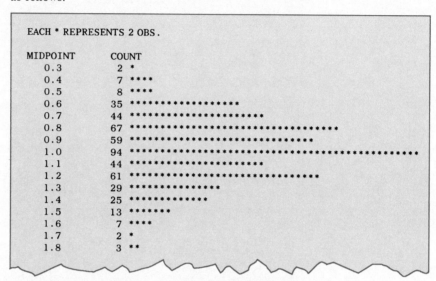

```
EACH * REPRESENTS 2 OBS.

MIDPOINT      COUNT
    0.3         2  *
    0.4         7  ****
    0.5         8  ****
    0.6        35  ******************
    0.7        44  **********************
    0.8        67  **********************************
    0.9        59  ******************************
    1.0        94  ***********************************************
    1.1        44  **********************
    1.2        61  ******************************
    1.3        29  ***************
    1.4        25  ************
    1.5        13  *******
    1.6         7  ****
    1.7         2  *
    1.8         3  **
```

Empirical Distribution	Theoretical Distribution
Mean = 0.9887	Mean = 0.9750
Standard error = 0.2656	Standard error = 0.2630

It is easy to use MINITAB to simulate an empirical sampling distribution of the mean. In the illustration that follows, 100 random samples of size 25 are generated from the population of integers from 1 to 10. Each of the 25 columns contains 100 integers. The first entries in each of the 25 columns make up the first sample. The second entries in each column make up the second sample, and so on. The RMEAN command is used to determine the mean of each sample (row), and these 100 means are stored in column C26. Computer display 8.3 contains the MINITAB commands and output.

Computer Display 8.3

```
MTB > RANDOM 100 C1-C25;
SUBC > INTEGER 1 10.
MTB > RMEAN C1-C25 C26
MTB > DESCRIBE C26
```

	N	MEAN	MEDIAN	TRMEAN	STDEV	SEMEAN
C26	100	5.4256	5.4800	5.4444	0.5679	0.0568

	MIN	MAX	Q1	Q3
C26	3.3200	6.8800	5.0500	5.8000

A histogram for the 100 sample means is constructed using the HISTOGRAM command. The results are contained in computer display 8.4.

Computer Display 8.4

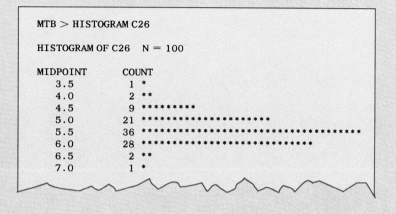

```
MTB > HISTOGRAM C26

HISTOGRAM OF C26   N = 100

MIDPOINT        COUNT
    3.5           1  *
    4.0           2  **
    4.5           9  *********
    5.0          21  *********************
    5.5          36  ************************************
    6.0          28  ****************************
    6.5           2  **
    7.0           1  *
```

EXAMPLE 8.23

As another example of the central limit theorem, suppose ordered samples of size 2 are drawn with replacement from a population consisting of the values 0, 2, 4, 6, and 8. The 25 samples listed in table 8.5 can be drawn.

TABLE 8.5

Samples of Size 2 from Population of Size 5, with Values of Sample Mean

Sample	$\bar{x}$	Sample	$\bar{x}$	Sample	$\bar{x}$
{0, 0}	0	{4, 0}	2	{8, 0}	4
{0, 2}	1	{4, 2}	3	{8, 2}	5
{0, 4}	2	{4, 4}	4	{8, 4}	6
{0, 6}	3	{4, 6}	5	{8, 6}	7
{0, 8}	4	{4, 8}	6	{8, 8}	8
{2, 0}	1	{6, 0}	3		
{2, 2}	2	{6, 2}	4		
{2, 4}	3	{6, 4}	5		
{2, 6}	4	{6, 6}	6		
{2, 8}	5	{6, 8}	7		

The mean of the population is

$$\mu = \frac{0 + 2 + 4 + 6 + 8}{4} = 4$$

The variance of the population is

$$\sigma^2 = \frac{(0 - 4)^2 + (2 - 4)^2 + (4 - 4)^2 + (6 - 4)^2 + (8 - 4)^2}{5}$$

$$= \frac{40}{5} = 8$$

and the standard deviation of the population is

$$\sigma = \sqrt{8} \approx 2.83$$

A graph of the frequency distribution for the population is shown in figure 8.13. The graph could not be considered bell shaped or normal. The 25 sample means can be grouped into the following frequency table:

$\overline{x}$	f
0	1
1	2
2	3
3	4
4	5
5	4
6	3
7	2
8	1

FIGURE 8.13

Frequency distribution for population of values 0, 2, 4, 6, 8

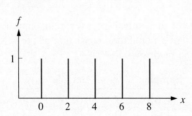

FIGURE 8.14

Frequency distribution for $\overline{x}$

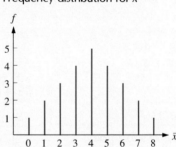

A graph of the frequency distribution for $\overline{x}$ is shown in figure 8.14. From the bell-shaped appearance of the distribution of means, we conclude that it is reasonable to approximate the sampling distribution of $\overline{x}$ by a normal distribution, once the mean and standard deviation of the sampling distribution are known. Table 8.6 makes it convenient to compute $\sigma_{\overline{x}}$, the standard error of the mean, using the formulas:

$$\mu_{\overline{x}} = \frac{\Sigma\,[\overline{x}f]}{N} \qquad \sigma_{\overline{x}} = \sqrt{\frac{\Sigma\,[(\overline{x} - \mu)^2 f]}{N}}$$

TABLE 8.6

Computations for Determining Standard Error of Mean

$\overline{x}$	f	$f\overline{x}$	$\overline{x} - \mu$	$(\overline{x} - \mu_{\overline{x}})^2$	$(\overline{x} - \mu_{\overline{x}})^2 f$
0	1	0	$0 - 4 = -4$	16	16
1	2	2	$1 - 4 = -3$	9	18
2	3	6	$2 - 4 = -2$	4	12
3	4	12	$3 - 4 = -1$	1	4
4	5	20	$4 - 4 = 0$	0	0
5	4	20	$5 - 4 = 1$	1	4
6	3	18	$6 - 4 = 2$	4	12
7	2	14	$7 - 4 = 3$	9	18
8	1	8	$8 - 4 = 4$	16	16
	$N = 25$	100			100

The mean of the sampling distribution of the mean is

$$\mu_{\bar{x}} = \frac{100}{25} = 4$$

and the variance of the sampling distribution of the mean is

$$\sigma_{\bar{x}}^2 = \frac{100}{25} = 4$$

Hence, the standard error of the mean is

$$\sigma_{\bar{x}} = \sqrt{4} = 2$$

Notice that the value of the standard error of the mean, found by using formula (8.2), agrees with the above result, since

$$\sigma_{\bar{x}} = \frac{\sigma}{\sqrt{n}}$$

$$= \frac{\sqrt{8}}{\sqrt{2}} = 2$$

APPLICATION 8.12

If in example 8.23, an ordered sample of size 60 is drawn with replacement from the population, what is the probability that the sample mean will be between 3.5 and 4.3? That is, determine $P(3.5 < \bar{X} < 4.3)$.

Solution: In this case, $\sigma_{\bar{x}} = \sigma/\sqrt{n} = \sqrt{8}/\sqrt{60} \approx 0.37$. Since $n \geq 30$ and as a consequence of the central limit theorem, the sampling distribution of the mean is well approximated by a normal distribution. The z scores for 3.5 and 4.3 are found by using the z score formula

$$z = \frac{\bar{x} - \mu_{\bar{x}}}{\sigma_{\bar{x}}}$$

The z score for 3.5 is

$$z = \frac{3.5 - 4}{0.37} = -1.35$$

and the z score for 4.3 is

$$z = \frac{4.3 - 4}{0.37} = 0.81$$

Thus,

$$P(3.5 < \bar{X} < 4.3) = P(-1.35 < Z < 0.81) = 0.2910 + 0.4115 = 0.7025 \quad ▪$$

It is important to note once again that as n increases, the variation among sample means decreases. This is due to the fact that $\sigma_{\bar{x}} = \sigma/\sqrt{n}$, which decreases as n increases. Recall that the area under a normal curve is equal to 1. So as the variance of the sampling distribution decreases, the sampling distribution of the mean narrows and becomes taller to contain an area of 1. This decrease in variance and the effects on the sampling distribution of the mean as n increases are illustrated in figure 8.15.

FIGURE 8.15

Effects of increasing *n* on sampling distribution of mean

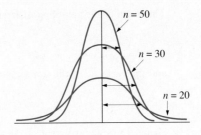

Sampling Distribution of Sample Sums

Since the sum of the measurements in a sample is directly related to the sample mean, one might suspect that the sampling distribution of the sample sums is approximately normal, and that the approximation improves as the sample size increases. This is indeed the case, and it can be shown that the mean of the sampling distribution of the sample sums has a mean equal to $n\mu$ and a standard deviation equal to $\sqrt{n}\sigma$, where μ and σ are the mean and standard deviation, respectively, of the sampled population and n is the sample size. That is

$$\mu_{\Sigma x} = n\mu$$
$$\sigma_{\Sigma x} = \sqrt{n}\sigma$$

EXAMPLE 8.24

Suppose a population consists of the integers from 1 to 10. The mean of the population is $\mu = 5.5$ and the standard deviation is $\sigma = 2.87228$. If random samples of size 35 are selected from this population and the sum of each sample is calculated, the mean of the sampling distribution of the sample sum is

$$\mu_{\Sigma x} = n\mu$$
$$= (35)(5.5) = 192.5$$

and the standard deviation of the distribution of the sample sum is

$$\sigma_{\Sigma x} = \sqrt{n}\sigma$$
$$= (\sqrt{35})(2.87228) = 16.99265$$

EXAMPLE 8.25

Let's calculate the standard deviation of the distribution of sample sums described in motivator 8. The standard deviation of the integers 1 to 6 is approximately equal to 1.871.

$$\sigma_{\Sigma x} = \sqrt{n}\sigma$$
$$= (\sqrt{10})(1.871) = 5.92$$

Notice that this value is close to 5.397, the standard deviation of the 1000 sums computed by MINITAB.

MINITAB can be used to simulate the sampling distribution of the sample sum. Computer display 8.5 contains the commands and a histogram displaying the 100 sample sums based on random samples of size 35 from the population of integers from 1 to 10. The sum of each sample is stored in column C36.

Computer Display 8.5

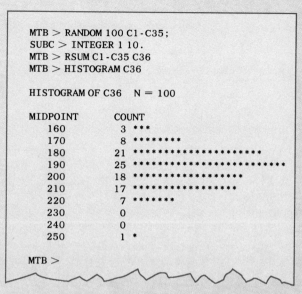

```
MTB > RANDOM 100 C1-C35;
SUBC > INTEGER 1 10.
MTB > RSUM C1-C35 C36
MTB > HISTOGRAM C36

HISTOGRAM OF C36   N = 100

MIDPOINT        COUNT
     160          3   ***
     170          8   ********
     180         21   *********************
     190         25   *************************
     200         18   ******************
     210         17   *****************
     220          7   *******
     230          0
     240          0
     250          1   *

MTB >
```

Notice that histogram for the empirical distribution suggests that the sampling distribution of the sample sum is approximately normal.

Computer display 8.6 contains the mean and standard deviation for the empirical distribution of 100 sample sums.

Computer Display 8.6

```
MTB > DESCRIBE C36

            N      MEAN    MEDIAN     TRMEAN    STDEV    SEMEAN
C36       100    192.41    191.00     192.24    15.53      1.55

           MIN       MAX        Q1        Q3
C36     158.00    248.00    181.25    204.75

MTB >
```

Notice that the mean of the empirical distribution of the sample sum is 192.41 and the mean of the theoretical distribution of the sample sum is 192.5, as stated above. And the standard deviation of the empirical distribution of the sample sum is 15.53, compared to 16.99265 for the theoretical distribution.

In chapter 6 we saw that the number of successes in a binomial experiment is a binomial random variable with $\mu = np$ and $\sigma = \sqrt{np(1 - p)}$. For a binomial experiment, if we let the random variable X_i take on the value 1 if a success is obtained on the ith trial and 0 if a failure is obtained on the ith trial, then ΣX_i represents the number of successes. By the preceding remarks, we know that the sampling distribution of ΣX_i, the number of successes, is approximately normal. This observation reinforces the result of section 7.3 that normal distributions can be used to approximate binomial distributions.

Computer display 8.7 contains a histogram of the results of a simulation for 100 replications of a binomial experiment for which $p = 0.5$ and $n = 25$.

Computer Display 8.7

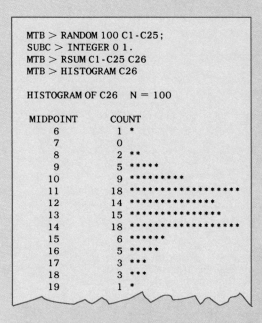

```
MTB > RANDOM 100 C1-C25;
SUBC > INTEGER 0 1.
MTB > RSUM C1-C25 C26
MTB > HISTOGRAM C26

HISTOGRAM OF C26   N = 100

MIDPOINT      COUNT
   6            1  *
   7            0
   8            2  **
   9            5  *****
  10            9  *********
  11           18  ******************
  12           14  **************
  13           15  ***************
  14           18  ******************
  15            6  ******
  16            5  *****
  17            3  ***
  18            3  ***
  19            1  *
```

Computer display 8.8 contains a description of this empirical distribution of the number of successes, which includes the mean and standard deviation. Note that the mean of the binomial distribution is $\mu = np = (25)(0.5) = 12.5$ and the standard deviation is $\sigma = \sqrt{np(1-p)} = \sqrt{(25)(0.5)(0.5)} = 2.5$.

Computer Display 8.8

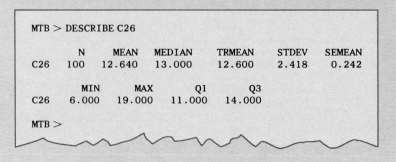

```
MTB > DESCRIBE C26

          N     MEAN    MEDIAN   TRMEAN    STDEV   SEMEAN
C26     100   12.640   13.000   12.600    2.418    0.242

          MIN      MAX       Q1       Q3
C26     6.000   19.000   11.000   14.000

MTB >
```

Applications of the Central Limit Theorem

APPLICATION 8.13   A traffic study shows that the average number of occupants in a car is 1.75 and the standard deviation is 0.65. In a sample of 50 cars, find the probability that the mean number of occupants is more than 2.

Solution: We are given that $\mu = 1.75$ and $\sigma = 0.65$. As a consequence of the central limit theorem, the sampling distribution of the mean is approximately normal. The approximation is considered good since $n = 50$.

The mean of the sampling distribution of the mean is $\mu_{\bar{x}} = \mu = 1.75$ (see figure 8.16).

FIGURE 8.16

Sampling distribution of mean for application 8.13

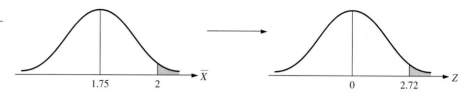

The standard error of the mean is $\sigma_{\bar{x}} = \sigma/\sqrt{n} = 0.65/\sqrt{50} = 0.092$.
The z-value for a mean of 2 is

$$z = \frac{2 - 1.75}{0.092} = 2.72$$

Thus, $P(\bar{X} > 2) = P(\bar{Z} > 2.72) = 0.5 - 0.4967 = 0.0033.$ ▪

APPLICATION 8.14

The manager of a credit bureau has calculated that the average amount of money borrowed by customers for a new car is $4685.54 and the standard deviation is $748.72. During the next month, if 40 people receive car loans, what is the probability that the average amount borrowed is between $4500 and $4800?

Solution: We are given that $\mu = \$4685.54$ and $\sigma = \$748.72$. As a consequence of the central limit theorem and the fact $n = 40 \geq 30$, a normal distribution provides a good approximation to the sampling distribution of the mean. The mean of the sampling distributon of the mean is

$$\mu_{\bar{x}} = \mu = \$4685.54$$

The standard error of the mean is

$$\sigma_{\bar{x}} = \frac{\sigma}{\sqrt{n}}$$

$$= \frac{\$748.72}{\sqrt{40}} = \$118.38$$

The z values for $4500 and $4800 (see figure 8.17) are

$$z = \frac{4500 - 4685.54}{118.38} = -1.57$$

$$z = \frac{4800 - 4685.54}{118.38} = 0.97$$

FIGURE 8.17

Sampling distribution of mean for application 8.14

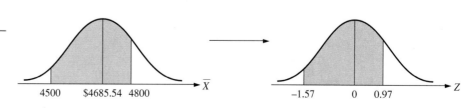

Thus, $P(4500 < \overline{X} < 4800) = P(-1.57 < Z < 0.97) = 0.4418 + 0.3340 = 0.7758$. Hence, the probability that 40 people borrow an average amount of money for a car between $4500 and $4800 is 0.7758. ■

APPLICATION 8.15

The mean breaking point for a batch of 1950 special fuses is 4.5 amperes and the standard deviation is 1.5 amperes. If 200 of these special fuses are sampled (without replacement), find the probability that the mean breaking point of the sample is less than 4.3 amperes.

Solution: We are given that $N = 1950$, $n = 200$, $\mu = 4.5$, and $\sigma = 1.5$. As a consequence of a central limit theorem, the sampling distribution of the mean is approximately normal. The mean of the sampling distribution of the mean is $\mu_{\overline{x}} = 4.5$. Since sampling is done without replacement from a finite population and $20n > N$, we should calculate $\sigma_{\overline{x}}$ using the following formula:

$$\sigma_{\overline{x}} = \frac{\sigma}{\sqrt{n}} \sqrt{\frac{N - n}{N - 1}}$$

Thus, the standard error of the mean is

$$\sigma_{\overline{x}} = \frac{\sigma}{\sqrt{n}} \sqrt{\frac{N - n}{N - 1}}$$
$$= \frac{1.5}{\sqrt{200}} \sqrt{\frac{1950 - 200}{1950 - 1}} = 0.10$$

The z value for a mean of 4.3 is

$$z = \frac{4.3 - 4.5}{0.10} = -2$$

Thus, $P(\overline{X} < 4.3) = P(\overline{Z} < -2) = 0.5 - 0.4772 = 0.0228$. Hence, the probability that the mean breaking point for a sample of 200 fuses is less than 4.3 amperes is 0.0228. ■

APPLICATION 8.16

A collection of 1000 test scores is bimodal with a mean of 76 and a standard deviation of 15. If a sample of 40 of these scores is selected without replacement, find the probability that the sample mean is larger than 79.

Solution: We are given that $N = 1000$, $n = 40$, $\mu = 76$, and $\sigma = 15$. The sampling distribution of the mean is approximately normal, as a consequence of the central limit theorem, with $\mu_{\overline{x}} = 76$. Since sampling is done without replacement from a finite population and $N > 20n$, we can caluclate $\sigma_{\overline{x}}$ by using

$$\sigma_{\overline{x}} = \frac{\sigma}{\sqrt{n}}$$

Thus, the standard error of the mean is

$$\sigma_{\overline{x}} = \frac{15}{\sqrt{40}} = 2.37$$

The z value for a mean of 79 is

$$z = \frac{79 - 76}{2.37} = 1.27$$

The probability that the sample mean is larger than 79 is thus
$$P(\overline{X} > 79) = P(Z > 1.27) = 0.5 - 0.3980 = 0.1020 \quad ▪$$

EXERCISE SET 8.4

Basic Skills

1. If ordered samples of size 30 are selected with replacement from the population with values 0, 3, 6, 9, and 12, find
   a. $\mu_{\overline{x}}$.
   b. $\sigma_{\overline{x}}$.
   c. the approximate shape of the sampling distribution of the mean.

2. If ordered samples of size 60 are selected with replacement from the population with values 2, 4, 6, 8, and 10, find
   a. $\mu_{\overline{x}}$.
   b. $\sigma_{\overline{x}}$.
   c. the approximate shape of the sampling distribution of the mean.

3. Suppose ordered samples of size 2 are selected with replacement from the population 1, 3, 5 and 7.
   a. Find μ and σ.
   b. Find $\mu_{\overline{x}}$ and $\sigma_{\overline{x}}$.
   c. Construct a frequency distribution for $\overline{x}$.
   d. Verify that the results of the central limit theorem hold.

4. Suppose ordered samples of size 2 are selected with replacement from the population 2, 4, 6 and 8.
   a. Find μ and σ.
   b. Find $\mu_{\overline{x}}$ and the standard error of the mean.
   c. Construct a frequency distribution for $\overline{x}$.
   d. Verify that the results of the central limit theorem hold.

5. Suppose ordered samples of size 3 are selected with replacement from the population 0, 3, 6, and 9. Find
   a. μ and σ.
   b. $\mu_{\overline{x}}$ and $\sigma_{\overline{x}}$.

6. Suppose ordered samples of size 3 are selected without replacement from the population 0, 9, 18, and 27. Find
   a. the mean and standard deviation of the population.
   b. the mean and standard deviation of the sampling distribution of the mean.

7. If in exercise 1, a sample is selected from the sampling distribution of the mean, what is the probability that the sample mean is
   a. less than 6.6?
   b. between 5 and 7?
   c. greater than 5.5?

8. If in exercise 2, a sample is selected from the sampling distribution of the mean, what is the probability that the sample mean is
   a. less than 5.6?
   b. between 5.5 and 6.5?
   c. greater than 6.8?

9. If the sampling distribution of the mean is normal for all sample sizes of samples n, what do you know about the population sampled? Explain.

10. Identify a practical situation where a variable of interest is probably not normal, but where the central limit theorem could be used for calculating probabilities associated with a sample mean.

11. Suppose samples of size 2 are drawn with replacement from the population having values 1, 7, and 13.
    a. Construct a frequency distribution for the nine sample sums.
    b. Verify that the mean of the sampling distribution of sample sums is given by $\mu_{\Sigma x} = n\mu$.
    c. Verify that the standard error of the sample sums is given by $\sigma_{\Sigma x} = \sqrt{n}\sigma$.

12. Suppose samples of size 2 are drawn with replacement from the population having values 3, 7, 11, and 15.
    a. Construct a frequency distribution for the 16 sample sums.
    b. Verify that the mean of the sampling distribution of sample sums is given by $\mu_{\Sigma x} = n\mu$.
    c. Verify that the standard error of the sample sums is given by $\sigma_{\Sigma x} = \sqrt{n}\sigma$.

Further Applications

13. A service station located in a large city has found that its gasoline sales average 12.4 gallons per customer and have a standard deviation of 2.9 gallons. For a random sample of 40 gasoline customers, find the probability that the average gasoline purchase
    a. is less than 13 gallons.
    b. exceeds 12 gallons.
    c. is between 11.5 gallons and 13.1 gallons.
    d. exceeds 12.6 gallons.

14. The average life of a certain brand of light bulbs is 1000 hours and the standard deviation is 50 hours. What is the

probability that the average life for a sample of 36 such light bulbs is
a. greater than 1010 hours?
b. less than 990 hours?
c. between 980 hours and 1010 hours?

15. An automatic machine fills bottles with root beer. The mean amount of fill is 16 ounces and the standard deviation is 0.5 ounce. What is the probability that a sample of 35 bottles with a mean
a. greater than 16.1 ounces?
b. between 15.9 ounces and 16.1 ounces?
c. less than 15.9 ounces?

16. The birth weights of 5000 newborn babies have a mean of 7.3 pounds and a standard deviation of 1.5 pounds. If a sample of 100 newborn babies is selected and weighed, what is the probability that the mean weight is
a. between 7 and 7.5 pounds?
b. less than 7.1 pounds?
c. greater than 7.2 pounds?

17. Five thousand students took a statistics examination. The mean score was 75 and the standard deviation was 10. If 300 random samples of size 40 are drawn from this population of test scores, find
a. $\mu_{\bar{x}}$ and $\sigma_{\bar{x}}$.
b. the approximate number of sample means that fall between 73 and 77.
c. the approximate number of sample means that fall above 72.

18. Four hundred students took a history examination. The mean score was 65 and the standard deviation was 20. If 100 random samples of size 35 are drawn from this population of test scores, find
a. the mean of the sampling distribution of the mean and the standard error of the mean.
b. the approximate number of sample means that fall between 62 and 67.
c. the approximate number of sample means that fall above 71.

Going Beyond

19. The sampling distribution of the sample variance s^2 is not normal. But if sampling is from a normal population with mean μ and variance σ^2, the sampling distribution of s^2 has a mean equal to σ^2 and a variance equal to $2\sigma^4/(n-1)$. For large values of n, the sampling distribution of s^2 is approximately normal. If a sample of size 75 is taken from a normal population with $\sigma^2 = 85$, find the probability that s^2 will be between 60 and 100.

20. The scores on the SAT mathematics examination in a recent year had the normal distribution with mean 445 and standard deviation of 95. Fifty seniors from Southern High School took the test. Illustrate the central limit theorem in this case using a computer program to simulate drawing 100 samples of size 50 and calculating the sample means. Use a histogram to summarize the results of the calculations. Does the histogram appear to be bell shaped? Is it likely that the 50 seniors would have an average greater than 472?

SECTION 8.5   *Sampling Distribution of Sample Proportions*

A **binomial** or **dichotomous population** is a population that consists of qualitative data that can be classified into one of two distinct classes, successes or failures. In figure 8.18, successes are labeled *a* and failures are labeled *b*.

FIGURE 8.18

Dichotomous or binomial population

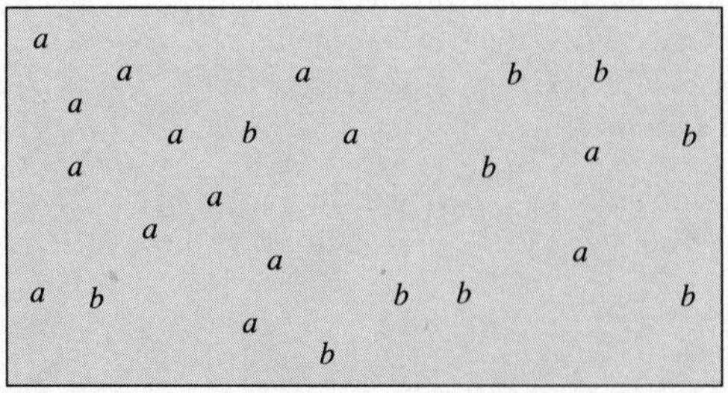

EXAMPLE 8.26

The following are examples of binomial populations:

Population	Classes
Gender classifications of 100 people	Males, females
Marital status of 40 adults	Single, married
70 whole numbers	Even numbers, odd numbers
100 tosses of a coin	Heads, tails
30 test scores	Passing scores, failing scores
Programs of study for 50 students math majors	Majors, not math majors
Classifications of 80 senior citizens	Retired, not retired
Classifications of 500 adult males	Employed, unemployed

A	A		1	1
A	A		1	1
B	B		0	0
B	B		0	0
B	B		0	0

Dichotomous population   Population of 0s and 1s

Any binomial population can be associated with a population of 0s and 1s. A population with a members labeled A and b members labeled B can be associated with a population having a 0s and b 1s or a 1s and b 0s. Let's assume that the observations in one of the two classes in a binomial population are assigned the value 1 and the observations in the other class are assigned the value 0. If X is a variable that can assume only the values 0 and 1, then ΣX is just the number of observations for which X equals 1 in the binomial population. For example, suppose we have a binomial population with 4 As and 6 Bs and each A is associated with 1 and each B is associated with 0 as shown in the diagram.

The proportion p of As in the binomial population is $p = 4/10$, and the mean of the population of 0s and 1s is

$$\mu = \frac{\Sigma x}{N}$$

$$= \frac{1 + 1 + 1 + 1 + 0 + 0 + 0 + 0 + 0 + 0}{10}$$

$$= \frac{4}{10} = \frac{2}{5}$$

Thus, if a binomial population consists of As and Bs, each A is associated with 1, and each B is associated with 0, then the proportion of As p in the population is equal to the mean of the population of 0s and 1s, μ. As a result, we have

$$\mu = p$$

The variance σ^2 of a population of 0s and 1s can be found by using the formula

Variance of a Population of 0s and 1s

$$\sigma^2 = p(1 - p) \tag{8.4}$$

where p is the proportion of 1s. For the above population of 4 As and 6 Bs, let's suppose X is a variable that assigns 1 to A and 0 to B.

EXAMPLE 8.27

The following is a frequency table for the binomial population:

Observation	x	f
B	0	6
A	1	4
		10

To find σ^2 for the binomial population of 0s and 1s, we complete the following table, sum the last column, and divide the sum by $N = 10$:

x	f	xf	$x - \mu$	$(x - \mu)^2$	$(x - \mu)^2 f$
0	6	0	-0.4	0.16	0.96
1	4	4	0.6	0.36	1.44
	10	4			2.40

The population variance is therefore equal to

$$\sigma^2 = \frac{\Sigma\,[f(x - \mu)^2]}{N}$$

$$= \frac{2.40}{10} = 0.24$$

This same result can be obtained by using formula (8.4):

$$\sigma^2 = p(1 - p)$$
$$= (0.6)(0.4) = 0.24$$

Estimating Population Proportions

EXAMPLE 8.28

Frequently we will have occasion to estimate or determine an unknown proportion for a binomial population. For example, we may want to estimate the proportion of college students who live off campus, or we may be interested in the percentage of women who favor abortion. To estimate the population proportion, we will use a random sample from the binomial population and use the sample proportion $\hat{p} = x/n$ to estimate p, where x is the number of successes (the observations of interest) and n is the size of the sample. For example, suppose a sample from a binomial population consists of 2 As and 3 Bs, as shown in figure 8.19. For the sample of As and Bs, the proportion of As is $\hat{p} = 2/5$ and the mean of the sample of 0s and 1s is $\bar{x} = 2/5$. Thus, we have $\hat{p} = \bar{x}$.

FIGURE 8.19

Binomial population of 2 As and 3 Bs

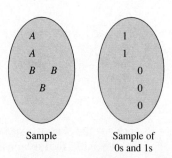

Sample   Sample of
                  0s and 1s

Sampling Distribution
of Sample Proportions

If random samples of size *n* are drawn from a large binomial population and the **sample proportion** $\hat{p}$ is computed for each sample, the collection of sample proportions is called the **sampling distribution of sample proportions.**

EXAMPLE 8.29

Suppose we have a population of five transistors, consisting of two defective ones (*D*) and three good ones (*G*). If we associate a 1 with a defective transistor, then we have the following population of 0s and 1s:

Transistor	Transistor Quality	Population Value
1	*D*	1
2	*G*	0
3	*G*	0
4	*D*	1
5	*G*	0

If unordered samples of size 3 are drawn without replacement from this population, there are ten possible samples. They are listed in table 8.7 along with the proportion of defectives $\hat{p}$ in each sample.

TABLE 8.7

Samples of Size 3 from
Binomial Population

Sample Number	Sample of Transistors	Proportion of Defectives	Sample Number	Sample of Transistors	Proportion of Defectives
1	{1, 2, 3}	1/3	6	{3, 4, 5}	1/3
2	{1, 2, 4}	2/3	7	{1, 4, 5}	2/3
3	{1, 2, 5}	1/3	8	{2, 4, 5}	1/3
4	{2, 3, 4}	1/3	9	{1, 3, 5}	1/3
5	{2, 3, 5}	0	10	{1, 3, 4}	2/3

The mean for the population of ten sample proportions is

$$\mu_{\hat{p}} = \frac{(6)(1/3) + (1)(0) + (3)(2/3)}{10} = \frac{2}{5}$$

Note that the proportion of defectives in the population is also 2/5. A frequency graph for the sampling distribution of $\hat{p}$ is as follows:

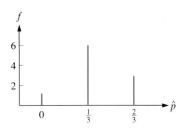

Since samples from a binomial population can be identified with samples of 0s and 1s and the proportion of successes in a sample from the binomial population is equal to the mean of the corresponding sample of 0s and 1s, the sampling distribution of the mean can assist us in answering the following three questions concerning the sampling distribution of sample proportions:

1. What distribution does the sampling distribution of $\hat{p}$ have?
2. What is the mean of the sampling distribution of $\hat{p}$?
3. What is the standard deviation of the sampling distribution of $\hat{p}$?

Recall the following facts concerning the sampling distribution of the mean:

1. The sampling distribution of the mean is approximately normal, the approximation is considered good for $n \geq 30$.
2. $\mu_{\bar{x}} = \mu$.
3. $\sigma_{\bar{x}} = \sigma/\sqrt{n}$.

As a consequence of these facts, we have the following properties concerning the sampling distribution of the sample proportion:

Properties of the Sampling Distribution of Sample Proportions

1. The sampling distribution of sample proportions is approximately normal, the approximation is considered good for $n \geq 30$.
2. $\mu_{\hat{p}} = \mu_{\bar{x}} = \mu = p$, where μ is the mean of the population of 0s and 1s and p is the proportion of 1s in the population.
3. $\sigma_{\hat{p}} = \sigma_{\bar{x}} = \sigma/\sqrt{n}$, where σ is the standard deviation of the population of 0s and 1s.

Because the standard deviation for a population of 0s and 1s is given by $\sigma = \sqrt{p(1 - p)}$, the standard error of $\hat{p}$ (the standard deviation of the distribution of sample proportions) is given by

$$\sigma_{\hat{p}} = \frac{\sigma}{\sqrt{n}}$$

$$= \frac{\sqrt{p(1 - p)}}{\sqrt{n}}$$

$$= \frac{\sqrt{p(1 - p)}}{n}$$

Hence, the standard error of the sample proportion is given by

Standard Error of the Sample Proportion

$$\sigma_{\hat{p}} = \sqrt{\frac{p(1 - p)}{n}} \tag{8.5}$$

If sampling is done from a finite binomial population with $20n > N$, then the standard error of $\hat{p}$ is given by

Standard Error of the Sample Proportion for Small Samples

$$\sigma_{\hat{p}} = \sqrt{\frac{p(1 - p)}{n}} \sqrt{\frac{N - n}{N - 1}} \tag{8.6}$$

APPLICATION 8.17

For the illustration in example 8.29 of selecting samples of size 3 from a population of five transistors, two of which are defective, find the standard deviation of the sampling distribution of $\hat{p}$

    a. without using formula (8.6).

    b. using formula (8.6).

Solution:

    a. We shall use the formula for calculating the standard deviation of a population. Recall that $\mu_{\hat{p}} = 2/5$. By the definition of population standard deviation, we have

$$\sigma_{\hat{p}} = \sqrt{\frac{\Sigma\,[(\hat{p} - \mu_{\hat{p}})^2 f]}{N}}$$

$$= \sqrt{\frac{(1/3 - 2/5)^2(6) + (0 - 2/5)^2(1) + (2/3 - 2/5)^2(3)}{10}} = 0.2$$

    b. By using formula (8.6):

$$\sigma_{\hat{p}} = \sqrt{\frac{p(1 - p)}{n}}\sqrt{\frac{N - n}{N - 1}}$$

$$= \sqrt{\frac{(2/5)(3/5)}{3}}\sqrt{\frac{5 - 3}{5 - 1}}$$

$$= \sqrt{0.08}\,\sqrt{0.5} = \sqrt{0.04} = 0.2 \quad ▪$$

    In summary, a sample proportion can be thought of as a special case of a sample mean for a binomial population consisting of 0s and 1s, where a success is identified with 1. As such, by using the population of 0s and 1s as the sampled population and as a consequence of the central limit theorem for the sampling distribution of the mean, we have the following properties for the sampling distribution of the sample proportion $\hat{p}$:

Properties of the Sampling Distribution of $\hat{p}$

1. The sampling distribution of $\hat{p}$ is approximately normal.
2. The mean of the sampling distribution of $\hat{p}$ is equal to p, the population proportion.
3. The standard deviation of $\hat{p}$ (the standard error of $\hat{p}$) is equal to $\sqrt{p(1 - p)/n}$, where n is the size of the sample and p is the proportion of successes in the population.

In addition, if $n \geq 30$, the above normal approximation is considered good.

    As a result of these properties for the sampling distribution of $\hat{p}$, it follows that the z score formula for $\hat{p}$ is given by

z score for $\hat{p}$

$$z = \frac{\hat{p} - p}{\sqrt{p(1 - p)/n}} \tag{8.7}$$

and the sampling distribution of z is approximately the standard normal.

A computer simulation using MINITAB can demonstrate that the sampling distribution of the sample proportion is approximately normal. One hundred random samples of size 35 are selected from a population having an equal number of 0s and 1s. The mean (proportion of 1s) in each sample is computed. The results are contained in computer display 8.9.

Computer Display 8.9

```
MTB > RANDOM 100 C1-C35;
SUBC > INTEGER 0 1.
MTB > RMEAN C1-C35 C36
MTB > NAME C36 'MEANS'
MTB > PRINT C36

MEANS
  0.428571   0.428571   0.485714   0.542857   0.400000   0.514286   0.485714
  0.514286   0.485714   0.628571   0.485714   0.428571   0.571429   0.457143
  0.485714   0.571429   0.514286   0.542857   0.542857   0.628571   0.485714
  0.457143   0.428571   0.485714   0.485714   0.371429   0.428571   0.342857
  0.514286   0.514286   0.542857   0.600000   0.628571   0.342857   0.428571
  0.400000   0.457143   0.600000   0.400000   0.600000   0.342857   0.571429
  0.457143   0.485714   0.657143   0.400000   0.457143   0.457143   0.514286
  0.571429   0.600000   0.485714   0.600000   0.457143   0.571429   0.485714
  0.542857   0.457143   0.457143   0.400000   0.657143   0.571429   0.514286
  0.600000   0.600000   0.485714   0.514286   0.457143   0.428571   0.428571
  0.400000   0.485714   0.314286   0.457143   0.457143   0.371429   0.685714
  0.485714   0.457143   0.514286   0.485714   0.571429   0.542857   0.571429
  0.600000   0.628571   0.571429   0.571429   0.600000   0.485714   0.428571
  0.571429   0.542857   0.514286   0.342857   0.428571   0.542857   0.485714
  0.514286   0.514286
```

The HISTOGRAM command was used to list the 100 proportions into the frequency distribution and histogram in computer display 8.10. Note that the output of this empirical distribution resembles the shape of a normal distribution.

Computer Display 8.10

```
MTB > HISTOGRAM C36

HISTOGRAM OF MEANS    N = 100

MIDPOINT      COUNT
  0.30          1  *
  0.35          6  ******
  0.40          6  ******
  0.45         23  ***********************
  0.50         29  *****************************
  0.55         19  *******************
  0.60          9  *********
  0.65          6  ******
  0.70          1  *
```

The mean of the sample proportions, the median, and other measures, such as the quartiles, are printed out by using the DESCRIBE command. The results are contained in computer display 8.11.

Computer Display 8.11

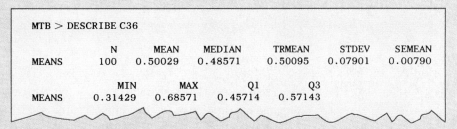

```
MTB > DESCRIBE C36

            N       MEAN     MEDIAN     TRMEAN     STDEV     SEMEAN
MEANS      100    0.50029    0.48571    0.50095   0.07901   0.00790

           MIN       MAX        Q1         Q3
MEANS    0.31429   0.68571   0.45714    0.57143
```

The proportion of 1s in a binomial population consisting of an equal number of 0s and 1s is 0.5. The result compares favorably with the mean of 0.50029 provided by MINITAB. The standard deviation of the number of 1s in a binomial population consisting of an equal number of 0s and 1s is:

$$\sigma = \sqrt{p(1 - p)}$$
$$= \sqrt{(0.5)(0.5)} = 0.5$$

By using formula (8.5) we determine the standard deviation of the population of sample proportions to be

$$\sigma_{\hat{p}} = \sqrt{\frac{p(1 - p)}{n}}$$

$$= \sqrt{\frac{0.25}{35}} \approx 0.0845$$

This result also compares favorably with the result of 0.07901 provided by MINITAB. The means and standard deviations for the two distributions differ because we have used only 100 of all the possible samples. Since 100 is a large number, we would expect the means and standard deviations to be close but not exactly equal.

APPLICATION 8.18   It has been determined that 60% of the students at a large university smoke cigarettes. If a random sample of 800 students polled indicate that 440 smoke cigarettes, locate the sample proportion $\hat{p}$ within the sampling distribution of $\hat{p}$ by finding what percentage of the sample proportions fall below it.

Solution: Since $n = 800$, the sampling distribution of $\hat{p}$ is approximately normal. The mean of the sampling distribution of $\hat{p}$ is

$$\mu_{\hat{p}} = p = 0.60$$

The standard error of $\hat{p}$ is

$$\sigma_{\hat{p}} = \sqrt{\frac{p(1 - p)}{n}}$$

$$= \sqrt{\frac{(0.6)(0.4)}{800}} = 0.0173$$

The value of the sample proportion $\hat{p}$ is

$$\hat{p} = \frac{x}{n}$$

$$= \frac{440}{800} = 0.55$$

Therefore, the z value for $\hat{p}$ is

$$z = \frac{\hat{p} - p}{\sqrt{p(1 - p)/n}}$$

$$= \frac{0.55 - 0.60}{0.0173} = -2.89$$

Thus, $P(\hat{p} < 0.55) = P(Z < -2.89) = 0.5 - 0.4981 = 0.0019 = 0.19\%$ of the sampling distribution of $\hat{p}$ falls to the left of $\hat{p} = 0.55$ (see figure 8.20).

FIGURE 8.20

Sampling distribution of $\hat{p}$ for application 8.18

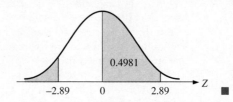

APPLICATION 8.19

A medication for an upset stomach carries a warning that some users may have an adverse reaction to it. Further, it is thought that about 3% of the users have such a reaction. If a random sample of $n = 150$ people with upset stomachs use the medication, find the probability that $\hat{p}$, the proportion of the users who actually have an adverse reaction, will be greater than 5%.

Solution: To determine the probability that more than 5% of the users will have an adverse reaction to the medication, we find the area to the right of $\hat{p} = 0.05$ under the following normal curve above the horizontal axis.

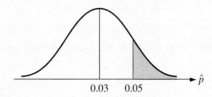

The z score for $\hat{p} = 0.05$ is

$$z = \frac{\hat{p} - p}{\sqrt{p(1 - p)/n}}$$

$$= \frac{0.05 - 0.03}{\sqrt{(0.03)(0.97)/150}}$$

$$= 1.44$$

By examining the z table (see the front endpaper), we determine the area under the standard normal curve between $z = 0$ and $z = 1.44$ and above the horizontal axis to be

$$P(0 < Z < 1.44) = 0.4251$$

Thus, the probability that more than 5% of the users will have an adverse reaction is

$$P(\hat{p} > 0.05) = P(Z > 1.44)$$
$$= 0.5 - 0.4251 = 0.0749. \quad \blacksquare$$

Binomial Probability Distributions

A binomial population is closely associated with a binomial probability distribution. A binomial population is a collection of successes and failures, whereas a binomial probability distribution contains the probabilities or proportions of all possible numbers of successes in a binomial experiment. As a consequence of this relationship, probability statements involving the sample proportion $\hat{p}$ can be evaluated by using the normal approximation to the binomial, provided that $np \geq 5$ and $n(1 - p) \geq 5$. In chapter 6 we worked with counts of successes, instead of proportions, when dealing with binomial experiments. Any count can be converted to a proportion by dividing the count by the number of trials n; that is, $x/n = \hat{p}$.

EXAMPLE 8.30

For application 8.19, let's calculate the probability that the proportion of users who actually have an adverse reaction will be greater than 5% by using the normal approximation to the binomial. Recall that $n = 150$ and $p = 0.03$. Since $\hat{p} = x/n$ and x represents a count and must be a whole number, we have

$$P(\hat{p} > 0.05) = P\left(\frac{X}{150} > 0.05\right)$$
$$= P(X > 7.5)$$
$$= P(X \geq 8)$$

The binomial distribution of the number of users who have an adverse reaction has a mean of

$$\mu = np$$
$$= (150)(0.03) = 4.5$$

and a standard deviation of

$$\sigma = \sqrt{np(1 - p)}$$
$$= \sqrt{(4.5)(0.97)} = 2.089$$

To approximate the area of all the bars labeled $x = 8$ or higher in a probability graph for the binomial distribution, we use the normal distribution with a mean of 4.5 and a standard deviation of 2.089. The z score for $x = 7.5$ is

$$z = \frac{7.5 - 4.5}{2.089} = 1.44$$

Hence, the probability that $\hat{p}$ is greater than 0.05 is

$$P(\hat{p} > 0.05) = P(X \geq 8)$$
$$\approx P(Z > 1.44)$$
$$= 0.5 - 0.4251 = 0.0749$$

which agrees with the result obtained in application 8.19. Note that $np \geq 5$ and $n(1 - p) \geq 5$.

EXERCISE SET 8.5

Basic Skills

1. Five percent of all diodes manufactured by a certain firm are defective. Suppose a random sample of 50 diodes is selected from 1500 newly manufactured diodes and $\hat{p}$ represents the percentage of defective diodes.

a. Describe the sampling distribution of $\hat{p}$ and find $\mu_{\hat{p}}$ and $\sigma_{\hat{p}}$.
b. Find $P(\hat{p} < 0.08)$.
c. Find $P(0.01 < \hat{p} < 0.10)$.
d. Find $P(\hat{p} > 0.04)$.

2. A particular county in Pennsylvania has a 12% unemployment rate. A survey of 500 individuals was made.
   a. If $\hat{p}$ represents the percentage unemployed in the sample, describe the sampling distribution of $\hat{p}$ and find $\mu_{\hat{p}}$ and $\sigma_{\hat{p}}$.
   b. Find $P(\hat{p} > 0.11)$.
   c. Find $P(0.11 < \hat{p} < 0.13)$.
   d. If $\hat{p} = 0.13$, locate $\hat{p}$ within the sampling distribution of $\hat{p}$ by finding what percentage of sample proportions will fall below 0.13.

Further Applications

3. If a drug is 80% effective in treating a certain disease and a random sample of 500 patients is treated with the drug, find the following probabilities if $\hat{p}$ represents the percentage of effective treatments:
   a. $P(\hat{p} > 0.81)$
   b. $P(0.70 < \hat{p} < 0.81)$
   c. $P(\hat{p} < 0.84)$

4. An executive officer for a large brokerage firm surveyed 120 of their clients and learned that 72 of them were extremely satisfied with the firm's service.
   a. Find $\hat{p}$, the proportion of customers in the sample who were extremely satisfied with the firm's service.
   b. Estimate the standard error of the sample proportion of customers who are extremely satisfied by using $\hat{p}$ as an estimate for p, the true proportion of customers who are extremely satisfied.

5. Refer to exercise 4.
   a. Find the percentage of sample proportions that fall within two standard deviations of the population proportion p.
   b. Find the percentage of sample proportions that fall beyond 1.5 standard deviations of the population proportion p.

6. Suppose that it is known that the true proportion of all manufactured components made by a firm that are defective is 4%. Find the probability that a random sample of size 60 will have
   a. fewer than 3% defective components.
   b. more than 1% but less than 5% defective components.

7. Refer to exercise 6. Suppose $\hat{p}$ represents the proportion of defective components in a sample drawn from a population of size 1000. Find
   a. $\sigma_{\hat{p}}$.
   b. $P(\hat{p} < 0.03)$.
   c. $P(0.02 < \hat{p} < 0.07)$.

8. Refer to exercise 3. Suppose the drug was 80% effective for a population of size 50,000. Find
   a. $\sigma_{\hat{p}}$.
   b. $P(\hat{p} > 0.80)$.
   c. $P(0.75 < \hat{p} < 0.80)$.
   d. $P(\hat{p} < 0.81)$.

9. According to a recent survey of 500 doctors in the United States, 75% approve of requiring second opinions before nonemergency surgery. Use this value as the true proportion of U.S. doctors who approve of requiring second opinions before nonemergency surgery. Suppose $\hat{p}$ represents the proportion of doctors in a sample of size 500 that approve of requiring second opinions. Find
   a. the standard error of the proportion.
   b. an interval centered at 0.75 that contains 80% of the sample proportions.

10. Refer to exercise 9. Suppose random samples of size 500 are drawn from a population of $N = 8000$ doctors. Find
    a. the standard error of the proportion.
    b. an interval centered at 0.75 that contains 90% of the sample proportions.

Going Beyond

11. Suppose an experiment consists of tossing a fair coin 50 times. If $\hat{p}$ represents the proportion of heads obtained, find
    a. $\mu_{\hat{p}}$ and $\sigma_{\hat{p}}$.
    b. $P(\hat{p} > 0.6)$.
    c. $P(\hat{p} < 0.04)$.
    d. $P(0.4 < \hat{p} < 0.6)$.

12. Show that formula (8.7) can be written as $z = (x - np)/\sqrt{np(1 - p)}$. What type of discrete random variable is x? Explain.

13. Suppose a large binomial population consists of 0s and 1s, where the proportion of 1s is p and the proportion of 0s is $1 - p$.
    a. Prove that the mean of the population is p.
    b. Prove that the variance of the population is $p(1 - p)$.

14. Refer to exercise 13 and recall that $\hat{p} = x/n$ can be regarded as the mean of a sample of size n drawn from a population of 0s and 1s.
    a. Prove that the mean of the sampling distribution of $\hat{p}$ is p.
    b. Prove that the standard error of $\hat{p}$ is $\sqrt{p(1 - p)/n}$.

CHAPTER SUMMARY

In this chapter we learned about sampling distributions of sample statistics, a concept that is basic to understanding inferential statistics. Random samples are used in order to ensure a representative subcollection from the whole population. We learned several methods for obtaining a random sample from a population of interest. We learned that the standard error of the mean depends on the size of the sample relative to the population and decreases as the sample size increases. The central limit theorem states that the sampling distribution of the mean is approximately normal when sampling is from non-normal populations. If the population is normal, then the sampling distribution of the mean is normal. Finally, we studied the sampling distribution of the sample proportion, where sampling is done from a binomial population. We learned that the distribution of a sample statistic has a certain shape, mean, variance, and standard deviation. When such information is known, as in the case for the sampling distributions of the mean and sample proportion, probability statements can be made about the sampled population and the sampling distribution.

CHAPTER REVIEW

▪ IMPORTANT TERMS ▪

The following chapter terms have been mixed in ordering to provide you better review practice. For each of the following terms, provide a definition in your own words. Then check your responses against the definitions given in the chapter.

central limit theorem
sampling distributions
binomial population
cluster sampling
degrees of freedom
random number table
randomization
sampling distribution of the sample
    standard deviation
sampling error
correction factor

simple random sample
standard error of a statistic
standard error of the mean
unbiased
stratified sampling
sampling bias
census
sample proportion
dichotomous population
sampling distribution of sample
    proportions

random sample
sampling distribution of the sample
    mean
systematic sample
nonsampling errors
t distribution
estimate
t statistic
zero bias
sampling distribution of a statistic

▪ IMPORTANT SYMBOLS ▪

e, sampling error

$\mu_{\bar{x}}$, mean of the sampling distribution of the mean

$\sigma_{\bar{x}}$, standard deviation of the sampling distribution of the mean, also the standard error of the mean

t, t statistic

df, degrees of freedom

p, population proportion

$\hat{p}$, sample proportion

$\mu_{\hat{p}}$, the mean of the sampling distribution of the sample proportion

$\sigma_{\hat{p}}$, the standard deviation of the sampling distribution of the sample proportion, also the standard error of the sample proportion.

■ **IMPORTANT FACTS AND FORMULAS** ■

Mean of sampling distribution of $\bar{x}$: $\mu_{\bar{x}} = \mu$. (8.1)

Standard error of the mean for large samples: $\sigma_{\bar{x}} = \dfrac{\sigma}{\sqrt{n}}$. (8.2)

If sampling is without replacement, then

a. $\sigma_{\bar{x}} = \dfrac{\sigma}{\sqrt{n}} \sqrt{\dfrac{N - n}{N - 1}}$. (8.3)

b. if $20n \leq N$ or sampling is from an infinite population, then $\sigma_{\bar{x}} \approx \sigma/\sqrt{n}$.

z value for $\bar{x}$: $z = (\bar{x} - \mu)/(\sigma/\sqrt{n})$.

t value for $\bar{x}$: $t = (\bar{x} - \mu)/(s/\sqrt{n})$.

Degrees of freedom for t: df $= n - 1$.

Central limit theorem: The sampling distribution of the mean is approximately normal. The approximation is considered good for $n \geq 30$.

Variance of a population of 0s and 1s: $\sigma^2 = p(1 - p)$. (8.4)

Sample proportion: $\hat{p} = x/n$.

Mean of the sampling distribution of $\hat{p}$: $\mu_{\hat{p}} = p$.

Standard error of $\hat{p}$: $\sigma_{\hat{p}} = \sqrt{p(1 - p)/n}$.

Standard error of $\hat{p}$ for small samples:

$\sigma_{\hat{p}} = \sqrt{\dfrac{p(p - p)}{n}} \sqrt{\dfrac{N - n}{N - 1}}$. (8.6)

z score for $\hat{p}$: $z = (\hat{p} - p)/\sqrt{p(1 - p)/n}$. (8.7)

REVIEW EXERCISES

1. Suppose we want to obtain a random sample of size 5 from the numbers 1, 2, 3, 4, 5, 6, 7, 8, 9, and 10. To select the sample a coin is tossed. If it lands heads up, then the numbers 1, 3, 5, 7 and 9 are chosen. If the coin lands tails up then the numbers 2, 4, 6, 8, and 10 are chosen.
   a. Does every number between 1 and 10 have the same chance of being chosen?
   b. Will the resulting sample be a simple random sample? Explain.

2. Use the random number table (table 8.1) to select a sample of size 5 from the 26 letters of the alphabet and explain your procedure.

3. Is the sample $\{1, 2\}$ a random sample from the population of values $\{1, 2, 3, 4, 5, 6, 7, 8, 9, 10\}$? Explain.

4. Suppose ordered samples of size 2 are drawn with replacement from the populaton of values 1, 3, 5, and 7. For the sampling distribution of the mean,
   a. find $\mu_{\bar{x}}$.
   b. find $\sigma_{\bar{x}}$.
   c. construct a graph for the frequency distribution.

5. Suppose ordered samples of size 2 are drawn without replacement from the population of values 1, 3, 5, and 7. For the sampling distribution of the mean,
   a. find $\mu_{\bar{x}}$.
   b. find $\sigma_{\bar{x}}$.
   c. construct a graph for the frequency distribution.

6. Suppose ordered samples of size 2 are drawn with replacement from the population of values 0, 3, 5, and 7. If the sample mean is computed for each sample, find
   a. the sampling error for each sample mean.
   b. the mean of the sampling errors.
   c. the standard deviation of the sampling errors.

7. Ordered samples of size 3 are selected with replacement from the population of values 1, 2, 3, and 6. The following is a frequency table for the sample medians:

Median	f
1	10
2	22
3	22
6	10

Suppose the median of a sample is used to measure or estimate the population mean.
   a. Find the sampling error for each median.

b. Construct a graph for the sampling distribution of the median.

c. Find the mean of the sampling errors.

d. Find the mean of the sampling distribution of the median.

e. Find the standard deviation of the sampling errors.

f. Find the standard error of the median.

g. Compare your results for parts e and f.

8. Consider the following population of six voters, where yes means that the voter will vote for candidate A and no indicates that the voter will not vote for candidate A.

Voter	Response
1	yes
2	no
3	yes
4	yes
5	yes
6	no

a. Selecting unordered samples of size 4 without replacement provides a total of 15 possible samples. List the 15 samples.

b. Compute the proportion of yes votes for each sample.

c. Find the mean of the sampling distribution of the proportion of yes votes.

d. Find the standard deviation of the sampling distribution of the proportion of yes votes.

e. Construct a graph for the sampling distribution of the proporton of yes votes.

9. Suppose an automatic machine used to fill cans of soup has $\mu = 16$ ounces and $\sigma = 0.5$ ounce. If a sample of 50 cans is obtained, find the probability that the sample mean fill $\bar{x}$ is

a. greater than 15.88 ounces.

b. greater than 15.9 ounces and less than 16.09 ounces.

c. less than 16.2 ounces.

10. Suppose a sample of size 50 is selected from a populaton of 5000 employees in order to estimate the average age for the population. If the standard deviation of the population is 7.5 years, find $\sigma_{\bar{x}}$, the standard error of the mean.

11. What is the probability that the sample mean age of the employees in exercise 10 will be within two years of the population mean age?

12. It is known that 10% of all items produced on an assembly line are defective. If a sample of 50 parts are tested and $\hat{p}$ denotes the percentage of tested items that are defective, find

a. $P(\hat{p} > 10.8\%)$.

b. $P(9.5\% < \hat{p} < 10.5\%)$.

c. $P(\hat{p} < 9.6\%)$.

13. It is claimed that a new drug is 80% effective in treating patients with a particular disease. If the drug is used on a sample of 64 patients and $\hat{p}$ denotes the sample proportion of patients effectively treated, find

a. $P(0.78 < \hat{p} < 0.82)$.

b. $P(0.82 < \hat{p} < 0.83)$.

c. $P(\hat{p} < 0.81)$.

14. Suppose the daily water usage by customers is normally distributed with a mean of 250 gallons. On a particular day, a sample of 20 meter readings showed a standard deviation of 45 gallons. If $P(\bar{X} > c) = 0.95$, find the value of c.

15. The number of passengers carried each day by a certain bus company is normally distributed with a mean of 220 passengers and a standard deviation of 50 passengers. If a sample of 12 days is used to estimate the average number of passengers using a company bus each day, find the probability that a 12-day sample will produce an average of less than 300 customers.

Computer Applications

1. Use a computer program to simulate an empirical t distribution with df $= 15$ having 500 values. Plot these values and find the mean, median, and standard deviation of the empirical distribution and compare these values with the theoretical values. (*Hint:* use commands similar to

```
RANDOM 500 C1;
T 15.
DESCRIBE C1
DOTPLOT C1.)
```

2. Refer to exercise 1. Find the 99th percentile of the distribution and compare this to the theoretical value obtained from the t table. (*Hint:* use commands similar to

```
SORT C1 C2
LET K1 = C2(495)
PRINT K1.)
```

3. Use a computer program to simulate a binomial population with $p = 0.3$ having 500 values. Plot these values and find the mean, median, and standard deviation of the empirical distribution and compare these values with the theoretical values. (*Hint:* In MINITAB the simulation can be accomplished with the RANDOM command and its DISCRETE subcommand:

```
RANDOM 500 C3;
DISCRETE C1 C2.
DESCRIBE C3
DOTPLOT C3.)
```

4. Use a computer program to simulate tossing a six-sided die 500 times. Then find the mean and standard deviation of the numbers showing on the 500 tosses. (*Hint:* use commands similar to

```
RANDOM 500 C1;
INTEGER 1 6.
DESCRIBE C1.)
```

5. Use a computer program to do exercise 19 of exercise set 8.2.

6. Use commands similar to the commands in computer display 8.2 to illustrate the central limit theorem.

7. **a.** Simulate drawing 100 samples of size 25 from the normal population having $\mu = 200$ and $\sigma = 10$. For each sample, compute the mean. Use these 100 sample means to form an empirical distribution of sample means.
   b. Determine the mean and standard deviation of the empirical distribution of sample means and construct a histogram for the 100 means.
   c. Compare your results to the theoretical results.

8. Simulate an empirical sampling distribution of sample proportions by drawing 100 samples of size 10 from a population of 0s and 1s and computing the proportion of 1s in each sample. Use these 100 sample proportions to form an empirical sampling distribution.
   a. Determine the mean and standard deviation of the empirical distribution and construct a histogram for the distribution.
   b. Compare the results with the theoretical results.

9. Determine c so that $P(t > c) = 0.20$ where df $= 15$.

10. Determine $P(-1.5 < t < 2.3)$ if df $= 10$.

EXPERIMENTS WITH REAL DATA

Treat the 720 subjects listed in the database in appendix C as if they constituted the population of all U.S. subjects.

1. Select a random sample of 40 subjects and calculate the values of the following statistics:
   a. mean age for males
   b. proportion of subjects that smoke
   c. mean diastolic blood pressure
   d. correlation between diastolic and systolic blood pressures
   e. correlation between age and diastolic blood pressure

2. Select a random sample of 30 males in the 20–29 year age group and calculate the values of the following statistics:
   a. proportion that smoke
   b. mean weight
   c. mean diastolic blood pressure
   d. standard deviation of diastolic blood pressures

3. Select a random sample of 30 females in the 20–29 year age group and calculate the values of the following statistics:
   a. proportion that smoke
   b. mean weight
   c. mean diastolic blood pressure
   d. standard deviation of diastolic blood pressures

4. Compare your results in experiments 2 and 3. What do they suggest, if anything, for the population as a whole?

▪ CHAPTER ACHIEVEMENT TEST ▪

1. If a random sample has a mean $\overline{x} = 20$ and the sampled population has a mean $\mu = 25$, determine the sampling error if $\overline{x}$ is used to estimate μ.

2. Describe how to obtain a random sample of three letters from the alphabet using a random number table.

3. If random samples of size 36 are obtained from a population with mean $\mu = 10$ and standard deviation $\sigma = 9$, determine $\sigma_{\overline{x}}$.

4. If a random sample with $\overline{x} = 7$, $s = 2.5$, and $n = 10$ was obtained from a normal population with $\mu = 5$ and $\sigma = 3$, find
   a. the value of the t statistic associated with $\overline{x} = 7$.
   b. the degrees of freedom associated with t.
   c. the value of the z statistic associated with $\overline{x} = 7$.

5. If a random sample of size 100 is obtained from a population with $\mu = 40$ and $\sigma = 12$, find
   a. $P(38 < \overline{X} < 41)$.
   b. $P(40 < \overline{X} < 55)$.
   c. the value x_0 so that $P(\overline{X} > x_0) = 0.40$.

6. Eighty percent of the students at a large university favor pass-fail grades for elective courses. If a random sample of 100 students is selected to determine the proportion who favor pass-fail grades and the sample proportion $\hat{p}$ is calculated, find
   a. $P(\hat{p} > 0.85)$.
   b. $P(0.78 < \hat{p} < 0.83)$.

7. If a random sample with $\overline{x} = 5$, $s = 2$, and $n = 10$ is selected from a normal population with $\mu = 8$ and $\sigma = 3$, find the value of the z statistic associated with $\overline{x} = 5$.

8. Sixty percent of a large number of people polled indicated that they prefer brand X toothpaste to other brands. If in a sample of 40 people, 65% indicated they prefer brand X toothpaste to other brands, find the value of the z statistic associated with $\hat{p} = 65$.

9. If ordered samples of size 4 are obtained with replacement from the population of values 2, 3, 4, 5, and 6, find
   a. $\mu_{\overline{x}}$.
   b. $\sigma_{\overline{x}}$.

9

Estimation

CHAPTER OBJECTIVES

In this chapter we will investigate

▷ *The difference between an estimate and an estimator.*

▷ *Two kinds of estimates for an unknown parameter.*

▷ *Biased estimators.*

▷ *Critical values.*

▷ *How to find critical values.*

▷ *Point estimates of μ, σ^2, σ, and p.*

▷ *What the maximum error of estimate is for a point estimate.*

▷ *How to make probability statements concerning the maximum error of estimate for a point estimate.*

▷ *How to construct confidence intervals for the mean of a normal population using large samples.*

▷ *How to construct confidence intervals for the mean of a normal population using small samples.*

▷ *How to determine sample sizes in order to be confident that the error of estimate is at most a specified value.*

▷ *The chi-square distributions.*

▷ *How to construct confidence intervals for the variance and standard deviation of a normal population.*

|||| MOTIVATOR 9 ⟩

*S*uppose you are in charge of training for a large corporation and have a target population of 2500 that you want to sample for a training needs assessment study. The target population contains the following makeup: 5% supervisors and managers, 10% administrative and clerical staff, 25% professional staff, 20% technical personnel, and 40% manufacturing and production workers. How many questionnaires will you need to send out in order to meet the minimum sample size required to make reliable inferences about the questionnaire data, assuming a 50% return rate? The minimum number of questionnaires to send out depends on the following three factors:

■ The expected response rate of the questionnaire (a 50% return rate is considered good).

■ The precision of the population estimate (for example, within plus or minus 5%).

▪ The confidence level (for example, a 95% confidence level means that 95 out of a 100 times a sample will provide the desired precision level).

According to Kenneth M. Nowack,[36] for a precision of 0.05 and a 95% confidence level, the minimum number of questionnaires to have returned is determined by the following formula:

$$\text{Minimum number returned} = \frac{(0.96)(\text{population size})}{(0.0025)(\text{population size}) + 0.96}$$

Assuming a 50% rate of return for the questionnaire, you would have to send out twice the number determined by Nowack's formula. By using Nowack's formula for a population size of 2500, you determine that a minimum of 333 questionnaires need to be returned. With a return rate of 50%, this means that 666 questionnaires need to be sent out.

In order to make sure that the sample is representative of the target audience, about $(0.05)(666) = 33$ supervisors and managers would need to be selected at random to receive the questionnaire, about $(0.10)(666) = 67$ of the administrative and clerical staff, and so on.

In section 9.4 we will examine formulas for selecting sample sizes in order to make valid inferences about certain population parameters.

Chapter Overview

Estimation is a major objective of inferential statistics. By studying one sample from a population, we want to generalize our findings to the entire population. As we have seen in chapter 8, statistics vary greatly within their sampling distributions, and the smaller the standard error of the statistic, the closer the values of the statistic are to each other. While studying **statistical inference** (generalizing from a sample to a population), we will concern ourselves with two elements, the inference and how good it is. For example, if we estimate the annual average income of all families living within 1 mile of a new shopping center to be $28,200, we would like to know how good this estimate is. Since the population mean is unknown, the sampling error $\bar{x} - \mu = 28,200 - \mu$ is unknown. In this chapter we will learn how to make probability statements concerning the size of the error.

There are two general processes for making inferences about unknown population parameters: *estimation* and *hypothesis testing*. Although the two processes share common elements (for example, both procedures involve probability theory), the process of estimation is more direct and conceptually easier to understand; we shall start with it. The logic of hypothesis testing involves an indirect method for making inferences and, as a result, is conceptually more involved; we save it for chapter 10. Keep in mind that the method used for making an inference is usually a matter of choice for the experimenter. Since both processes are used in experimental research, it is important to understand both approaches.

Point Estimates for μ

As an illustration of the ideas involved in estimation, suppose a biologist is interested in determining the average number of eggs laid per nest per season for the Eastern Phoebe bird. A random sample of 50 nests was examined and the following results were obtained:

Number of Eggs/Nest	1	2	3	4	5	6
Frequency f	2	1	1	8	36	2

The mean number of eggs for the sample of 50 nests is 4.62. This figure can serve as an **estimate** of μ, the true average number of eggs laid per nest by the Eastern Phoebe bird during one season. If this experiment were to be repeated a large number of times, values of the sample means would vary, providing different estimates. As a result of the central limit theorem, we know the sampling distribution of these means is approximately normal with $\mu_{\bar{x}} = \mu$ and $\sigma_{\bar{x}} = \sigma/\sqrt{50}$. By using these facts, we will be able to determine the goodness of our estimate.

There are two types of estimates for parameters: point and interval estimates. A **point estimate** is a single value of a statistic that is used to estimate a parameter. The statistic that we use is called an **estimator.** For our example above, the sample mean $\bar{x}$ is the estimator and the value 4.62 is the point estimate. An **interval estimate,** on the other hand, is an interval (usually of finite width) that is expected to contain the parameter. For example, the interval (4.57, 4.67) could be an interval estimate for the true average number of eggs per nest. The width of this interval is $4.67 - 4.57 = 0.10$. This interval either contains the populatin mean μ or it does not.

For estimation purposes, we can think of the sample mean $\bar{x}$ as a measurement of the value of the population mean μ. Any measurement has an associated degree of precision, which indicates how accurate it is. If the interval (4.57, 4.67) is an interval estimate for μ, based on $\bar{x} = 4.62$, we feel reasonably confident that $4.57 < \mu < 4.67$. Stated differently, we have a certain degree of confidence that the value of μ is strictly within 4.62 ± 0.05. The value 0.05 indicates a probable degree of precision involved in using $\bar{x} = 4.62$ to estimate μ. Knowing that a measurement is precise is of limited value if it is not reliable. **Reliability** is the probability than an estimate is correct. As a result of the central limit theorem, the sampling distribution of $\bar{x}$ is approximately normal for large samples. Therefore, when we estimate μ using $\bar{x}$, we can increase the reliability and precision by using larger samples. Since $\mu_{\bar{x}} = \mu$ and $\sigma_{\bar{x}} = \sigma/\sqrt{n}$, the values for $\bar{x}$ will cluster more tightly around μ as the size of the sample increases.

Population values are commonly estimated by their corresponding sample values. A sample mean $\bar{x}$ is an estimate of the population mean μ. A sample proportion $\hat{p} = x/n$ is an estimate of the population proportion p. The population variance σ^2 is estimated by s^2, the sample variance, whereas the standard deviation σ of a population is estimated by the standard deviation s of a sample drawn from the population.

An **unbiased estimator** is one where the mean of the sampling distribution is the parameter being estimated. If the sample mean $\bar{x}$ is used to estimate the population mean μ, we learned in chapter 8 that $\bar{x}$ is an unbiased estimator, that is, $\mu_{\bar{x}} = \mu$. Similarly, we learned that $\hat{p}$ is an unbiased estimator of p.

The following is a list of parameters and corresponding estimators:

Parameter	Estimator	Unbiased
μ	$\bar{x}$	yes
σ^2	s^2	yes
p	$\hat{p} = x/n$	yes
σ	s	no

Although $\bar{x}$, s^2, and $\hat{p}$ are unbiased estimators of μ, σ^2, and p, respectively, s is a biased estimator of σ. But for most practical applications, the error resulting is negligible, provided the size of the sample is at least 30.

It should be clear that the precision of an estimator increases with the size of the sample. If the sample is the entire population, then $\bar{x} = \mu$. The absolute value of the difference between an estimate and the parameter being estimated is called the **error of estimate**. When $\bar{x}$ is used to estimate μ, $|\bar{x} - \mu|$ is the error of estimate. Recall from section 8.1 that $\bar{x} - \mu$ is called the *sampling error*. The difference between the *sampling error* and the *error of estimate* is that the error of estimate is always greater than or equal to 0. The average error of estimate decreases as the sample size increases. Unfortunately, we never know the error of estimate, since we do not know the value of the parameter. But we can make probability statements concerning the error of estimate.

EXAMPLE 9.1

Suppose we use $\bar{x}$ to estimate μ, the mean number of eggs per nest for the Eastern Phoebe bird data, and want to determine the probability that the error of estimate is less than 0.05. That is, we want to determine the probability $P(-0.05 < \bar{X} - \mu < 0.05)$. Further, suppose we use the sample standard deviation $s = 0.99$ to estimate the true standard deviation s. Since $n = 50$, the standard error of the mean is given by

$$\sigma_{\bar{x}} = \frac{\sigma}{\sqrt{n}}$$

$$\approx \frac{s}{\sqrt{n}}$$

$$= \frac{0.99}{\sqrt{50}} = 0.14$$

Note that since

$$z = \frac{\bar{x} - \mu}{\sigma/\sqrt{n}}$$

it follows that

$$P(|\bar{X} - \mu| < 0.05) = P(-0.05 < \bar{X} - \mu < 0.05)$$

$$= P\left(\frac{-0.05}{\sigma/\sqrt{n}} < \frac{\bar{X} - \mu}{\sigma/\sqrt{n}} < \frac{0.05}{\sigma/\sqrt{n}}\right)$$

$$= P\left(\frac{-0.05}{0.14} < Z < \frac{0.05}{0.14}\right)$$

$$= P(-0.36 < Z < 0.36)$$

$$= 2P(0 < Z < 0.36) = 2(0.1406) = 0.2812$$

If $\bar{x}$ is used to estimate μ, we have determined that the probability of the error of estimate being less than 0.05 is 0.2812.

The probability value 0.2812 in example 9.1 is called the reliability or **confidence level.** Thus, if the sample mean is used to estimate the average number of eggs laid per nest per season by the Eastern Phoebe bird, we can be 28.12% confident that the error of estimate is less than 0.05. Notice that in order to evaluate the probability that the error of estimate $|\bar{x} - \mu|$ is less than some value E, we need know only $\sigma_{\bar{x}}$, the standard error of the mean.

APPLICATION 9.1

A study was conducted to determine the mean amount of cola dispensed from a cola machine. In a previous study, it was determined that the standard deviation of the amounts dispensed is $s = 0.40$ ounce. If a sample of size 40 is used to estimate μ, find the confidence level for the error of estimate to be less than 0.01 ounce.

Solution: We want to determine the probability $P(|\bar{X} - \mu| < 0.01)$. As a consequence of the central limit theorem and the fact that $n = 40$, the sampling distribution of $\bar{X}$ is approximately normal. The standard error of the mean is given by

$$\sigma_{\bar{x}} = \frac{\sigma}{\sqrt{n}}$$

$$= \frac{0.40}{\sqrt{40}} = 0.063$$

Thus,
$$P(|\bar{X} - \mu| < 0.01) = P\left(|Z| < \frac{0.01}{0.063}\right)$$

$$= P(|Z| < 0.16)$$

$$= 2P(0 < Z < 0.16) = 2(0.0636) = 0.1272$$

Hence, we can be 12.72% confident that the error of estimate is less than 0.01. ■

For each of the situations thus far, the maximum error of estimate was stipulated in advance, and the confidence level was found. We now want to develop the ideas involved in determining the maximum error of estimate if the confidence level has been stipulated in advance. But in order to make probability statements about the error of estimate, we must first examine the concept of critical value. In figure 9.1, the value z_α is called a **critical value.** The critical value z_α is that standard normal value such that the area under the standard normal curve to the right of it is α.

For example, $z_{0.025}$ is the z value that has an area of 0.025 to the right of it under the standard normal curve and above the horizontal axis (see figure 9.2). Since the standard normal table (z table; see front endpaper) provides areas under the standard normal curve and above the horizontal axis between 0 and specified z values, to determine the value of $z_{0.025}$ we need to identify the z value that corresponds to a tabled

FIGURE 9.1

Critical value z_α

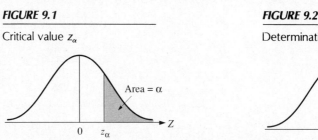

FIGURE 9.2

Determination of $z_{0.025}$

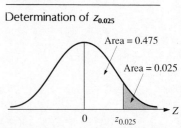

FIGURE 9.3

Critical values of confidence level $1 - \alpha$

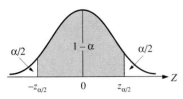

area of $0.5 - 0.025 = 0.475$. (If the exact area can not be found, we will determine the area that comes closest to it and identify its corresponding z value as the critical value.) Checking the z table, we can locate 0.475 in the table; its corresponding z value is 1.96. Hence, the desired critical value is $z_{0.025} = 1.96$.

A distribution can have two critical values. In figure 9.3, $-z_{\alpha/2}$ and $z_{\alpha/2}$ are both called critical values; one is the negative of the other. The area under the standard normal curve and between the two critical values is called the confidence level. The confidence level is equal to $1 - \alpha$, since the sum of three areas, $\alpha/2$, $1 - \alpha$, and $\alpha/2$, is 1. Applications 9.2 and 9.3 demonstrate the concepts of confidence level and critical value.

APPLICATION 9.2

If the confidence level is 85%, find the positive critical value.

Solution: Since $1 - \alpha = 85\%$, $\alpha = 0.15$ and $\alpha/2 = 0.075$. To find the critical value $z_{0.075}$ we use the standard normal table. Since the area under the curve and above the horizontal axis to the right of $z_{0.075}$ is 0.075 (see figure 9.4), the area between 0 and

FIGURE 9.4

Critical value $z_{0.075}$ for appliction 9.2

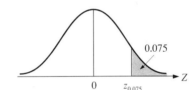

Area	z Value
0.4236	1.43
0.4250	
→0.4251	1.44

$z_{0.075}$ is $0.5 - 0.075 = 0.425$. We look in the z table (see front endpaper) for the area that comes closest to 0.425 and find the corresponding z value. Since the area 0.4251 comes closest to the area 0.425, the corresponding z value is 1.44. Thus, the positive critical value $z_{0.075}$ is 1.44. ▪

APPLICATION 9.3

If the positive critical value is 1.65, find the confidence level $1 - \alpha$.

Solution: The area under the standard normal curve and above the horizontal axis between 0 and 1.65 is 0.4505 as shown in the accompanying figure. We double this to obtain the confidence level. Thus, the confidence level is $1 - \alpha = 0.901$.

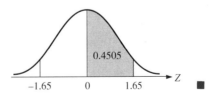

▪

Point Estimates for μ Using Large Samples

Suppose we want our estimate of μ to be $(1 - \alpha)100\%$ reliable. What can we say about the level of precision or error associated with the estimate? The maximum error E that corresponds to a confidence level of $1 - \alpha$ can be found as follows. Since

$$P(-z_{\alpha/2} < Z < z_{\alpha/2}) = 1 - \alpha \tag{9.1}$$

and

$$z = \frac{\overline{x} - \mu}{\sigma/\sqrt{n}} \tag{9.2}$$

we substitute the value of z given in formula (9.2) into the left-hand side of equation (9.1) to get

$$P\left(-z_{\alpha/2} < \frac{\overline{X} - \mu}{\sigma/\sqrt{n}} < z_{\alpha/2}\right) = 1 - \alpha$$

If we now apply some relatively simple algebra, we can rewrite this last statement as

$$P\left(|\overline{X} - \mu| < z_{\alpha/2}\frac{\sigma}{\sqrt{n}}\right) = 1 - \alpha \tag{9.3}$$

Thus, we have the following formula for the **maximum error of estimate E:**

Maximum Error of Estimate for Large Samples

$$E = z_{\alpha/2}\frac{\sigma}{\sqrt{n}} \tag{9.4}$$

One should be careful not to conclude that E is the maximum *possible* error. The error of estimate $|\overline{x} - \mu|$ can exceed the "maximum" error E. What the probability statement (9.3) indicates is that in repeated sampling, $(1 - \alpha)100\%$ of the sample means will fall no more than a distance of E from the population mean μ. Since our estimate $\overline{x}$ of μ is based on a single random sample, we feel $(1 - \alpha)100\%$ confident that the error of estimate $|\overline{x} - \mu|$ is less than the maximum error of estimate E. In other words, the reliability of our estimate $\overline{x}$ having precision E is $(1 - \alpha)100\%$. Equation (9.4) defines the maximum error as the product of the critical value and the standard error of the mean. Consider applications 9.4 and 9.5. The maximum error E depends, in part, on the confidence level. If we want to be more confident, then the maximum error E must increase, as illustrated in applications 9.6 and 9.7.

APPLICATION 9.4

In an effort to estimate the average resting pulse rate of adults 40 years of age, a random sample of pulse-rate measurements from 50 40-year-old individuals was used. If $\overline{x}$ is used to estimate μ and it is desired to be 95% confident that the population mean μ differs by no more than E from $\overline{x}$, find the maximum error E, assuming that $\sigma = 10$ beats per minute.

Solution: Since $1 - \alpha = 0.95$, $\alpha = 0.05$ and $\alpha/2 = 0.025$. From the z table (see front endpaper) we find $z_{0.025} = 1.96$. By using equation (9.4) we have,

$$E = z_{\alpha/2}\frac{\sigma}{\sqrt{n}}$$

$$= (1.96)\left(\frac{10}{\sqrt{50}}\right) = 2.77$$

Thus, if $\overline{x}$ is used to estimate μ, we can be 95% confident that the maximum error of estimate is less than 2.77 beats per minute. ■

APPLICATION 9.5

Suppose a random sample of 40 newborn baby boys is to be used to estimate the mean weight of all newborn baby boys. If it is found that $\bar{x} = 7.5$ pounds and $x = 1.2$ pounds, find the maximum error E so that we can be 92% confident that the error of estimate is less than E.

Solution: Note that $\alpha = 0.08$ and $\alpha/2 = 0.04$. Since $n > 30$ and σ is unknown, we can use s to estimate σ. Thus, by using formula (9.4) we have

$$E = z_{\alpha/2}\frac{\sigma}{\sqrt{n}}$$

$$\approx z_{\alpha/2}\frac{s}{\sqrt{n}}$$

$$= 1.75\left(\frac{1.2}{\sqrt{40}}\right) = 0.33 \text{ pound}$$

Thus, we can be 92% sure that μ is within 0.33 pound of $\bar{x} = 7.5$ pounds. ■

APPLICATION 9.6

For application 9.5, find the maximum error E so that we can be 99% confident that the error of estimate is less than E.

Solution: Since $1 - \alpha = 0.99$, $\alpha = 0.01$ and $\alpha/2 = 0.005$. We find that $z_{0.005} = 2.58$ by locating the z score in the z table corresponding to the area that comes closest to the area 0.495. The value 2.58 is chosen because 0.495 is equidistant from areas 0.4949 and 0.4951 and we choose the largest corresponding z value. By applying formula (9.4), we obtain the maximum error of estimate:

$$E = z_{\alpha/2}\frac{\sigma}{\sqrt{n}}$$

$$= (2.58)\left(\frac{1.2}{\sqrt{40}}\right) = 0.49$$

Thus, we can be 99% confident that the error of estimate $|\bar{x} - \mu|$ is less than 0.49. ■

APPLICATION 9.7

For application 9.5, find E so that we can be 90% confident that the maximum error of estimate is less than E.

Solution: Since $1 - \alpha = 0.90$, $\alpha = 0.10$ and $\alpha/2 = 0.05$. By examining the z table (see the front endpaper), we determine the positive critical value to be 1.65, since 0.45 is equidistant from 0.4495 and 0.4505 and 1.65 is the largest corresponding z value. By using formula (9.4), we have

$$E = z_{\alpha/2}\frac{\sigma}{\sqrt{n}}$$

$$= 1.65\left(\frac{1.2}{\sqrt{40}}\right) = 0.31 \quad ■$$

The results of applications 9.5, 9.6, and 9.7, are summarized in the following table:

Confidence Level	Maximum Error E
0.90	0.31
0.95	0.37
0.99	0.49

Notice that as the confidence level increases, the maximum error also increases.

There is a direct relationship between the confidence level and the maximum error of estimate. The maximum error of estimate varies directly with the confidence level. When one increases so does the other, and when one decreases so does the other. Note also that the maximum error E varies directly with the standard error of the mean $\sigma_{\bar{x}}$, since $E = z_{\alpha/2}\sigma_{\bar{x}}$. This means that E gets smaller as the standard error $\sigma_{\bar{x}}$ gets smaller and E gets larger when $\sigma_{\bar{x}}$ gets larger. Since $\sigma_{\bar{x}} = \sigma/\sqrt{n}$, $\sigma_{\bar{x}}$ decreases as n increases. Consequently, the maximum error of estimate E will also decrease as the sample size increases. It is usually desirable to choose as large a sample as possible when using the sample mean $\bar{x}$ to estimate the population mean μ.

Point Estimates for μ Using Small Samples

If $\bar{x}$ is used to estimate μ and a random sample of size $n < 30$ is selected from a normal population with σ unknown, then the sampling distribution of the sample mean is normal, but we cannot compute a z score for $\bar{x}$. Recall from chapter 8 that the sampling distribution of the statistic $(\bar{x} - \mu)/(s/\sqrt{n})$ is a t distribution with df $= n - 1$.

FIGURE 9.5

Critical values for t distribution

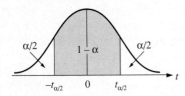

Critical values for t are identified with the t distributions in a fashion similar to the standard normal distribution. The values $-t_{\alpha/2}$ and $t_{\alpha/2}$ in figure 9.5 are *critical values*. The value $t_{\alpha/2}$ is such that the area under the t curve (with df $= n - 1$), above the horizontal axis, and to the right of it is $\alpha/2$.

The following probability statement can be made for a t distribution with df degrees of freedom:

$$P(-t_{\alpha/2} < t < t_{\alpha/2}) = 1 - \alpha$$

Since the t value for $\bar{x}$ is given by

$$t = \frac{\bar{x} - \mu}{s/\sqrt{n}}$$

we can substitute this value for t in the above probability statement and use some simple algebra to deduce the maximum error of estimate E when $\bar{x}$ is used to estimate μ for a small sample ($n < 30$) drawn from a normal population with unknown variance:

Maximum Error of Estimate for Small Samples

$$E = t_{\alpha/2}\frac{s}{\sqrt{n}} \tag{9.5}$$

APPLICATION 9.8

In an effort to determine the average noise level for large trucks, the Environmental Protection Agency obtained a random sample of noise-level readings (in decibels) for eight large trucks. The sample mean and standard deviation were found to be 85.6 decibels and 0.65 decibel, respectively. If the sample mean is used to estimate μ, determine the maximum error E so that the EPA can be 95% confident that μ is within E of $\bar{x}$. Assume that the truck noise levels are normally distributed.

Solution: By reading from the t table (see back endpaper) under the column labeled 0.025 for df $= 7$, we find that $t_{0.025} = 2.365$. Hence, using formula (9.5) we have

$$E = t_{\alpha/2}\frac{s}{\sqrt{n}}$$

$$= 2.365\left(\frac{0.65}{\sqrt{8}}\right) = 0.54$$

Thus, if $\bar{x} = 85.6$ is used to estimate μ, we can be 95% confident that μ is within 0.54 decibels of $\bar{x} = 85.6$ decibels. Stated differently, we can be 95% confident that the maximum error of estimate is less than 0.54 when $\bar{x} = 85.6$ is used to estimate μ. ■

APPLICATION 9.9

It is known that the time (in minutes) required to hand assemble a certain electronic module is normally distributed. If the following sample data were obtained, find the maximum error of estimate E so that we can be 99% confident that the population mean assembly time μ is within E of the sample mean $\bar{x}$.

$$6.2 \quad 7.1 \quad 5.7 \quad 6.8 \quad 5.4$$

Solution: We first find $\bar{x}$ and s for the sample. The table organizes the computations.

	x	x^2
	6.2	38.44
	7.1	50.41
	5.7	32.49
	6.8	46.24
	5.4	29.16
Sums	31.2	196.74

The sample mean is

$$\bar{x} = \frac{\Sigma x}{n}$$

$$= \frac{31.2}{5} = 6.24$$

The sum of squares is

$$SS = \Sigma x^2 - \frac{(\Sigma x)^2}{n}$$

$$= 196.74 - \frac{(31.2)^2}{5} = 2.052$$

Therefore, the sample standard deviation is

$$s = \sqrt{\frac{SS}{n-1}}$$

$$= \sqrt{\frac{2.052}{4}} = 0.716$$

The confidence level is $1 - \alpha = 0.99$. Thus, $\alpha = 0.1$ and $\alpha/2 = 0.005$.

The critical value $t_{0.005}$ is found in the t table (see back endpaper) with df $= n - 1 = 5 - 1 = 4$. Thus, $t_{0.005} = 4.604$.

We use formula (9.5) to find E:

$$E = t_{\alpha/2} \frac{s}{\sqrt{n}}$$

$$= 4.604 \left(\frac{0.716}{\sqrt{5}} \right) = 1.47$$

Thus, we can be 99% confident that μ is within 1.47 of $\bar{x}$. In other words, we can be 99% confident that the error of estimate is less than 1.47 minutes when $\bar{x} = 6.24$ is used to estimate μ. ■

EXERCISE SET 9.1

Basic Skills

1. Find the confidence level for the following critical values:
   a. $z_{\alpha/2} = 1.28$
   b. $-z_{\alpha/2} = -1.44$
   c. $z_{\alpha/2} = 2.58$
   d. $z = 2.76$

2. Find the confidence level for the following critical values:
   a. $z = 2.34$
   b. $z = -1.75$
   c. $z = -0.98$
   d. $z = 1.75$

3. Find the positive critical z value if
   a. the confidence level is 0.94.
   b. $\alpha = 10\%$.
   c. the confidence level is 88%.

4. Find the positive critical z value if
   a. the confidence level is 0.80.
   b. $\alpha = 15\%$
   c. the confidence level is 92%.

5. If the confidence level is 95%, find the critical value $t_{\alpha/2}$ for $n = 15$.

6. If the confidence level is 90%, find $t_{\alpha/2}$ for $n = 20$.

7. If $1 - \alpha = 0.98$, find $t_{\alpha/2}$ for df = 29.

8. Find the value of t_α for df = 20 so that $P(t > t_\alpha) = 0.10$.

Further Applications

9. In an effort to determine the average amount spent per customer for lunch at a large Chicago-area restaurant, data were collected for 64 customers over a 1-month period. If $\sigma = \$2.25$ and the sample mean $\bar{x}$ is used to estimate μ, find the maximum error E so that one can be 95% confident that the error of estimate is less than E.

10. Fifteen employees of a large manufacturing company were involved in a study to test a new production method. The mean production rate for the sample of 15 employees was 63 components per hour and the standard deviation was 8 components per hour. If $\bar{x}$ is used to estimate μ, determine the maximum error E in order to be 99% confident that μ is within E of $\bar{x}$. Assume that the production rates are normally distributed.

11. A random sample of 50 entrance examination scores at a large university is used to estimate the true mean of all entrance examination scores. If $\bar{x} = 98.2$, $s = 17$, and the confidence level is 99%, find the maximum error if $\bar{x}$ is used to estimate μ and interpret your result.

12. Each member of a random sample of 60 college students was asked how many hours he or she studied per week. If $\bar{x} = 18$ and $s = 2$, and $\bar{x}$ is used to estimate μ, find the maximum error so that the probability that the error of estimate is less than E is 0.95.

13. If a random sample of 50 cups of cola dispensed from an automatic vending machine showed that the mean amount of cola dispensed was 7.9 ounces with $s = 0.35$ ounce, find the confidence level if $E = 0.06$.

14. The drying times (in hours) for newly painted parts are normally distributed. Using the following sample data, find the maximum error of estimate E so that we can be 95% confident that the mean drying time of all newly painted parts μ is within E of $\bar{x}$:

    $$6.4 \quad 7.2 \quad 5.9 \quad 6.8 \quad 7.1 \quad 5.5$$

15. A car manufacturer tested 100 cars to determine the mileage traveled before a motor overhaul was needed. He obtained $\bar{x} = 81{,}250$ miles and $s = 6325$ miles. Find the maximum error of estimate so that we can be 95% confident that μ is within E of $\bar{x}$.

Going Beyond

16. Using the statement $P(-z_{\alpha/2} < Z < z_{\alpha/2}) = 1 - \alpha$ and the fact that $z = (\bar{x} - \mu)/(\sigma/\sqrt{n})$, prove that $P(|\bar{x} - \mu| < z_{\alpha/2}\, \sigma/\sqrt{n}) = 1 - \alpha$ and $E = z_{\alpha/2}\, \sigma/\sqrt{n}$.

17. Using the statement $P(-t_{\alpha/2} < t < t_{\alpha/2}) = 1 - \alpha$ and the fact that $t = (\bar{x} - \mu)/(s/\sqrt{n})$, prove that $P(|\bar{X} - \mu| < t_{\alpha/2}\, s/\sqrt{n}) = 1 - \alpha$ and $E = t_{\alpha/2}\, s/\sqrt{n}$.

18. If μ is within E of $\bar{x}$, show that $\bar{x}$ is within E of μ. That is, if $\bar{x}$ is close to μ, then μ is close to $\bar{x}$.

19. Suppose we wanted to estimate the mean number of children per family. There are at least two methods we could use to arrive at this estimate.

    Method 1. Select a random sample of families, count the number of children per family, and determine the average number of children per family.
    Method 2. Select a random sample of people, count the number of children each has in his or her family, and determine the average number of children per family.

    Do these methods provide unbiased estimates of μ, the average number of children per family? Explain.

20. Which of the two methods listed in exercise 19 would you use to estimate the mean number of passengers per car? Explain.

21. Suppose we want to use a mail survey to sample work attitudes from a population of 700 employees. If we assume a return rate of 45% for the survey, a precision is 0.05, and a level of confidence of 0.95, use Nowack's formula, discussed in motivator 9, to determine the minimum size of the sample to survey in order to make valid inferences about the work attitudes of the 700 employees.

SECTION 9.2   *Confidence Intervals for μ*

Suppose we want to estimate a population mean μ by using an interval estimate and that either the population is normal or the sample size n is large ($n \geq 30$). For the present, we shall assume that sampling is done without replacement. If the population is normal, then the sampling distribution of the sample mean is normal; if the sample size is large, then, as a consequence of the central limit theorem, the sampling distribution of the sample mean is approximately normal. In section 9.1 we saw that if $\bar{x}$ is used to estimate μ, then the maximum error E can be determined so that sampling can be $(1 - \alpha)100\%$ confident that μ is within E of $\bar{x}$ or that $\bar{x} - E < \mu < \bar{x} + E$. That is, we can determine the maximum error E so that $P(\bar{x} - E < \mu < \bar{x} + E) = 1 - \alpha$. The interval extending from $\bar{x} - E$ to $\bar{x} + E$ is called a **$(1 - \alpha)100\%$ confidence interval for μ.** The value $\bar{x} - E$ is called the **lower confidence limit** and is denoted by L_1 and the value $\bar{x} + E$ is called the **upper confidence limit** and is denoted by L_2. The confidence interval is written using interval notation as (L_1, L_2). Consequently, the limits for a $(1 - \alpha)100\%$ confidence interval for μ are given by one of the following three equivalent forms:

$$
\begin{array}{c}
\textbf{Limits for Confidence Interval} \\
1.\ \bar{x} \pm E \\
2.\ \bar{x} \pm z_{\alpha/2}\, \sigma/\sqrt{n} \qquad\qquad (9.6)\\
3.\ \bar{x} \pm z_{\alpha/2}\, \sigma_{\bar{x}}
\end{array}
$$

The maximum error E represents the precision when we use $\bar{x}$ to estimate μ. That is, as a measurement of μ, we can be $(1 - \alpha)100\%$ confident that μ is somewhere between $\bar{x} - E$ and $\bar{x} + E$.

$$
\underset{\underset{\text{Measurement}}{\uparrow}}{\bar{x}} \qquad \pm \qquad \underset{\underset{\text{Precision}}{\uparrow}}{E}
$$

APPLICATION 9.10   Type-B light bulbs supplied by a firm to a theater for use in its display signs have useful lives that have a standard deviation of 35 hours. If a random sample of 45 type-B bulbs has an average life of 750 hours, find a 95% confidence interval for the population mean life of all type-B light bulbs manufactured by the firm, and find the width of the confidence interval.

Solution: We use formula (9.4). Since $1 - \alpha = 0.95$, $\alpha = 0.05$ and $\alpha/2 = 0.025$. The positive critical value is $z_{0.025} = 1.96$, and the maximum error E is found by using formula (9.4):

$$E = z_{\alpha/2} \frac{\sigma}{\sqrt{n}}$$

$$= 1.96 \frac{35}{\sqrt{45}} = 10.23$$

From formulas (9.6), the confidence interval limits are

$$\overline{x} \pm E$$

$$750 \pm 10.23$$

The 95% confidence interval for μ is $(750 - 10.23, 750 + 10.23)$ or $(739.77, 760.23)$. The width of this confidence interval is found by subtracting the lower confidence limit from the upper confidence limit:

$$L_2 - L_1 = 760.23 - 739.77 = 20.46 \quad ▪$$

MINITAB can be used to determine this confidence interval for the light bulb data. The ZINTERVAL command is used. Computer display 9.1 contains the commands and output.

Computer Display 9.1

```
MTB > SET C1
DATA> 45(750)
DATA> END
MTB > ZINTERVAL 95 PERCENT SIGMA=35 C1

THE ASSUMED SIGMA =35.0

        N      MEAN    STDEV    SE MEAN    95.0 PERCENT C.I.
C1     45    750.00     0.00      5.22    ( 739.76, 760.24)
    * NOTE * ALL VALUES IN COLUMN ARE IDENTICAL
```

The confidence interval limits using MINITAB differ slightly from those we calculated because of the increased precision involved in the MINITAB calculations.

It is important to realize that the confidence interval specified by formulas (9.6) has L_1 and L_2 as variables. Their values depend on the values for $\overline{x}$ and α once the sample size n is determined. Once L_1 and L_2 are fixed, the confidence interval (L_1, L_2) either contains μ, with a probability of 1, or it doesn't, with a probability of 0. That is, if μ is contained in (L_1, L_2), then the probability that the interval contains μ is 1, and if μ is not contained in the interval (L_1, L_2), then the probability that the interval contains μ is 0.

EXAMPLE 9.2

In application 9.10, we obtained $(739.77, 760.23)$ as a 95% confidence interval for μ. How should we interpret this interval? It either contains μ or it does not. If it does, the probability that the interval contains μ is 1, and if it doesn't the probability that the interval contains μ is 0. The statement, "the probability that the interval $(739.77, 760.23)$ contains μ is 95%" is *false*.

Since the values of L_1 and L_2 depend on the sample mean $\overline{x}$ once α and n are specified, we can think of the variable confidence interval (L_1, L_2) as generating a collection of confidence intervals. Every confidence interval is centered at a sample mean $\bar{x}$ and has a width w equal to

$$w = L_2 - L_1$$
$$= (\bar{x} + E) - (\bar{x} - E)$$
$$= 2E$$
$$= 2z_{\alpha/2}\frac{\sigma}{\sqrt{n}}$$

which is a fixed value. Some of the intervals generated in this manner contain μ and some don't. The value $1 - \alpha$ is the proportion of the intervals that contain μ, while α is the proportion of the intervals that do not contain μ.

EXAMPLE 9.3

For application 9.10, since 95% of the confidence intervals constructed with a fixed width of 20.46 contain μ, we can say we are 95% confident that our interval (739.77, 760.23) contains μ. We do not know whether our sample is one of the lucky 95% that provide an interval that contains μ. All we know is we used a procedure that provides correct results 95% of the time.

Remember being 95% confident that a confidence interval contains μ can not be interpreted to mean that the probability the interval contains μ is 95%. Rather, it means that the procedure used to arrive at the confidence interval produces correct results 95% of the time.

Computer display 9.2 contains twenty 95% confidence intervals for μ and the commands used to obtain them. MINITAB was used to draw 20 random samples of size 50, $n = 50$, from a normal population with $\mu = 100$ and $\sigma = 10$. For each sample, a 95% confidence interval was constructed using formulas (9.6).

Computer Display 9.2

```
MTB > RANDOM 50 C1-C20;
SUBC>NORMAL 100 10.
MTB > ZINTERVAL 95% SIGMA = 10 C1-C20

THE ASSUMED SIGMA =10.0
```

	N	MEAN	STDEV	SE MEAN	95.0 PERCENT C.I.
C1	50	100.24	10.14	1.41	(97.46, 103.01)
C2	50	100.60	11.38	1.41	(97.83, 103.38)
C3	50	101.40	9.63	1.41	(98.62, 104.18)
C4	50	99.25	8.25	1.41	(96.48, 102.03)
C5	50	97.99	9.52	1.41	(95.21, 100.76)
C6	50	100.40	9.20	1.41	(97.62, 103.18)
C7	50	99.52	10.42	1.41	(96.74, 102.29)
C8	50	101.68	10.30	1.41	(98.90, 104.45)
C9	50	100.18	10.14	1.41	(97.40, 102.96)
C10	50	101.70	10.82	1.41	(98.93, 104.48)
C11	50	102.23	8.88	1.41	(99.45, 105.00)
C12	50	101.32	10.06	1.41	(98.54, 104.10)
C13	50	98.34	10.09	1.41	(95.57, 101.12)
C14	50	99.52	10.57	1.41	(96.74, 102.30)
C15	50	98.94	11.52	1.41	(96.17, 101.72)
C16	50	96.98	8.56	1.41	(94.21, 99.76)
C17	50	102.27	8.59	1.41	(99.49, 105.04)
C18	50	99.08	11.38	1.41	(96.31, 101.86)
C19	50	99.90	11.21	1.41	(97.13, 102.68)
C20	50	100.07	11.17	1.41	(97.29, 102.85)

Annotations above table columns: $\frac{\sigma}{\sqrt{n}}$ over STDEV; $\frac{\sigma}{\sqrt{n}}$ over SE MEAN; $\bar{x} - z_{0.025}\frac{\sigma}{\sqrt{n}},\ \bar{x} + z_{0.025}\frac{\sigma}{\sqrt{n}}$ over 95.0 PERCENT C.I.

Notice that 19 intervals contain μ. We would expect 95% of *all* such intervals constructed to contain μ. Of course, if another 20 samples were selected and the 95% confidence intervals for the mean were computed, there is no guarantee that exactly 19 of them would contain μ.

Recall from section 9.1 that the maximum error of estimate varies directly with the confidence level; as the confidence level increases, the maximum error increases, and as the confidence level decreases, the maximum error decreases. Since the width of a confidence interval is $w = L_2 - L_1 = 2E$, we know the confidence level also varies directly with the width of the interval. A larger confidence level produces a wider interval, and a smaller confidence level produces a shorter interval.

The user usually chooses a confidence level and the width of the interval follows from this choice and the sample selected. A short interval with a high confidence level is always desirable. For a given set of sample data, there is a trade-off between the confidence level and the width of the interval; to decrease the width of the interval means we have to decrease the confidence, and to increase the confidence, we have to increase the width.

Since the maximum error of estimate is given by

$$E = z_{\alpha/2}\frac{\sigma}{\sqrt{n}}$$

for a fixed confidence level, we can always shorten the width of the interval by choosing a larger sample. We will examine this point further in section 9.4.

Sampling from Small Populations without Replacement

Sometimes it is desirable to construct a confidence interval for μ when the population is not large and sampling is done without replacement. Recall from chapter 8 that whenever 20 times the sample size is greater than or equal to the population size ($20n \geq N$), then the standard error of the mean $\sigma_{\bar{x}}$ is found by using the following formula:

> **Standard Error of the Mean when Sampling without Replacement**
>
> $$\sigma_{\bar{x}} = \frac{\sigma}{\sqrt{n}}\sqrt{\frac{N-n}{N-1}} \qquad (9.7)$$

The maximum error of estimate E is then

$$E = z_{\alpha/2}\sigma_{\bar{x}}$$

And the confidence interval limits are then found by using formulas (9.6):

$$\bar{x} \pm z_{\alpha/2}\sigma_{\bar{x}} \quad \text{or} \quad \bar{x} \pm E$$

Application 9.11 illustrates the process.

APPLICATION 9.11   A sample of 45 students is selected from a population of 800 students and given a test to determine their reaction times to respond to a given stimulus. If the mean reaction time is determined to be $\bar{x} = 0.75$ second and the standard deviation is $x = 0.15$

second, find a 95% confidence interval for μ, the mean reaction time for the population of 800 students.

Solution: Since $20n = (20)(45) = 900 > N = 800$, we use formula (9.7) to find $\sigma_{\bar{x}}$. Also, since $n \geq 30$ and σ is unknown, we can use s to estimate σ.

$$\sigma_{\bar{x}} = \frac{\sigma}{\sqrt{n}} \sqrt{\frac{N - n}{N - 1}}$$

$$= \frac{0.15}{\sqrt{45}} \sqrt{\frac{800 - 45}{800 - 1}} = 0.022$$

The positive critical value is $z_{0.025} = 1.96$, and the limits for the 95% confidence interval are given by

$$\bar{x} \pm z_{\alpha/2}\sigma_{\bar{x}}$$
$$0.75 \pm (1.96)(0.022)$$
$$0.75 \pm 0.04$$

Thus, the 95% confidence interval is

$$(0.75 - 0.04, 0.75 + 0.04)$$
$$(0.71, 0.79)$$

We can be 95% confident that the interval (0.71, 0.79) contains μ; 95% of all such intervals constructed contain μ and 5% of them do not. ▪

Confidence Intervals Using Small Samples

If we want to construct a $(1 - \alpha)100\%$ confidence interval for μ using a sample of size $n < 30$ selected from a normal population with an unknown standard deviation σ, the maximum error E is determined by using a t distribution. Recall from section 9.1 that the maximum error of estimate E is given by $E = t_{\alpha/2}(s/\sqrt{n})$ for df $= n - 1$. As a result, the limits for a $(1 - \alpha)100\%$ confidence interval for μ are given by $\bar{x} \pm E$ or

Limits for Confidence Interval Using Small Samples

$$\bar{x} \pm t_{\alpha/2}\frac{s}{\sqrt{n}} \tag{9.8}$$

APPLICATION 9.12

Five similar bags of potatoes weigh 14.8, 16.1, 15.3, 15.2, and 15.4 pounds. Find a 95% confidence interval for the mean weight of all such bags of potatoes. Assume that the population of weights is approximately normal.

Solution: The sample mean and standard deviation are found to be $\bar{x} = 15.36$ and $s = 0.472$, respectively. For df $= 5 - 1 = 4$, we find $t_{0.025} = 2.776$ from the t table (see the back endpaper). Thus, the maximum error E is given by

$$E = t_{0.025}\frac{s}{\sqrt{n}}$$

$$= (2.776)\left(\frac{0.472}{\sqrt{5}}\right) = 0.59$$

and the limits for the 95% confidence interval are

$$\bar{x} \pm E$$
$$15.36 \pm 0.59$$

Thus, the required interval is $(15.36 - 0.59, 15.36 + 0.59)$ or $(14.77, 15.95)$. With this result we feel 95% confident that the interval contains μ, since 95% of all such intervals constructed contain μ. In addition, we are 95% confident that μ is within 0.58 of $\bar{x} = 15.36$. ■

MINITAB can also be used to determine this confidence interval. The TINTERVAL command is used. Computer display 9.3 contains the commands and output.

Computer Display 9.3

```
MTB > SET C1
DATA> 14.8 16.1 15.3 15.2 15.4
DATA> END
MTB > TINTERVAL 95 PERCENT C1

         N     MEAN    STDEV    SE MEAN    95.0 PERCENT C.I.
C1       5    15.360   0.472    0.211     ( 14.773, 15.947)

MTB >
```

If (L_1, L_2) is a $(1 - \alpha)100\%$ confidence interval for μ found by using formula (9.8), then the width of the interval is given by

$$
\begin{aligned}
w &= L_2 - L_1 \\
&= \bar{x} + E - (\bar{x} - E) \\
&= 2E \\
&= 2t_{\alpha/2}\frac{s}{\sqrt{n}}
\end{aligned}
$$

Note that the width of the intervals constructed in repeated sampling is not constant since each is dependent on the value of s, which varies from sample to sample. Recall that this was not the case when we constructed a confidence interval for μ using a random sample drawn from a large population with a known standard deviation; when μ is known, all $(1 - \alpha)100\%$ confidence intervals constructed using a fixed sample size have the same width.

EXERCISE SET 9.2

Basic Skills

1. A sample of 100 observations is taken from a population with unknown mean μ and standard deviation $\sigma = 4.5$. If the mean of the sample is 28.3, construct confidence intervals for μ for each of the following confidence levels. In addition, for each interval find the width and compare your results for parts a–d.

   a. 90%   b. 95%   c. 99%   d. 99.7%

2. A sample of 144 observations is taken from a population with unknown mean μ and standard deviation $\sigma = 3.2$. If the mean of the sample is 45.2, construct confidence intervals for μ for each of the following confidence levels. In addition, for each interval find the width and compare your results for parts a–d.

   a. 90%   b. 95%   c. 99%   d. 99.7%

3. a. For df = 11, find a value for c such that
$P(-c < t < c) = 0.90$.
   b. Find a value for α such that
$P(-z_\alpha < Z < z_\alpha) = 0.40$.
   c. Find $P(-z_{0.05} < Z < z_{0.10})$.

4. a. For df = 15, find a value for c such that
$P(-c < t < c) = 0.80$.
   b. Find a value for α such that
$P(-z_\alpha < Z < z_\alpha) = 0.60$.
   c. Find $P(-z_{0.15} < Z < z_{0.25})$.

Further Applications

5. A mathematical skills test given to 12 randomly selected eighth-grade students showed an average score of 77.8 and a standard deviation of 11.1. Assuming that the population of scores is normally distributed, find a 95% confidence interval for the mean mathematical skills score for eighth graders.

6. A college student measured the heights of 35 male colleagues in her statistics class. Assuming this sample to be representative of all males attending college, find a 95% confidence interval for the true average height of all male college students, if the sample yielded a mean of $\bar{x} = 70.4$ inches and a standard deviation of $s = 2.45$ inches.

7. The management of a large national chain of 20 motels, each with 100 rooms, decided to estimate the mean cost per room of repairing damages made by its customers. A random sample of 150 vacated rooms was inspected by the management and indicated a mean repair cost of $\bar{x} = \$28.10$ and a sample standard deviation of $s = \$12.40$. Construct a 95% confidence interval for the mean repair cost μ for the 2000 motel rooms.

8. Repeat exercise 7 if the national chain consists of 50 motels, each with 100 rooms. Compare the width of this interval to the width of the interval obtained in exercise 7.

9. A study of the health records of a large group of deceased males who smoked at least one pack of cigarettes daily over a 5-year period was conducted to determine the mean life span for all such individuals. A random sample of 16 health records for deceased smokers indicated an average life span of 65.7 years and a standard deviation of 3.4 years. Using these statistics, construct a 99% confidence interval for the true average life span μ for the population of male smokers who smoke at least one pack of cigarettes daily over a 5-year period. Assume that the life spans for such males are normally distributed.

10. Assume that the useful lifetimes (in minutes) of a certain brand of type-D batteries are normally distributed. A

random sample of seven batteries was tested to determine how long the batteries would adequately function in a certain electronic device. The sample yielded a mean of $\bar{x} = 152$ minutes and a standard deviation of $s = 5$ minutes. Construct a 95% confidence interval for the true mean lifetime of all type-D batteries operating the electronic device.

11. Assuming that the useful lifetimes (in minutes) of type-C batteries are normally distributed, use the following sample of battery lives to construct a 90% confidence interval for the true mean lifetime of all type-C batteries: 150, 162, 178, 158, 162, and 171.

12. A health-care organization conducted a study to estimate the average number of days a surgical patient is hospitalized. A random sample of 150 surgical patients yielded an average stay of 4.8 days and a standard deviation of 2.6 days. Find a 90% confidence interval for the mean number of days a surgical patient remains in the hospital.

13. Repeated assessments on a chemical determination of human blood during a laboratory analysis are known to be normally distributed. Ten assessments on a given sample of blood yielded the values 1.002, 0.958, 1.014, 1.009, 1.041, 0.962, 1.058, 1.024, 1.019, and 1.020. Find a 99% confidence interval for the true chemical determination in the blood for repeated assessments on the sample.

14. A special type of hybrid corn was planted on eight different plots. The plots produced yield values (in bushels) of 140, 70, 39, 110, 134, 104, 100, and 125. Assuming the yields follow a normal distribution, find a 95% confidence interval for the true average yield of this type of hybrid corn.

Going Beyond

15. Refer to exercise 6. If $L_1 = \bar{x} - z_{0.02}(s/\sqrt{n})$ and $L_2 = \bar{x} + z_{0.03}(s/\sqrt{n})$, is (L_1, L_2) a 95% confidence interval for μ? Why? How does the width of this interval compare to the width of the interval found in exercise 6?

16. Prove that the inequalities $\bar{x} - E < \mu < \bar{x} + E$ are equivalent to the inequality $|\bar{x} - \mu| < E$.

17. If $P(-z_{\alpha/2} < Z < z_{\alpha/2}) = 1 - \alpha$ and $z = (\bar{x} - \mu)/(\sigma/\sqrt{n})$, prove that
$$P\left(\bar{x} - z_{\alpha/2}\frac{\sigma}{\sqrt{n}} < \mu < \bar{x} + z_{\alpha/2}\frac{\sigma}{\sqrt{n}}\right) = 1 - \alpha$$

18. If $P(-t_{\alpha/2} < t < t_{\alpha/2}) = 1 - \alpha$ and $t = (\bar{x} - \mu)/(s/\sqrt{n})$, prove that
$$P\left(\bar{x} - t_{\alpha/2}\frac{s}{\sqrt{n}} < \mu < \bar{x} + t_{\alpha/2}\frac{s}{\sqrt{n}}\right) = 1 - \alpha$$

19. A one-sided $(1 - \alpha)100\%$ confidence interval for μ is an interval with an area equal to α located either in the left tail or the right tail of the distribution, but not both. A $(1 - \alpha)100\%$ one-sided confidence interval for μ has one of the following two forms:

$$\mu > \bar{x} - z_\alpha \frac{\sigma}{\sqrt{n}} \quad \text{or} \quad \mu < \bar{x} + z_\alpha \frac{\sigma}{\sqrt{n}}$$

A random sample of 30 statistics scores was chosen from a large group of scores to determine the mean μ for the entire group. From previous tests, s was determined to be 13. If the sample mean was $\bar{x} = 72$, construct an appropriate one-sided 95% confidence interval for μ.

20. Based on a random sample of 80 Jersey cows, the confidence interval $(38.6, 42.3)$ was constructed for estimating the true average yield (in ounces) per milking for Jersey cows. If the amounts of milk from Jersey cows per milking are normally distributed and $\sigma = 6$, determine the level of confidence for the interval. Also, determine the mean of the sample used to construct the interval.

SECTION 9.3   *Estimates of Population Proportions*

In section 9.1 we learned that the sample proportion $\hat{p}$ can be used to estimate the population proportion p.

Example 9.4

In 1984 a Baltimore newspaper, the *Sun,* was interested in determining, among other things, the proportion of Maryland residents who smoke. It hired a Baltimore agency to conduct a survey of Maryland adults. The survey revealed a total of 250 smokers from a random sample of 806 adults.[37] By using this information, the polling agency determined the proportion of smokers in the sample to be 0.31. The estimate $\hat{p} = 0.31$ serves as a point estimate for the true proportion of all Maryland adults who smoke. The maximum error of estimate was reported by the *Sun* to be no more than 3.5% with a confidence level of 95%. This means that if $\hat{p} = 0.31$ is used to estimate p, the true proportion of Maryland residents who smoke, then the polling agency is 95% confident that p is within 0.035 of $\hat{p} = 0.31$.

Given a certain level of confidence, we would like to determine the *maximum error of estimate E* when the sample proportion $\hat{p}$ is used as a point estimate for the population proportion p. Toward this goal, recall the following facts from section 8.5 concerning the sampling distribution of $\hat{p}$:

1. When the sample size n is large, the sampling distribution of $\hat{p}$ is approximately normal.
2. The mean of the sampling distribution of $\hat{p}$ is $\mu_{\hat{p}} = p$.
3. The standard error of $\hat{p}$ is $\sigma_{\hat{p}} = \sqrt{p(1 - p)/n}$.

For those applications where p is to be estimated, the standard error of $\hat{p}$ will be unknown. Fortunately it can be estimated by using the sample proportion $\hat{p}$. The estimate $\hat{\sigma}_{\hat{p}}$ of $\sigma_{\hat{p}}$ is given by

$$\hat{\sigma}_{\hat{p}} = \sqrt{\frac{\hat{p}(1 - \hat{p})}{n}}$$

EXAMPLE 9.5

For the *Sun's* poll mentioned above, the standard error of $\hat{p}$ can be estimated as

$$\hat{\sigma}_{\hat{p}} = \sqrt{\frac{\hat{p}(1-\hat{p})}{n}}$$

$$= \sqrt{\frac{(0.31)(0.69)}{806}} = 0.0163$$

The sampling distribution of the statistic

$$\frac{\hat{p} - \mu_{\hat{p}}}{\sigma_{\hat{p}}}$$

is approximately the standard normal. Thus, we have

$$z \approx \frac{\hat{p} - p}{\sqrt{p(1-p)/n}} \tag{9.9}$$

Toward finding the maximum error of estimate E, we begin with the following probability statement:

$$P(-z_{\alpha/2} < Z < z_{\alpha/2}) = 1 - \alpha$$

By substituting the z value from formula (9.9) into this probability statement and using some simple algebra, we can obtain the following probability statement:

$$P\left(|\hat{p} - p| < z_{\alpha/2}\sqrt{\frac{p(1-p)}{n}}\right) \approx 1 - \alpha \tag{9.10}$$

$$\uparrow \qquad\qquad\qquad \uparrow$$

error of estimate E confidence level

Therefore, the maximum error of estimate E can be approximated by

$$E = z_{\alpha/2}\sqrt{\frac{\hat{p}(1-\hat{p})}{n}} = z_{\alpha/2}\hat{\sigma}_{\hat{p}}$$

Thus, if $\hat{p}$ is used to estimate p with a confidence level of $1 - \alpha$, the maximum error of estimate E can be approximated by using

> **Approximate Maximum Error of Estimate if $\hat{p}$ is Used to Estimate p**
>
> $$E = z_{\alpha/2}\sqrt{\frac{\hat{p}(1-\hat{p})}{n}} = z_{\alpha/2}\hat{\sigma}_{\hat{p}} \tag{9.11}$$

EXAMPLE 9.6

For the *Sun's* poll of Maryland adults, if $\hat{p} = 0.31$ is used to estimate the true proportion of Maryland adults who smoke, the *Sun* can be 95% confident that the error of estimate is approximately

$$E \approx z_{\alpha/2}\hat{\sigma}_{\hat{p}}$$

$$= (1.96)(0.0163) = 0.032$$

Hence, the *Sun* can be 95% confident that p is within approximately 3.2% of $\hat{p} = 0.31$. Notice that our maximum error of estimate differs from the *Sun's* by 0.3% or 0.003.

APPLICATION 9.13

In a sample of 400 type-B batteries manufactured by the Everlast Company, 20 defective batteries were found. If the proportion $\hat{p}$ of defective batteries in the sample is used to estimate p, the true proportion of all defective type-B batteries manufactured by the Everlast Company, find the maximum error of estimate E so that one can be 95% confident that p is within E of $\hat{p}$.

Solution: Since $1 - \alpha = 0.95$, $\alpha = 0.05$ and $\alpha/2 = 0.025$. Hence, $z_{\alpha/2} = 1.96$. Since $\hat{p} = x/n = 20/400 = 0.05$, an estimate for $\sigma_{\hat{p}}$ is given by

$$\hat{\sigma}_{\hat{p}} = \sqrt{\frac{\hat{p}(1 - \hat{p})}{n}}$$

$$= \sqrt{\frac{(0.05)(0.95)}{400}} = 0.0109$$

Hence, by using formula (9.11), we have

$$E = z_{\alpha/2} \sqrt{\frac{\hat{p}(1 - \hat{p})}{n}}$$

$$= (1.96)(0.0109) = 0.021$$

If $\hat{p} = 0.05$ is used to estimate p, we can be 95% confident that p is within approximately 0.021 of $\hat{p}$. In other words, if $\hat{p} = 0.05$ is used to estimate p, the maximum error of estimate will be approximately 0.021 with a confidence level of 95%. ▪

EXAMPLE 9.7

If $\hat{p}$ is used to estimate p, and we are 95% confident that p is within approximately E of $\hat{p}$, then the expression $\hat{p} \pm E$ can be used to establish an approximate 95% confidence interval for p. For application 9.13, an approximate 95% confidence interval for the true proportion of defective type-B batteries is found by using the limits

$$\hat{p} \pm E$$
$$0.05 \pm 0.021$$

Thus, an approximate 95% confidence interval for p, the true proportion of defective type-B batteries, is

$$(0.029, 0.071)$$

In summary, the limits for an approximate $(1 - \alpha)100\%$ confidence interval for the population proportion p are given by any one of the following three equivalent expressions:

> **Limits for an Approximate Confidence Interval for p**
>
> $$\hat{p} \pm E$$
>
> $$\hat{p} \pm z_{\alpha/2} \sqrt{\frac{\hat{p}(1 - \hat{p})}{n}} \qquad (9.12)$$
>
> $$\hat{p} \pm z_{\alpha/2}\hat{\sigma}_{\hat{p}}$$

APPLICATION 9.14

In a study of 300 automobile accidents in a particular city, 60 resulted in fatalities. Based on this sample, construct an approximate 90% confidence interval for the proportion of all auto accidents in the city that result in fatalities.

Solution: We first note that the best point estimate of p is $\hat{p} = 60/300 = 0.2$. The critical value $z_{0.05}$ is found in the standard normal table (z table; see front endpaper) to be 1.65. Since an approximation for the standard error of $\hat{p}$ can be found by using

$$\hat{\sigma}_{\hat{p}} = \sqrt{\frac{\hat{p}(1 - \hat{p})}{n}}$$

we have

$$\hat{\sigma}_{\hat{p}} = \sqrt{\frac{(0.2)(0.8)}{300}} = 0.0231$$

Approximate confidence interval limits are found from expressions (9.12) as follows:

$$\hat{p} \pm z_{\alpha/2}\hat{\sigma}_{\hat{p}}$$
$$0.20 \pm (1.65)(0.0231)$$
$$0.20 \pm 0.038$$

Thus, an approximate 90% confidence interval for p is (0.162, 0.238). In addition, using $\hat{p} = 0.2$ to estimate p, we can be 90% confident that our error of estimate is approximately $E = 0.038$. ▪

MINITAB can also be used to determine the confidence interval for application 9.14. The TINTERVAL command is used. Computer display 9.4 contains the commands used and the output.

Computer Display 9.4

```
MTB > SET C1
DATA> 60(1) 240(0)
DATA> END
MTB > TINTERVAL 90 PERCENT C1

            N      MEAN     STDEV     SE MEAN     90.0 PERCENT C.I.
C1        300    0.2000    0.4007     0.0231     ( 0.1618, 0.2382)

MTB >
```

When sampling without replacement from a small population (relative to the sample size) and estimating p, the standard error of $\sigma_{\hat{p}}$ can be approximated by using the following formula

Approximate Standard Error of $\hat{p}$ when Sampling Without Replacement

$$\hat{\sigma}_{\hat{p}} = \sqrt{\frac{\hat{p}(1 - \hat{p})}{n}}\sqrt{\frac{N - n}{N - 1}} \tag{9.13}$$

where $\hat{\sigma}_{\hat{p}} \approx \sigma_{\hat{p}}$. Note that formula (9.13) is used whenever $N < 20n$, as illustrated in application 9.15.

APPLICATION 9.15

A political candidate is planning his campaign strategy and would like to determine the extent to which he is known. In a random sample of 3000 of the country's 25,000 registered voters 1800 indicated that they recognized the candidate's name. Construct a 95% confidence interval for the true proportion of voters in the country who are familiar with the candidate.

Solution: The critical value is $z_{0.025} = 1.96$. Since $25,000 < (20)(3000)$, formula (9.13) should be used to compute $\sigma_{\hat{p}}$ (note that $\hat{p} = 1800/3000 = 0.6$):

$$\hat{\sigma}_{\hat{p}} = \sqrt{\frac{\hat{p}(1 - \hat{p})}{n}} \sqrt{\frac{N - n}{N - 1}}$$

$$= \sqrt{\frac{(0.6)(0.4)}{3000}} \sqrt{\frac{25,000 - 3000}{25,000 - 1}} = 0.00839$$

Hence, by using expressions (9.12), we obtain the approximate confidence limits:

$$\hat{p} \pm z_{\alpha/2} \hat{\sigma}_{\hat{p}}$$
$$0.6 \pm (1.96)(0.00839)$$
$$0.6 \pm 0.02$$

Thus, an approximate 95% confidence interval for p is (0.58, 0.62). We can be approximately 95% confident that the interval (0.58, 0.62) contains p, since approximately 95% of all such intervals constructed contain p. ▪

EXERCISE SET 9.3

Basic Skills

1. Construct the appropriate confidence intervals for p using the given information.
   a. $n = 400$, $x = 100$, $\alpha = 0.05$
   b. $n = 500$, $x = 125$, $1 - \alpha = 0.90$
   c. $n = 1500$, $x = 900$, $1 - \alpha = 0.80$
   d. $n = 250$, $N = 4000$, $x = 100$, $1 - \alpha = 0.95$
   e. $n = 45$, $N = 1000$, $x = 10$, $\alpha = 0.01$.

2. Construct the appropriate confidence intervals for p using the given information.
   a. $n = 500$, $x = 200$, $\alpha = 0.05$
   b. $n = 125$, $x = 25$, $1 - \alpha = 0.99$
   c. $n = 25$, $x = 40$, $1 - \alpha = 0.85$
   d. $n = 30$, $N = 500$, $x = 5$, $1 - \alpha = 0.94$

3. A fair coin was tossed 500 times and 255 heads appeared.
   a. Find E so that we can be 99% sure that p is within approximately E of $\hat{p}$.
   b. Construct an approximate 99% confidence interval for the probability of getting a head on one toss and interpret your result.

4. A fair coin was tossed 750 times and 371 heads appeared.
   a. Find E so that we can be 95% sure that p is within approximately E of $\hat{p}$.

   b. Construct an approximate 95% confidence interval for the probability of getting a head on one toss and interpret your result.

Further Applications

5. A survey of 672 audited tax returns showed that 448 resulted in additional payments. Construct an approximate 95% confidence interval for the true proportion of all audited tax returns that result in additional payments to the Internal Revenue Service. Also, find E so that we can be 95% confident that the true proportion p is approximately within E of $\hat{p}$.

6. An intrauterine device used by 500 women for the purpose of preventing pregnancy failed in 20 women. Construct an approximate 99% confidence interval for the true proportion of failures for this device and interpret your result. Also, find E so that we can be 95% confident that the true proportion p is approximately within E of $\hat{p}$.

7. If the device in exercise 6 is experimental and used by only 4000 women, construct an approximate 95% confidence interval for the true proportion of failures out of 500 and interpret your result. Compare this interval to the one obtained in exercise 4.

8. Suppose a random sample of 25 mining caps (explosive devices for setting off dynamite) was tested from a population of 300 mining caps and 20 exploded properly. Construct an approximate 95% confidence interval for the proportion of mining caps that will explode properly and interpret your result.

9. When sprayed with a certain type of insecticide, 38 of 60 Japanese beetles died. Construct an approximate 99% confidence interval for the true proportion of beetles that will die when exposed to the insecticide.

10. In a random poll taken in a large city, 428 of 975 people indicated they drink at least one cup of coffee a day. Construct an approximate 80% confidence interval for the true proportion of people who drink at least one cup of coffee per day.

11. A sample of voters was polled to determine the support for candidate A. Out of 140 voters surveyed, 74 expressed plans to vote for candidate A during the election. Construct an approximate 90% confidence interval for the proportion of the voters in the population who will vote for candidate A.

12. In a city of size 25,000, a random sample of 700 voters revealed that 420 opposed the reelection of their mayor. Construct an approximate 92% confidence interval for the proportion of all city voters who oppose the mayor's reelection.

13. In a sample of 350 students on a university campus who were questioned regarding their participation in athletics, 161 said they participate. Construct an approximate 96% confidence interval for the true proportion of students who participate in athletics.

14. A sample of 1168 rabbis revealed that 315 experience their work as very stressful. Construct an approximate 99% confidence interval for the true proportion of rabbis who find their work very stressful.

15. In a random sample of 249 Maryland smokers, it was found that 87% believe smoking is harmful to them. Construct an approximate 92% confidence interval for the true proportion of Maryland smokers who believe smoking is harmful to them.

16. In a random sample of 249 Maryland smokers, it was found that 147 believe smoking is hazardous to nonsmokers. Construct an approximate 96% confidence interval for the true proportion of Maryland smokers who believe that smoking is hazardous to nonsmokers.

17. On a certain day at the New York Stock Exchange, 2500 different stocks were traded. Of a random sample of 100 stocks, 70 declined in price. Construct an approximate 99% confidence interval for the proportion of all stocks that declined.

18. Researchers at the Centers for Disease Control (CDC) believe that the flu vaccine does not work as well for nursing-home residents as it does for younger, healthier people. The flue vaccine is usually 50% to 80% effective in warding off disease in the general population. Reasons for the lower effectiveness in the nursing home population include a decline with age in the body's immunity levels and the close contact of residents in a nursing home. In an effort to estimate the effectiveness of the flu vaccine in nursing-home residents, CDC researchers conducted a survey of 1068 nursing-home residents who received the flu vaccine and found that 269 came down with flu or flulike illness. Construct a 95% confidence interval for the effectiveness of the flu vaccine in nursing-home residents.

19. An unconventional new treatment for advanced liver cancer, a generally fatal disease, shows promise in shrinking tumors and has even produced a handful of apparent cures. Dr. Stanley Order of Johns Hopkins Hospital has indicated that the treatment, which involves injections of antibodies carrying radioactive isotopes, has significantly shrunk inoperable tumors in 50 of 104 patients.[38] Construct a 99% confidence interval for the true proportion of inoperable tumors that are significantly shrunk by Dr. Order's new treatment.

20. As one of the criteria for job employment as a teacher in Baltimore City, newly hired teachers must take a writing test to determine a teacher's ability to spell, punctuate, and construct a sentence. Of 158 newly-hired teachers in 1985, 32 were unable to score a passing grade on the test.[39] Construct a 95% confidence interval for the true proportion of all newly hired teachers who are unable to pass the test.

Going Beyond

21. By using the statement $P(-z_{\alpha/2} < Z < z_{\alpha/2}) = 1 - \alpha$ and the fact that $z = (\hat{p} - p)/\sigma_{\hat{p}}$, prove that $P(\hat{p} - z_{\alpha/2}\sigma_{\hat{p}} < p < \hat{p} + z_{\alpha/2}\sigma_{\hat{p}}) = 1 - \alpha$.

22. Suppose $P(|(p - \hat{p})/\sigma_{\hat{p}}| < z_{\alpha/2}) = 1 - \alpha$. The inequality $|p - \hat{p}| < z_{\alpha/2}\sigma_{\hat{p}}$ can be solved for p to provide a $(1 - \alpha)100\%$ confidence interval for p. Let z represent $z_{\alpha/2}$.
    a. Show that if $|p - \hat{p}| < z_{\alpha/2}\sigma_{\hat{p}}$, then $(n + z^2)p^2 - (2n\hat{p} + z^2)p + n\hat{p}^2 < 0$.
    b. If $f(p) = (n + z^2)p^2 - (2n\hat{p} + z^2)p + n\hat{p}^2$, show that the zeros p_0 of the quadratic function f are given by

$$p_0 = \frac{2n\hat{p} + z^2 + z\sqrt{4n\hat{p}(1 - \hat{p}) + z^2}}{2n + 2z^2}$$

c. Show that the two zeros found in part b determine a $(1 - \alpha)100\%$ confidence interval for p.

d. If n is large, show that the interval obtained in part c is approximately equal to the interval determined by expressions (9.12). (*Hint:* Divide the numerator and denominator of the expression in part b by $2n$ and observe the limiting value for p_0 as n gets large.)

23. Using the results in exercise 21, construct a 95% confidence interval for the true proportion of all defective type-B batteries in application 9.13 and compare your result with that obtained in application 9.13.

SECTION 9.4 — Determining Sampling Sizes for Estimates

Population Mean

How large should a sample be if the sample mean is to be used to estimate the population mean? The answer depends on the standard error of the mean. If the standard error of the mean is 0, then only one measurement would be needed; this measurement must necessarily equal the unknown mean μ, since $\sigma = 0$. This extreme case is not encountered in practice, but it supports the fact that the smaller the standard error of the mean, the smaller the sample size needed to achieve a desired degree of accuracy.

It was stated earlier that one way to decrease the error of estimate is to increase the size of the sample. If the sample included the whole population, then $\bar{x}$ would equal μ and the error of estimate $|\bar{x} - \mu|$ would equal 0. With this in mind, it seems reasonable that for a fixed confidence level, it should be possible to determine a sample size so that the error of estimate is as small as we want. To be more precise, given a fixed confidence level $1 - \alpha$ and a fixed error of estimate E, we can choose a sample size n so that $P\left(|\bar{x} - \mu| < E\right) = 1 - \alpha$. Toward the task of determining n, recall from section 9.1 that the maximum error of estimate is given by

$$E = z_{\alpha/2}\frac{\sigma}{\sqrt{n}}$$

If we square both sides of this equation and solve the resulting equation for n, we have

> **Sample Size in Order to Be $(1 - \alpha)100\%$ Confident that the Error of Estimate is at Most E**
>
> $$n = \left(\frac{z_{\alpha/2}\sigma}{E}\right)^2 \tag{9.14}$$

Since n must be a whole number, we round up all fractional parts.

EXAMPLE 9.8

If formula (9.14) produced a value of $n = 89.13$, we would round this value to $n = 90$. If instead of rounding up to 90, we used $n = 89$, we could not be $(1 - \alpha)100\%$ confident that μ is within E of $\bar{x}$. With the value $n = 90$, we can be *at least* $(1 - \alpha)100\%$ confident that μ is within E of $\bar{x}$.

Formula (9.14) requires that the population standard deviation σ be known. Since this is seldom the case, a preliminary sampling must be conducted with $n \geq 30$ in order to obtain an estimate s for σ. Application 9.16 illustrates the use of formula (9.14).

APPLICATION 9.16

A biologist wants to estimate the mean weight of deer killed in th
A preliminary study of ten deer killed showed the standard devia
to be 12.2 pounds. How large a sample should be taken so that
95% confident that the error of estimate is at most 4 pounds.

Solution: By using formula (9.14) we have

$$n = \left(\frac{z_{\alpha/2}\sigma}{E}\right)^2$$

$$= \left(\frac{(1.96)(12.2)}{4}\right)^2 = 35.736$$

Thus, if the size of the sample is $n = 36$, we can be 95% confident that μ is within 4
of $\bar{x}$. ∎

Population Proportion

We next want to determine the sample size n that is necessary in order to be
$(1 - \alpha)100\%$ confident that the error of estimate is at most E when $\hat{p}$ is used to estimate p. Stated more precisely, given the maximum error E that can be tolerated
and a fixed confidence level $1 - \alpha$, we want to find the sample size n so that
$P(|\hat{p} - p| < E) = 1 - \alpha$. Recall from section 9.3 that the maximum error of estimate
E is given by

$$E = z_{\alpha/2}\sqrt{\frac{p(1 - p)}{n}}$$

Squaring both sides and solving for n, we have

$$n = \left(\frac{z_{\alpha/2}}{E}\right)^2 p(1 - p)$$

Since we are trying to estimate p, which is unknown, we need a preliminary estimate
$\hat{p}$ for p in order to use the above formula. If a preliminary estimate $\hat{p}$ for p is available,
we can use the following formula to determine the sample size n:

> **Sample Size in Order to Be $(1 - \alpha)100\%$ Confident that the Error
> of Estimate is at Most E if a Prior Estimate of p is Known**
>
> $$n = \left(\frac{z_{\alpha/2}}{E}\right)^2 \hat{p}(1 - \hat{p}) \tag{9.15}$$

Again, we round up the value for n obtained by using (9.15) to the next larger integer
value.

   If a preliminary estimate $\hat{p}$ for p is not available, we could substitute the maximum
possible value of $p(1 - p)$ in formula (9.15) to obtain the sample size n. In most cases
this would probably make the sample size unnecessarily large for a given level of confidence. The value of n would then be a conservative estimate in the sense that it is
larger than what it needs to be; in this case, we can say that we are at least
$(1 - \alpha)100\%$ confident that p is within E of $\hat{p}$. The maximum value of $p(1 - p)$ is

easily found to be 1/4 (see exercise 14 at the end of this section). By using this maximum value, we obtain the following sample size formula:

> **Sample Size in Order to Be $(1 - \alpha)100\%$ Confident that the Error of Estimate is at Most E if a Prior Estimate of p is Unknown**
>
> $$n = \left(\frac{z_{\alpha/2}}{2E}\right)^2 \tag{9.16}$$

EXAMPLE 9.9

For the *Sun's* poll to determine the proportion of adult smokers in Maryland (discussed in section 9.3), we can use formula (9.16) to determine the size of sample to be polled in order to be 95% confident that the error of estimate is no more than 3.5 percentage points. The size of the random sample to be polled is

$$n = \left(\frac{z_{\alpha/2}}{2E}\right)^2$$
$$= \left(\frac{1.96}{0.07}\right)^2 = 784$$

Since the *Sun's* poll involved 806 adults, which is 22 more than the required sample size we calculated, we can be at least 95% confident that the true proportion of adult smokers in Maryland is within 0.035 of $\hat{p} = 0.31$.

APPLICATION 9.17

A large teachers' organization wants to estimate the percentage of its membership who are in favor of collective bargaining. The union wants to be certain that the error of estimate is at most 1.5% with a confidence level of 95%.

a. If no preliminary estimate for p is available, how large a sample must be polled?

b. If a preliminary sample of 200 teachers indicated that 65% favored collective bargaining, how many more teachers should be polled?

Solution:

a. When no preliminary estimate of p is available, we use formula (9.16). Thus,

$$n = \left(\frac{z_{\alpha/2}}{2E}\right)^2$$
$$= \left(\frac{1.96}{(2)(0.015)}\right)^2 \approx 4268.44 = 4269$$

b. In this case since we have a preliminary estimate, $\hat{p} = 0.65$, we can use formula (9.15). Hence, the sample size is

$$n = \left(\frac{z_{\alpha/2}}{E}\right)^2 \hat{p}(1 - \hat{p})$$
$$= \left(\frac{1.96}{0.015}\right)^2 (0.65)(0.35) \approx 3884.28 = 3885$$

Thus, we need to poll $3885 - 200 = 3685$ more teachers in order to achieve the desired accuracy. Note that by having a preliminary estimate for p, we can sample 384 fewer teachers than required for the same nominal level of confidence when no preliminary estimate for p is available. ∎

EXERCISE SET 9.4

Further Applications

1. A biometrician wants to estimate the percentage of oxygen in the blood of all newborn babies, and prior research has suggested that this value is $p = 68\%$.
   a. If it is desired to be off by no more than 2 percentage points with a confidence level of 95%, find the sample size needed.
   b. If no prior estimate of p is available, how large should be sample be?

2. In an area where water is not fluoridated a random sample of citizens is to be surveyed to estimate the percentage of citizens who favor having the water fluoridated.
   a. How large a sample is needed to be 99% confident that the true percentage is within 1 percentage point of the sample percentage?
   b. If a prior estimate of p is known to be equal to 0.85, how large a sample is needed?

3. We want to determine the mean weight of a certain type of fish. If we also want to estimate μ to within 0.05 gram with a confidence level of 0.80, how large must the sample be if σ is approximately 2.5?

4. If we want in exercise 3 to estimate μ so that the width of the confidence interval is no more than 0.12 grams with a confidence level of 96%, how large must be sample be?

5. The amount of time that a doctor spends with patients has a standard deviation of approximately 7.8 minutes. If we want to estimate the mean time doctors spend with their patients, find the sample size that is needed to be 88% confident that the true mean is within 2.5 minutes of the estimate.

6. A random sample of 50 lawyers indicated that the average length of law experience is 9.6 years and the standard deviation is 4.4 years. How many more lawyers should be included in the sample to be 99% confident that the true mean work experience is within 6 months of the sample mean work experience?

7. A student group wants to estimate the percentage of students living in dormitories who are dissatisfied with cafeteria food. Determine the number to be surveyed if the student group wants to be 92% confident that the true proportion p is within 1 percentage point of the estimate $\hat{p}$.

8. A seed company wants to estimate the proportion of its cucumber seeds that will germinate. How large a sample of cucumber seeds should be used if the seed company wants to be 94% confident that the error of estimate is at most 0.05?

9. A random sample of 75 medical doctors in Chicago indicated they worked an average of 51.2 hours per week; the standard deviation of the number of hours worked was 6.8 hours. If we want to gather another random sample of doctors in Chicago to estimate the true average number of hours worked per week, how large a sample is needed in order to be 93% confident that the true mean is within 0.5 hour of the sample mean?

10. Doctors at Sinai Hospital in Baltimore have been studying the effects of doses of vitamin E in the treatment of the ailment of noncancerous breast lumps in females. When treating 26 patients with 600 international units (i.u.) of vitamin E daily for eight weeks, they found that 22 were good or fair responders to the treatment and 4 were nonresponders. How many more patients should be tested in order for the doctors to be 91% confident that the sample proportion of female patients taking 600 i.u. of vitamin E daily for eight weeks who are good or fair responders to the treatment differs no more than 0.10 from the true proportion?

Going Beyond

11. An opinion survey conducted by a college reported that 1067 students were surveyed and that the error factor was 3%. Verify that an error factor of 3% is approximately correct. (*Hint:* Use $\alpha = 0.5$ and find E.)

12. Suppose sampling is done without replacement from a finite population of size N. Show that in order to be $(1 - \alpha)100\%$ confident that p will differ from $\hat{p}$ by at most E, the sample size should be at least $n = N z^2_{\alpha/2}/[z^2_{\alpha/2} + 4(N - 1)E^2]$ when no prior information concerning p is available.

13. A labor organization that has 4000 members wants to conduct a survey to estimate the percentage of its members favoring a strike. What is the minimum number of members that should be included in a sample so that the organization can be 98% confident that the true proportion p will be within 4 percentage points of the sample estimate $\hat{p}$?

14. Show that the maximum value of $p(1 - p)$ is $1/4$.

SECTION 9.5   *Chi-Square Distributions*

In order to estimate the population variance or standard deviation, we need to be familiar with the χ^2 statistic. If a sample of size n is selected from a normal population with variance σ^2, the statistic

$$\frac{(n-1)s^2}{\sigma^2}$$

has a sampling distribution that is a **chi-square distribution** with **degrees of freedom** df $= n - 1$ and is denoted by χ^2 (χ is the lowercase Greek letter chi). The **chi-square statistic** is given by

$$\chi^2 = \frac{(n-1)s^2}{\sigma^2}$$

where n is the sample size, s^2 is the sample variance, and σ^2 is the variance of the sampled population. Note that since $s^2 = SS/(n-1) = \Sigma(x - \bar{x})^2/(n-1)$, the chi-square statistic can also be written as either of the following two expressions:

$$\chi^2 = \frac{SS}{\sigma^2} \quad \text{and} \quad \chi^2 = \frac{\Sigma(x - \bar{x})^2}{\sigma^2}$$

The chi-square distributions have the following properties:

Properties of the Chi-Square Distributions

1. The values of χ^2 are greater than or equal to 0.
2. The shapes of the χ^2 distributions depend on df $= n - 1$. As a result, there are an infinite number of χ^2 distributions.
3. The area under a chi-square curve and above the horizontal axis is 1.
4. The χ^2 distributions are not symmetric. They have narrow tails extending to the right; that is, they are skewed to the right.
5. When $n > 2$, the mean of a χ^2 distribution is $n - 1$ and the variance is $2(n - 1)$.
6. The modal value of a chi-square distribution occurs at the value $(n - 3)$.

Figure 9.6 illustrates three χ^2 distributions. Notice that the modal value for a χ^2 distribution occurs at the value $(n - 3) = (df - 2)$.

FIGURE 9.6

Graphs of chi-square
distributions

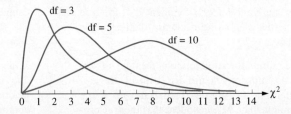

MINITAB can be used to simulate an empirical chi-square distribution. Computer display 9.5 contains the commands used and the corresponding output. One thousand random samples of size 11 were selected from a normal population having a mean of 40 and a standard deviation of 2. For each sample, the value of $(11 - 1)s^2/4 = SS^2/2$ was computed, and these 1000 values were stored in C14. The values stored in C14 constitute an empirical sampling distribution of the statistic $SS^2/2$. The empirical sampling distribution has a mean of 9.9317 and a variance of 19.5542. These values compare quite favorably with the corresponding parameters of the theoretical sampling distribution. By using property 5 above, the mean of the chi-square distribution is

$$\mu = n - 1$$
$$= 11 - 1 = 10$$

and the variance is

$$\sigma^2 = 2(n - 1)$$
$$= 2(10) = 20$$

Notice also that the midpoint of the modal class of the histogram is equal to $(n - 3) = 8$. This also agrees with the theoretical result.

Computer Display 9.5

```
MTB > RANDOM 1000 C1-C11;
SUBC> NORMAL 40 2.
MTB > RSTDEV C1-C11 C12
MTB > LET C13 = C12**2
MTB > LET C14 = 10*C13/4
MTB > MEAN C14
 MEAN = 9.9317
MTB > LET K1 = STDEV (C14)**2
MTB > PRINT K1
K1   19.5542
MTB > HISTOGRAM C14

HISTOGRAM OF C14  N = 1000
EACH * REPRESENTS 5 OBS.

MIDPOINT         COUNT
    2             20 ****
    4             88 ******************
    6            166 *********************************
    8            205 *******************************************
   10            167 **********************************
   12            146 *****************************
   14             89 ******************
   16             47 *********
   18             35 *******
   20             15 ***
   22              8 **
   24              8 **
   26              2 *
   28              4 *
```

Table 5 in appendix **B** gives critical values $\chi^2_\alpha(\mathrm{df})$ for ten selected values of α. The symbol $\chi^2_\alpha(\mathrm{df})$ is used to denote the critical value for a chi-square distribution with degrees of freedom df. This critical value specifies an area of α under the χ^2 curve above the horizontal axis and to the right of it (see figure 9.7). For example, to find $\chi^2_{0.05}(6)$ in table 5, we locate df = 6 at the left side of the table and $\alpha = 0.05$ across the top of the table. The critical value is found where the row labeled by df = 6 intersects the column labeled by α = 0.05, the critical value is found, as illustrated in the following diagram:

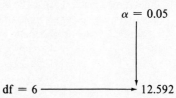

$$df = 6 \longrightarrow 12.592$$

Thus, the critical value is $\chi^2_{0.05}(6) = 12.592$.

FIGURE 9.7

Critical values of chi-square distribution

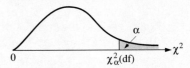

APPLICATION 9.18   Find the critical value $\chi^2_{0.10}(13)$ in table 5 of appendix B.

Solution: Locate the row corresponding to df = 13 and the column corresponding to α = 0.10. Their intersection in the table determines the critical value; it is found to be $\chi^2_{0.10}(13) = 19.812$. We have the following diagram for the χ^2 distribution:

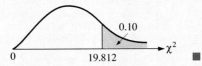

APPLICATION 9.19   Suppose the times required for a particular bus to arrive at one of its destinations in a large city form a normal distribution with a standard deviation of σ = 1 minute. If a sample of 17 bus arrival times is selected at random, find the probability that the sample variance is greater than 2; that is, find $P(s^2 > 2)$.

Solution: We find the chi-square value corresponding to $s^2 = 2$ as follows:

$$\chi^2 = \frac{(n-1)s^2}{\sigma^2}$$

$$= \frac{(17-1)(2)}{(1)^2} = 32$$

Thus,

$$P(s^2 > 2) = P(\chi^2 > 32)$$

Using table 5 and reading across the row labeled by df = 17 − 1 = 16, we locate 32.000. Then we look up to find the corresponding value of α, which is 0.01. Hence, the probability value is $P(s^2 > 2) = 0.01$.   ■

Confidence Intervals for σ^2 and σ

Often we need to estimate the variance of a given population of measurements. A company that manufactures nuts and bolts must have close tolerances on all produced parts. A bolt that is 8.4 mm in diameter will not screw into a 8-mm hole.

EXAMPLE 9.10

Suppose a bolt manufacturer is producing 8-mm diameter bolts and the bolt diameters are normally distributed. For quality control purposes, a random sample of 25 bolts is obtained from a production line to estimate the variance of all 8-mm bolt diameters. The variance for the sample of bolt diameters is $s^2 = 0.009$ square mm. The sample variance $s^2 = 0.009$ provides a point estimate for σ^2, the variance of the diameters of all 8-mm bolts produced. Toward obtaining a 95% confidence interval for σ^2, consider figure 9.8.

FIGURE 9.8

Critical values for 95% confidence interval for σ^2

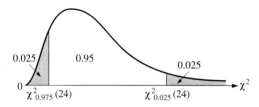

We want to determine two critical values so that the area between them is 0.95 and the "tail" regions each contain an area of 0.025. Note that the critical value $\chi^2_{0.9754}(24)$ has an area of 0.975 to its right and an area of 0.025 to its left ($0.975 + 0.025 = 1$). In general, if $\chi^2_{\alpha/2}(\text{df})$ denotes the right-tail critical value, $\chi^2_{1-\alpha/2}(\text{df})$ will denote the left-tailed critical value. By using table 5, we find the critical values to be

$$\chi^2_{0.975}(24) = 12.401 \quad \text{and} \quad \chi^2_{0.025}(24) = 39.364$$

We will use these values to construct a 95% confidence interval for σ^2. The confidence interval (L_1, L_2) will be such that

$$P(L_1 < \sigma^2 < L_2) = 0.95$$

L_1 represents the smallest value that σ^2 can be and L_2 represents the largest value that σ^2 can be. Since the χ^2 statistic is defined by

$$\chi^2 = \frac{(n-1)s^2}{\sigma^2}$$

we can solve this equation for σ^2 in order to obtain the confidence limits L_1 and L_2. By doing so, we have

$$\sigma^2 = \frac{(n-1)s^2}{\chi^2} \qquad (9.17)$$

To find the confidence limits L_1 and L_2, we substitute the chi-square critical values for χ^2 in formula (9.17). The limits become

$$\sigma^2 = \frac{(n-1)s^2}{\chi^2}$$

$$= \frac{(25-1)(0.009)}{39.364} = 0.0055$$

and

$$\sigma^2 = \frac{(n-1)s^2}{\chi^2}$$

$$= \frac{(25-1)(0.009)}{12.401} = 0.0174$$

The smaller value obtained by using equation (9.17) is the lower confidence limit L_1 and the larger value is the upper confidence limit L_2. Thus, $L_1 = 0.006$, $L_2 = 0.017$, and a 95% confidence interval for σ^2 is (0.006, 0.017). If this is acceptable to the plant engineer and if at some future date a sample of bolt diameters produces a sample variance of $s^2 = 0.020$, this may be an indication that something is wrong, since the value 0.020 is not contained in the 95% confidence interval (0.006, 0.017).

In general, to find a $(1 - \alpha)100\%$ confidence interval (L_1, L_2) for the variance σ^2 of a normal population, the confidence limits are found by using L_1 and L_2 given by the following expressions:

> **Limits of $(1 - \alpha)\%$ Confidence Interval for σ^2**
>
> $$L_1 = \frac{(n-1)s^2}{\chi^2_{\alpha/2}(\text{df})}$$
>
> $$L_2 = \frac{(n-1)s^2}{\chi^2_{1-\alpha/2}(\text{df})}$$
>
> where n is the sample size and s^2 is the sample variance.

(9.18)

To obtain a confidence interval for the population standard deviation σ of a normal population, we use the positive square roots of the limits found in (9.18). A confidence interval for σ is given by

> **Confidence Interval for σ**
>
> $$(\sqrt{L_1}, \sqrt{L_2})$$
>
> where L_1 and L_2 are defined by equations (9.18)

(9.19)

APPLICATION 9.20

The variation of the potency of a drug must be small; otherwise, the drug could be harmful or ineffective. A pharmaceutical company was interested in determining the variance of potency measurements for a new batch of a certain drug. Toward this end, a random sample of 20 cc vials of the drug produced a variance equal to 0.0018. Construct a 99% confidence interval for the variance of the drug's potency measurements.

Solution: Since $1 - \alpha = 0.99$, $\alpha = 0.01$, $\alpha/2 = 0.005$, and $1 - \alpha/2 = 0.995$. In addition, df $= n - 1 = 20 - 1 = 19$. The critical values $\chi^2_{0.005}(19)$ and $\chi^2_{0.995}(19)$ are found in table 5 of appendix B.

$$\chi^2_{0.005}(19) = 38.582 \quad \text{and} \quad \chi^2_{0.995}(19) = 6.844$$

By dividing the value $(n - 1)s^2$ by the critical values, we obtain the confidence interval limits L_1 and L_2. Thus, we have

$$\frac{(n-1)s^2}{\chi^2} = \frac{(19)(0.0018)}{38.58} = 0.0008864$$

and

$$\frac{(n-1)s_2}{\chi^2} = \frac{(19)(0.0018)}{6.844} = 0.0050$$

Hence, $L_1 = 0.00089$ and $L_2 = 0.005$. Thus, a 99% confidence interval for σ^2 is (0.00089, 0.005).

The confidence-interval limits for σ are found by taking the positive square roots of the limits L_1 and L_2:

$$(\sqrt{L_1}, \sqrt{L_2}) = (\sqrt{0.00089}, \sqrt{0.005}) = (0.030, 0.071)$$

Thus, $(0.030, 0.071)$ is a 99% confidence for σ. ▪

EXERCISE SET 9.5

Basic Skills

1. Find the critical values $\chi^2_{1-\alpha/2}(df)$ and $\chi^2_{\alpha/2}(df)$ if
   a. $n = 17$ and $1 - \alpha = 99\%$.
   b. $n = 25$ and $1 - \alpha = 95\%$.

2. Find the critical values $\chi^2_{1-\alpha/2}(df)$ and $\chi^2_{\alpha/2}(df)$ if
   a. $n = 11$ and $1 - \alpha = 95\%$.
   b. $n = 18$ and $1 - \alpha = 90\%$.

3. Construct 90% confidence intervals for σ^2 and σ in each of the following cases where the statistics resulted from samples drawn from normal populations.
   a. $s = 0.15$, $n = 25$ b. $s = 28.7$, $n = 14$
   c. $s = 1.01$, $n = 38$ d. $s = 842$, $n = 29$

4. Construct 95% confidence intervals for σ^2 and σ in each of the following cases where the statistics resulted from samples drawn from normal populations.
   a. $s = 0.20$, $n = 17$ b. $s = 2.03$, $n = 21$
   c. $s^2 = 5.07$, $n = 24$ d. $s^2 = 56.4$, $n = 22$

5. Assume that $n = 41$, $\bar{x} = 72.4$, and $s^2 = 2.6$ result from a random sample taken from a normal population.
   a. What is the best point estimate for σ?
   b. Construct a 90% confidence interval for σ.

6. Assume that $n = 35$, $\bar{x} = 12.2$, and $s^2 = 1.2$ result from a random sample taken from a normal population.
   a. What is the best point estimate for σ?
   b. Construct a 95% confidence interval for σ.

7. A sample of 20 castings is taken from a production run known to produce flaws that are normally distributed with a standard deviation of 5 flaws. If the sample has a mean of 4 flaws and a standard deviation of 2.83 flaws, find the value of the χ^2 statistic.

8. A sample of 35 tires is taken from a production run known to produce flaws that are normally distributed with a standard deviation of 1.2 flaws. If the sample has a mean of 1.5 flaws and a standard deviation of 0.9 flaws, find the value of the χ^2 statistic.

Further Applications

9. A sample of 12 cans of soup produced by the XYZ Soup Company yielded the following net weights, measured in ounces:

11.9	12.2	11.6	12.1	12.1	11.8
11.9	11.8	12.0	12.3	11.8	12.0

   Assuming the net weights are normally distributed, construct 95% confidence intervals for the variance and standard deviation of the population of net weights of all cans of soup produced by the company.

10. The daily arrival times for a certain train at one of its destinations are normally distributed. A sample of 12 train arrival times indicated $s = 1.789$ minutes. Construct a 99% confidence interval for the variance of the population of arrival times.

11. The weights of a certain species of fish are known to be normally distributed with a standard deviation of 2 grams. If a sample of 17 fish is selected from the population, find the probability that the sample variance is less than 8.

12. Suppose the number of hours that teenagers spend watching television per week has a normal distribution with a variance of 3. If a sample of 17 teenagers is selected, and the number of hours per week of viewing television is recorded for each, find the probability that the sample standard deviation of viewing times is greater than $\sqrt{6}$.

13. A precision meter is guaranteed to be accurate within 2% of full scale. A sample of five meter readings on the same object yielded the following measurements: 350, 348, 348, 352, and 351. Construct a 95% confidence interval for σ, assuming the population of measurements is normally distributed.

14. The concentrations of artificial food coloring in six lots were recorded as follows: 0.010, 0.013, 0.018, 0.014, 0.015, and 0.013. If the food coloring concentrations are assumed to

be normally distributed, construct a 95% confidence interval for the variance of concentrations for the population of lots.

15. For the data in exercise 14, construct a 90% confidence interval for the standard deviation of the concentrations for the population of lots.

Going Beyond

16. If $P(\chi^2_{1-\alpha/2}(df) < \chi^2 < \chi^2_{\alpha/2}(df)) = 1 - \alpha$ and $\chi^2 = (n - 1)s^2/\sigma^2$, prove that $P((n - 1)s^2/\chi^2_{\alpha/2}(df) < \sigma^2 < (n - 1)s^2/\chi^2_{1-\alpha/2}(df)) = 1 - \alpha$.

17. When $df > 30$, the statistic $\sqrt{2\chi^2} - \sqrt{2df - 1}$ has a sampling distribution that is approximately the standard normal. That is,

$$z \approx \sqrt{2\chi^2} - \sqrt{2\ df - 1}$$

Use this result to find $\chi^2_{0.05}(40)$ and compare the result to that contained in the chi-square table. (*Hint:* Let $z = 1.65$, $df = 40$, and solve for the value of χ^2.)

18. For large values of n, the sampling distribution of

$$\frac{\chi^2 - (n - 1)}{\sqrt{2(n - 1)}}$$

is approximately the standard normal. Use this fact to find $\chi^2_{0.05}(40)$ and compare the result to that contained in the chi-square table.

19. For samples of size $n \geq 30$, the sampling distribution of s is approximately normal with $\mu_s = \sigma$ and $\sigma_s = \sigma/\sqrt{2n}$. Use this result to solve exercise 5, part b, and compare the results obtained by using the two procedures.

20. Refer to exercise 14. For this sample of data, how many 95% confidence intervals for σ^2 can be constructed? Is the 95% confidence interval constructed in exercise 14 of minimum width?

CHAPTER SUMMARY

In this chapter we learned how to construct point and interval estimates for μ, σ^2, σ, and p. The estimates result from random samples taken from populations and are used to estimate characteristics of the populations, such as the mean and variance. A confidence interval shows a range of values within which the population parameter might fall. A given parameter is either contained in a confidence interval or it is not. Being 95% confident that a confidence interval contains the population mean does not mean that the probability the interval contains μ is 95%; rather, it means that 95% of all such intervals con-

structed will contain μ. All confidence intervals for μ and p have the following form:

estimate $\pm$ [$z_{\alpha/2}$ or $t_{\alpha/2}$] (standard error of the estimate)

We learned also how to determine the sample size needed to construct confidence intervals of a given width. Finally, we learned how to construct confidence intervals for the variance and standard deviation of a normal population. The following table summarizes the ideas used in the chapter.

To Estimate	Employ Sampling Distribution	Conditions	Use	Formula Number
μ	$\bar{x}$	$n \geq 30$ or the population is normal	*assume σ^2 known* ↓ $\bar{x} \pm z_{\alpha/2}(\sigma/\sqrt{n})$ or $\bar{x} \pm z_{\alpha/2}(s/\sqrt{n})$ ←don't know σ^2	(9.6)
		$n < 30$, σ is unknown and the population is normal	$\bar{x} \pm t_{\alpha/2}(s/\sqrt{n})$	(9.8)
p *(proportions)*	$\hat{p}$	The population is large	$\hat{p} \pm z_{\alpha/2}\sqrt{\hat{p}(1 - \hat{p})/n}$	(9.12)
σ^2	χ^2	The population is normal	(L_1, L_2) where $L_1 = (n - 1)s^2/\chi^2_{\alpha/2}(df)$ and $L_2 = (n - 1)s^2/\chi^2_{1-\alpha/2}(df)$	(9.18)
σ	χ^2	The population is normal	$(\sqrt{L_1}, \sqrt{L_2})$	(9.19)

To Determine	Conditions	Use	Formula Number
Sample size n so that $P(\lvert x - \mu \rvert < E) = 1 - \alpha$	σ is known	$n = (z_{\alpha/2}\sigma/E)^2$	(9.14)
	σ is unknown, $n \geq 30$, and have an estimate for σ	$n = (z_{\alpha/2}s/E)^2$	
Sample size n so that $P(\lvert \hat{p} - p \rvert < E) = 1 - \alpha$	Have an estimate for p	$n = (z_{\alpha/2}/E)^2\,\hat{p}(1 - \hat{p})$	(9.15)
	Have no estimate for p	$n = (z_{\alpha/2}/2E)^2$	(9.16)

CHAPTER REVIEW

■ IMPORTANT TERMS ■

The following chapter terms have been mixed in ordering to provide you better review practice. For each of the following terms, provide a definition in your own words. Then check your responses against the definitions given in the chapter.

reliability
chi-square distribution
confidence level
critical value
degrees of freedom
$(1 - \alpha)100\%$ confidence interval for μ

error of estimate
estimate
estimation
estimator
interval estimate
lower confidence limit

maximum error of estimate
point estimate
statistical inference
unbiased estimator
upper confidence limit

■ IMPORTANT SYMBOLS ■

$\lvert \bar{x} - \mu \rvert$, error of estimate when estimating μ

$z_{\alpha/2}$, critical value for z

$1 - \alpha$, confidence level

E, maximum error of estimate

$\sigma_{\bar{x}}$, the standard error of $\bar{x}$

$t_{\alpha/2}$, critical value for t

L_1, lower confidence limit

L_2, upper confidence limit

$\sigma_{\hat{p}}$, the standard error of $\hat{p}$

$\hat{\sigma}_{\hat{p}}$, an estimate of $\sigma_{\hat{p}}$

$\lvert \hat{p} - p \rvert$, error of estimate when estimating p

df, degrees of freedom

χ^2, chi-square statistic

$\chi^2_{1-\alpha/2}(\mathrm{df})$, left-tail critical value for χ^2

$\chi^2_{\alpha/2}(\mathrm{df})$, right-tail critical value for χ^2

■ IMPORTANT FACTS AND FORMULAS ■

Maximum error of estimate for large samples when $\bar{x}$ is used to estimate μ: $E = z_{\alpha/2}\dfrac{\sigma}{\sqrt{n}}$.

Maximum error of estimate for small samples when $\bar{x}$ is used to estimate μ: $E = t_{\alpha/2}\dfrac{s}{\sqrt{n}}$.

Confidence interval limits for μ (σ is known):

$$\bar{x} \pm z_{\alpha/2}\frac{\sigma}{\sqrt{n}}. \quad (9.6)$$

Confidence interval limits for μ (σ is unknown):

$$\bar{x} \pm t_{\alpha/2}\frac{s}{\sqrt{n}}. \quad (9.8)$$

Approximate maximum error of estimate if $\hat{p}$ is used to estimate p:

$$E = z_{\alpha/2}\sqrt{\frac{\hat{p}(1 - \hat{p})}{n}}. \quad (9.11)$$

Approximate confidence interval limits for p:
$$\hat{p} \pm z_{\alpha/2}\sqrt{\hat{p}(1 - \hat{p})/n}. \quad (9.12)$$

Approximate standard error of estimate of $\hat{p}$ when sampling without replacement:

$$\hat{v}_{\hat{p}} \approx \sqrt{\frac{\hat{p}(1 - \hat{p})}{n}}\sqrt{\frac{N - n}{N - 1}}. \quad (9.13)$$

Sample size needed in order to estimate μ:
$$n = (z_{\alpha/2}\sigma/E)^2. \quad (9.14)$$

Sample size needed in order to estimate p when a prior estimate of p is available: $n = (z_{\alpha/2}/E)^2\hat{p}(1 - \hat{p})$. (9.15)

Sample size needed in order to estimate p when no prior estimate of p is available: $n = (z_{\alpha/2}/2E)^2$. (9.16)

Chi-square statistic: $\chi^2 = \dfrac{(n - 1)s^2}{\sigma^2}$, df $= n - 1$.

Confidence interval limits for σ^2: $L_1 = \dfrac{(n - 1)s^2}{\chi^2_{\alpha/2}(\text{df})}$

$$L_2 = \frac{(n - 1)s^2}{\chi^2_{1-\alpha/2}(\text{df})} \quad (9.18)$$

Confidence interval for σ: $(\sqrt{L_1}, \sqrt{L_2})$. (9.19)

REVIEW EXERCISES

1. **a.** In a study to estimate the true proportion of U.S. citizens who are in favor of gun control legislation, a random sample of 5000 people yielded 2800 in favor of gun-control legislation. Construct an approximate 90% confidence interval for the true proportion of U.S. citizens who are in favor of gun control legislation.

   b. How large a sample must be selected if we want to be 95% confident that the error of estimate is less than 0.01? Use the value of $\hat{p}$ obtained in part a to estimate p.

2. A random sample of 125 college students was taken to determine the percentage of students who receive financial aid at a certain college. It was found that 88 received some form of financial aid. Construct an approximate 99% confidence interval for the true proportion of college students receiving financial aid.

3. Refer to review exercise 2. How large a sample is needed if we want to estimate the true proportion to within 0.05 with a confidence level of 99% and no preliminary estimate for p is available?

4. In order to determine if ball bearings are conforming to manufacturing specifications, a sample of 25 ball-bearing diameters was inspected, yielding a variance of 0.0015. Construct a 95% confidence interval for the true variance of all ball-bearing diameters manufactured. Assume that the ball-bearing diameters are normally distributed.

5. In order to determine the proportion of students at a large university who are in default on their government loans, administrators selected a sample of 380 former students. The investigation indicated that 150 loans were in default. Construct an approximate 95% confidence interval for the true proportion of university students who are in default on their loans. What is the maximum error of estimate?

6. A new cigarette was just marketed. A study to determine the mean nicotine content of 35 cigarettes yielded $\bar{x} = 25.4$ mg and $s = 1.9$ mg.
   a. Find E so that we can be 95% confident that μ is within E of $\bar{x}$.
   b. Construct a 95% confidence interval for the true average nicotine content.
   c. Using $s = 1.9$ to estimate σ, find the sample size needed in order to be 95% confident that the error of estimate is less than 0.01 mg.

7. In order to estimate the variance in the diameters of screw-top lids for medicine bottles, a sample of 20 lids was examined, yielding a sample variance of 0.18 square mm. Assuming the lid diameters are normally distributed, construct a 99% confidence interval for the true variance of lid diameters.

8. A large university wants to estimate the mean number of days that students are sick each school year. A sample of 500 students indicated that $\bar{x} = 2.3$ days and $s = 10.2$ days.
   a. Find E so that we can be 90% confident that when $\bar{x}$ is used to estimate μ, the error of estimate is less than E.
   b. Estimate μ by constructing a 90% confidence interval for μ.
   c. By using $s = 10.2$ as a point estimate for σ, find the sample size needed in order to be 90% confident that the maximum error of estimate is 0.25 day.

9. On a final examination for a calculus course at a large university, a sample of 45 tests was graded early to estimate the average grade. From previous experience, it had been determined that $\sigma = 14.5$. If the 45 tests produced a mean of 62 and a standard deviation of 17.1, construct a 95% confidence interval for the true average test grade for all calculus students at the university.

10. A sample of 75 customers at a certain gasoline station indicated that the mean number of gallons purchased was 14.3 gallons and the standard deviation was 2.7 gallons.
    a. By using $\bar{x}$ to estimate μ, find the maximum error of estimate E so that μ is within E of $\bar{x}$ with a confidence level of 95%.
    b. Construct a 95% confidence interval for the true average number of gallons of gasoline purchased.
    c. By using $s = 2.7$ to estimate σ, find the sample size needed in order to be 95% confident that the maximum error of estimate is less than 0.2 gallon.

11. Shown below are the durations (in minutes) for a sample of telephone orders at a large department store. Assuming the lengths of telephone calls to place orders are normally distributed, construct a 95% confidence interval for the true mean time needed to place telephone orders.

    10.1   5.0   4.3   4.2   3.6   6.7   3.1   2.8   4.2

12. Sales personnel at a large company are required to submit weekly reports concerning the number of customer contacts. A random sample of 78 weekly reports showed an average of 27.2 customer contacts per week; the standard deviation was 7.2 contacts.
    a. By using the value 27.2 to estimate the true average, find E so that we can be 90% confident that μ is within E of $\bar{x} = 27.2$.
    b. Construct a 90% confidence interval for μ.
    c. By using $s = 7.2$ to estimate σ, find the sample size necessary to be 90% confident that E is less than 0.5 customers.

13. The diameters of a certain type of bolt produced by a large firm are normally distributed. Suppose a sample of 20 bolts yields $\bar{x} = 8.02$ mm and $s = 0.06$ mm.
    a. Find E so that we can be 95% confident that μ is within E of $\bar{x}$.
    b. Estimate μ by constructing a 95% confidence interval.

14. A random sample of monthly rents for one bedroom apartments in a large city was taken to estimate the average rent. Suppose $n = 800$, $\bar{x} = \$235.10$, and $s = \$47.55$.
    a. Find E so that we can be 85% confident that μ is within E of $\bar{x}$.
    b. Estimate μ by constructing a 85% confidence interval.
    c. If s is used as a point estimate for σ, find the sample size needed in order to be 85% confident that E is less than \$5.00.

15. In a survey to determine the proportion of viewers for a new television program, a sample of 575 households indicated that 238 were watching the program.
    a. Construct an approximate 99% confidence for the true proportion of households watching the program.
    b. If no preliminary estimates are available, how large a sample must be polled in order to be 95% confident that E is less than 3%?

16. Of 125 people who were given a vaccine for type-A flu virus, 100 developed immunity to the virus. Construct an approximate 95% confidence interval for the true proportion of those vaccinated who develop immunity.

Computer Applications

1. The accompanying data represent the pulse rates (in beat per minute) of a random sample of 100 college students.

```
68  80  68  89  82  61  67  80  76    74  63
74  90  74  78  73  64  71  93  60    74  87
74  83  73  88  90  80  87  81  68  100  79
70  91  83  68  79  74  79  65  84    71  74
62  69  95  76  97  62  71  71  80    85  67
78  57  66  76  75  84  84  62  71    82  82
67  66  86  75  60  60  85  76  80    95  79
79  74  71  79  91  74  62  75  80    75  77
70  71  72  74  73  61  64  67  91    70  79
77
```

Use a computer program to construct a 95% confidence interval for the population mean. Based on this confidence interval, would it be very likely that $\mu = 75$ beats per minute?

2. The accompanying data represent the systolic blood pressure measurements of a random sample of 100 college students.

```
 93   96  141   95  106  105  108  100  140  121  178
112  119   95  126  137  106  143  127  114  123  118
147  136   98  126  159  110  166  130  135  116   95
124  119  125   96  146  139  128  105  114  117  145
149  113  135  101  107  130   98  118   94  100   94
 97   90  127  104  126  133  122  128  119  133   96
134   91  130  111  135  119  144   62  149  102   87
117  146  113  147  144  133  101  109  117  108  122
110  136  130  115   73   91  112  170  151  110  120
135
```

Use a computer program to construct a 90% confidence interval for the population mean. Is it very likely that $\mu = 120$?

3. Use a computer program to simulate drawing 100 random samples of size 35 from a normal population having a mean of 100 and a standard deviation of 10. For each sample, construct a 90% confidence interval for the population mean. How many of these confidence intervals contain $\mu = 10$? How many intervals out of 100 intervals would you expect to contain the population mean?

4. Use a computer program to simulate drawing 100 random samples of size 10 from a normal population having a mean of 100. For each sample, construct a 90%

confidence interval for the population mean. How many of these confidence intervals contain $\mu = 10$? How many intervals out of 100 intervals would you expect to contain the population mean?

5. Use a computer software package to simulate an empirical chi-square distribution with df $= 10$ having 1000 elements. Construct a histogram and compare it to the shape of the theoretical distribution by comparing the modal values of each. Then determine the 95th percentile of the empirical distribution, and compare this value to the theoretical value obtained from table 5 of appendix B.

6. **a.** Simulate an empirical χ^2 distribution having df $= 12$ by drawing 100 random samples of size 5 from the normal distribution having a mean of 50 and a variance of 3 and calculating the value of $(n - 1)s^2/\sigma^2$ for each sample.
   b. Determine the mean, mode, and standard deviation of the 100 values and construct a histogram of the results.
   c. Compare your results with the theoretical results.

7. **a.** Repeat computer application 6, part a, for samples of samples of size 20. Construct a histogram for the results.
   b. Repeat computer application 6, part a, for samples of size 30. Construct a histogram for the results.
   c. Repeat computer application 6, part a, for samples of size 40. Construct a histogram for the results.
   d. By examining the three histograms, what appears to be true concerning the shapes as the sample size increases.

8. **a.** Simulate drawing 20 random samples of size 50 from the normal population having $\mu = 60$ and $\sigma = 10$. For each sample, construct a 90% confidence interval for μ.
   b. Determine the number of intervals that contain μ.
   c. Are your results consistent with the theory? Explain.

9. Determine $P(5 < \chi^2 < 15)$ if df $= 15$.

10. Determine C so that $P(\chi^2 > C) = 0.40$ if df $= 12$.

11. Demonstrate that s is a biased estimator for σ.

12. Demonstrate that SS/n is a biased estimator for σ^2.

13. Consider the experiment of tossing a fair coin five times and observing the number of heads that occur. Use a

computer program to simulate 200 repititions of the experiment.

a. Determine the mean and standard deviation of the 200 results.

b. Construct a relative frequency table for the six possible outcomes of the experiment.

c. Determine the theoretical probabilities for each of the six possible outcomes of the experiment.

d. Compare your results to parts b and c.

e. Determine the mean and standard deviation of the theoretical distribution of outcomes and compare your results to those obtained in part a.

(*Hint:* Use commands similar to

```
RANDOM 200 C1-C5;
INTEGER 01.
RSUM C1-C5 C6
HISTOGRAM C6
DESCRIBE C6).
```

14. Demonstrate that s^2 is an unbiased estimator of σ^2.

15. A match company wants to determine a 90% confidence interval for the standard deviation of the time (in seconds) its matches stay lit when held in an upright position. A random sample of 27 matches indicated that $s = 2.24$ minutes. Use a trial-and-error method to determine the shortest such 90% confidence interval for this random sample of data.

EXPERIMENTS WITH REAL DATA

Treat the 720 subjects listed in the database in appendix C as if they constituted the population of all U.S. subjects.

1. Select a random sample of 30 males in the 30–39 year age group and estimate the values of the following parameters:
  a. proportion that smoke
  b. mean weight
  c. mean diastolic blood pressure
  d. standard deviation of diastolic blood pressures

2. Refer to experiment 1.
  a. Construct a 95% confidence interval for the mean diastolic blood pressure of all males in the 30–39 year age group.
  b. Construct a 95% confidence interval for the proportion of males in the 30–39 year age group that smoke.
  c. Construct a 90% confidence interval for the variance of diastolic blood pressure of all males in the 30–39 year age group.

3. Select a random sample of 30 females in the 30–39 year age group and estimate the values of the following parameters:
  a. proportion that smoke
  b. mean weight
  c. mean diastolic blood pressure
  d. standard deviation of diastolic blood pressures

4. Refer to experiment 3.
  a. Construct a 95% confidence interval for the mean diastolic blood pressure of all females in the 30–39 year age group.
  b. Construct a 95% confidence interval for the proportion of females in the 30–39 year age group that smoke.
  c. Construct a 90% confidence interval for the variance of diastolic blood pressure of all females in the 30–39 year age group.

5. Select a random sample of 20 males in the 40–49 year age group. Use this sample to construct a
  a. 95% confidence interval for the mean systolic blood pressure of all males in the 40–49 year age group.
  b. 90% confidence interval for the variance of systolic blood pressure of all males in the 40–49 year age group.

1. Find the following critical values:
   a. $t_{0.05}$ in $n = 18$
   b. $\chi^2_{0.01}(df)$ if $n = 25$

2. Find the value of the following statistics:
   a. z if $\bar{x} = 17$, $s = 4.2$, $n = 80$, and $\mu = 15$
   b. t if $\bar{x} = 75$, $s = 1.8$, $n = 12$, and $\mu = 80$
   c. χ^2 if $n = 16$, $s = 18$, and $\sigma^2 = 9$

3. Construct a 95% confidence interval for μ given $\bar{x} = 17.2$, $s = 2.5$, and $n = 45$.

4. Construct an approximate 90% confidence interval for p given $x = 25$ and $n = 150$.

5. Construct a 99% confidence interval for σ given that a sample from a normal population yielded $n = 18$, $\bar{x} = 2.1$, and $s = 1.4$.

6. If $\bar{x}$ is used to estimate μ and $\sigma = 1.7$, find the value of n so that we can be 90% confident that E is less than 0.12.

7. If $\hat{p}$ is used to estimate p, find the value of n so that we can be 95% confident that $\hat{p}$ is within 0.11 of p.

8. Find the following:
   a. $P(\chi^2_{0.70} < \chi^2 < \chi^2_{0.10})$
   b. $P(t_{0.90} < t < t_{0.05})$
   c. $P(-z_{0.05} < Z < z_{0.15})$
   d. the confidence level $(1 - \alpha)$ if $\chi^2_{\alpha/2} = 7.564$ and df $= 17$.
   e. the confidence level $(1 - \alpha)$ if $t_{\alpha/2} = 2.552$ and df $= 18$.

10

Hypothesis Testing

CHAPTER OBJECTIVES

In this chapter we will investigate

▷ *Hypothesis testing.*

▷ *The logic and language of hypothesis testing.*

▷ *The types of errors associated with hypothesis testing.*

▷ *The steps in the hypothesis-testing procedure.*

▷ *How to test hypotheses concerning a population mean.*

▷ *How to test hypotheses concerning a population proportion.*

▷ *How to test hypotheses concerning a variance or standard deviation of a normal population.*

MOTIVATOR 10

*M*uch experimental research has been conducted to support the use of microcomputers in college science laboratories. Albert E. Cordes carried out a study involving a microcomputer-based laboratory (MBL) in a college general physics course.[40] Students in two sections of the course were involved in the study. One section was randomly selected to be the MBL group, and the other the control group. Students from each section attended three hours of lecture and one two-hour laboratory session per week. The course content was identical for each group. The MBL group used computers in the laboratory, and the other group (control) explored the same laboratory concepts using traditional methods. The computers in the MBL group were used for data analysis and data collection. Data collection on two laboratory activities involving motion and heat of fusion was accomplished by interfacing temperature probes and timing devices with the computer. Tests were administered to both groups following each of these activities. Students in the MBL group were allowed to do their data analysis immediately after they collected their data. Students in the control group computed means, percentage error, and estimated lines of best fit without the benefit of the computer. At the end of the semester, both groups were administered a test designed to measure students' knowledge of statistical principles. The following table contains the results of the three tests:

	Computer Group			Control Group				
Test	N	Mean	SD	N	Mean	SD	t	F
Motion	15	8.93	1.67	12	8.33	1.87	0.88	1.25
Heat	14	7.96	2.37	14	6.79	2.67	1.12	1.27
Statistics	16	4.0	1.26	12	2.83	1.03	2.62	1.03

The *F* test was used to test for homogeneity of variance for all three tests. These tests showed no significant differences ($p < 0.05$) between the variances of the experimental group and the control group for the three tests. Two *t* tests were used to analyze the results of the tests given at the end of the motion and heat laboratories. Results indicated that there were no significant differences ($p > 0.05$) between the mean test scores of the experimental group and the control group. A *t* test was also used to analyze the results of the statistics test; it was found that the mean score of the experimental group was significantly higher ($p < 0.01$) than the control group. Many of these ideas will be examined in this chapter. And after studying chapter 11, you should be able to verify the values listed in the last two columns of the above table.

Chapter Overview

In chapter 9 we studied estimation procedures for making inferences about unknown population parameters, such as μ, σ, and p, by using single values and intervals. In this chapter we want to develop an alternate procedure for making inferences—hypothesis testing. With the hypothesis-testing procedure, we follow a formal set of steps or ordered procedures that lead to decisions concerning statements (called statistical hypotheses) about the value of a parameter.

SECTION 10.1   *Logic of Hypothesis Testing*

Unlike estimation procedures, hypothesis testing is not an exploratory procedure. With hypothesis testing, we are more interested in confirming the relationship of the parameter of interest to a fixed, known value, than in exploring its unknown value. Consider examples 10.1, 10.2, and 10.3.

EXAMPLE 10.1

According to the 1978 Energy Act passed by Congress, a tax is imposed on the manufacturer of any new car that fails to average at least 22.5 miles per gallon of gasoline. (This cost gets passed on, of course, to the consumer.) As a result, a new-car manufacturer may not want to estimate the average gas mileage, but instead would want to determine if the average mileage exceeds 22.5 miles per gallon. The car manufacturer would want to test the hypothesis

> A: The mean gas mileage does not exceed 22.5 miles per gallon.

against the hypothesis

> B: The mean gas mileage exceeds 22.5 miles per gallon.

with the hope that it can gather sufficient information to support hypothesis B.

EXAMPLE 10.2

Some time ago it was accidentally discovered that minoxidil, a drug manufactured by Upjohn Pharmaceutical Company and prescribed for severe high blood pressure, promotes hair growth. The drug is usually given in tablet form to control blood pressure. It has been estimated that 80% of patients taking minoxidil experience thickening, elongation, and darkening of body hair within three to six weeks after starting the drug. As a result of these side effects of the drug, Upjohn has been researching the possibilities of using minoxidil in topical form (a drug called Rogaine) to treat male baldness, a condition that is widespread in the population and is generally untreatable by drugs. The benefits of the use of Rogaine to prevent baldness would be enormous, especially to Upjohn. An experiment was conducted by a researcher to test the effects of Rogaine on baldness. The experiment was conducted over a six-month period to test the hypothesis

A: Rogaine has no therapeutic benefits for preventing baldness.

against the hypothesis

B: Rogaine has therapeutic benefits for preventing baldness.

The experiment was conducted under the assumption that hypothesis A was correct, but with the hope of gathering evidence to the contrary. Two groups of bald people were used. One group (the treatment group) received a fixed dosage of Rogaine, and the other group (the control group), received a placebo—a substance having an appearance identical to Rogaine but having no known effects for treating baldness. The medical staff administering the drug was not made aware of which subjects were in the treatment and control groups; this was done to control any possible effects that the persons administering the drug might have. (An experiment such as this one, in which neither the experimental subject nor the person administering the treatment knows who is in the treatment or control groups is called a *double-blind experiment*.) After the experimental period of six months, the researcher found evidence to suggest that Rogaine has real benefits for treating baldness in males. She thus rejected hypothesis A in favor of hypothesis B. The process that the researcher used to arrive at her decision is called hypothesis testing.

EXAMPLE 10.3

A leading drug in the treatment of hypertension has an advertised therapeutic success rate of 84%. A medical researcher believes he has found a new drug for treating hypertensive patients that has a higher therapeutic success rate than the leading drug, but with fewer side effects. To test his assertion, the researcher gets approval from the Food and Drug Administration to conduct an experimental study involving a random sample of 60 hypertensive patients. The proportion of patients in the sample receiving therapeutic success will be used to determine if the success rate, p, for the population of hypertension patients is higher than 0.84. The researcher arbitrarily decides that he will conclude that the therapeutic success rate of the new drug is higher than 0.84 if the proportion of hypertensive patients in the sample having therapeutic success after taking the drug is 0.86 or higher. Otherwise, he will conclude that his drug is no more effective than the currently used drug. It appears, at first glance, that the decision should be clear-cut—just compute the sample proportion, compare it to 0.84, then make a decision. But the decision rests on the outcome of a single sample proportion, which will vary from sample to sample. If the true proportion in the population receiving therapeutic success is less than or equal to 0.84 (i.e., if $p \leq 0.84$), there is the possibility that $\hat{p}$ could be greater than 0.86, by chance; similarly, there is the possibility that the sample proportion could be less than 0.84 even if the true proportion is, say, 0.87. The uncertainty involving the researcher's decision can be attributed, in part, to the standard error of the sample proportion.

The two statements

$$p \leq 0.84 \quad \text{and} \quad p > 0.84$$

are called **statistical hypotheses.** A statistical hypothesis is an assertion concerning one or more parameters involving one or more populations.

Null Hypothesis and Alternative Hypothesis

The logic of hypothesis testing is based on the simple fact that a statement and its negation cannot both be true (or false) at the same time. Consider the following two statements concerning the population proportion p:

A: $p \leq 0.84$
B: $p > 0.84$

If statement A is true, then statement B is false, and if statement B is true, then statement A is false. Both statements cannot be true (or false) at the same time.

Hypothesis testing involves a pair of statistical hypotheses, such that acceptance of one means rejection of the other. The hypothesis that the researcher would like to establish is called the **experimental hypothesis** or **alternative hypothesis** and is denoted by H_1, whereas the other hypothesis is called the **null hypothesis** and is denoted by H_0. *Null* means none, so it is sometimes convenient to think of the null hypothesis as the hypothesis of no difference. The null hypothesis is commonly written as a statement involving the equality relation when a population parameter is being compared with a numerical value. Note that the relation expressed by the null hypothesis must include one of the three symbols $\leq$, $\geq$, or $=$. The statistical hypothesis $p < 0.84$ is not an acceptable null hypothesis, but it is an acceptable alternative hypothesis.

EXAMPLE 10.4

H_0: $\mu = 20$, H_0: $\mu \geq 20$, and H_0: $\mu \leq 20$ could be null hypotheses.

EXAMPLE 10.5

For example 10.3, the null hypothesis is H_0: $p \leq 0.84$ and the alternative hypothesis is H_1: $p > 0.84$. If the researcher rejects the null hypothesis, he will have support for his drug's having superior therapeutic benefits over the leading drug that has widespread use.

EXAMPLE 10.6

The following demonstrate pairs of statistical hypotheses. For each, H_0 is under test; if H_0 is rejected, the researcher has produced evidence that the alternative hypothesis H_1 is true.

   a. H_0: $\mu = 25$ (the population mean equals 25).
   H_1: $\mu \neq 25$ (the population mean does not equal 25).

   b. H_0: $\sigma \leq 4$ (the population standard deviation is less than or equal to 4).
   H_1: $\sigma > 4$ (the population standard deviation is greater than 4).

   c. H_0: $p \geq 0.5$ (the population proportion is greater than or equal to 0.5).
   H_1: $p < 0.5$ (the population proportion is less than 0.5).

Note that in each pair, H_1 is the negation (opposite) of H_0.

In order for an experimental or alternative hypothesis to be established, the null hypothesis must be rejected. For the hypotheses to be tested in this chapter, this occurs when the evidence obtained from one sample is inconsistent with the statement of the null hypothesis. If the sample evidence is not inconsistent with the null hypothesis, it does not necessarily imply that the null hypothesis is true; it simply means we do not have sufficient evidence to reject it. The statistician or researcher generally labels the statistical hypothesis that he hopes to reject as the null hypothesis. The truth of a statistical hypothesis is never known with certainty. The only way to determine the unknown value of the parameter is to examine the whole population. Since this is not always practical or possible, a decision involving the null hypothesis could involve an error. The process of choosing between H_0 and H_1 is called **hypothesis testing.** Consider examples 10.7–10.13.

EXAMPLE 10.7

A doctor believes that drug A is better than drug B in treating a certain disease. This becomes the alternative hypothesis H_1. The null hypothesis H_0 is that drug A is not better than drug B. If the doctor gathers sufficient evidence to reject H_0, there is a strong likelihood that H_1 is true.

EXAMPLE 10.8

A consumer who purchased a new car four months ago does not believe the advertised claim that the average gas mileage is at least 45 miles per gallon on the highway; he suspects it is less than 45 miles per gallon. Thus, the statement of the alternate hypothesis becomes H_1: $\mu < 45$, and the statement of the null hypothesis is H_0: $\mu \geq 45$. Note that if H_0 is rejected, then the consumer will have support for H_1 being true.

EXAMPLE 10.9

A prosecuting attorney believes the accused is guilty. His job is to convict the accused party. The null hypothesis is the statement that the accused is innocent, whereas the alternative hypothesis is the statement that the accused is guilty. Failure to reject the null hypothesis does not mean the accused is innocent; it means only that there is insufficient evidence to convict.

EXAMPLE 10.10

A business executive believes sales for this year will be higher than last year. The null hypothesis becomes H_0: sales for this year will be no higher than sales last year. The alternative hypothesis becomes H_1: sales this year will be higher than sales last year.

EXAMPLE 10.11

An agronomist would like to determine if a new brand of fertilizer will increase grain production. The null hypothesis is H_0: the new fertilizer will not increase grain production. The alternative hypothesis is H_1: the new brand of fertilizer will increase grain production.

EXAMPLE 10.12

A pollster would like to determine if the proportion of voters in Maryland who are Democrats is different from the proportion of voters in Pennsylvania who are Democrats. The null hypothesis is H_0: the proportion of Democrats in Pennsylvania is equal to the proportion of Democrats in Maryland. The alternative hypothesis is H_1: the proportion of Democrats in Pennsylvania is different from the proportion of Democrats in Maryland.

EXAMPLE 10.13

Suppose a drug company conducted an experiment to determine which of two hypertension drugs, A or B, is more effective on patients with mild hypertension. Drug A was given to one group of patients and drug B was given to another group of patients. On the whole, the drug company found that the patients receiving drug B had more therapeutic benefits than those receiving drug A. There are two possible explanations for the observed difference:

1. Drug B is superior to drug A in terms of therapeutic benefits.

2. The observed difference in therapeutic benefits is due to chance factors or sampling error.

Suppose the drug company was able to calculate the plausibility of the second explanation and found it to be 5%. The company might then conclude, based on the evidence at hand, that it is reasonably certain that drug B is superior to drug A.

Types of Errors with Hypothesis Testing

When testing any null hypothesis H_0, there are four possible outcomes that can occur. The following diagram contains the four possibilities:

	H_0 is true	H_0 is false
Reject H_0	Incorrect decision (Type I error)	Correct decision
Do not reject H_0	Correct decision	Incorrect decision (Type II error)

Two of the outcomes involve correct decisions and two involve incorrect decisions. Rejecting H_0 when it is true and not rejecting H_0 when it is false are incorrect decisions. Rejecting the null hypothesis when it is true is called a **Type I error,** and not rejecting the null hypothesis when false is called a **Type II error.** The probability of committing a Type I error is denoted by the Greek letter α, and is called the **level of significance.**

EXAMPLE 10.14

For the hypertension drug study of example 10.3, deciding on the basis of the test data that the new drug is more effective than the old one (i.e., deciding there is greater than 0.84 therapeutic success rate) when in fact the drug is not more effective for the population as a whole would be a Type I error. The probability of making this error can be arrived at using techniques we'll develop shortly. The probability of committing a Type II error is denoted by the Greek letter β. In the hypertension study of example 10.3, deciding from the test data that the new drug is only as effective or less effective than the old one (i.e., deciding that the therapeutic success rate is not greater than 84%), when in reality it is more effective for the entire population would be a Type II error.

EXAMPLE 10.15

Testing a null hypothesis H_0 is similar to trying an accused individual under our judicial system. The defendant is on trial and H_0 is the statement that the accused is innocent. H_1 is the statement that the defendant is guilty. The prosecution lawyers try to prove H_1 is true (or H_0 is false). In arriving at their verdict, the jury may render a correct decision or an incorrect decision. The correct decisions are the jury votes not guilty when the accused is innocent or votes guilty when the accused is guilty. The incorrect decisions are the jury votes to convict the accused when the accused is innocent (Type I error) or votes to acquit the accused when the accused is guilty (Type II error).

EXAMPLE 10.16

Suppose a new and more expensive diagnostic procedure for detecting breast cancer in women is being tested for superiority over the currently used method. The statistical hypotheses are

H_0: The new method is no better than the currently used method.

H_1: The new method is better than the currently used method.

A Type I error results when the null hypothesis H_0 is rejected when it is true. The consequences of this error would be increased medical costs. A Type II error would result when H_0 is not rejected when it is false. Consequences of this error would be inferior testing and possibly a higher death rate from breast cancer. The more serious type of error in this case is the Type II error. If H_0 is true and it is not rejected or if H_0 is false and it is rejected, then no errors are committed.

The probability associated with making a correct decision—not rejecting H_0 when true or rejecting H_0 when it is false—can be determined. The probability of not rejecting H_0 when it is true is equal to $1 - \alpha$. This can be shown by noting that

$$P(\text{rejecting } H_0 \text{ when it is true}) + P(\text{not rejecting } H_0 \text{ when it is true}) = 1$$

Since $P(\text{rejecting } H_0 \text{ when it is true}) = \alpha$, we have

$$P(\text{not rejecting } H_0 \text{ when } H_0 \text{ is true}) = 1 - \alpha$$

Note that the probability of not rejecting H_0 when it is true is the confidence level $1 - \alpha$ studied in chapter 9.

The probability of rejecting H_0 when it is false is equal to $1 - \beta$. This can be shown by noting that

$$P(\text{rejecting } H_0 \text{ when it is false}) + P(\text{not rejecting } H_0 \text{ when it is false}) = 1$$

But since $P(\text{not rejecting } H_0 \text{ when it is false}) = \beta$, we have

$$P(\text{rejecting } H_0 \text{ when it is false}) = 1 - \beta$$

The probability of rejecting the null hypothesis H_0 when it is false is called the **power of the test.** The probabilities associated with the four possible outcomes of a hypothesis test are summarized in table 10.1.

	Probability Symbol	Definition
TABLE 10.1	α	Level of significance: probability of a Type I error
Outcomes and Probabilities	β	Probability of a Type II error
for Hypothesis Testing	$1 - \alpha$	Level of confidence: probability of not rejecting H_0 when true
	$1 - \beta$	Power of the test: probability of rejecting H_0 when false

Proof-by-contradiction in mathematics provides a reasonable parallel to hypothesis testing. The only difference is that proof-by-contradiction does constitute a proof in the mathematical sense, but hypothesis testing does not. With hypothesis testing a statement is rejected when probabilities derived from the data based on the statement of the null hypothesis cast serious doubt on the truth of the null hypothesis. A mathematician does not conclude that $A = B$ just because he cannot show otherwise. The reasoning with hypothesis testing is no less logical in that failure to reject H_0 does not constitute a proof that H_0 is true or that it has been verified. A good point to remember about hypothesis testing is

Hypothesis testing can never be used to "establish" absolute truths. With any decision, there is always the possibility of error. When the null hypothesis is rejected, we have statistical evidence that indicates that the alternative hypothesis is plausible, but not necessarily true. In addition, failure to reject H_0 should not imply that one should accept H_0. Rather, judgment should be reserved unless the probability of committing a Type II error is known. If β is small, then one might conclude that H_0 is plausible, but not necessarily true.

Types of Hypothesis Tests

Hypothesis tests are classified as directional or nondirectional, depending on whether the alternative hypothesis H_1 involves the unequal sign ($\neq$). If the statement of H_1 involves $\neq$, then we call the test a **nondirectional test,** whereas if the statement of H_1 does not involve $\neq$ (that is, if it involves $<$ or $>$), then we call the test a **directional test.** Nondirectional tests are also called **two-tailed tests,** and directional tests are also called **one-tailed tests.** For tests involving one sample of data, if the statement of H_1 involves the symbol $<$, then the test is called a **left-tailed test,** and if the statement of H_1 involves the symbol $>$, then the test is called a **right-tailed test.** The following table summarizes the language used. Terms listed in each column are frequently used interchangeably.

Sign in the Statement of H_1		
$<$	$\neq$	$>$
Directional	Nondirectional	Directional
One-tailed	Two-tailed	One-tailed
Left-tailed		Right-tailed

EXAMPLE 10.17

The following demonstrate the different types of hypothesis tests:

Directional	Nondirectional	Left-Tailed	Right-Tailed
$H_1: \mu > 2$	$H_1: \mu \neq 2$	$H_1: \mu < 2$	$H_1: \mu > 2$
$H_1: p < 0.2$	$H_1: p \neq 0.2$	$H_1: p < 0.2$	$H_1: p > 0.2$
$H_1: \sigma < 4$	$H_1: \sigma \neq 4$	$H_1: \sigma < 4$	$H_1: \sigma > 4$

Determining H₁

The form of the alternative hypothesis H_1 will be dictated by the practical aspects of the problem. It will reflect the direction of the desired result as formulated by the experimenter. Frequently it is possible to select H_0 so that an erroneous rejection of H_0 is the more serious consequence. As we shall see later, this is due to the fact that the probability of committing a Type I error can be controlled by the experimenter (since it is the level of significance and should be stipulated in advance).

EXAMPLE 10.18

Suppose a medical researcher is interested in testing the effects of a new drug. Consider the following two pairs of statistical hypotheses:

a. H_0: The drug is safe.
   H_1: The drug is harmful.

b. H_0: The drug is harmful.
   H_1: The drug is safe.

For situation a, a Type I error is to conclude that a safe drug is harmful, and a Type II error is to conclude that a harmful drug is safe. For situation b, a Type I error is to conclude that a harmful drug is safe, and a Type II error is to conclude that a safe drug is harmful. Since the Type I error for situation b (concluding that a harmful drug is safe) is more serious and α can be controlled, the researcher would select the hypotheses in situation b for the experiment and would formulate the null hypothesis to reflect that the drug is harmful.

In verbal problems written to provide experience in using the hypothesis testing procedure (such as the exercises in this chapter) the type of test (one-tailed or two-tailed) can be determined by careful reading of the problem. A one-tailed test can usually be identified by use of such terms as *more than, less than, better than, worse than, at least, at most, too much,* and the like. Two-tailed tests are usually identified in verbal problems by such terms as *not equal to, different from, changed for better or worse, unequal, a difference,* and the like.

APPLICATION 10.1

For each of the following situations, identify the alternative hypothesis H_1, the type of test (one-tailed or two-tailed), and the identifying term(s).

   a. The national average salary for state and local government employees was $19,044 for the year 1982. Employees of the state of Maryland suspected that they were below the national average in terms of salary.

   b. The mean yield of cotton grown by a particular farmer was 1225 pounds per acre for a recent year. In the next year, the farmer tried a new type of insecticide on a random sample of plots and was interested in determining if the average yield of cotton per acre would increase.

   c. A sample of Camel cigarettes shows the average tar content to be 12 mg. Does this indicate that the average tar content for Camel cigarettes exceeds 11 mg?

   d. A pharmaceutical company claims that its best-selling painkiller has a mean effective period of at least 3 hours. Do the sample data indicate that the company's claim is too high?

   e. In 1979, the average age of students enrolled in college was 19.8 years. Based on a sample taken last year, has the average age of college students changed?

f. John Smith, a Democrat, is running against Jane Jones, a Republican, for the office of governor. Last month's poll showed that 65% of the voters supported Smith, but then Jones started mudslinging and now Smith is interested in whether the proportion who support him has dropped.

Solution:

Alternative Hypothesis	Type of Test	Identifying Terms
a. $H_1: \mu < \$19,044$	left-tailed test	below
b. $H_1: \mu > 1225$	right-tailed test	increase
c. $H_1: \mu > 11$	right-tailed test	exceeds
d. $H_1: \mu < 3$	left-tailed test	too high
e. $H_1: \mu \neq 19.8$	two-tailed test	changed
f. $H_1; p < 0.65$	left-tailed test	dropped

▪

EXERCISE SET 10.1

Basic Skills

1. For each pair of statistical hypotheses, indicate which one is the null hypothesis, H_0.
   a. A: $\mu > 21$
      B: $\mu \leq 21$
   b. A: $p = 0.7$
      B: $p \neq 0.7$
   c. A: $\sigma \neq 1.2$
      B: $\sigma = 1.2$
   d. A: $\sigma^2 < 8.1$
      B: $\sigma^2 \geq 8.1$

2. For each pair of statistical hypotheses, indicate which one is the null hypothesis, H_0.
   a. A: $\mu \geq -235$
      B: $\mu < -235$
   b. A: $p \neq 0.9$
      B: $p = 0.9$
   c. A: $\sigma^2 = 32$
      B: $\sigma^2 \neq 32$
   d. A: $\sigma > 8$
      B: $\sigma \leq 8$

3. Identify which of the hypotheses in exercise 1 are associated with one-tailed tests.

4. Identify which of the hypotheses in exercise 2 are associated with one-tailed tests.

5. For each of the following situations, identify the hypotheses and the type of error, when appropriate.
   a. The drug cibenzoline is being compared to propranolol for possible use in controlling cardiac arrhythmia. The claim is that cibenzoline is better than propranolol. In fact, it is not the case. However, research concluded that cibenzoline is better.
   b. Two schools are equally effective in training students. An evaluation team concluded that there was no significant difference between the effectiveness of the two schools in training students.

6. For each of the following situations, identify the hypotheses and the type of error, when appropriate.

   a. A greater proportion of RCA televisions need repairs than Zenith televisions. A study done by a consumer advocate group concluded the same thing.
   b. Two people are compared for efficiency in performing a particular task. Individual B is really more efficient than individual A. An evaluation concluded that there is no difference in efficiency ratings between A and B.

7. For each of the following research situations, indicate whether a one-tailed or two-tailed test is appropriate.
   a. The Food and Drug Administration wants to test a new prescription drug manufactured by Upjohn to determine if it contains 5 mg of codeine as claimed by Upjohn.
   b. A survey was done by mechanics to determine if there is a difference between the proportion of Fords and the proportion of Chevrolets that need repairs. They think there is a difference, but are not sure.
   c. An accrediting association was convinced that college A's grading of students was different from college B's; however, the association was not sure from which college the students got better grades.

8. For each of the following research situations, indicate whether a one-tailed or two-tailed test is appropriate.
   a. Dr. Jones believes that the discovery method of teaching is more effective than the expository method. She is interested in showing this is the case.
   b. The National Academy of Sciences has recommended that the mean daily sodium intake for a person should be between 1100 mg and 3300 mg. A medical researcher thinks the mean daily sodium intake for the average person exceeds 4500 mg.

c. A biologist believes the average number of eggs laid per nest for the Eastern Phoebe bird is four. He is interested in showing this is the case.

9. Suppose the Food and Drug Administration wants to test a new prescription drug manufactured by Upjohn to determine if it contains 5 mg of codeine as claimed by Upjohn.
   a. What null hypothesis should they test?
   b. Why did you choose your answer over the other two possibilities for stating the alternative hypothesis?
   c. Describe Type I and Type II errors associated with your null hypothesis and discuss the possible consequences of making each type of error.

10. Suppose that the EPA wants to test a claim by an automobile manufacturer that their compact car gets at least 55 miles per gallon of gasoline.
    a. What null hypothesis should they test?
    b. Why did you choose your answer over the other two possibilities for stating the alternative hypothesis?
    c. Describe Type I and Type II errors associated with your null hypothesis and discuss the possible consequences of making each type of error.

11. For a statistical test, what do $1 - \alpha$ and $1 - \beta$ represent?

12. For a statistical test, what do α and β represent?

13. Suppose we want to test the following null hypothesis:

    H_0: Smoking is not harmful to your health.

    In terms of the null hypothesis, state in words what is represented by
    a. a Type I error.
    b. a Type II error.
    c. a good decision.

    Which type of error is more serious?

14. Suppose we want to test the following null hypothesis:

    H_0: Potatoes grown with fertilizer A are at least as large as potatoes grown with fertilizer B.

    In terms of the null hypothesis, state in words what is represented by
    a. a Type I error.
    b. a Type II error.
    c. a good decision.

    Which type of error is more serious?

15. For each of the following null hypotheses, state what actions would constitute Type I and Type II errors.
    a. H_0: The discovery method of teaching statistics is at least as good as the expository method.

b. H_0: At most 2% of the machines are defective.
c. H_0: At least 95% of Americans are against war.

16. For each of the following null hypotheses, state what actions would constitute Type I and Type II errors.
    a. H_0: The new production process is at least as good as the old process.
    b. H_0: 90% of doctors recommend preparation A.

17. Suppose we conduct a hypothesis test with a level of significance equal to 0.05, determine whether the following statements are true or false.
    a. The probability of committing a Type I error is 0.95.
    b. The probability of rejecting H_0 is 0.05.
    c. The probability of committing a Type I error is 0.05.

18. Suppose we conduct a hypothesis test with a level of significance equal to 0.01, determine whether the following statements are true or false.
    a. The probability of not rejecting H_0 is 0.99.
    b. The probability of not rejecting H_0 when true is 0.99.
    c. The probability of rejecting H_0 when false is 0.01.

19. When we fail to reject H_0, why do we not automatically accept H_0? Explain.

20. Is the probability of making a Type I error equal to 1 minus the probability of making a Type II error? Explain.

21. Consider each of the following situations. For each decision, state whether it is a correct decision and (if applicable) indicate the type of error.

    H_0: The lifetime of battery A does not exceed the lifetime of battery B.
    a. Change to A when B lasts as long or longer.
    b. Keep B when A lasts longer.
    c. Keep B when B lasts at least as long.
    d. Change to A when A lasts at least as long.

22. Consider each of the following situations. For each decision, state whether it is a correct decision and (if applicable) indicate the type of error.

    H_0: The new drug is safe.
    a. Approve the drug when it is unsafe.
    b. Disapprove the drug when it is unsafe.
    c. Disapprove the drug when it is safe.
    d. Approve the drug when it is safe.

Going Beyond

23. If you reject H_0, is it possible that you have made a Type II error?

24. If it has been determined to reject the null hypothesis H_0 when $\alpha = 0.05$, does $\beta = 0$? Explain.

SECTION 10.2   *Introduction to the Hypothesis-Testing Procedure*

The hypothesis-testing procedure involves an indirect approach for making an inference about an unknown population parameter. The procedure is based on the assertion that the null hypothesis H_0 is true. A statistical test is then conducted to determine the plausibility of H_0. A null hypothesis is plausible if a researcher finds insufficient evidence to suggest that it is false. If the results of the test are judged unlikely to occur (or indicate a rare event), then the original assumption that H_0 is true is rejected; in such cases there is support for the alternative hypothesis H_1 to be plausible. Of course, with any decision, we may make an error. By using the hypothesis-testing procedure, we will make generalizations about population parameters based on the evidence gathered from only one sample (for the tests in this chapter). The **hypothesis-testing procedure** involves the following steps:

To Test a Null Hypothesis

1. Formulate and state H_0.
2. Formulate and state H_1.
3. Choose an appropriate test statistic.
4. Formulate and state a decision rule.
5. Compute a statistic from a test sample.
6. Make a decision—either (a) fail to reject H_0 or (b) reject H_0.

The hypothesis-testing procedure involves gathering a random sample of test data relating to the hypotheses and computing a statistic, called the **test statistic.** This test statistic belongs to a sampling distribution with a particular shape, mean, and standard deviation. If a value of the test statistic (such as the sample mean $\bar{x}$) is located too far to the left of the mean of its sampling distribution (in the left tail of the sampling distribution) or too far to the right of the mean (in the right tail of its sampling distribution), then the null hypothesis H_0 is rejected. How far is too far is determined by a decision rule. The **decision rule** is stated in terms of α, the level of significance. Recall that the level of significance is the probability of committing a Type I error, that is, rejecting H_0 when it is true. The decision rule specifies those values in the sampling distribution of the test statistic that result in H_0 being rejected. These values are the values of the test statistic that are unlikely to occur if H_0 is true, and are commonly referred to as the **rejection region.** Consider example 10.19.

EXAMPLE 10.19

Suppose a certain coin-operated soft-drink machine was designed to dispense, on the average, 8 ounces of beverage per cup. Over a long period of use, we suspect that the machine is dispensing an average of less than 8 ounces per cup. As a result, we decide to test the following statistical hypotheses:

Step 1 H_0: $\mu \geq 8$.

Step 2 H_1: $\mu < 8$.

If we can reject H_0, we conclude that our suspicions are correct.

Step 3 Next we decide to obtain a random sample of 30 cups of beverage and compute the sample mean, $\bar{x}$. Suppose $\bar{x} = 7.6$ ounces and $s = 0.75$ ounce. The sample mean is chosen as the *test statistic*.

Step 4 We next stipulate the level of significance α. This will assist us in formulating our decision rule. The level of significance can be any value between 0 and 1, but is usually 0.05 or 0.01. Let's use $\alpha = 0.05$. This means that the probability of rejecting H_0 when it true is 0.05. Since we are interested in producing evidence that supports H_1 being true, we shall assume that H_0 is true and hope to reject H_0. The statement that H_0 is true is only an assumption and must be tested for truth or plausibility. As a result, we proceed under the assumption that H_0 is true.

Since the level of significance is $\alpha = 0.05$ and the test is a left-tailed test (since $H_1: \mu < 8$) a rejection region can be established. This is done by locating the value in the sampling distribution of the mean that has 5% of the sample means falling below it. In other words, a value can be determined that cuts off an area of 0.05 in the left tail of the sampling distribution. This value is called the **critical value,** and it separates the sampling distribution into two regions, the rejection region and the nonrejection region. We shall denote a critical value by the symbol C. All values to the left of the critical value make up the rejection region. If $\bar{x} = 7.6$ falls in the rejection region, we reject H_0, otherwise, we fail to reject H_0, as indicated in the figure.

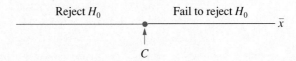

Since the sampling distribution of the mean is approximately normal, to find the critical value C, we find the z value for C and refer to the standard normal distribution (see fig. 10.1). Since σ is unknown and $n \geq 30$, s is an adequate point estimate for σ. Thus,

$$z = \frac{C - \mu}{\sigma / \sqrt{n}}$$

$$-1.65 = \frac{C - 8}{0.75 / \sqrt{30}}$$

The value 1.65 was found in the z table (see the front endpaper) corresponding to an area of $0.5 - 0.05 = 0.45$. By solving the last equation for C, the critical value is

$$C = 7.77$$

FIGURE 10.1

Critical value for soft-drink example

Step 5 We want to locate the value of the sample mean $\bar{x} = 7.6$ within its sampling distribution. As a consequence of the central limit theorem and the fact that $n = 30$, the sampling distribution of the mean is approximately normal. Its mean is μ, the mean of the sampled population. Since we assumed H_0 to be true, the mean of the sampling distribution is greater than or equal to 8 ounces. Which value for μ should we use? We always use the equality value expressed by H_0 (called the **null value**). Thus, $\mu = 8$ is the null value. If a value of μ greater than 8 is used, it can be shown that the probability of rejecting H_0 when H_0 is true is less than the stipulated level of significance $\alpha = 0.05$. That is, the null value presents the worst possible case in terms of providing the largest possible probability of rejecting H_0 when it is assumed to be true. As a result, throughout the remainder of the text, we shall always employ the null value as the mean of the sampling distribution of the test statistic.

Step 6 Since $\bar{x} = 7.6 < 7.77$ we reject H_0 in favor of H_1. With this decision we risk committing a Type I error, rejecting H_0 when it is true. The probability of this type of error is $\alpha = 0.05$. When a test results in rejecting a null hypothesis under a specified level of significance, the test is said to be *significant* and the result is referred to as a **significant finding.** For our example, the result $\mu < 8$ is classified as a significant finding. Thus, we have significant statistical evidence to indicate the machine is dispensing on the average less than 8 ounces per cup of soft-drink beverage.

A test procedure is considered good if both its Type I and Type II error probabilities are small. We have control over the Type I error probability α, since it is stipulated before the data are gathered. We generally have little control over β, the probability of a Type II error. The probability of committing a Type II error varies, depending on the true value of the population parameter. For example, if we want to test the null hypothesis H_0: $\mu = 25$ and it is asserted that the true value of μ is 26, then a value for β can be calculated. The value for β (for a fixed sample size) depends on the difference between the asserted true value and the hypothesized value.

It is desirable to minimize the probabilities for both types of errors; for a fixed significance level α, β can generally be kept to a minimum by choosing the sample size as large as possible. Recall that the probability of rejecting the null hypothesis when it is false is called the power of the test and is denoted by $1 - \beta$. The power can be increased by increasing the sample size n. Let's examine the logic underlying the hypothesis test introduced in this section. The following events occurred in the order listed:

1. The statistical hypotheses H_0 and H_1 were formulated such that rejection of the null hypothesis H_0: $\mu \geq 8$ ounces would establish that the alternative hypothesis H_1: $\mu < 8$ ounces is plausible.

2. The null hypothesis was assumed to be true and the null value $\mu = 8$ was chosen to be the mean of the sampling distribution of the sample mean.

3. To test the null hypothesis, a random sample of 30 cups of beverage was obtained and the sample mean and standard deviation were computed ($\bar{x} = 7.6$ and $s = 0.75$). The sample mean $\bar{x}$ is the test statistic and 7.6 is the value of the test statistic.

4. The maximum risk of committing a Type I error, that of rejecting H_0 when it is true, was stipulated to be $\alpha = 0.05$.

5. Next, we located $\bar{x} = 7.6$ in its sampling distribution to determine the likelihood of $\bar{x}$ being at most as large as 7.6 under the assumption that H_0: $\mu = 8$ was true. Since the level of significance was stipulated to be 0.05, a cutoff point C (called the critical value) was established in the left tail of the distribution such that $P(\bar{x} < C) = 0.05$.

All values in the sampling distribution to the left of the critical value were identified as the rejection region. If the value of the test statistic falls within the rejection region, we reject H_0, otherwise, we fail to reject H_0.

6. For our example, $\bar{x} = 7.6$ fell in the rejection region, so H_0 was rejected in favor of H_1 being true. If H_0 is true and our sample is random, we would not expect the sample mean to be located so far into the left tail of its sampling distribution. A sample mean falling in the lower left portion of the distribution would signal a rare or unlikely event, in which case the null hypothesis is not plausible and should be rejected.

EXERCISE SET 10.2

Basic Skills

1. The mean of SAT math scores is 500 and the standard deviation is 100. A random sample of 64 students received an average SAT math score of 480. To determine if the population mean score is less than 500, a hypothesis test with $\alpha = 0.05$ is to be conducted.
   a. Determine the critical value.
   b. Determine the rejection region.
   c. What is the null hypothesis?
   d. What is the alternative hypothesis?
   e. What is the test statistic and corresponding value?
   f. What is the significance level?
   g. What is the decision?
   h. What type of error is associated with the decision in part g?

2. A soft-drink machine is set to dispense 7.0 ounces per cup. A random sample of 35 cups yielded $\bar{x} = 7.23$ ounces and $s = 0.14$ ounces. To determine if the machine is overfilling its cups, a hypothesis test is to be conducted with $\alpha = 0.01$.
   a. Determine the critical value.
   b. Determine the rejection region.
   c. What is the null hypothesis?
   d. What is the alternative hypothesis?
   e. What is the test statistic and corresponding value?
   f. What is the significance level?
   g. What is the decision?
   h. What type of error is associated with the decision in part g?

3. A new type of tire is tested to determine if it can average 30,000 miles under normal highway driving conditions. A sample of 100 tire sets was tested and averaged 30,200 miles; the standard deviation was 850 miles. Do the data indicate that the tires wear significantly more than 30,000 miles? Use $\alpha = 0.05$.
   a. Determine the critical value.
   b. Determine the rejection region.
   c. What is the null hypothesis?

   d. What is the alternative hypothesis?
   e. What is the test statistic and corresponding value?
   f. What is the significance level?
   g. What is the decision?
   h. What type of error is associated with the decision in part g?

4. A toy company manufactures snow sleds having a pull rope. The mean breaking strength of the rope is advertised to be at least 150 pounds. Suspecting that the mean breaking strength of the rope is less than 150 pounds, company engineers chose a random sample of 50 ropes from a new shipment and found a mean breaking strength of 145 pounds and a standard deviation equal to 4 pounds. Should the toy manufacturer accept the shipment? Use $\alpha = 0.05$.
   a. Determine the critical value.
   b. Determine the rejection region.
   c. What is the null hypothesis?
   d. What is the alternative hypothesis?
   e. What is the test statistic and corresponding value?
   f. What is the significance level?
   g. What is the decision?
   h. What type of error is associated with the decision in part g?

Going Beyond

5. In nine test runs, a new large diesel truck was operated with 1 gallon of special fuel oil, and obtained the following mileages: 7, 9, 9, 6, 8, 11, 9, 7, and 6. Is this evidence that the truck is not operating at an average of 10.4 miles per gallon of fuel oil? Use $\alpha = 0.05$. (*Hint:* use the t statistic.)
   a. Determine the critical value.
   b. Determine the rejection region.
   c. What is the null hypothesis?
   d. What is the alternative hypothesis?
   e. What is the test statistic and corresponding value?

f. What is the significance level?

g. What is the decision?

h. What type of error is associated with the decision in part g?

6. For large sample sizes, small differences tend to be significant, but not always of practical importance. Occasionally, a small sample yields a difference that is of practical importance. For example, a weight gain of 5 pounds would not be considered of practical importance for a 20-year-old male football player, but it would be for a year-old baby. The following situations illustrate that small differences become significant for large sample sizes. In which of the following situations would you reject H_0 for a one-tailed test?

a. $N = 30$, $\sigma = 5$, $\alpha = 0.05$, H_0: $\mu = 70$, and $\overline{x} = 71$

b. $n = 100$, $\sigma = 5$, $\alpha = 0.05$, H_0: $\mu = 70$, and $\overline{x} = 71$

c. $n = 1000$, $\sigma = 0.5$, $\alpha = 0.05$, H_0: $\mu = 70$, and $\overline{x} = 71$

d. $n = 2500$, $\sigma = 2$, $\alpha = 0.05$, H_0: $\mu = 1.70$, and $\overline{x} = 1.80$

e. $n = 3600$, $\sigma = 0.5$, $\alpha = 0.05$, H_0: $\mu = 0.60$, and $\overline{x} = 0.61$

7. Suppose it is desired to test the null hypothesis H_0: $\mu \leq 10$ versus the alternative hypothesis H_1: $\mu > 10$. If the decision rule is to reject H_0 if $\mu > 10$, what is the probability of committing a Type I error?

8. For exercise 7, state a different rule that has a probability of 0 of leading to a Type I error. With such a rule, what is the probability of committing a Type II error?

9. Can a null hypothesis be of the form H_0: $\mu > \mu_0$, where μ_0 is a fixed value? Explain.

10. If H_0 cannot be rejected at $\alpha = 0.05$, can it be rejected at $\alpha = 0.01$? Explain.

SECTION 10.3 Hypothesis Tests Involving μ

EXAMPLE 10.20

Suppose, as members of a consumer protection group, we are interested in determining if the average weight of a certain brand of chocolate chip morsels, packaged in 15-ounce packages, is less than the advertised weight of 15 ounces. We formulate the following statistical hypotheses and decide to test the null hypotheses H_0 against the alternative hypothesis H_1:

$$H_0: \mu \geq 15$$
$$H_1: \mu < 15$$

As the level of significance, we decide on $\alpha = 0.05$. Since we are dealing with the parameter μ and desire information concerning its magnitude, we choose a random sample of 50 bags and compute the value of the sample mean $\overline{x}$. Suppose we find that $\overline{x} = 14.4$ ounces and $s = 1.2$ ounces. Our best point estimate of μ is then 14.4. Recall from chapter 8 that $\overline{x}$ can be represented as a sum of two components, the population mean μ and sampling error e:

$$\overline{x} = \mu + e$$

Testing the null hypothesis H_0: $\mu \geq 15$ is equivalent to determining whether the observed difference between $\overline{x}$ and μ can be attributed to sampling error alone if we assume that $\mu = 15$. To aid us in our determination, we use the sampling distribution of the mean.

If $\mu = 15$ and the sample is random then we would expect the sample mean ($\overline{x} = 14.4$) to be representative of the population for which $\mu = 15$ and to fall somewhere near the center of the sampling distribution of the mean. On the other hand, if $\overline{x} = 14.4$ falls too far into the lower tail of the distribution, this would indicate an unusual (or rare) event and signal that perhaps something was wrong.

How can we determine if $\overline{x} = 14.4$ is located "too far" into the left-tail area? As we learned in section 10.2, our level of significance $\alpha = 0.05$ assists us in this determination (see figure 10.2). Recall that the sample mean C that has 5% of the sample means to the left of it is called the *critical value* for the test and determines the rejection region. Thus, if $\overline{x} = 14.4$ is less than or equal to C, we will reject H_0; otherwise, we will fail to reject H_0. What remains is to

FIGURE 10.2

Critical value for hypothesis test

determine the critical value C, located within the normal distribution having $\mu_{\bar{x}} = 15$ and $\sigma_{\bar{x}} = 1.2/\sqrt{50} = 0.17$.

The critical value C can be found by using the z score formula:

$$z = \frac{C - \mu_{\bar{x}}}{\sigma_{\bar{x}}}$$

$$= \frac{C - 15}{1.2/\sqrt{50}} = \frac{C - 15}{0.17}$$

By examining the standard normal table, we find $-z_{0.05} = -1.65$. Thus, we have the following equation:

$$-1.65 = \frac{C - 15}{0.17}$$

Multiplying both sides of the above equation by 0.17 and then adding 15 to both sides, we get $C = 14.72$. The rejection region becomes all values of $\bar{x}$ less than or equal to 14.72.

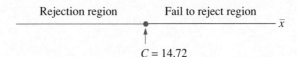

Since the value of our test statistic, $\bar{x} = 14.4$, falls in the rejection region, we reject H_0: $\mu \geq 15$ in favor of H_1: $\mu < 15$ being true. It thus follows that we have statistically significant information that supports the alternative hypothesis H_1: $\mu < 15$ as being the true state of affairs. As a result, we can conclude that the average weight of a 15-ounce package of chocolate chips is less than 15 ounces. Of course, with this decision we may have committed a Type I error, the error of rejecting H_0 when it is true. But the chance of doing this is only 5%, or 1 chance in 20.

If the test is predicated on the assumptions that H_0 is true and the sample is random, then to have the test statistic $\bar{x} = 14.4$ fall in the rejection region signals that something may be wrong—either the sample is not random or H_0 is not true. If we are confident concerning the randomness of the sample, then it is probably true that H_0: $\mu \geq 15$ is false. That is, $\mu \geq 15$ is false and $\mu < 15$ is true. If H_0 is true, a sample mean as small as 14.4 (i.e., 14.4 or smaller) would occur only 5% of the time.

In the hypothesis test of example 10.20, if H_0 is true and $n \geq 30$, then the sampling distribution of the mean is approximately normal and has a mean equal to 15. In figure 10.3, this distribution is identified with the normal curve drawn with the solid line. If the sample were random, we would expect the value of $\bar{x}$ to fall near 15, the mean of its distribution. For the test in example 10.20, the observed value of the sample mean was $\bar{x} = 14.4$, which suggests that $\mu_{\bar{x}} \leq 15$ in order for the observed sample mean $\bar{x} = 14.4$ to be located near the mean of its sampling distribution. Such a distribution is indicated by the dashed normal curve in figure 10.3. Reflecting back on the error term $e = \bar{x} - \mu = -0.6$ (if H_0 is true), we see that the data suggest this difference cannot be attributed solely to sampling error alone.

FIGURE 10.3

Sampling distribution of mean for hypothesis test

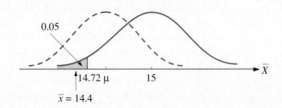

In summary, the following important terminology was used in our hypothesis testing procedure for example 10.20:

- Null hypothesis, H_0: $\mu \geq 15$
- Alternative hypothesis, H_1: $\mu < 15$
- Value of test statistic, $\bar{x} = 14.4$
- Level of significance, $\alpha = 0.05$
- Critical value, $C = 14.72$
- Rejection region, all values of $\bar{x}$ less than or equal to $C = 14.72$
- Sampling distribution of the mean, the normal distribution with $\mu_{\bar{x}} = 15$ and $\sigma_{\bar{x}} = 1.2/\sqrt{50} = 0.17$

Equivalent Testing Procedures

There are three equivalent procedures for deciding whether the null hypothesis H_0 is true or plausible. They are described here in terms of the test in example 10.20.

Procedure 1 Compare the value of the test statistic $\bar{x}$ (calculated from the sample) to the critical value obtained from the sampling distribution of the mean with $\mu_{\bar{x}} = 15$ and $\sigma_{\bar{x}} = 0.17$.

Procedure 2 Find the z-value for the value of the test statistic and compare it with $-z_{0.05}$ (the z value corresponding to an area equal to $\alpha = 0.05$ in the left tail of the standard normal distribution).

Procedure 3 Find the probability of obtaining a value at most as large as the test statistic $\bar{x} = 14.4$ and compare this with the significance level $\alpha = 0.05$. If this probability is less than or equal to $\alpha = 0.05$, we will reject H_0; otherwise, we will fail to reject H_0.

We have already used the first procedure with the illustration in this section. We now demonstrate the second procedure, in example 10.21, which will be our preference in later work dealing with hypothesis testing. Then, before we examine the third procedure for testing a null hypothesis, let's consider applications 10.2–10.4, showing the details of the second procedure.

EXAMPLE 10.21

For the second procedure, we find the z value for $\bar{x} = 14.4$ and compare it with $-z_{0.05}$. The z value for $\bar{x} = 14.4$ is

$$z = \frac{\bar{x} - \mu_{\bar{x}}}{\sigma_{\bar{x}}}$$

$$= \frac{14.4 - 15}{0.17} = -3.53$$

Within the standard normal distribution, $-z_{0.05}$ is found to be -1.65. Since $z = -3.53 < -z_{0.05} = -1.65$, we reject H_0 in favor of H_1 being true. When we use the second procedure to arrive at a decision regarding H_0, the z value for the sample mean $\bar{x}$ becomes the *test statistic* and the tabulated value of z corresponding to α in the standard normal table becomes the *critical value*. The only difference between the first and second approaches is the normal distribution that is used—the standard normal or some other normal distribution. See figure 10.4, comparing the normal distributions used for procedures 1 and 2.

FIGURE 10.4

Normal distributions of test
statistics

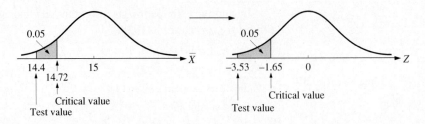

APPLICATION 10.2

A cold tablet is supposed to contain 10 grains of aspirin. A random sample of 100
tablets yielded a mean of 10.2 grains and a standard deviation of 1.4 grains. Can we
conclude at the 0.05 level of significance that μ is different from 10?

Solution: Note that in this case, the alternative hypothesis indicates a nondirectional
test. As such, $\mu < 10$ or $\mu > 10$ and the rejection region consists of values in the left
and right tails of the distribution. Since the probability of committing a Type I error,
rejecting H_0 when it is true, is 0.05 and both tails constitute the rejection region, we
place $\alpha/2 = 0.025$ of the distribution in each of the two tail regions, as indicated in
the accompanying figure.

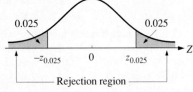

1. H_0: $\mu = 10$.
2. H_1: $\mu \neq 10$ (two-tailed test).
3. Sampling distribution: The sampling distribution of the mean. The test statistic is
   the z value for $\bar{x}$. Since σ is unknown and $n = 100$, the sample standard deviation s
   provides a good estimate for σ.

$$z = \frac{\bar{x} - \mu}{\sigma/\sqrt{n}}$$
$$= \frac{10.2 - 10}{1.4/\sqrt{100}} = 1.43$$

4. Decision rule: Reject H_0 if $z < -z_{0.025}$ or $z > z_{0.025}$.
5. The critical values are $\pm z_{0.025} = \pm 1.96$, as shown in the figure.

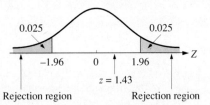

6. Decision: We fail to reject H_0. We thus conclude that there is no statistical support
   for μ being different from 10. Our failure to reject H_0: $\mu = 10$ is interpreted to mean
   that our finding is not statistically significant at $\alpha = 0.05$.
7. Type of error possible: Type II, since H_0 may be false and we fail to reject it. The
   probability β in this case is unknown. As a result, the experimenter should reserve

judgment on H_0 until more data are gathered. In this case, the decision is fail to reject H_0. As we pointed out earlier, this statement does *not* imply that H_0 is accepted as being true or plausible. ▪

APPLICATION 10.3

It is known that the voltages of a certain brand of size C batteries are normally distributed. A random sample of 15 batteries was tested and it was found that $\bar{x} = 1.4$ volts and $s = 0.21$ volt. At the 0.01 level of significance, does this indicate that $\mu < 1.5$ volts?

Solution:

1. H_0: $\mu \geq 1.5$.
2. H_1: $\mu < 1.5$ (one-tailed test, left tail).
3. Sampling distribution: Distribution of the sample mean. The test statistic is the t value for $\bar{x}$.

$$t = \frac{\bar{x} - \mu}{s/\sqrt{n}}, \quad df = 14$$

4. Decision rule: Reject H_0 if $t < -t_{0.01}$.

   Note that t is used instead of z, since $n < 30$ and σ is unknown. In addition, we use the normality assumption here, since the t statistic requires sampling from a normal population. The value of the test statistic is

$$t = \frac{1.4 - 1.5}{0.21/\sqrt{15}} = -1.84$$

5. Critical value: $-t_{0.01} = -2.624$. The value of 2.624 was obtained from the t table (see the back endpaper) and represents the t value corresponding to $\alpha = 0.01$ and $df = 14$ (see the accompanying figure).

6. Decision: Since $t = -1.84 > -2.624 = t_{0.01}$, we cannot reject H_0. Hence, there is no evidence to support that $\mu < 1.5$. In this case the difference $e = \bar{x} - \mu_{\bar{x}} = 1.4 - 1.5 = -0.1$ can be attributed to sampling error alone.

7. Type of error possible: Type II; the value of β is unknown. ▪

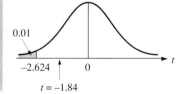

0.01

-2.624 0 t

$t = -1.84$

APPLICATION 10.4

A commercial school advertises that its students can type an average of 80 words per minute (wpm) upon graduation. A sample of 60 recent graduates was tested and the results showed $\bar{x} = 78$ words per minute and $s = 6.2$ wpm. Can this difference of $78 - 80 = -2$ be explained by sampling error alone, or is $\mu < 80$? Test at the 0.01 level of significance.

Solution:

1. H_0: $\mu \geq 80$.
2. H_1: $\mu < 80$ (one-tailed test, left tail).
3. Sampling distribution: Distribution of the mean. The test statistic is the z value for $\bar{x}$.

$$z = \frac{\bar{x} - \mu_{\bar{x}}}{\sigma_{\bar{x}}}$$

$$= \frac{78 - 80}{6.2/\sqrt{60}} = -2.50$$

4. Decision rule: Reject H_0 if $z < -z_{0.05}$.

5. Critical value: $-z_{0.05} = -1.65$.

6. Decision: Reject H_0, since $-2.50 < -1.65$. The difference -2 is not likely to be due to sampling error alone. We conclude that $\mu < 80$. The finding is significant at the 0.05 level.

7. Type of error possible: Type I. The probability of rejecting H_0 when true is 0.05. ■

p-Values

The third procedure for deciding the fate of a null hypothesis H_0 involves the concept of *p*-value. When computer programs involving library packages (such as MINITAB, SYSTAT, SPSSx, BMDP, and SAS) are used to analyze statistical data, it is common for the resulting printouts to contain *p*-values. The **p-value** (or *probability value*) associated with a test statistic is the smallest level of significance that would have resulted in H_0 being rejected. Thus, for a right-tailed test, the *p*-value is the probability that the test statistic is at least as large as the value of the statistic calculated from the sample under the assumption that H_0 is true. For a left-tailed test, the *p*-value corresponds to the probability that the test statistic is at most as large as the value of the statistic calculated from the sample when H_0 is true. And for a two-tailed test, the *p*-value is the probability of getting a sample result at least as extreme in either direction as the one obtained when H_0 is true. If the level of significance of a test is greater than the reported *p*-value, then H_0 is rejected. Thus, the following rule can be used for deciding whether to reject the null hypothesis:

$$\text{Reject } H_0 \text{ if } p\text{-value} \le \alpha$$

The *p*-value provides an indication of how strongly the data disagree with H_0. Consider the following illustrations. In each case, the *p*-value is indicated by the shaded area in the figure.

1. Left-tailed test:

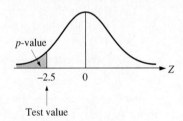

We would reject H_0 if *p*-value $\le \alpha$.

2. Right-tailed test:

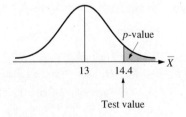

We would reject H_0 if *p*-value $\le \alpha$.

3. Two-tailed test:

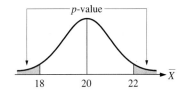

We would reject H_0 if p-value $\leq \alpha$.

APPLICATION 10.5

For the left-tailed test in application 10.4, compute the p-value.

Solution: To compute the p-value, we need to find $P(Z < -2.5)$ or the area under the standard normal curve to the left of the test statistic, as indicated in the accompanying figure. By examining the normal table, we find the p-value to be $0.5 - 0.4938 = 0.0062$. Thus, p-value $= 0.0062$. Since p-value $< \alpha = 0.05$, we reject H_0. This test would have been significant at any level of significance less than or equal to 0.0062.

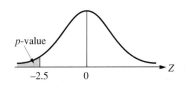

MINITAB has a ZTEST command that can be used to test the null hypothesis of application 10.4. Computer display 10.1 contains the commands used and the output.

Computer Display 10.1

```
MTB > SET C1
DATA> 60(78)
DATA> END
MTB > ZTEST MU=80 SIGMA=6.2 C1;
SUBC> ALTERNATIVE −1.

TEST OF MU = 80.000 VS MU L.T. 80.000
THE ASSUMED SIGMA = 6.20


         N      MEAN      STDEV    SE MEAN       Z     P VALUE
C1      60    78.000      0.000      0.800   −2.50      0.0063

* NOTE * ALL VALUES IN COLUMN ARE IDENTICAL
```

Notice that p-value is 0.0063. This value differs from our value of 0.0062 because of the increased accuracy that MINITAB affords. It is common practice among some researchers to report that the hypothesis test is significant at the p-value. Recall that if the p-value is greater than the significance level α, then the null hypothesis cannot be rejected.

APPLICATION 10.6   For the left-tailed test in application 10.3, compute the p-value.

Solution: The t-value for the test statistic is -1.84. We need to find the area under the t curve with df $= 14$, above the horizontal axis, and to the left of $t = -1.84$ (see the accompanying figure). We scan along the df $= 14$ row of the t table (see back endpaper) and find that $t = 1.84$ lies between $t_{0.05} = 1.761$ and $t_{0.025} = 2.145$. Thus, $0.025 < p$-value < 0.05. Since $\alpha = 0.01$ and p-value $> \alpha$, we fail to reject H_0.

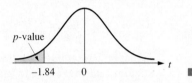

APPLICATION 10.7   For the two-tailed test in application 10.2, compute the p-value.

Solution: Since this test is two-tailed, we need to compute the p-value using both tails. Since the p-value is the smallest level of significance that would have resulted in H_0 being rejected and since the distribution is symmetric, we just double the p-value for one side. The value of the test statistic was $z = 1.43$. We thus need to find the area under the z curve to the right of $z = 1.43$; this value will represent $1/2$ (p-value), as illustrated in the figure. From the z table, we find $1/2$ (p-value) $= 0.5 - 0.4236 = 0.0764$. Hence, p-value $= 2(0.0764) = 0.1528$. This test would have been significant at the 15.3% significance level. Since p-value $> \alpha$, we cannot reject H_0.

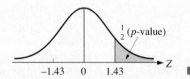

MINITAB was used to test the null hypothesis H_0: $\mu = 10$ versus the alternative hypothesis H_1: $\mu \neq 10$ of application 10.2 by using the ZTEST command. Computer display 10.2 contains the MINITAB commands and output.

Computer Display 10.2

```
MTB > SET C1
DATA> 100(10.2)
DATA> END
MTB > ZTEST MU=10 SIGMA=1.4 C1

TEST OF MU = 10.000 VS MU N.E. 10.000
THE ASSUMED SIGMA = 1.40


        N       MEAN      STDEV     SE MEAN      Z       P VALUE
C1     100     10.200     0.000      0.140     1.43       0.15

* NOTE * ALL VALUES IN COLUMN ARE IDENTICAL
```

Note that the p-value is 0.15, which, to two decimal places, agrees with our answer.

Comparing Confidence Intervals and Two-Tailed Hypothesis Tests

There are two standard techniques for making an inference about the value of an unknown parameter: estimation and hypothesis testing. We discussed estimation in chapter 9. A comparison of an unknown parameter with a known constant involving a two-tailed test at a significance level equal to α can be made by constructing a $(1 - \alpha)100\%$ confidence interval for the parameter. If the hypothesized value of the parameter is contained in the confidence interval, then we cannot conclude that the parameter is different from the known constant.

EXAMPLE 10.22

Let's refer again to the cold-tablet data in application 10.2, and construct a 95% confidence interval for the average amount of aspirin contained in the cold tablet. Recall that the confidence interval limits are found by using

$$\bar{x} \pm z_{\alpha/2} \frac{\sigma}{\sqrt{n}}$$

The critical value is $z_{0.025} = 1.96$. Since $n = 100$ and σ is unknown, s provides a good estimate of σ. Hence, the limits are

$$10.2 \pm 1.96 \frac{1.4}{\sqrt{100}} = 10.2 \pm 0.27$$

A 95% confidence interval for μ is (9.93, 10.47). Since the hypothesized value 10 is contained in the interval, we cannot conclude that $\mu \neq 10$. Note that this result is the same conclusion we arrived at by using the hypothesis-testing procedure in application 10.2.

EXAMPLE 10.23

A confidence interval provides more information than does a hypothesis test. Based on the data, we could reject the null hypothesis and find that the result is of no practical significance. But by using the corresponding confidence interval and some common sense, we can determine if the significant results of the hypothesis test are of any practical importance. Let's consider a hypothetical situation in which a new study guide for increasing SAT mathematics scores is being tested. The mean and standard deviation for the SAT mathematics scores are known to be 500 and 100, respectively. A sample of 1,000,000 students is used to determine whether the new study guide produces a mean SAT mathematics score different from 500. The null hypothesis is that the mean SAT mathematics test score for students using the new study guide is 500, the same as the group not using the new study guide. That is, the null hypothesis is

$$H_0: \mu = 500$$

Let's suppose that the group using the new study guide has a mean SAT mathematics score of 500.4. By using the formula

$$\bar{x} \pm z_{0.05} \frac{\sigma}{\sqrt{n}}$$

we find that a 95% confidence interval for μ is

$$500.4 \pm 1.96 \frac{100}{\sqrt{1,000,000}} = 500.4 \pm (1.96)(0.1) = (500.2, 500.6)$$

Since 500 is not contained in the interval, we would reject the null hypothesis and conclude that the new study guide does have a statistically significant effect on the SAT mathematics test scores. The average test score for the group, $\bar{x} = 500.4$, is four standard deviations above the mean of its sampling distribution. But does this significant finding have any practical importance? Surely, few people (if any) would use the new study guide in order to raise their SAT mathematics score by 0.4 point.

Example 10.23, while hypothetical, does serve to illustrate the following two points:

1. A hypothesis test may yield a significant finding but have no *practical* importance.
2. A large sample size increases the chance of rejecting the null hypothesis.

We learned in chapter 9 that as the sample size increases, the width of the confidence interval shrinks to 0. When the sample is the entire population, $\bar{x} = \mu$ and the width of the confidence interval is 0. In this situation, any null hypothesis $H_0: \mu = \mu_0$ would be rejected, except for the case where μ_0 is the true value of μ. From a theoretical point of view, any null hypothesis can be rejected by choosing a large enough sample. One might then conclude that failure to reject a null hypothesis is a result of the sample not being large enough. Of course, in most practical applications involving hypothesis tests, the amount of data collected is based on economic considerations, as well as the nature of the experiment. For some experiments, such as in the study of rare diseases, it is not possible to gather a large amount of data.

EXERCISE SET 10.3

Further Applications

For exercises 1–11, conduct a hypothesis test and determine the following information:

a. H_0
b. H_1
c. α and type of test (one-tailed or two-tailed)
d. sampling distribution
e. value of the test statistic
f. critical value(s)
g. decision
h. type of error possible and associated probability (if a Type I error)
i. *p*-value

1. A newspaper article stated that college students at a state university spend an average of $95 a year on beer. A student investigator who believed the state average was too high polled a random sample of 50 students and found that $\bar{x} = \$92.25$ and $s = \$10$. Use these results to test at the 0.05 level of significance the statement made by the newspaper.

2. A survey of 50 homemakers selected at random showed that they watch television an average of 15 hours per week; the standard deviation was 12.5 hours. Test $H_0: \mu \geq 20$ against $H_1: \mu < 20$ at the 0.02 level of significance.

3. Truck loads of coal arriving at a power plant are contracted to carry 10 tons of coal per load. A sample of 15 loads showed $\bar{x} = 9.5$ tons and $s = 0.9$ tons. If the distribution of weights is assumed to be normal, test the null

hypothesis $H_0: \mu \geq 10$ versus the alternative hypothesis $H_1: \mu < 10$ by using a level of significance of 0.01.

4. A random sample of ten high school seniors took a standardized mathematics test and made the scores 88, 86, 90, 84, 85, 89, 91, 86, 83, and 87. Past scores at the same high school have been normally distributed with $\sigma = 4$ and $\mu = 85$. Test $H_0: \mu \leq 85$ against $H_1: \mu > 85$. Use a level of significance of 0.05.

5. The scores of ten students on a statistics examination were 43, 61, 67, 70, 74, 76, 79, 85, 94, and 81. Assuming these scores are from a normal population, test $H_0: \mu = 70$ against $H_1: \mu \neq 70$ at the 0.05 level of significance.

6. A certain restaurant advertises that it puts a quarter pound (0.25 pound) of beef in its burgers. A customer who frequents the restaurant thinks that the burgers contain less than 0.25 pound of beef. With permission from the owner, the customer selected a random sample of 50 burgers and found that $\bar{x} = 0.23$ and $s = 0.12$. Test the restaurant's claim using a level of significance of 0.10.

7. A dairy advertises that a tub of its ice cream produces an average of 84 scoops. An ice-cream parlor that buys wholesale from the dairy found that its clerks obtained an average of 83.7 scoops per tub from 72 tubs. The standard deviation was 11.43 scoops. Test the null hypothesis $H_0: \mu \geq 84$ against the alternative hypothesis $H_1: \mu < 84$ by using a level of significance of 0.05. To what do you attribute the results?

8. The lengths (in centimeters) of a random sample of 30 trout caught at Piney Dam were determined; it was found that $\bar{x} = 37.8$ and $s^2 = 27.04$. Do the data indicate that $\mu > 35$? Test by using the 0.05 level of significance.

9. The weights (in pounds) of a random sample of six-month-old babies are 14.6, 12.5, 15.3, 16.1, 14.4, 12.9, 13.7, and 14.9. Test at the 5% level of significance to determine if the average weight of all six-month-old babies is different from 14 pounds. Assume that the weights of six-month-old babies are normally distributed.

10. A sugar manufacturer uses an automatic machine to fill 5-pound bags with sugar. The machine was initially set to dispense an average of 5 pounds and to have a standard deviation of 0.12 pound. To determine if the machine is out of adjustment, a sample of 35 bags was weighed, and it was found that the average weight was 4.9 pounds. Use the 5% level of significance to determine if the machine needs readjusting.

11. The administrator of a large hospital is concerned about the number of days spent in the hospital by surgery patients. She suspects that there may be a tendency for doctors to keep patients in the hospital longer than necessary from a medical viewpoint because virtually all patients are covered by some type of hospital insurance. A random sample of 48 surgery patients is chosen and the number of days spent in the hospital by each is recorded. The data are recorded here.

10	10	4	1	4	10	10	10	8	13	7	10
7	3	4	7	10	11	12	15	3	5	8	6
6	8	7	8	15	11	12	11	10	9	8	5
12	5	6	12	8	9	3	4	12	7	10	7

By using $\alpha = 0.05$, test to determine if the average hospital stay for surgery patients exceeds seven days.

12. For exercise 5, construct a 95% confidence interval for μ. Does the interval contain $\mu = 70$? Compare your results with those obtained in exercise 5.

13. For exercise 9, construct a 95% confidence interval for μ. Does the interval contain $\mu = 14$? Compare your results with those obtained in exercise 9.

14. For exercise 10, construct a 95% confidence interval for μ. Does the interval contain $\mu = 5$? Compare your results with those obtained in exercise 10.

Going Beyond

15. Refer to exercise 8. If the true value of μ is 37, find β, the probability of committing a Type II error.

16. Refer to exercise 6. If the true value of μ is 0.24, find the probability of committing a Type II error.

SECTION 10.4   Testing Proportions and Variances

The hypothesis-testing procedure for testing proportions is similar to the procedure for testing means. The logic is the same; the only difference is that we work with sampling distributions of proportions rather than means. Recall that if n is large, the sampling distribution of $\hat{p} = x/n$ has the following properties:

1. It is approximately normal.
2. $\mu_{\hat{p}} = p$.
3. $\sigma_{\hat{p}} = \sqrt{p(1(-p)/n}$.

A sample is considered large if both $np \geq 5$ and $n(1 - p) \geq 5$. Care should be taken not to confuse the *p*-value (introduced in section 10.3) with *p,* the population proportion, since both are numbers between 0 and 1, inclusive. Consider applications 10.8 and 10.9.

APPLICATION 10.8   A doctor claims that 12% of all his appointments are canceled. Over a six-week period, 21 of the doctor's 200 appointments were canceled. Test at $\alpha = 0.05$ to determine if the true proportion of all appointments that are canceled is different from 12%.

Solution:

1. H_0: $p = 0.12$.

2. H_1: $p \neq 0.12$ (two-tailed test).

3. Sampling distribution: distribution of sample proportions. The test statistic is the z value for p, where $\hat{p} = 21/200 = 0.105$

4. The significance level is $\alpha = 0.05$. Note that $np = (200)(0.12) = 24$ and $n(1 - p) = (200)(0.88) = 176$. Thus, n can be considered large and the sampling distribution of $\hat{p}$ is approximately normal. The z value for $\hat{p}$ is

$$z = \frac{\hat{p} - \mu_{\hat{p}}}{\sigma_{\hat{p}}}$$

$$= \frac{\hat{p} - p}{\sqrt{p(1 - p)/n}}$$

$$= \frac{0.105 - 0.12}{\sqrt{(0.12)(0.88)/200}} = -0.65$$

Note that we used the hypothesized value of p in the denominator of z since we assumed that H_0: $p = 0.12$ is true. In chapter 9, we used $\hat{p}$ in the construction of a confidence interval for p.

5. Critical values: $\pm z_{0.025} = \pm 1.96$ (see the accompanying figure).

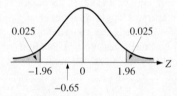

6. Decision: We fail to reject H_0. Thus, we have no statistical evidence to reject the doctor's claim.

7. Type of error possible: Type II, the probability β is unknown.

8. p-value: We have the situation illustrated in the figure. Thus, the p-value is $2(0.2578) = 0.5156$.

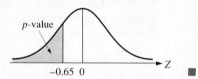

APPLICATION 10.9   Suppose it has been suggested that pregnant women who once took birth-control pills are more likely to have girls than boys. A random sample of 30 mothers who once used the pill and subsequently gave birth to one child produced 7 boys and 23 girls. Do the data indicate that girls are more likely to be born than boys to mothers who once used the pill? Use the 0.05 level of significance.

Solution: Suppose p represents the proportion of girls born to mothers who once used the pill. The value of the sample proportion is $\hat{p} = 23/30 = 0.767$.

1. $H_0: p \le 0.5$.
2. $H_1: p > 0.5$ (one-tailed, right tail).
3. Sampling distribution: distribution of $\hat{p}$. Note that $np = n(1 - p) = 30(0.5) = 15 > 5$. The test statistic is the z value for $\hat{p}$:

$$z = \frac{\hat{p} - p}{\sqrt{p(1 - p)/n}}$$
$$= \frac{0.767 - 0.5}{\sqrt{(0.5)(0.5)/30}} = 2.92$$

4. The significance level is $\alpha = 0.05$.
5. Critical value: $z_{0.05} = 1.65$.
6. Decision: Since $2.92 > 1.65$, we reject H_0. We thus conclude that the proportion of girls born to mothers who once used the pill is greater than 0.5. In other words, we can conclude that more girls than boys are born to mothers who once used the pill.
7. Type of error possible: Type I. The probability of a Type I error is $\alpha = 0.05$.
8. p-value: We have the situation shown in the figure. Therefore, the p-value $= 0.5 - 0.4982 = 0.0018 = 0.18\%$. Thus, the finding is significant at the 0.2% level. ■

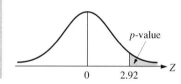

MINITAB can be used to test the null hypothesis $H_0: p \le 0.5$ for the birth-control data in application 10.9 by employing the ZTEST command. Computer display 10.3 contains the commands used and the output.

Computer Display 10.3

```
MTB > SET C1
DATA> 23(1) 7(0)
DATA> END
MTB > ZTEST MU=0.5 SIGMA=0.5 C1;
SUBC> ALTERNATIVE=1.

TEST OF MU = 0.5000 VS MU G.T. 0.5000
THE ASSUMED SIGMA = 0.500

        N      MEAN     STDEV    SE MEAN      Z     P VALUE
C1     30    0.7667   0.4302    0.0913     2.92    0.0018
```

Testing Variances

EXAMPLE 10.24

Many practical applications require making inferences about population variances. In quality control studies, the concern is on consistency of the product. A bolt-manufacturing company, for example, may be interested in testing its manufacturing process to determine whether bolt diameters have a variance smaller than an allowable specification.

Recall that if a random sample is drawn from a normal population, then the statistic $(n - 1)s^2/\sigma^2$ has a sampling distribution that is a chi-square distribution with df $= n - 1$. That is,

$$\chi^2 = \frac{(n - 1)s^2}{\sigma^2}$$

serves as our test statistic, and the critical values are read directly from the chi-square table of probabilities (table 5 of appendix B). Consider application 10.10. On the other hand, sometimes we are interested in testing a null hypothesis involving the population standard deviation. Since the variance is equal to the square of the standard deviation, the procedure used to test a null hypothesis involving the variance can be used. Consider application 10.11.

APPLICATION 10.10

An engineer believes that the variance in waiting times (in seconds), by machinists at a tool cage is greater than 25. To test this, he selected a random sample of 30 machinists' times logged in at the cage and found that $s^2 = 41.4$ square seconds. Assuming that the waiting times are normally distributed, test the null hypothesis H_0: $\sigma^2 \leq 25$ by using the 0.05 level of significance.

Solution:

1. H_0: $\sigma^2 \leq 25$.

2. H_1: $\sigma^2 > 25$ (one-tailed, right tail).

3. Sampling distribution: χ^2, $df = 30 - 1 = 29$. The value of the χ^2 test statistic is

$$\chi^2 = \frac{(n-1)s^2}{\sigma^2}$$

$$= \frac{(29)(41.4)}{25} = 48.024$$

4. The significance level is $\alpha = 0.05$.

5. Critical value: $\chi^2_{0.05}(29) = 42.557$ (found in table 5).

6. Decision: Since $48.024 > 42.557$, we reject H_0 (see the accompanying figure). The engineer can thus conclude that the variance in waiting times is greater than 25.

7. Type of error possible: Type I, $\alpha = 0.05$

8. *p*-value: Reading across the row labeled df = 29 in table 5, we see that the chi-square test statistic 48.024 falls between $\chi^2_{0.05}(29) = 45.772$ and $\chi^2_{0.01}(29) = 49.558$. Thus, $0.01 < p\text{-value} < 0.025$. ■

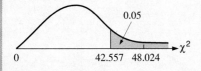

APPLICATION 10.11

An automatic sugar-packaging machine is used to fill 5-pound bags of sugar. A random sample of 15 bags indicated that $\bar{x} = 4.94$ pounds and $s = 0.02$ pound. If the weights are assumed to be normal, and from past experience it is known that the standard deviation of the weights is 0.015 pound, can the apparent increase in variability be explained by sampling error alone? Use the 0.05 level of significance.

Solution:

1. H_0: $\sigma \leq 0.015$.

2. H_1: $\sigma > 0.015$ (right-tailed test).

3. Note that we assume that the null hypothesis H_0: $\sigma = 0.015$ is true, then the hypothesis H_0: $\sigma^2 = (0.015)^2$ is also true, and we can use the chi-square distribution. The test statistic is χ^2. Its value is

$$\chi^2 = \frac{(n-1)s^2}{\sigma^2}$$

$$= \frac{(15-1)(0.02)^2}{(0.015)^2} = 24.89$$

4. The significance level is $\alpha = 0.05$.

5. Critical value: $\chi^2_{0.05}(14) = 23.685$.

6. Decision: Reject H_0, since $\chi^2 = 24.89 > \chi^2_{0.05}(14) = 23.685$.

7. Type of error possible: Type I, $\alpha = 0.05$.

8. p-value: By reading across the row labeled by df $= 14$ in table 5, we locate $\chi^2 = 24.89$ between $\chi^2_{0.05}(14) = 23.685$ and $\chi^2_{0.025}(14) = 26.119$. Thus, $0.025 < p$-value < 0.05.

Hence, the increase in variability should not be attributed to sampling error alone, but to some other factor; perhaps the machine is not being adequately calibrated. ▪

MINITAB can be used to find the p-value for the test in application 10.11. Computer display 10.4 contains the commands used and corresponding output.

Computer Display 10.4

```
MTB > CDF 24.89;
SUBC> CHISQUARE 14.
  24.8900  0.9643
MTB > LET K1 = 1 - 0.9643
MTB > PRINT K1
K1   0.0357000
```

The p-value of 0.0357 is stored in constant K1. Notice that this result agrees with the result obtained in application 10.11, but is more precise.

EXERCISE SET 10.4

Basic Skills

1. Conduct each of the following hypothesis tests at $\alpha = 0.01$:

   a. H_0: $p = 0.5$
   H_1: $p \neq 0.5$
   $\hat{p} = 0.45$
   $n = 500$

   b. H_0: $p = 0.5$
   H_1: $p \neq 0.5$
   $\hat{p} = 0.6$
   $n = 100$

   c. H_0: $p \geq 0.6$
   H_1: $p < 0.6$
   $\hat{p} = 0.5$
   $n = 200$

   d. H_0: $p \leq 0.3$
   H_1: $p > 0.3$
   $\hat{p} = 0.35$
   $n = 50$

2. Conduct each of the following hypothesis tests at $\alpha = 0.05$:

   a. H_0: $p = 0.5$
   H_1: $p \neq 0.5$
   $\hat{p} = 0.55$
   $n = 400$

   b. H_0: $p = 0.5$
   H_1: $p \neq 0.5$
   $\hat{p} = 0.4$
   $n = 75$

   c. H_0: $p \geq 0.55$
   H_1: $p < 0.55$
   $\hat{p} = 0.5$
   $n = 150$

   d. H_0: $p \leq 0.2$
   H_1: $p > 0.3$
   $\hat{p} = 0.25$
   $n = 40$

3. Conduct the following hypothesis tests:

   a. H_0: $\sigma^2 \geq 100$
   H_1: $\sigma^2 < 100$
   $s^2 = 50$
   $n = 28$
   $\alpha = 0.05$

   b. H_0: $\sigma^2 \leq 200$
   H_1: $\sigma^2 > 200$
   $s^2 = 250$
   $n = 25$
   $\alpha = 0.01$

   c. H_0: $\sigma^2 = 150$
   H_1: $\sigma^2 \neq 150$
   $s^2 = 120$
   $n = 37$
   $\alpha = 0.05$

4. Conduct the following hypothesis tests:

   a. H_0: $\sigma^2 \geq 90$
   H_1: $\sigma^2 < 90$
   $s^2 = 80$
   $n = 30$
   $\alpha = 0.01$

   b. H_0: $\sigma^2 \leq 100$
   H_1: $\sigma^2 > 100$
   $s^2 = 150$
   $n = 41$
   $\alpha = 0.05$

c. H_0: $\sigma^2 = 25$
   H_1: $\sigma^2 \neq 25$
   $s^2 = 12$
   $n = 38$
   $\alpha = 0.01$

5. The following sample was taken from a normal population: 12, 10, 13, 12, 11, 13, 14, 13, 14, 12, and 10. Test H_0: $\sigma \geq 2.5$ against H_1: $\sigma < 2.5$ using the 0.05 level of significance.

6. The following sample was drawn from a normal population: 41, 10, 25, 5, 10, 10, 30, 19, 6, 10, 14, 14, 41, 25, 14, 30, 25, 14, 30, and 25. By using $\alpha = 0.01$, test H_0: $\sigma^2 = 121$ versus H_1: $\sigma^2 \neq 121$.

Further Applications

7. From a random sample of 500 males interviewed, 125 indicated that they watch professional football on Monday night television. Does this evidence indicate that more than 20% of the male TV viewers watch professional football on Monday evenings? Use the 0.01 level of significance, and find the *p*-value.

8. A random sample of 300 shoppers in a shopping mall is selected and 182 are found to favor longer shopping hours. Is this sufficient evidence to conclude that less than 65% of the shoppers favor longer shopping hours? Use $\alpha = 0.05$ and find the *p*-value.

9. A candidate for the local board of education thinks that at least 55% of the voters will vote for her. If a random sample of 50 voters indicated that 42% would vote for her, test H_0: $p \geq 0.55$ against H_1: $p < 0.55$ at $\alpha = 0.05$. Find the *p*-value.

10. If 60% of a sample of 100 new-car buyers indicated that they want air conditioning, does this indicate that less than 70% of all new-car buyers want air conditioning? Use $\alpha = 0.02$ and find the *p*-value.

11. A national survey indicated that 60% of all new homes constructed have four bedrooms. If a random sample of 100 newly constructed homes indicated 52 with four bedrooms, does this suggest that the national survey is wrong? Use $\alpha = 0.05$ and find the *p*-value.

12. The following sample is thought to have come from a normal population with variance $\sigma^2 = 25$: 2.1, 3.6, 3.8, 4.2,

4.7, and 15.3. Test at the 0.05 level of significance to determine if $\sigma^2 \neq 25$.

13. The Metro Bus Company in a large city claims to have a variance in bus-arrival times (arrival times measured in minutes) at its various bus stops of no more than 5. A bus company executive ordered that arrival data be collected at various bus stops in order to determine if bus drivers are maintaining consistent schedules. If a sample of 12 bus arrivals at a particular bus stop produced a variance of 5.7 and the population of arrival times is assumed to be normal, test the null hypothesis H_0: $\sigma^2 \leq 5$ versus H_1: $\sigma^2 > 5$. Use $\alpha = 0.05$.

14. The variance in the diameters of roller bearings during production is of critical importance. Large variances in bearing diameters promote wear and bearing failure. Industry standards call for a variance of no more than 0.0001 when the bearing diameters are measured in inches. A bearing manufacturer selected a random sample of 25 bearings and found that $s = 0.015$ inch. Does this indicate that $\sigma^2 > 0.0001$? Use $\alpha = 0.01$ and assume that the bearing diameters are normally distributed.

15. For exercise 11, construct a 95% confidence interval for the true proportion of all new homes constructed that have four bedrooms. Compare your results with those obtained in exercise 11.

16. For exercise 12, construct a 95% confidence interval for the true population variance σ^2. Compare your results with those obtained in exercise 12.

17. A large dairy continually monitors the level of butterfat content in its milk. In its 2% milk, the percentage of butterfat should not deviate much from this percentage. A standard deviation of 10% is acceptable. A sample of 20 cartons of milk was obtained and the percentage of butterfat in each carton was recorded. The results are given below.

1.85	2.25	2.01	1.90	1.97	1.80	2.05
2.23	1.65	1.86	2.02	2.09	2.04	2.07
2.14	1.93	2.08	2.17	1.91	1.93	

a. Construct a 95% confidence interval for the variance of butterfat percentages of 2% milk.

b. By using $\alpha = 0.05$ test to determine if the variance in butterfat percentages exceeds 1%.

CHAPTER SUMMARY

In this chapter we discussed hypothesis testing, an alternative to estimation for making inferences. When using hypothesis testing, we are interested in determining whether a parameter is different from a specified value. The logic of hypothesis testing is indirect; that is, we assume the null hypothesis is true, and based on this assumption, we determine if our test statistic is likely to be from the hypothesized population. If not, we conclude that the null hypothesis should be rejected in favor of the alternative hypothesis. Since for each type of decision—

reject H_0 or fail to reject H_0—there is a probability of committing an error, we never "prove" anything using hypothesis testing. Failure to reject the null hypothesis does not necessarily mean that we will accept the null hypothesis. Instead, we shall reserve judgment on H_0 since the probability of committing a Type II error is unknown. We discussed tests for μ, p, σ^2, and σ. The test procedures are summarized in the following table, where the values μ_0, p_0, σ_0^2, and σ_0 are considered to be fixed values.

Parameter	Null Hypothesis	Sampling Distribution	Test Statistic	Critical Values One-tailed test	Critical Values Two-tailed test	Assumptions
μ	$H_0: \mu = \mu_0$ $H_0: \mu \leq \mu_0$ $H_0: \mu \geq \mu_0$	Normal distribution	$z = \dfrac{\bar{x} - \mu_0}{\sigma/\sqrt{n}}$ $z = \dfrac{\bar{x} - \mu_0}{s/\sqrt{n}}$	z_α or $-z_\alpha$	$\pm z_{\alpha/2}$	Normal population or $n \geq 30$
μ	$H_0: \mu = \mu_0$ $H_0: \mu \leq \mu_0$ $H_0: \mu \geq \mu_0$	t distribution	$t = \dfrac{\bar{x} - \mu_0}{s/\sqrt{n}}$	t_α or $-t_\alpha$ df $= n - 1$	$\pm t_{\alpha/2}$ df $= n - 1$	Normal population or $n < 30$ σ unknown
p	$H_0: p = p_0$ $H_0: p \leq p_0$ $H_0: p \geq p_0$	Normal distribution $np \geq 5$ and $n(1 - p) \geq 5$	$z = \dfrac{\hat{p} - p_0}{\sqrt{p_0(1 - p_0)/n}}$	z_α or $-z_\alpha$	$z_{\alpha/2}$	$n \geq 30$
σ^2, σ	$H_0: \sigma^2 = \sigma_0^2$ $H_0: \sigma^2 \leq \sigma_0^2$ $H_0: \sigma^2 \geq \sigma_0^2$	χ^2 distribution	$\chi^2 = \dfrac{(n - 1)s^2}{\sigma_0^2}$	χ_α^2 (df) or $\chi_{1-\alpha}^2$ (df)	$\chi_{\alpha/2}^2$ (df) or $\chi_{1-\alpha/2}^2$ (df)	Normal population

CHAPTER REVIEW

■ **IMPORTANT TERMS** ■

The following chapter terms have been mixed in ordering to provide you better review practice. For each of the following terms, provide a definition in your own words. Then check your responses against the definitions given in the chapter.

decision rule
alternative hypothesis
right-tailed test
critical value
directional test

statistical hypotheses
experimental hypothesis
hypothesis-testing procedure
nondirectional test
null hypothesis

left-tailed test
null value
one-tailed test
power of a test
p-value

rejection region   significant finding   two-tailed test

level of significance   statistical hypothesis   Type I error

hypothesis testing   test statistic   Type II error

■   *IMPORTANT SYMBOLS*   ■

H_1, alternative hypothesis

H_0, null hypothesis

α, probability of Type I error, level of significance

β, probability of Type II error

C, critical value

χ^2, chi-square test statistic

■   *IMPORTANT FACTS AND FORMULAS*   ■

If the null hypothesis cannot be rejected, judgment should be reserved concerning the truth of the null hypothesis. One should conclude that the data offer no significant evidence to warrant that the null hypothesis be rejected in favor of the alternative hypothesis.

REVIEW EXERCISES

1. A large television repair service claims that its average repair charge is $24. Suspecting that the average repair charge is higher, a consumer group randomly chose a sample of 35 statements for television repairs done by the repair service and found $\overline{x} = 25.50$ and $s = 2.25$. Test the claim at the 5% level of significance and find the p-value.

2. A basketball team claims that their opponents scored an average of at most 87 points against them over the past ten years. Suspecting that the true average was somewhat higher, a sports reporter gathered a random sample of 30 game summaries for the period. He found $\overline{x} = 89.6$ points and $s = 6.2$ points. Test the team's claim at $\alpha = 0.05$.

3. A newspaper in a large city advertised that 62% of the registered voters were opposed to abortions. A social service agency, believing the estimate was too high, polled a random sample of 500 registered voters and found that 290 were opposed to abortions. Using the 5% level of significance, test to determine if the newspaper's estimate is too high and find the p-value.

4. A coffee dispenser at a local cafeteria is supposed to dispense an average of 7 ounces of coffee. Suspecting that the average amount dispensed is somewhat lower, a customer obtained a random sample of 15 cups over a period of two weeks and found $\overline{x} = 6.4$ ounces and $s = 0.71$ ounce. Assuming that the amounts of coffee dispensed are normally distributed, test at $\alpha = 0.01$ to determine if the average amount of coffee dispensed is less than 7 ounces.

5. The weights (in pounds) of a random sample of 16-year-old male high school students are 146, 149, 137, 153, 125, 219, and 161. Assuming that the weights of 16-year-old boys are normally distributed, test the null hypothesis that the average weight of 16-year-old boys is equal to 140 pounds. Use the 0.05 level of significance.

6. The statistics department at a certain university has never been able to achieve a failure rate less than 11% for its introductory statistics course. During an experimental semester, all students enrolled in introductory statistics were required to attend a 1-hour laboratory in addition to classes with the hope of lowering the failure rate. At the end of the semester, 171 students out of 1800 students failed. Test using the 0.05 level of significance to determine if there has been a significant decrease in failure rate since the lab was instituted.

7. A beer company uses dispensing machines to fill beer cans that provide a maximum variance of 0.05 (amount of beer is measured in ounces) so that cans are not overfilled or underfilled. A sample of fills for 25 cans yielded $s^2 = 0.07$. If the amounts of beer dispensed are normally distributed, test the null hypothesis $\sigma^2 \leq 0.05$ against the alternative hypothesis $\sigma^2 > 0.05$. Use $\alpha = 0.05$.

8. A moped manufacturer advertises that its moped gets an average of at least 127.6 miles per gallon of gasoline. The standard deviation σ is known to be 9.8 miles per gallon and the gas mileage ratings are normally distributed. Sus-

pecting the manufacturer's claim is too high, a consumer took a sample of ten mopeds and found a sample mean of 120.6 miles per gallon and a sample standard deviation of 8.4 miles per gallon. Test the manufacturer's claim using $\alpha = 0.05$.

9. A new drug, Cyclosporin A, is claimed to have been 86% successful in increasing the success rate in 30 organ transplant operations. Before this new drug was available, a success rate of 60% had been obtained with organ-transplant patients. By using $\alpha = 0.05$, determine if the success rate has improved as a result of the new drug.

10. For exercise 8, test using $\alpha = 0.05$ to determine if $\sigma \neq 9.8$.

11. One-pound cans of nuts are to contain a net weight of 16 ounces, but there is considerable variability. A random sample of six cans of brand A nuts revealed the following net weights (in ounces): 16.1, 15.8, 15.1, 15.4, 16.1, and 16.2. Using $\alpha = 0.01$, determine if the true net weight is different from 16 ounces.

12. A new alloy is claimed to have a tensile strength of 120 pounds. A sample of seven independent tests provided the following readings (in pounds): 116.5, 118.7, 122.3, 118.7, 122.3, 122.6, and 121.6. At $\alpha = 0.05$, do the data indicate that $\mu \neq 120$?

Computer Applications

1. A study involving a random sample of adult men was conducted to test the claim that the average weight of adult men exceeds 82 kilograms. Listed below are the weights (in kilograms) for the adult men.

84	84	84	82	77	71	81	75	66	90	92
89	84	89	77	68	76	83	97	76	85	89
83	85	94	68	85	76	82	80	87	87	81
84	85	85	86	76	77	100	72	83	66	89
79	74	83	86	98	79	92	76	78	81	82
72	80	85	89	97	97	78	79	85	75	81
72	75	88	85	93	103	88	81	75	87	82
87	76	93	84	88	77	74	71	77	79	69
78	78	85	85	84	86	78	79	68	94	85
81										

Use $\alpha = 0.01$ to test the claim.

2. A study was conducted to determine if a new method of teaching statistics is more effective than the traditional method. A random sample of 100 students was involved in the study. A standardized examination having a mean of 74 was used to access the achievement. The results of the final examination are as follows:

68	80	68	89	82	61	67	80	76	74	63
74	90	74	78	73	64	71	93	60	74	87
74	83	73	88	90	80	87	81	68	100	79
70	91	83	68	79	74	79	65	84	71	74
62	69	95	76	97	62	71	71	80	85	67
78	57	66	76	75	84	84	62	71	82	82
67	66	86	75	60	60	85	76	80	95	79
79	74	71	79	91	74	62	75	80	75	77
70	71	72	74	73	61	64	67	91	70	79
77										

Use $\alpha = 0.05$ to determine if the new method is more effective than the traditional method.

3. A ninth-grade teacher suspects that the students at her school have an average IQ lower than 100. The IQ scores for a random sample of 24 students are as follows:

132	103	94	78	108	105	98	114	89	112	95	80
82	86	124	118	120	87	120	107	95	104	100	99

Use $\alpha = 0.05$ to determine if $\mu < 100$.

4. **a.** Simulate drawing 20 random samples from a normal population with mean 50 and standard deviation 2. For each sample, test the null hypothesis H_0: $\mu = 50$ by using $\alpha = 0.10$.
   b. How many hypotheses are rejected?
   c. Are your results consistent with the theory? Explain.

EXPERIMENTS WITH REAL DATA

Treat the 720 subjects listed in the database in appendix C as if they constituted the population of all U.S. subjects.

1. Conduct a hypothesis test to determine if the mean systolic blood pressure of smokers aged 30–39 years is greater than 120. Draw a random sample of 40 subjects, use the sample mean as the test statistic, and use the 0.05 level of significance.

2. Conduct a hypothesis test to determine if the proportion of smokers is different than 0.40. Use a sample of size 50 and the 0.01 level of significance. Determine the *p*-value of the test.

3. Conduct a hypothesis test to determine if the standard deviation of the diastolic blood pressures for subjects aged 40–59 is greater than 10. Use a sample size of 35 and the 0.05 level of significance.

■ CHAPTER ACHIEVEMENT TEST ■

1. A claim is made that the mean height of male college teachers is 71 inches. In an investigation to test the claim, a random sample of 12 male teachers yielded $\bar{x} = 72$ inches and $s = 3$ inches. Assuming that the heights of male teachers are normally distributed, test the claim at the 0.05 level of significance and provide the following information:
   a. value of the test statistic
   b. critical value(s)
   c. decision
   d. *p*-value

2. A particular medicine is claimed to be 85% effective in alleviating a certain type of allergic reaction. A consumer group believes the medicine is less than 85% effective and gathered a sample of 60 people who experience the type of allergic reaction. Of this group who used the medicine, 48 people got relief. Test the claim at the 0.01 level of significance and provide the following information:
   a. value of the test statistic
   b. critical value(s)

   c. decision
   d. *p*-value

3. A manufacturer claims that the diameters of its 8-mm bolts have a variance of at most 0.02. The diameters of a sample of 20 8-mm bolts yielded a variance of 0.025. Test at $\alpha = 0.05$ to determine if $\sigma^2 > 0.02$, and provide the following information:
   a. value of the test statistic
   b. critical value
   c. decision

4. A new type of radial tire is tested to determine if it can average at least 60,000 miles of road wear. A sample of 35 tires was experimentally tested and it was determined that $\bar{x} = 59,600$ miles and $s = 968$ miles. At $\alpha = 0.05$, do the data indicate that $\mu < 60,000$ miles?

5. In exercise 4, assume that the tread wear is normally distributed and test, using $\alpha = 0.01$, to determine if $\sigma < 1200$ miles.

11 Inferences Comparing Two Parameters

CHAPTER OBJECTIVES

In this chapter we will investigate

▷ *The two types of samples used in making inferences about two populations.*

▷ *The properties of the sampling distribution of the differences between sample means.*

▷ *How to construct confidence intervals for $\mu_1 - \mu_2$.*

▷ *How to test the null hypothesis $H_0: \mu_1 - \mu_2 = 0$.*

▷ *The properties of the sampling distribution of the differences between sample proportions.*

▷ *How to construct confidence intervals for $p_1 - p_2$.*

▷ *How to test the null hypothesis $H_0: p_1 - p_2 = 0$.*

▷ *The properties of the F distributions.*

▷ *How to construct confidence intervals for σ_1^2 / σ_2^2.*

▷ *How to test the null hypothesis $H_0: \sigma_1^2 = \sigma_2^2$.*

||||| **MOTIVATOR 11**

*C*ourses in basic computer literacy are very common in colleges and universities, and students entering these courses come with a full range of attitudes and varying degrees of anxiety regarding their abilities to use computers. Members of the Department of General Education at Towson State University in Maryland conducted a study to investigate the relationship between computer anxiety and achievement in an introductory undergraduate computer course, Computers in Society.[41] The Computers in Society course may be used to satisfy a general university requirement and may also be substituted for a mathematics requirement at Towson. The course met three hours per week for approximately 16 weeks. The course included information about the history of computers, basic hardware and software concepts, and practical experience working with personal computers. About one-half of the class time was spent in a lab setting in which students worked with Logo and Appleworks.

Students from nine sections of the course were subjects for the study. Participation in the study was voluntary and not related to course grade. Twenty-seven percent of the students were males; 73 percent were females. The median age was 22. Sixty-three percent of the students had taken no previous course with computers, and of those who reported having had a course, 80 percent had one previous course.

The procedures for the study included administering a Computer Anxiety Scale at the beginning and end of the course. Pre- and post-test scores were analyzed by using a t test for dependent samples to answer the question "As a result of the course, Computers in Society, are there reductions in students' level of anxiety?" No significant ($\alpha = 0.05$) difference between mean pretest scores and mean posttest scores were found. Table 11.1 contains summary information for the study.

TABLE 11.1

Summary Information for Computers In Society Study

N	Variable	Mean	t Value	Probability
	Pre-Anxiety	63.48		
	Post-Anxiety	60.34		
89			1.95	0.055

The researchers also found that age was significantly related to computer anxiety. Pearson's correlation coefficient for age and pre-anxiety score was -0.241 ($p < 0.01$) and for age and post-anxiety score was -0.218($p < 0.05$). Unlike previous studies, the researchers did not find a significant relationship between gender and computer anxiety. In addition, no significant relationship was found between anxiety and students' academic achievement in the computer course as measured by their course grade. The t test for dependent samples will be examined in section 11.6.

Chapter Overview

Many practical applications involve the comparison of two populations. We commonly compare two populations by comparing their corresponding parameters, such as μ_1 and μ_2, p_1 and p_2, σ_1 and σ_2, or σ_1^2 and σ_2^2. For example, population means μ_1 and μ_2 might be compared when deciding which brand of tooth paste, A or B, is more effective in preventing cavities. Or population percentages p_1 and p_2 might be compared when trying to decide if a greater proportion of males than females favor lowering the minimum wage. Two population variances σ_1^2 and σ_2^2 might be compared when a drug manufacturer is interested in comparing the consistencies of two different methods of producing a heart drug containing 25 mgs of digoxin. In this chapter we will use hypothesis testing and estimation to compare two population parameters.

SECTION 11.1   *Independent and Dependent Samples*

When researchers compare two parameters, such as μ_1 and μ_2, it is common practice to consider the difference of the parameters $\mu_1 - \mu_2$. By determining whether the difference $\mu_1 - \mu_2$ equals zero, we can make comparisons concerning μ_1 and μ_2. That is, if $\mu_1 - \mu_2 = 0$, then $\mu_1 = \mu_2$; if $\mu_1 - \mu_2 > 0$ then $\mu_1 > \mu_2$; and if $\mu_1 - \mu_2 < 0$, then $\mu_1 < \mu_2$. To estimate the difference $\mu_1 - \mu_2$, we usually use the point estimator $\bar{x}_1 - \bar{x}_2$; to estimate $p_1 - p_2$, we usually use the point estimator $\hat{p}_1 - \hat{p}_2$, the difference between sample proportions. When certain assumptions are satisfied, inferences comparing two parameters can be made by considering the sampling distribution of the difference between sample statistics, such as $\bar{x}_1 - \bar{x}_2$ and $\hat{p}_1 - \hat{p}_2$. We begin by drawing a distinction between two basic types of samples used in comparing two population parameters.

In order to make statistical inferences about two populations, we need to have a sample from each population. The two samples will be independent or dependent, according to how they are selected. If the selection of sample data from one population is unrelated to the selection of sample data from the other population, the samples are called **independent samples.** If the samples are chosen in such a way that each measurement in one sample can be naturally paired with a measurement in the other sample, the samples are called **dependent samples.** Each piece of data results from some source. A **source** is anything, a person or an object, that produces a piece of data. If two measurements result from the same source, then the measurements can be thought of as being paired. As a result, two samples resulting from the same set of sources are dependent. Note that if two samples are dependent, then they necessarily have the same size. Therefore if two samples have different sizes, they cannot be dependent. Examples 11.1–11.6 and applications 11.1 and 11.2 will help clarify the two types of samples.

EXAMPLE 11.1

Ten overweight adults were randomly selected to evaluate a particular diet. Each person was weighed before beginning the diet and again after being on the diet 12 weeks. The sample of weights before dieting and the sample of weights after dieting are dependent samples. A source is a person, who provides two measurements, one for each sample. The paired observations provide the sample measurements.

EXAMPLE 11.2

A farmer from the Midwest conducted an experiment to determine if the use of a special chemical additive with the fertilizer he has been using to grow soybeans will accelerate plant growth. Fifteen locations were randomly chosen for the study. At each location, two soybean plants located close to one another were treated, one with the standard fertilizer and one with the standard fertilizer with the chemical additive. Plant growth was measured in inches for each plant after a four-week period. The plant measurements associated with each type of fertilizer constitute two dependent samples. A source is a location; each location produces a pair of measurements.

EXAMPLE 11.3

A medical researcher compared two flu vaccines, A and B, for localized side effects. Ten subjects were randomly selected and each subject was injected twice, with vaccine A in the left arm and vaccine B in the right arm. Sufficient time was allowed between injections to serve as a *washout* period for vaccine A. The side effects of each vaccine were measured using a special numerical index. The set of numerical indices for each vaccine constitutes a sample, and the two samples are dependent. A source is a person.

EXAMPLE 11.4

In an experiment to determine whether persons afflicted with glaucoma have abnormally thick corneas, eight persons with glaucoma in only one eye were examined. The cornea thickness (in microns) was measured for each person's eyes. The eight thickness measurements for the glaucomatous eyes constitute one sample, and the eight other eye measurements constitute the other sample. The samples are dependent, and each person serves as a source.

EXAMPLE 11.5

In an experiment to determine whether persons afflicted with glaucoma have abnormally thick corneas, 16 subjects were involved. Eight had glaucoma and 8 did not. Cornea thickness was measured (in microns) for each subject. The cornea measurements for the glaucoma patients comprise one sample and the cornea measurements for the subjects not afflicted with glaucoma comprise the other sample. The two samples are independent; no pairing is involved.

EXAMPLE 11.6

Twelve one-week-old white male infants from middle- and upper-middle-class families were involved in an experiment to determine whether special walking exercises in the newborn can lower the average age at which infants first walk alone. Two groups were randomly chosen from the twelve infants. Group A received special training for an eight-week period, while group B received no special training. After the eight-week period, the ages (in months) when the children first walked were recorded. The ages for the two groups comprise independent samples.

APPLICATION 11.1

A medical researcher wants to determine whether drug therapy can improve the IQ of children with learning and behavioral problems. An experiment will be conducted in which two groups of children, A and B, are used. Both groups are to be the same size. Group A will receive a placebo for six weeks, whereas group B will receive a widely used anticonvulsant drug for six weeks. A verbal IQ test will be used to assess results at the end of the six-week period.

    a. Describe how you would obtain two independent samples of IQ scores to evaluate.

    b. Describe how you would obtain two dependent samples of IQ scores to evaluate.

Solution:

    a. Randomly divide 20 children with learning and behavioral problems into two groups of ten. One group will receive the placebo (no therapy) and the other group will receive the drug therapy. The two samples of IQ test scores will be independent.

    b. Choose ten children with learning and behavioral difficulties. Each child will be administered the placebo for three weeks and the drug for three weeks. After each three-week period, all children will be administered an IQ test. Because a child might do better the second time he or she takes the IQ test, the order in which the placebo and drug are administered will be randomized; some children will be given the placebo first and some will be given the drug first. IQ scores for each child will be recorded following the three-week placebo period and following the three-week drug period. The two samples of ten IQ scores form dependent samples. A source in each case is a student. ▪

APPLICATION 11.2

Ten pairs of identical twins are used in an experiment to determine which of two experimental methods is better for teaching statistics. One method involves a small-group discovery method and the other method involves a lecture-discussion format with computer assisted instruction. One member of each pair of twins is assigned to each method. Following the instruction, an examination is administered to each group. Are the two samples of test scores independent? Explain.

Solution: Identical twins come from the same fertilized egg and therefore have the same set of genes, which determine their physical and mental characteristics. As a result, the sample can be considered dependent. The pairing is natural. ▪

Why Use Dependent Samples

The procedures for making inferences about two population means involve the sample means calculated from two dependent or two independent samples. Dependent samples are used to control the effects of certain "nuisances," thus reducing the effects of unwanted factors, as we see in example 11.7. Inferences concerning two populations using dependent samples are covered in section 11.6. The remaining sections of this chapter deal with the methods for making inferences about population parameters using independent samples.

EXAMPLE 11.7

Suppose we were given the task of determining the wearing quality of two different brands of tires, A and B. To do this, we must road test the two brands of tires. Automobiles must be selected, equipped with both brands of tires, and then driven for a fixed distance under similar conditions. Tread wear will then be measured (in thousandths of an inch) for each brand of tire. If we want the tread-wear measurements to reflect only the quality of the tires, then we want to control as many of the following factors as possible that could also affect tread wear:

1. Type of car, including size and weight

2. Mechanical condition of the car

3. Driver's driving habits

4. Type of roads, including terrain

5. Distance traveled

6. Location of tires on the car

7. Weather conditions

If we randomly choose, say ten cars, randomly place one tire of each brand on the rear wheels of each car, and then drive the cars a fixed distance under similar conditions, we can control or limit the influence that the above factors can have on the results. Those factors we can not control, such as driver habits and course traveled, should have an equal effect on both brands of tires due to the way we designed the experiment. Thus, the tread-wear measurements should reflect the wearing qualities of the two brands of tires.

EXERCISE SET 11.1

Further Applications

1. An experiment to test a new variety of sugarcane against an old variety was conducted on ten different 1-acre plots. The new variety was expected to yield higher sugar content than the variety currently used. The ten different 1-acre plots were chosen where soil and general climate conditions vary. Each acre was divided into two half-acre plots containing similar soil, and the new variety was planted in one half-acre plot and the old type was planted in the other half-acre plot. After harvest, the average sugar content for each half-acre plot was determined. Do the samples of data for the two varieties of sugarcane represent dependent or independent samples? Explain.

2. To test whether honeybees show a preference for stinging objects that have already been stung, an experiment was repeated ten times. Each time eight cotton balls wrapped in muslin were dangled up and down in front of a beehive entrance. Four of the balls were exposed to a swarm of angry bees and were filled with stingers, while the other four were fresh. After a specified length of time, the number of new stingers in each ball was counted. Are the two sets of bee-sting counts independent or dependent? Explain.

3. An experiment was conducted in Milwaukee, Wisconsin, to test the effectiveness of artificial food flavoring in rat poison. For each survey, approximately 1600 poison baits were placed around garbage storage bins; half of the poison baits were plain cornmeal and half were butter-vanilla flavored cornmeal. To ensure that a rat would have equal access to both kinds of poison baits, the baits were always placed in pairs. After two weeks, the sites were inspected and the percentage of poison baits that were gone was recorded. A different set of locations in the same general vicinity was selected and the experiment was repeated again for another two-week period. This same procedure was repeated three more times. For each of the five surveys, the percentage of baits accepted was recorded for each flavoring. Are the two samples of five percentages independent or dependent? Explain.

4. An experiment was designed to determine if fluoride helps to prevent tooth cavities. Fifteen children had their teeth cleaned and treated with fluoride. Another 15 had their teeth cleaned, but received no fluoride. Six months later, the number of cavities was recorded for each child. Are the two sets of cavity counts independent or dependent? Explain.

5. Describe how you would redesign the experiment of exercise 3 to obtain samples of the opposite type.

6. Describe how you would redesign the experiment of exercise 4 to obtain samples of the opposite type.

7. A manufacturer of waterproofing for footwear claims its product is superior to the leading brand. Ten pairs of shoes are available for a test.
   a. Explain how you would conduct a test using dependent samples. How would you make your assignments of waterproofing?
   b. Explain how you would conduct a test using independent samples.

8. An experiment is to be designed to compare two competing headache remedies, A and B, for fast-acting relief. Sixteen sets of identical twins are available for the study. Eight subjects are to be assigned to each of two groups. Each subject in one group is to be administered 500 mg of remedy A and each subject in the other group is to be administered 500 mg of remedy B. The length of time in minutes for each drug to reach a specified level in the blood will be recorded for each individual.
   a. Describe how you could obtain two independent samples of absorption times to evaluate.
   b. Describe how you could obtain two dependent samples of absorption times to evaluate.

9. To determine what effect a rat's environment early in life has on his behavior late in life, 16 rats from the same litter were separated from their mothers when they were three weeks old. One half (Group A) were put into individual cages and the other half (Group B) were put into the same cage. After living under these conditions for a period of approximately eight months, each rat was individually put through a series of swimming trials to determine the length of time required to swim a specified distance underwater. Based on the swimming times the rats within each group were ranked, and the rats in the two groups were paired according to respective ranks: the rat ranking number 1 in group A was paired with the rat ranking number 1 in group B, and so on. Rats in each pair had approximately the same swimming rates. Pair by pair, the rats were subjected to competitive underwater swimming competitions. Average times were recorded for each rat under competition. Each of the eight pairs of rats competed 60 times. The average times for each group constitute a sample. Are the two samples independent or dependent? Explain.

SECTION 11.2   *Inferences Concerning $\mu_1 - \mu_2$ Using Large, Independent Samples*

In each of the following situations, we might be interested in comparing two population means, μ_1 and μ_2:

■ Medicine—Determining which of two treatments is most effective in treating a certain disease

■ Education—Comparing test scores resulting from two modes of instruction to determine the better mode

■ Manufacturing—Comparing two automobiles for the best average gas mileage

■ Sports—Comparing two defensive strategies in basketball by examining opponents' scores

■ Agriculture—Comparing two types of fertilizer on crop yields

■ Law—Comparing two municipal courts in a large city for the average length of time it takes to try a case

■ Science—Comparing a new synthetic material for strength against an old material

■ Business—Comparing two methods of marketing a product

■ Mining—Comparing two methods of mining coal

■ Religion—Comparing attitudes between youth and adults concerning the Bible

There are few new ideas covered in this section that were not presented in chapters 9 and 10; apart from involving a new sampling distribution, the techniques and logic involved in comparing two parameters are the same as those presented in chapters 9 and 10. Before we can begin to make inferences concerning the difference between two population means, however, we need to develop the background needed for making such comparisons. We begin by examining the sampling distribution of the differences between sample means.

Sampling Distribution of the Differences between Sample Means

EXAMPLE 11.8

Suppose we want to decide which of two brands of toothpaste, A or B, is more effective in preventing cavities in teenagers. How do we decide whether a difference exists, or how can we estimate the difference? We would agree that if there is a difference in therapeutic benefits, then this difference should be reflected in the difference in the average numbers of cavities for teenagers using the two brands. A starting point is to gather some data from two different groups of teenagers, one using brand A and the other using brand B. The best estimate for the difference between the population means is the corresponding difference between the means for the two samples. The difference in sample means, $\bar{x}_1 - \bar{x}_2$, becomes our estimator for the difference in population means, $\mu_1 - \mu_2$. Therefore, we need to examine and understand the sampling distribution of $\bar{x}_1 - \bar{x}_2$; we will need to use it to determine if there is a difference in the corresponding population parameters by using a hypothesis test or estimation.

In general, suppose we have two distinct populations, the first with mean μ_1 and standard deviation σ_1, and the second with mean μ_2 and standard deviation σ_2. Further, suppose a random sample of size n_1 is selected from the first population and an independent random sample of size n_2 is selected from the second population. The sample mean is computed for each sample and the difference between sample means is calculated. The collection of all such differences is called the **sampling distribution of the differences between means** or the **sampling distribution of the statistic $\bar{x}_1 - \bar{x}_2$**. Figure 11.1 illustrates the sampling distribution of $\bar{x}_1 - \bar{x}_2$, and the sampling distribution of the differences between sample means is shown in applications 11.3–11.4. The sampling distribution of $\bar{x}_1 - \bar{x}_2$ has the following properties:

FIGURE 11.1

Sampling distribution of $\bar{x}_1 - \bar{x}_2$

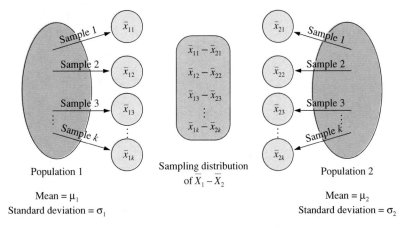

Population 1
Mean = μ_1
Standard deviation = σ_1

Sampling distribution of $\bar{X}_1 - \bar{X}_2$

Population 2
Mean = μ_2
Standard deviation = σ_2

Properties of the Sampling Distribution of $\bar{x} - \bar{x}_2$

1. The distribution is approximately normal for $n_1 \geq 30$ and $n_2 \geq 30$. If the populations are normal, then the sampling distribution is normal regardless of the sample sizes.

2. $\mu_{\bar{x}_1 - \bar{x}_2} = \mu_1 - \mu_2$ (11.1)

3. $\sigma_{\bar{x}_1 - \bar{x}_2} = \sqrt{\dfrac{\sigma_1^2}{n_1} + \dfrac{\sigma_2^2}{n_2}}$

MINITAB can be used to simulate the sampling distribution of the difference in sample means. Fifty random samples of size 20 are drawn from a normal population having a mean of 50 and a standard deviation of 5. Means of the 50 samples are stored in C46. In addition, 50 random samples of size 25 are drawn from a normal population having a mean of 75 and a standard deviation of 8. Means of the 50 samples are stored in C47. The differences of the sample means are stored in C48. The mean of the 50 differences of sample means is stored in constant K1, and the standard deviation of the 50 differences in sample means is stored in constant K2. A histogram of the 50 differences in sample means is constructed. Computer display 11.1 contains the commands and corresponding output.

Computer Display 11.1

```
MTB > RANDOM 50 C1-C20;
SUBC> NORMAL 50 5.
MTB > RANDOM 50 C21-C45;
SUBC> NORMAL 75 8.
MTB > RMEAN C1-C20 C46
MTB > RMEAN C21-C45 C47
MTB > LET C48 = C47 - C46
MTB > LET K1 = MEAN(C48)
MTB > LET K2 = STDEV(C48)
MTB > PRINT K1 K2
K1   24.6727
K2   2.16680
MTB > HISTOGRAM C48

HISTOGRAM OF C48   N = 50

MIDPOINT     COUNT
    20         1  *
    21         2  **
    22         5  *****
    23         5  *****
    24        13  *************
    25         7  *******
    26         6  ******
    27         5  *****
    28         3  ***
    29         3  ***
```

The mean and standard deviation of the empirical distribution of 50 differences in sample means and the mean and standard deviation of the theoretical distribution of the sampling distribution of the differences in sample means can be compared. From computer display 11.1 we can determine that the empirical distribution of 50 differences in sample means has a mean of 24.6727 and a standard deviation of 2.1668. And by using formula (11.1) we can determine that the mean of the theoretical distribution of differences in sample means is

$$\mu_{\bar{x}_1-\bar{x}_2} = \mu_1 - \mu_2$$
$$= 75 - 50 = 25$$

and the standard deviation of the theoretical distribution of differences in sample means is

$$\sigma_{\bar{x}_1-\bar{x}_2} = \sqrt{\frac{\sigma_1^2}{n_1} + \frac{\sigma_2^2}{n_2}} = \sqrt{\frac{25}{20} + \frac{64}{25}} = 1.95$$

The means for the two distributions are approximately equal (25 and 24.67) and the standard deviations for the two distributions are approximately equal (1.95 and 2.17). The histogram for the empirical distribution contained in computer display 11.1 suggests that the sampling distribution of the differences in sample means is normal.

APPLICATION 11.3

In a study to compare the average weights of sixth-grade boys and girls at a large middle school, a random sample of 20 boys and a random sample of 25 girls are to be used. It is known that the weights are normally distributed for both boys and girls. All sixth-grade boys at the school have an average weight of 100 pounds and a standard deviation of 14.142 pounds, whereas all sixth-grade girls have an average weight of 85 pounds and a standard deviation of 12.247 pounds. If $\bar{X}_1$ represents the average weight of a sample of 20 boys and $\bar{X}_2$ represents the average weight of a sample of 25 girls, find $P(\bar{X}_1 - \bar{X}_2 > 20)$, the probability that the average weight for the 20 males is at least 20 pounds more than the average weight for the 25 females.

Solution: A value for the difference between sample means $\bar{x}_1 - \bar{x}_2$ is an element in the sampling distribution of the differences between sample means. As a result of formula (11.1), we have

1. The sampling distribution of $\bar{x}_1 - \bar{x}_2$ is approximately normal.
2. The mean of the sampling distribution of $\bar{x}_1 - \bar{x}_2$ is $\mu_{\bar{x}_1-\bar{x}_2} = \mu_1 - \mu_2$
   $= 100 - 85 = 15$.
3. The standard deviation of the sampling distribution of $\bar{x}_1 - \bar{x}_2$ is

$$\sigma_{\bar{x}_1-\bar{x}_2} = \sqrt{\frac{\sigma_1^2}{n_1} + \frac{\sigma_2^2}{n_2}}$$
$$\sqrt{\frac{(14.142)^2}{20} + \frac{(12.247)^2}{25}} = 4.0$$

Thus, to determine $P(\bar{X}_1 - \bar{X}_2 > 20)$, we find the shaded area under the following standard normal curve:

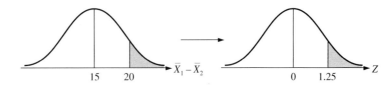

The z value for a difference of 20 is given by

$$z = \frac{20 - \mu_{\bar{x}_1 - \bar{x}_2}}{\sigma_{\bar{x}_1 - \bar{x}_2}}$$

$$= \frac{20 - 15}{4} = 1.25$$

Hence, we have

$$P(\overline{X}_1 - \overline{X}_2 \geq 20) = P(Z \geq 1.25)$$

$$= 0.5 - 0.3944 = 0.1056$$

Therefore, the probability that the average weight of the sample of boys is at least 20 pounds greater than the average weight of the sample of girls is 0.1056. ■

APPLICATION 11.4

A leading television manufacturer purchases picture tubes from two companies, A and B. The tubes from company A have a mean lifetime of 7.2 years and a standard deviation of 0.8 year, whereas those from company B have a mean lifetime of 6.7 years and a standard deviation of 0.7 year. Determine the probability that a random sample of 34 picture tubes from company A will have a mean lifetime that is at least 1 year more than the mean lifetime of a random sample of 40 picture tubes from company B.

Solution: We summarize the given information as shown in the table. As a consequence of formula (11.1), the mean of the sampling distribution of differences between sample means is

Company A	Company B
$\mu_1 = 7.2$	$\mu_2 = 6.7$
$\sigma_1 = 0.8$	$\sigma_2 = 0.7$
$n_1 = 34$	$n_2 = 40$

$$\mu_{\bar{x}_1 - \bar{x}_2} = \mu_1 - \mu_2$$

$$= 7.2 - 6.7 = 0.5$$

and the standard error of the difference between sample means is

$$\sigma_{\bar{x}_1 - \bar{x}_2} = \sqrt{\frac{\sigma_1^2}{n_1} + \frac{\sigma_2^2}{n_2}}$$

$$= \sqrt{\frac{0.8^2}{34} + \frac{0.7^2}{40}}$$

$$= \sqrt{\frac{0.64}{34} + \frac{0.49}{40}}$$

$$= \sqrt{0.01882 + 0.01225} = 0.176$$

If the mean of the sample from company A is at least one year more than the mean of the sample from company B, then $\bar{x}_1 - \bar{x}_2 \geq 1$. The z value for $\bar{x}_1 - \bar{x}_2 = 1$ is given by

$$z = \frac{1 - 0.5}{0.176} = 2.84$$

Thus, the probability is

$$P(\overline{X}_1 - \overline{X}_2 > 1) = P(Z > 2.84)$$

$$= 0.5 - 0.4977 = 0.0023$$

Hence, the probability that the mean lifetime of a random sample of 34 pictures tubes from company A will exceed the mean lifetime of a random sample of 40 picture tubes from company B by at least one year is 0.0023. ▪

It is interesting to note that if independent samples of sizes n_1 and n_2 are drawn from the same normal population with mean μ and variance σ, then by formula (11.1) we have

1. The sampling distribution of $\bar{x}_1 - \bar{x}_2$ is normal.
2. $\mu_{\bar{x}_1 - \bar{x}_2} = \mu - \mu = 0.$
3. $\sigma_{\bar{x}_1 - \bar{x}_2} = \sqrt{\sigma^2/n_1 + \sigma^2/n_2} = \sigma\sqrt{1/n_1 + 1/n_2}.$

(11.2)

Application 11.5 illustrates sampling from the same normal population.

APPLICATION 11.5 Suppose two independent samples of size 20 are drawn from a normal population with mean 10 and variance 2.5. Find $P(\bar{X}_1 - \bar{X}_2 < 1)$.

Solution: Since the sampling distribution of the differences between sample means is normal, we will find the z score for 1 and then determine the area under the standard normal curve below this z value. From formula (11.2) we know that the mean of the sampling distribution of the differences between sample means is

$$\mu_{\bar{x}_1 - \bar{x}_2} = 10 - 10 = 0$$

and the standard deviation of the sampling distribution of the differences between means is

$$\sigma_{\bar{x}_1 - \bar{x}_2} = \sqrt{2.5}\ \sqrt{\frac{1}{20} + \frac{1}{20}}$$
$$= \sqrt{(2.5)(0.10)} = 0.5$$

The z score for 1 thus becomes

$$z = \frac{1 - \mu_{\bar{x}_1 - \bar{x}_2}}{\sigma_{\bar{x}_1 - \bar{x}_2}}$$
$$= \frac{1 - 0}{0.5} = 2$$

Thus, $P(\bar{X}_1 - \bar{X}_2 < 1) = P(Z < 2) = 0.5 + 0.4772 = 0.9772.$ ▪

If two independent samples of the same size are selected from the same population, one would expect the difference between any pair of sample means to be close to 0 for large sample sizes. This can be seen by examining the variance of the sampling distribution of the difference between sample means $\sigma^2_{\bar{x}_1 - \bar{x}_2}$.

$$\sigma^2_{\bar{x}_1 - \bar{x}_2} = \frac{\sigma^2}{n} + \frac{\sigma^2}{n} = \frac{2\sigma^2}{n}$$

As n gets large, the values of $2\sigma^2/n$ will approach 0. Thus, for large values of n, the variability of the differences between means is small, and the differences between sample means are close to their mean, which is 0.

Confidence Intervals
for $\mu_1 - \mu_2$

Confidence intervals can be used to compare two population means, μ_1 and μ_2. To do so, a $(1 - \alpha)100\%$ confidence interval for $\mu_1 - \mu_2$ is constructed. If the interval contains 0, then μ_1 could equal μ_2, μ_1 could be smaller than μ_2, or μ_1 could be larger than μ_2. If the confidence interval contains only negative values, then one can be $(1 - \alpha)100\%$ confident that $\mu_1 - \mu_2 < 0$ or $\mu_1 < \mu_2$. Or, if the confidence interval contains only positive values, one can be $(1 - \alpha)100\%$ confident that $\mu_1 - \mu_2 > 0$ or $\mu_1 > \mu_2$. Recall from chapter 9 that the limits for a confidence interval can be obtained by employing a point estimator and its standard error. Consequently, the limits for a $(1 - \alpha)100\%$ confidence interval for $\mu_1 - \mu_2$ can be obtained by using the following limits:

> **Limits for $(1 - \alpha)100\%$ Confidence Interval for $\mu_1 - \mu_2$**
>
> $$\overline{x}_1 - \overline{x}_2 \pm z_{\alpha/2}\sigma_{\overline{x}_1 - \overline{x}_2}$$
>
> $\uparrow$ point estimate $\uparrow$ standard error

(11.3)

When a confidence interval is constructed for the difference between population means, there are two possibilities for expressing the difference in means: $\mu_1 - \mu_2$ and $\mu_2 - \mu_1$. The confidence intervals for $\mu_1 - \mu_2$ and $\mu_2 - \mu_1$ are not identical, but they are related in a special way. If (L_1, L_2) is a $(1 - \alpha)100\%$ confidence interval for $\mu_1 - \mu_2$, then $(-L_2, -L_1)$ is a $(1 - \alpha)100\%$ confidence interval for $\mu_2 - \mu_1$. This can be seen by observing that if

$$L_1 < \mu_1 - \mu_2 < L_2$$

then by multiplying both inequalities by -1, we have $-L_1 > \mu_2 - \mu_1 > -L_2$. And rewriting these inequalities, we have

$$-L_2 < \mu_2 - \mu_1 < -L_1$$

The choice of which interval to construct is arbitrary, but we need to be consistent when interpreting the interval.

As a general rule, when we are requested to determine a confidence interval for the difference between population means and the difference is not specified, we shall use the interval corresponding to a positive difference between sample means.

APPLICATION 11.6

In an effort to determine if more college males watch television than college females, independent random samples of 50 males and 40 females were used. The number of hours each student spent watching television during a specific week was recorded. The results (in hours) were as shown in the table.

Males	Females
$n_1 = 50$	$n_2 = 40$
$\overline{x}_1 = 10.25$	$\overline{x}_2 = 8.50$
$s_1 = 5.2$	$s_2 = 4.8$

Construct a 95% confidence interval for $\mu_1 - \mu_2$ and interpret the result.

Solution: We will use $\bar{x}_1 - \bar{x}_2 = 10.25 - 8.50 = 1.75$ to estimate $\mu_1 - \mu_2$. Since both samples are large (greater than 30), we can use s_1 and s_2 to estimate σ_1 and σ_2, respectively. Also, since $1 - \alpha = 0.95$, $\alpha = 0.05$ and $\alpha/2 = 0.025$. The critical value $z_{0.025}$ is found in the z table (see front endpaper) to be 1.96. The standard error of $\bar{x}_1 - \bar{x}_2$ is

$$\sigma_{\bar{x}_1 - \bar{x}_2} = \sqrt{\frac{\sigma_1^2}{n_1} + \frac{\sigma_2^2}{n_2}}$$

$$= \sqrt{\frac{5.2^2}{50} + \frac{4.8^2}{40}} = 1.06$$

By using expression (11.3) to construct a 95% confidence interval for $\mu_1 - \mu_2$, we have

$$\bar{x}_1 - \bar{x}_2 \pm z_{\alpha/2}\,\sigma_{\bar{x}_1 - \bar{x}_2} = 1.75 \pm (1.96)(1.06) = 1.75 \pm 2.08$$

Hence, a 95% confidence interval for $\mu_1 - \mu_2$ is $(-0.33, 3.83)$. Since this interval contains 0 we can not be 95% confident that $\mu_1 \neq \mu_2$ or that more male students watch television than female students. ▪

Hypothesis Tests for $\mu_1 - \mu_2$

Hypothesis tests can also be used to compare μ_1 and μ_2. The logic and procedures for testing are identical to those presented in chapter 10. Two samples of data are collected and the value of the test statistic $\bar{x}_1 - \bar{x}_2$ is located in its sampling distribution under the assumption that H_0 is true. If the value of the test statistic falls in the tail area(s) determined by α, then the value of the statistic is determined to be unlikely from the assumed normal population, and the null hypothesis H_0 is rejected in favor of the alternative hypothesis H_1. And if the value of the statistic falls near the center of its sampling distribution, there is no statistical evidence to suggest that H_0 is not true. When testing H_0: $\mu_1 - \mu_2 = 0$, the z value for the test statistic $\bar{x}_1 - \bar{x}_2$ is computed using the following formula:

Z value for $\bar{x}_1 - \bar{x}_2$

$$z = \frac{\bar{x}_1 - \bar{x}_2}{\sqrt{(\sigma_1^2/n_1) + (\sigma_2^2/n_2)}} \qquad (11.4)$$

When directional hypotheses (e.g., H_1: $\mu_1 - \mu_2 < 0$) are used, an upper-tailed test for one researcher might be a lower-tailed test for another. For example, if one is interested in establishing that $\mu_1 > \mu_2$, there are two possibilities for stating H_1:

$$H_1 \text{:} \ \mu_1 - \mu_2 > 0 \quad \text{and} \quad H_1 \text{:} \ \mu_2 - \mu_1 < 0$$

This is because the inequality $\mu_1 > \mu_2$ is equivalent to the inequality $\mu_2 < \mu_1$; hence, either $\mu_1 - \mu_2 > 0$ or $\mu_2 - \mu_1 < 0$ is correct. If the alternative hypothesis is H_1: $\mu_1 - \mu_2 > 0$, then the test statistic becomes $\bar{x}_1 - \bar{x}_2$, and if the alternative hypothesis is H_1: $\mu_2 - \mu_1 < 0$, then the test statistic becomes $\bar{x}_2 - \bar{x}_1$. These observations are summarized in the table.

H_1	Test Statistic
$\mu_1 - \mu_2 < 0$	$\bar{x}_1 - \bar{x}_2$
$\mu_2 - \mu_1 > 0$	$\bar{x}_2 - \bar{x}_1$
$\mu_1 - \mu_2 \neq 0$	$\bar{x}_1 - \bar{x}_2$ or $\bar{x}_2 - \bar{x}_1$

It is generally a good practice to use the test statistic that preserves uniformity of "match" between the subscripts used for the sample means and the subscripts used for the corresponding population means. For example, we would choose $\overline{x}_1 - \overline{x}_2$ as our test statistic for the one-tailed test involving H_1: $\mu_1 - \mu_2 > 0$. It is best not to label directional tests involving two parameters as right tailed or left tailed.

Applications 11.7 and 11.8 illustrate the procedures of estimation and hypothesis testing for making inferences concerning a comparison of two population means, μ_1 and μ_2. In practice, a researcher would choose only one procedure, based on personal preference, for drawing an inference concerning $\mu_1 - \mu_2$. We will use both procedures, side by side, to illustrate that both procedures lead to consistent results.

APPLICATION 11.7 The state department of education in a southern state compared high school seniors' knowledge of the basic skills in mathematics at two different high schools, one located in the northern part of the state and one located in the southern part of the state. Random samples of 50 seniors from each high school were obtained and given a standardized mathematics achievement examination. An analysis of the examination scores yielded the results shown in the table.

Northern School	Southern School
$n_1 = 50$	$n_2 = 50$
$\overline{x}_1 = 81.4$	$\overline{x}_2 = 84.5$
$s_1 = 4.6$	$s_2 = 4.0$

Determine whether μ_1 is significantly different from μ_2 using $\alpha = 0.05$.

Solution:

Estimation Procedure Since $\alpha = 0.05$, $1 - \alpha = 0.95$. Also, since $\overline{x}_2 > \overline{x}_1$, a 95% confidence interval will be constructed for $\mu_2 - \mu_1$. The best point estimator for the difference between population means $\mu_2 - \mu_1$ is the difference between sample means $\overline{x}_2 - \overline{x}_1$. The value of $\overline{x}_2 - \overline{x}_1$ is $84.5 - 81.4 = 3.1$. Since σ_1 and σ_2 are unknown and $n_1 = n_2 = 50$, the values of s_1 and s_2 can be used as point estimates for σ_1 and σ_2, respectively. In addition, since n_1 and n_2 are both greater than 30, the sampling distribution of the differences between sample means is approximately normal. The standard error of the difference between means is found as follows:

$$\sigma_{\overline{x}_1 - \overline{x}_2} = \sqrt{\frac{\sigma_1^2}{n_1} + \frac{\sigma_2^2}{n_2}}$$

$$= \sqrt{\frac{(4.0)^2}{50} + \frac{(4.6)^2}{50}} = 0.862$$

The positive critical value is $z_{0.025} = 1.96$. Limits for the confidence interval are found using (11.3):

$$\overline{x}_2 - \overline{x}_1 \pm z_{\alpha/2}\sigma_{\overline{x}_1 - \overline{x}_2} = 3.1 \pm (1.96)(0.862) = 3.1 \pm 1.69$$

The 95% confidence interval for $\mu_2 - \mu_1$ is (1.41, 4.79). Thus, we can be 95% confident that $\mu_2 - \mu_1$ is contained in the interval (1.41, 4.79); 95% of all such intervals constructed will contain $\mu_2 - \mu_1$. Since the confidence interval (1.41, 4.79) contains only positive values, we can be 95% confident that $\mu_2 - \mu_1 > 0$ or $\mu_2 > \mu_1$.

Hypothesis-Testing Procedure We carry out the following steps:

1. The null hypothesis is H_0: $\mu_1 - \mu_2 = 0$.
2. The alternative hypothesis is H_1: $\mu_1 - \mu_2 \neq 0$ (two-tailed test).
3. The sampling distribution is the distribution of the differences in sample means. The standard error of the difference was found above to be 0.862. The z value for -3.1 is found by using formula (11.4).

$$z = \frac{\bar{x}_1 - \bar{x}_2}{\sqrt{\sigma_1^2/n_1 + \sigma_2^2/n_2}}$$

$$= \frac{-3.1}{0.862} = -3.60$$

4. The level of significance is $\alpha = 0.05$.
5. The critical values for z are $\pm z_{0.025}$ or ± 1.96.
6. Decision: Since $-3.6 < -1.96$, we reject H_0.
7. Type of error possible: Type I, $\alpha = 0.05$.
8. p-value: $p = 2(0.5 - 0.49984) = 0.00032$ (a value very close to 0). ▪

Note, with hypothesis testing we cannot conclude that $\mu_1 - \mu_2 < 0$ or $\mu_1 < \mu_2$, since the alternative hypothesis H_1: $\mu_1 - \mu_2 \neq 0$ is a two-tailed test. We can conclude only that $\mu_1 \neq \mu_2$. Another directional hypothesis test would have to be used to statistically establish (at $\alpha = 0.05$) that $\mu_1 < \mu_2$. Although one-sided confidence intervals can be constructed to parallel directional hypothesis tests, we shall not develop these concepts since they are not consistent with the general objectives of this text.

APPLICATION 11.8

A large company wants to hire a secretary. Two private secretarial schools are available to recruiters. The personnel officer of the company gave a typing test to independent random samples of 50 recent graduates from each school. The test was developed to determine the correct number of words typed per minute. The results obtained are shown in the table. Based on the data, can we conclude at the 95% level of confidence that there is a significant difference between the average scores of students from the two schools? Which school should be chosen to recruit from, if all other factors are equal?

School A	School B
$n_1 = 50$	$n_2 = 50$
$\bar{x}_1 = 67$	$\bar{x}_2 = 70$
$s_1 = 15$	$s_2 = 11$

Solution: A personnel officer not trained in statistics might think that school B is better, since it has the higher sample average score. But do the mean scores differ because of sampling error or do the means differ because graduates of school B actually have a greater average typing score than graduates of school A? Inferential statistics can help us answer this question. The difference between means $\bar{x}_2 - \bar{x}_1$ is $70 - 67 = 3$. Can this difference of 3 be attributed to random error alone? If the schools were identically effective and if two more samples of size 50 were obtained, we would not expect to obtain a difference between sample means of exactly 3. Again, we shall use both procedures for drawing inferences to illustrate how the procedures compare. In practice, a researcher would use only one procedure.

Hypothesis-Testing Procedure

1. The null hypothesis is H_0: $\mu_2 - \mu_1 = 0$.
2. The alternative hypothesis H_1: $\mu_2 - \mu_1 \neq 0$.

3. The sampling distribution is the distribution of differences in sample means, and the test statistic is the z value for $\bar{x}_2 - \bar{x}_1$.

$$z = \frac{\bar{x}_2 - \bar{x}_1}{\sqrt{\sigma_1^2/n_1 + \sigma_2^2/n_2}}$$

$$= \frac{3}{\sqrt{15^2/50 + 11^2/50}}$$

$$= \frac{3}{2.63} = 1.14$$

4. The level of significance is $\alpha = 0.05$ (two-tailed test).

5. The critical values are $\pm z_{0.025}$ or ± 1.96.

6. Decision: Since $-1.96 < 1.14 < 1.96$, we fail to reject H_0.

7. Type of error possible: Type II, β is unknown.

8. p-value: $p = 2(0.5 - 0.3729) = 2(0.1271) = 0.2542$.

As a result of this test, we can conclude that the observed difference of 3 between sample means can be accounted for by random error in sampling; there is no statistical evidence that one school is better than the other. Note that we should not conclude that there is no difference between school averages, since β is unknown.

Estimation Procedure We shall construct a $(1 - \alpha)100\% = (1 - 0.05)100\% = 95\%$ confidence interval for $\mu_2 - \mu_1$ (note that $\bar{x}_2 - \bar{x}_1$ is positive). We find the limits for the confidence interval by using (11.3):

$$\bar{x}_2 - \bar{x}_1 \pm z_{\alpha/2}\,\sigma_{\bar{x}_2 - \bar{x}_1} = 3 \pm (1.96)(2.63) = 3 \pm 5.15$$

Hence, we find that $(-2.15, 8.15)$ is a 95% confidence interval for $\mu_2 - \mu_1$. Since the interval contains 0, any of the three relationships may be true: $\mu_1 = \mu_2$, $\mu_1 < \mu_2$, or $\mu_1 > \mu_2$. Thus, we would not want to conclude that the null hypothesis H_0: $\mu_2 - \mu_1 = 0$ is true. ■

EXERCISE SET 11.2

Basic Skills

1. Two brands of golf balls A and B are to be compared with respect to driving distance. The balls are tested on an automatic driving device known to give normally distributed distances with a standard deviation of 15 yards. It is known that $\mu_a = 285$ yards and $\mu_b = 280$ yards. If random samples of 25 of each type of ball are hit, determine the probability that $\bar{x}_a - \bar{x}_b$ is greater than 11 yards.

2. At a certain eastern college, the average score for freshmen on an entrance examination is 450 and the standard deviation is 40. If two groups of freshmen students are selected at random, one of size 40 and the other of size 45, what is the probability that the two groups of students will differ in their mean scores by more than 10 points?

Further Applications

3. Independent random samples of grades for males and females were selected from the student population of a large

university in an effort to determine which gender had the higher grade-point average (GPA). The results are as follows:

Males	Females
$n_1 = 50$	$n_2 = 75$
$\bar{x}_1 = 2.1$	$\bar{x}_2 = 2.3$
$s_1 = 0.8$	$s_2 = 0.7$

a. By using hypothesis testing, determine if there is a difference between average college male and female GPAs. Use $\alpha = 0.01$.

b. Determine if there is a difference between average male and female GPAs by constructing a 99% confidence interval for the difference between averages.

4. A study was conducted to determine the difference between salaries of college science teachers and industrial

employees who were once college science teachers. Two random samples of salary information were gathered and the results follow:

Science Teachers	Industrial Employees
$n_1 = 50$	$n_2 = 60$
$\bar{x}_1 = \$34,960$	$\bar{x}_2 = \$35,440$
$s_1 = \$1,200$	$s_2 = \$1,000$

a. Construct a 90% confidence interval for $\mu_2 - \mu_1$ and interpret it.
b. At the 0.05 level of significance, do the data support that $\mu_2 > \mu_1$? Find the p-value.

5. Two brands of cigarettes, C and D, are compared for their nicotine contents (in milligrams). Random samples of 40 brand C cigarettes and 50 brand D cigarettes yielded the following results:

Brand C	Brand D
$n_1 = 40$	$n_2 = 50$
$\bar{x}_1 = 14.3$	$\bar{x}_2 = 15.7$
$s_1 = 2.9$	$s_2 = 3.8$

a. At the 1% level of significance, do the two brands of cigarettes differ in their mean nicotine contents?
b. Construct a 99% confidence interval for the difference between mean nicotine contents for the two brands of cigarettes.

6. The length of time (in days) to complete recovery for hernia patients randomly assigned to two different surgical procedures are as follows:

Procedure 1	Procedure 2
$n_1 = 30$	$n_2 = 35$
$\bar{x}_1 = 7.50$	$\bar{x}_2 = 8.25$
$s_1 = 1.12$	$s_2 = 1.38$

At $\alpha = 0.05$, do the data indicate a difference between many recovery times for the two surgical procedures? Use estimation and hypothesis testing to arrive at your conclusion. What is the p-value for the test?

7. Samples of hourly wages of truck drivers in cities A and B yielded the following data:

City A	City B
$n_1 = 30$	$n_2 = 30$
$\bar{x}_1 = \$5.30$	$\bar{x}_2 = \$5.40$
$s_1 = \$0.16$	$s_2 = \$0.15$

Test the null hypothesis $H_0: \mu_2 - \mu_1 \leq 0$ against the alternative hypothesis $H_1: \mu_2 - \mu_1 > 0$ using $\alpha = 0.01$. Find the p-value.

8. The director of athletics at a large university was interested in determining whether male students who participate in college athletics are taller than other male students. Two independent random samples of height data (in inches) were collected and the following results were obtained:

Participants	Nonparticipants
$n_1 = 50$	$n_2 = 75$
$\bar{x}_1 = 68.2$	$\bar{x}_2 = 67.4$
$s_1 = 5.2$	$s_2 = 2.9$

Test the hypothesis at the 0.01 significance level that male students who participate in college athletics are taller than nonparticipants. Find the p-value.

9. Of 80 recently hired employees for a large firm, half were assigned to a special one-day orientation class and half received no special orientation. After three months, on-the-job evaluations were conducted, producing the following information:

Received Orientation	No Orientation
$n_1 = 40$	$n_2 = 40$
$\bar{x}_1 = 84.1$	$\bar{x}_2 = 81.8$
$s_1 = 3.6$	$s_2 = 4.1$

At the 0.05 level of significance, do the data indicate that employees receiving special orientation perform better on the job than those who do not? Find the p-value.

10. Two drugs, A and B, are compared for duration of pain relief in postoperative patients. Records are kept on the number of hours of pain relief for 40 randomly selected patients using drug A and 50 randomly selected patients using drug B. The results are as follows:

Drug A	Drug B
$n = 40$	$n = 50$
$\bar{x} = 5.14$	$\bar{x} = 4.53$
$s = 1.20$	$s = 1.79$

By using $\alpha = 0.05$, determine if drug A has a significantly longer duration of pain relief than drug B.

Going Beyond

11. Two random samples of the same size are taken from normal populations with variances of 5 and 10. If $\bar{x}_1 - \bar{x}_2$ is used to estimate $\mu_1 - \mu_2$, determine the sample size needed in order to be 95% confident that $\bar{x}_1 - \bar{x}_2$ differs by no more than one unit from $\mu_1 - \mu_2$.

12. Two methods of teaching ninth-grade science are being compared, an old method and a new method. The following data on test scores resulted:

Old Method	New Method
$n_1 = 60$	$n_2 = 75$
$\bar{x}_1 = 68.1$	$\bar{x}_2 = 72.9$
$s_1 = 5.1$	$s_2 = 5.5$

At the 0.05 level of significance, test the hypothesis that the mean science score of students taught under the new method is over three points above the mean science score of students taught by the old method. Find the *p*-value.

SECTION 11.3   Inferences Comparing Population Proportions or Percentages

Many applications involve populations of qualitative data that need to be compared by using proportions or percentages. The following are examples:

■ Politics—Is there a difference between the percentages of Democrats and Republicans favoring SALT talks?

■ Education—Is the proportion of students who pass mathematics greater than the proportion who pass English?

■ Medicine—Is the percentage of users of drug A who have an adverse reaction less than the percentage of users of drug B who have an adverse reaction?

■ Management—Is there a difference between the percentages of men and women in management positions?

■ Marketing—Is the percentage of beer drinkers who prefer Budweiser greater than the percentage who prefer Miller's?

Both estimation and hypothesis-testing procedures can be used to draw inferences comparing two population proportions or percentages. The techniques are similar to those involved with comparing population means using large samples. The major exception is that a sampling distribution of the differences between sample proportions is used instead of a sampling distribution of differences between sample means.

Sampling Distribution of $\hat{p}_1 - \hat{p}_2$

Many practical applications involve qualitative data that can be placed into one of two categories. In the case of one-sample problems encountered in chapter 10 we saw that, as a consequence of the central limit theorem, the sampling distribution of the sample proportion $\hat{p} = x/n$ is approximately normal with $\mu_{\hat{p}} = p$ and $\sigma_{\hat{p}} = \sqrt{p(1 - p)/n}$. When sampling is from two binomial populations and two sample proportions are involved, the sampling distribution of $\hat{p}_1 - \hat{p}_2$ is approximately normal for large sample sizes [$n_1 p_1 \geq 5$, $n_1(1 - p_1) \geq 5$, $n_2 p_2 \geq 5$, and $n_2(1 - p_2) \geq 5$]. Then $\hat{p}_1$ and $\hat{p}_2$ have approximately normal sampling distributions, so their difference $\hat{p}_1 - \hat{p}_2$ also has an approximately normal sampling distribution. The following properties of the sampling distribution of $\hat{p}_1 - \hat{p}_2$ hold.

> **Properties of the Sampling Distribution of $\hat{p}_1 - \hat{p}_2$**
>
> 1. The distribution is approximately normal
> 2. $\mu_{\hat{p}_1 - \hat{p}_2} = p_1 - p_2$ $\qquad$ (11.5)
> 3. $\sigma_{\hat{p}_1 - \hat{p}_2} = \sqrt{\dfrac{p_1(1 - p_1)}{n_1} + \dfrac{p_2(1 - p_2)}{n_2}}$
>
> where p_1 is the first population proportion, p_2 is the second population proportion, n_1 is the size of the first sample, and n_2 is the size of the second sample.

Consider application 11.9, showing the sampling distribution of the difference between sample proportions.

APPLICATION 11.9   Adult males and females living in a large northern city differ in their views concerning the issue of the death penalty for persons found guilty of murder. It is believed that 12% of the adult males favor the death penalty, whereas only 10% of the adult females favor the death penalty. If a random sample of 150 males and a random sample of 100 females are polled concerning their views on the issue of the death penalty for persons found guilty of murder, determine the probability that the percentage of males who favor the death penalty is at least 3% higher than the percentage of females who favor the death penalty.

Solution: Let p_1 represent the percentage of males who favor the death penalty and p_2 represent the percentage of females who favor the death penalty. As a consequence of (11.5), the mean of the sampling distribution of the differences between sample proportions is

$$\mu_{\hat{p}_1 - \hat{p}_2} = p_1 - p_2 = 0.12 - 0.10 = 0.02$$

and the standard error of the differences between sample proportions is

$$\sigma_{\hat{p}_1 - \hat{p}_2} = \sqrt{\frac{p_1(1 - p_1)}{n_1} + \frac{p_2(1 - p_2)}{n_2}}$$

$$= \sqrt{\frac{(0.12)(0.88)}{150} + \frac{(0.10)(0.90)}{100}} = 0.04$$

Thus, the z value for $\hat{p}_1 - \hat{p}_2 = 0.03$ is given by

$$z = \frac{\hat{p}_1 - \hat{p}_2 - \mu_{\hat{p}_1 - \hat{p}_2}}{\sigma_{\hat{p}_1 - \hat{p}_2}}$$

$$= \frac{0.03 - 0.02}{0.04} = \frac{0.01}{0.04} = 0.25$$

Hence,

$$P(\hat{p}_1 - \hat{p}_2 \geq 0.03) = P(Z \geq 0.25)$$

$$= 0.5 - 0.0987 = 0.4013$$

Thus, the probability that the percentage of males who favor the death penalty for persons found guilty of murder is at least 3% higher than the percentage of females who favor the death penalty is 0.4013. ■

Confidence Intervals for $p_1 - p_2$

By properties (11.5), the standard error of the difference between sample proportions is given by

$$\sigma_{\hat{p}_1-\hat{p}_2} = \sqrt{\frac{p_1(1 - p_1)}{n_1} + \frac{p_2(1 - p_2)}{n_2}}$$

Since p_1 and p_2 are unknown, we can use $\hat{p}_1$ and $\hat{p}_2$ to estimate p_1 and p_2, respectively. As a result, $\hat{\sigma}_{\hat{p}_1-\hat{p}_2}$ is used to estimate $\sigma_{\hat{p}_1-\hat{p}_2}$, where

$$\hat{\sigma}_{\hat{p}_1-\hat{p}_2} = \sqrt{\frac{\hat{p}_1(1 - \hat{p}_1)}{n_1} + \frac{\hat{p}_2(1 - \hat{p}_2)}{n_2}}$$

Hence, the limits for an approximate $(1 - \alpha)100\%$ confidence interval for $(p_1 - p_2)$ can be found by using the following rule:

> **Limits for a $(1 - \alpha)\%$ Confidence Interval for $p_1 - p_2$**
>
> $$\hat{p}_1 - \hat{p}_2 \pm z_{\alpha/2}\hat{\sigma}_{\hat{p}_1-\hat{p}_2} \qquad (11.6)$$
>
> where $n_1p_1 \geq 5$, $n_1(1 - p_1) \geq 5$, $n_2p_2 \geq 5$, and $n_2(1 - p_2) \geq 5$.

Hypothesis Tests for $p_1 - p_2$

Whenever the null hypothesis is of the form H_0: $p_1 - p_2 = 0$ and is assumed to be true ($p_1 = p_2 = p$), we will not know the common value p. In order to compute an approximate value of $\sigma_{\hat{p}_1-\hat{p}_2}$, we must have an estimate for p, since

$$\sigma_{\hat{p}_1-\hat{p}_2} = \sqrt{\frac{p(1 - p)}{n_1} + \frac{p(1 - p)}{n_2}}$$

$$= \sqrt{p(1 - p)\left(\frac{1}{n_1} + \frac{1}{n_2}\right)}$$

Which estimate for p should we use, $\hat{p}_1$ or $\hat{p}_2$ or neither? An average $(\hat{p}_1 + \hat{p}_2)/2$ would be appropriate only when $n_1 = n_2$. Since $\hat{p}_1 = x_1/n_1$ and $\hat{p}_2 = x_2/n_2$, we can form a **pooled** or weighted **estimate for p** defined by

> **Pooled Estimate of p**
>
> $$\hat{p} = \frac{x_1 + x_2}{n_1 + n_2} \qquad (11.7)$$

The value of $\hat{p}$ must necessarily lie between $\hat{p}_1$ and $\hat{p}_2$. Then the standard error of the difference between sample proportions can be estimated by using the following formula:

> **Approximate Standard Error of the Difference in Sample Proportions**
>
> $$\sigma_{\hat{p}_1-\hat{p}_2} \approx \sqrt{\hat{p}(1 - \hat{p})\left(\frac{1}{n_1} + \frac{1}{n_2}\right)} \qquad (11.8)$$
>
> where $\hat{p}$ is given by formula (11.7).

The above techniques are demonstrated in applications 11.10–11.12. Note that if we want to test a null hypothesis of the form H_0: $p_1 - p_2 = p_0$, where p_0 is a nonzero value, there is no need to obtain a pooled estimate for p, since $p_1 \neq p_2$. In this case $\hat{p}_1$ is used to estimate p_1 and $\hat{p}_2$ is used to estimate p_2.

APPLICATION 11.10   A college student has heard that professor X gives a higher percentage of A grades in math 101 than professor Y. Suspecting that Professor X is the easier "A grader," the student compares published grades for both professors from the past semester. Assuming that students are randomly assigned to math 101 sections by the registrar, do the following data indicate that professor X gives a higher percentage of A grades than professor Y?

Professor X	Professor Y
$x_1 = 60$	$x_2 = 70$
$n_1 = 150$	$n_2 = 250$

Use $\alpha = 0.01$ and find the *p*-value.

Solution: The sample proportion of A grades given by professor X is $\hat{p}_1 = 60/150 = 0.40$, and the sample proportion of A grades given by professor Y is $\hat{p}_2 = 70/250 = 0.28$. From formula (11.7), the pooled estimate for p is

$$\hat{p} = \frac{x_1 + x_2}{n_1 + n_2}$$

$$= \frac{60 + 70}{150 + 250} = 0.325$$

The estimated standard error of $\hat{p}_1 - \hat{p}_2$ is found using formula (11.8):

$$\sigma_{\hat{p}_1 - \hat{p}_2} \approx \sqrt{\hat{p}(1 - \hat{p})\left(\frac{1}{n_1} + \frac{1}{n_2}\right)}$$

$$= \sqrt{(0.325)(0.675)\left(\frac{1}{150} + \frac{1}{250}\right)}$$

$$= 0.0484$$

By using the hypothesis testing procedure, we have the following:

1. The null hypothesis is H_0: $p_1 - p_2 \leq 0$.
2. The alternative hypothesis is H_1: $p_1 - p_2 > 0$ (one-tailed test).
3. The sampling distribution is the distribution of the differences between sample proportions. Note that
   a. $n_1 p_1 = x_1 = 60 \geq 5$.
   b. $n_1(1 - p_1) = n_1 - x_1 = 90 \geq 5$.
   c. $n_2 p_2 = x_2 = 70 \geq 5$.
   d. $n_2(1 - p_2) = n_2 - x_2 = 180 \geq 5$.
   Hence, the sampling distribution of $\hat{p}_1 - \hat{p}_2$ is approximately normal. The test statistic is the z value for $\hat{p}_1 - \hat{p}_2$.

$$z = \frac{\hat{p}_1 - \hat{p}_2 - 0}{\sigma_{\hat{p}_1 - \hat{p}_2}}$$

$$= \frac{0.40 - 0.28}{0.0484} = 2.48$$

4. The level of significance is $\alpha = 0.01$.

5. The critical value is $z_{0.01}$ is 2.33.

6. We reject H_0, since the value of the test statistic falls in the rejection region, that is, $z = 2.48 \geq 2.33$.

7. Type of error possible: Type I; $\alpha = 0.01$.

8. The p-value is $0.5 - 0.4934 = 0.0066$.

Thus, the observed difference $\hat{p}_1 - \hat{p}_2 = 0.012$ can not be attributed to sampling error alone, and one can conclude that professor X gives a higher percentage of A grades in math 101 than professor Y. ▪

APPLICATION 11.11

To landscape its campus, a large university experimented with two varieties of plants. Four hundred plants of variety A and 600 plants of variety B were initially planted. If 42 of the variety A plants failed to grow, whereas 48 of the variety B plants failed to grow, test at the 0.05 level of significance whether there is a difference between the percentages of varieties A and B that will fail to grow on campus.

Solution: We will illustrate hypothesis testing and estimation procedures for solving the problem. Let p_1 represent the proportion of variety A plants that fail to grow on campus and let p_2 represent the proportion of variety B plants that fail to grow.

Hypothesis-Testing Procedure We first compute the sample proportions and the pooled estimate for p. The sample proportions are

$$\hat{p}_1 = \frac{42}{400} = 0.105$$

and

$$\hat{p}_2 = \frac{48}{600} = 0.08$$

The pooled estimate for p is

$$\hat{p} = \frac{x_1 + x_2}{n_1 + n_2} = \frac{42 + 48}{1000} = 0.09$$

The approximate standard error of $\sigma_{\hat{p}_1 - \hat{p}_2}$ is

$$\sigma_{\hat{p}_1 - \hat{p}_2} \approx \sqrt{\hat{p}(1 - \hat{p})\left[\frac{1}{n_1} + \frac{1}{n_2}\right]}$$

$$= \sqrt{(0.09)(0.91)\left[\frac{1}{400} + \frac{1}{600}\right]} = 0.018$$

1. The null hypothesis is H_0: $p_1 - p_2 = 0$.

2. The alternative hypothesis is H_1: $p_1 - p_2 \neq 0$ (two tailed test).

3. The sampling distribution is the distribution of differences between sample percentages. The test statistic is the z value for $\hat{p}_1 - \hat{p}_2$.

$$z = \frac{\hat{p}_1 - \hat{p}_2}{\sigma_{\hat{p}_1 - \hat{p}_2}} = \frac{0.105 - 0.08}{0.018} = 1.39$$

4. The level of significance is $\alpha = 0.05$.
5. The critical values are $\pm z_{0.025} = \pm 1.96$.
6. The decision is we fail to reject H_0.
7. Type of error possible: Type II; β is unknown.
8. The p-value is $2(0.5 - 0.4177) = 2(0.0823) = 0.1646$.

Thus, the difference $\hat{p}_1 - \hat{p}_2 = 0.025$ can be accounted for as sampling error and not due to the variety of the plant.

Estimation Procedure We first find the estimated value of $\hat{\sigma}_{\hat{p}_1-\hat{p}_2}$:

$$\hat{\sigma}_{\hat{p}_1-\hat{p}_2} = \sqrt{\frac{\hat{p}_1(1-\hat{p}_1)}{n_1} + \frac{\hat{p}_2(1-\hat{p}_2)}{n_2}}$$

$$= \sqrt{\frac{(0.105)(0.895)}{400} + \frac{(0.08)(0.92)}{600}} = 0.019$$

Note that we do not use the pooled estimate, since we have not assumed that $p_1 = p_2$ is true. By (11.6), the limits for the 95% confidence interval for $p_1 - p_2$ are

$$(\hat{p}_1 - \hat{p}_2) \pm z_{0.025}\,\hat{\sigma}_{\hat{p}_1-\hat{p}_2} = 0.025 \pm (1.96)(0.019) = 0.025 \pm 0.037$$

Thus, a 95% confidence interval for $p_1 - p_2$ is $(-0.012, 0.062)$. Since 0 is contained the interval, $p_1 - p_2$ may equal 0. Hence, we have no statistical evidence to suggest that there is a difference in the percentages of varieties A and B that fail to grow on campus. ▪

APPLICATION 11.12

An administrator of a large college claims that at least 10% more male students than female students have an automobile on campus. A statistics professor takes issue with the claim and randomly polls 100 males and 100 females. He found that 34 males have cars on campus and 27 females have cars on campus. Can he conclude at the 5% level of significance that the administrator's claim is false?

Solution: Let p_1 represent the proportion of male students who have cars on campus and let p_2 represent the proportion of female students who have cars on campus.

1. The null hypothesis is H_0: $p_1 - p_2 \geq 0.1$.
2. The alternative hypothesis is H_1: $p_1 - p_2 < 0.1$ (one-tailed test).
3. The sampling distribution is the sampling distribution of $\hat{p}_1 - \hat{p}_2$. The test statistic is the z value for $\hat{p}_1 - \hat{p}_2 = 0.34 - 0.27 = 0.07$. The z value for 0.07 is

$$z = \frac{(\hat{p}_1 - \hat{p}_2) - (p_1 - p_2)}{\sqrt{[\hat{p}_1(1-\hat{p}_1)/n_1] + [\hat{p}_2(1-\hat{p}_2)/n_2]}}$$

$$= \frac{0.07 - 0.1}{\sqrt{[(0.34)(0.66)/100] + [(0.27)(0.73)/100]}} = -0.46$$

4. The level of significance is $\alpha = 0.05$.
5. The critical value is $-z_{0.05} = -1.65$.
6. We fail to reject H_0, since $-0.46 > -1.65$.

7. Type error possible: Type II; β is unknown.

8. The *p*-value is $P(Z < -0.46) = 0.5 - 0.1772 = 0.3228$.

Hence, there is no statistical evidence to reject the administrator's claim that at least 10% more male students have cars on campus than female students. ■

EXERCISE SET 11.3

Basic Skills

1. In a study to estimate the proportions of residents in a small city and its suburbs who subscribe to the newspaper, it was found that 48 of 120 city residents subscribe, whereas only 30 of 125 suburban residents subscribe to the newspaper. If the true proportion of the city residents who subscribe to the newspaper is 0.5 and if the true proportion of the suburbanites who subscribe is 0.3, find

a. $\mu_{\hat{p}_r - \hat{p}_s}$.
b. $\sigma_{\hat{p}_r - \hat{p}_s}$.
c. the corresponding z score for $\hat{p}_r - \hat{p}_s$.

2. Adults males and females living in a large northern city differ in their views concerning the issue of the death penalty for persons found guilty of murder. It is believed that 15% of the adult males favor the death penalty, whereas only 11% of the adult females favor the death penalty. A random sample of 100 males and a random sample of 125 females are polled concerning their views on the issue of the death penalty for persons found guilty of murder, and it was found that 18 males favored the death penalty and 12 females favored the death penalty. Find

a. $\mu_{\hat{p}_m - \hat{p}_f}$.
b. $\sigma_{\hat{p}_m - \hat{p}_f}$.
c. the corresponding z score for $\hat{p}_m - \hat{p}_f$.

Further Applications

3. A study was conducted to determine whether homemakers from Boston have the same preference as homemakers from St. Louis for one of two brands of floor wax, A and B. It was found that among 300 randomly selected homemakers from Boston, 171 preferred brand A to brand B, whereas among 400 homemakers from St. Louis, 236 preferred A to B.

a. At $\alpha = 0.05$ determine if there is a significant difference between the percentages of homemakers who prefer brand A for the two groups. Find the *p*-value.
b. Construct a 95% confidence interval for the difference in percentages who prefer brand A for the two groups. Does the interval contain 0?

4. In a study to determine the effects of color in commercial television advertising, two groups of 500 persons were ex-

posed to a program, including commercials. One group watched the program in color and the other group watched the identical program in black and white. Two hours after the program, each person was asked which products were advertised. Two hundred of those who watched the program in color remembered the products, whereas 180 of those who watched the program without color remembered the products.

a. At $\alpha = 0.05$ test the null hypothesis that there is no difference in retention between persons who watch television in color as opposed to those who watch television in black and white. Find the *p*-value.
b. Construct a 95% confidence interval for the true difference between proportions. Is 0 in the interval?

5. Independent random samples of 50 males and 75 females were selected to study GPAs of students at a large university. The group of males included 15 who had GPAs less than 2.0, whereas the group of females had 24 with GPAs below 2.0.

a. Construct a 95% confidence interval for the true difference between proportions of all male and female students with GPAs below 2.0.
b. Test the null hypothesis that there is no difference between the true proportion of university males with GPAs below 2.0 and the true proportion of university females with GPAs below 2.0. Use $\alpha = 0.05$.

6. A manufacturer of microprocessors buys its processor chips from two suppliers. A sample of 300 chips from supplier A resulted in 50 defective chips, whereas a sample of 400 chips from supplier B resulted in 70 defective chips.

a. Construct a 95% confidence interval for the difference between the proportions of defective chips received from the two suppliers.
b. Test the null hypothesis that there is no difference between the proportions of defective chips for the two suppliers. Use $\alpha = 0.05$.

7. It is common knowledge that not everyone cooperates with answering questions from door-to-door poll takers. For an experiment to determine whether women are more cooperative than men, the following results indicate the number of each gender who cooperated with an interviewer:

Men	Women
$n_1 = 175$	$n_2 = 250$
$x_1 = 97$	$x_2 = 143$

a. Construct a 99% confidence interval for the difference in the true proportions of men and women who cooperate with poll takers.

b. Test whether there is a difference between the proportions of men and women who cooperate with door-to-door poll takers. Use $\alpha = 0.01$ and find the *p*-value.

8. Consider the following data from two random samples:

Sample 1	Sample 2
$n_1 = 200$	$n_2 = 300$
$x_1 = 120$	$x_2 = 168$

a. Test the null hypothesis $H_0: p_1 - p_2 = 0.01$ versus the alternative hypothesis $H_1: p_1 - p_2 \neq 0.01$ using $\alpha = 0.05$. Also find the *p*-value.

b. Construct a 95% confidence interval for $p_1 - p_2$. Is 0.01 contained in the interval?

9. The western Maryland branch of the American Heart Association studied cardiovascular disease-related deaths in Allegany and Garrett counties for 1983. It reported that of the 969 deaths in Allegany County, 510 were attributed to cardiovascular disease, whereas of the 250 deaths reported in Garrett County, 150 were due to cardiovascular disease. Do the data suggest that the percentages of deaths

resulting from cardiovascular disease differ for the two western Maryland counties? Use a $= 0.05$.

10. Diet drinks have become very popular as more people are concerned about controlling their weight. Much of the advertising has been aimed at females, reflecting a feeling that females are more likely than males to purchase diet drinks. A study was designed to investigate possible differences between males and females in preferences for regular cola or diet cola. In a sample of 400 males, 256 choose the regular cola and 144 chose diet cola. In a sample of 400 females, 192 chose the regular cola and the remaining 208 chose the diet cola.

a. Use $\alpha = 0.05$ to determine if the percentage of females preferring diet cola is greater than the percentage of males preferring diet cola.

b. Construct a 95% confidence interval for the true difference in the proportion of females preferring diet cola and the proportion of males preferring diet cola.

Going Beyond

11. Refer to exercise 6. Test the null hypothesis $H_0: p_2 - p_1 = 0.5\%$ against $H_1: p_2 - p_1 \neq 0.5\%$ using the 5% level of significance. Find the *p*-value.

12. A study is to be conducted to estimate the difference between the proportions of defective parts shipped by two suppliers, A and B. If samples of the same size are to be used, how large must the samples be in order to be 95% confident that the maximum error of estimate is less than 0.01?

SECTION 11.4   *Comparing Population Variances*

Another method of comparing two populations is to compare their variances. Many statistical applications arise in which population variances must be compared. In industrial applications involving two different methods or machines for producing the same product, variances are often used and compared for quality-control purposes. Many statistical tests, such as those to be presented in section 11.5, require population variances to be equal in order for the tests to be appropriate. In order to be able to compare two population variances, we need to be familiar with a new sampling distribution, the *F* distribution.

F Distributions

If two independent random samples are taken from normal populations with equal population variances ($\sigma_1^2 = \sigma_2^2$), then the sampling distribution of the statistic s_1^2/s_2^2 is an *F* **distribution**. Thus, the *F* **statistic** is defined by the following rule.

F Statistic

$$F = \frac{s_1^2}{s_2^2}$$

The exact shape of an F distribution depends on two degrees-of-freedom parameters, df_1 and df_2. These are defined by:

Degrees of Freedom for the F Statistic

$df_1 = n_1 - 1$, where n_1 is the size of the sample yielding s_1^2, and $df_2 = n_2 - 1$, where n_2 is the size of the sample yielding s_2^2.

We will identify the degrees of freedom as an ordered pair (df_1, df_2). The F distributions, like the χ^2 distributions, are not symmetric distributions, but all elements of both distributions are greater than or equal to 0 (see fig. 11.2).

FIGURE 11.2

Graph of F distributions

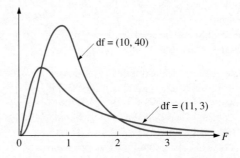

APPLICATION 11.13

In a test of the effectiveness of two different kinds of sleeping pills, A and B, two independent groups of insomniacs are to be used. One group of size 40 will be administered pill A and the other group of size 60 will be administered pill B. The number of hours of sleep will be recorded for each individual used in the study. If the number of hours of sleep for individuals using each pill is assumed to be normally distributed and $\sigma_a^2 = \sigma_b^2$, calculate the value of the F statistic and determine df_a and df_b if $s_a^2 = 9$, $s_b^2 = 5$, $n_a = 40$, and $n_b = 60$.

Solution: The value of the F statistic is

$$F = \frac{s_a^2}{s_b^2} = \frac{9}{5} = 1.8$$

The degrees of freedom are

$$df_a = n_a - 1 = 40 - 1 = 39$$
$$df_b = n_b - 1 = 60 - 1 = 59 \quad ■$$

Note that if the statistic s_1^2/s_2^2 has a sampling distribution that is an F distribution with $df_1 = n_1 - 1$ and $df_2 = n_2 - 1$, then the statistic s_2^2/s_1^2 (the reciprocal of F) has a sampling distribution that is an F distribution with $df_1 = n_2 - 1$ and $df_2 = n_1 - 1$. This is summarized in the following relationship:

Relationship Between F and $1/F$

If F has an F distribution with $df = (n_1 - 1, n_2 - 1)$, then $1/F$ has an F distribution with $df = (n_2 - 1, n_1 - 1)$.

Since both statistics have F distributions, it is common practice to place the larger sample variance in the numerator of the F ratio.

Hypothesis Tests Comparing σ_1^2 and σ_2^2

To compare population variances we need to use the quotient of the variances, rather than the difference of the variances. If the quotient of the population variances is equal to 1, then we can say that the variances are equal; if the quotient of the population variances is not equal to 1, then the population variances are not equal. That is,

$$\frac{\sigma_1^2}{\sigma_2^2} = 1 \quad \text{is equivalent to} \quad \sigma_1^2 = \sigma_2^2$$

If $\sigma_1^2 = \sigma_2^2$, we would expect the value of the F statistic to be close to 1; the difference from 1 may be attributed to sampling error. The more removed the value of F is from 1, the more unlikely it is that the F value belongs to a particular F distribution.

As we have previously noted, the F distributions are not symmetric; as a result, to form the rejection region for a two-tailed test, we can simplify the computations by making certain that the right tail of the F distribution is used. A right-tailed test is used because only these areas are given in the *F* **tables.** If the larger sample variance is placed in the numerator of the F statistic, then a right-tailed test will always be indicated. By doing this, we are in effect doubling the table value for α by making the test a two-tailed test and using only the right tail. Thus, for a two-tailed test employing a significance level of α, a right-tailed test is used that has a tail area of $\alpha/2$. The F statistic becomes

$$F = \frac{\text{larger sample variance}}{\text{smaller sample variance}}$$

For directional hypotheses, we place the sample variance corresponding to the larger population variance, as expressed by the alternative hypothesis H_1, in the numerator of the F ratio. For example, if the alternative hypothesis is $\sigma_1^2 > \sigma_2^2$, then the sample variance corresponding to the population variance σ_1^2 is placed in the numerator of the F ratio. One should be careful that df_1 represents the degrees of freedom corresponding to the variance in the numerator of the F ratio.

To determine a critical value in the F table (tables 6a and 6b of appendix B), we locate the correct table for $\alpha(0.01$ or $0.05)$ and intersect the row identified by df_2 with the *column* identified by df_1. We shall denote the F distribution with $df_1 = n_1 - 1$ and $df_2 = n_2 - 1$ as $F(n_1 - 1, n_2 - 1)$. As a result, the F statistic has a sampling distribution that is an $F(n_1 - 1, n_2 - 1)$ distribution. Note that the df corresponding to the numerator of the F ratio is always located at the top of the F table, and the df corresponding to the denominator of the F ratio is always located at the left side of the F table. Applications 11.14 and 11.15 demonstrate testing hypotheses to determine if two population variances differ.

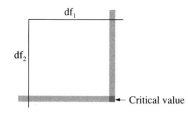

df$_1$

df$_2$

← Critical value

A P P L I C A T I O N 11.14

Two different processes are used to produce 100-ohm resistors with a tolerance of 1% (the true resistance is somewhere between 99 and 101 ohms). A random sample of resistors is selected for each process. Assume that the distribution of resistor values for each process is normally distributed. If $n_1 = 21$, $s_1^2 = 39$, $n_2 = 31$, and $s_2^2 = 33$, test using $\alpha = 0.05$ to determine if the variance of the resistors manufactured by process 1 is greater than the variance of the resistors manufactured by process 2.

Solution:

1. The null hypothesis is H_0: $\sigma_1^2 \leq \sigma_2^2$.
2. The alternative hypothesis is H_1: $\sigma_1^2 > \sigma_2^2$ (one-tailed test).
3. The sampling distribution is $F(20, 30)$. (The F distribution with $df_1 = 20$ and $df_2 = 30$).

$$F = \frac{s_1^2}{s_2^2} = \frac{39}{33} = 1.18$$

4. The level of significance is $\alpha = 0.05$.
5. The critical value is $F_{0.05}(20, 30) = 1.93$. This value was found by locating the F distribution table corresponding to $\alpha = 0.05$ (table 6b of appendix B) and determining the value where the column labeled by 20 intersects the row labeled by 30.
6. The decision is that we cannot reject H_0 (see the accompanying figure).
7. With this decision, we may have committed a type II error; β is unknown.

The value of the F statistic differs from 1 by 0.18. We conclude that this difference can be explained as sampling error. We have no statistical evidence to indicate that the variance of process 1 is greater than the variance of process 2. ▪

MINITAB can be used to determine the critical value for the F distribution in application 11.14. The INVCDF command and the accompanying F subcommand are used. Computer display 11.2 contains the commands used and the output.

Computer Display 11.2

```
MTB > INVCDF 0.95;
SUBC> F 20 30.
   0.9500  1.9317
MTB >
```

The output from MINITAB (1.9317) contains more precision than the value found in table 6b, but to two decimal places they both agree.

A P P L I C A T I O N 11.15

Assume that the weights of items produced by two different manufacturing processes are normally distributed. If the weights of a sample of $n_1 = 15$ items produced by one process have a variance of 33 and the weights of a sample of $n_2 = 41$ items produced by the other process have a variance of 15, test using $\alpha = 0.10$ to determine if $\sigma_1^2 \neq \sigma_2^2$.

Solution:

1. The null hypothesis is H_0: $\sigma_1^2 = \sigma_2^2$.
2. The alternative hypothesis is H_1: $\sigma_1^2 \neq \sigma_2^2$.
3. The sampling distribution is $F(14, 40)$. The test statistic is F. Note that the larger variance is placed in the numerator of the F statistic. The value of the F statistic is

$$F = \frac{33}{15} = 2.2$$

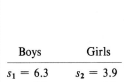

4. The level of significance is $\alpha = 0.10$ (two-tailed test).
5. The critical value is $F_{0.05}(14, 40) = 1.95$, as indicated in the figure.

Note that it is not necessary to obtain a critical value in the lower tail of the F distribution, since we placed the larger sample variance in the numerator of the F statistic. The larger critical value corresponds to a right-tailed area of $\alpha/2$ for a two-tailed test.

6. The decision is to reject H_0.
7. With this decision we may have committed a Type I error; $\alpha = 0.10$.

Thus, we can conclude that the population variances are significantly different at the 10% significance level. ■

Comparing Population Standard Deviations

To determine whether $\sigma_1 \neq \sigma_2$, we determine whether $\sigma_1^2 \neq \sigma_2^2$. That is, $\sigma_1 \neq \sigma_2$ if $\sigma_1^2 \neq \sigma_2^2$. Thus, to test the null hypothesis H_0: $\sigma_1 = \sigma_2$, we test the equivalent hypothesis H_0: $\sigma_1^2 = \sigma_2^2$.

APPLICATION 11.16

In an attempt to determine whether the standard deviation in weights among first-grade boys is greater than the standard deviation in weights among first-grade girls, two random samples of weights were obtained. The data recorded are shown in the table. Do the data indicate that $\sigma_1 > \sigma_2$? Use $\alpha = 0.05$.

Solution: Note that $\sigma_1 > \sigma_2$ if $\sigma_1^2 > \sigma_2^2$. Thus, we test to determine if $\sigma_1^2 > \sigma_2^2$.

Boys	Girls
$s_1 = 6.3$	$s_2 = 3.9$
$n_1 = 25$	$n_2 = 10$

1. The null hypothesis is H_0: $\sigma_1^2 \leq \sigma_2^2$.
2. The alternative hypothesis is H_1: $\sigma_1^2 > \sigma_2^2$.
3. The sampling distribution is $F(24, 9)$. The value of the test statistic is

$$F = \frac{6.3^2}{3.9^2} = 2.61$$

4. The level of significance is $\alpha = 0.05$.
5. The critical value is $F_{0.05}(24, 9)$. Since $F_{0.05}(24, 9)$ is between 2.86 and 2.93 we shall use 2.93 as our critical value.
6. Since $2.61 \leq 2.93$, we fail to reject H_0.
7. A Type II error is possible with this decision, and β is unknown.

Hence, we have no statistical evidence to suggest that the standard deviation of the weights of first-grade boys is greater than the standard deviation of the weights of first-grade girls. ■

Left-Tailed Critical Values for F

In order to construct a confidence interval for σ_1^2/σ_2^2, we need to be able to find both critical values associated with the F statistic. To find the critical value for the left tail in an F distribution, we use the following fact:

Left-Tailed Critical Value for an F Distribution

$$F_{1-\alpha/2}(n_2 - 1, n_1 - 1) = \frac{1}{F_{\alpha/2}(n_1 - 1, n_2 - 1)}$$

(11.9)

EXAMPLE 11.9

Let's determine $F_{0.95}(40, 14)$ for the F distribution in application 11.15. We use formula (11.9).

$$F_{0.95}(40, 14) = \frac{1}{F_{0.05}(14, 40)}$$

$$= \frac{1}{1.95} = 0.513$$

Recall that the critical value 1.95 represents the right-tailed critical value $F_{0.05}(14, 40)$, which was found in the F table (appendix B).

Confidence Intervals for the Quotient of Two Population Variances

When constructing a confidence interval for the quotient of two population variances, we shall follow the convention adopted earlier of placing the larger sample variance in the numerator of the F statistic. In the notation that follows, s_1^2 will represent the larger sample variance. The confidence intervals for σ_1^2/σ_2^2 and σ_2^2/σ_1^2 are not identical, but they are related to one another in a special way. If (L_1, L_2) is a $(1 - \alpha)100\%$ **confidence interval for σ_1^2/σ_2^2**, then $(1/L_2, 1/L_1)$ is a $(1 - \alpha)100\%$ confidence interval for σ_2^2/σ_1^2.

A $(1 - \alpha)100\%$ confidence interval for σ_1^2/σ_2^2 is given by (L_1, L_2) where the limits L_1 and L_2 are given by the formulas (11.10).

Limits for a $(1 - \alpha)100\%$ Confidence Interval for σ_1^2/σ_2^2

$$L_1 = \frac{s_1^2}{s_2^2}F_{1-\alpha/2}(n_2 - 1, n_1 - 1)$$

$$L_2 = \frac{s_1^2}{s_2^2}F_{\alpha/2}(n_2 - 1, n_1 - 1)$$

(11.10)

If the confidence interval for σ_1^2/σ_2^2 does not contain 1, then we can be $(1 - \alpha)100\%$ confident that $\sigma_1^2 \neq \sigma_2^2$. And we can reject $H_0: \sigma_1^2 = \sigma_2^2$ using a level of significance equal to α.

Limits for a $(1 - \alpha)100\%$ **confidence interval for σ_1/σ_2** are given by the following rule:

Confidence Limits for σ_1/σ_2

A $(1 - \alpha)100\%$ confidence interval for σ_1/σ_2 is given by $(\sqrt{L_1}, \sqrt{L_2})$ where L_1 and L_2 are given by (11.10).

The following four steps can be used to construct a $(1 - \alpha)100\%$ confidence interval for σ_1^2/σ_2^2:

Four Steps to Construct a $(1 - \alpha)100\%$ Confidence Interval for σ_1^2/σ_1^2

1. Find the value of s_1^2/s_2^2 making sure the larger sample variance is placed in the numerator of the F ratio.
2. Find the value of $F_{1-\alpha/2}(n_2 - 1, n_1 - 1)$ by using formula (11.9). Be sure that n_2 represents the size of the sample from the population having variance σ_2^2.
3. Find the value of $F_{\alpha/2}(n_2 - 1, n_1 - 1)$, making sure n_2 represents the size of the sample from the population having variance σ_2^2.
4. Find the confidence interval limits L_1 and L_2 by using (11.10).

APPLICATION 11.17

For application 11.15 construct a 90% confidence interval for σ_1^2/σ_2^2. Recall that $n_1 = 15$, $s_1^2 = 33$, $n_2 = 41$, and $s_2^2 = 15$.

Solution:

1. The value of the F statistic is
$$F = \frac{s_1^2}{s_2^2} = \frac{33}{15} = 2.2$$

Note that the larger sample variance is placed in the numerator of the F statistic.

2. The left-tailed critical value for F is

$$F_{1-\alpha/2}(n_2 - 1, n_1 - 1) = F_{0.95}(40, 14)$$
$$= \frac{1}{F_{0.05}(14, 40)}$$
$$= \frac{1}{1.95} = 0.51$$

3. The right-tailed critical value for F is

$$F_{\alpha/2}(n_2 - 1, n_1 - 1) = F_{0.05}(40, 14) = 2.27$$

From (11.10), the limits for the 90% confidence interval are

$$L_1 = \frac{s_1^2}{s_2^2}F_{1-\alpha/2}(n_2 - 1, n_1 - 1)$$
$$= (2.2)(0.51) = 1.122$$

and

$$L_2 = \frac{s_1^2}{s_2^2}F_{\alpha/2}(n_2 - 1, n_1 - 1)$$
$$= (2.2)(2.27) = 4.994$$

Thus, a 90% confidence interval for σ_1^2/σ_2^2 is (1.12, 4.99). Note that if a $(1 - \alpha)100\%$ confidence interval for σ_1^2/σ_2^2 contains 1, then we can not be $(1 - \alpha)100\%$ confident that the population variances are different. In this example, we can be 90% confident that the population variances are different, since the interval (1.12, 4.99) does not contain 1. Note that the 90% confidence interval for σ_2^2/σ_1^2 is $(1/4.99, 1/1.12) = (0.20, 0.89)$. ■

APPLICATION 11.18

Two new motor assembly methods are tested by an automobile manufacturer for variance in assembly times (in minutes). The results are shown in the table.

Method 1	Method 2
$n_1 = 31$	$n_2 = 25$
$s_1^2 = 50$	$s_2^2 = 24$

a. Construct a 90% confidence interval for σ_1^2/σ_2^2.

b. Conduct the corresponding hypothesis test to determine if there is a difference in the variances for the assembly times for the two methods.

Solution:

a. Construction of a 90% Confidence Interval

1. The value of s_1^2/s_2^2 is $50/24 = 2.08$.

2. The left-tailed critical value is

$$F_{1-\alpha/2}(n_2 - 1, n_1 - 1) = F_{0.95}(24, 30)$$

$$= \frac{1}{F_{0.05}(30, 24)}$$

$$= \frac{1}{1.94} = 0.52$$

3. The right-tailed critical value is

$$F_{\alpha/2}(n_2 - 1, n_1 - 1) = F_{0.05}(24, 30)$$

Note that the F table does not have a column label for df $= 24$, but it does contain df $= 20$ and df $= 30$. For these adjacent values, the critical values for F are $F_{0.05}(20, 30) = 1.93$ and $F_{0.05}(30, 30) = 1.84$. Let's use the larger and more conservative value of 1.93. Hence, $F_{0.05}(24, 30) \approx 1.93$. (In this text, whenever the tables do not contain a df value entry, we shall follow the practice of choosing the larger F value corresponding to the adjacent values of df.)

The limits for the 90% confidence interval for σ_1^2/σ_2^2 are given by (11.10):

$$L_1 = \frac{s_1^2}{s_2^2}F_{1-\alpha/2}(n_2 - 1, n_1 - 1)$$

$$= \left(\frac{50}{24}\right)(F_{0.95}(24, 30))$$

$$= (2.08)(0.52) = 1.08$$

and

$$L_2 = \frac{s_1^2}{s_2^2}F_{\alpha/2}(n_2 - 1, n_1 - 1)$$

$$= \left(\frac{50}{24}\right)(F_{0.05}(24, 30))$$

$$= (2.08)(1.93) = 4.01$$

Thus, a 90% confidence interval for σ_1^2/σ_2^2 is (1.08, 4.01). Since this interval does not contain 1, we can be 90% confident that $\sigma_1^2 \neq \sigma_2^2$.

b. Hypothesis Test

1. The null hypothesis is $\sigma_1^2 = \sigma_2^2$.
2. The alternative hypothesis is $\sigma_1^2 \neq \sigma_2^2$ (two-tailed test).
3. The sampling distribution is the F distribution with $df_1 = 30$ and $df_2 = 24$, $F(30, 24)$. The value of the F statistic is

$$F = \frac{50}{24} = 2.08$$

4. The level of significance is 10%.
5. The critical value is $F_{.05}(30, 24) = 1.94$. Since we put the larger sample variance in the numerator of the F statistic, we only need to compare the value of the F statistic against the larger critical value, $F_{0.05}(30, 24)$.
6. The decision is to reject H_0, since $F = 2.08 > F_{0.05}(30, 24) = 1.94$. Note that this result is consistent with our interpretation of the confidence interval found in part *a*. ▪

EXERCISE SET 11.4

Basic Skills

1. To determine whether grades of instructor A are more variable than those of instructor B, two independent random samples of grades on equivalent examinations were selected. The following results were obtained:

Instructor A	Instructor B
$s_a^2 = 81.24$	$s_b^2 = 97.86$
$n_a = 15$	$n_b = 20$

If it is assumed that the $\sigma_a = \sigma_b$, calculate the value of the F statistic and find its associated df values.

2. To determine whether the weights of 5-ounce packages of candy produced by machine 1 are consistent with the weights of the 5-ounce packages produced by machine 2, two independent random samples of items were produced by the two machines. The following results were obtained:

Machine 1	Machine 2
$s_1^2 = 4.68$	$s_2^2 = 6.34$
$n_1 = 25$	$n_2 = 30$

If it is assumed that $\sigma_1 = \sigma_2$, calculate the value of the F statistic and find its associated df values.

3. Find the following critical values for F:
   a. $F_{0.95}(21, 31)$ b. $F_{0.99}(31, 21)$ c. $F_{0.99}(10, 10)$

4. Find the following critical values for F:
   a. $F_{0.95}(15, 10)$ b. $F_{0.99}(10, 15)$ c. $F_{0.95}(10, 10)$

5. Find 90% confidence intervals for σ_1^2/σ_2^2 if
   a. $n_1 = 15$, $s_1^2 = 15.1$, $n_2 = 31$, and $s_2^2 = 6.8$.
   b. $n_1 = 21$, $s_1^2 = 4.8$, $n_2 = 11$, and $s_2^2 = 10.4$.

6. Refer to exercise 5. Find 98% confidence intervals for σ_1/σ_2.

7. Perform the following hypothesis tests. Assume sampling is from normal populations.

   a. $H_0: \sigma_1^2 \geq \sigma_2^2$
      $H_1: \sigma_1^2 < \sigma_2^2$
      $\alpha = 0.05$
      $n_1 = 16$
      $s_1^2 = 12.83$
      $n_2 = 26$
      $s_2^2 = 21.75$

   b. $H_0: \sigma_1^2 = \sigma_2^2$
      $H_1: \sigma_1^2 \neq \sigma_2^2$
      $\alpha = 0.10$
      $n_1 = 21$
      $s_1^2 = 10.89$
      $n_2 = 11$
      $s_2^2 = 22.87$

8. Perform the following hypothesis tests. Assume sampling is from normal populations.

   a. $H_0: \sigma_1 = \sigma_2$
      $H_1: \sigma_1 \neq \sigma_2$
      $\alpha = 0.10$
      $n_1 = 26$
      $s_1^2 = 3.84$
      $n_2 = 11$
      $s_2^2 = 16.4$

   b. $H_0: \sigma_1^2 \geq \sigma_2^2$
      $H_1: \sigma_1^2 < \sigma_2^2$
      $\alpha = 0.05$
      $n_1 = 31$
      $s_1 = 10.65$
      $n_2 = 12$
      $s_2 = 15.32$

9. The following data represent statistics from independent samples from two normal populations with unknown parameters. For each, test at $\alpha = 0.02$ to determine whether $\sigma_1^2 \neq \sigma_2^2$.
   a. $n_1 = 25$, $s_1 = 14$, $n_2 = 31$, $s_2 = 18$
   b. $n_1 = 21$, $s_1 = 5.2$, $n_2 = 24$, $s_2 = 3.2$

10. The following data represent statistics from independent samples from two normal populations with unknown parameters. For each, test at $\alpha = 0.02$ to determine whether $\sigma_1^2 \neq \sigma_2^2$.

a. $n_1 = 10, s_1 = 4.1, n_2 = 12, s_2 = 2.1$

b. $n_1 = 13, s_1 = 3.1, n_2 = 14, s_2 = 1.6$

Further Applications

11. Assume that an industrial plant produces 70-mm bolts on two different shifts. Samples of bolts are taken from those made by both shifts and their diameters are measured in mm. The results follow:

First Shift	Second Shift
$n_1 = 31$	$n_2 = 21$
$s_1^2 = 0.045$	$s_2^2 = 0.080$

a. By using $\alpha = 0.10$ test the null hypothesis H_0: $\sigma_1 = \sigma_2$ against the alternative hypothesis H_1: $\sigma_1 \neq \sigma_2$.

b. Construct a 90% confidence interval for σ_1/σ_2 to determine whether $\sigma_1 \neq \sigma_2$.

12. The following data represent two samples from different normal populations:

Sample 1		Sample 2	
9	9	13	14
9	7	12	16
11	8	11	10
11	6	13	8
12	5	12	9

a. Construct a 90% confidence interval for σ_1^2/σ_2^2.

b. Test the null hypothesis H_0: $\sigma_1^2 = \sigma_2^2$ against the alternative hypothesis H_1: $\sigma_1^2 \neq \sigma_2^2$ using $\alpha = 0.10$.

13. A new machine is being considered by a sugar packaging firm to replace its present machine. The weights of a sample of 21 5-pound bags of sugar packaged by the old machine produced a variance of $s_1^2 = 0.16$, whereas the weights for 20 5-pound bags of sugar packaged by the new machine produced a variance of $s_2^2 = 0.09$.

a. Use $\alpha = 0.05$ to test the null hypothesis H_0: $\sigma_2 \geq \sigma_1$.

b. Construct a 90% confidence interval for σ_1/σ_2.

c. Based on the data, would you advise management to consider the new machine as being superior to the old machine?

14. A criminologist is interested in comparing the consistency of the lengths of sentences given to persons convicted of robbery by two county judges. A random sample of 17 people convicted of robbery by one judge showed a standard deviation of $s_1 = 1.34$ years, whereas a random sample of 21 persons convicted by the other judge showed a standard deviation of $s_2 = 2.53$ years.

a. Do the data suggest that the variances of the lengths of sentences for the two judges differ? Use $\alpha = 0.10$.

b. Construct a 90% confidence interval for σ_1/σ_2.

c. Construct a 90% confidence interval for σ_2/σ_1.

Going Beyond

15. Verify the results of the F tests listed in motivator 10.

16. If χ_1^2 and χ_2^2 are independent random variables, then $(\chi_1^2/df_1)/(\chi_2^2/df_2)$ has an F distribution with df = (df_1, df_2). Use this fact to derive formulas (11.10).

17. A variance-reduction process is being considered by a plant to cut rising costs. The new process will not be implemented unless it can be verified statistically at $\alpha = 0.01$ that the new process will reduce the standard deviation by more than 20%. Suppose a study resulted in the following data:

Old Process	New Process
$n_1 = 17$	$n_2 = 11$
$s_1 = 3.61$	$s_2 = 1.72$

Should the new process be implemented? Explain.

SECTION 11.5   *Inferences Concerning $\mu_1 - \mu_2$ Using Small, Independent Samples*

If the sample sizes are large enough ($n_1 \geq 30$ and $n_2 \geq 30$), the techniques presented in section 11.2 can be used to compare almost any two populations. This is a consequence of the central limit theorem. In situations for which the population variances are unknown, sample variances provide satisfactory substitutes for the population variances in most situations when large samples are used. In this section, we shall develop methods for comparing two populations by using the difference between sample means obtained from small samples ($n_1 < 30$ and/or $n_2 < 30$) to estimate $\mu_1 - \mu_2$. The samples will be independent of one another. We shall treat the case of dependent samples in section 11.6.

Independent Samples

The assumptions underlying the methods for making inferences concerning $\mu_1 - \mu_2$ based on small, independent samples from two populations are

1. Both populations are normal.

2. The samples are independent and random.

3. The population variances are equal ($\sigma_1^2 = \sigma_2^2$).

Assumption 3 can be tested using the F statistic, explained in section 11.4. If we fail to reject the null hypothesis H_0: $\sigma_1^2 = \sigma_2^2$, we cannot necessarily conclude that $\sigma_1^2 = \sigma_2^2$. Nevertheless, whenever we fail to reject the null hypothesis of equal population variances, we will proceed under the assumption that the population variances are equal.

Inferences comparing μ_1 and μ_2 using small samples require that certain parameters must be estimated. If the population variances are equal, we will denote the common population variance by σ^2. Point estimates for μ_1, μ_2, and σ must be known. The best point estimate for μ_1 is $\bar{x}_1$, and the best estimate for μ_2 is $\bar{x}_2$. The difference between sample means $\bar{x}_1 - \bar{x}_2$ provides an estimate for the difference between population means $\mu_1 - \mu_2$. How should we estimate σ? If the sample sizes are equal ($n_1 = n_2$), then the average sample variance $(\sigma_1^2 + \sigma_2^2)/2$ is appropriate. But if $n_1 \neq n_2$, we need a weighted average that takes into account both sample sizes. This weighted average (sometimes called the **pooled estimate for** σ), denoted by s_p, is defined as follows:

Pooled Estimate for σ

$$s_p = \sqrt{\frac{(n_1 - 1)s_1^2 + (n_2 - 1)s_2^2}{n_1 + n_2 - 2}}$$

(11.11)

Recall that the standard error of the difference between sample means is defined by

$$\sigma_{\bar{x}_1 - \bar{x}_2} = \sqrt{\frac{\sigma_1^2}{n_1} + \frac{\sigma_2^2}{n_2}}$$

(11.12)

Since $\sigma_1^2 = \sigma_2^2 \ (= \sigma^2)$, we can rewrite equation (11.12) as

$$\sigma_{\bar{x}_1 - \bar{x}_2} = \sqrt{\frac{\sigma^2}{n_1} + \frac{\sigma^2}{n_2}}$$

$$= \sqrt{\sigma^2 \left(\frac{1}{n_1} + \frac{1}{n_2}\right)} = \sigma \sqrt{\frac{1}{n_1} + \frac{1}{n_2}}$$

If σ were known, the following statistic would have the standard normal as its sampling distribution:

$$\frac{(\bar{x}_1 - \bar{x}_2) - (\mu_1 - \mu_2)}{\sigma \sqrt{(1/n_1) + (1/n_2)}}$$

But since σ is unknown, we can use s_p to estimate σ, and it can be shown that the sampling distribution of $[(\bar{x}_1 - \bar{x}_2) - (\mu_1 - \mu_2)]/[s_p\sqrt{(1/n_1) + (1/n_2)}]$ is a t distribution with df $= n_1 + n_2 - 2$:

$$t \text{ Statistic for } \bar{x}_1 - \bar{x}_2$$

$$t = \frac{(\bar{x}_1 - \bar{x}_2) - (\mu_1 - \mu_2)}{s_p\sqrt{(1/n_1) + (1/n_2)}} \tag{11.13}$$

Application 11.19 illustrates the use of formula (11.13) for calculating the value of the t statistic.

APPLICATION 11.19

Two methods for teaching calculus, A and B, were compared using two random groups of students. At the conclusion of the experimental instruction each group was given the same achievement examination. The scores for each group comprise random samples from two normal populations with equal variances. Find the value of the t statistic and its associated degrees of freedom if we assume that $\mu_1 = \mu_2$ and $n_1 = 16$, $n_2 = 12$, $\bar{x}_1 = 80.7$, $\bar{x}_2 = 73.2$, $s_1^2 = 12.8$, and $s_2^2 = 8.5$.

Solution: We first compute the value of s_p using formula (11.11):

$$s_p = \sqrt{\frac{(n_1 - 1)s_1^2 + (n_2 - 1)s_2^2}{n_1 + n_2 - 2}}$$

$$= \sqrt{\frac{(16 - 1)(12.8) + (12 - 1)(8.5)}{16 + 12 - 2}} \approx 3.31$$

From formula (11.13), the value of the t statistic is

$$t = \frac{(\bar{x}_1 - \bar{x}_2) - (\mu_1 - \mu_2)}{s_p\sqrt{(1/n_1) + (1/n_2)}}$$

$$= \frac{(80.7 - 73.2) - 0}{3.31\sqrt{(1/16) + (1/12)}} \approx 5.93$$

The df associated with t is

$$\text{df} = n_1 + n_2 - 2$$
$$= 16 + 12 - 2 = 26 \quad ■$$

Inferences Concerning $\mu_1 - \mu_2$ Using Small, Independent Samples

Hypothesis-testing and estimation procedures can be used to draw inferences concerning $\mu_1 - \mu_2$ using small samples. If we are testing the null hypothesis H_0: $\mu_1 - \mu_2 = 0$ under the assumptions of normality and equal variances, then the test statistic becomes

$$t = \frac{\bar{x}_1 - \bar{x}_2}{s_p\sqrt{(1/n_1) + (1/n_2)}} \tag{11.14}$$

with df $= n_1 + n_2 - 2$.

The limits for a $(1 - \alpha)100\%$ confidence interval for $\mu_1 - \mu_2$ are given by

$$\text{Limits for a } (1 - \alpha)100\% \text{ Confidence Interval for } \mu_1 - \mu_2$$

$$\bar{x}_1 - \bar{x}_2 \pm t_{\alpha/2}s_p\sqrt{\frac{1}{n_1} + \frac{1}{n_2}} \tag{11.15}$$

$$\text{where df} = n_1 + n_2 - 2$$

Applications 11.20 and 11.21 illustrate the methods used to draw inferences concerning $\mu_1 - \mu_2$ for small, independent samples from two normal populations with unknown, but equal variances.

APPLICATION 11.20

Drug A	Drug B
$n_1 = 12$	$n_2 = 12$
$\bar{x}_1 = 26.8$	$\bar{x}_2 = 32.6$
$s_1^2 = 15.57$	$s_2^2 = 17.54$

An experiment was conducted to compare the mean length of time required for the human body to absorb two drugs, A and B. Assume that the length of time it takes either drug to reach a specified level in the bloodstream is normally distributed. Twelve people were randomly assigned to test each drug, and the length of time (in minutes) for each drug to reach a specified level in the blood was recorded. The results shown in the table were obtained. Determine using $\alpha = 0.05$ if $\mu_2 - \mu_1 \neq 0$.

Solution:

Hypothesis Testing Procedure We first test the assumption $\sigma_1^2 = \sigma_2^2$ by using hypothesis testing with $\alpha = 0.10$.

1. The null hypothesis is H_0: $\sigma_1^2 = \sigma_2^2$.
2. The alternative hypothesis is H_1: $\sigma_1^2 \neq \sigma_2^2$ (two-tailed test).
3. The sampling distribution is $F(11, 11)$. The value of the test statistic is

$$F = \frac{17.54}{15.57} = 1.13$$

4. The level of significance is $\alpha = 0.10$.
5. Recall that we can use the right-tailed critical value if we put the larger sample variance in the numerator of the F ratio. Although we cannot find $F_{0.05}(11, 11)$ in table 6b of appendix B, the following values can be found:

$$F_{0.05}(10, 11) = 2.86 \quad \text{and} \quad F_{0.05}(12, 11) = 2.79$$

The critical value for F is some value between 2.86 and 2.79. We shall follow our convention and choose the larger value. Hence, $F_{0.05}(11, 11) \approx 2.86$.

6. The decision is we fail to reject H_0, since $F = 1.13 < 2.86$. Thus, there is no statistical evidence to conclude that the population variances are different.

We next determine if $\mu_2 - \mu_1 \neq 0$.

1. The null hypothesis is H_0: $\mu_2 - \mu_1 = 0$.
2. The alternative hypothesis is H_1: $\mu_2 - \mu_1 \neq 0$ (two-tailed test).
3. The sampling distribution is the distribution of the differences between sample means. The test statistic is the t value for $\bar{x}_2 - \bar{x}_1$.
   We first find the value for s_p from (11.11):

$$s_p = \sqrt{\frac{(n_1 - 1)s_1^2 + (n_2 - 1)s_2^2}{n_1 + n_2 - 2}}$$

$$= \sqrt{\frac{(11)(15.57) + (11)(17.54)}{22}} = 4.07$$

Note that since $n_1 = n_2$, s_p^2 represents the average variance:

$$s_p^2 = \frac{15.57 + 17.54}{2} = 16.555$$

Since the standard deviation is equal to the positive square root of the variance, we have

$$s_p = \sqrt{16.555} = 4.07$$

which agrees with the above result. From (11.13), the value of the t statistic is

$$t = \frac{\bar{x}_2 - \bar{x}_1}{s_p \sqrt{(1/n_1) + (1/n_2)}}$$

$$= \frac{32.6 - 26.8}{4.07 \sqrt{(1/12) + (1/12)}} = 3.49$$

The degrees of freedom is

$$df = n_1 + n_2 - 2$$
$$= 12 + 12 - 2 = 22$$

4. The level of significance is $\alpha = 0.05$.
5. The critical values are $\pm t_{0.025}$ (df = 22) = ± 2.074.
6. The decision is to reject H_0, since $t = 3.49 > 2.074$.
7. With this decision, we risk committing a Type I error, and $\alpha = 0.05$.
8. The p-value associated with this test is less than 0.005.

Hence, we can conclude that the mean length of time for drug A to reach a specified level in the bloodstream is different from the mean length of time it takes drug B to reach the specified level in the bloodstream.

Estimation Procedure We next demonstrate the use of estimation to determine if $\mu_1 - \mu_2 \neq 0$. Since $\alpha = 0.05$, $1 - \alpha = 0.95$. The limits for a 95% confidence interval for $\mu_1 - \mu_2$ are found using (11.15) to be:

$$\bar{x}_1 - \bar{x}_2 \pm t_{\alpha/2} s_p \sqrt{\frac{1}{n_1} + \frac{1}{n_2}}$$

$$5.8 \pm (2.074)(4.07) \sqrt{\frac{1}{12} + \frac{1}{12}}$$

$$5.8 \pm 3.45 = (2.35, 9.25)$$

Since 0 is not contained in this interval, we can be 95% confident that $\mu_1 - \mu_2 \neq 0$, or that $\mu_1 \neq \mu_2$. ■

APPLICATION 11.21

Above Town	Below Town
4.7	5.1
5.1	4.6
4.9	4.8
4.8	4.9
5.2	5.0
5.0	

A common method to check for pollution in a river is to measure the dissolved oxygen content in the river water at different locations. A reduction of the oxygen content in the water from one location to another indicates the presence of pollution somewhere between the two locations. The Environmental Protection Agency (EPA) suspected that a small town was dumping untreated sewage into a river running through the town. Six random specimens of river water were taken at a location above town and five random specimens of river water were taken at a location below the town. The dissolved oxygen readings, in parts per million, are as shown in the table. Determine whether there is sufficient evidence to indicate that the mean oxygen content above the town is greater than the mean oxygen content below the town. Use $\alpha = 0.05$.

Solution: The statistics calculated are shown in the table. Note that since the two sample variances are equal, there is no reason to suspect that the two population variances are unequal. In addition, with an F value of 1, we cannot reject a null hypothesis of equal population variances, since all the tabulated critical values of F are greater than 1.

Above Town	Below Town
$n_1 = 6$	$n_2 = 5$
$\bar{x}_1 = 4.95$	$\bar{x}_2 = 4.88$
$s_1 = 0.19$	$s_2 = 0.19$

We now proceed to test the hypothesis of interest to the EPA.

1. The null hypothesis is H_0: $\mu_1 - \mu_2 \leq 0$.
2. The alternative hypothesis is H_1: $\mu_1 - \mu_2 > 0$ (one-tailed test).
3. The sampling distribution is the distribution of the differences between sample means. The test statistic is the t value for $\bar{x}_1 - \bar{x}_2$. In order to calculate the value t, we first calculate the value of s_p:

$$s_p = \sqrt{\frac{(n_1 - 1)s_1^2 + (n_2 - 1)s_2^2}{n_1 + n_2 - 2}}$$

$$= \sqrt{\frac{5(0.19)^2 + 4(0.19)^2}{9}} = 0.19$$

The value of the t statistic is

$$t = \frac{\bar{x}_1 - \bar{x}_2}{s_p \sqrt{(1/n_1) + (1/n_2)}}$$

$$= \frac{4.95 - 4.88}{0.19\sqrt{(1/6) + (1/5)}} = 0.61$$

The value of df is

$$df = n_1 + n_2 - 2$$
$$= 6 + 5 - 2 = 9$$

4. The level of significance is $\alpha = 0.05$.
5. The critical value is $t_{0.05} = 1.833$ for df $= 9$.
6. The decision is fail to reject H_0, since $t = 0.61 < 1.833$. Thus, there is no statistical evidence to indicate that the community is polluting the stream. The difference between sample means $\bar{x}_1 - \bar{x}_2 = 0.07$ can be explained by sampling error and should not be attributed to pollution.
7. With this decision we may be committing a Type II error; β is unknown.
8. The p-value for the test is greater than 0.1.

For illustrative purposes, let's construct a 95% confidence interval for $\mu_1 - \mu_2$. The limits are given by (11.15).

$$\bar{x}_1 - \bar{x}_2 \pm t_{\alpha/2} s_p \sqrt{\frac{1}{n_1} + \frac{1}{n_2}}$$

$$0.07 \pm (2.262)(0.19)\sqrt{\frac{1}{6} + \frac{1}{5}}$$

$$0.07 \pm 0.26$$

Thus, a 95% confidence interval for $\mu_1 - \mu_2$ is $(-0.19, 0.33)$. Since this interval contains 0, we cannot conclude that there is a difference between mean oxygen contents for the two locations. ■

MINITAB can be used to construct a confidence interval for the EPA data of application 11.21. The TWOSAMPLE command is used to test the hypothesis and construct a 95% confidence interval. Computer display 11.3 contains the commands used and corresponding output.

Computer Display 11.3

```
MTB > SET C1
DATA> 4.7 5.1 4.9 4.8 5.2 5.0
DATA> END
MTB > SET C2
DATA> 5.1 4.6 4.8 4.9 5.0
DATA> END
MTB > TWOSAMPLE .95 C1 C2;
SUBC> ALTERNATIVE=+1.
SUBC> POOLED.


TWO SAMPLE T FOR C1 VS C2
         N     MEAN    STDEV    SEMEAN
  C1     6    4.950    0.187    0.076
  C2     5    4.880    0.192    0.086

95 PCT CI FOR MU C1 − MU C2: (−0.190, 0.330)


TTEST MU C1 = MU C2 (VS GT): T= 0.61 P=0.28 DF = 9
```

Notice that the MINITAB output agrees with our calculations.

EXERCISE SET 11.5

Basic Skills

1. The following data resulted from independent random samples taken from two normal populations with equal variances:

Sample 1	Sample 2
$n_1 = 11$	$n_2 = 9$
$\bar{x}_1 = 14$	$\bar{x}_2 = 17$
$s_1 = 6$	$s_2 = 8$

   a. At $\alpha = 0.05$, do the data indicate a difference between population means?
   b. Construct a 95% confidence interval for $\mu_1 - \mu_2$.

2. Independent random samples were taken from different normal populations with equal variances. The resulting statistics are as follows:

Sample 1	Sample 2
$n_1 = 20$	$n_2 = 25$
$\bar{x}_1 = 212$	$\bar{x}_2 = 231$
$s_1 = 8.5$	$s_2 = 9.9$

   a. Find 95% confidence intervals for $\mu_1 - \mu_2$ and $\mu_2 - \mu_1$.
   b. At $\alpha = 0.05$ test $H_0: \mu_1 - \mu_2 = 0$ against $H_1: \mu_1 - \mu_2 \neq 0.02$.

3. The following summary information was obtained from two independent random samples taken from two normal populations with equal variances:

Sample 1	Sample 2
$n_1 = 12$	$n_2 = 10$
$\bar{x}_1 = 13$	$\bar{x}_2 = 16$
$s_1 = 3.9$	$s_2 = 4.5$

   Determine if the difference $\bar{x}_1 - \bar{x}_2 = -3$ can be attributed to sampling error or to the possibility that $\mu_1 \neq \mu_2$ by using each of the following procedures:
   a. estimation with $1 - \alpha = 0.95$
   b. hypothesis testing with $\alpha = 0.05$

4. The following summary information was obtained from two independent random samples taken from two normal populations with equal variances:

Sample 1	Sample 2
$n_1 = 15$	$n_2 = 11$
$\bar{x}_1 = 19$	$\bar{x}_2 = 16$
$s_1 = 3.8$	$s_2 = 4.3$

   Determine if the difference $\bar{x}_1 - \bar{x}_2 = -3$ can be attributed to sampling error or to the possibility that $\mu_1 \neq \mu_2$

by using each of the following procedures:
a. estimation with $1 - \alpha = 0.90$
b. hypothesis testing with $\alpha = 0.10$

Further Applications

5. Two methods for teaching calculus were applied to two randomly selected groups of college students and compared by means of a standardized calculus test given at the end of the course. The results were as follows:

Method A	Method B
$n_1 = 10$	$n_2 = 13$
$\bar{x}_1 = 63$	$\bar{x}_2 = 68$
$s_1 = 7.2$	$s_2 = 8.4$

a. Do the data present sufficient evidence to indicate a difference between mean population scores? Use $\alpha = 0.01$.
b. Construct a 99% confidence interval for the difference between population means $\mu_a - \mu_b$.

6. Two different surgical procedures were being compared with respect to the length of time in days to recovery. The following data resulted:

Procedure A	Procedure B
$n_1 = 15$	$n_2 = 16$
$\bar{x}_1 = 6.8$	$\bar{x}_2 = 7.5$
$s_1 = 1.25$	$s_2 = 1.49$

a. At $\alpha = 0.05$, do the data indicate a difference between mean recovery times for the two procedures?
b. Construct a 95% confidence interval for the difference between mean recovery times.

7. A study was undertaken to determine if undercoating late-model automobiles inhibits the development of body rust. A random sample of 20 cars of a recent make and model were used in the study. Of these, 12 were undercoated and 8 were not. At the end of a four-year period of use, the amount of rust in square inches was determined for each car. The following are the results:

Undercoated	Not Undercoated
$\bar{x} = 30.7$	$\bar{x}_2 = 38.5$
$s_1 = 4.9$	$s_2 = 4.1$
$n_1 = 12$	$n_2 = 8$

Assume that the populations are normal and have equal variances.

a. By using $\alpha = 0.05$, determine whether the average amount of rust for cars that are undercoated is less than the average amount of rust for cars that are not undercoated.
b. Construct a 95% confidence for $\mu_2 - \mu_1$.

8. Two drugs were compared for their effects on blood cholesterol levels. Two groups of patients were randomly selected to compare the two drugs, A and B. The results (serum cholesterol in milligrams per deciliter) are as follows:

Drug A	Drug B
$n_1 = 18$	$n_2 = 13$
$\bar{x}_1 = 175$	$\bar{x}_2 = 201$
$s_1 = 14.07$	$s_2 = 20.86$

a. Test the variance assumption $H_0: \sigma_1^2 = \sigma_2^2$ at $\alpha = 10\%$.
b. Do the data provide sufficient evidence to indicate that the mean cholesterol level associated with drug A is lower than the corresponding mean for drug B? Use $\alpha = 0.01$ and assume the blood cholesterol levels are normally distributed.
c. Construct a 90% confidence interval for $\mu_b - \mu_a$.

9. In an effort to determine which river, A or B, has brook trout of the greater average length, two samples of trout were compared. The following data were obtained (lengths were measured in inches):

River A	River B
$\bar{x} = 11.2$	$\bar{x}_2 = 12.4$
$s_1 = 2.2$	$s_2 = 1.9$
$n_1 = 14$	$n_2 = 11$

By using $\alpha = 0.05$, determine if $\mu_a \neq \mu_b$ by using each of the following procedures:
a. estimation
b. hypothesis testing

10. Type A fertilizer was applied to six plots and type B fertilizer was applied to seven plots. The following represent the yields of corn (in bushels) per plot:

Type A	50	59	50	58	48	44	
Type B	57	50	60	51	63	49	55

a. Construct a 95% confidence interval for $\mu_b - \mu_a$.
b. Test the null hypothesis $H_0: \mu_b - \mu_a = 0$ against the alternative hypothesis $H_1: \mu_b - \mu_a \neq 0$ by using $\alpha = 0.05$.

11. A union organization conducted a study of the ages of workers in two large industries, A and B. A random sample of 35 workers from industry A had a mean age of 36.1 years and a standard deviation of 5.5 years, whereas a random sample of 40 workers from industry B had a mean age of 32.4 years and a standard deviation of 6.3 years. Assume that the populations of ages are normal and that the variances are equal. Construct a 95% confidence interval for the difference in mean ages for employees in industries A and B to determine if there is a difference between the mean ages.

12. Verify the t values given in the table of motivator 10.

Going Beyond

13. For small independent samples from normal populations with unknown and unequal population variances, the sampling distribution of the statistic

$$\frac{\bar{x}_1 - \bar{x}_2}{\sqrt{(s_1^2/n_1) + (s_2^2/n_2)}}$$

is approximately a t distribution with df found to be the closest integer to

$$\frac{(s_1^2/n_1) + (s_2^2/n_2)}{[(s_1^2/n_1)/(n_1 - 1)] + [(s_2^2/n_2)/(n_2 - 1)]}$$

Use this result to solve the following problem. The data resulted from two independent random samples taken from normal populations with unequal variances.

Sample 1	Sample 2
$n_1 = 11$	$n_2 = 9$
$\bar{x}_1 = 14$	$\bar{x}_2 = 17$
$s_1 = 6$	$s_2 = 8$

a. At $\alpha = 0.05$, do the data indicate a difference between population means?
b. Construct a 95% confidence interval for $\mu_2 - \mu_1$.

SECTION 11.6 **Inferences Concerning $\mu_1 - \mu_2$ Using Small, Dependent Samples**

EXAMPLE 11.10

Many practical applications involve making comparisons between two populations based on paired data, or dependent samples. Applications that could involve dependent samples include

■ Medicine—Testing the effects of a diet by obtaining weight measurements on the same people before and after dieting.

■ Teaching—Testing the effectiveness of a teaching strategy by giving pretests and posttests to the same individuals.

■ Agriculture—Testing the effects of two fertilizers on soy bean production by comparing the yields from similar plots under similar conditions.

■ Business—Comparing repair estimates of two different auto-body shops for the same wrecked automobiles.

■ Industry—Testing two brands of tires for tread wear by placing one of each brand on the rear wheels of a sample of the same type of car.

Reducing Two Samples of Data to One Sample

If we have two dependent random samples of size n, where each element in the first sample is paired with exactly one element in the second sample, then these two samples give rise to a sample of pairs or a sample of differences, as indicated in figure 11.3. The sample of differences $d = x_1 - x_2$ can be thought of as a sample from the population of the differences of paired data from two populations. The mean of the population of differences has a mean equal to the difference of the population means. If a

FIGURE 11.3

Reducing two samples of data
to a sample of differences

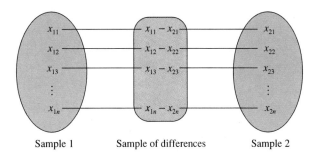

Sample 1   Sample of differences   Sample 2

difference score is represented by $d = x_1 - x_2$ and μ_1 and μ_2 represent the two population means, then we can say

$$\mu_{x_1-x_2} = \mu_d = \mu_1 - \mu_2$$

EXAMPLE 11.11

That the mean of the differences is the difference of the means can be demonstrated by considering the following two populations whose elements have been paired:

	Population 1	Population 2	Difference d
	2	5	$2 - 5 = -3$
	4	6	$4 - 6 = -2$
	6	2	$6 - 2 = 4$
	8	4	$8 - 4 = 4$
	10	8	$10 - 8 = 2$
Sum	30	25	5
Mean	6	5	1

The difference between population means is

$$\mu_1 - \mu_2 = 6 - 5 = 1$$

and the mean of the population of differences is

$$\mu_d = \frac{\Sigma d}{N} = \frac{5}{5} = 1$$

Hence, we see that the mean of the population of differences is equal to the difference between the population means. By following the same line of reasoning, we can show that for two dependent samples, the mean of the sample of differences is equal to the difference between the sample means. That is, if $x_1 - x_2 = d$, then $\bar{x}_1 - \bar{x}_2 = \bar{d}$.

If we have a random sample of n data pairs and if the differences d are normally distributed, then the statistic

$$\frac{\bar{d} - \mu_d}{s_d/\sqrt{n}}$$

has a sampling distribution that is a t distribution with df $= n - 1$, where s_d represents the standard deviation of the sample of difference scores. Hence, the t value for $\bar{d}$ is given by the following formula.

$$\boxed{\begin{array}{c} t \text{ Statistics for } \bar{d} \\[2mm] t = \dfrac{\bar{d} - \mu_d}{s_d/\sqrt{n}} \\[2mm] \text{with df} = n - 1. \end{array}} \qquad (11.16)$$

As a result of the above remarks, if two dependent samples are used to compare two population means, then the one-sample t statistic presented in chapter 10 can be used. Even though two populations are involved, we can interpret the problem of comparing μ_1 and μ_2 as comparing $\mu_{\bar{x}_1 - \bar{x}_2}$ with some fixed value (which is usually 0), based on one sample of difference scores from one population of differences.

When testing the null hypothesis that $\mu_1 - \mu_2 = 0$ and paired data are involved, we can write the null hypothesis as $H_0\colon \mu_d = 0$, where d represents the difference scores. If the population of difference scores is normal, then the test statistic is given by

$$t = \frac{\bar{d}}{s_d/\sqrt{n}}$$

where $\bar{d}$ is the mean of the sample of difference scores, s_d is the standard deviation of the sample of difference scores, and n is the size of the sample of difference scores. The t statistic has $n - 1$ degrees of freedom associated with it. Limits for a $(1 - \alpha)100\%$ confidence interval for $\mu_1 - \mu_2$ are given by

$$\boxed{\begin{array}{c} \textbf{Confidence Interval Limits for } \mu_1 - \mu_2 \textbf{ Using Dependent Samples} \\[2mm] \bar{d} \pm t_{\alpha/2}\dfrac{s_d}{\sqrt{n}} \\[2mm] \text{where df} = n - 1. \end{array}} \qquad (11.17)$$

Applications 11.22–11.24 illustrate the methods involved in making inferences about $\mu_1 - \mu_2$ using paired data.

APPLICATION 11.22   The agricultural departments of five different universities were awarded grants to study the potential yields of two new varieties of corn. Three acres of each variety were planted at each university. The following data (yields in bushels) were collected:

University	Variety 1	Variety 2
1	58	60
2	61	64
3	52	52
4	60	65
5	71	75

If it is known that the population of difference scores d is normal, determine, using $\alpha = 0.05$, if there is a difference between the mean yields for the two varieties.

Solution: The two samples of measurements are dependent, since the sources are universities, and each university produces two measurements. By planting both varieties at each university, we can control for the differences in soil, climate, and weather that could influence the experiment and bias the results. Let x_1 represent the yields for variety 1 and x_2 represent the yields for variety 2. We first calculate the value of s_d. The following table is used to facilitate calculation:

x_1	x_2	d	d^2
58	60	−2	4
61	64	−3	9
52	52	0	0
60	65	−5	25
71	75	−4	16
		−14	54

The mean of the differences is

$$\bar{d} = \frac{14}{5} = -2.8$$

In order to calculate s_d, we first compute the sum of squares for the difference scores:

$$SS = \Sigma d^2 - \frac{(\Sigma d)^2}{n} = 54 - \frac{(-14)^2}{5} = 14.8$$

Thus, the standard deviation of the difference scores is

$$s_d = \sqrt{\frac{SS}{n-1}} = \sqrt{\frac{14.8}{4}} \approx 1.92$$

Therefore, by (11.16), the value of the t statistic is

$$t = \frac{\bar{d}}{s_d/\sqrt{n}} = \frac{-2.8}{1.92/\sqrt{5}} \approx -3.26$$

In addition, the value for df is

$$df = n - 1 = 5 - 1 = 4$$

The statistical hypotheses are:

$$H_0: \mu_1 - \mu_2 = 0$$
$$H_1: \mu_2 - \mu_2 \neq 0$$

For df $= 4$, the critical values for t are $\pm t_{0.025} = \pm 2.776$. Since the test statistic $t = -3.26$ is less than -2.776, we reject H_0, and conclude that the yields of the two varieties of corn are different. ■

APPLICATION 11.23 A study was conducted to determine if jogging reduces a person's at-rest heart rate. Ten volunteers were examined before and following a six-month jogging program. Their at-rest heart rates (beats per minute) were recorded as follows:

Before	73	77	68	62	72	80	76	64	70	72
After	68	72	64	60	71	77	74	60	64	68

By using $\alpha = 0.05$, determine if jogging reduces at-rest heart rates.

Solution: The difference scores are 5, 5, 4, 2, 1, 3, 2, 4, 6, and 4. The mean difference score is $\overline{d} = 3.6$ and the standard deviation of the sample of difference scores is $s_d = 1.58$.

1. The null hypothesis is H_0: $\mu_1 - \mu_2 \leq 0$.
2. The alternative hypothesis is H_1: $\mu_1 - \mu_2 > 0$ (one-tailed test).
3. The test statistic is the t value for $\overline{d}$. The t value for $\overline{d} = 3.6$ is

$$t = \frac{3.6 - 0}{1.58/\sqrt{10}} = 7.20$$

4. The level of significance is $\alpha = 0.05$.
5. For df $= 9$, the critical value is $t_{0.05} = 1.833$.
6. The decision is to reject H_0, since $t = 7.20 > 1.833$. Thus, one can conclude that the data indicate that jogging significantly lowers at-rest heart rates.
7. With this decision, we may have committed a Type I error; $\alpha = 0.05$.
8. The p-value is less than 0.005. ■

APPLICATION 11.24   Ten men were placed on a special diet; their weights were recorded before starting the diet and after one month on the dieting program. The results of weighings (in pounds) before and after are as follows:

Male	A	B	C	D	E	F	G	H	I	J
Before	181	172	190	186	210	202	166	173	183	184
After	178	175	185	184	207	201	160	168	180	189

Test using $\alpha = 0.05$ to determine whether dieting makes a difference (positive or negative).

Solution: The difference scores d are 3, -3, 5, 2, 3, 1, 6, 5, 3, and -5. The mean and standard deviation are $\overline{d} = 2$ pounds and $s_d = 3.53$ pounds, respectively. For illustrative purposes, we will determine if the diet is effective for a change in weight by using hypothesis testing and estimation procedures.

Hypothesis-Testing Procedure

1. The null hypothesis is H_0: $\mu_{\text{before}} - \mu_{\text{after}} = 0$.
2. The alternative hypothesis is H_1: $\mu_{\text{before}} - \mu_{\text{after}} \neq 0$ (two-tailed test).
3. The test statistic is the t value for $\overline{d}$. The t value for $\overline{d} = 2$ is

$$t = \frac{2}{3.53/\sqrt{10}} = 1.79$$

4. The level of significance is $\alpha = 0.05$.
5. For df $= 9$, the critical values are $\pm t_{0.025} = \pm 2.262$.
6. We fail to reject H_0, since $t = 1.79 < 2.262$.
7. With this decision, we may be committing a Type II error; β is unknown. Hence, we have no statistical evidence to support that the diet is effective in changing weight.
8. The p-value is between 0.10 and 0.20.

Estimation Procedure The limits for a 95% confidence interval for $\mu_1 - \mu_2$ are given by (11.17).

$$\bar{d} \pm t_{\alpha/2} \frac{s_d}{\sqrt{n}} = 2 \pm 2.262 \frac{3.53}{\sqrt{10}} = 2 \pm 2.53$$

The 95% confidence interval is $(-0.53, 4.53)$. Since this interval contains 0, we cannot conclude that the diet is effective in changing weight. ■

MINITAB can be used to determine if the diet is effective in application 11.23. Computer display 11.4 contains the read command to input the data.

Computer Display 11.4

```
MTB > READ C1 C2
DATA> 181 178
DATA> 172 175
DATA> 190 185
DATA> 186 184
DATA> 210 207
DATA> 202 201
DATA> 166 160
DATA> 173 168
DATA> 183 180
DATA> 184 189
DATA> END
   10 ROWS READ
```

The TTEST command is used to test the null hypothesis. Computer display 11.5 contains the commands and accompanying output.

Computer Display 11.5

```
MTB > LET C3 = C1 − C2
MTB > TTEST MU=0 C3

TEST OF MU = 0.000 VS MU N.E. 0.000

            N      MEAN    STDEV    SE MEAN     T     P VALUE
C3         10     2.000    3.528     1.116     1.79     0.11
```

Notice that the output agrees with our calculations. The TINTERVAL command is used to determine the 95% confidence interval. Computer display 11.6 contains the command and accompanying output.

Computer Display 11.6

```
MTB > TINTERVAL 95 PERCENT C3

            N      MEAN    STDEV    SE MEAN    95.0 PERCENT C.I.
C3         10     2.00     3.53      1.12      ( −0.52, 4.52)
```

Notice that the endpoints of the confidence interval obtained by MINITAB varies slightly from the limits we obtained. This difference is due to the greater precision of calculations done by MINITAB.

When dealing with small independent samples, we should check the normality assumption and the assumption of equal variances whenever possible. These assumptions, along with the assumption that H_0 is true, make it possible to reject H_0 whenever the test statistic lands in the rejection region. If both the normality and variance assumptions were tenuous and the test statistic fell in the rejection region, we would not know whether the unlikely event of rejection should be attributed to violation of the assumptions underlying the test or to the possibility that H_0 could be false. Of course, nonrandom samples could also lead to false conclusions, and both types of error can be affected when the assumptions are violated.

The t tests are called **robust tests,** meaning that they work reasonably well in the face of minor violations of their theoretical assumptions. This is particularly so when two-sample tests involve samples of equal sizes. When there are major violations of the assumptions, a nonparametric test should be investigated. These will be discussed in chapter 15.

There are many methods available to test the normality assumption underlying the χ^2 and t tests. Graphical procedures, such as the z score plot presented in section 7.2, are easy to use. Other tests, such as the chi-square goodness-of-fit test and Lilliefor's test,[42] are also available. Both of these tests are beyond the scope of this text and will not be treated.

EXERCISE SET 11.6

Basic Skills

1. Given the following paired-sample data:

x	4	2	6	4
y	3	1	3	5

  a. Determine the value of the t statistic.
  b. How many degrees of freedom does this t have?

2. Given the following paired-sample data:

x	5	4	5	3	8
y	2	3	4	9	7

  a. Determine the value of the t statistic.
  b. How many degrees of freedom does this t have?

3. A new method of study was developed to increase SAT mathematics scores. The method was used on ten randomly chosen students who had already taken the SAT test once. To test the effectiveness of the method, an SAT test was again administered to the ten students. Their before and after scores were as shown in the accompanying table.

Student	A	B	C	D	E
Before	596	610	598	613	588
After	599	612	607	610	588

Student	F	G	H	I	J
Before	592	606	619	600	597
After	610	607	623	591	599

If the difference scores (after − before) are assumed to be normally distributed with $\mu_d = 0$, calculate the value of the t statistic and determine its degrees of freedom.

4. Eleven men were placed on a special diet; their weights were recorded before starting the diet and after one month on the dieting program. The results of weighings (in pounds) before and after are as follows:

Male	A	B	C	D	E	F
Before	183	192	195	184	205	212
After	178	178	184	184	207	201

Male	G	H	I	J	K
Before	186	153	175	169	175
After	160	155	170	159	170

If the difference scores (before − after) are assumed to be normally distributed with $\mu_d = 0$, calculate the value of the t statistic and determine its degrees of freedom.

Further Applications

5. A study was conducted on the effectiveness of an industrial safety program to reduce lost-time accidents. The results (expressed in mean man-hours lost per month over a period of one year) were taken at six plants before and after an industrial safety program was put into effect.
   a. Do the data in the accompanying table provide sufficient evidence (at $\alpha = 0.05$) to indicate that the program was effective in reducing time lost due to industrial accidents?
   b. Use the data in the accompanying table to construct a 95% confidence interval for $\mu_{before} - \mu_{after}$.

Plant	1	2	3	4	5	6
Before	40	66	44	72	60	32
After	33	60	45	67	54	31

6. A tire manufacturer wanted to compare the wearing qualities of two types of tires, A and B. Six cars of similar weight and design were randomly chosen for the experiment. For each car, a tire of type A and a tire of type B were randomly placed on the rear wheels. The cars were then operated for 5000 miles over similar terrain under similar driving conditions, and the amount of tread wear (in thousandths of an inch) was recorded, with the following results:

Car	1	2	3	4	5	6
Tire A	10.7	9.8	11.3	9.5	8.6	9.4
Tire B	10.3	9.4	11.5	9.0	8.2	9.0

   a. Do the data provide sufficient evidence to indicate a difference between average tread wear for tires A and B? Use $\alpha = 0.05$. Assume that the difference scores are normally distributed.
   b. Construct a 95% confidence interval for the difference between average tread wear for tires A and B.

7. Fourteen heart patients were placed on a special diet to lost weight. Their weights (in kilograms) were recorded before starting the diet and after one month on the diet. The following data were obtained:

Patient	1	2	3	4	5	6	7
Weight before	62	62	65	88	76	57	60
Weight after	59	60	63	78	75	58	60

Patient	8	9	10	11	12	13	14
Weight before	59	54	68	65	63	60	56
Weight after	52	52	65	66	59	58	55

   Assume that the differences in weights are normally distributed.
   a. Use $\alpha = 0.05$ and test to determine if the diet is effective.
   b. Construct a 95% confidence interval for $\mu_{before} - \mu_{after}$.

8. Calculate the value of s_d for motivator 11.

Going Beyond

9. For two dependent samples, the standard error of the difference between sample means $\sigma_{\bar{x}_1 - \bar{x}_2}$ is sometimes approximated by $\hat{\sigma}_{\bar{x}_1 - \bar{x}_2}$ where

$$\hat{\sigma}_{\bar{x}_1 - \bar{x}_2} = \sqrt{\frac{s_1^2}{n} + \frac{s_2^2}{n} - \frac{2rs_1s_2}{n}}$$
$$= \sqrt{\frac{s_1^2 + s_2^2 - 2rs_1s_2}{n}}$$

where r is the correlation coefficient between x_1 and x_2. For the data in exercise 3, show that

$$s_d = \sqrt{s_1^2 + s_2^2 - 2rs_1s_2}$$

10. Consider the following computer printout for computing the t statistic for a sample of nine bivariate pairs. Using the statistics in the printout, find s_d and determine whether the population means are significantly different at $\alpha = 0.01$.

```
              T-TEST RESULTS

VARIABLE X: BEFORE          VARIABLE Y: AFTER
MEAN OF X = 7.27667         MEAN OF Y = 11.2722
S.D. OF X = 2.60702         S.D. OF Y = 5.24589
S.E. MEAN = 0.921721        S.E. MEAN = 1.8547


NUMBER OF PAIRS (N) = 9
CORRELATION OF X WITH Y (R) = 0.622
DIFFERENCE (MEAN X = MEAN Y) = -3.99556
DEGREES OF FREEDOM (DF) = 8
T-RATIO FOR THE DIFFERENCE = -2.71684
PROBABILITY (2 TAILED TEST) = 0.025
```

CHAPTER SUMMARY

In this chapter we discussed estimation and hypothesis testing involving two populations. The methods are based on the statistics from two random samples. For independent samples, the two-sample z statistic is used to compare population means when both samples are large, the two-sample t statistic is used to compare population means when sampling involves small samples from normal populations with equal variances, and the F statistic is used to compare population variances when samples are from normal populations. For dependent samples, the one-sample t statistic is used whenever the population of difference scores is normal and the variance of difference scores is unknown. The statistical procedures for analyzing the differences between population proportions are similar to the procedures for analyzing the differences between population means, both involve sampling distributions that are approximately normal for large samples. Confidence intervals for $\mu_1 - \mu_2$ and $p_1 - p_2$ have the following form:

(test statistic) $\pm [z_{\alpha/2} \text{ or } t_{\alpha/2}]$ (standard error of test statistic)

The following table summarizes the methods discussed in this chapter for drawing inferences concerning the comparison of two population parameters.

Samples	Parameter Comparison	Assumptions	Test Statistic and Sampling Distribution	t or z Value for Test Statistic	Confidence Interval Limits
Small, dependent	$\mu_1 - \mu_2$	1. Samples are random. 2. Sampling distribution of d is normal.	$\bar{x}_1 - \bar{x}_2 = \bar{d}$	$t = \dfrac{\bar{d}}{s_d/\sqrt{n}}$, df $= n - 1$	$\bar{d} \pm t_{\alpha/2}\dfrac{s_d}{\sqrt{n}}$
Small, independent	$\mu_1 - \mu_2$	1. Samples are random. 2. Both populations are normal. 3. Samples are independent. 4. $\sigma_1^2 = \sigma_2^2$.	$\bar{x}_1 - \bar{x}_2$	$t = \dfrac{\bar{x}_1 - \bar{x}_2}{s_p\sqrt{(1/n_1) + (1/n_2)}}$ $s_p = \sqrt{\dfrac{(n_1 - 1)s_1^2 + (n_2 - 1)s_2^2}{n_1 + n_2 - 2}}$ df $= n_1 + n_2 - 2$	$\bar{x}_1 - \bar{x}_2 \pm t_{x/2}s_p\sqrt{\dfrac{1}{n_1} + \dfrac{1}{n_2}}$
Large, independent	$\mu_1 - \mu_2$	1. Samples are random. 2. Samples are independent.	$\bar{x}_1 - \bar{x}_2$	$z = \dfrac{\bar{x}_1 - \bar{x}_2}{\sqrt{(\sigma_1^2/n_1) + (\sigma_2^2/n_2)}}$	$\bar{x}_1 - \bar{x}_2 \pm z\sqrt{\dfrac{\sigma_1^2}{n_1} + \dfrac{\sigma_2^2}{n_2}}$
Large, independent	$p_1 - p_2$	1. Samples are random. 2. Samples are independent. 3. $n_1p_1 \geq 5$, $n_1(1 - p_1) \geq 5$, $n_2p_2 \geq 5$, $n_2(1 - p_2) \geq 5$.	$\hat{p}_1 - \hat{p}_2$	$z = \dfrac{\hat{p}_1 - \hat{p}_2}{\sqrt{\hat{p}(1 - \hat{p})(1/n_1 + 1/n_2)}}$ $\hat{p} = \dfrac{x_1 + x_2}{n_1 + n_2}$	$\hat{p}_1 - \hat{p}_2 \pm z_{\alpha/2}\sqrt{\dfrac{\hat{p}_1(1 - \hat{p}_1)}{n_1} + \dfrac{\hat{p}_2(1 - \hat{p}_2)}{n_2}}$
Any size, independent	$\dfrac{\sigma_1^2}{\sigma_2^2}$ or $\dfrac{\sigma_1}{\sigma_2}$	1. Samples are random. 2. Samples are independent. 3. Populations are normal.	$\dfrac{s_1^2}{s_2^2}$	$F = \dfrac{s_1^2}{s_2^2}$	$\dfrac{s_1^2}{s_2^2}F_{1-\alpha/2}(n_2 - 1, n_1 - 1) = L_1$ $\dfrac{s_1^2}{s_2^2}F_{\alpha/2}(n_2 - 1, n_1 - 1) = L_2$

CHAPTER REVIEW

∎ **IMPORTANT TERMS** ∎

The following chapter terms have been mixed in ordering to provide you better review practice. For each of the following terms, provide a definition in your own words. Then check your responses against the definitions given in the chapter.

robust tests
F distribution
dependent samples
F statistic
sampling distribution of the
 difference between means

F tables
source
pooled estimate for p
independent samples
pooled estimate for σ

$(1 - \alpha)100\%$ confidence interval for σ_1^2/σ_2^2
$(1 - \alpha)100\%$ confidence interval for σ_1/σ_2
sampling distribution of the statistic $\bar{x}_1 - \bar{x}_2$

∎ **IMPORTANT SYMBOLS** ∎

$\mu_{\bar{x}_1-\bar{x}_2}$, the mean of the sampling distribution of $\bar{x}_1 - \bar{x}_2$

$\sigma_{\bar{x}_1-\bar{x}_2}$, the standard error of $\bar{x}_1 - \bar{x}_2$

$\mu_{\hat{p}_1-\hat{p}_2}$ mean of the sampling distribution of $\hat{p}_1 - \hat{p}_2$

$\sigma_{\hat{p}_1-\hat{p}_2}$, the standard error of $\hat{p}_1 - \hat{p}_2$

$\hat{p}$, pooled estimate for p

F, F statistic

$F(n_1 - 1, n_2 - 1)$, F distribution with df $= (n_1 - 1, n_2 - 1)$

$F_{1-\alpha/2}(n_2 - 1, n_1 - 1)$, left-tailed critical value for F distribution

$F_{\alpha/2}(n_1 - 1, n_2 - 1)$, right-tailed critical value for F distribution

s_p, pooled estimate for σ

d, difference score

$\bar{d}$, the mean of the sample of difference scores

s_d, the standard deviation of the sample of difference scores

∎ **IMPORTANT FACTS AND FORMULAS** ∎

Mean of the sampling distribution of $\bar{x}_1 - \bar{x}_2$:
$\mu_{\bar{x}_1-\bar{x}_2} = \mu_1 - \mu_2.$ (11.1)

Standard error of $\bar{x}_1 - \bar{x}_2$:
$\sigma_{\bar{x}_1-\bar{x}_2} = \sqrt{(\sigma_1^2/n_1) + (\sigma_2^2/n_2)}.$ (11.1)

Confidence interval limits for $\mu_1 - \mu_2$:
$\bar{x}_1 - \bar{x}_2 \pm z_{\alpha/2}\sigma_{\bar{x}_1-\bar{x}_2}.$ (11.3)

Mean of the sampling distribution of $\hat{p}_1 - \hat{p}_2$:
$\mu_{\hat{p}_1-\hat{p}_2} = p_1 - p_2.$ (11.5)

Approximate standard error of $\hat{p}_1 - \hat{p}_2$:
$\sigma_{\hat{p}_1-\hat{p}_2} = \sqrt{\dfrac{\hat{p}_1(1 - \hat{p}_1)}{n_1} + \dfrac{\hat{p}_2(1 - \hat{p}_2)}{n_2}}.$

Approximate confidence interval limits for $p_1 - p_2$:
$\hat{p}_1 - \hat{p}_2 \pm z_{\alpha/2}\hat{\sigma}_{\hat{p}_1-\hat{p}_2}.$ (11.6)

Pooled estimate for p: $\hat{p} = (x_1 + x_2)/(n_1 + n_2).$ (11.7)

Approximate standard error of $\hat{p}_1 - \hat{p}_2$ when $p_1 = p_2$:
$\sigma_{\hat{p}_1-\hat{p}_2} \approx \sqrt{\hat{p}(1 - \hat{p})\left(\dfrac{1}{n_1} + \dfrac{1}{n_2}\right)}.$ (11.8)

F statistic used to test equality of population variances:
$F = s_1^2/s_2^2$, df $= (n_1 - 1, n_2 - 1)$.

$F_{1-\alpha/2}(n_2 - 1, n_1 - 1) = \dfrac{1}{F_{\alpha/2}(n_1 - 1, n_2 - 1)}.$
(11.9)

Confidence interval limits for σ_1^2/σ_2^2:
$L_1 = \dfrac{s_1^2}{s_2^2}F_{1-\alpha/2}(n_2 - 1, n_1 - 1)$ and

$L_2 = \dfrac{s_1^2}{s_2^2}F_{\alpha/2}(n_2 - 1, n_1 - 1).$ (11.10)

Pooled estimate for σ:
$s_p = \sqrt{\dfrac{(n_1 - 1)s_1^2 + (n_2 - 1)s_2^2}{n_1 + n_2 - 2}}.$ (11.11)

t statistic for $\bar{x}_1 - \bar{x}_2$: $t = \dfrac{(\bar{x}_1 - \bar{x}_2) - (\mu_1 - \mu_2)}{s_p\sqrt{(1/n_1) + (1/n_2)}}$.

(11.13)

Confidence interval limits for $\mu_1 - \mu_2$ using small,

independent samples: $\bar{x}_1 - \bar{x}_2 \pm t_{\alpha/2}s_p\sqrt{\dfrac{1}{n_1} + \dfrac{1}{n_2}}$.

(11.15)

t statistic for $\bar{d}$: $t = (\bar{d} - \mu_{\bar{d}})/(s_d/\sqrt{n})$.   (11.16)

Confidence interval limits for $\mu_1 - \mu_2$ using small,

dependent samples: $\bar{d} \pm t_{\alpha/2}\dfrac{s_d}{\sqrt{n}}$.   (11.17)

REVIEW EXERCISES

1. Two different types of brake lining were tested for differences in wear. Twenty-four cars were used. A sample of each brand was tested with the results (listed in hundreds of miles) given in the table.

Brand A	42 58 64 40 47 50 62 54 42 38 66 52
Brand B	48 40 30 44 54 38 32 42 40 62 50 34

   a. Assuming the populations are normal and the population standard deviations are equal to 8, test for a difference in average brake lining wear using $\alpha = 0.05$. Find the p-value.
   b. Construct a 95% confidence interval for $\mu_a - \mu_b$.

2. Two groups of males are polled concerning their interest in a new electric razor that has four cutting edges. A sample of 64 males under age 40 indicated that only 12 were interested, whereas in a sample of 36 males over age 40, only 8 indicated an interest. Test whether there is a difference between the proportion of the two age groups who indicate an interest in the new razor. Use $\alpha = 0.05$ and determine the p-value.

3. For review exercise 2, construct a 95% confidence interval for the difference between age-group proportions and determine if there is a difference between population proportions.

4. Eleven workers performed a task using two different methods. The completion times (in minutes) for each task are given below:

Worker	Method A	Method B
1	15.2	14.5
2	14.6	14.8
3	14.2	13.8
4	15.6	15.6
5	14.9	15.3
6	15.2	14.3

Worker	Method A	Method B
7	15.6	15.5
8	15.0	15.0
9	16.2	15.6
10	15.7	15.2
11	15.6	14.8

If the completion times are normally distributed for each method, test to determine whether $\mu_a \neq \mu_b$ using $\alpha = 0.05$.

5. Before the primary elections, candidate A showed the following voter support for president in polls taken in Maryland and Pennsylvania:

Maryland	Pennsylvania
$n_1 = 1000$	$n_2 = 540$
$x_1 = 720$	$x_2 = 324$

Let p_1 represent the proportion of Maryland voters who support candidate A and let p_2 represent the proportion of Pennsylvania voters who support candidate A. Test at the 0.05 level of significance the null hypothesis H_0: $p_1 - p_2 = 0$ against the alternative hypothesis H_1: $p_1 - p_1 \neq 0$.

6. For review exercise 5, construct a 95% confidence interval for $p_1 - p_2$, the difference between voter proportions favoring candidate A in the two states.

7. Independent samples of male and female employees selected from a large industrial plant yielded the following hourly-wage results.

Males	Females
$n_1 = 45$	$n_2 = 32$
$\bar{x}_1 = \$6.00$	$\bar{x}_2 = \$5.75$
$s_2 = \$0.95$	$s_2 = \$0.75$

Test the statistical hypothesis that the mean hourly wage for male employees exceeds that for female employees. Use $\alpha = 0.01$ and determine the p-value.

8. A certain precision part is manufactured by two different companies and its length is critical. Random samples of parts were obtained from the two manufacturers and measured for length (in millimeters), with the following results:

Company A	Company B
$n_1 = 15$	$n_2 = 31$
$s_1 = 0.008$	$s_2 = 0.012$

Test using the 0.02 level of significance to determine whether there is a difference between the variances of the lengths of the parts manufactured by the two companies. Assume that the lengths of the parts are normally distributed.

9. For review exercise 8, construct a 98% confidence interval for σ_2/σ_1 to determine if there is a difference between the variances of the lengths of the parts made by the two companies.

10. In order to compare the nicotine contents (in milligrams) of two brands of cigars, A and B, the following results were obtained:

Brand A	Brand B
$\bar{x}_1 = 15.3$	$\bar{x}_2 = 16.7$
$s_1 = 1.7$	$s_2 = 2$
$n_1 = 60$	$n_2 = 40$

a. Test using $\alpha = 0.05$ to determine if the mean nicotine contents for the two brands of cigars are unequal.
b. Construct a 95% confidence interval for $\mu_a - \mu_b$. Does it contain 0?

11. Two varieties of apples, A and B, were analyzed for their potassium content in milligrams. The following information was obtained:

Variety A	Variety B
$n_1 = 100$	$n_2 = 150$
$\bar{x}_1 = 0.30$	$\bar{x}_2 = 0.27$
$s_1 = 0.07$	$s_2 = 0.05$

Do the data indicate that the average potassium contents differ for the two varieties? Use $\alpha = 0.01$. Can we conclude that $\mu_b < \mu_a$?

12. Answer review exercise 11 by constructing a 99% confidence interval for the difference between the average potassium contents for the two varieties of apples.

13. Two different types of brake lining were tested for differences in wear. Twelve cars were used; each car had both brands randomly placed on the rear wheels. A sample of each brand was tested, with the results (listed in hundreds of miles) given in the table.

Car	1 2 3 4 5 6 7 8 9 10 11 12
Brand A	42 58 64 40 47 50 62 54 42 38 66 52
Brand B	48 40 30 44 54 38 32 42 40 62 50 34

a. Assuming the population of difference scores are normal, test for a difference between average brake lining wear using $\alpha = 0.05$. Find the p-value.
b. Construct a 95% confidence interval for $\mu_a - \mu_b$.

14. In order to evaluate two methods of teaching German, students were randomly assigned to two groups. Students in one group were taught by method A and students in the other group were taught by method B. At the end of the instructional units, both groups were given the same achievement test. The scores for the two groups are given in the accompanying table.

Group A	84 86 91 93 84 88
Group B	90 88 92 94 84 85 92

At the 0.05 level of significance determine whether method B produces higher average results than method A using the t test.

15. Eleven workers performed a task using two different methods. The completion times (in minutes) for each task are given below:

Worker	Method A	Method B
1	15.2	24.5
2	14.6	14.8
3	14.2	13.8
4	15.6	15.6
5	14.9	15.3
6	15.2	14.3
7	15.6	15.5
8	15.0	15.0
9	16.2	15.6
10	15.7	15.2
11	15.6	14.8

a. Use the t test to test to determine whether there is a difference in average completion times for the two methods. Use $\alpha = 0.01$.

b. Construct a 99% confidence interval for the difference in average completion times.

16. An auto manufacturing plant plans to institute a new employee incentive plan. To evaluate the new plan, five employees use the incentive plan for an experimental period. Their work outputs before and after implementation of the new plan are as follows:

Employee	Output Before	Output After
A	20	23
B	17	19
C	23	24
D	20	23
E	21	23

At $\alpha = 0.05$, use the t test to determine whether the new incentive plan results in greater average output.

17. For review exercise 16, construct a 95% confidence interval for $\mu_1 - \mu_2$, where μ_1 denotes the population mean output before the new plan was implemented and μ_2 denotes the population mean output after the plan was implemented. Does this result contradict the result found in review exercise 16? Explain.

18. The military conducted a study to determine the accuracy that can be obtained with two types of rifles. A random sample of equally proficient soldiers was chosen and divided into two groups. Each group shot only one type of rifle. Accuracy ratings are given in the table where higher values signify greater accuracy.

Rifle A	88 84 88 90 86 92 88
Rifle B	90 94 92 90 88 86

Use the t test at the 0.01 level of significance to determine whether rifle B is more accurate, on the average, than rifle A.

19. Two groups of overweight teachers were randomly selected to test two diets. One group followed diet A for one month and the other group followed diet B for one month. Their weight losses (in pounds) were recorded at the end of the one-month period, with the results following:

Diet A	Diet B
$n_1 = 50$	$n_2 = 40$
$\bar{x}_1 = 10.2$	$\bar{x}_2 = 8.4$
$s_1 = 2.6$	$s_2 = 1.8$

Test at $\alpha = 0.05$ to determine if there is a difference in the average weight losses for the two diets.

20. For review exercise 19, determine if there is a difference between the average weight losses for the two groups of teachers by constructing a 90% confidence interval.

21. For review exercise 18, test at $\alpha = 0.10$ to determine whether the population variances are different. Does this result affect the use of the t test? Explain.

22. Two brands of feed are being compared to determine if chickens fed one brand will weigh more than chickens fed the other brand. A random sample of 10,000 chickens was chosen and fed with brand A feed and another random sample of 10,000 chickens was chosen and fed with brand B. The following results were obtained (weights were measured in pounds):

Brand A	Brand B
$\bar{x}_1 = 9.072$	$\bar{x}_2 = 9.023$
$s_1 = 1.05$	$s_2 = 1.14$

Determine if there is a difference between the average weights of chickens fed with feeds A and B. Use $\alpha = 0.05$.

Computer Applications

1. A candy manufacturer has two electronic filling machines that measure 5 ounces of candy for each package. As part of its quality control efforts, the following weights (in ounces) were taken for each machine.

Machine A: 5.20 5.46 3.98 4.23 5.30 5.91
 6.27 4.71 4.86 4.59 4.08 4.66
 3.58 3.64 4.65 5.27 4.92 4.71
 5.62 3.73 3.65 3.97

Machine B: 4.70 5.23 2.05 4.75 4.62 5.85
 4.90 5.80 3.42 6.79 5.32

a. Use a computer program and $\alpha = 0.05$ to determine if $\sigma_a^2 \neq \sigma_b^2$.

b. Use a dotplot to determine if there is any information to suggest that the samples did not come from normal populations.

c. Use a computer program and $\alpha = 0.01$ to determine if $\mu_a \neq \mu_b$.

d. Should either or both machines be recalibrated?

2. Do computer application 1, but use the following data:

Machine A:
5.20	5.46	3.98	4.23	5.30	5.91
6.27	4.71	4.86	4.59	4.08	4.66
3.58	3.64	4.65	5.27	4.92	4.71
5.62	3.73	3.65	3.97	4.70	5.23
2.05	4.75	4.62	5.85	4.90	5.80
3.42	6.79	5.32	5.43	4.86	5.49
5.26	5.58	5.82	4.61	4.60	4.12
3.55	3.71	3.48	4.13	4.18	5.32
4.95	5.97	5.90	4.18	3.62	6.45
4.97	5.35	4.85	4.73	5.09	4.60
5.95	4.41	4.94	4.88	3.70	5.04

Machine B:
5.77	3.86	3.73	5.72	6.50	5.73
4.25	5.28	7.33	4.20	4.29	4.95
6.10	5.82	6.51	5.14	5.69	5.79
4.32	5.80	5.81	4.75	5.34	5.17
5.87	4.86	4.65	6.61	4.68	5.89
7.37	4.13	4.20	4.80		

3. a. Simulate an empirical sampling distribution of the difference in sample means by drawing 100 random samples of size 35 from the normal distribution having a mean of 75 and a standard deviation of 3 and 100 random samples of size 40 from the normal distribution having a mean of 60 and a standard deviation of 5. For each sample compute the sample mean. Then form the empirical distribution of the differences in sample means.

b. Determine the mean and standard deviation of the empirical distribution and construct a histogram for the distribution.

c. Compare your results with the theoretical results.

4. a. Simulate an empirical sampling distribution of the difference in sample proportions by drawing 100 random samples of size 35 from a large population having an equal number of 0s and 1s and 100 random samples of size 40 from a large population of 0s and 1s having the same number of 0s and 1s. For each sample, compute the proportion of 1s in the sample. Then form the empirical distribution of the differences in sample proportions.

b. Determine the mean and standard deviation of the empirical distribution and construct a histogram for the distribution.

c. Compare your results with the theoretical results.

EXPERIMENTS WITH REAL DATA

Treat the 720 subjects listed in the database in appendix C as if they constituted the population of all U.S. subjects.

1. For subjects in the 20–29 year age group,

a. test the null hypothesis that the proportion of females that smoke is greater than or equal to the proportion of males who smoke. Use two random samples of size 35 and the 0.05 level of significance.

b. use two random samples of size 35 to construct a 95% confidence interval for the difference in proportions of males and females who smoke.

2. For subjects in the 30–39 year age group, use samples of size 40 to

a. test the hypothesis that the mean systolic blood pressure of males is different than the mean systolic blood pressure of females. Use $\alpha = 0.05$.

b. construct a 95% confidence interval for the difference in mean systolic blood pressures for males and females.

3. For subjects in the 20–29 year age group, use samples of size 40 to

a. test the null hypothesis that the mean systolic blood pressure of smokers is less than or equal to the mean systolic blood pressure of nonsmokers. Use $\alpha = 0.05$.

b. construct a 95% confidence interval for the difference in mean systolic blood pressures for smokers and nonsmokers.

4. For subjects in the 20–29 year age group, use samples of size 35 to construct a 95% confidence interval for the difference in weights of female smokers and female nonsmokers.

■ CHAPTER ACHIEVEMENT TEST ■

1. The following information is based on random samples selected from two normal populations:

Sample 1	Sample 2
$n_1 = 31$	$n_2 = 31$
$\bar{x}_1 = 82.4$	$\bar{x}_2 = 85.5$
$s_1^2 = 21.2$	$s_2^2 = 16$

   a. Test $H_0: \sigma_1^2 \leq \sigma_2^2$ against $H_1: \sigma_1^2 > \sigma_2^2$ using $\alpha = 0.05$.
   b. Test $H_0: \mu_2 - \mu_1 \leq 0$ against $H_1: \mu_2 - \mu_1 > 0$ using $\alpha = 0.05$.
   c. Construct a 90% confidence interval for $\mu_2 - \mu_1$.

2. In order to test $H_0: p_1 - p_2 = 0$, the following information was obtained:

Sample 1	Sample 2
$n_1 = 50$	$n_2 = 60$
$x_1 = 29$	$x_2 = 33$

   a. Find $\hat{p}$, the pooled estimate for p.
   b. Find the z value for $\hat{p}_1 - \hat{p}_2$ under the assumption that H_0 is true.
   c. Test $H_0: p_1 - p_2 \geq 0$ versus $H_1: p_1 - p_2 < 0$ using $\alpha = 0.05$.
   d. Construct a 95% confidence interval for $p_1 - p_2$.

3. For test question 1,
   a. Construct a 90% confidence interval for σ_1^2/σ_2^2.
   b. Construct a 90% confidence interval for σ_2/σ_1.

4. The following information was obtained from two random samples selected from large populations:

Sample 1	Sample 2
$n_1 = 49$	$n_2 = 64$
$\bar{x}_1 = 61$	$\bar{x}_2 = 54$
$s_1 = 14$	$s_2 = 16$

   a. Find $\sigma_{\bar{x}_1 - \bar{x}_2}$.
   b. Find the z value for $\bar{x}_1 - \bar{x}_2$ if $\mu_1 - \mu_2 = 1$.
   c. Test $H_0: \mu_1 - \mu_2 \leq 0$ against $H_1: \mu_1 - \mu_2 > 0$ at $\alpha = 0.05$.
   d. Find the p-value for the test conducted in part c.
   e. Construct a 90% confidence interval for $\mu_1 - \mu_2$.

5. The following information was obtained from random samples selected from two normal populations:

Sample 1	Sample 2
$n_1 = 15$	$n_2 = 15$
$\bar{x}_1 = 82.4$	$\bar{x}_2 = 85.5$
$s_1^2 = 21.2$	$s_2^2 = 16$

   a. Find s_p, the pooled estimate of σ.
   b. Test $H_0: \sigma_1^2 \leq \sigma_2^2$ against $H_1: \sigma_1^2 > \sigma_2^2$ using $\alpha = 0.05$.
   c. Test $H_0: \mu_2 - \mu_1 \leq 0$ against $H_1: \mu_2 - \mu_1 > 0$ using $\alpha = 0.05$.
   d. Construct a 90% confidence interval for $\mu_2 - \mu_1$.
   e. Construct a 90% confidence interval for $\mu_1 - \mu_2$.

12

Analyses of Count Data

CHAPTER OBJECTIVES

In this chapter we will investigate:

▷ *Contingency tables.*

▷ *How to test the equality of two or more population proportions using the chi-square test statistic.*

▷ *How to conduct multinomial tests using the chi-square test statistic.*

▷ *How to conduct chi-square goodness-of-fit tests.*

▷ *How to test for independence using the chi-square test statistic.*

▷ *How to test for homogeneity using the chi-square test statistic.*

|||| MOTIVATOR 12 ▶

*I*t has become increasingly important for companies to prepare employees to function in international markets. According to McEnery and DesHarnais of Eastern Michigan University, in 1980, more than 200 Fortune 500 companies generated at least 20 percent of their sales abroad.[43] In an effort to gauge businesspeople's perception about the necessary preparation for international work, McEnery and DesHarnais conducted a survey. Table 12.1 shows how 40 random respondents ranked the importance of different types of skills to success in international work. Respondents replied on a five-point scale ranging from 1 (definitely not important) to 5 (definitely important).

TABLE 12.1

Importance of Skills in International Positions

Type of Skill	Mean	Standard Deviation	Chi Square	*P*-Value
Technical skills	4.6	0.87	41.0	0.000
Knowledge of business practices in relevant country	4.3	1.00	18.2	0.000
Human relations	4.2	0.98	18.2	0.009
Foreign relations	4.0	0.99	8.7	0.034
Cultural knowledge	3.9	1.00	4.8	0.183

Technical or functional skills were considered to be most important with a mean of 4.6 on a five-point scale. The chi-square test was used to determine if the five responses to each type of skill were equally likely to be chosen. By using the 0.05 level of significance for the first four skill areas, the data indicated that the five responses were not all equally likely to be chosen. No statistical evidence (with $\alpha = 0.05$) was found to suggest that the responses to the cultural knowledge item were not equally likely to be chosen. The mean response for this area was 3.9.

The above analysis of survey results uses the chi-square goodness-of-fit test, which is presented in section 12.3. For each skill area, if the five choices were equally likely to occur, we would expect eight respondents to check each choice. The chi-square test measures the extent to which the actual frequencies for each choice differ from eight.

Chapter Overview

In this chapter we will examine and illustrate additional statistical methods for dealing with data in the form of frequencies. All these methods are based on the chi-square distributions.

SECTION 12.1   *Introduction*

Frequencies or count data often arise when variables of classification, each with several levels, are used to analyze relationships. Such would be the case when each observation is measured by two qualitative variables. For example, suppose a researcher is interested in the distribution of hypertension (high blood pressure) in 200 individuals. The two variables of classification might be degree of hypertension and amount of smoking. Let's suppose the hypertension variable has three levels: none, mild, and severe. And suppose the smoking variable also has three levels: none, moderate, and heavy. The two variables, each with three levels, give rise to table 12.2, called a **three-by-three contingency table.**

TABLE 12.2

3×3 Contingency Table

		Amount of Smoking		
		None	Moderate	Heavy
Degree of hypertension	Severe	10	14	20
	Mild	20	18	31
	None	40	22	25

If a contingency table has r rows and c columns, then it is called a **r-by-c contingency table.** The nine nonoverlapping groups that make up the contingency table in table 12.2 are called **cells.** An $r \times c$ contingency table has $(r)(c)$ cells. The group of individuals who have mild hypertension and are heavy smokers consists of 31 people. Note also that these 31 people are counted only in this one cell. The entries in each cell consist of count data, called **observed frequencies** and are denoted by O_{ij}. Note that O_{ij} represents the observed frequency for the cell located at the ith row and the jth column. Also note that the sum of the cell frequencies is the size of the sample, n:

$$\Sigma\, O_{ij} = 200 = n$$

where n represents the total frequency count or the size of the sample. Each cell in a contingency table will have two kinds of frequencies associated with it, observed and expected. The **expected frequencies** are denoted by E_{ij} and must be calculated. We will see how this is done shortly. As with the observed frequencies,

$$\Sigma\, E_{ij} = n$$

For example, suppose we are interested in determining whether a certain coin is fair. To decide, we toss it 200 times and record the number of heads and tails. Assume that we observed 90 heads and 110 tails. If the coin is fair, we would *except* 100 heads and 100 tails; that is, the expected frequency for each of the two outcomes is 100. This information can be summarized in a 1×2 contingency table. We need to determine whether the difference between O_{ij} and E_{ij} for each of the two cells is due to sampling error or to the coin being biased. For, if we were to toss the coin another 200 times, we would not expect to get 90 heads and 110 tails again. Thus, we want to determine how well the observed values fit the expected values in the above table. Toward this end, we consider the difference $O_{ij} - E_{ij}$ for each cell. The closer O_{ij} is to E_{ij}, the smaller the value of the difference $O_{ij} - E_{ij}$ and, the further apart they are, the larger the value of the difference $O_{ij} - E_{ij}$. For cell (1, 1) we have $O_{11} - E_{11} = -10$, and for cell (1, 2), $O_{12} - E_{12} = 10$. Notice that the differences sum to 0:

$$\Sigma\, (O_{ij} - E_{ij}) = 0$$

Position of Coin	
Heads	Tails
$O_{11} = 90$	$O_{12} = 110$
$E_{11} = 100$	$E_{12} = 100$

Since the total of the differences is 0, we square them first before we add, just as we did for the deviation scores when we studied the concept of variance in chapter 3. Since large E_{ij} values generally correspond to large squared differences $(O_{ij} - E_{ij})^2$, we write each cell's squared difference value $(O_{ij} - E_{ij})^2$ as a percentage of its expected frequency E_{ij}; as a result, we get a more accurate measure of the difference for each cell.

If the coin is fair, an individual cell's contribution to lack of fit is determined by the value of $(O_{ij} - E_{ij})^2/E_{ij}$, and the contribution of all cells is measured by the statistic

$$\Sigma\, \frac{(O_{ij} - E_{ij})^2}{E_{ij}}$$

For our example, the value of this statistic is found to be 2, since

$$\frac{(90 - 100)^2}{100} + \frac{(110 - 100)^2}{100} = 2$$

It can be shown that the statistic $\Sigma\, (O_{ij} - E_{ij})^2/E_{ij}$ has a sampling distribution that is approximately a χ^2 distribution. Thus, we have

Chi-Square Statistic

$$\chi^2 = \Sigma\, \frac{(O_{ij} - E_{ij})^2}{E_{ij}} \tag{12.1}$$

For our example, the value of the chi-square statistic is 2. To determine if the coin is biased, we ask whether $\chi^2 = 2$ has a larger value than can reasonably be attributed to sampling error. The minimum value that χ^2 can assume and not be attributed to chance is found in the chi-square table. As we shall learn later, the critical value for our illustration is 3.841 for $\alpha = 0.05$. Thus, our value of $\chi^2 = 2$ can reasonably be accounted for by sampling error; as a result, we have no evidence to suggest that the coin is biased.

SECTION 12.2   *Testing Two or More Population Proportions*

In section 11.3 we learned how to test to determine whether two population proportions are different by using two random samples drawn from dichotomous or binomial populations. Each population consists of two categories, called successes and failures. The null hypothesis has the form H_0: $p_1 = p_2$, where p_1 is the proportion of successes in one population and p_2 is the proportion of successes in the other population. The probability of obtaining a success remains constant (property of a binomial experiment). We now want to extend the z test learned in section 11.3 to more than two binomial populations. This extension, when restricted to two populations, will be consistent with the z test covered in section 11.3.

In this section we are interested in testing null hypotheses of the form

$$H_0: p_1 = p_2 = p_3 = \cdots = p_k$$

where $k > 2$ and p_i represents the proportion of successes in the ith binomial population. For each such null hypothesis, the alternative hypothesis is

$$H_1: \text{At least two population proportions are unequal.}$$

The two variables of classification for problems of this type are the outcome category (success or failure) and sample number (with k levels). The $2 \times k$ contingency table takes on the following form:

	Sample 1	Sample 2	Sample 3	$\cdots$	Sample k
Success					
Failure					

EXAMPLE 12.1

Suppose a certain large university is proposing a three-hour statistics requirement for graduation. To determine student opinion, random samples of sophomores, juniors, and seniors are polled. The two outcomes in the dichotomous population are favoring the requirement and opposing the requirement. The results of the poll are contained in the 2×3 contingency table shown in table 12.3.

TABLE 12.3

Results of Student
Opinion Poll

		Category		
		Sophomore	Junior	Senior
Outcome	Favor	12	5	13
	Oppose	13	15	17

If we let p_1, p_2, and p_3 represent the proportions of sophomores, juniors, and seniors, respectively, who favor the new graduation requirement, we are interested in testing the following null hypothesis at $\alpha = 0.05$:

$$H_0: p_1 = p_2 = p_3$$

Continuing with our example, we first find the row and column totals (called **marginal frequencies**), which are shown in table 12.4.

TABLE 12.4

Marginal Frequencies for
Student Opinion Poll

		Sample			
		Sophomore	Junior	Senior	TOTAL
	Favor	12	5	13	30
Outcome	Oppose	13	15	17	45
	TOTAL	25	20	30	$75 = n$

The column totals give us the size of each random sample, and the row totals give us the total number of students who favor the new requirement and the total number of students who oppose the new requirement. Note that the sum of the column totals equals the sum of the row totals:

$$25 + 20 + 30 = 30 + 45 = 75$$

The total number of responses is $n = 75$.

Before we can calculate the value of the test statistic, we must calculate the expected frequency for each cell. The expected frequencies are found by assuming the null hypotheses $H_0: p_1 = p_2 = p_3$ is true. Since each cell will have two frequencies (observed and expected) associated with it, we will place parentheses around the expected frequencies in each cell for identification purposes, as depicted here:

		Category			
		Sophomore	Junior	Senior	TOTAL
	Favor	12 ( )	5 ( )	13 ( )	30
Outcome	Oppose	13 ( )	15 ( )	17 ( )	45
	TOTAL	25	20	30	$75 = n$

Since $H_0: p_1 = p_2 = p_3$ is assumed to be true, we shall obtain a pooled estimate for $p_1 = p_2 = p_3 \, (=p)$. We obtain the pooled estimate by combining the three samples and determining the proportion $\hat{p}$ of this combined group who are in favor of the new graduation requirement. Thus, we have

$$\hat{p} = \frac{12 + 5 + 13}{25 + 20 + 30} = \frac{30}{75} = \frac{2}{5}$$

To obtain E_{11} the expected number of sophomores in favor of the new requirement, we multiply $\hat{p} = 2/5$ by 25, the number of sophomores, to get $E_{11} = (2/5)(25) = 10$. Thus, we would expect $E_{11} = 10$ sophomores to be in favor of the new requirement and $E_{21} = 15$ sophomores to be opposed ($E_{21} = 25 - 10$). Similarly, to get E_{12}, the expected number of juniors in favor of the requirement, we multiply $\hat{p}$ by 20, the number of juniors, to get $E_{12} = (2/5)(20) = 8$. The expected number of juniors opposed to the requirement is $E_{22} = 20 - 8 = 12$. Since $E_{11} + E_{12} + E_{13} = 30$, it follows that $E_{13} = 12$. And since $E_{13} + E_{23} = 30$, we have $E_{23} = 30 - 12 = 18$. The complete 2×3 contingency table showing observed and expected frequencies is given in table 12.5.

TABLE 12.5

2 × 3 Contingency Table
for Student Opinion Poll

		Sample			
		Sophomore	Junior	Senior	TOTAL
Outcome	Favor	12 (10)	5 (8)	13 (12)	30
	Oppose	13 (15)	15 (12)	17 (18)	45
	TOTAL	25	20	30	75 = n

There is a more general procedure for obtaining the expected frequencies in a contingency table. For the student opinion poll example, we would first calculate the probability that one of the 75 responses is favorable and made by a sophomore under the assumption that the null hypothesis H_0: $p_1 = p_2 = p_3$ is true. Note that if the population proportions are equal, then the proportion of favorable responses is independent of the class category. We learned in chapter 5 that E and F are independent events if $P(E \cap F) = P(E)P(F)$. Thus, the probability that a response is favorable and made by a sophomore is

$$P(\text{Favor and Sophomore}) = P(\text{Favor}) \cdot P(\text{Sophomore})$$
$$= \left(\frac{30}{75}\right)\left(\frac{25}{75}\right)$$

We multiply this probability by the total number of students n to obtain E_{11}, the expected number of students who are sophomores and favor the new graduation requirement. Hence, we have

$$E_{11} = P(\text{Favor and Sophomore}) \cdot n$$
$$= \left(\frac{30}{75}\right)\left(\frac{25}{75}\right)(75)$$
$$= \frac{30 \times 25}{75} = 10$$

Since $E_{11} + E_{21} = 25$, by subtraction we have

$$E_{21} = 25 - E_{11}$$
$$= 25 - 10 = 15$$

Note that this agrees with the results found previously. Note also that

$$E_{11} = \frac{(30)(25)}{75}$$
$$= \frac{(\text{Row total}) \times (\text{Column total})}{n}$$

It can be shown in general that any expected frequency E_{ij} can be found by using the following formula:

Expected Frequency for Cell (i, j)

$$E_{ij} = \frac{(i\text{th row total}) \times (j\text{th colum total})}{n} \qquad (12.2)$$

We next compute the value of the χ^2 test statistic using formula (12.1):

$$\chi^2 = \Sigma \frac{(O_{ij} - E_{ij})^2}{E_{ij}}$$

Table 12.6 helps to organize our computations for the χ^2 statistic. From table 12.6 we see that the value of the χ^2 test statistic is $\chi^2 = 2.6806$. Note that if the null hypothesis were true and there were no sampling error involved, then the expected frequency and observed frequency for each cell would be equal. As a result, the value of the χ^2 statistic would equal zero. The question we need to answer is: How large must the value of χ^2 be in order to conclude that it no longer reflects sampling or random error?

TABLE 12.6

Computations Required for χ^2 Statistic

Cell	O_{ij}	E_{ij}	$(O_{ij} - E_{ij})^2$	$\dfrac{(O_{ij} - E_{ij})^2}{E_{ij}}$
(1, 1)	12	10	4	0.400
(1, 2)	5	8	9	1.125
(1, 3)	13	12	1	0.0833
(2, 1)	13	15	4	0.2667
(2, 2)	15	12	9	0.75
(2, 3)	17	18	1	0.0556
				2.6806

The critical value of χ^2 that is tabulated in the chi-square table (table 5 of appendix B) determines the rejection region; at or beyond this value, the test statistic reflects that H_0 is false and that at least two population proportions are unequal. As a consequence, all χ^2 tests for equality of population proportions are right-tailed tests.

Before we can find the critical value from the chi-square table, we must know the degrees of freedom for χ^2. The number of degrees of freedom for an $r \times c$ contingency table is found by multiplying $(r - 1)$ by $(c - 1)$, where r is the number of rows in the contingency table and c is the number of columns.

> **Degrees of Freedom for Chi-Square Test of Proportions**
>
> $$\text{df} = (r - 1)(c - 1) \tag{12.3}$$

Consequently, the number of degrees of freedom associated with a 2×3 contingency table is

$$\text{df} = (2 - 1)(3 - 1) = 2$$

Recall that $\chi^2(\text{df})$ is used to indicate the chi-square distribution with df degrees of freedom. For our example, the chi-square critical value for $\alpha = 0.05$ and $\text{df} = 2$ is denoted by $\chi^2_{0.05}(2)$. It is found in table 5 to be 5.991, as indicated in figure 12.1. The value of the test statistic was found to be $\chi^2 = 2.68$. Since this statistic is not located in the rejection region, we cannot reject H_0, and we can conclude that $\chi^2 = 2.68$ reflects only sampling error. As a result, we have no statistical evidence to indicate that the population proportions are unequal. Of course, with this decision, we may have committed a Type II error. But the probability of a Type II error, β, is unknown.

FIGURE 12.1

Rejection region for chi-square test

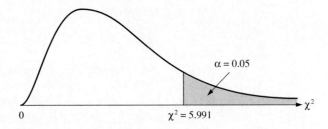

APPLICATION 12.1

Let's examine the concept of degrees of freedom more closely. How many degrees of freedom are there in choosing the measures of the angles in a triangle?

Solution: Consider triangle ABC shown in the accompanying figure.

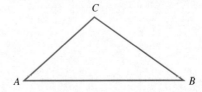

Since $\angle A + \angle B + \angle C = 180°$, we are free to choose the measures of any two angles, since the third angle can be found by subtraction. Hence, there are two degrees of freedom associated with choosing the measures of the three angles in a triangle. ▪

EXAMPLE 12.2

For the student opinion poll of example 12.1 involving a 2×3 contingency table (table 12.5), we can determine the number of degrees of freedom we have when choosing the cell frequencies once the column and row totals are known. Initially, we are free to choose any cell frequency. Suppose we choose cell $(2, 2)$. So far we have enjoyed one degree of freedom. We have no freedom to choose cell $(1, 2)$, since $O_{22} + O_{12} = 20$. Thus, we are free to choose any of the four remaining cells $(1, 1)$, $(1, 3)$, $(2, 1)$, or $(2, 3)$. Let's choose cell $(2, 1)$. So far, we have enjoyed two degrees of freedom. We have no freedom with cell $(2, 3)$, since $O_{21} + O_{22} + O_{23} = 45$. Similarly, we have no freedom with cell $(1, 3)$, or with cell $(1, 1)$. We have indicated freedom in our choice of cells by using a down arrow ($\downarrow$) in the following diagram. Cells marked $\times$ were not free to be chosen. Thus, for our example, df = 2. Note that this result is the same as that obtained by using formula (12.3).

$\times$	$\times$	$\times$
$\downarrow$	$\downarrow$	$\times$

Computational Formula for χ^2

There is a computational formula for computing the value of the chi-square test statistic given by formula (12.1). It is as follows:

Computational Formula for Chi-Square

$$\chi^2 = \Sigma \frac{O_{ij}^2}{E_{ij}} - n \qquad (12.4)$$

EXAMPLE 12.3

Let's use formula (12.4) to compute the value of the chi-square test statistic for the student poll application used in this section. The calculations are organized in table 12.7.

TABLE 12.7

Calculation of χ^2 Using Computational Formula

Cell	O_{ij}	E_{ij}	$\dfrac{O_{ij}^2}{E_{ij}}$
(1, 1)	12	10	14.4
(1, 2)	5	8	3.125
(1, 3)	13	12	14.0833
(2, 1)	13	15	11.2667
(2, 2)	15	12	18.75
(2, 3)	17	18	16.0556
			77.6806

The value of the chi-square test statistic is

$$\chi^2 = \Sigma \frac{O_{ij}^2}{E_{ij}} - n$$
$$= 77.68 - 75 = 2.68$$

which agrees with the result found previously by using formula (12.1).

In order for the sampling distribution of the statistic $\Sigma (O_{ij} - E_{ij})^2/E_{ij}$ to be approximately a χ^2 distribution with df $= (r - 1)(c - 1)$, the following assumptions must be met:

Chi-Square Assumptions

1. At least 80% of the expected cell frequencies should be 5 or more.
2. No expected frequencies less than 1 should occur.

A number less than 1 in the denominator of $(O_{ij} - E_{ij})^2/E_{ij}$ tends to inflate the value of the χ^2 statistic for that cell and, as a result, leads to erroneous decisions.

APPLICATION 12.2

A voter running for public office in a three-county area is interested in knowing if the proportion of voters who favor her is the same for each county. To determine this she took a random sample of voter opinions in each county and obtained the results shown in the table.

County A	County B	County C
$x_1 = 46$	$x_2 = 48$	$x_3 = 42$
$n_1 = 120$	$n_2 = 125$	$n_3 = 110$

Do the data indicate that the true proportions of voters who favor her differ among the counties? Use $\alpha = 0.05$.

Solution:

1. The null hypothesis is H_0: $p_1 = p_2 = p_3$.

2. The alternative hypothesis is H_1: At least two proportions are unequal.

3. The sampling distribution is the distribution of $\Sigma\,[(O_{ij} - E_{ij})^2/E_{ij}]$, which is approximately $\chi^2(2)$. In order to calculate the value of the test statistic, we first calculate the expected frequencies under the assumption that H_0 is true. The expected and observed frequencies are listed in the 2×3 contingency table (table 12.8).

TABLE 12.8

2×3 Contingency Table for Application 12.2

		County			TOTAL
		A	B	C	
	Yes	46 (45.97)	48 (47.89)	42 (42.14)	136
Favor	No	74 (74.03)	77 (77.11)	68 (67.86)	219
	TOTAL	120	125	110	355 = n

The first two expected frequencies in the first row were found by using formula (12.2):

$$E_{11} = \frac{(120)(136)}{355} = 45.97$$

$$E_{12} = \frac{(125)(136)}{355} = 47.89$$

The remaining four expected frequencies were found by subtraction. In fact, for any $r \times c$ contingency table, formula (12.2) need be used to find only df $= (r - 1)(c - 1)$ expected frequencies; the remaining $(r + c - 1)$ expected frequencies can then be found by subtraction. Note that each expected frequency in this example is at least 5.

Next, we compute the value of the chi-square test statistic by using formula (12.4). Computations are organized in the following table:

Cell	O_{ij}	E_{ij}	$\dfrac{O_{ij}^2}{E_{ij}}$
(1, 1)	46	45.97	46.0300
(1, 2)	48	47.89	48.1103
(1, 3)	42	42.14	41.8605
(2, 1)	74	74.03	73.9700
(2, 2)	77	77.11	76.8902
(2, 3)	68	67.86	68.1403
			355.0013

Thus,

$$\chi^2 = \Sigma\,\frac{O_{ij}^2}{E_{ij}} - n$$

$$\chi^2 = 355.0013 - 355 = 0.0013$$

The degrees of freedom are found by using formula (12.3):

$$df = (r - 1)(c - 1)$$
$$= (2 - 1)(3 - 1) = 2$$

4. The level of significance is $\alpha = 0.05$.
5. The critical value is $\chi^2_{0.05}(2) = 5.991$.
6. The decision is we fail to reject H_0, since $\chi^2 = 0.0013 < 5.991$.
7. With this decision, we may have committed a Type II error, but β is unknown. Hence, there is no statistical evidence to suggest that the true proportions of the voters who favor the candidate differ among the counties. ▪

MINITAB can be used to test the null hypothesis of application 12.2. The commands used and the accompanying output are contained in computer display 12.1.

Computer Display 12.1

```
MTB > READ C1 C2 C3
DATA> 46 48 42
DATA> 120 125 110
DATA> END
   2 ROWS READ
MTB > CHISQUARE C1 C2 C3

EXPECTED COUNTS ARE PRINTED BELOW OBSERVED COUNTS

              C1       C2       C3    TOTAL
    1         46       48       42      136
           45.98    47.92    42.10

    2        120      125      110      355
          120.02   125.08   109.90

TOTAL       166      173      152      491

CHISQ = 0.000 + 0.000 + 0.000 + 0.000 + 0.000 + 0.000 = 0.001
DF = 2
```

The critical value for the test is found by using the INVCDF command and its CHISQUARE subcommand. The commands and accompanying output are contained in computer display 12.2.

Computer Display 12.2

```
MTB > INVCDF .95;
SUBC> CHISQUARE DF=2.
   0.9500 5.9915
```

In the case of testing the null hypothesis H_0: $p_1 = p_2$ against the alternative hypothesis H_1: $p_1 \neq p_2$, the χ^2 test is equivalent to the z test presented in chapter 10. This is because z^2 has a sampling distribution that is a χ^2 distribution with one degree of freedom. That is,

$$z^2 = \chi^2(1)$$

Since the z test for this application is nondirectional and the χ^2 test is directional, it can be shown that the two critical values have the following relationship:

$$z_{\alpha/2}^2 = \chi_\alpha^2(1)$$

If $\alpha = 0.05$, we have $z_{0.025} = 1.96$ and $z^2 = (1.96)^2 = 3.8416$. Notice that this value agrees to two decimal places with the value $\chi_{0.05}^2(1)$ found in table 5 of appendix **B**. The difference in the two values is due to rounding error.

APPLICATION 12.3

A study was conducted to determine the extent to which adults drink alcoholic beverages. Two random samples of data were gathered, one from males and the other from females. Of 150 females polled, 72 indicated they drink alcohol, whereas of 200 males polled, 104 indicated they drink alcohol. At $\alpha = 0.05$, test to determine if the proportions of adult males and adult females who drink alcohol are different.

Solution: Let p_1 denote the proportion of adult males who drink alcohol and p_2 denote the proportion of adult females who drink alcohol.

1. The null hypothesis is H_0: $p_1 = p_2$.
2. The alternative hypothesis is H_1: $p_1 \neq p_2$.

For illustrative purposes, we shall test the null hypothesis by using two procedures, the z test and the χ^2 test.

Procedure 1 Using the z Test

3. The sampling distribution is the distribution of the differences between sample proportions. The test statistic is the z value for $\hat{p}_1 - \hat{p}_2$. The sample proportions are

$$\hat{p}_1 = \frac{104}{200} = 0.52 \quad \text{and} \quad \hat{p}_2 = \frac{72}{150} = 0.48$$

and the pooled estimate for p is

$$\hat{p} = \frac{104 + 72}{200 + 150} = 0.502$$

The standard error of $\hat{p}_1 - \hat{p}_2$ is approximately equal to

$$\hat{\sigma}_{\hat{p}_1 - \hat{p}_2} = \sqrt{(0.5)(0.5)\left(\frac{1}{200} + \frac{1}{150}\right)} = 0.054$$

The z value is

$$z = \frac{\hat{p}_1 - \hat{p}_2}{\hat{\sigma}_{\hat{p}_1 - \hat{p}_2}}$$

$$= \frac{0.52 - 0.48}{0.054} = 0.74$$

4. The level of significance is $\alpha = 0.05$.
5. The critical values are $\pm z_{0.025} = \pm 1.96$.
6. We fail to reject H_0, since $-1.96 < z = 0.74 < 1.96$.
7. With this decision we may have committed a Type II error; the probability of doing this is unknown. Hence, there is no statistical evidence to suggest that the proportions of adult males and adult females who drink alcohol are different.

Procedure 2: Using a χ^2 Test Assuming H_0: $p_1 = p_2$ is true, we compute the expected frequencies. They are shown with the observed frequencies in the 2×2 contingency table (table 12.9).

TABLE 12.9

2×2 Contingency Table for Application 12.3

	Males	Females	TOTAL
Alcohol	104 (100.57)	72 (75.43)	176
No alcohol	96 (99.43)	78 (74.57)	174
TOTAL	200	150	$350 = n$

The expected frequency for cell $(1, 1)$ is

$$E_{11} = \frac{(200)(176)}{350} = 100.57$$

Expected frequencies E_{12}, E_{21}, and E_{22} are found by subtraction:

$$E_{12} = 176 - E_{11} = 176 - 100.57 = 75.43$$
$$E_{21} = 200 - E_{11} = 200 - 100.57 = 99.43$$
$$E_{22} = 174 - E_{21} = 174 - 99.43 = 74.57$$

The following table organizes the calculations for finding the value of the χ^2 test statistic using formula (12.4).

Cell	O_{ij}	E_{ij}	$\dfrac{O_{ij}^2}{E_{ij}}$
$(1, 1)$	104	100.57	107.5470
$(1, 2)$	72	75.43	68.7260
$(2, 1)$	96	99.43	92.6883
$(2, 2)$	78	74.57	81.5878
			350.5491

Hence, the value of the chi-square test statistic is

$$\chi^2 = \Sigma \frac{O_{ij}^2}{E_{ij}} - n = 350.55 - 350 = 0.55$$

The critical value obtained from table 5 is $\chi_{0.05}^2(1) = 3.841$. Since $\chi^2 = 0.55 < \chi_{0.05}^2(1) = 3.841$, we cannot reject H_0. We thus conclude that there is no real evidence to suggest that the proportions of adult males and adult females who drink alcohol are different. Also note the following:

1. $z^2 = (0.74)^2 \approx 0.55 = \chi^2$.
2. $\chi_{0.05}^2(1) = 3.843$ and $(z_{0.025})^2 = (1.96)^2 \approx 3.84 \approx \chi_{0.05}^2(1)$. ▪

The interested student might wonder why we are using a χ^2 test to test the null hypothesis H_0: $p_1 = p_2 = p_3$ instead of the three pairwise z tests associated with the following null hypotheses:

1. H_0: $p_1 = p_2$.
2. H_0: $p_2 = p_3$.
3. H_0: $p_1 = p_3$.

There are several good reasons. The method involving three separate tests is too time consuming and not very efficient. However, the most important reason is that sizable and unknown error rates result. For example, if each of the above three tests uses $\alpha = 0.05$ as the significance level and if the tests are independent of each other (in fact, they are not), then *the probability of committing at least one Type I error in three independent tests* is found to be:

$$1 - \text{Probability of making no Type I error in three tests} = 1 - (0.95)^3$$
$$\approx 0.14$$

Thus, the error rate for the experiment is somewhere between 0.05 and 0.14, instead of 0.05, the error rate for one χ^2 test. As the number of tests increases, the error rate for the experiment increases.

EXERCISE SET 12.2

Basic Skills

1. For contingency tables having the following dimensions, find the degrees of freedom and the appropriate critical value of χ^2 at the indicated significance level.
   a. 6×2, $\alpha = 0.05$
   b. 5×6, $\alpha = 0.01$

2. For contingency tables having the following dimensions, find the degrees of freedom and the appropriate critical value of χ^2 at the indicated significance level.
   a. 4×7, $\alpha = 0.05$
   b. 4×4, $\alpha = 0.01$

Further Applications

3. The accompanying data represent the results of a college survey to study the graduation rate for male and female students admitted as first-time freshmen students and graduating within five years.

	Graduated	Did Not Graduate
Male	16	28
Female	18	19

   a. By using the 0.05 level of significance and the z test, determine if the true proportions of males and females who graduate within five years differ.
   b. Determine by using the χ^2 test whether there is a difference between the proportions of males and females who graduate. Use $\alpha = 0.05$.
   c. Show that the value of the χ^2 test statistic is equal to the square of the value of the z statistic, that is, $\chi^2 = z^2$.

4. In an effort to gauge the success of a candidate's speech, the campaign manager compared the proportion of voters who indicated support for the candidate before and after the speech. The results are summarized as follows:

	Before	After
Supporters	52	76
Nonsupporters	98	125

   a. By using the 0.05 level of significance and the z test, determine if the true proportion of supporters has changed following the speech.
   b. Determine by using the χ^2 test whether there is a difference between the proportions of supporters before and after the speech.
   c. Show that the value of the χ^2 test statistic is equal to the square of the value of the z statistic, that is, $\chi^2 = z^2$.

5. At the end of a semester, the grades for Mathematics 101 were tabulated in the following 3×2 contingency table to study the relationship between class attendance and grade received.

	Grade Received	
Number Days Absent	Pass	Fail
0–3	135	110
4–6	36	4
7–45	9	6

   At $\alpha = 0.05$, do the data indicate that the proportions of students who pass differ among the three absence categories?

6. A survey was undertaken to determine if the proportions

of voters in Frostburg, LaVale, and Cumberland who favor an elected county school board are different. The results of the survey are given in the accompanying table.

	Frostburg	LaVale	Cumberland
Favor	125	150	133
Oppose	130	160	102

Determine if the true proportions for the three areas differ by using $\alpha = 0.05$.

7. A electronic supply center wants to determine if there are any differences in the proportions of service cells for four major brands of television sold by them in a certain city. The following data were collected during a two-year period:

	Brand A	Brand B	Brand C	Brand D
Service	20	30	55	45
No Service	280	289	350	89

Can the supply center conclude using $\alpha = 0.01$ that the proportions of defective televisions differ among brands?

8. A study was undertaken to determine if the proportions of male and female babies are the same for mothers in three different areas: the United States, Europe, and Africa. The results shown in the table were obtained.

	U.S.	Europe	Africa
Male	261	207	50
Female	174	169	139

By using $\alpha = 0.05$, can we conclude that the proportions of male babies differ by country?

9. Women in four sections of the United States were polled and asked if they watch Monday-night football on television. The results are summarized in the accompanying 2×4 contingency table.

	East	West	North	South
Yes	90	57	70	97
No	110	93	105	128

Test using $\alpha = 0.05$ to determine whether the proportions of women who watch Monday-night football on television differ by section of the United States.

10. A certain fraternity on a large college campus conducted a study to determine the proportion of students failed by three statistics instructors in math 209. The results are summarized in the following 2×3 contingency table:

	Instructor		
	A	B	C
Failed	8	12	10
Passed	67	88	115

By using the 0.01 level of significance, test to determine whether there is a difference among the proportions of students failed by each instructor.

11. A study was conducted to determine if nest location has any effect on the proportion of young birds of a certain species that survive during the nesting period. Four nest-site locations identified for study were bridges, buildings, natural rock formations, and road cuts. At each site, the nest histories were recorded for one season, with the results shown in the table.

	Nest Location			
	Bridges	Buildings	Rock Formations	Road Cuts
Number eggs hatched	15	40	32	28
Number eggs in nest	20	42	44	35

By using $\alpha = 0.05$ test the null hypothesis of equal survival proportions for the four nest locations.

12. As part of a health inventory questionnaire, samples of third-, sixth-, and ninth-grade students in a Missouri school district were asked the following question: "Do you eat breakfast?" Their responses were as follows:

Responses	Grade		
	Third	Sixth	Ninth
Yes	132	168	126
No	52	33	46

At significance level $\alpha = 0.05$, do the data provide sufficient evidence to reject the null hypothesis that the proportions of positive responses for the three grades are equal?

Going Beyond

13. Prove that $\Sigma (O_{ij} - E_{ij}) = 0$.

14. The value of χ^2 for the following 2×2 contingency table is given by

$$\chi^2 = \frac{(ad - bc)^2 m}{efgh}$$

(Assume that none of the totals is equal to zero.)

	Category 2		TOTAL
	a	*b*	*g*
Category 1	*c*	*d*	*h*
TOTAL	*e*	*f*	*m*

a. Use this result to find the value of χ^2 for the following 2×2 contingency table.

18	16
22	44

b. Determine the value of χ^2 by using the computational formula.

15. Prove that $\Sigma\,[(O_{ij} - E_{ij})^2/E_{ij}] = \Sigma\,(O_{ij}^2/E_{ij}) - n.$

16. Calculate the χ^2 statistics for the following 2×2 contingency tables and compare your results (assume that the observed frequencies are different from zero):

Table 1

a	*b*
c	*d*

Table 2

ax	*bx*
cx	*dx*

17. Prove the formula for χ^2 given in exercise 14.

18. Discuss the use of chi-square tests used in the case discussed in motivator 11. How did the researchers arrive at the values of the chi-square test statistic? Why did they report standard deviations and what significance do they play in the analysis of the results?

SECTION 12.3   Multinomial Tests

In section 12.2 we discussed testing hypotheses of the form H_0: $p_1 = p_2 = p_3 = \cdots = p_k$ for k dichotomous (binomial) populations. For these applications, qualitative data were classified into one of two distinct outcome categories, which we labeled as success or failure. The purpose of this section is to generalize these tests to **multinomial experiments** involving populations of qualitative data having more than two distinct outcome categories. For such applications, we draw one sample from a multinomial population having m distinct categories to aid us in determining whether the population categories have proportions equal to specified values.

Multinomial experiments were discussed in chapter 6. Recall that a multinomial experiment has properties closely resembling those of a binomial experiment. The properties of a multinomial experiment include the following:

1. The experiment consists of n trials.

2. Each trial results in exactly one of m outcomes.

3. The trials are independent.

4. The probability of any outcome remains constant from trial to trial, and the sum of the m probabilities is 1.

By examining these four properties, we can see that a binomial experiment is a special case of a multinomial experiment with $m = 2$ outcomes. For all of our applications, we will assume that sampling occurs from large (or infinite) populations.

EXAMPLE 12.4

The following are examples of multinomial populations:

1. The outcomes from tossing a six-sided die

2. The responses to a survey question by answering yes, no, or undecided

3. The results of achievement in terms of letter grades A, B, C, D, or F

4. The classifications of religious affiliation such as Catholic, Protestant, Jewish, or other

Possible null hypotheses for each might be

    1. H_0: $p_1 = p_2 = p_3 = p_4 = p_5 = p_6 = 1/6$.

    2. H_0: $p_y = 0.2$, $p_n = 0.7$, p_u 0.1.

    3. H_0: $p_A = 0.1$, $p_B = 0.2$, $p_C = 0.3$, $p_D = 0.2$, $p_F = 0.2$.

    4. H_0: $p_C = 0.3$, $p_P = 0.4$, $p_J = 0.2$, $p_o = 0.1$.

Note that for each null hypothesis, $\Sigma\, p = 1$.

The statistical calculations for testing multinomial parameters are similar to those involved in section 12.1. The only difference is in the calculation of the degrees of freedom for the χ^2 test statistic. For testing multinomial parameters, df is equal to 1 less than the number of possible categories of distinct outcomes in a multinomial experiment. That is,

> **Degrees of Freedom for Multinomial Experiments**
>
> $$df = (\text{number of outcome categories}) - 1 \qquad (12.5)$$

Applications 12.4 and 12.5 show testing parameters for multinomial experiments.

APPLICATION 12.4

A sportmen's association stocked its lake ten years ago with trout, bass, bluegill, and catfish in the following respective percentages: 20, 15, 40, and 25. Has the original distribution of fish changed over ten years if a recent sample of fish provided the following numbers? Use $\alpha = 0.05$.

Trout	Bass	Bluegill	Catfish
132	100	200	168

Solution: Let p_1 = percentage of trout, p_2 = percentage of bass, p_3 = percentage of bluegill, and p_4 = percentage of catfish.

    1. H_0: $p_1 = 0.20$, $p_2 = 0.15$, $p_3 = 0.4$, $p_4 = 0.25$.

    2. H_1: At least one percentage is incorrect.

    3. The total number of fish caught is $n = 132 + 100 + 200 + 168 = 600$. For this application, we have the 1×4 contingency table shown in table 12.10.

TABLE 12.10

Contingency Table for Application 12.4

Trout	Bass	Bluegill	Catfish
132	100	200	168
(120)	(90)	(240)	(150)

To find the expected frequencies, we multiply the hypothesized percentages by the total $n = 600$. Thus, we have

$$E_{11} = (600)(0.20) = 120$$
$$E_{12} = (600)(0.15) = 90$$
$$E_{13} = (600)(0.40) = 240$$
$$E_{14} = (600)(0.25) = 150$$

Note that E_{14} could also have been found by subtraction:

$$E_{14} = 600 - 120 - 90 - 240 = 150$$

The value of the test statistic is computed using the computational formula:

$$\chi^2 = \Sigma \frac{O_{ij}^2}{E_{ij}} - n$$

The following table organizes the computations:

Cell	O_{ij}	E_{ij}	$\dfrac{O_{ij}^2}{E_{ij}}$
1	132	120	145.2
2	100	90	111.1111
3	200	240	166.6667
4	168	150	188.16
			611.1378

Hence, the value of the chi-square test statistic is

$$\chi^2 = 611.1378 - 600 \approx 11.14$$

The degrees of freedom are found by using formula (12.5).

$$df = (\text{number of outcome categories}) - 1 = 4 - 1 = 3$$

4. The level of significance is $\alpha = 0.05$.

5. The critical value is $\chi_{0.05}^2(3) = 7.815$.

6. Since $\chi^2 = 11.14 > \chi_{0.05}^2(3) = 7.815$, we reject H_0 and conclude that the original distribution of fish has changed over the ten-year period.

7. With this decision we may be committing a Type I error; the probability of this is $\alpha = 0.05$. ■

APPLICATION 12.5

A six-sided die is tossed with the results shown in table 12.11.

TABLE 12.11

Observed Frequencies for Application 12.5

Face	1	2	3	4	5	6
Frequency	45	38	37	40	37	43

At $\alpha = 0.05$, do these results indicate that the die is biased?

Solution:

1. The null hypothesis is that the die is fair; that is, H_0: $P(1) = P(2) = P(3) = P(4) = P(5) = P(6) = 1/6$.

2. The alternative hypothesis is that the die is not fair; that is, H_i: For some i, $P(i) \neq 1/6$.

3. The die was tossed $n = 45 + 38 + 37 + 40 + 37 + 43 = 240$ times. The resulting 1×6 contingency table is as follows:

	Face Showing					
Frequency	1	2	3	4	5	6
Observed	45	38	37	40	37	43
Expected	(40)	(40)	(40)	(40)	(40)	(40)

The expected frequencies are all found to be $(1/6)(240) = 40$.
The value of the test statistic is found by using

$$\chi^2 = \Sigma \frac{O_{ij}^2}{E_{ij}} - n$$

The computations are organized in the following table:

Cell	O_{ij}	E_{ij}	$\dfrac{O_{ij}^2}{E_{ij}}$
1	45	40	50.625
2	38	40	36.100
3	37	40	34.225
4	40	40	40.000
5	37	40	34.225
6	43	40	46.225
			241.400

Thus, the value of χ^2 is

$$\chi^2 = 241.4 - 240 = 1.4$$

The number of degrees of freedom is

$$df = (\text{number of outcome categories}) - 1$$
$$= 6 - 1 = 5$$

4. The level of significance is $\alpha = 0.05$.
5. The critical value is $\chi_{0.05}^2(5) = 11.070$.
6. The decision is that we cannot reject H_0, since $\chi^2 = 1.4 < \chi_{0.05}^2(5) = 11.070$. Thus, we have no statistical evidence that the die is biased.
7. With this decision we may have committed a Type II error; β is unknown. ▪

Goodness-of-Fit Tests

A P P L I C A T I O N 12.6

The chi-square **goodness-of-fit test** can be used to test whether a given sample belongs to a specified population. For example, table 12.12 contains the observed frequencies for the first 100 digits in a random number table.

TABLE 12.12

Observed Frequencies for Application 12.6

Digit	0	1	2	3	4	5	6	7	8	9
f	11	9	15	11	9	11	6	6	13	9

Test to determine whether the digits in the random number table occur with different probabilities using $\alpha = 0.05$.

Solution: Let p_i be the percentage of occurrences of digit i.

1. H_0: $p_i = 0.1$ for $i = 0, 1, 2, \ldots , 9$.
2. H_1: At least one p_i differs from 0.1.
3. We compute the expected frequencies under the assumption that H_0 is true. If H_0 is true, we would expect $(0.1)(100) = 10$ occurrences of each digit.

We next calculate the value of the χ^2 statistic. The following table organizes the computations:

Cell	O_{ij}	E_{ij}	$\dfrac{O_{ij}^2}{E_{ij}}$
1	11	10	12.1
2	9	10	8.1
3	15	10	22.5
4	11	10	12.1
5	9	10	8.1
6	11	10	12.1
7	6	10	3.6
8	6	10	3.6
9	13	10	16.9
10	9	10	8.1
			107.2

By applying formula (12.4), we obtain the value of the χ^2 statistic:

$$\chi^2 = \Sigma \frac{O_{ij}^2}{E_{ij}} - n$$
$$= 107.2 - 100 = 7.2$$

4. Level of significance: $\alpha = 0.05$.

5. Critical value: $\chi_{0.05}^2(9) = 16.919$.

6. Decision: We fail to reject H_0, since $\chi^2 = 7.2 < \chi_{0.05}^2(9) = 16.919$. Thus, we have no statistical evidence to suggest that the frequencies are different. For a random number table we would expect this to be the case.

7. Type of error possible: Type II; β is unknown. ■

MINITAB can be used to test the randomness of the digits of application 12.6. The commands and output are contained in computer display 12.3. Note that SUM = 7.2000 represents the value of the chi-square statistic using formula (12.1).

Computer Display 12.3

```
MTB > SET C1
DATA> 11 9 15 11 9 11 6 6 13 9
DATA> END
MTB > LET C2 = SUM(C1)/COUNT(C1)
MTB > LET C3 = C1 − C2
MTB > LET C4 = C3**2/C2
MTB > SUM C4
  SUM   =   7.2000
MTB > INVCDF .95;
SUBC> CHISQUARE DF=9.
  0.9500   16.9190
MTB >
```

A P P L I C A T I O N 12.7

TABLE 12.13

Frequency Table for
Application 12.7

Three coins were tossed 80 times and the number of heads X occurring each time was recorded. The resulting data are provided in table 12.13.

x	0	1	2	3
f	20	38	18	4

Test the null hypothesis that X is binomial with $n = 3$ and $p = 0.5$. Use $\alpha = 0.05$.

Solution: Let p be the probability of getting a head.

1. H_0: X is binomial with $n = 3$ and $p = 0.5$.
2. H_1: X is not binomial with $n = 3$ and $p = 0.5$.
3. We assume H_0 is true and determine the expected frequencies. The binomial probability table (table 1 of appendix B) is used to find the binomial probabilities $P(x)$ for $n = 3$ and $p = 0.5$. The expected frequencies are found by multiplying $P(x)$ by 80, as indicated in the table.

x	$P(x)$	$80 \cdot P(x)$
0	0.125	10
1	0.375	30
2	0.375	30
3	0.125	10

We next calculate the value of the χ^2 statistic. The following table organizes the results:

O_{ij}	E_{ij}	$\dfrac{O_{ij}^2}{E_{ij}}$
20	10	40.0000
38	30	48.1333
18	30	10.8000
4	10	1.6000
		100.5333

From formula (12.4), the value of the χ^2 statistic is

$$\chi^2 = \Sigma \frac{O_{ij}^2}{E_{ij}} - n$$

$$= 100.53 - 80 = 20.53$$

4. Significance level: $\alpha = 0.05$.
5. Critical value: $\chi^2_{0.05}(3) = 7.815$.
6. Decision: Since $20.53 > \chi^2_{0.05}(3) = 7.815$, we reject H_0. Hence, we conclude that X is not binomial with $n = 3$ and $p = 0.05$.
7. Type of error possible: Type I; $\alpha = 0.05$. ▪

A P P L I C A T I O N 12.8

In a study to determine the percentage of television viewers who watch the 11:00 P.M. news, a random sample of viewers was obtained. It was found that out of 500 viewers, 190 watched the late news on television. Use $\alpha = 0.05$ to determine whether p, the true percentage of viewers who watch the late news on television, differs from 40%.

Solution:

1. H_0: $p = 0.4$

2. H_1: $p \neq 0.4$

3. If the null hypothesis H_0: $p = 0.4$ is true, then $(0.4)(500) = 200$ of the sampled viewers would be expected to watch the late news, and 300 viewers would not be expected to watch the late news. The value of the χ^2 test-statistic is then

$$\chi^2 = \Sigma \frac{(O_{ij} - E_{ij})^2}{E_{ij}}$$

$$= \frac{(190 - 200)^2}{200} + \frac{(310 - 300)^2}{300} = 0.83$$

4. Significance level: $\alpha = 0.05$.

5. Critical value: $\chi^2_{0.05}(1) = 3.841$.

6. Decision: We cannot reject H_0, since $\chi^2 = 0.83 < \chi^2_{0.05}(1) = 3.841$. Hence, we have no statistical evidence to suggest that the percentage of viewers who watch the late news differs from 0.40.

7. Type of error possible: Type II; β is unknown.

Note that a z test could also have been used to test H_0: $p = 0.4$. The value of the z statistic for $\hat{p} = 190/500 = 0.38$ is

$$z = \frac{\hat{p} - p}{\sqrt{p(1 - p)/n}} = \frac{0.38 - 0.40}{\sqrt{(0.4)(0.6)/500}} = 0.913$$

The critical values are ± 1.96. Since $-1.96 < -0.91 < 1.96$, the null hypothesis cannot be rejected, which agrees with the χ^2 test. Note that $z^2 = (-0.913)^2 \approx 0.83 = \chi^2$. ■

EXERCISE SET 12.3

Further Applications

1. For the past five years, the percentages of A's, B's, C's, D's, and F's in a statistics class taught by Dr. Smith have been 11%, 18%, 35%, 25%, and 11%, respectively. Dr. Henry taught the course last year and he give 7 A's, 13 B's, 30 C's, 16 D's, and 9 F's. Can we conclude that Dr. Henry's grades differ from Dr. Jone's grades? Use $\alpha = 0.01$.

2. An elementary school teacher is supposed to spend an equal amount of time on each subject taught. Do the accompanying data collected for a four-week period indicate that the teacher does not spend the same amount of time on each subject? Use $\alpha = 0.01$.

Subject	Science	Reading	Writing	Spelling	Arithmetic	History
Time spent (hours)	15	30	10	27	13	25

3. In order to plan his staffing assignments, a city police chief assumes that automobile accidents occur daily with equal frequencies during the summer months. To test the assumption, the number of automobile accidents was recorded for a random sample of ten summer days. The results were 4, 9, 6, 15, 10, 2, 20, 8, 14, and 12. Do the data indicate that the frequencies of accidents differ during the summer months? Use $\alpha = 0.05$.

4. Families with four children each were involved in a study to determine the gender distribution. The results listed in the table were obtained.

Number of boys	0	1	2	3	4
Number of families	3	12	11	9	5

Test at the 0.01 level of significance to determine if gender distribution follows a binomial distribution with $p = 0.5$.

5. Color preferences for new cars are indicated by a random sample of potential customers. The information in the accompanying table was obtained.

Color	Red	Yellow	Blue	Green	Brown
Frequency	40	64	46	36	14

Test using $\alpha = 0.05$ to determine if color preferences are different.

6. Grades are normally distributed if the following percentages are followed: 2% A's, 14% B's, 68% C's, 14% D's, and 2% F's. A sample of 120 statistics grades yielded the following results: 5 A's, 24 B's, 60 C's, 10 D's, and 21 F's. Using $\alpha = 0.05$, test to determine if the statistics grades are normally distributed.

7. In an experiment of tossing three coins 200 times, the observed frequencies were recorded as shown in the table. At the 0.05 level of significance, determine whether the results fit a binomial distribution with $n = 3$ and $p = 0.5$.

Number of heads	0	1	2	3
f	17	63	82	38

8. In an experiment, five balls were drawn, one at a time with replacement, from a box containing an equal number of red and white balls. The experiment was repeated 800 times and the following distribution of the number of white balls resulted:

Number of white balls	0	1	2	3	4	5
f	30	121	270	220	132	27

Test at $\alpha = 0.05$ to determine if the observed frequencies fit a binomial distribution.

9. The following is a calculation of π that has been carried out to 200 decimal places.

$\pi \approx 3.14159$	26535	89793	23846	26433
58209	74944	59230	78164	06286
82148	08651	32823	06647	09384
48111	74502	84102	70193	85211
50288	49171	69399	37510	83279
86280	34825	34211	70679	20899
50582	23172	53594	08128	46095
64462	29489	54930	38196	05559

Test to determine whether the digits of π to the right of the decimal point occur with unequal frequencies. Use $\alpha = 0.05$.

10. A biologist has collected information regarding the age and gender of 646 birds. He wants to determine if the gender ratios among adult and juvenile birds are the same this year as they were in the previous year. Last year there were 3 adult males to 7 adult females. In this year's sample there are 121 adult males, 166 adult females, 179 juvenile males, and 180 juvenile females. At the 0.05 significance level, does the gender ratio for adult birds differ from 3:7 this year? Does the gender ratio for juvenile birds differ from 3:7?

11. By using $\alpha = 0.05$ and the digits in rows 6 to 9, inclusive, of the random number table (table 8.1 of section 8.1) determine if the randomization process produces digits of unequal frequencies.

Going Beyond

12. A machine was found to be producing 10% defective parts. After the machine was repaired, it produced 7 defective parts in the first batch of 100 parts produced.
    a. By using the z test at $\alpha = 0.025$, determine if the proportion of defective parts has been reduced.
    b. Can we arrive at the same result by using the χ^2 statistic? Explain.

13. Both the χ^2 test and the z test can be used to test the null hypothesis $H_0: p = p_0$, provided $np \geq 5$ and $n(1 - p) \geq 5$. Prove that $\chi^2 = z^2$.

SECTION 12.4   *Chi-Square Tests for Independence*

EXAMPLE 12.5

Frequently in statistical applications we are interested in determining whether two variables of classification (either quantitative or qualitative) are independent or associated. The following are situations where one might be interested in determining whether two variables are related:

▪ Are reading habits associated with the gender of a reader?

▪ Are grades received in a course related to the number of days absent?

- Is opinion on foreign policy independent of political party?
- Is a person's gender independent of color preference?
- Is gender associated with having a college education?
- Are heart disease and smoking associated?
- Are family size and the level of education of parents independent?
- Is grade distribution independent of major subject for college students?
- Is unemployment associated with an increase in crime?
- Are faculty rank and the amount of time spent working with college students related?
- Are class size and the number of students who drop a college course associated?

Another way to express the fact that two variables are independent is to say that the two variables have nothing to do with one another; they are not related or associated. Care should be taken not to conclude that two variables are correlated if they are not independent. Two variables can be uncorrelated without being independent, but all independent variables are uncorrelated.

If we know that two quantitative variables are related, we can estimate the strength of the linear relationship using the correlation coefficient, r; and using regression analysis, we can estimate the values of one variable knowing the values of the other. Of course, regression analysis can be used only when dealing with quantitative data.

For all tests of independence, the hypotheses are

H_0: The two variables of classification are independent.

H_1: The two variables of classification are dependent.

The methods for testing H_0 against H_1 are identical to those used for testing the differences in population proportions based on the χ^2 test presented in section 12.1. Again, we will compare the observed and expected frequencies (derived under the assumption that H_0 is true) to determine how large a departure can be tolerated before the hypothesis of independence can be rejected. If the value of the χ^2 test statistic is greater than or equal to the tabulated critical value, then we no longer can assume that this value could have resulted from two independent variables of classification. This is why all χ^2 tests for independence are right-tailed tests. Recall that the value of the test statistic is found by

$$\chi^2 = \Sigma \frac{(O_{ij} - E_{ij})^2}{E_{ij}}$$

$$= \Sigma \frac{O_{ij}^2}{E_{ij}} - n$$

and the corresponding degrees of freedom are found by

$$df = (r - 1)(c - 1)$$

APPLICATION 12.9   An association of university professors wants to determine whether faculty satisfaction is independent of faculty rank. The association conducted a national survey of university faculty and found the results shown in table 12.14. At $\alpha = 0.05$ test to determine whether job satisfaction and rank are dependent based on the 3×4 contingency table.

TABLE 12.14

Contingency Table for Application 12.9

		Rank			
		Instructor	Assistant professor	Associate professor	Professor
Job Satisfaction	High	40	60	52	63
	Medium	78	87	82	88
	Low	57	63	66	64

Solution:

1. H_0: Job satisfaction and rank are independent.

2. H_1: Job satisfaction and rank are dependent.

3. We next find the marginal frequencies (totals), as shown in the following table:

	Rank				
Satisfaction	Instructor	Assistant professor	Associate professor	Professor	Total
High	40	60	52	63	215
Medium	78	87	82	88	335
Low	57	63	66	64	250
Total	175	210	200	215	$800 = n$

The df is found to be df $= (3 - 1)(4 - 1) = 6$.

The next step is to find the expected frequencies using formula (12.2):

$$E_{ij} = \frac{(i\text{th Row total})(j\text{th Column total})}{n}$$

Since df $= 6$, we need to compute only six expected frequencies, say, E_{11}, E_{12}, E_{13}, E_{21}, E_{22}, and E_{23}; the remaining expected frequencies are found by subtraction. The observed and expected frequencies are shown in the accompanying table.

	Rank				
Satisfaction	Instructor	Assistant professor	Associate professor	Professor	Total
High	40 (47.03)	60 (56.44)	52 (53.75)	63 (57.78)	215
Medium	78 (73.28)	87 (87.94)	82 (83.75)	88 (90.03)	335
Low	57 (54.69)	63 (65.62)	66 (62.50)	64 (67.19)	250
Total	175	210	200	215	800

The value of the χ^2 statistic is found next. The calculations are summarized in the following table:

Cell	O_{ij}	E_{ij}	$\dfrac{O_{ij}^2}{E_{ij}}$
(1, 1)	40	47.03	34.0208
(1, 2)	60	56.44	63.7846
(1, 3)	52	53.75	50.3070
(1, 4)	63	57.78	68.6916
(2, 1)	78	73.28	83.0240
(2, 2)	87	87.94	86.0700
(2, 3)	82	83.75	80.2866
(2, 4)	88	90.03	86.0158
(3, 1)	57	54.69	59.4076
(3, 2)	63	65.62	60.4846
(3, 3)	66	62.50	69.6960
(3, 4)	64	67.19	60.9615
			802.7501

Thus we have

$$\chi^2 = \Sigma \frac{O_{ij}^2}{E_{ij}} - n = 802.75 - 800 = 2.75$$

4. The level of significance is $\alpha = 0.05$.
5. The critical value is $\chi_{0.05}^2(6) = 12.592$.
6. Decision: Since $\chi^2 = 2.75 < \chi_{0.05}^2(6) = 12.592$, we fail to reject H_0.
7. Type of error possible: Type II, but β is unknown. Thus, we have no statistical evidence to conclude that job satisfaction and faculty rank are related (or associated). ▪

APPLICATION 12.10

Consider table 12.15, a hypothetical 3 × 4 contingency table representing four populations, each classified into three categories. By using the chi-square test for independence, determine if the two variables are dependent.

TABLE 12.15

3 × 4 Contingency Table for Application 12.10

	Population				
	I	II	III	IV	Total
A	100	150	200	50	500
B	40	60	80	20	200
C	60	90	120	30	300
Total	200	300	400	100	$N = 1000$

Solution: Note that for each cell, the expected cell frequency equals the observed cell frequency; hence, the value of the χ^2 statistic is

$$\chi^2 = \Sigma \frac{(O_{ij} - E_{ij})^2}{E_{ij}}$$

$$= \Sigma \frac{0}{E_{ij}}$$

$$= 0$$

The two variables of classification are purely independent. Note the following relationships between rows and columns:

1. For any two rows, the column entries are proportional; for example,

$$\frac{\text{Row A}}{\text{Row B}}: \frac{100}{40} = \frac{150}{60} = \frac{200}{80} = \frac{50}{20}$$

$$\frac{\text{Row B}}{\text{Row C}}: \frac{40}{60} = \frac{60}{90} = \frac{80}{120} = \frac{20}{30}$$

2. For any two columns, the row entries are proportional; for example,

$$\frac{\text{Column I}}{\text{Column II}}: \frac{100}{150} = \frac{40}{60} = \frac{60}{90}$$

$$\frac{\text{Column II}}{\text{Column IV}}: \frac{150}{50} = \frac{60}{20} = \frac{90}{30}$$

Although this application is hypothetical, it does illustrate the relationships that the cell frequencies must exhibit in order for the two variables of classification to be independent. For a contingency table involving sample data, minor deviations due to sampling error can be tolerated. Of course, any deviation from the contingency table (table 12.15) would indicate dependency. For our hypothetical example, there would be no need to use a χ^2 test for independence, since we already have all the population frequencies. ▪

APPLICATION 12.11

In a study to determine if opinions on school closings are related to profession, the 3×3 contingency table of data shown in table 12.16 was obtained.

TABLE 12.16

Contingency Table for Application 12.11

Opinion	Teachers	Lawyers	Doctors	Total
Favor	3	67	15	85
Oppose	105	50	75	230
No opinion	4	3	8	15
Total	112	120	98	330

Profession

Test using $\alpha = 0.05$ to determine if opinion is related to profession.

Solution:

1. H_0: Opinion and profession are independent.
2. H_1: Opinion and profession are dependent.
3. We find the expected frequencies to two decimal places under the assumption that H_0 is true. The expected frequencies are as follows:

$$E_{11} = \frac{(112)(85)}{330} = 28.85$$

$$E_{12} = \frac{(120)(85)}{330} = 30.91$$

$$E_{13} = 85 - 28.85 - 30.91 = 25.24$$

$$E_{21} = \frac{(112)(230)}{330} = 78.06$$

$$E_{22} = \frac{(120)(230)}{330} = 83.64$$

$$E_{23} = 230 - 78.06 - 83.64 = 68.30$$

$$E_{31} = 112 - 28.85 - 78.06 = 5.09$$

$$E_{32} = 120 - 30.91 - 83.64 = 5.45$$

$$E_{33} = 15 - 5.09 - 5.45 = 4.46$$

The observed and expected frequencies are shown in the accompanying table.

	Teachers	Lawyers	Doctors	Total
Favor	3 (28.85)	67 (30.91)	15 (25.24)	85
Oppose	105 (78.06)	50 (83.64)	75 (68.30)	230
No opinion	4 (5.09)	3 (5.45)	8 (4.46)	15
Total	112	120	98	$330 = n$

We next compute the value of the χ^2 statistic. We shall use

$$\chi^2 = \Sigma \frac{(O_{ij} - E_{ij})^2}{E_{ij}}$$

The following table is used to organize the computations:

Cell	O_{ij}	E_{ij}	$\dfrac{(O_{ij} - E_{ij})^2}{E_{ij}}$
(1, 1)	3	28.85	23.1620
(1, 2)	67	30.91	42.1381
(1, 3)	15	25.24	4.1544
(2, 1)	105	78.06	9.2975
(2, 2)	50	83.64	13.5300
(2, 3)	75	68.30	0.6572
(3, 1)	4	5.09	0.2334
(3, 2)	3	5.45	1.1014
(3, 3)	8	4.46	2.8098
			97.0838

Thus, the value of the chi-square statistic is $\chi^2 = 97.0838 \approx 97.08$.

$$df = (r - 1)(c - 1) = (3 - 1)(3 - 1) = 4.$$

4. Significance level: $\alpha = 0.05$.

5. Critical value: $\chi^2_{0.05}(4) = 9.488$.

6. Decision: Since $\chi^2 = 97.08 > 9.488$, we reject H_0. Hence, we have statistical evidence that opinion on school closings is related to profession.

7. Type of error possible: Type I, $\alpha = 0.05$.

8. *p*-value: *p*-value < 0.005. ▪

Test for Homogeneity

When one of the two variables of classification in a contingency table is controlled by the researcher so that either the row totals or column totals are predetermined or fixed before the data are collected, the χ^2 test is called a **test for homogeneity,** instead of a test for independence. Suppose there are four fixed column totals and three row totals in a 3×4 contingency table. The four columns of data correspond to samples from four populations. Each sample of data is then classified into three categories or cells. The objective of a test for homogeneity is to determine whether the populations that correspond to the samples of predetermined sizes are homogeneous, or alike, with respect to the cell probabilities. Beyond the design of the experiment, the computations for a test of homogeneity are identical to those required for a test of independence. The test statistic is

$$\chi^2 = \Sigma \frac{(O_{ij} - E_{ij})^2}{E_{ij}}$$

with df $= (r - 1)(c - 1)$.

EXAMPLE 12.6

Suppose 200 teachers, 300 lawyers, and 400 doctors are involved in a study to determine the extent of alcohol consumption within the three professions. The frequency counts would be listed in a contingency table similar to the following:

Consumption	Teachers	Lawyers	Doctors	Total
Light				
Moderate				
Heavy				
Total	200	300	400	$900 = n$

Suppose that a survey of 200 teachers, 300 lawyers, and 400 doctors yielded the following information:

Consumption	Teachers	Lawyers	Doctors	Total
Light	100	50	100	250
Moderate	50	150	200	400
Heavy	50	100	100	250
Total	200	300	400	$900 = n$

The null hypothesis is that the three populations are homogeneous with respect to the three categories of alcohol consumption. That is,

1. The population percentage of light drinking is the same for all three professions.

2. The population percentage of moderate drinking is the same for all three professions.

3. The population percentage of heavy drinking is the same for all three professions.

For our sample, the population percentages are

	Teachers	Lawyers	Doctors
Light	$\dfrac{100}{200} = \dfrac{1}{2}$	$\dfrac{50}{300} = \dfrac{1}{6}$	$\dfrac{100}{400} = \dfrac{1}{4}$
Moderate	$\dfrac{50}{200} = \dfrac{1}{4}$	$\dfrac{150}{300} = \dfrac{1}{2}$	$\dfrac{200}{400} = \dfrac{1}{2}$
Heavy	$\dfrac{50}{200} = \dfrac{1}{4}$	$\dfrac{100}{300} = \dfrac{1}{3}$	$\dfrac{100}{400} = \dfrac{1}{4}$

The alternative hypothesis is that the populations are not homogeneous with respect to the three categories of drinking.

The value of the χ^2 test statistic is

$$\chi^2 = 76.875$$

For df $= 4$, the critical value corresponding to a level of significance of $\alpha = 0.05$ is

$$\chi_{0.05}^2(4) = 9.488$$

Since the value of the test statistic falls in the rejection region, we reject the null hypothesis that the three populations are homogeneous with respect to the three categories of alcohol consumption.

Summary

The chi-square applications presented in this chapter are often classified as **nonparametric methods.** The only restrictions in their use are that at least 80% of the expected cell frequencies should be at least 5 and no expected frequencies less than 1 should occur. In cases where these conditions are not satisfied, it may be possible to combine rows (or columns) by adding expected frequencies so that the limitations can be overcome, or the size of the sample can be increased. Of course, the nature of the experimental study will determine if it is reasonable to combine rows or columns.

The chi-square goodness-of-fit test can be used to test the normality assumption underlying the t and F tests. This topic is left to a more advanced treatment of the subject and will be omitted here.

EXERCISE SET 12.4

Further Applications

1. In a study to determine if gender of student and interest in mathematics are related, a sample of sixth grade students yielded the information given in the table. Using $\alpha = 0.05$, test to determine if gender and interest in mathematics are related.

	Interest		
Gender	Low	Average	High
Male	15	50	25
Female	10	35	15

2. In a study to determine the relationship between English grades and favorite type of book for sixth-grade students, the results were categorized in the accompanying 2 × 5 contingency table.

	English Grades				
	A	B	C	D	F
Fiction	30	20	15	10	5
Nonfiction	20	15	8	5	4

Determine at $\alpha = 0.05$ if type of book and grade in English are dependent.

3. In a study to determine the relationship between ability in

science and interest in science, the results given in the table were obtained from a random sample of high school students.

	Interest		
Ability	Low	Average	High
High	10	15	20
Average	5	20	15
Low	25	30	10

Test to determine whether interest and ability in science are dependent. Use $\alpha = 0.05$.

4. The accompanying contingency table contains the results of a random sample taken to determine if IQ scores are independent of salaries for high school graduates.

	Salary		
IQ Score	< $25,000	$25,000–50,000	> $50,000
High	5	50	30
Average	20	90	15
Low	15	70	5

Test using $\alpha = 0.05$ to determine if IQ score and salary are dependent variables.

5. In an effort to determine if family status and high school study program are related, the data in the contingency table shown here were gathered from a random sample of high school students in a large school district.

High School Program	Social Status		
	Lower	Middle	Upper
Academic	25	30	60
Commercial	40	60	70
General	35	70	40
Vocational	10	15	20

Test using $\alpha = 0.01$ to determine if social status and high school program are dependent.

6. A popular claim among teachers is that teacher ratings by students are related to grades received by students. To test this claim, an administrator collected the accompanying data:

Rating	Grades				
	A	B	C	D	F
Truly outstanding	13	17	15	3	12
Effective and competent	20	38	60	16	10
Needs improvement	20	30	45	12	5

Test the claim using $\alpha = 0.01$.

7. Use the data in the table to test the claim that the number of cigarettes smoked per day and blood pressure level are dependent. Use $\alpha = 0.05$.

Systolic Blood Pressure	Number of Cigarettes Smoked					
	Under 5	6–10	11–15	16–20	21–25	Over 25
High	12	6	3	15	15	14
Slightly elevated	13	4	7	8	6	7
Normal-to-low	15	11	10	10	2	0

8. A sociology class studied the types of crime occurring in the four quadrants of a large city. The accompanying table displays the frequencies for various types of crimes committed during random periods in each quadrant for the past year.

Quadrant	Type Crime			
	Homicide	Larceny	Assault	Burglary
1	30	120	450	150
2	25	200	1000	300
3	15	190	450	260
4	10	90	100	90

At $\alpha = 0.05$, can we conclude that type of crime is related to city quadrant?

9. As part of a health inventory questionnaire, random samples of sixth- and ninth-grade students in a Missouri school district were asked the following question: "Do you use seat belts?" The results are given in the table.

Response	Grade	
	Sixth	Ninth
Always	5	4
Sometimes	70	45
Never	121	169

Do the results indicate that student responses are related to grade level? Use $\alpha = 0.05$.

10. In a large school district, a study was conducted to determine whether the grades for four academic subjects are related. The results are summarized in the 5 × 4 contingency table shown here.

Grade	Subject			
	Math	Science	English	History
A	15	8	7	11
B	18	17	10	20
C	15	17	10	20
D	10	7	7	7
E	11	8	11	8

Test for dependence of academic subject and grade at the 0.01 level of significance.

Going Beyond

11. Find the values of the χ^2 statistic for the following two 2 × 3 contingency tables. How do the values compare? Generalize the results.

Table 1

5	10	15
7	8	5

Table 2

(3)(5)	(3)(10)	(3)(15)
(3)(7)	(3)(8)	(3)(5)

12. Prove that if the value of the chi-square statistic for a contingency table is χ_0^2, and each entry in the table is multiplied by a constant c, then the new value of the chi-square statistic is $c \cdot \chi_0^2$.

CHAPTER SUMMARY

In this chapter we presented various chi-square tests. These tests measure the differences between observed and expected frequencies under the assumption that the null hypothesis is true. The larger the chi-square statistic, the stronger the evidence that the null hypothesis should be rejected. As a result, all tests studied in this chapter are right-tailed tests. Tests were presented for the equality of two or more population proportions, multinomial parameters, goodness-of-fit, independence, and homogeneity.

The purpose of goodness-of-fit tests is to determine whether a hypothesized population fits a particular population of interest. A test for independence is used to determine if the two qualitative variables of classification used in the contingency table are related or dependent. Tests for homogeneity of cell probabilities in the population are used whenever the row totals or column totals (but not both) in a contingency table are determined before the sample data are gathered. The test is similar to a test for independence; only the experimental design is different.

In order to use the chi-square tests in this chapter, 80% of the expected cell frequencies should be at least 5 and no cell frequency should be less than 1.

CHAPTER REVIEW

■ IMPORTANT TERMS ■

The following chapter terms have been mixed in ordering to provide you better review practice. For each of the following terms, provide a definition in your own words. Then check your responses against the definitions given in the chapter.

marginal frequencies	expected frequencies	multinomial experiments
tests for homogeneity	goodness-of-fit test	nonparametric methods
cells	observed frequencies	r-by-c contingency table
contingency table	three-by-three contingency table	

■ IMPORTANT SYMBOLS ■

c, number of columns	O_{ij}, observed frequency	E_{ij}, expected frequency
r, number of rows (note that r also denotes the correlation coefficient)	n, total number of counts	χ^2, chi-square statistic

---

■ *IMPORTANT FACTS AND FORMULAS* ■

Value of the chi-square test statistic:
$\chi^2 = \Sigma\,[(O_{ij} - E_{ij})^2/E_{ij}]$.   (12.1)

Expected cell frequencies:
$$E_{ij} = \frac{(i\text{th row total})(j\text{th column total})}{n}.$$ (12.2)

Degrees of freedom for testing population proportions:
$df = (r - 1)(c - 1)$.   (12.3)

Degrees of freedom for tests of independence:
$df = (r - 1)(c - 1)$.

Computational formula for chi-square:
$\chi^2 = \Sigma\,(O_{ij}^2/E_{ij}) - n$.   (12.4)

Degrees of freedom for multinomial experiments:
$df = (\text{the number of outcome categories}) - 1$.   (12.5)

At least 80% of the expected cell frequencies should be at least 5 and no expected frequency should be less than 1.

REVIEW EXERCISES

1. In a study to determine the preferred brand of coffee for coffee drinkers, the following results were obtained:

Brand	A	B	C	D	E
Number Preferred	185	205	235	195	180

Test at $\alpha = 0.05$ to determine if preferences for the five brands differ.

2. A sports preference study yielded the results given in the table.

	Sports Preference			
Gender	Baseball	Tennis	Basketball	Football
Male	38	25	30	48
Female	32	15	36	38

Test using $\alpha = 0.01$ to determine if gender and sport preference are dependent.

3. In a sample of 200 people with colds, a new drug was used to relieve the symptoms. The results indicated that 155 people obtained relief from their symptoms as a result of the drug. Use a chi-square test and $\alpha = 0.05$ to determine if the drug was 85% effective.

4. The numbers of cars sold by three salesmen at a large GM dealership over a period of six months are shown in the table.

		Cars		
		Pontiacs	Buicks	Chevrolets
	A	28	24	8
Salesmen	B	42	32	16
	C	30	10	20

At $\alpha = 0.01$, test for dependence of salesman and type of GM car sold.

5. A homemade six-sided die was tossed 240 times. The following results were obtained:

Face	1	2	3	4	5	6
Frequency	54	43	35	37	39	32

Using $\alpha = 0.05$, determine if the die is biased.

6. Voter support for a candidate for mayor in a large city was determined by sampling the four city quadrants. The results shown in the accompanying table were obtained.

	Quadrant			
	I	II	III	IV
Favor	110	90	96	38
Oppose	90	60	79	37

At $\alpha = 0.05$, test the null hypothesis $H_0\colon p_{\mathrm{I}} = p_{\mathrm{II}} = p_{\mathrm{III}} = p_{\mathrm{IV}}$.

7. The number of orders filled by two major suppliers over a one-month period are contained in the accompanying 2 × 2 contingency table.

	Supplier	
	ABC	XYZ
Back ordered	35	28
Not back ordered	165	172

Determine if the two suppliers have different percentages of orders that are back ordered. Use $\alpha = 0.05$.

8. Do review exercise 7 by using a z test. Then show that $z^2 = \chi^2(1)$.

9. In a certain city last year, 75% of the drivers had no accidents, 15% had one accident, and 10% had more than one accident. This year a random sample of drivers produced the following information:

Number of accidents	0	1	More than 1
Frequency	280	72	48

At $\alpha = 0.05$, does this sample indicate that the accident percentages have changed?

10. A study was conducted by an independent consulting firm to investigate the reliability of automobile brakes. By studying a sample of automobile accident reports involving brake failure, the firm obtained the data given in the table.

Cause of Failure

Type of Brake	Wheel cylinder	Master cylinder	Scoring	Worn lining
Disk	70	30	5	20
Drum	80	20	20	30

Test to determine if type of brake and cause of failure are dependent. Use $\alpha = 0.05$.

11. In an effort to determine if time of day and the number of incorrectly assembled automobiles are related, an industrial engineer collected the data given here.

Time of Day

Assembled	8–10 A.M.	10–12 A.M.	12–2 P.M.	2–4 P.M.
Correctly	44	40	52	40
Incorrectly	4	5	9	7

At $\alpha = 0.01$, test the null hypothesis that the proportions of incorrectly assembled units differ among the four time periods.

12. The accompanying data illustrate employment in various occupations by race.

Occupation

Race	White color	Blue color	Farm works	Household services
White	560	360	76	80
Black	30	50	24	15

At $\alpha = 0.05$, determine if race and occupation are dependent.

13. A small-town newspaper publisher conducted a study to determine whether social class and newspaper subscription are related. The data listed in the table were collected.

Social Class

	Poor	Middle class	Rich
Subscribe	20	48	16
Do not subscribe	85	90	2

Test at $\alpha = 0.05$ to determine if subscriptions and social class are dependent.

14. According to Mendel's law, a certain pea plant should produce offspring that have white, red, and pink flowers in the ratios 1:1:2. A sample of 500 offspring of the pea plant were colored in the ratios 23:25:52. At $\alpha = 0.01$, can Mendel's law be rejected?

15. A study was made of the opinions of undergraduate students at a large university concerning a new proposed alcoholic drinking policy for the university. Five hundred males and 500 females were polled concerning their opinions regarding the proposed policy. The results are provided in the table.

	Favor	Indifferent	Oppose
Male	55	225	220
Female	95	360	45

Do the data indicate that for the three response categories, the probabilities differ for the populations of males and females? Use $\alpha = 0.05$.

16. In a study to investigate the relationship between price and perceived quality, 180 adults were asked to purchase beer from three presumably different brands of beer that were in fact identical. Subjects received a 5-cent discount per bottle for choosing the low-priced beer, a 2-cent discount per bottle for choosing the medium-priced beer, and no discount for choosing the highest-priced beer. After purchasing and consuming the beer, the subjects were asked to rate the beers. The results are given in the accompanying table.

Price Level

Rating	High	Medium	Low
Undrinkable	4	1	4
Poor	8	21	20
Fair	26	22	23
Good	15	12	0
Very pleasant	7	4	4

Use $\alpha = 0.05$ to determine if there is a relationship between price and perceived product quality.

Computer Applications

1. One of the qualities of a good random number generator is that each number should be equally likely to be generated. Evaluate the random number generator for your computer by generating 100 random numbers between 0 and 1 and grouping them into the following classes:

   Class 1: Numbers at least 0.0 but less than 0.2

   Class 2: Numbers at least 0.2 but less than 0.4

   Class 3: Numbers at least 0.4 but less than 0.6

   Class 4: Numbers at least 0.6 but less than 0.8

   Class 5: Numbers at least 0.8 but less than 1.0

   Then use a computer program and a goodness-of-fit test to determine if the random numbers are uniformly distributed.

2. Use a computer program to do the following: Generate 100 random values from the standard normal distribution. Square each one of these z values and construct a histogram for the 100 values of z^2. We would expect the histogram to be positively skewed. Use the chi-square goodness-of-fit test and five classes to determine if the squared z values follow a chi-square distribution with one degree of freedom. Use $\alpha = 0.05$.

3. Use a computer program to determine if there is a relationship between price and perceived product quality for review exercise 16.

EXPERIMENTS WITH REAL DATA

Treat the 720 subjects listed in the database in appendix C as if they constituted the population of all U.S. subjects.

1. Are diastolic blood pressures and smoking related for subjects aged 20–39? Systolic pressures are rated as following:
   Low: Less than 75
   Normal: Greater than or equal to 75 but less than 85
   Moderately high: Greater than or equal to 85 but less than 100
   High: Greater than or equal to 100

Use a random sample of 100 subjects and $\alpha = 0.05$ to determine if smoking and blood pressure are related.

2. Are gender and smoking related? Use a 2×2 contingency table, a random sample of 150, and $\alpha = 0.05$ to determine if there is a relationship between smoking and gender.

3. Use a random sample of 150 subjects to determine if there are differences in the proportions of smokers aged in their 20s, 30s, 40, and 50s. Use $\alpha = 0.01$.

■ CHAPTER ACHIEVEMENT TEST ■

1. If $\chi^2(1) = 15$, find the positive value of the z statistic.

2. A random sample of voters were polled and asked two questions:
   a. Do you have children attending public schools?
   b. Are you in favor of increased tax support for public education?
   The results are classified in the following 2×2 contingency table:

	Children Attending	No Children Attending
Favor	125	35
Opposed	25	140

   Let p_A represent the percentage of voters in favor who have children in school and let p_N represent the percentage of voters in favor who do not have children in school. Use $\alpha = 0.01$ and test $H_0: p_A = 0.90$, $p_N = 0.10$ and provide the following information:
   a. H_1
   b. value of the chi-square test statistic
   c. critical value
   d. decision

3. Four coins were tossed 200 times. Do the following observed frequencies follow a binomial distribution with $p = 0.5$? Test at $\alpha = 0.1$ and provide the following information.

Number of heads	0	1	2	3	4
Frequency	10	43	82	60	5

a. value of the test statistic
b. critical value
c. decision

4. If two variables of classification yield the contingency table given here, are the two variables dependent? Use $\alpha = 0.05$

and provide the following information:

	I	II	III
A	26	24	50
B	174	176	550

a. value of the test statistic
b. critical value
c. decision

13 *Analysis of Variance*

CHAPTER OBJECTIVES

In this chapter we will investigate

▷ *The methodology involved with analysis of variance.*

▷ *Two different approaches to viewing analysis of variance.*

▷ *The procedures for using analysis of variance to test the null hypothesis of equality of two or more population means.*

▷ *The fact that one-way analysis of variance is a generalization of the two-sample t test for comparing two population means.*

▷ *The methods and procedures for two-way ANOVA involving a blocking factor.*

▷ *The methods and procedures involved with two-way factorial ANOVA.*

▷ *How to test for differences in population means following a significant F test.*

▷ *How to estimate the strength of association following a significant F test.*

MOTIVATOR 13

South Carolina requires that prospective teachers pass the EEE (Education Entrance Examination), a basic skills test in reading, mathematics, and writing, as standard practice for admission to approved teacher education programs. All three portions of the test must be passed in order to pass the EEE, and the EEE test can be taken no more than three times. Throughout the state of South Carolina minority students have had difficulty passing the mathematics basic skills portion of the EEE. In general, minority students tend to score low on standardized tests, especially mathematics. A research study was conducted at South Carolina State College, a predominantly black institution in a rural setting, to determine if computer-assisted remediation instruction (CAI) could help minority students pass the EEE.[44] Prior to this study, seminars had been used to prepare students to take the EEE. Two types of instruction were investigated (CAI versus non-CAI) along with gender of the subjects. Forty-nine EEE seminar students were randomly assigned to an experimental group and a control group. The students in the control group participated in six weeks (18 60-minute sessions) of EEE instruction. Students in the experimental group also participated in six weeks of instruction (18 sessions of CAI plus EEE seminar instruction in mathematics). Thirty minutes were CAI instruction in mathematics and

30 minutes were EEE instruction in mathematics. Upon termination of the six-weeks instructional period, a test of basic mathematics was administered. Two-way ANOVA was used to analyze the results. An ANOVA summary table for student measures of mathematics basic skills is contained in table 13.1.

TABLE 13.1

ANOVA Summary Table

Source of Variation	df	SS	MS	F value	Significance of F
Group	1	1061.40	1061.40	13.58	0.001
Sex	1	127.68	127.68	1.63	0.21
Group $\times$ Sex	1	170.24	170.24	2.18	0.15

The findings indicated that there were statistically significant differences with regards to type of instruction ($F = 13.58$, $p = 0.001$). There were no statistically significant differences in mathematics basic skills scores with regard to gender ($F = 1.63$, $p = 0.21$) and the interaction between gender and type of instruction ($F = 2.18$, $p = 0.15$). The findings indicated that mathematics scores were significantly increased by the CAI but not affected by gender of the student. This is an example of two-way ANOVA for a **factorial design**. After studying section 13.5 you should be able to understand the above results.

Chapter Overview

Analysis of variance (ANOVA) is a statistical method used to test the equality of two or more population means; it is an extension or generalization of the two-sample *t* test to more than two independent samples.

SECTION 13.1 Introduction to One-Way ANOVA

Suppose we are interested in determining which of three new mid-sized, four-cylinder automobiles (Citation, LeBaron, or Reliant) is most fuel efficient in terms of miles traveled per gallon of gasoline. All three cars have approximately the same size engine and total weight. To determine the most fuel efficient car, we will conduct an experiment involving fifteen cars, five of each make. Each car will be tested once. All cars will be supplied with a special gas tank containing 1 gallon of gasoline and driven over a test course until it runs out of gasoline. The mileage driven will then be recorded. Fifteen professional test car drivers will be randomly assigned to drive the fifteen cars. Suppose the following results were obtained. We will consider three possibilities, A, B, and C.

Possibility A

	Mileages		
	Citation	LeBaron	Reliant
	23	30	27
	24	31	27
	24	32	26
	24	32	26
	25	30	24
$\bar{x}$	24	31	26

Possibility B

	Mileages		
	Citation	LeBaron	Reliant
	22	30	29
	28	29	25
	24	33	26
	25	36	26
	21	27	24
$\bar{x}$	24	31	26

Possibility C

	Mileages		
	Citation	LeBaron	Reliant
	25	23	32
	17	32	34
	28	40	23
	20	34	21
	30	26	20
$\bar{x}$	24	31	26

Our task is to determine if the cars differ in average gas consumption. That is, are the sample averages (24, 31, and 26) different because the underlying population means are unequal (the cars differ in fuel efficiency)? Or can the differences be attributed to random error or chance variation? Possibility A suggests that the cars may differ in average miles-per-gallon ratings. Possibility B is not as suggestive of a difference in population means as possibility A, whereas possibility C depicts an erratic situation. Note that for all three possibilities the sample means are the same. The three possibilities differ in the extent to which the mileage ratings vary within each sample. As a result, it is appropriate to take into account the extent to which the sample means vary from one another (between-sample variance) and the variability of the measurements within each sample (within-sample variance). If our experiment is well designed and uses random assignments to control for any factors that could influence the outcome, then we would not expect the within-sample variation ratings to be sizable, and would attribute this variability to sampling error. Consider the following diagrams illustrating the between-sample variance and within-sample variance for each of the possibilities A, B, and C.

Possibility A

Car	Range
Citation	$25 - 23 = 2$
LeBaron	$32 - 30 = 2$
Reliant	$27 - 24 = 3$
Average Range $= 2\ 1/3$	

The variation within each sample appears to be small compared to the variation between the samples. By using the range as a measure of variation we have the results shown in the table. The range of the sample means is $R = 31 - 24 = 7$. Thus, by using the range as a measure of variation, we see the between-sample variation is three times the within-sample variation. That is,

$$\frac{\text{Between-sample variation}}{\text{Within-sample variation}} = \frac{7}{2\ 1/3} = 3$$

Thus, the data for possibility A suggest that the average gas mileage ratings for the three cars are different.

Possibility B

Car	Range
Citation	$28 - 21 = 7$
LeBaron	$36 - 27 = 9$
Reliant	$29 - 24 = 5$
Average Range $= 7$	

The variation within each sample appears to be approximately equal to the variation between the samples. By using the range as a measure of variation, we have the results shown in the table. The range of the sample means is $R = 31 - 24 = 7$. Thus we see that

$$\frac{\text{Between-sample variation}}{\text{Within-sample variation}} = \frac{7}{7} = 1.0$$

The data for possibility B do not suggest that the average miles-per-gallon fuel ratings are different.

Possibility C

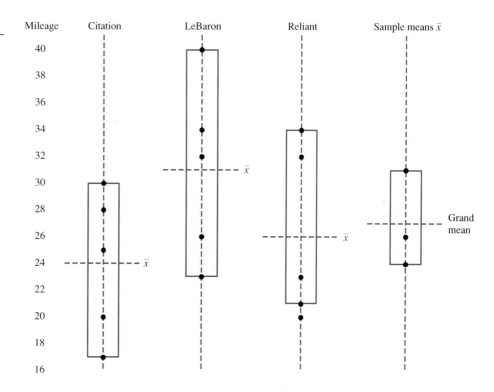

Car	Range
Citation	30 − 17 = 13
LeBaron	40 − 23 = 17
Reliant	34 − 20 = 14
Average Range	= 14 2/3

Compared to the between-sample variation, the within-sample variation appears to be greater. Again, using the range as a measure of variation, we have the results shown in the table. The range of the sample means is $R = 31 - 24 = 7$; thus, we have

$$\frac{\text{Between-sample variation}}{\text{Within-sample variation}} = \frac{7}{14\ 2/3} = \frac{21}{44} \approx 0.5$$

The within-sample variation is greater than the between-sample variation for each make of car. This erratic behavior seemingly could have occurred by chance.

In summary, if the between-sample variation is large relative to the within-sample variation, we can conclude that the average miles-per-gallon ratings for the three cars are not equal.

One-way analysis of variance (one-way ANOVA) is a methodology for analyzing the variation between samples and the variation within samples using variances, rather than ranges. As such, it is a useful statistical method for comparing two or more population means. One-way ANOVA enables us to test hypotheses such as

$$H_0: \mu_1 = \mu_2 = \mu_3 = \cdots = \mu_k.$$

H_1: At least two population means are unequal.

Recall the assumptions underlying the two-sample t test involving independent samples:

1. Both populations are normal.
2. The population variances are equal, that is, $\sigma_1^2 = \sigma_2^2$.

Since one-way ANOVA is a generalization of the two-sample t test, the assumptions for one-way ANOVA are:

1. All k populations are normal.
2. $\sigma_1^2 = \sigma_2^2 = \sigma_3^2 = \cdots = \sigma_k^2 (= \sigma^2)$.

The one-way ANOVA method involves calculating two independent estimates of σ^2, the common population variance. These two estimates are denoted by s_b^2 and s_w^2. s_b^2 is called the **between-samples variance estimate** and s_w^2 is called the **within-samples variance estimate**. The test statistic then becomes s_b^2/s_w^2 and has a sampling distribution that is an F distribution.

F Statistic for One-Way ANOVA

$$F = \frac{s_b^2}{s_w^2}$$

(13.1)

For the two-sample t-test, we had two samples of data from which to compute the t statistic. For one-way ANOVA, we have k samples of data as shown here.

	Sample 1	Sample 2	Sample 3	. . .	Sample k
Sample mean	$\bar{x}_1$	$\bar{x}_2$	$\bar{x}_3$		$\bar{x}_k$
Sample standard deviation	s_1	s_2	s_3		s_k
Sample size	n_1	n_2	n_3		n_k

To simplify computations, we will assume that all samples are of the same size n. That is,

$$n_1 = n_2 = n_3 = \cdots = n_k (= n)$$

Recall that for the two-sample t test, s_p^2 is called the *pooled estimate for* σ^2 and is found by using

$$s_p^2 = \frac{(n_1 - 1)s_1^2 + (n_2 - 1)s_2^2}{n_1 + n_2 - 2}$$

If $n_1 = n_2 (= n)$, then s_p^2 can be written as

$$s_p^2 = \frac{(n - 1)s_1^2 + (n - 1)s_2^2}{n + n - 2} = \frac{(n - 1)(s_1^2 + s_2^2)}{2(n - 1)} = \frac{s_1^2 + s_2^2}{2}$$

Thus, the average variance can be used as an estimate for σ^2 when samples of the same size are used. Generalizing this result to k samples, we see that the average variance can be used to estimate σ^2. We denote this estimate by s_w^2, and call it the *within-*

samples variance estimate. Thus,

Within Samples Variance Estimate

$$s_w^2 = \frac{s_1^2 + s_2^2 + s_3^2 + \cdots + s_k^2}{k} = \frac{\Sigma s_i^2}{k} \qquad (13.2)$$

Since this estimate is based on k samples, each with size n, and each sample has $(n-1)$ degrees of freedom associated with it (see fig. 13.1), there are $k(n-1)$ total degrees of freedom associated with the variance estimate s_w^2. This can be shown as follows:

$$\text{df}_w = (n-1) + (n-1) + (n-1) + \cdots + (n-1) = k(n-1)$$

Thus, the degrees of freedom associated with s_w^2 is given by the following formula:

Degrees of Freedom for s_w^2

$$\text{df}_w = k(n-1) \qquad (13.3)$$

where k is the number of samples, and n is the size of each sample

FIGURE 13.1

Degrees of freedom associated with s_w^2

   ···

Sample 1
size n
df = $n - 1$

Sample 2
size n
df = $n - 1$

Sample 3
size n
df = $n - 1$

Sample k
size n
df = $n - 1$

Recall that the sampling distribution of the mean has a variance $\sigma_{\bar{x}}^2$ defined by

$$\sigma_{\bar{x}}^2 = \frac{\sigma^2}{n}$$

By multiplying both sides of the above equation by n, we have

$$\sigma^2 = n\sigma_{\bar{x}}^2$$

If we had an estimate for $\sigma_{\bar{x}}^2$, we could obtain an estimate for σ^2 by multiplying this estimate by n. Since we have k samples, each with a sample mean $\bar{x}$ (see fig. 13.2), the variance of the sample of k sample means can be used to estimate the variance of the sampling distribution of the mean, $\sigma_{\bar{x}}^2$. We will then multiply this estimate for $\sigma_{\bar{x}}^2$ by n, thereby obtaining our second estimate for σ^2. The mean of the sample of k sample means is called the **grand mean** and denoted by $\bar{x}$.

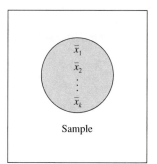

FIGURE 13.2

Sampling distribution of $\bar{x}$

Grand Mean

$$\bar{x} = \frac{\bar{x}_1 + \bar{x}_2 + \bar{x}_3 + \cdots + \bar{x}_k}{k} = \frac{\Sigma \bar{x}_i}{k} \qquad (13.4)$$

Consider the following table to organize the computations needed to help find the variance of the sample of k sample means:

Sample Mean	Sample Mean $- \bar{x}$	(Sample Mean $- \bar{x})^2$
$\bar{x}_1$	$\bar{x}_1 - \bar{x}$	$(\bar{x}_1 - \bar{x})^2$
$\bar{x}_2$	$\bar{x}_2 - \bar{x}$	$(\bar{x}_2 - \bar{x})^2$
$\bar{x}_3$	$\bar{x}_3 - \bar{x}$	$(\bar{x}_3 - \bar{x})^2$
.	.	.
.	.	.
.	.	.
$\bar{x}_k$	$\bar{x}_k - \bar{x}$	$(\bar{x}_k - \bar{x})^2$

The variance of the sample of k sample means is the sum of the last column in the table divided by $k - 1$. Hence, the estimator s_b^2 is given by

$$s_b^2 = \frac{n[(\bar{x}_1 - \bar{x})^2 + (\bar{x}_2 - \bar{x})^2 + (\bar{x}_3 - \bar{x})^2 + \cdots + (\bar{x}_k - \bar{x})^2]}{k - 1}$$

or

Variance of k Sample Means

$$s_b^2 = \frac{n \, \Sigma \, (\bar{x}_i - \bar{x})^2}{(k - 1)} \tag{13.5}$$

Since the sample of sample means has k elements, the between-samples variance estimate s_b^2 has $k - 1$ degrees of freedom associated with it. The symbol df_b denotes the number of degrees of freedom associated with s_b^2.

Degrees of Freedom of s_b^2

$$df_b = k - 1 \tag{13.6}$$

where k is the number of samples

Recall from section 11.4 that to determine whether two sample variances estimate the same population variance, we used the F statistic. Since both s_b^2 and s_w^2 are estimates for σ^2, the test statistic $F = s_b^2/s_w^2$ has a sampling distribution that is an F distribution with $df_b = (k - 1)$ and $df_w = k(n - 1)$. The critical value, obtained from the F tables (tables 6a and 6b of appendix B), is denoted by $F_\alpha(k - 1, k(n - 1))$.

Critical Value for F Test

$$F_\alpha(k - 1, k(n - 1))$$

where the degrees of freedom for the numerator is $k - 1$ and the degrees of freedom for the denominator is $k(n - 1)$ and α is the level of significance

---

APPLICATION 13.1

For the data from possibility A of the gasoline-mileage illustration involving the Citation, the LeBaron, and the Reliant (repeated in table 13.2), determine using $\alpha = 0.05$ if the mean gasoline mileage ratings are different.

	Citation	LeBaron	Reliant
TABLE 13.2	23	30	27
Data for Application 13.1	24	31	27
	24	32	26
	24	32	26
	25	30	24

Solution: If we let μ_1, μ_2, and μ_3 denote the population mean mileage ratings for the Citation, LeBaron, and Reliant, respectively, then we can write the statistical hypotheses as

$$H_0: \mu_1 = \mu_2 = \mu_3.$$

H_1: At least two populations means are unequal.

The sample statistics are as shown in the table.

Citation	LeBaron	Reliant
$n = 5$	$n = 5$	$n = 5$
$\bar{x} = 24$	$\bar{x} = 31$	$\bar{x} = 26$
$s^2 = 0.5$	$s^2 = 1$	$s^2 = 1.5$

Calculate s_w^2 using formula (13.2):

$$s_w^2 = \frac{s_1^2 + s_2^2 + s_3^2}{3} = \frac{0.5 + 1 + 1.5}{3} = \frac{3}{3} = 1$$

Calculate df_w using formula (13.3):

$$df_w = k(n - 1) = 3(5 - 1) = 12$$

Calculate s_b^2 using formula (13.5). The grand mean is first computed by using formula (13.4) to be

$$\bar{x} = \frac{\bar{x}_1 + \bar{x}_2 + \bar{x}_3}{3} = \frac{24 + 31 + 26}{3} = 27$$

Then

$$s_b^2 = \frac{n\Sigma\,(\bar{x}_i - \bar{x})^2}{k - 1}$$
$$= \frac{5\,[(24 - 27)^2 + (31 - 27)^2 + (26 - 27)^2]}{2} = 65$$

Calculate df_b using formula (13.6):

$$df_b = k - 1 = 3 - 1 = 2$$

Calculate the value of F using formula (13.1):

$$F = \frac{s_b^2}{s_w^2} = \frac{65}{1} = 65$$

Find the critical value $F_{0.05}(2, 12)$ in table 6b:

$$F_{0.05}(2, 12) = 3.89$$

Since our test statistic F exceeds the tabulated critical value, we reject the null hypothesis $H_0: \mu_1 = \mu_2 = \mu_3$ and conclude that the population mean gas mileage ratings are unequal. ▪

If the null hypothesis is true and the two one-way ANOVA assumptions hold, then the k samples can be thought of as being selected from the same normal population. If H_0 is true, the between-samples variance estimate s_b^2 indicates the variance of the sample means and reflects sampling error (or error variance). The within-samples variance estimate, being an average variance, reflects sampling error (or error variance) and does not depend on the null hypothesis being true or false. The measurements within each sample come from a single population and are not affected by a difference between population means. As a result, the values of s_w^2 and s_b^2 in this case should be close. If $H_0: \mu_1 = \mu_2 = \mu_3 = \cdots = \mu_k$ is false, then the between-samples variance estimate s_b^2 should be greater than the within-samples variance estimate s_w^2, because s_w^2 reflects only error variance and does not depend on whether the population means are equal. But if H_0 is false, the variance of the sample of sample means will reflect more than sampling error, since the population means are not equal. The tabulated critical value for F determines how much larger than 1 the test statistic can be and still reflect only sampling error. If the test statistic F is greater than or equal to the critical value (and thus falls in the rejection region), then the null hypothesis H_0 is rejected and we conclude that the test statistic reflects more than sampling error. This explains why one-way ANOVA is a right-tailed test procedure.

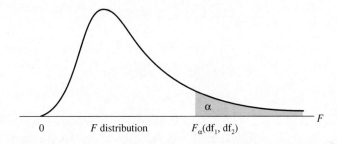

APPLICATION 13.2 Three different makes of gasoline engines were tested to determine the useful lifetimes before an overhaul was needed. If the engine lifetimes for each make are normally distributed and have the same variance, test using $\alpha = 0.05$ to determine if the mean useful lifetimes before an overhaul differ. The useful engine lifetimes (in tens of thousands of miles) for each engine are given in table 13.3.

TABLE 13.3

Data for Application 13.2

A	B	C
6	8	3
2	7	2
4	7	5
1	2	4
7	6	1

Solution: If we let μ_1, μ_2, and μ_3 denote the population mean useful lifetimes for makes A, B, and C, respectively, then we can write the statistical hypotheses as

$$H_0: \mu_1 = \mu_2 = \mu_3.$$

H_1: At least two populations means are unequal.

We shall proceed according to the following steps:

1. Find the sample means and variances.
2. Find the within-samples variance estimate s_w^2 and its associated degrees of freedom df_w.
3. Find the grand mean $\bar{x}$ for the sample of sample means.
4. Find the between-samples variance estimate s_b^2 and its associated degrees of freedom df_b.
5. Find the value of the F test statistic.
6. Find the critical value for F based on df_b and df_w.
7. Decide whether to reject H_0.

Step 1 The following table displays Σx and Σx^2, which are used to find the sample means and variances for the three groups A, B, and C.

	A	B	C
	6	8	3
	2	7	2
	4	7	5
	1	2	4
	7	6	1
Σx	20	30	15
Σx^2	106	202	55

By using a hand-held calculator, we compute the sample means to be $\bar{x}_1 = 4$, $\bar{x}_2 = 6$, and $\bar{x}_3 = 3$. We then apply the formulas

$$SS = \Sigma x^2 - \frac{(\Sigma x)^2}{n} \quad \text{and} \quad s^2 = \frac{SS}{n-1}$$

with the help of a calculator to compute the sample variances, $s_1^2 = 6.5$, $s_2^2 = 5.5$, and $s_3^2 = 2.5$.

Step 2 By using formula (13.2), we find s_w^2:

$$s_w^2 = \frac{s_1^2 + s_2^2 + s_3^2}{3} = \frac{6.5 + 5.5 + 2.5}{3} = 4.83$$

By using formula (13.3), we find df_w:

$$df_w = k(n-1) = 3(5-1) = 12$$

Note that each sample has four degrees of freedom and that there are three samples; thus, $df = (4)(3) = 12$.

Step 3 By using formula (13.4), we find $\bar{x}$:

$$\bar{x} = \frac{\bar{x}_1 + \bar{x}_2 + \bar{x}_3}{3} = \frac{4 + 6 + 3}{3} = 4.33$$

Step 4 By using formula (13.5), we find s_b^2:

$$s_b^2 = \frac{n[\bar{x}_1 - \bar{x})^2 + (\bar{x}_2 - \bar{x})^2 + (\bar{x}_3 - \bar{x})^2]}{k - 1}$$

$$= \frac{5\left[(4 - 4.33)^2 + (6 - 4.33)^2 + (3 - 4.33)^2\right]}{2}$$

$$= \frac{5\,(4.667)}{2} = 11.67$$

The degrees of freedom for s_b^2 is found by using formula (13.6):

$$df_b = k - 1 = 3 - 1 = 2$$

Step 5 The value for the F test statistic is given by formula (13.1):

$$F = \frac{s_b^2}{s_w^2} = \frac{11.67}{4.83} = 2.42$$

Step 6 The critical value for the test statistic is $F_{0.05}(2, 12)$ and is found in the F table:

$$F_{0.05}(2, 12) = 3.89$$

Step 7 Since $F = 2.42 < F_{0.05}(2, 12) = 3.8853$, we fail to reject H_0.

We have no statistical evidence to support that the useful engine lifetimes before overhaul differ. The differences in sample means can be attributed to sampling error. ◼

By examining the entries in the F table (tables 6a and 6b of appendix B), we note that all values are greater than 1. As a result, there is no need to find the critical value for the F statistic when the value for the F test statistic is less than 1. That is, there is no chance for rejecting the null hypothesis H_0 when $F < 1$.

We noted earlier that the one-way ANOVA method generalizes the nondirectional two-sample t test for independent samples having the same size. This is because when only $k = 2$ samples are involved, the square of the t statistic is identical to the F statistic. That is, we have

$$t^2 = F$$

where the t distribution has df $= k(n - 1) = 2(n - 1)$ and the F distribution has $df_b = (k - 1) = 1$ and $df_w = k(n - 1) = 2(n - 1)$. Consider application 13.3.

A P P L I C A T I O N   13.3

As part of a readability study conducted to determine the clarity of two different texts, 20 random pages from each text were selected. A readability score for each page was determined. Assume that the readability scores are normal for each text and have equal variances. The data were obtained as shown in the table. Test at $\alpha = 0.05$ to determine if the population average readability scores are different.

Text 1	Text 2
$n_1 = 20$	$n_2 = 20$
$\bar{x}_1 = 76$	$\bar{x}_2 = 70$
$s_1^2 = 60$	$s_2^2 = 100$

Solution:

> **Method 1: t-test**
>
> 1. $H_0: \mu_1 = \mu_2$.
> 2. $H_1: \mu_1 \neq \mu_2$.

3. The difference between sample means is

$$\bar{x}_1 - \bar{x}_2 = 76 - 70 = 6$$

The pooled estimate for σ is

$$s_p = \sqrt{\frac{(n_1 - 1)s_1^2 + (n_2 - 1)s_2^2}{n_1 + n_2 - 2}}$$

$$= \sqrt{\frac{(19)(60) + (19)(100)}{38}}$$

$$= \sqrt{80} = 8.94$$

The value of the test statistic is

$$t = \frac{\bar{x}_1 - \bar{x}_2}{s_p \sqrt{(1/n_1) + (1/n_2)}}$$

$$= \frac{6}{8.94 \sqrt{(1/20) + (1/20)}} = 2.122$$

The degrees of freedom for the test statistic is

$$df = n_1 + n_2 - 2 = 20 + 20 - 2 = 38$$

4. The level of significance is $\alpha = 0.05$.
5. The critical values are $\pm t_{0.025} = \pm 2.021$ for $df = 38$.
6. Decision: Reject H_0, since $t = 2.122 > t_{0.025} = 2.021$ for $df = 38$. Hence, we have statistical evidence to suggest that the average clarity scores for the texts are unequal.
7. Type of error possible: Type I, $\alpha = 0.05$.
8. p-value: $0.025 < p$-value < 0.05.

Method 2: One-way ANOVA

1. The sample means and variances are given above.
2. The within samples variance estimate is

$$s_w^2 = \frac{s_1^2 + s_2^2}{2} = \frac{60 + 100}{2} = 80$$

The df for s_w^2 is

$$df_w = k(n - 1) = 2(20 - 1) = 38$$

3. The grand mean is

$$\bar{x} = \frac{\bar{x}_1 + \bar{x}_2}{2} = \frac{76 + 70}{2} = 73$$

4. The between samples variance estimate is

$$s_b^2 = \frac{n\,[(\bar{x}_1 - \bar{x})^2 + (\bar{x}_2 - \bar{x})^2]}{k - 1}$$

$$= \frac{20[(76 - 73)^2 + (70 - 73)^2]}{2 - 1}$$

$$= 20(9 + 9) = 360$$

The degrees of freedom for s_b^2 is

$$df_b = k - 1 = 2 - 1 = 1$$

5. The value of the F statistic is

$$F = \frac{s_b^2}{s_w^2} = \frac{360}{80} = 4.50$$

6. The critical value $F_{0.05}(1, 38)$ is not contained in table 6b of appendix B. We shall approximate it by using $F_{0.05}(1, 30) = 4.17$, which is contained in table 6b. Note that since $t^2 = F$ and $t_{0.025} = 2.021$ we can find $F_{0.05}(1, 30)$: $F_{0.05}(1, 38) = (2.021)^2 = 4.08$.

7. Since $F = 4.5 > F_{0.05}(1, 38)$, we reject H_0. As with the t test, we conclude that the average clarity scores are different. Note also that $t^2 = (2.122)^2 = 4.502884 \approx 4.50 = F$. ■

The F test has been determined by statisticians to be a robust test, particularly when equal sample sizes are involved. A test is called **robust** if it is relatively insensitive to minor violations of the normality and equal-variance assumptions. Determining which violations are minor is open to speculation and depends on the judgment of the individual using the test. Knowing that the F test is robust should not lead one to believe that the assumptions do not have to be checked. They should always be checked, even for equal sample sizes.

Analysis of variance is a mathematical model used to solve practical problems. As a result, it never fits a given situation exactly, and there are times when one-way ANOVA should not be used to model a given situation. These times are when the assumptions are very tenuous, at best. And in such cases, a nonparametric method may be appropriate. Nonparametric methods are presented in chapter 15.

EXERCISE SET 13.1

Further Applications

For exercises 1–10, assume that the assumptions for one-way ANOVA hold.

1. A readability study was done on two textbooks to investigate their clarity. Five independent determinations of readability were made for each textbook. The following data represent the readability scores for the two books:

Text A	50	49	54	57	58
Text B	48	57	61	55	60

   a. At $\alpha = 0.05$ use the t-test to determine if the two texts have different average readability scores.
   b. At $\alpha = 0.05$, use one-way ANOVA to answer the same question.
   c. Show that $t^2 = F$.

2. An experiment was conducted to determine the best of two methods to train the newly hired in a large industry. Ten trainees were randomly assigned to method 1 and ten trainees were randomly assigned to method 2. Following the training session, both groups took a standardized achievement test. The scores are as follows:

Method 1	90	80	90	94	87	86	95	99	94	99
Method 2	80	86	83	80	78	85	81	77	87	83

   a. At $\alpha = 0.05$ use the t-test to determine if the two methods have different average achievement scores.
   b. At $\alpha = 0.05$, use one-way ANOVA to answer the same question.
   c. Show that $t^2 = F$.

3. Five different methods are used to teach a basic unit on one-way ANOVA to college business majors. A different teaching method was used in each of five classes and 13 students were randomly assigned to each class. The results, based on an examination given at the end of the unit,

yielded the following data:

Method A	Method B	Method C	Method D	Method E
$n_1 = 13$	$n_2 = 13$	$n_3 = 13$	$n_4 = 13$	$n_5 = 13$
$\bar{x}_1 = 75$	$\bar{x}_2 = 76$	$\bar{x}_3 = 70$	$\bar{x}_4 = 74$	$\bar{x}_5 = 76$
$s_1 = 7.7$	$s_2 = 7.1$	$s_3 = 9.9$	$s_4 = 6$	$s_5 = 6.3$

By using $\alpha = 0.05$, test to determine if there is a difference among the five population means.

4. Three laboratories are used by a chemical firm for performing analyses. For quality control purposes, it is decided to submit five samples of the same material to each lab and to compare their analyses to determine whether they give, on the average, the same results. The analytical results from the three labs are as follows:

$$\begin{array}{llllll} \text{Lab A:} & 58.6 & 60.7 & 59.4 & 59.6 & 60.5 \\ \text{Lab B:} & 61.6 & 64.8 & 62.8 & 59.2 & 60.4 \\ \text{Lab C:} & 60.7 & 55.6 & 57.3 & 55.2 & 60.2 \end{array}$$

By using $\alpha = 0.05$, determine whether the labs produce, on the average, the same results.

5. The accompanying table contains the number of words typed per minute by four college secretaries at five different times using the same typewriter.

A	B	C	D
82	55	69	87
79	67	72	61
75	84	78	82
68	77	83	61
65	71	74	72

By using $\alpha = 0.05$, determine if the typing speeds of the four secretaries differ.

6. Basic skills tests are given to ninth-grade students in four high schools in a certain county. Random samples of their scores are shown in the table.

School A	School B	School C	School D
20	24	16	19
21	21	21	20
22	22	18	21
24	25	13	20

By using $\alpha = 0.05$, test to determine if the mean scores for each school differ.

7. A study was conducted to study the length of time (in seconds), required for students majoring in art, music, and physical education to complete a particular task involving certain motor skills. Seven people were randomly chosen from each of the three disciplines. The results are listed in the table.

Art	Music	Physical Education
17	24	25
21	18	24
25	19	25
16	22	21
19	23	24
22	20	28
18	21	19

By using the 0.01 level of significance, test to determine if there is a difference in the population mean completion times for the three disciplines.

8. Tourists were polled and asked which of three activities was most important in influencing their choice of resort and the number of hours per day they spent participating in the activity. The results of the survey for activity participation times are as follows:

	Activity	
Tennis	Swimming	Golf
$n = 20$	$n = 20$	$n = 20$
$\bar{x} = 3.05$	$\bar{x} = 2.15$	$\bar{x} = 3.82$
$s = 2.51$	$s = 3.26$	$s = 1.17$

At $\alpha = 0.05$ do the data indicate different population mean participation times for the three activities?

9. A study was conducted to investigate the contaminant level of coal gases in the atmosphere from four different utility sources that use coal to generate energy. The results given in the accompanying table were produced by taking five readings from each utility at different times.

		Utility	
A	B	C	D
0.047	0.037	0.019	0.041
0.039	0.041	0.021	0.037
0.051	0.036	0.018	0.038
0.048	0.035	0.022	0.047
0.046	0.040	0.017	0.048

By using $\alpha = 0.05$, determine if there are differences among the average contaminant levels for the four utility plants.

10. In a study of the effect of diet on blood pressure, 18 adults ages 25–30 were randomly assigned to one of three diets. After subjects had been on the diets for two months, blood pressure measurements were recorded for each individual, with the results shown in the table.

Diet

A	B	C
122	117	128
130	123	124
118	116	119
115	112	132
128	119	135
118	115	118

At $\alpha = 0.01$, is there a difference in mean blood pressure measurements for subjects on the three diets?

Going Beyond

11. Suppose two random samples of size n are selected from normal populations with equal means and variances and yield the following information:

Sample 1	Sample 2
Size $= n$	Size $= n$
Mean $= \bar{x}_1$	Mean $= \bar{x}_2$
Variance $= s_1^2$	Variance $= s_2^2$

Prove that $t^2 = F$.

SECTION 13.2 — Computational Formulas for One-Way ANOVA

Although the approach to one-way ANOVA used in section 13.1 is reasonably good from a developmental standpoint, it does not lend itself to short-cut computations that minimize the amount of work involved. Nor does the approach generalize easily to ANOVA techniques involving more than one independent variable (called a **factor**) or to applications requiring samples with unequal sizes. In this section we will approach one-way ANOVA from a different point of view and will present computational formulas that will facilitate the calculation of the F statistic. The methods presented can be extended or generalized to the more advanced techniques of ANOVA, such as two-factor ANOVA discussed in sections 13.4 and 13.5.

Notation

Let's assume k samples of data are used to test the statistical hypotheses

$$H_0: \mu_1 = \mu_2 = \mu_3 = \cdots = \mu_k.$$

H_1: At least two population means are unequal.

The following notation will be used. The jth measurement in the ith sample will be represented by x_{ij}, as depicted here.

$$x_{i\,j}$$
$$\uparrow\uparrow$$
$$\text{Sample Measurement}$$

For example, x_{24} represents the fourth measurement in sample 2.

TABLE 13.4

Experimental Procedure for
Single-Factor ANOVA

	Sample 1	Sample 2	Sample 3	. . .	Sample k
	x_{11}	x_{21}	x_{31}	. . .	x_{k1}
	x_{12}	x_{22}	x_{32}	. . .	x_{k2}
	x_{13}	x_{23}	x_{33}	. . .	x_{k3}
	.	.	.	.	.
	.	.	.	.	.
	.	.	.	.	.
	x_{1n_1}	x_{2n_2}	x_{3n_3}	. . .	x_{kn_k}
Totals	C_1	C_2	C_3	. . .	$C_k = T$

The experimental procedure may be summarized as in table 13.4. There are k samples and the size of each sample is n_i. The sum of all the column totals C_i is called the **grand total** and is represented by T. Thus, $T = \Sigma x_{ij} = \Sigma C_i$. We let the total number of measurements be represented by $N = \Sigma n_i$. If the grand total T is divided by the total number of measurements, then the quotient T/N is called the grand mean and is denoted by $\overline{T}$. Thus, the grand mean is represented by

> **Grand Mean**
>
> $$\overline{T} = \frac{T}{N}$$

Formulas

Recall from chapter 3 that the sum of squares SS represents the sum of the squared deviations from the mean. That is,

$$SS = \Sigma (x - \overline{x})^2$$

Also recall the following computational formula for SS:

$$SS = \Sigma x^2 - \frac{(\Sigma x)^2}{n}$$

The sum of squared deviations about the grand mean is called the **sum of squares for total** and is denoted by SST. That is,

> **Sum of Squares for Total**
>
> $$SST = \Sigma (x_{ij} - \overline{T})^2 \tag{13.7}$$

To find SST, we first find the deviation of each measurement from the grand mean. These deviations are then squared and totaled.

EXAMPLE 13.1

Consider the three samples of data shown in table 13.5.

TABLE 13.5

Data to Illustrate Calculation
of SST

Sample 1	Sample 2	Sample 3
3	2	4
5	2	3
	4	3
	1	

The column totals are

$$C_1 = 8 \qquad C_2 = 9 \qquad C_3 = 10$$

The grand total is $T = 8 + 9 + 10 = 27$, and the total number of measurements is $N = 2 + 4 + 3 = 9$. Hence, the grand mean $\overline{T}$ is

$$\overline{T} = \frac{T}{\Sigma n_i} = \frac{27}{9} = 3$$

The total sum of squares SST is

$$\begin{aligned}
\text{SST} &= (3 - 3)^2 + (5 - 3)^2 + (2 - 3)^2 + (2 - 3)^2 + (4 - 3)^2 \\
&\quad + (1 - 3)^2 + (4 - 3)^2 + (3 - 3)^2 + (3 - 3)^2 \\
&= 0 + 4 + 1 + 1 + 1 + 4 + 1 + 0 + 0 = 12
\end{aligned}$$

A computational formula for SST is given by

> **Computational Formula for SST**
>
> $$\text{SST} = \Sigma x_{ij}^2 - \frac{T^2}{N}$$
>
> where $N = \Sigma n_i$

(13.8)

EXAMPLE 13.2

For the data in table 13.5, we have

$$\begin{aligned}
T &= C_1 + C_2 + C_3 = 8 + 9 + 10 = 27 \\
N &= \Sigma n_i = 2 + 4 + 3 = 9 \\
\Sigma x_{ij}^2 &= 3^2 + 5^2 + 2^2 + 2^2 + 4^2 + 1^2 + 4^2 + 3^2 + 3^2 = 93
\end{aligned}$$

By using formula (13.8), we have

$$\text{SST} = \Sigma x_{ij}^2 - \frac{T^2}{N} = 93 - \frac{27^2}{9} = 12$$

This result agrees with the answer found by using formula (13.7).

The sum of squares for total SST can be partitioned (or separated) into two component parts, called **sum of squares between samples (SSB)** and **sum of squares within samples (SSW)**. That is,

> **Partitioning SST**
>
> $$\text{SST} = \text{SSB} + \text{SSW}$$

(13.9)

SSB is defined by the formula

> **Sum of Squares for Between Samples**
>
> $$\text{SSB} = \Sigma n_i (\overline{C}_i - \overline{T})^2$$
>
> where $\overline{C}_i$ is the mean of the ith column or sample

and SSW is defined by the formula

> **Sum of Squares for Within Samples**
> $$SSW = \Sigma (x_{ij} - \overline{C}_i)^2$$

Hence, the sum of squares for total can be expressed as

$$SST = \Sigma (x_{ij} - \overline{T})^2 = \Sigma n_i(\overline{C}_i - \overline{T})^2 + \Sigma (x_{ij} - \overline{C}_i)^2$$

A computational formula for SSB is given by

> **Computational Formula for SSB**
> $$SSB = \Sigma \frac{C_i^2}{n_i} - \frac{T^2}{N} \qquad (13.10)$$

EXAMPLE 13.3

Let's find the values of SSB and SSW for the data in table 13.5. By using formula (13.10),

$$SSB = \frac{C_1^2}{n_1} + \frac{C_2^2}{n_2} + \frac{C_3^2}{n_3} - \frac{T^2}{n} = \frac{8^2}{2} + \frac{9^2}{4} + \frac{10^2}{3} - \frac{27^2}{9} = 4.58$$

As a result of formula (13.9), we have

$$SSW = SST - SSB = 12 - 4.58 = 7.42$$

The normal procedure (computational formula) for finding SSW is first to find SST and SSB and then to subtract SSB from SST. This involves less work than finding SSW by using the formula $SSW = \Sigma (x_{ij} - \overline{C}_i)^2$.

> **Computational Formula for SSW**
> $$SSW = SST - SSB$$

Each of the three sum of squares, SST, SSB, and SSW, has an associated degrees of freedom. These are given by

> **Degrees of Freedom**
> SST: $df_t = N - 1$
> SSB: $df_b = k - 1$ $\qquad (13.11)$
> SSW: $df_w = N - k$

The degrees of freedom for SST is 1 less than the total number of measurements. The degrees of freedom for SSB is 1 less than the number of samples. Since the ith sample has $n_i - 1$ degrees of freedom associated with it, the k samples have a total number of degrees of freedom given by

$$\begin{aligned} df_w &= (n_1 - 1) + (n_2 - 1) + (n_3 - 1) + \cdots + (n_k - 1) \\ &= (n_1 + n_2 + n_3 + \cdots + n_k) - k \\ &= N - k \end{aligned}$$

Note that since $N - 1 = (k - 1) + (N - k)$ we have

$$df_t = df_b + df_w$$

Hence, the following two relations must hold:

$$SST = SSB + SSW$$
$$df_t = df_b + df_w$$

EXAMPLE 13.4

Let's compute the degrees of freedom for the data in example 13.1. By using formulas (13.11) we have

$$df_t = N - 1 = 9 - 1 = 8$$
$$df_b = k - 1 = 3 - 1 = 2$$
$$df_w = N - k = 9 - 3 = 6$$

As a computational check, we note that

$$df_t = df_b + df_w$$
$$8 = 2 + 6$$

In statistics, a sum of squares divided by its associated degrees of freedom is called a **mean square** and represents a variance estimate. Recall from chapter 3, that

$$s^2 = \frac{SS}{n - 1} = \frac{SS}{df}$$

Since a sample of size n has $(n - 1)$ degrees of freedom associated with it, the *mean square for SSB,* denoted by MSB, is defined as

$$MSB = \frac{SSB}{df_b} = \frac{SSB}{k - 1}$$

Thus, the **mean square for between samples** is given by

Mean Square for Between Samples
$$MSB = \frac{SSB}{k - 1}$$

(13.12)

The mean square for SSW is denoted by MSW and defined as

$$MSW = \frac{SSW}{df_w} = \frac{SSW}{N - k}$$

Thus, the **mean square for within samples** is given by

Mean Square for Within Samples
$$MSW = \frac{SSW}{N - k}$$

(13.13)

EXAMPLE 13.5

For the data in example 13.1, we have

$$MSB = \frac{SSB}{k-1} = \frac{4.58}{2} = 2.29$$

$$MSW = \frac{SSW}{N-k} = \frac{7.42}{6} = 1.24$$

As we have already pointed out, MSB and MSW are both variance estimates. MSB reflects the variation between samples and MSW reflects the variation within samples. Since each observation within a given sample comes from a population with a given mean, the differences between observations within a random sample are explained by sampling error. As a result, MSW is typically referred to as **mean square for error.** The value of the F test statistic is given by

Value of F Statistic
$$F = \frac{MSB}{MSW}$$

(13.14)

EXAMPLE 13.6

For the data in example 13.1, the value of F is

$$F = \frac{MSB}{MSW} = \frac{2.29}{1.24} = 1.85$$

The computations we have just presented are usually displayed in a special summary table, called an **ANOVA summary table,** the general form of which is shown in table 13.6.

TABLE 13.6

General Form of ANOVA Summary Table

Source of Variation	SS	df	MS	F
Between samples	SSB	$k-1$	$\dfrac{SSB}{k-1}$	$\dfrac{MSB}{MSW}$
Within samples	SSW	$N-k$	$\dfrac{SSW}{N-k}$	
Total	SST	$N-1$		

EXAMPLE 13.7

For the data of example 13.1, the ANOVA summary table is given in table 13.7.

TABLE 13.7

ANOVA Summary Table for Data of Table 13.5

Source of Variation	SS	df	MS	F
Between samples	4.58	2	2.29	1.85
Within samples	7.42	6	1.24	
Total	12.00	8		

Summarizing, we use the following steps to find the value for the F test statistic:

1. Find the column totals and the grand total.
2. Find SST by using formula (13.8).
3. Find SSB by using formula (13.10).
4. Find SSW by subtracting SSB from SST.
5. Find df_b and df_w by using formulas (13.11).
6. Find MSB and MSW by using formulas (13.12) and (13.13).
7. Find F by using formula (13.14).

The computations are usually summarized in an ANOVA summary table. Consider the application 13.4.

APPLICATION 13.4

For application 13.2, find the value of F by using the methods presented in this section. The data are repeated for convenience in table 13.8.

TABLE 13.8

Data for Application 13.4

A	B	C
6	8	3
2	7	2
4	7	5
1	2	4
7	6	1

Solution:

1. We first find the column totals C_i and the grand total T, as shown in the following table:

	A	B	C	
	6	8	3	
	2	7	2	
	4	7	5	
	1	2	4	
	7	6	1	
Totals	$C_1 = 20$	$C_2 = 30$	$C_3 = 15$	$T = 65$

2. Find SST using formula (13.8):

$$\text{SST} = \Sigma x_{ij}^2 - \frac{T^2}{N}$$

$$= 6^2 + 2^2 + 4^2 + \cdots + 5^2 + 4^2 + 1^2 - \frac{65^2}{15} = 81.33$$

3. Find SSB using formula (13.10):

$$\text{SSB} = \Sigma \frac{C_i^2}{n_i} - \frac{T^2}{N}$$

$$= \frac{20^2}{5} + \frac{30^2}{5} + \frac{15^2}{5} - \frac{65^2}{15} = 23.33$$

4. Find SSW by subtracting SSB from SST:

$$\text{SSW} = \text{SST} - \text{SSB} = 81.33 - 23.33 = 58$$

5. Find df_b and df_w by using formulas (13.11):

$$df_b = k - 1 = 3 - 1 = 2$$
$$df_w = N - k = 15 - 3 = 12$$

6. Find MSB and MSW using formulas (13.12) and (13.13):

$$\text{MSB} = \frac{\text{SSB}}{k - 1} = \frac{23.33}{2} = 11.67$$

$$\text{MSW} = \frac{\text{SSW}}{N - k} = \frac{58}{12} = 4.83$$

Note that for application 13.2, $s_b^2 = 11.67$ and $s_w^2 = 4.83$.

7. Find F by using formula (13.14):

$$F = \frac{\text{MSB}}{\text{MSW}} = \frac{11.67}{4.83} = 2.42$$

Note that this value of F agrees with the value found in application 13.2.

The summary table is given in table 13.9.

TABLE 13.9

ANOVA Summary Table
for Application 13.4

Source of Variation	SS	df	MS	F
Between samples	23.33	2	11.67	
Within samples	58	12	4.83	2.42
Total	81.33	14		

At $\alpha = 0.05$, the critical value is found to be $F_{0.05}(2, 12) = 3.89$. Since the value of our test statistic F is less than 3.89, we fail to reject H_0. ▪

We can use MINITAB to do application 13.4. Computer display 13.1 contains the data and commands. The MINITAB command AOVONEWAY is used. The output is contained in computer display 13.2.

Computer Display 13.1

```
MTB > READ C1 C2 C3
DATA> 6 8 3
DATA> 2 7 2
DATA> 4 7 5
DATA> 1 2 4
DATA> 7 6 1
DATA> END
    5 ROWS READ
MTB > AOVONEWAY C1 C2 C3
```

Computer Display 13.2

```
ANALYSIS OF VARIANCE
SOURCE      DF      SS        MS       F       P
FACTOR       2    23.33     11.67    2.41    0.132
ERROR       12    58.00      4.83
TOTAL       14    81.33
                                 INDIVIDUAL 95 PCT CI'S FOR MEAN
                                 BASED ON POOLED STDEV
LEVEL        N    MEAN      STDEV
C1           5    4.000     2.550    ------+----------+----------+---------+
C2           5    6.000     2.345       (----------*----------)
C3           5    3.000     1.581    (----------*----------)
POOLED STDEV =    2.198
MTB >                                    2.0       4.0       6.0       8.0
```

APPLICATION 13.5

In a study to determine if the rates of unemployment are different for eastern, central, and western cities, random samples of cities were selected from the three areas of the United States for study. The data in table 13.10 represent the extent of unemployment (in percentages).

TABLE 13.10

Data for Application 13.5

East	Central	West
5.2	11.4	7.2
11.5	9.1	15.9
6.3	6.6	10.3
6.6	10.5	9.5
7.7	3.6	
3.8		
7.6		

By assuming that the assumptions for ANOVA hold and by using $\alpha = 0.05$, test the null hypothesis $H_0: \mu_E = \mu_C = \mu_W$.

Solution:

1. The following table is used to find Σx and Σx^2 for the measurements of each area:

East		Central		West	
x	x^2	x	x^2	x	x^2
52	27.04	11.4	129.96	7.2	51.84
11.5	132.25	9.1	82.81	15.9	252.81
6.3	39.69	6.6	43.56	10.3	106.09
6.6	43.56	10.5	110.25	9.5	90.25
7.7	59.29	3.6	12.96	42.9	500.99
3.8	14.44	41.2	379.54		
7.6	57.76				
48.7	374.03				

We perform the necessary preliminary calculations:

$$T = 48.7 + 41.2 + 42.9 = 132.8$$
$$\Sigma x_{ij}^2 = 374.03 + 379.54 + 500.99 = 1254.56$$
$$N = 7 + 5 + 4 = 16$$

2. Find SST by using formula (13.8):

$$\text{SST} = \Sigma x_{ij}^2 - \frac{T^2}{N} = 1254.56 - \frac{132.8^2}{16} = 152.32$$

3. Find SSB using formula (13.10):

$$\text{SSB} = \Sigma \frac{C_i^2}{n_i} - \frac{T^2}{N}$$

$$= \frac{48.7^2}{7} + \frac{41.2^2}{5} + \frac{42.9^2}{4} - \frac{132.8^2}{16} = 36.16$$

4. Find SSW by subtracting SSB from SST:

$$\text{SSW} = \text{SST} - \text{SSB} = 152.32 - 36.16 = 116.16$$

5. By using formulas (13.11), find df_b and df_w:

$$df_b = k - 1 = 3 - 1 = 2$$
$$df_w = N - k = 16 - 3 = 13$$

6. Find MSB and MSW using formulas (13.12) and (13.13):

$$\text{MSB} = \frac{\text{SSB}}{k - 1} = \frac{36.16}{2} = 18.08$$

$$\text{MSW} = \frac{\text{SSW}}{N - k} = \frac{116.16}{13} = 8.94$$

7. Find the value of the test statistic F using formula (13.14):

$$F = \frac{\text{MSB}}{\text{MSW}} = \frac{18.08}{8.94} = 2.02$$

8. Find the critical value for F:

$$F_{0.05}(2, 13) = 3.81$$

Since $F = 2.02 < 3.81$, we cannot reject H_0. Thus, we have no statistical evidence to suggest that the unemployment rates are different for the three sections of the United States. An ANOVA summary table for our calculations is given in table 13.11.

TABLE 13.11

ANOVA Summary Table for Application 13.5

Source of Variance	SS	df	MS	F
Between samples	36.16	2	18.08	2.02
Within samples	116.16	13	8.94	
Total	152.32	15		

As a final remark, note that the F test for ANOVA tells us whether we can reject the null hypothesis that the population means are equal. If the decision is to reject

$H_0: \mu_1 = \mu_2 = \mu_3 = \cdots = \mu_k$, ANOVA does not tell us where the differences are. To find which population means are different, further tests, called **post hoc tests,** are used. We will discuss one popular post hoc test in section 13.3.

EXERCISE SET 13.2

Basic Skills

In exercises 1–4, use the computational formulas to find the value for F and summarize your computations using an ANOVA summary table.

1. Readability scores from exercise 1 of exercise set 13.1.

Test A	50	49	54	57	58
Test B	48	57	61	55	60

2. Laboratory analysis results from exercise 4 of exercise set 13.1.

    Lab A: 58.6 60.7 59.4 59.6 60.5
    Lab B: 61.6 64.8 62.8 59.2 60.4
    Lab C: 60.7 55.6 57.3 55.2 60.2

3. Typing scores from exercise 5 of exercise set 13.1.

A	B	C	D
82	55	69	87
79	67	72	61
75	84	78	82
68	77	83	61
65	71	74	72

4. Basic skills test scores from exercise 6 of exercise set 13.1.

School A	School B	School C	School D
20	24	16	19
21	21	21	20
22	22	18	21
24	25	13	20

Further Applications

5. Three sections of elementary statistics were taught by different teachers. A common final examination was given. The test scores are given in the table.

Teacher A	Teacher B	Teacher C
75	90	17
91	80	81
83	50	55
45	93	70
82	53	61
75	87	43

Teacher A	Teacher B	Teacher C
68	76	89
62	58	73
47	82	73
95	98	93
38	78	58
79	64	81
	80	70
	33	
	79	

Assume that the populations of test scores are normally distributed with equal variances. By using the 0.05 level of significance, test to determine if there is a difference in the average grades given by the three instructors. Summarize your calculations using an ANOVA summary table.

6. An experiment was conducted to compare the effects of four different diets on reduction of serum cholesterol. Twenty-one men were randomly assigned to each diet group, and after three months, their serum cholesterol was taken. The results are contained in the accompanying table.

Diet I	Diet II	Diet III	Diet IV
205	245	185	265
265	280	225	265
225	265	250	285
240	240	200	235
245	220		250
240			230

By using the 0.05 level of significance, test to determine if there is a difference in the average serum cholesterol for the four diets. Summarize your calculations using an ANOVA summary table.

7. Test scores from exercise 3 of exercise set 13.1:

Test Scores

Method A	Method B	Method C	Method D	Method E
$n_1 = 13$	$n_2 = 13$	$n_3 = 13$	$n_4 = 13$	$n_5 = 13$
$\bar{x}_1 = 75$	$\bar{x}_2 = 76$	$\bar{x}_3 = 70$	$\bar{x}_4 = 74$	$\bar{x}_5 = 76$
$s_1 = 7.7$	$s_2 = 7.1$	$s_3 = 9.9$	$s_4 = 6$	$s_5 = 6.3$

Test using $\alpha = 0.01$ to determine if there is a difference among the five methods.

Going Beyond

8. Prove that SST $= \Sigma (x_{ij} - \overline{T})^2 = \Sigma x_{ij}^2 - T^2/N$.

9. Prove that SSB $= \Sigma n_i(\overline{C}_i - \overline{T})^2 = \Sigma (C_i^2/n_i) - T^2/N$.

10. Prove that SST $=$ SSB $+$ SSW.

11. Show that $s_b^2 =$ MSB for equal sample sizes.

12. Show that $s_w^2 =$ MSW for equal sample sizes.

13. Consider the following data:

Class 1	Class 2	Class 3
3	2	4
5	3	3
1	1	5

a. Find the value of the F statistic.

b. If each of the above observations is multiplied by 5, find the value of the F statistic for the transformed data.

c. If 4 is added to each observation above, find the value of the F statistic for the transformed data.

d. Transform each of the above observations by using the formula $y = 5x + 4$. Then find the value of the F statistic for the transformed data.

e. Generalize the results of parts a–d.

14. Suppose k samples of size n are randomly drawn from normal populations having equal variances and one-way ANOVA is used to determine if the k population means are different.

a. If a constant is added to each measurement in the k samples, what affect does this have on the value of the F statistic?

b. If each measurement in the k samples is multiplied by a nonzero constant, what affect does this have on the value of the F statistic?

15. Each value x_{ij} satisfies the following relation:

$$x_{ij} = T + (C_i - T) + (x_{ij} - C_i)$$

Use the data in exercise 13 and the given relation to do the following:

a. For each x_{ij}, express $x_{ij} - T$ as a sum of two quantities.

b. Show that $\Sigma (x_{ij} - T) = \Sigma (C_i - T) + \Sigma (x_{ij} - C_i)$.

c. Show that $\Sigma (x_{ij} - T)^2 = \Sigma (C_i - T)^2 + \Sigma (x_{ij} - C_i)^2$.

16. In an ANOVA problem involving samples of the same size, a $(1 - \alpha)100\%$ confidence interval for the difference between two population means μ_i and μ_j can be formed using MSW to estimate σ^2. In this case, $\sigma_{\overline{x}_i - \overline{x}_j} \approx \sqrt{(MSW)/n} + (MSW/n)$. The limits for the confidence interval become

$$\overline{x}_i - \overline{x}_j \pm t_{\alpha/2} \sqrt{\frac{2MSW}{n}}$$

where df $= 2n - 2$. Use this result on the data in exercise 2 to find a 95% confidence interval for

a. $\mu_A - \mu_C$.

b. $\mu_B - \mu_C$.

c. Can you be 95% confident of both results simultaneously? Explain.

SECTION 13.3   *Follow-Up Procedures for a Significant F*

For the gasoline mileage illustration of section 13.1, we conclude that the Citation, the LeBaron, and the Reliant have different mean gas mileage ratings. We did not determine where the differences in mean gasoline mileages exist among the three automobile styles. To investigate further we might use three t tests, each at the 0.05 significance level. The trouble with doing this is that we have three opportunities for making a Type I error. If the three tests were independent (in fact, they are not), the probability of making at least one Type I error would be

$$1 - P(\text{no Type I error}) = 1 - 0.95^3$$
$$\approx 1 - 0.86 = 0.14$$

That is, the actual error rate for the experiment is somewhere between 0.05 and 0.14, not the chosen error rate of 0.05.

There are many procedures designed to test for differences among population means following the rejection of a null hypothesis of equal population means that attempt to control the overall error rate for the experiment. These procedures are commonly known as **multiple-comparison procedures** or **post hoc procedures.** One such is **Bonferroni's procedure.** It is one of the easiest procedures to apply. In addition, it requires only the use of the t distributions and is applicable to ANOVA tests involving unequal sample sizes. This procedure attempts to control the overall error rate for the experiment by reducing the level of significance for each comparison in means that is conducted following a significant F test. Bonferroni's procedure can be applied to either hypothesis tests of pairwise differences in population means or to constructing confidence intervals for the difference in population means. We shall first examine Bonferroni's procedure applied to hypothesis tests of differences between population means.

Bonferroni's Procedure Applied to Hypothesis Tests of Pairwise Differences between Population Means

Bonferroni's procedure is based on the same assumptions as ANOVA: the sampled populations are normal and possess a common variance. If there are k treatments (samples), then there are $m = \binom{k}{2} = k(k-1)/2$ **pairwise comparisons** of means to be performed. The error rate for each comparison is set to α/m, where α is the overall level of significance level for the experiment and m is the number of comparisons to be made.

Recall from section 11.5 that the test statistic for comparing two population means to determine if $\mu_i = \mu_j$ is:

$$t = \frac{|\bar{x}_i - \bar{x}_j|}{s_p \sqrt{(1/n_i) + (1/n_j)}}$$

If we multiply both sides of this expression by $s_p \sqrt{(1/n_i) + (1/n_j)}$, we get the following expression:

$$|\bar{x}_i - \bar{x}_j| = t s_p \sqrt{\frac{1}{n_i} + \frac{1}{n_j}}$$

And if on the right-hand side of this expression we replace s_p by $\sqrt{MSW}$ to estimate the common population standard deviation and t by the critical value, $t_{\alpha/(2m)}$ (for df $=$ df$_w$), then the resulting expression is called the **critical difference** (CD).

When Bonferroni's procedure is used, the critical difference that a positive difference of sample means, $|\bar{x}_i - \bar{x}_j|$, must exceed to be declared significant is given by $t_{\alpha/(2m)} \sqrt{MSW[(1/n_i) + (1/n_j)]}$, where α is the overall significance level required and m is the number of pairwise comparisons of means to be performed.

Bonferroni's Critical Difference

$$CD = t_{\alpha/(2m)} \sqrt{MSW\left(\frac{1}{n_i} + \frac{1}{n_j}\right)}$$

$$\text{where } m = \binom{k}{2}$$

(13.15)

If all samples are of the same size, then the expression for CD in formula (13.15) can be simplified to the following formula:

$$CD = t_{\alpha/(2m)} \sqrt{\frac{2MSW}{n}} \tag{13.16}$$

where $m = \binom{k}{2}$ and n is the common sample size. The values of $t_{\alpha/(2m)}$ in formula (13.16) can be found by using MINITAB or table 4 in appendix B.

EXAMPLE 13.8

If ANOVA involved $k = 5$ treatments, then there would be $m = \binom{5}{2} = \frac{5 \cdot 4}{2} = 10$ pairs of means to be compared.

EXAMPLE 13.9

For the gas mileage data of application 13.1, the critical difference CD can be found using formula (13.16). Note that MSW $= 1$ and $n = 5$. Since $k = 3$, the number of pairwise comparisons of means is $m = \binom{3}{2} = \frac{3 \cdot 2}{2} = 3$. The significance level for each of the three pairwise comparisons must be $\alpha/(2m)$ in order that the overall significance level of α for the experiment isn't exceeded. Since

$$\frac{\alpha}{2m} = \frac{0.05}{2 \cdot 3} = \frac{0.05}{6} = 0.0083$$

the critical difference that a pair of means must meet or exceed in order to be declared significant is given by

$$CD = t_{0.008} \sqrt{\frac{2 \cdot 1}{5}}$$

for df $= 12$. Unless we round the value 0.008 to 0.01, we can't use the t table given in the text to determine the critical difference. At this point, if we want a more exact tabled t value than $t_{0.01}$ (for df $= 12$), we must use table 4 in appendix B to determine the critical value $t_{0.0083}$ for df $= 12$ or use a statistical package, such as MINITAB. Computer display 13.3 contains the commands and corresponding output for determining the critical value $t_{0.0083}$ for df $= 12$.

Computer Display 13.3

```
MTB > INVCDF .0083;
SUBC> T 12.
   0.008  -2.7816
```

Since the t distributions are symmetrical about their mean of 0, the t critical value for df $= 12$ is $t_{0.0083} = 2.7816 \approx 2.78$.

Table 4 in appendix B provides the t critical values for each of m comparisons with an overall significance level of $\alpha = 0.05$. To determine the critical value for an overall significance level of 0.05 using table 4 in appendix B, we locate the column containing the number of comparisons m along the top of the table and the row containing the number of degrees of freedom along the left side of the table. Where this row and column intersect is the critical value. For $m = 3$ and df $= 12$, we determine that $t_{0.0083} = 2.78$. Note that to two decimal places, this agrees with the value supplied by MINITAB.

Thus, in order for two population means to be declared different, the positive difference in the two corresponding sample means must meet or exceed the critical difference of

$$CD = t_{0.0083} \sqrt{\frac{2 \cdot 1}{5}} = 2.78 \sqrt{\frac{2}{5}} = 1.7582$$

The three sample means are $\bar{x}_1 = 24$, $\bar{x}_2 = 31$, and $\bar{x}_3 = 26$. The three positive pairwise differences are

$$\text{LeBaron–Citation:} \quad \bar{x}_3 - \bar{x}_1 = 31 - 24 = 7$$
$$\text{LeBaron–Reliant:} \quad \bar{x}_3 - \bar{x}_2 = 31 - 26 = 5$$
$$\text{Reliant–Citation:} \quad \bar{x}_2 - \bar{x}_1 = 26 - 24 = 2$$

Since all three of the positive differences in sample means are greater than the critical difference $CD = 1.7582$, we can conclude that any two models differ significantly in average gasoline mileages. And with these three tests, we can be confident that the probability of committing at least one Type I error for the three comparisons is at most 0.05.

APPLICATION 13.6

Now let's see how to use Bonferroni's procedure for unequal sample sizes. An appliance company is considering the purchase of a large quantity of paint from one of five paint companies with nearly equal bid prices. Since drying time (in minutes) is a critical factor for the appliance company, it was decided to paint 35 appliances, 7 appliances with each of the five different brands of paint. The sample sizes and mean drying times are given in table 13.12.

TABLE 13.12

Data for Application 13.6

Company	Sample Size	Mean (in minutes)
A	7	29.0
B	5	25.8
C	6	25.7
D	6	26.8
E	7	31.0

The sample sizes are unequal because it was discovered too late that 4 appliances had been inadequately prepared for painting.

By using one-way ANOVA it was determined that the mean drying times for the paint from the five paint companies were significantly different ($\alpha = 0.05$). The following is the ANOVA summary table for the analysis:

Source of Variance	SS	df	MS	F
Between samples	134.52	4	33.63	4.03
Within samples	216.97	26	8.34	
Total	351.48	30		

Use Bonferroni's procedure and $\alpha = 0.05$ to test for pairwise differences between the average drying times for paint from the five paint companies.

Solution: Since $k = 5$, there are $m = \binom{5}{2} = 10$ pairwise comparisons to be made. To organize our results, we shall order the sample means from largest to smallest and construct the following table of positive differences in sample means:

		Sample				
		E 31	A 29	D 26.8	B 25.8	C 25.7
E	31		2	4.2	5.2	5.3
A	29			2.2	3.2	3.7
D	26.8				1.0	1.1
B	25.8					0.1
C	25.7					

The critical difference for assessing the positive difference $|\bar{x}_i - \bar{x}_j|$ is

$$CD = t_{\alpha/(2m)} \sqrt{MSW\left(\frac{1}{n_i} + \frac{1}{n_j}\right)}$$

$$= t_{0.05/(2 \cdot 10)} \sqrt{MSW\left(\frac{1}{n_i} + \frac{1}{n_j}\right)}$$

$$= t_{0.0025} \sqrt{MSW\left(\frac{1}{n_i} + \frac{1}{n_j}\right)}$$

$$= (3.07)\sqrt{MSW} \sqrt{\frac{1}{n_i} + \frac{1}{n_j}}$$

$$= (3.07)\sqrt{8.34} \sqrt{\frac{1}{n_i} + \frac{1}{n_j}}$$

$$= (8.8659) \sqrt{\frac{1}{n_i} + \frac{1}{n_j}}$$

where df $= 26$.

The critical difference for comparing A versus E is

$$CD = (8.8659) \sqrt{\frac{1}{7} + \frac{1}{7}} = 4.7390$$

Similarly, the critical differences for the other comparisons are

C versus D}

$$CD = (8.8659) \sqrt{\frac{1}{6} + \frac{1}{6}} = 5.1187$$

A versus B
B versus E}

$$CD = (8.8659) \sqrt{\frac{1}{5} + \frac{1}{7}} = 5.1913$$

A versus C
A versus D
E versus C
E versus D}

$$CD = (8.8659) \sqrt{\frac{1}{7} + \frac{1}{6}} = 4.9325$$

B versus C
B versus D}

$$CD = (8.8659) \sqrt{\frac{1}{5} + \frac{1}{6}} = 5.3686$$

By using the above table of differences in sample means, we see in the cases E versus C and E versus B that the positive difference between sample means is greater than the critical difference. Thus, by using Bonferroni's procedure, we can detect differences in mean drying times between paints manufactured by companies E and C and companies E and B. Note that we can not conclude, for example, that the drying time for brand E paint is longer than the drying time for brand C paint. Because the tests are all two-tailed, we can only conclude, for example, that the drying time for brand E paint is different from the drying time for brand C paint. ■

A convenient summary diagram of the results of Bonferroni's procedure is a listing of the sample means from highest to lowest, with those means not significantly different being underscored. For the brands of paint in application 13.7, the summary diagram is

$$\text{E} \quad \text{A} \quad \text{D} \quad \underline{\text{B} \quad \text{C}}$$

Simultaneous Confidence Intervals for Pairwise Differences of Means

Pairwise comparisons following a significant one-way ANOVA can be conducted by using hypothesis tests or constructing confidence intervals. The preceding Bonferroni procedure used multiple hypothesis tests. The main disadvantage to using multiple hypothesis tests is that the tests are all two-tailed; one can not ascertain that one population mean is greater than another. This will not be the case with constructing simultaneous confidence intervals for differences in population means.

Bonferroni's procedure applied to pairwise hypothesis tests attempts to control the overall experimentwise error rate by performing each hypothesis tests of pairwise differences in population means at a reduced significance level of α/m for the m pairwise hypothesis tests. By doing this the probability of committing at least one Type I error for the m tests is at most α.

Bonferroni's procedure applied to constructing confidence intervals attempts to control the overall confidence level for the experiment by increasing the confidence level for each of the m confidence intervals. Recall that a confidence interval becomes wider as the confidence level increases. By using Bonferroni's procedure, if we want to be at least $(1 - \alpha)100\%$ confident concerning the m confidence intervals, we construct a $(1 - \alpha/m)100\%$ confidence interval for each of the m differences in population means.

Let's now illustrate how pairwise comparisons of treatment means can be made by constructing a set of simultaneous confidence intervals for the $m = \binom{k}{2}$ differences in population means. Each confidence interval will have a level of confidence equal to $(1 - \alpha/m)100\%$, and the overall confidence level for the experiment will be $(1 - \alpha)100\%$. The formula for constructing each $(1 - \alpha/m)100\%$ confidence interval is

> **Bonferroni's Simultaneous Confidence Interval Limits**
> $$\bar{x}_i - \bar{x}_j \pm \text{CD} \qquad (13.17)$$
> where CD is given by formula (13.15) or formula (13.16).

EXAMPLE 13.10

Let's return to the gas mileage data of application 13.1 and construct the three simultaneous $(1 - 0.05/3)100\% = 98.33\%$ confidence intervals for the differences in population means. With these three confidence intervals, we can be 95% confident that they all three will contain the true difference. Recall from example 13.9 that CD = 1.7582. By using formula (13.17), the three 98.33% confidence intervals are:

LeBaron–Citation:

$$\bar{x}_2 - \bar{x}_1 \pm CD$$
$$(31 - 24) \pm 1.7582$$
$$7 \pm 1.7582$$

or

$$(5.2418, 8.7582)$$

Notice that this interval does not contain zero. Hence, $\mu_2 - \mu_1 > 0$ and $\mu_2 > \mu_1$, and we can conclude that we can be 98.33% confident that the mean gasoline mileage for the LeBaron is greater than the mean gasoline mileage for the Citation.

LeBaron–Reliant:

$$\bar{x}_2 - \bar{x}_3 \pm CD$$
$$(31 - 26) \pm 1.7582$$
$$5 \pm 1.7582$$

or

$$(3.2418, 6.7582)$$

Notice that this interval does not contain zero. We can be 98.33% confident that the mean gasoline mileage for the LeBaron is greater than the mean gasoline mileage for the Reliant.

Reliant–Citation:

$$\bar{x}_3 - \bar{x}_1 \pm CD$$
$$(26 - 24) \pm 1.7582$$
$$2 \pm 1.7582$$

or

$$(0.2418, 3.7582)$$

Notice that this interval does not contain zero. We can be 98.33% confident that the mean gasoline mileage for the Reliant is greater than the mean gasoline mileage for the Citation.

The results arrived at by using confidence intervals are consistent with the results arrived at by using hypothesis tests. But with using confidence intervals we can detect the larger population mean for each comparison, if a difference exists. This is not the case with hypothesis tests.

Bonferroni's procedure is known to be conservative for most applications. This means that the overall significance level is usually smaller (in the case of hypothesis tests) or the overall confidence level is usually greater (in the case of confidence intervals). With confidence intervals, this means that each of the m individual intervals is somewhat wider than they need be in order to have an overall specified confidence level. Multiple comparison procedures that are less conservative and more exact tend to be more complicated and involve special tables.

There are many other multiple-comparison procedures used by researchers for testing pairwise differences following a significant F test using ANOVA. Six such tests are *Fisher's LSD test, Scheffé's test, Duncan's multiple range test, the Waller-Duncan k-ratio test,* the *Newman-Keuls test,* and *Tukey's HSD test.* All except Scheffé's test and Fisher's LSD test involve special tables and a deeper study of statistics beyond the scope of this text. We shall leave them to a more advanced course in statistics. Scheffé's test is introduced in exercises 7 and 8 and Fisher's LSD test is introduced in exercise 12 at the end of this section.

A Measure of Association

A significant *F* value for ANOVA indicates that there is an association between two variables, the treatment variable and the dependent variable. The treatment variable for the gasoline mileage data of application 13.1 is the type of car and the dependent variable is the gas-mileage rating. Since the *F* statistic was significant, we conclude that there is a relationship between type of car and gas mileage rating. The *F* statistic does not, however, indicate the **strength of association** between these two variables. Such information is important in evaluating the outcome of an experiment since it is possible to have a small or trivial association between two variables although the association is statistically significant because sufficiently large samples are involved. In section 10.3 we saw an illustration involving a new study guide for improving quantitative SAT scores in which a difference of 0.4 on the SAT was significant because a sample of 1,000,000 was involved. In this case the association is trivial, but statistically significant.

A useful estimator of the strength of association between the treatment variable and the dependent variable in a one-way ANOVA is **Hays' omega-square statistic, $\hat{\omega}^2$.** The $\hat{\omega}^2$ statistic is given by

> **Hays' Omega-Square Statistic**
>
> $$\hat{\omega}^2 = \frac{\text{SSB} - (k - 1)\text{MSW}}{\text{SST} + \text{MSW}}$$

(13.18)

where SSB is the sum of squares for between groups, SST is the total sum of squares, MSW is the mean square for within groups or error, and k is the number of samples. Hays' $\hat{\omega}^2$ statistic provides a **measure of association** between the treatment variable and the dependent variable.

EXAMPLE 13.11

For the gas mileage illustration, the value of $\hat{\omega}^2$ is

$$\hat{\omega}^2 = \frac{\text{SSB} - (k - 1)\text{MSW}}{\text{SST} + \text{MSW}} = \frac{130 - 2(1)}{142 + 1} = 0.90$$

We can conclude that the treatment variable, type of car, accounts for 90% of the variance in the dependent variable, gas mileage rating. Not only is the association between car type and gas mileage rating statistically significant, but it is also very strong.

A P P L I C A T I O N 13.7

Calculate the value of Hays' omega-square statistic for the data given in application 13.6.

Solution: The value of Hays' omega-square statistic is

$$\hat{\omega}^2 = \frac{\text{SSB} - (k - 1)\text{MSW}}{\text{SST} + \text{MSW}} = \frac{134.52 - (5 - 1)(8.34)}{351.48 + 8.34} = \frac{101.16}{359.82} = 0.28$$

We can conclude that type of paint accounts for 28% of the variance in the drying times of the paints. Although there is a relationship between type of paint and drying time, it cannot be considered strong. ▪

EXERCISE SET 13.3

Further Applications

For exercises 1–6 assume that the assumptions for ANOVA hold.

1. Four diets were compared for controlling blood sugar in diabetic patients. Nine patients were randomly assigned to each diet. The data in the accompanying table represent the blood glucose readings of the patients following two weeks on the special diets.

A	B	C	D
140	160	200	200
120	110	220	180
80	110	220	180
120	130	190	170
130	150	150	130
120	100	210	210
120	140	230	150
120	100	190	170
120	130	180	170

a. By using one-way ANOVA and $\alpha = 0.05$, determine if the four diets produce different mean effects on blood glucose readings for diabetic patients.
b. For a significant F test, use Bonferroni's procedure and multiple hypothesis tests to test for pairwise differences. Use an overall significance level of $\alpha = 0.05$.
c. For a significant F test, use Bonferroni's procedure and simultaneous confidence intervals to test for pairwise differences. Use an overall significance level of $\alpha = 0.05$.
d. For a significant F statistic calculate the value of Hays' $\hat{\omega}^2$ statistic and interpret the results.

2. Four different assessments of acidity of high-sulfur coal at a power plant produced the pH data readings listed in the table.

	Assessment			
	I	II	III	IV
n	10	10	10	10
$\bar{x}$	6.6	6.9	3.0	5.3
s	0.38	0.50	0.43	0.47

a. By using one-way ANOVA and $\alpha = 0.05$, determine if there are differences in the mean pH values for the different assessments.
b. For a significant F test, use Bonferroni's procedure and multiple hypothesis tests to test for pairwise differences. Use an overall significance level of $\alpha = 0.05$.
c. For a significant F test, use Bonferroni's procedure and simultaneous confidence intervals to test for pairwise differences. Use an overall significance level of $\alpha = 0.05$.
d. For a significant F statistic, calculate the value of Hays' $\hat{\omega}^2$ statistic and interpret the results.

3. A large oil-drilling firm developed a new high-speed drill bit in an attempt to reduce drilling costs. The new high-speed drill bit, called the DB3, is believed to be superior to the two fastest drill bits known, the DB1 and the DB2. In an experiment to test the new drill bit, four drilling sites were randomly assigned to use each bit. The accompanying data indicate the rate of penetration (in feet per hour) after drilling 2000 feet at each site.

DB3	DB1	DB2
37.4	28.0	16.9
32.3	31.9	31.1
39.8	28.8	25.5
36.5	32.3	18.4

a. By using ANOVA and $\alpha = 0.05$, determine if the mean rates of penetration differ for the three drill bits.
b. For a significant F test, use Bonferroni's procedure and multiple hypothesis tests to test for pairwise differences. Use an overall significance level of $\alpha = 0.05$.
c. For a significant F test, use Bonferroni's procedure and simultaneous confidence intervals to test for pairwise differences. Use an overall significance level of $\alpha = 0.05$.
d. For a significant F statistic calculate the value of Hays' $\hat{\omega}^2$ statistic and interpret the results.

4. A national engineering society is interested in comparing the starting hourly rates of engineering graduates of three

large universities, A, B, and C. Independent random samples of six engineering graduates of each university were selected for the study. The table shows the starting hourly salary (in dollars) for each graduate.

University

A	B	C
11.25	12.50	11.75
11.25	13.05	12.00
12.35	13.12	10.85
12.25	13.35	11.61
12.00	12.55	12.10
11.85	12.60	12.15

a. By using ANOVA and $\alpha = 0.05$, determine if there is a difference among the mean starting salaries of engineering graduates at the three universities.
b. For a significant F test, use Bonferroni's procedure and multiple hypothesis tests to test for pairwise differences. Use an overall significance level of $\alpha = 0.05$.
c. For a significant F test, use Bonferroni's procedure and simultaneous confidence intervals to test for pairwise differences. Use an overall significance level of $\alpha = 0.05$.
d. For a significant F statistic calculate the value of Hays' $\hat{\omega}^2$ statistic and interpret the results.

5. The accompanying data represent yields per plot for tomatoes grown with three different fertilizers. Seven plots were grown with fertilizer A, eight with fertilizer B, and six with fertilizer C.

Fertilizer

A	B	C
31.0	40.0	41.4
31.8	39.6	42.5
28.3	35.3	36.0
29.7	33.0	36.4
28.0	35.7	36.8
27.1	33.7	38.1
32.6	37.4	
	38.6	

a. Use ANOVA and $\alpha = 0.05$ to determine if there are differences in tomato yields for the three fertilizers.
b. For a significant F test, use Bonferroni's procedure to test for pairwise differences.
c. For a significant F statistic, calculate the value of Hays' $\hat{\omega}^2$ statistic and interpret the results.

6. Five brands of cigarettes were tested for tar content. The

following table gives the tar contents (in milligrams) for five packs of cigarettes for each of the five brands:

Brand A	Brand B	Brand C	Brand D	Brand E
339	357	334	357	330
322	319	317	344	315
318	339	329	329	325
329	359	319	349	316
334	337	337	339	340

The following is a partial ANOVA summary table for the data:

Source of Variance	SS	df	MS
Between samples	1560.75	4	390.1875
Within samples	2560.75	20	128.0375
Total	4121.50	24	

a. Use ANOVA and $\alpha = 0.10$ to determine if there are differences in the mean tar contents for the five brands of cigarettes. [Since the value $F_{0.10}(4, 20)$ isn't contained in the F table, table 6a in appendix B, use $F_{0.10}(4, 20) = 2.25$.]
b. For a significant F test, use Bonferroni's procedure to test for pairwise differences.
c. For a significant F statistic calculate the value of Hays' $\hat{\omega}^2$ statistic and interpret the results.

Going Beyond

7. *Scheffé's procedure* is another multiple comparison procedure that also tends to be conservative. This method consists of testing for pairwise differences in population means $\mu_i - \mu_j$ in much the same way as we did with Bonferroni's procedure. The critical difference is given by

$$CD = \sqrt{(k-1)F_\alpha(df_1, df_2)} \sqrt{MSW\left(\frac{1}{n_i} + \frac{1}{n_j}\right)}$$

where MSW represents the mean square within-groups-variance estimate, n_i is the size of the ith sample, and k is the number of treatments (samples).

a. Use the Scheffé's procedure and an overall level of significance of $\alpha = 0.05$ to test for pairwise differences for the gasoline-mileage data in application 13.1.
b. Compare the critical differences (CD) for Bonferroni's t procedure and Scheffé's procedure.

8. Following Bonferroni's t procedure, we can construct a set of $(1 - \alpha)100\%$ simultaneous confidence intervals using Scheffé's procedure for pairwise differences $\mu_i - \mu_j$ by using the following expression:

$$\bar{x}_i - \bar{x}_j \pm CD$$

where CD is given by the expression in exercise 7. With this procedure, the probability that all m confidence intervals contain the differences in population means is at least $1 - \alpha$.

a. Use this procedure to construct three 95% simultaneous confidence intervals for the gasoline-mileage data in application 13.1. The data are as follows:

Citation	LeBaron	Reliant
23	30	27
24	31	27
24	32	26
24	32	26
25	30	24

b. Compare the lengths of these intervals to the corresponding lengths of those found in example 13.10. Which method tends to be the most conservative?

9. For the data given in exercise 6 on the five brands of cigarettes that were tested for tar content, use Scheffé's procedure and an overall significance level of $\alpha = 0.10$ to test for pairwise differences in the population mean tar contents for the five brands of cigarettes.

10. Refer to the data in exercise 6.
   a. Test for pairwise differences using Bonferroni's procedure and an overall significance level of $\alpha = 0.10$.
   b. Use Hays' omega-square statistic to estimate the strength of association between brand of cigarettes and tar content.

11. Refer to the data in exercise 6.
   a. Construct simultaneous 90% confidence intervals using Bonferroni's t procedure for the pairwise comparisons of the population mean tar contents for the five brands of cigarettes.
   b. Construct simultaneous 90% confidence intervals using Scheffé's procedure for the pairwise comparisons of the population mean tar contents for the five brands of cigarettes.
   c. Compare the lengths of the corresponding confidence intervals for the two procedures. For which procedure are the intervals shorter? Explain.

12. Fisher's restricted least significant difference (LSD) test does not attempt to control the overall significance level for the experiment. The critical difference is given by the following formula:

$$CD = t_{\alpha/2} \sqrt{MSW \left(\frac{1}{n_i} + \frac{1}{n_j} \right)}$$

where α is the significance level chosen for the experiment, n_i is the size of the ith sample, and $df = df_w$.

a. Use Fisher's procedure and an overall level of significance of $\alpha = 0.05$ to test for pairwise differences for the gasoline-mileage data in application 13.1.
b. Compare the critical differences (CD) for Fisher's LSD procedure, Bonferroni's t procedure, and Scheffé's procedure.
c. Of the three procedures, which procedure appears to be most conservative? Least conservative?

13. Show that Hays' $\hat{\omega}^2$ statistic can be written as

$$\hat{\omega}^2 = \frac{(k - 1)(F - 1)}{(k - 1)(F - 1) + nk}$$

14. Another useful strength-of-association statistic is eta-square ($\hat{\eta}^2$). It is defined by

$$\hat{\eta}^2 = \frac{SSB}{SST}$$

It can be shown that in a one-way ANOVA context with k treatment groups, the square of the simple correlation coefficient between each measurement and its group mean is equal to $\hat{\eta}^2$. Consider the following three sets of data:

A:	1	5	6	7	8	9
B:	9	10	11	18		
C:	3	4	5	7	8	9

a. For this data, determine the square of the simple correlation coefficient between each measurement and its group mean.
b. Determine SSB/SST for the three groups and show that $\hat{\eta}^2 = SSB/SST$.
c. By using the three groups of data, show that $F = (df_2/df_1)[\hat{\eta}^2/(1 - \hat{\eta}^2)]$.
d. Plot a scatter diagram of the data against their group means. Verify that the means of the three groups lie on a straight line.

15. Use the eta-square statistic to estimate the strength of association for the gasoline mileage data of example 13.1. Compare this estimate with the value of the omega-square statistic.

16. Can a value of the omega-square statistic be negative? (See exercise 13.)

In section 11.1 we made the distinction between independent samples of data and dependent samples of data. In experiments to compare two population means, samples are either independent or dependent (paired). Where paired data are used, an attempt is made by the experimenter to control some unwanted or "nuisance" variable. In both cases data are collected to compare the mean of one treatment with the mean of another treatment. How the measurements are recorded is different, and this difference depends on whether we use independent or dependent samples. These two different methods of analyzing the data can have a profound effect on whether the null hypothesis is rejected or not.

When more than two dependent samples of data are involved, we say the data are *blocked,* rather than paired. The blocks play the same role in experiments involving more than two samples as paired data do in experiments involving two samples. Measurements within a given **block** are relatively homogeneous with respect to some condition. The basic design of an experiment involving blocks is called a **block design.** If the treatments are applied randomly within each block, as is the case with example 13.12, the experimental design is called a **randomized block design.**

EXAMPLE 13.12

An experiment was conducted to determine if the mean numbers of out-patient surgeries performed at three hospitals (Memorial, Sacred Heart, and Mercy) during a week are different. The accompanying data show the number of outpatient surgeries performed by the three hospitals during a particular week.

| | Number of Surgeries Performed | | | |
Day	Memorial	Sacred Heart	Mercy	
Monday	19	25	25	Block 1
Tuesday	19	23	23	Block 2
Wednesday	18	22	23	Block 3
Thursday	14	21	13	Block 4
Friday	12	22	14	Block 5
Sample means	16.4	22.6	19.6	

In this experiment the blocking factor (variable) is day of the week. The measurements within each block are dependent, since they are related by the time factor. All treatments (hospital variable) are applied exactly once in each block. This experiment uses a block design.

The sample means suggest that the mean weekly numbers of surgeries performed at the three hospitals in example 13.12 differ. Let's use MINITAB to analyze the data by using one-way ANOVA studied in the section 13.2. Computer display 13.4 contains the commands used and part of the corresponding output.

Computer Display 13.4

```
MTB > READ C1 C2 C3
DATA> 19 25 25
DATA> 19 23 23
DATA> 18 22 23
DATA> 14 21 13
DATA> 12 22 14
DATA> END
   5 ROWS READ
MTB > AOVONEWAY C1 C2 C3

ANALYSIS OF VARIANCE
SOURCE    DF      SS      MS      F       P
FACTOR     2     96.1    48.1   3.25   0.075
ERROR     12    177.6    14.8
TOTAL     14    273.7
```

At the 0.05 level of significance, we fail to reject the null hypothesis of equal population means.

Recall that one-way ANOVA involves partitioning the total sum of squares into two components: sum of squares for between samples and sum of squares for within samples. The sum of squares for between samples reflects the variation of the sample means about the grand mean; as such, it is often called **sum of squares for treatment** and is written as SSTr. The sum of squares for within samples reflects the variation of the individual sample measurements about their sample mean and is often called the **sum of squares for error** and is written as SSE.

$$SST = SSTr + SSE$$

The F statistic is used as the test statistic to test the null hypothesis. For equal sample sizes, the F statistic is defined as

$$F = \frac{SSTr/(k - 1)}{SSE/k(n - 1)}$$

The smaller the value of SSE, the larger the value of F and the greater the chances of rejecting the null hypothesis.

When a blocking factor is used, ANOVA involves partitioning the total sum of squares into three components: sum of squares for treatments (SSTr), sum of squares for blocks (SSBl), and sum of squares for error (SSE).

$$SST = SSTr + SSBl + SSE$$

Since this analysis involves two factors, the sum of squares for between samples is called sum of squares for treatments; that is, SSB = SSTr. The **two-way analysis of variance** is called **two-way ANOVA** or **two-factor ANOVA**. The purpose of using a blocking factor is to partition SSE (for a one-way ANOVA) into two new components: SSBl and SSE. By using a two-way ANOVA, rather than a one-way ANOVA, the error term is smaller and thus increases the likelihood of rejecting the null hypothesis using the F statistic (see fig. 13.3).

FIGURE 13.3

Partitioning SST for one-way ANOVA and two-way ANOVA using a blocking factor

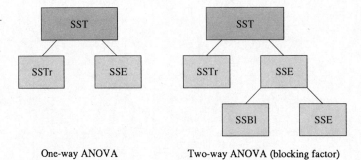

One-way ANOVA   Two-way ANOVA (blocking factor)

For equal sample sizes, we learned in section 13.2 that SSTr is found by using the following formula:

$$SSTr = \Sigma \frac{C_i^2}{n} - \frac{T^2}{N} \tag{13.19}$$

The sum of squares for treatment SSTr has $(k - 1)$ degrees of freedom associated with it. The formula for SSBl is found in an analogous way; the row totals are used instead of the column totals:

$$SSBl = \Sigma \frac{R_j^2}{k} - \frac{T^2}{N} \tag{13.20}$$

The sum of squares for blocks has $(n - 1)$ degrees of freedom associated with it. Since SST = SSTr + SSBl + SSE, the sum of squares for error is found by subtraction:

$$SSE = SST - SSTr - SSBl \tag{13.21}$$

The sum of squares for error SSE has $(k - 1)(n - 1)$ degrees of freedom associated with it. A mean square is found by dividing a sum of squares by its degrees of freedom. Table 13.13 displays the two-way ANOVA summary table including the **mean square for treatments** (MSTr) and the **mean square for blocks** (MSBl).

TABLE 13.13

Summary Table for Two-way ANOVA Involving Randomized Blocks

Source of Variation	Degrees of Freedom	Sum of Squares	Mean square	F
Treatments	$k - 1$	SSTr	$MSTr = \dfrac{SSTr}{k - 1}$	$\dfrac{MSTr}{MSE}$
Blocks	$n - 1$	SSBl	$MSBl = \dfrac{SSBl}{n - 1}$	$\dfrac{MSBl}{MSE}$
Error	$(k - 1)(n - 1)$	SSE	$MSE = \dfrac{SSE}{(k - 1)(n - 1)}$	
Total	$kn - 1$	SST		

APPLICATION 13.8

Analyze the hospital data by using $\alpha = 0.05$ and two-way ANOVA to determine if the differences among the three hospital means are significant.

Number of Surgeries Performed

Day	Memorial	Sacred Heart	Mercy
Monday	19	25	25
Tuesday	19	23	23
Wednesday	18	22	23
Thursday	14	21	13
Friday	12	22	14

Solution: We first find the row totals (denoted by R) and column totals (denoted by C):

Number of Surgeries Performed

Day	Memorial	Sacred Heart	Mercy	R
Monday	19	25	25	69
Tuesday	19	23	23	65
Wednesday	18	22	23	63
Thursday	14	21	13	48
Friday	12	22	14	48
C	82	113	98	293

Find SST by using formula (13.8):

$$SST = \Sigma x_{ij}^2 - \frac{T^2}{N}$$

$$= 19^2 + 25^2 + 25^2 + \cdots + 14^2 - \frac{293^2}{15}$$

$$= 5997 - 5723.27 = 273.73$$

Find SSTr by using formula (13.19):

$$SSTr = \Sigma \frac{C_i^2}{n} - \frac{T^2}{N}$$

$$= \frac{82^2 + 113^2 + 98^2}{5} - \frac{293^2}{15}$$

$$= \frac{29,097}{5} - 5723.27 = 96.13$$

Find SSBl by using formula (13.20):

$$SSBl = \Sigma \frac{R_j^2}{k} - \frac{T^2}{N}$$

$$= \frac{69^2 + 65^2 + 63^2 + 48^2 + 48^2}{3} - \frac{293^2}{15}$$

$$= \frac{17,563}{3} - 5723.27 = 131.07$$

Find SSE by using formula (13.21):

$$SSE = SST - SSTr - SSBl$$
$$= 273.73 - 96.13 - 131.07 = 46.53$$

Find degrees of freedom using table 13.13:

$$df_{Tr} = k - 1 = 3 - 1 = 2$$
$$df_{Bl} = n - 1 = 5 - 1 = 4$$
$$df_E = (k - 1)(n - 1) = (2)(4) = 8$$

Find the mean squares using table 13.13:

$$MSTr = \frac{SSTr}{k - 1} = \frac{96.13}{2} = 48.07$$

$$MSBl = \frac{SSBl}{n - 1} = \frac{131.07}{4} = 32.77$$

$$MSE = \frac{SSE}{(k - 1)(n - 1)} = \frac{46.53}{8} = 5.82$$

The statistical hypotheses are

Hospital factor:

$$H_0: \mu_{Mem} = \mu_{SH} = \mu_M.$$
H_1: At least two population means are different.

Day factor:

$$H_0: \mu_{mon} = \mu_{tues} = \mu_{wed} = \mu_{thurs} = \mu_{fri}.$$
H_1: At least two block means are different.

The value of the F statistic to test for treatments (difference in hospital means) is

$$F = \frac{MSTr}{MSE} = \frac{48.07}{5.82} = 8.26$$

The critical value of F is $F_{0.05}(2, 8) = 4.46$. Since $8.26 > 4.46$, we can conclude that at least two hospitals differ in average number of surgeries per week.

It should be noted that the blocking factor can be examined like any other factor in ANOVA. The value of the F statistic to test for blocks (days) is

$$F = \frac{MSBl}{MSE} = \frac{32.77}{5.82} = 5.63$$

The critical value is $F_{0.05}(4, 8) = 3.84$. Since $5.63 > 3.84$, we can conclude that the average number of surgeries per day differs for at least two days. ■

MINITAB can be used to analyze the data in application 13.8. Computer display 13.5 contains the commands used and a partial printout of the results.

Computer Display 13.5

```
MTB > READ CUTS C1 DAY C2 HOSPITAL C3
DATA> 19 1 1
DATA> 25 1 2
DATA> 25 1 3
DATA> 19 2 1
DATA> 23 2 2
DATA> 23 2 3
DATA> 18 3 1
DATA> 22 3 2
DATA> 23 3 3
DATA> 14 4 1
DATA> 21 4 2
DATA> 13 4 3
DATA> 12 5 1
DATA> 22 5 2
DATA> 14 5 3
DATA> END
   15 ROWS READ

MTB > NAME C1 'CUTS' C2 'DAYS' C3 'HOSPITAL'
MTB > TWOWAY C1 C2 C3

ANALYSIS OF VARIANCE CUTS

SOURCE        DF      SS       MS
DAY            4    131.07    32.77
HOSPITAL       2     96.13    48.07
ERROR          8     46.53     5.82
TOTAL         14    273.73
```

The data are entered by identifying the treatment and block numbers. For the data entered first, "19 1 1" means 19 surgeries for day 1 (Monday) and hospital 1 (Memorial Hospital); for the data entered second, "25 1 2" means 25 surgeries for day 1 (Monday) in hospital 2 (Sacred Heart Hospital), and so on.

Notice that the mean-square-for-error term in the one-way ANOVA is 14.8, and that the mean-square-for-error term in the two-way ANOVA is 5.82. Also note that for the one-way ANOVA, SSE = 177.6. For the two-way ANOVA, SSBl + SSE = 131.07 + 46.53 = 177.6, the value of SSE in the one-way ANOVA.

Bonferroni's Procedure
for Detecting Pairwise
Differences

The following formulas are used to determine the critical difference when using Bonferroni's procedure to detect differences between population means following a significant *F* test in a randomized block design.

Critical Difference for Treatment Factor

$$CD = t_{\alpha/(2m)} \sqrt{\frac{2MSE}{n}} \qquad (13.22)$$

where $m = \binom{k}{2}$, k = number of treatments, and n = number of blocks

Critical Difference for Blocking Factor

$$CD = t_{\alpha/(2m)} \sqrt{\frac{2MSE}{k}} \qquad (13.23)$$

where $m = \binom{n}{2}$, n = the number of blocks, and k = the number of treatments

EXAMPLE 13.13

For the data in application 13.8 let's use Bonferroni's procedure to test for differences in the hospital or treatment factor. Since there are $k = 3$ treatments, the number of comparisons to be made is $m = \binom{3}{2} = 3$. By using formula (13.22), the value of the critical difference is

$$CD = t_{\alpha/(2m)} \sqrt{\frac{2MSE}{n}}$$

$$= t_{0.05/(2 \cdot 3)} \sqrt{\frac{(2)(5.82)}{5}}$$

$$= t_{0.0083} \sqrt{2.328} \quad \text{(for df = 8)}$$

$$= (3.02) \sqrt{2.328} = 4.60785$$

The t critical value $t_{0.0083}$ (for df = 8) was found by using table 4 in appendix B.

The three $(1 - \alpha/m)100\% = (1 - 0.05/3)100\% = 98.33\%$ confidence intervals for the difference in population means are found by using the following formula:

$$\bar{x}_i - \bar{x}_j \pm CD \qquad (13.24)$$

Sacred Heart Versus Memorial	**Sacred Heart Versus Mercy**	**Mercy Versus Memorial**
$\bar{x}_i - \bar{x}_j \pm CD$	$\bar{x}_i - \bar{x}_j \pm CD$	$\bar{x}_i - \bar{x}_j \pm CD$
$(22.6 - 16.4) \pm 4.6079$	$(22.6 - 19.6) \pm 4.6079$	$(19.6 - 16.4) \pm 4.6079$
6.2 ± 4.6079	3 ± 4.6079	3.2 ± 4.6079
$(1.59, 10.81)$	$(-1.61, 7.61)$	$(-1.41, 7.81)$

The only confidence interval that does not contain zero is the interval for the difference in the mean number of surgeries for Memorial and Sacred Heart Hospitals. Thus, the mean number of surgeries at Sacred Heart Hospital is greater than the mean number of surgeries performed at Memorial Hospital.

Hays' Omega-Square Statistic

Hays' $\hat{\omega}^2$ statistic can be used to measure the strength of association for factors that are found to be significant in a randomized block design.

Strength of Blocking Factor

$$\hat{\omega}^2 = \frac{SSBl - (n - 1)MSE}{SST + MSE} \qquad (13.25)$$

Strength of Treatment Factor

$$\hat{\omega}^2 = \frac{\text{SSTr} - (k - 1)\text{MSE}}{\text{SST} + \text{MSE}}$$

(13.26)

EXAMPLE 13.14

Let's calculate the value of $\hat{\omega}^2$ for the hospital variable of application 13.8. We shall use formula (13.26) since the hospital variable serves as the treatment variable.

$$\hat{\omega}^2 = \frac{\text{SSTr} - (k - 1)\text{MSE}}{\text{SST} + \text{MSE}} = \frac{96.13 - (3 - 1)(5.82)}{273.73 + 5.82} = \frac{84.49}{279.55} = 30.22\%$$

Hence, 30.22% of the variance in the number of surgeries can be accounted for by the hospital variable.

There are two different ways to analyze two-factor experiments using analysis of variance. Which method is used depends on whether the two factors are independent or not. When the factors are related, two-way ANOVA involves an **interaction component,** called **sum of squares for interaction,** and the total sum of squares is partitioned into four components:

$$\text{SST} = \text{SS(Tr 1)} + \text{SS(Tr 2)} + \text{SS(Interaction)} + \text{SSE}$$

Three F tests are performed, one for the first factor (Tr 1), one for the second factor (Tr 2), and one for the interaction component. In the interaction component is significant, caution must be exercised in interpreting significant differences for the two main factors. A detailed examination of two-way ANOVA is carried out in section 13.5.

EXERCISE SET 13.4

Basic Skills

1. Consider the accompanying randomized block data.

Block	Treatments		
	1	2	3
Block 1	9	7	7
Block 2	5	6	9
Block 3	2	3	1
Block 4	6	8	7

Complete a two-way ANOVA summary table.

2. Consider the accompanying randomized block data.

Block	Treatments			
	1	2	3	4
Block 1	8	7	6	6
Block 2	3	4	3	2
Block 3	7	5	8	6

Complete a two-way ANOVA summary table.

Further Applications

3. Three laboratories, A, B, and C, are used by food manufacturing companies for making nutrition analyses of its products. The following data are the fat contents (in grams) of two tablespoons of three similar types of peanut butter:

Peanut Butter	Laboratory			
	A	B	C	D
Brand 1	16.6	17.7	16	16.3
Brand 2	16	15.5	15.6	15.9
Brand 3	16.4	16.3	15.9	16.2

Analyze the data by performing a two-way ANOVA; use $\alpha = 0.05$ for both tests. If the laboratory variable is significant, use an overall significance level of $\alpha = 0.05$ and Bonferroni's procedure to test for pairwise differences.

4. In an effort to expand its services, a regional transit authority conducted an experiment to determine which of four routes to take from an airport to the center of the business district of the city. The following data indicate the travel times (in minutes) along each of the four routes.

Day	Route 1	Route 2	Route 3	Route 4
Monday	20	22	22	24
Tuesday	23	24	26	26
Wednesday	22	25	27	25
Thursday	27	23	30	27
Friday	28	26	30	27

a. Analyze the data by performing a two-way ANOVA; use $\alpha = 0.05$ for both tests.
b. If the route variable is significant, use an overall significance level of $\alpha = 0.05$ and Bonferroni's procedure to test for pairwise differences.

5. An electronics manufacturing firm operates 24 hours a day, five days a week. Three 8-hour shifts are used, and the workers rotate shifts each week. A management team conducted a study to determine whether there is a difference in the mean number of 14-inch video monitors produced when employees work on the various shifts. A random sample of five workers is selected and the number of 14-inch video monitors they produce for each shift is recorded as here.

Employee	Monitors Produced Morning	Afternoon	Night
Jones	10	4	14
Miller	12	5	12
Phillips	7	3	9
Ross	9	8	7
Stevens	7	5	6

a. By using the 0.05 level of significance and two-way ANOVA, can we conclude there is a difference in the mean production by shift and in the mean production by worker?
b. If the shifts variable is significant, use an overall significance level of $\alpha = 0.05$ and Bonferroni's procedure to test for pairwise differences.
c. Calculate the value of $\hat{\omega}^2$ for both factors.

6. The manager of a large department store conducted an experiment to determine if there is a difference in the average weekly sales of three salespersons in the store's application department. The following data indicate sales (in $100) for seven consecutive weeks for the three employees.

Week	Salesperson A	Salesperson B	Salesperson C
1	27.6	28.7	26.4
2	31.2	29.3	30.3
3	28.8	28.4	28.0
4	30.6	29.8	28.7
5	30.0	31.0	32.3
6	28.4	29.9	29.6
7	30.9	29.5	31.1

a. By using the 0.05 level of significance and two-way ANOVA, can we conclude there is a difference in the mean sales for the three salespersons and in the mean sales by week?
b. If the salesperson variable is significant, use an overall significance level of $\alpha = 0.05$ and Bonferroni's procedure to test for pairwise differences.
c. Calculate the value of $\hat{\omega}^2$ for both factors.

7. Listed below are the average sentences (in years) given felons by three judicial processes and by crime.[45]

Crime	Judicial Process Trial by jury	Trial by judge	Guilty plea
Murder	28	21	14
Robbery	24	15	10
Rape	18	14	11
Aggravated assault	14	9	7
Burglary	10	5	6
Drug trafficking	8	10	5
Larceny	4	4	4

a. By using the 0.05 level of significance and two-way ANOVA, can we conclude that there is a difference in the mean sentence for the three judicial processes and in the mean numbers of crime?
b. If the judicial process variable is significant, use an overall significance level of $\alpha = 0.05$ and Bonferroni's procedure to test for pairwise differences.

8. An experiment was conducted to compare automobile gasoline mileages for four brands of gasoline, A, B, C, and D. Three automobiles of the same make and model used each of four brands of gasoline for the experiment. The data (in miles per gallon) are recorded in the accompanying table.

	Gas Brand			
Auto Type	A	B	C	D
I	18.8	20.1	20.4	19.2
II	20.3	21.2	21.0	20.8
III	19.2	20.6	19.9	20.9

a. Use two-way ANOVA and the 0.05 level of significance to determine if there is a difference in mean mileage per gallon for the four gasoline brands and a difference in mean mileages for the three automobiles.

b. If the gasoline variable is significant, use an overall significance level of $\alpha = 0.05$ and Bonferroni's procedure to test for pairwise differences.

9. Four different kinds of fertilizer, A, B, C, and D, are used to study the yield of soy beans. The soil is divided into twelve plots of the same size. The soil is divided into three blocks, each containing four homogeneous plots. Fertilizers were randomly assigned to the plots within each block. The yields (in pounds per acre) are as follows.

	Fertilizer			
Block	A	B	C	D
1	52.8	49.4	58.6	42.9
2	60.1	48.1	61.0	50.3
3	62.0	56.4	63.6	61.2

a. Conduct a two-way ANOVA using $\alpha = 0.05$ to determine if the fertilizers and blocks are effective.

b. For a significant F statistic for blocks, use $\alpha = 0.05$ to test if there are any differences of mean yields between blocks. Also calculate the value of Hays' $\hat{\omega}^2$ statistic and interpret the result.

c. For a significant F statistic for fertilizers, test for any pairwise differences in fertilizers by using $\alpha = 0.05$. Also calculate the value of Hays' $\hat{\omega}^2$ statistic and interpret the result.

10. Four different machines are under consideration for the assembly of a product. A study is conducted and six different operators are selected to operate the machines and are given initial instruction and practice on their use. The machines are to be compared in time (in seconds) to assemble a product. The results are as follows:

	Operator					
Machine	I	II	III	IV	V	VI
A	32.4	29.2	29.5	29.8	32.8	33.5
B	29.7	30.0	30.4	32.2	32.4	33.0
C	30.1	30.4	31.2	33.3	34.8	35.0
D	31.2	32.1	33.4	34.1	35.8	32.2

a. Use $\alpha = 0.05$ and a two-way ANOVA to test if the machines perform at the same mean rate of speed and the operators perform at the same mean rate of speed.

b. For a significant F statistic for machines, use $\alpha = 0.05$ to test for any pairwise differences in machines rates. Also calculate the value of Hays' $\hat{\omega}^2$ statistic and interpret the result.

c. For a significant F statistic for operators, use $\alpha = 0.05$ to test for any pairwise differences in operator rates. Also calculate the value of Hays' $\hat{\omega}^2$ statistic and interpret the result.

SECTION 13.5 Two-Way ANOVA: Factorial Designs

So far we have discussed the use of ANOVA for dealing with one-factor experiments, experiments involving two or more treatments. A *factor* is a an independent variable whose effect on the dependent variable (also called the **response variable**) is of interest to the experimenter. The **levels of a factor** are the values of the factor used in the experiment. In this section we are interested in examining two-factor experiments where each factor has two or more levels.

EXAMPLE 13.15

For the gas-mileage illustration of application 13.1, make of automobile is the factor whose effect on the dependent variable, mean gasoline-mileage ratings, was of interest. This factor has three levels: Citation, LeBaron, and Reliant. We have previously referred to these levels as treatments of the experiment.

EXAMPLE 13.16

Suppose we were interested in determining which method of teaching basic chemistry (discovery, expository, inductive, or deductive) is most effective in terms of achievement as measured by a standardized final examination. Method of teaching is the factor, and the factor has four levels: discovery, expository, inductive, and deductive. The dependent or response variable is the final examination score.

EXAMPLE 13.17

Suppose, in addition to make of automobile, we were also interested in studying the effect of four levels of octane for a particular brand of gasoline on the mean gasoline mileage ratings. Octane rating is a second factor having four levels: low, medium, high, and premium. The three makes of automobile and four types of gasoline provide for 12 treatment combinations, as the following layout suggests:

		Citation	LeBaron	Reliant
	Low			
	Medium			
Octane	High			
	Premium			

Type of Automobile

Suppose an experiment has two factors, *A* and *B*, so that *A* has *a* levels and *B* has *b* levels. If the experiment has a total of *ab* treatment combinations, then it is called an **a × b factorial experiment.** The number of treatments in the experiment is always equal to the product of the numbers of levels of the two factors.

The randomized block experiments studied in section 13.4 contained two factors, a treatment factor and a blocking factor, and are special cases of two-factor experiments. The blocking factor is usually not of experimental interest. Its main purpose in an experiment is to control some variable that has an influence on the dependent variable. By controlling this "nuisance" variable, we are in a better position to examine the treatment variable for differences in the response (dependent) variable.

The two factors in a randomized block design are independent in the sense that they do not interact with each other. That is, the treatment variable does not combine with the blocking variable to produce any effects. Each block is affected equally by the treatments. In general, however, the two variables in a two-factor experiment need not be independent; they can interact. Examples 13.18 and 13.19 explain the concept of interaction of two factors.

EXAMPLE 13.18

Suppose we are examining the effects of two kinds of medication, A and B, and two types of diet, I and II, on the recovery time of gall bladder surgery patients. If the recovery time for patients taking medication A is shortest for patients on diet I, but for patients taking medication B, the recovery time is shortest for patients on diet II, then medication and diet are said to interact. The effect of medication changes with the two levels of diet. Figure 13.4 contains four possible **treatment mean curve plots** for the 2 × 2 factorial design.

FIGURE 13.4

Treatment mean curve plots for the experiment in example 13.18

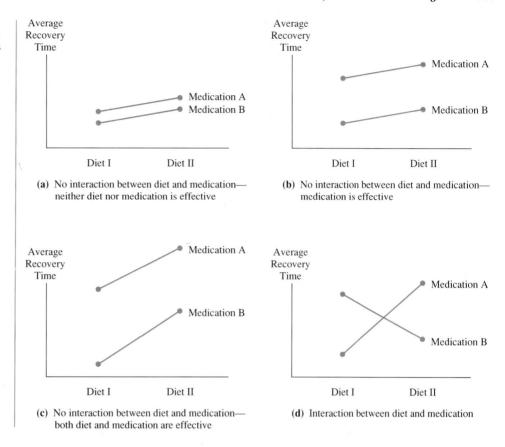

(a) No interaction between diet and medication—neither diet nor medication is effective

(b) No interaction between diet and medication—medication is effective

(c) No interaction between diet and medication—both diet and medication are effective

(d) Interaction between diet and medication

EXAMPLE 13.19

Suppose a marketing specialist for a leading candy manufacturer is concerned with studying the effect on its candy sales of package design (package size, color, lettering) and type of advertising (television, radio, billboard). A 3×3 factorial design is used for the study. For each of the nine treatment combinations, 20 retail stores are selected at random from appropriate regions of the country to participate in the study. The sales (in dollars) over a period of three months is computed for each treatment. If the sales from the effects of packaging change with the type of advertising, or if the sales from the type of advertising change with the type of packaging, then we have evidence of an interaction between the two factors. Figure 13.5 indicates a plot of means for which there is interaction between package design and type of advertising.

FIGURE 13.5

Interaction is present for the study of example 13.19

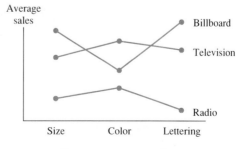

Treatment mean curve plots

For every two-way factorial experimental design involving factors A and B, there are three main questions that the researcher is interested in answering:

1. Is there a treatment effect on the response variable due to the interaction of factors A and B?
2. Is there a treatment effect on the response variable due to factor A?
3. Is there a treatment effect on the response variable due to factor B?

The statistical hypotheses are as follows:

Test for Interaction of Factors A and B:

H_0: Factors A and B do not interact to affect the response variable.
H_1: Factors A and B interact to affect the response variable.

Test for the Effects of Factor A:

H_0: There is no difference between the a mean levels of factor A.
H_1: At least two mean levels of factor A differ.

Test for the Effects of Factor B:

H_0: There is no difference between the b mean levels of factor B.
H_1: At least two mean levels of factor B differ.

Two-way ANOVA can be used to analyze the data resulting from a factorial design if each treatment has at least two measurements associated with it and the following assumptions are met:

1. The ab samples of n measurements are random and independent.
2. The ab samples come from normal populations.
3. The sampled populations have equal variances.

The two-way ANOVA technique involves partitioning the total sum of squares for the response variable y, denoted by SST, into four sum-of-squares components: the sum of squares for factor A, denoted by SSA, the sum of squares for factor B, denoted by SSB, the sum of squares for interaction, denoted by SSAB, and the sum of squares for error, denoted by SSE.

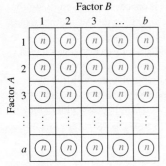

ab treatments with *n* measurements per treatment

FIGURE 13.6

The layout of an $a \times b$ factorial design with n replications per treatment

> **SS Relationship for Two-Way ANOVA**
>
> $$SST = SSA + SSB + SSAB + SSE \qquad (13.27)$$

Suppose factor A has a levels, factor B has b levels, and each of the ab treatment groups have n measurements associated with it (called **n replications**) as shown in figure 13.6.

A computer program, such as MINITAB, or the following calculation formulas can be used to determine the five sum of squares components in formula (13.27).

SS Formulas for Two-Way ANOVA

$$SST = \Sigma\, x_{ij}^2 - \frac{(\Sigma\, x_{ij})^2}{abn} \tag{13.28}$$

$$SSA = \frac{\Sigma\, A_i^2}{bn} - \frac{(\Sigma\, x_{ij})^2}{abn} \tag{13.29}$$

$$SSB = \frac{\Sigma\, B_j^2}{an} - \frac{(\Sigma\, x_{ij})^2}{abn} \tag{13.30}$$

$$SSAB = \frac{\Sigma\, (AB_{ij})^2}{n} - SSA - SSB - \frac{(\Sigma\, x_{ij})^2}{abn} \tag{13.31}$$

$$SSE = SST - SSA - SSB - SSAB \tag{13.32}$$

where

$a =$ Number of levels of factor A
$b =$ Number of levels of factor B
$n =$ Number of measurements (replicates) per treatment
$A_i =$ Total for level i of factor A ($i = 1, 2, \ldots, a$)
$B_j =$ Total for level j of factor B ($j = 1, 2, \ldots, b$)
$AB_{ij} =$ Total for treatment cell (i, j)

APPLICATION 13.9

The data shown in the accompanying table are for a 3×2 factorial experiment with two replications per treatment. Compute the five sum-of-squares components contained in formula (13.27).

		Factor A		
		Level a_1	Level a_2	Level a_3
Factor B	*Level b_1*	8	2	7
		9	1	9
	Level b_2	3	6	2
		4	4	3

Solution: We first compute the totals A_i, B_j, and AB_{ij}. These are contained in table 13.14.

TABLE 13.14

Sums for the 2×3 Layout of Example 13.13

		Factor A			
		a_1	a_2	a_3	B_j
Factor B	b_1	17	3	16	36
	b_2	7	10	5	22
	A_i	24	13	21	58

We use formula (13.28) to calculate the total sum of squares, SST. Note that $a = 3$, $b = 2$, $n = 2$, and $\Sigma\, x_{ij} = 58$.

$$SST = \Sigma x_{ij}^2 - \frac{(\Sigma x_{ij})^2}{abn}$$

$$= 8^2 + 9^2 + 2^2 + 1^2 + 7^2 + 9^2 + \cdots + 2^2 + 3^2 - \frac{58^2}{(3)(2)(2)}$$

$$= 370 - \frac{58^2}{12}$$

$$= 370 - 280.33 = 89.67$$

Note also that $(\Sigma x_{ij})^2/abn = 280.33$. We will use this result to determine SSA, SSB, and SSAB. We use formula (13.29) to compute the sum of squares for factor A, SSA.

$$SSA = \frac{\Sigma A_i^2}{bn} - \frac{(\Sigma x_{ij})^2}{abn}$$

$$= \frac{24^2 + 13^2 + 21^2}{(2)(2)} - 280.33$$

$$= \frac{1186}{4} - 280.33 = 16.17$$

We use formula (13.30) to compute the sum of squares for factor B, SSB.

$$SSB = \frac{\Sigma B_j^2}{an} - \frac{(\Sigma x_{ij})^2}{abn}$$

$$= \frac{36^2 + 22^2}{(3)(2)} - 280.33$$

$$= \frac{1780}{6} - 280.33$$

$$= 16.34$$

We use formula (13.31) to compute the sum of squares for interaction, SSAB.

$$SSAB = \frac{\Sigma (AB_{ij})^2}{n} - SSA - SSB - \frac{(\Sigma x_{ij})^2}{abn}$$

$$= \frac{17^2 + 7^2 + 3^2 + 10^2 + 16^2 + 5^2}{2} - 16.17 - 16.34 - 280.33$$

$$= \frac{728}{2} - 312.84 = 51.16$$

We use formula (13.32) to find the sum of squares for error, SSE.

$$SSE = SST - SSA - SSB - SSAB$$

$$= 89.67 - 16.17 - 16.34 - 51.16 = 6 \quad ■$$

Each of the sum-of-squares components in equation (13.27) has a number of degrees of freedom associated with it. Since the total number of measurements is abn, the number of degrees of freedom associated with SST is $(abn - 1)$.

Degrees of Freedom for SST

$$df_T = abn - 1 \tag{13.33}$$

The number of degrees associated with SSA is $(a - 1)$, since factor A has a levels.

> **Degrees of Freedom for SSA**
>
> $$\mathrm{df_A} = a - 1$$ (13.34)

And the number of degrees of freedom associated with factor B is $(b - 1)$, since factor B has b levels.

> **Degrees of Freedom for SSB**
>
> $$\mathrm{df_B} = b - 1$$ (13.35)

Since there are ab treatment groups and each group has $n - 1$ degrees of freedom associated with it, there are $ab(n - 1)$ degrees of freedom associated with the sum of squares for error.

> **Degrees of Freedom for SSE**
>
> $$\mathrm{df_E} = ab(n - 1)$$ (13.36)

The degrees of freedom for interaction can be found by subtraction:

$$
\begin{aligned}
\mathrm{df_{AB}} &= \mathrm{df_T} - \mathrm{df_A} - \mathrm{df_B} - \mathrm{df_E} \\
&= (abn - 1) - (a - 1) - (b - 1) - ab(n - 1) \\
&= abn - 1 - a + 1 - b + 1 - abn + ab \\
&= ab - a - b + 1 \\
&= (a - 1)(b - 1)
\end{aligned}
$$

Hence, the degrees of freedom associated with SSAB is

> **Degrees of Freedom for Interaction**
>
> $$\mathrm{df_{AB}} = (a - 1)(b - 1)$$ (13.37)

Mean squares are associated with each sum-of-squares component by dividing the sum of squares by its degrees of freedom:

> **Mean Squares for Two-way ANOVA**
>
> $$\mathrm{MSA} = \frac{\mathrm{SSA}}{a - 1}$$
>
> $$\mathrm{MSB} = \frac{\mathrm{SSB}}{b - 1}$$
>
> $$\mathrm{MSAB} = \frac{\mathrm{SSAB}}{(a - 1)(b - 1)}$$
>
> $$\mathrm{MSE} = \frac{\mathrm{SSE}}{ab(n - 1)}$$

F tests are then used to determine if factors A and B and the interaction of factors A and B have any effects on the response variable. The effects of factors A and B are called **main effects.** To test for interaction and main effects, the following F statistics and corresponding df's are used:

<table>
<tr><td colspan="3">F Statistics and Corresponding Degrees of Freedom for Two-way ANOVA</td></tr>
<tr><td>Treatment Effect</td><td>F statistic</td><td>df</td></tr>
<tr><td>A</td><td>$F = \dfrac{MSA}{MSE}$</td><td>$(a - 1,\ ab(n - 1))$</td></tr>
<tr><td>B</td><td>$F = \dfrac{MSB}{MSE}$</td><td>$(b - 1,\ ab(n - 1))$</td></tr>
<tr><td>AB</td><td>$F = \dfrac{MSAB}{MSE}$</td><td>$((a - 1)(b - 1),\ ab(n - 1))$</td></tr>
</table>

APPLICATION 13.10

Complete an ANOVA summary table for the data contained in application 13.91 and determine the critical values for the three F tests.

Solution: We compute mean squares for factor A, factor B, and the interaction of the factors A and B. By using formulas (13.34), (13.35), (13.36), and (13.37), we determine the following degrees of freedom:

$$df_A = a - 1 = 3 - 1 = 2$$
$$df_B = b - 1 = 2 - 1 = 1$$
$$df_{AB} = (a - 1)(b - 1) = (3 - 1)(2 - 1) = 2$$
$$df_E = ab(n - 1) = (3)(2)(2 - 1) = 6$$

The mean squares are found by dividing each sum-of-squares component by its corresponding degrees of freedom.

$$MSA = \frac{SSA}{a - 1} = \frac{16.17}{2} = 8.09$$

$$MSB = \frac{SSB}{b - 1} = \frac{16.34}{1} = 16.34$$

$$MSAB = \frac{SSAB}{(a - 1)(b - 1)} = \frac{51.16}{2} = 25.58$$

$$MSE = \frac{SSE}{ab(n - 1)} = \frac{6}{6} = 1$$

To test the main effect of factor A we use the following F ratio:

$$F = \frac{MSA}{MSE} = \frac{8.09}{1} = 8.09$$

To test the main effect of factor B we use the following F ratio:

$$F = \frac{MSB}{MSE} = \frac{16.34}{1} = 16.34$$

At the 0.05 level of significance, the critical values used to test for interaction and main effects are as shown in the table.

Effects	Critical Value
Interaction of A and B	$F_{0.05}(2, 6) = 5.14$
Factor A	$F_{0.05}(2, 6) = 5.14$
Factor B	$F_{0.05}(1, 6) = 5.99$

And to test for interaction between factors A and B we use the following F ra

$$F = \frac{MSAB}{MSE} = \frac{25.58}{1} = 25.58$$

These results are summarized in the following ANOVA summary table.

Source of Variance	SS	df	MS	F
A	16.17	2	8.09	8.09*
B	16.34	1	16.34	16.34*
AB	51.16	2	25.58	25.58*
Error	6	6	1	
Total	89.67	11		

*p-value < 0.05

Note that the ANOVA summary table of motivator 13 does not contain rows for error or total. This information can be obtained from the abbreviated table given (see exercise 16 at the end of the section). ▪

MINITAB can be used to determine the SS, df, and MS entries in the ANOVA summary table of application 13.10. Computer display 13.6 contains the commands used and corresponding output.

Computer Display 13.6

```
MTB > NAME C1 'Y' C2 'A' C3 'B'
MTB > SET 'Y'
DATA> 8 9 3 4
DATA> 2 1 6 4
DATA> 7 9 2 3
DATA> END
MTB > SET 'A'
DATA> (1:3)4
DATA> END
MTB > SET 'B'
DATA> 3(1:2)2
DATA> END
MTB > TWOWAY 'Y' 'B' 'A'

ANALYSIS OF VARIANCE Y

SOURCE          DF      SS      MS
B                1   16.33   16.33
A                2   16.17    8.08
INTERACTION      2   51.17   25.58
ERROR            6    6.00    1.00
TOTAL           11   89.67

MTB >
```

To enter the data for the A subscripts, the command (1:3)4 was used. This repeats each of the integers from 1 to 3 four times. The numbers entered into C2 with this command are 1 1 1 1 2 2 2 2 3 3 3 3. To enter the data for the B subscripts, the command 3(1:2)2 was used. The numbers entered into C3 with this command are 1 1 2 2 1 1 2 2 1 1 2 2. Notice that the data in the MINITAB output differ only slightly from the values we calculated. This is due to the increased accuracy of the computations performed by MINITAB.

When testing the null hypotheses for a two-way factorial design, the interaction effect is typically tested first. If the interaction is not significant, then the main effects of factors A and B are tested. If the interaction of factors A and B is significant, then Bonferroni's procedure can be used to compare pairs of treatment means of factor A for each level of factor B or of factor B for each level of factor A. If there is evidence of interaction between factors A and B and if one of the main factors is significant, it is usually not wise to test for differences between pairs of means for this factor, since the factors combine to affect the response variable.

APPLICATION 13.11   A study was conducted to examine the effects of two methods of teaching (discovery and expository) and three class periods (9:00 A.M., 11:00 A.M., and 1:00 P.M.) on achievement in general mathematics. A 2 × 3 factorial design was used for the study. All other variables that could influence the experiment, such as the teacher variable, were controlled as much as possible. Four students were randomly assigned to each of the six treatment combinations. Following a semester of study, all students were given a standardized achievement examination. The results of the examination are recorded in the accompanying table.

	Class Period		
Method	Period 1 9:00 A.M.	Period 3 11:00 A.M.	Period 5 1:00 A.M.
Expository	74	86	82
	77	78	89
	73	80	91
	80	88	86
Discovery	92	83	91
	83	95	84
	87	86	83
	90	92	90

a. Plot the means of the treatments to assist in your interpretation of the experiment.

b. Perform a statistical analysis of the experiment.

c. What assumptions must hold in order to use two-way ANOVA for the analysis?

Solution:

a. Figure 13.7 contains treatment mean curve plots for the 2 × 3 factorial experiment. The plots suggests that method and class period interact.

FIGURE 13.7

Treatment mean curve plots
for the data in application 13.1

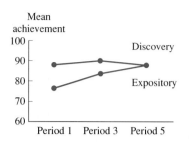

b. MINITAB is used to analyze the data. The commands used and the output are contained in computer display 13.7.

Computer Display 13.7

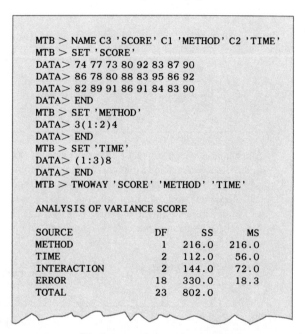

```
MTB > NAME C3 'SCORE' C1 'METHOD' C2 'TIME'
MTB > SET 'SCORE'
DATA> 74 77 73 80 92 83 87 90
DATA> 86 78 80 88 83 95 86 92
DATA> 82 89 91 86 91 84 83 90
DATA> END
MTB > SET 'METHOD'
DATA> 3(1:2)4
DATA> END
MTB > SET 'TIME'
DATA> (1:3)8
DATA> END
MTB > TWOWAY 'SCORE' 'METHOD' 'TIME'

ANALYSIS OF VARIANCE SCORE

SOURCE           DF      SS      MS
METHOD            1   216.0   216.0
TIME              2   112.0    56.0
INTERACTION       2   144.0    72.0
ERROR            18   330.0    18.3
TOTAL            23   802.0
```

We first test for any interaction between method and class period. The statistical hypotheses are:

H_0: Method and class period do not interact to affect mean achievement scores.

H_1: Method and class period interact to affect mean achievement scores.

The value of the test statistic is

$$F = \frac{\text{MSAB}}{\text{MSE}} = \frac{72}{18.3} = 3.93$$

The critical value for the test is $F_{0.05}(2, 18) = 3.55$. Since $3.93 > 3.55$, we can reject H_0 and conclude that method and class period interact to affect mean achievement. Because the factors interact, we do not proceed to test the main effects for method and class period. Instead, we shall compare treatment means to determine the nature of the interaction. Since the mean plots in figure 13.7 suggest that the two methods of teaching differ, let's first compare the differences in class-period achievement for each method. Since there are three periods, there are

$$m = \binom{3}{2} = 3 \text{ pairs of means that need to be tested for each teaching method. We}$$

shall use Bonferroni's method with an overall significance level of $\alpha = 0.05$ to test for pairwise differences in means. The critical difference is

$$CD = t_{\alpha/(2m)} \sqrt{\frac{2MSW}{n}} \quad \text{(for df} = 18)$$

$$= t_{0.05/(2)(3)} \sqrt{\frac{(2)(18.3)}{4}}$$

$$= t_{0.0083} \sqrt{\frac{18.3}{2}}$$

$$= (2.64)(3.02489) = 7.9857$$

Next, we order the differences in means for the three class periods for students taught by the discovery method and construct a table of positive differences in means.

		Time		
		Period 5	Period 3	Period 1
Time		89	88	87
Period 5	89	—	1	2
Period 3	88	—	—	1
Period 1	87	—	—	—

Since none of the positive differences is greater than the critical difference CD = 7.9857, we can conclude that there are no differences in mean achievement for class period for students taught by the discovery method.

Next, let's test for mean differences in achievement for students taught by the expository method during the three class periods. The table of ordered positive differences in mean achievement scores for students taught by the expository method is given below.

		Time		
		Period 5	Period 3	Period 1
Time		87	83	76
Period 5	87	—	4	11
Period 3	83	—	—	7
Period 1	76	—	—	—

The only positive difference in means that exceeds the critical difference of 7.9857 is 11. We therefore can conclude that there is a difference in average achievement scores for students taught by the expository method during the first period and the fifth period. The expository method appears to be most effective earlier in the day.

By examining the mean plots in figure 13.7, it appears that there are differences in mean achievement for students taught by the two methods during the first and third periods. Since there are only two means to compare for each of these periods, we can use either the t test or the F test to check for differences, since $t^2 = F$. The critical value for each positive difference in means is $F = F_{0.05}(1, 18) = 4.41$. The difference in means for the two groups taught during period 1 $(88 - 76 = 12)$ exceeds 4.41 and the difference in means of the two groups taught during period 3 $(89 - 83 = 6)$ exceeds 4.41. We can conclude that during periods 1 and 3, there is a difference in mean achievement for the two teaching methods. During the fifth period, there is obviously no difference in achievement for students taught by the two methods.

   c. The assumptions that must hold in order to conduct a two-way ANOVA are
     1. The six sampled populations are normal.
     2. The six populations have equal variances.
     3. The students are randomly assigned to the six treatment groups.
     4. The achievement test scores are independent.

The *F* test is *robust*. As long as equal sample sizes are used in factorial experiments, there is empirical evidence that minor violations in the normality and variance assumptions do not seriously affect the results of the analyses. ▪

APPLICATION 13.12

An automobile manufacturer conducted a controlled study to examine the effects of various octane ratings (87, 89, or 92) of gasoline and the type of driving (city, freeway, or mountainous) on the mean gasoline mileage ratings of one of its new models of automobiles. A 3×3 factorial design with two automobiles per treatment was used for the study. Each automobile was driven for a distance of 5000 miles and the miles-per-gallon ratings were recorded. The results are contained in the accompanying table. Use $\alpha = 0.05$ and a two-way ANOVA to analyze the data.

Type of Driving	Octane Rating 87	89	92	
City	18	19	21	$\bar{X}_{\cdot 1 \cdot}$
	20	21	25	
Freeway	25	27	30	$\bar{X}_{\cdot 2 \cdot}$
	23	31	26	
Mountainous	15	19	23	$\bar{X}_{\cdot 3 \cdot}$
	13	23	19	
	$\bar{X}_{1 \cdot \cdot}$	$\bar{X}_{2 \cdot \cdot}$	$\bar{X}_{3 \cdot \cdot}$	$\bar{X}_{\cdots}$

Solution: MINITAB is used to determine the mean squares for the analyses. Computer display 13.8 contains the commands used and the corresponding output.

Computer Display 13.8

```
MTB > NAME C3 'MILEAGE' C1 'DRIVING' C2 'OCTANE'
MTB > SET 'MILEAGE'
DATA> 18 20 25 23 15 13
DATA> 19 21 27 31 19 23
DATA> 21 25 30 26 23 19
DATA> END
MTB > SET 'DRIVING'
DATA> 3(1:3)2
DATA> END
MTB > SET 'OCTANE'
DATA> (1:3)6
DATA> END
MTB > TWOWAY 'MILEAGE' 'DRIVING' 'OCTANE'

ANALYSIS OF VARIANCE MILEAGE

SOURCE          DF        SS         MS
DRIVING          2     227.11     113.56
OCTANE           2      88.44      44.22
INTERACTION      4      22.22       5.56
ERROR            9      48.00       5.33
TOTAL           17     385.78
```

We first test for interaction of octane rating and type of driving. The value of the test statistic is

$$F = \frac{\text{MSAB}}{\text{MSE}} = \frac{5.56}{5.33} = 1.04$$

The critical value for the test is $F_{0.05}(4, 9) = 3.63$. Since $1.04 < 3.63$, we cannot reject the null hypothesis that there is no interaction between octane rating and type of driving. We can now test for the main effects of octane rating and type of driving

The statistical hypotheses for the octane factor are

H_0: Mean gasoline mileage ratings for the three octane ratings are equal.

H_1: Mean gasoline mileage ratings for the three octane ratings are different.

The value of the test statistic is

$$F = \frac{\text{MSA}}{\text{MSE}} = \frac{44.22}{5.33} = 8.30$$

The critical value for the test is $F_{0.05}(2, 9) = 4.26$. Since $8.30 > 4.26$, we can reject H_0 and conclude that there is a difference in gasoline mileage ratings for three octane ratings. We shall use Bonferroni's procedure to test for pairwise differences in means. The means for the three octane (levels) treatments are:

87 octane: $\dfrac{18 + 20 + 25 + 23 + 15 + 13}{6} = 19$

89 octane: $\dfrac{19 + 21 + 27 + 31 + 19 + 23}{6} = 23.33$

92 octane: $\dfrac{21 + 25 + 30 + 26 + 23 + 19}{6} = 24$

The table of ordered positive differences in mean gasoline mileage ratings is given below.

		Gasoline		
		92 octane	89 octane	87 octane
Gasoline		24	23.33	19
92 octane	24	—	0.67	5
89 octane	23.33	—	—	4.33
87 octane	19	—	—	—

Bonferonni's critical difference is

$$\text{CD} = t_{\alpha/(2m)} \sqrt{\frac{2\text{MSW}}{n}} \quad (\text{df} = 9)$$

$$= t_{0.05/(2\cdot 3)} \sqrt{\frac{(2)(5.33)}{6}}$$

$$= t_{0.0083} (1.33292)$$

$$= (2.93)(1.33292) = 3.9055$$

Note that $t_{0.0083} = 2.93$ was obtained from table 4 of appendix B for df $= 9$ and $m = 3$. There are two pairwise differences that are significant. Since $5 > 3.9055$, there is a difference in mean mileages for cars using 87 octane and 92 octane gasoline, and since $4.33 > 3.9055$, there is a difference in mean mileages for cars using 87 octane and 89 octane gasoline.

We next test for the main effects of type of driving. The statistical hypotheses for the type-of-driving factor are

H_0: The mean gasoline mileage ratings for the three types of driving are equal.

H_1: The mean gasoline mileage ratings for the three types of driving are different.

The value of the test statistic is

$$F = \frac{\text{MSB}}{\text{MSE}} = \frac{113.56}{5.33} = 21.31$$

The critical value for the test is $F_{0.05}(2, 9) = 4.26$. Since $21.31 > 4.26$, we can conclude that at least two levels of driving have different mean gasoline mileage ratings. We shall use Bonferroni's method to test for pairwise differences. The means for the three types of driving are

$$\text{City:} \quad \frac{18 + 20 + 19 + 21 + 21 + 25}{6} = 20.67$$

$$\text{Freeway:} \quad \frac{25 + 23 + 27 + 31 + 30 + 26}{6} = 27$$

$$\text{Mountainous:} \quad \frac{15 + 13 + 19 + 23 + 23 + 19}{6} = 18.67$$

The table of ordered positive differences in mean gasoline mileage ratings is given below.

Type of Driving		Type of Driving		
		Freeway	City	Mountainous
		27	20.67	18.67
Freeway	27	—	6.33	8.33
City	20.67	—	—	2
Mountainous	18.67	—	—	—

Bonferroni's critical difference was found above to be 3.9055. Since $8.33 > 3.9055$, we can conclude that freeway driving and mountainous driving have different mean gasoline mileage ratings for the particular automobile involved in the study. Also, since $6.33 > 3.9055$, we can conclude that freeway and city driving have different mean gasoline mileage ratings. ▪

EXERCISE SET 13.5

Basic Skills

1. Consider the following tabulation of the measurements for a two-way ANOVA:

		Factor B		
		b_1	b_2	b_3
Factor A	a_1	5	10	25
		10	25	35
	a_2	15	15	65
		25	45	25

a. Construct an ANOVA summary table for the data.
b. Test for interaction of factors A and B and for any main effects of factors A and B. Use $\alpha = 0.05$.

2. Consider the following tabulation of the measurements for a two-way ANOVA.

		Factor B		
		b_1	b_2	b_3
Factor A	a_1	20	10	45
		18	16	26
	a_2	10	25	20
		35	20	25
	a_3	20	24	36
		22	30	48

a. Construct an ANOVA summary table for the data.
b. Test for interaction of factors A and B and for any main effects of factors A and B. Use $\alpha = 0.05$.
c. Where appropriate, test for pairwise differences using $\alpha = 0.05$.

3. The individual treatment means for two 4×2 factorial designs are given below. For each determine by using mean plots whether or not there is interaction of the two factors.

a.

	a_1	a_2	a_3	a_4
b_1	10	12	14	16
b_2	8	10	12	14

b.

	a_1	a_2	a_3	a_4
b_1	10	12	14	16
b_2	8	9	10	11

4. The individual treatment means for two 4×2 factorial designs are given below. For each determine by using mean plots whether or not there is interaction of the two factors.

a.

	a_1	a_2	a_3	a_4
b_1	10	14	12	16
b_2	7	11	9	13

b.

	a_1	a_2	a_3	a_4
b_1	12	12	8	12
b_2	8	8	12	8

5. A two-way ANOVA is used to analyze the data for a factorial experiment. The MINITAB printout for the data is shown here.

SOURCE	DF	SS	MS
A	1	533	533
B	2	1138	569
INTERACTION	2	4	2
ERROR	6	1475	246
TOTAL	11	3150	

a. How many treatments are involved in the experiment?
b. How may replications are there for each treatment?
c. How many levels of factor A are there?
d. How many levels of factor B are there?
e. Calculate the value of the F statistic to test for interaction between factors A and B. How many degrees of freedom are associated with this statistic?
f. Calculate the values of the F statistics to test for main effects. How many degrees of freedom are associated with each statistic?

6. A two-way ANOVA is used to analyze the data for a factorial experiment. The MINITAB printout for the data is shown here.

SOURCE	DF	SS	MS
A	1	58.594	58.594
B	3	25.865	8.622
INTERACTION	3	4.865	1.622
ERROR	16	9.167	0.573
TOTAL	23	98.490	

a. How many treatments are involved in the experiment?
b. How may replications are there for each treatment?
c. How many levels of factor A are there?
d. How many levels of factor B are there?
e. Calculate the value of the F statistic to test for interaction between factors A and B. How many degrees of freedom are associated with this statistic?
f. Calculate the values of the F statistics to test for main effects. How many degrees of freedom are associated with each statistic?

7. Complete the following ANOVA summary table for a 3×5 factorial experiment.

Source	SS	DF	MS	F
A		2		
B	5			
AB	1.2			
Error	3			
Total	23	44		

8. Complete the following ANOVA summary table for a factorial experiment.

Source	SS	DF	MS	F
A	6	3	2	
B		4		
AB			1.5	
Error	5			
Total	30	69		

Further Applications

9. Four fertilizers and three varieties of corn were studied in a factorial experiment. Two subplots were exposed to each of the 12 treatment combinations. The yield (in bushels per acre) were recorded for each subplot. The results are contained in the following table.

		Corn Variety		
		A	B	C
Fertilizer	1	53 44	23 32	36 39
	2	64 67	42 57	54 58
	3	92 87	89 85	88 89
	4	89 82	93 96	95 90

a. Construct an ANOVA summary table for the data.
b. Test for interaction of corn variety and fertilizer and for any main effects of corn variety and fertilizer on the average yields. Use $\alpha = 0.05$ for all tests.

10. An experiment was conducted to test the effects of three feeds and two vitamin supplements on the weight gain of chicks. Eighteen chicks were randomly assigned to the six treatment combinations in a factorial design. The resulting weight gains (in grams) over a specified time period are given in the accompanying table.

		Feed		
		I	II	III
Vitamin	A	51 80 64	125 99 127	41 54 55
	B	97 92 90	121 101 93	38 52 51

a. Construct an ANOVA summary table for the data.
b. Test for interaction of the vitamin and feed factors and for any main effects of vitamins and feed factors. Use $\alpha = 0.05$ for all tests.
c. Where appropriate, use Bonferroni's procedure and $\alpha = 0.05$ to test for pairwise differences.

11. An automobile manufacturer conducted a study to determine the effects of three different gear differential ratios and three different air/fuel mixtures for its carburation system on gasoline mileage ratings. For each of the nine treatment combinations in the 3×3 factorial experiment, two cars were driven a fixed distance on a test track by trained drivers and the following miles-per-gallon ratings were recorded.

		Air/Fuel Mixture		
		I	II	III
Gear	A	26 28 27	24 27 23	22 21 26
	B	29 27 29	26 22 21	24 28 22
	C	32 28 29	30 26 29	26 22 23

a. Construct an ANOVA summary table for the data.
b. Test for interaction of gear differential ratios and air/fuel mixtures and for any main effects of gear differential ratios and air/fuel mixtures. Use $\alpha = 0.05$ for all tests.

12. Four varieties of corn were subjected to three different fertilizers in a 3×4 factorial design to produce the resulting yields (in bushels per plot) contained in the accompanying table. The plots were as identical as possible and two were used for each of the 12 possible treatments.

		Corn			
		Variety I	Variety II	Variety III	Variety IV
Fertilizer	A	85 95	115 125	95 85	105 95
	B	110 105	130 140	105 115	125 185
	C	105 95	105 115	85 95	115 125

a. Construct an ANOVA summary table.
b. Use $\alpha = 0.05$ to test for interaction and any main effects.
c. Where appropriate, analyze significant main effects using $\alpha = 0.05$ and Bonferroni's procedure.

13. In the formation of a certain synthetic material an undesirable by-product is formed. A 2×3 factorial design was used to examine these yields (in percentages) for three catalysts and two pressures. The data is contained in the following table.

		Catalyst		
		I	II	III
Pressure	High	0.35 0.41 0.25	0.24 0.41 0.53	0.29 0.64 0.27
	Low	0.22 0.50 0.56	0.06 0.34 0.08	0.74 0.74 0.81

Use $\alpha = 0.05$ and two-way ANOVA to determine whether the catalysts or pressure have a significant effect on the yield of the undesirable by-product. Is there any interaction between pressure and catalyst?

14. Three types of adhesive (I, II, and III) are used by a company to bond glass to glass in the assembly of a certain product. Three different types of assemblies are used, A, B, and C, in the assembly process. A tensile strength test is performed to determine the bond strength of the glass to glass assemblies of 45 of these glass products. A 3 × 3 factorial design is used to analyze the results of the tests. The following are the results:

		Assembly		
		A	B	C
Adhesive	I	20 18 23 22 23	21 27 24 20 18	17 23 18 21 25
	II	27 22 25 24 25	28 24 16 25 21	28 25 29 33 28
	III	31 32 18 30 21	18 30 18 32 31	21 22 17 20 22

a. Construct an ANOVA summary table for the data.
b. Test for interaction of adhesive and glass-to-glass assembly process and any main effects of adhesive and assembly process on bonding strength. Use $\alpha = 0.05$ for all tests.
c. Where appropriate, use $\alpha = 0.05$ and Bonferroni's procedure to test for pairwise differences.

15. A study was conducted to examine the effects of two methods of teaching (inductive and deductive) and three class periods (9:00 A.M., 11:00 A.M., and 1:00 P.M.) on achievement in general physics. A 2 × 3 factorial design was used for the study. All other variables that could influence the experiment, such as the teacher variable, were controlled as much as possible. Four students were randomly assigned to each of the six treatment combinations. Following six weeks of study, all students were given a standardized achievement test. The results of the examination are recorded in the accompanying table.

		Class Period		
		1	3	5
Method of teaching	Inductive	56 59 55 62	68 60 62 70	64 71 73 68
	Deductive	70 61 65 68	61 73 64 70	69 62 61 68

a. Plot the means of the treatments to assist in your interpretation of the experiment.
b. Perform a statistical analysis of the experiment. Use $\alpha = 0.05$.

16. Refer to motivator 13. Determine
   a. MSE. b. df_E. c. SSE. d. SST.

Going Beyond

17. For an $a \times b$ factorial design with n replicates per treatment, show that MSE is the average of the ab sample variances.

18. In order to conserve space, research journals sometimes do not publish ANOVA summary tables, but do publish tables of means and obtained F values. Suppose we have been given the following table of means for a 2 × 3 factorial design.

	a_1	a_2	a_3
b_1	11	12	10
b_2	3	10	14

and have been told that each of the six treatments had four replications and only the AB interaction is significant, $F = 3.93$, $P < 0.05$. Construct the ANOVA summary table from this information.

19. Explain why it is necessary to have at least two replications per treatment in a two-way factorial design in order to use ANOVA.

20. Generalize formula (13.27) to a three factor factorial design. (*Hint:* SST is portioned into eight sum-of-squares components.)

CHAPTER SUMMARY

In this chapter we presented two different views of one-way analysis of variance (ANOVA). One-way ANOVA is used to determine if k population means are different. The assumptions underlying the test are that the k populations are normal and the k population variances are equal. For equal sample sizes, the F test is robust, which means that minor violations in the assumptions underlying the test do not seriously affect the outcome of the test. We showed that the two-sample t test for determining whether two population means differ is a special case of one-way ANOVA for two samples. The simultaneous use of multiple t tests instead of ANOVA is not recommended to test for differences in more than two population means because the probability of making at least one Type I error for the experiment increases as the number of pairwise tests increases. Bonferroni's procedure was examined for determining which population means are significantly different following a significant F test, and we learned how to estimate the strength of association by using Hays' omega-square statistic. We also examined two-factor experimental designs. The randomized block design contained a blocking factor that did not interact with the treatment factor. The purpose of the blocking factor is to control the influence of a variable on the response variable. The factorial design allows the experimenter to study the effects of more than one variable on the response variable; these effects include those caused by the interaction of the two factors.

CHAPTER REVIEW

■ IMPORTANT TERMS ■

The following chapter terms have been mixed in ordering to provide you better review practice. For each of the following terms, provide a definition in your own words. Then check your responses against the definitions given in the chapter.

$a \times b$ factorial experiment	sum of squares for total	strength of association
one-way analysis of variance	block design	randomized block design
two-way analysis of variance	between-sample variance estimate	response variable
factorial design	n replications	robust
factor	two-factor ANOVA	sum of squares between samples
levels of a factor	two-way ANOVA	sum of squares for treatment
main effects	block	sum of squares within samples
grand mean	between-groups variance	sum of squares for interaction
mean square	mean square for between samples	sum of squares for error
one-way ANOVA	mean square for treatments	within samples variance
multiple-comparison procedures	mean square for within samples	interaction component
mean square for blocks	mean square for error	treatment mean plots
Bonferroni's procedure	measure of association	critical difference
post hoc tests	pairwise comparison	within-samples variance estimate
ANOVA summary table	post hoc comparisons	

▪ **IMPORTANT SYMBOLS** ▪

s_b^2, between samples variance estimate

s_w^2, within samples variance estimate

F, F statistic

$F_\alpha(\mathrm{df}_1, \mathrm{df}_2)$, critical value for the F test

df_b, degrees of freedom for s_b^2

df_w, degrees of freedom for s_w^2

df_{AB}, degrees of freedom for SSAB

df_E, degrees of freedom for SSE

n, sample size (number of blocks in a two-way design)

n_i, the size of the ith sample

k, number of samples

N, total number

SSA, sum of squares for factor A

SSB, sum of squares between samples (also sum of squares for factor B)

SSTr, sum of squares for treatments (two-way ANOVA)

SSBl, sum of squares for blocks (two-way ANOVA)

SSAB, sum of squares for interaction (two-way ANOVA)

SSW, sum of squares within samples

SSE, sum of squares for error (two-way ANOVA)

SST, total sum of squares

T, grand total

C_i, sum of measurements of ith sample

x_{ij}, the jth measurement in the ith sample

$\overline{T}$, the grand mean

$\overline{C}_i$, the mean of ith sample or column

R_j, sum of the measurements in the jth block (two-way ANOVA)

$\hat{\omega}^2$, Hays' omega-square statistic

▪ **IMPORTANT FACTS AND FORMULAS** ▪

F statistic (one-way ANOVA): $F = s_b^2 / s_w^2 = \dfrac{\mathrm{MSB}}{\mathrm{MSW}}$ (13.1)

Within samples variance estimate when sample sizes are equal: $s_w^2 = \dfrac{s_1^2 + s_2^2 + s_3^2 + \cdots + s_k^2}{k} = \dfrac{\Sigma s_i^2}{k}$. (13.2)

Degrees of freedom for within samples variance estimate (equal sample sizes): $\mathrm{df}_w = k(n - 1)$. (13.3)

Grand mean when sample sizes are equal:
$\overline{x} = \dfrac{\overline{x}_1 + \overline{x}_2 + \overline{x}_3 + \cdots + \overline{x}_k}{k} = \dfrac{\Sigma x_i}{k}$. (13.4)

Between-samples variance estimate when sample sizes are equal: $s_b^2 = \dfrac{n\Sigma (\overline{x}_i - \overline{x})^2}{k - 1}$.

Degrees of freedom for between samples estimate: $\mathrm{df}_b = k - 1$. (13.6)

Sum of squares for total (one-way ANOVA): $\mathrm{SST} = \Sigma (x_{ij} - T)^2$. (13.7)

Computational formula for SST (one-way ANOVA): $\mathrm{SST} = \Sigma x_{ij}^2 - \dfrac{T^2}{N}$. (13.8)

Sum of squares for total (one-way ANOVA): $\mathrm{SST} = \mathrm{SSB} + \mathrm{SSW}$. (13.9)

Sum of squares for between samples (one-way ANOVA): $\mathrm{SSB} = \Sigma\, n_i(\overline{C}_i - \overline{T})^2$

Computational formula for SSB: $\mathrm{SSB} = \Sigma \dfrac{C_i^2}{n_i} - \dfrac{T^2}{N}$. (13.10)

Degrees of freedom for within samples (unequal sample sizes): $\mathrm{df}_w = N - k$. (13.11)

Mean square for between samples (one-way ANOVA): $\mathrm{MSB} = \dfrac{\mathrm{SSB}}{k - 1}$. (13.12)

Mean square for within samples (one-way ANOVA): $\mathrm{MSW} = \dfrac{\mathrm{SSW}}{N - k}$. (13.13)

F statistic (one-way ANOVA): $F = \dfrac{\mathrm{MSB}}{\mathrm{MSW}}$. (13.14)

Bonferroni's critical difference for one-way ANOVA (unequal sample sizes): $\mathrm{CD} = t_{\alpha/(2m)} \sqrt{\mathrm{MSE}\left(\dfrac{1}{n_i} + \dfrac{1}{n_j}\right)}$. (13.15)

Bonferroni's critical difference for one-way ANOVA (equal sample sizes): $\mathrm{CD} = t_{\alpha/(2m)} \sqrt{\dfrac{2\,\mathrm{MSE}}{n}}$. (13.16)

Simultaneous confidence interval limits: $\bar{x}_i - \bar{x}_j \pm CD$. (13.17)

Hays' omega-square statistic (one-way ANOVA):
$$\hat{\omega}^2 = \frac{SSB - (k-1)MSW}{SST + MSW}. \quad (13.18)$$

Total sum of squares relationship (randomized block design): $SST = SSTr + SSBl + SSE$.

Sum of squares for treatment (randomized block design):
$$SSTr = \Sigma \frac{C_i^2}{n} - \frac{T^2}{N}. \quad (13.19)$$

Sum of squares for blocks (randomized block design):
$$SSBl = \Sigma \frac{R_j^2}{k} - \frac{T^2}{N}. \quad (13.20)$$

Sum of squares for error (randomized block design):
$SSE = SST - SSTr - SSBl$. (13.21)

Degrees of freedom for treatments (randomized block design): $df_{Tr} = k - 1$.

Mean square for blocks (randomized block design):
$$MSBl = \frac{SSBl}{n-1}.$$

Degrees of freedom for blocks (randomized block design): $df_{Bl} = n - 1$.

Mean square for treatments (randomized block design):
$$MSTr = \frac{SSTr}{k-1}.$$

Mean square for error (randomized block design):
$$MSE = \frac{SSE}{(k-1)(n-1)}.$$

F statistic for treatment factor (randomized block design): $F = \dfrac{MSTr}{MSE}$.

F statistic for blocks factor (randomized block design):
$$F = \frac{MSBl}{MSE}.$$

Bonferroni's critical difference for treatment factor (randomized block design): $CD = t_{\alpha/(2m)} \sqrt{\dfrac{2\,MSE}{n}}$. (13.22)

Bonferroni's critical difference for blocking factor (randomized block design): $CD = t_{\alpha/(2m)} \sqrt{\dfrac{2\,MSE}{k}}$. (13.23)

Hays' omega-square statistic for blocking factor (randomized block design): $\hat{\omega}^2 = \dfrac{SSBl - (k-1)MSE}{SST + MSE}$. (13.25)

Hays' omega-square statistic for treatment factor (randomized block design): $\hat{\omega}^2 = \dfrac{SSTr - (k-1)MSE}{SST + MSE}$. (13.26)

Sum of squares relationship for two-way factorial design: $SST = SSA + SSB + SSAB + SSE$. (13.27)

Sum of squares for total (two-way factorial design):
$$SST = \Sigma x_{ij}^2 - \frac{(\Sigma x_{ij})^2}{abn}. \quad (13.28)$$

Sum of squares for factor A (two-way factorial design):
$$SSA = \frac{\Sigma A_i^2}{bn} - \frac{(\Sigma x_{ij})^2}{abn}. \quad (13.29)$$

Sum of squares for factor B (two-way factorial design):
$$SSB = \frac{\Sigma B_j^2}{an} - \frac{(\Sigma x_{ij})^2}{abn}. \quad (13.30)$$

Sum of squares for interaction (two-way factorial design):
$$SSAB = \frac{\Sigma (AB_{ij})^2}{n} - SSA - SSB - \frac{(\Sigma x_{ij})^2}{abn}. \quad (13.31)$$

Sum of squares for error (two-way factorial design):
$SSE = SST - SSA - SSAB$. (13.32)

Degrees of freedom for SST (two-way factorial design): $df_T = abn - 1$. (13.33)

Degrees of freedom for SSA (two-way factorial design): $df_A = a - 1$. (13.34)

Degrees of freedom for SSB (two-way factorial design): $df_B = b - 1$. (13.35)

Degrees of freedom for error (two-way factorial design): $df_E = ab(n - 1)$. (13.36)

Degrees of freedom for interaction (two-way factorial design): $df_{AB} = (a - 1)(b - 1)$. (13.37)

Mean square for factor A (two-way factorial design):
$$MSA = \frac{SSA}{a-1}.$$

Mean square for factor B (two-way factorial design):
$$MSB = \frac{SSB}{b-1}.$$

Mean square for interaction (two-way factorial design):
$$MSAB = \frac{SSAB}{(a-1)(b-1)}.$$

Mean square for error (two-way factorial design):

$$MSE = \frac{SSE}{ab(n-1)}.$$

F statistic for factor A (two-way factorial design):

$$F = \frac{MSA}{MSE}.$$

F statistic for factor B (two-way factorial design):

$$F = \frac{MSB}{MSE}.$$

F statistic for interaction (two-way factorial design):

$$F = \frac{MSAB}{MSE}.$$

REVIEW EXERCISES

1. Three random groups of students were exposed to different types of training, after which each group was administered the same standardized task to perform. The completion times (in minutes) for each group are given in the table.

Types of Training

A	B	C
19	16	15
17	14	11
21	12	16
18	22	19
15	12	15
20	15	13
20	21	13
25	19	18
18	16	18
17	13	14

By using ANOVA test at $\alpha = 0.05$ to determine if the three population mean task completion times are different.

2. A comparison of the mean recovery times (in days to recovery) of patients suffering from a certain disease and using three different treatments was made. The data shown in the table were obtained.

Treatment

A	B	C
4	8	5
9	7	4
7	10	6
10	6	3
8	6	7
6		4
5		

Use $\alpha = 0.05$ to determine if there is a difference in the mean recovery times for the three treatments. Assume the assumptions for ANOVA hold.

3. Twelve patients with a particular disease were randomly assigned to four groups. Each group received a different dosage of a new experimental drug. The accompanying data indicate the percentage absorption of the new drug for the four groups of patients.

Group 1	Group 2	Group 3	Group 4
71.0	48.6	52.2	51.1
71.1	61.1	47.1	58.5
61.8	49.5	54.0	49.2

a. By using ANOVA and $\alpha = 0.05$, test to determine if the average absorption rates for different dosages of the new drug are different. Summarize your results in an ANOVA summary table.

b. Test for pairwise differences using Bonferroni's procedure and an overall significance level of $\alpha = 0.05$.

c. Use Hays' omega-square statistic to estimate the strength of association between the amount of the experimental drug and the average absorption rate. Interpret your finding.

4. The accompanying data represent samples of examination scores for four different calculus classes. Use ANOVA and $\alpha = 0.01$ to test H_0: $\mu_1 = \mu_2 = \mu_3 = \mu_4$.

Class 1	Class 2	Class 3	Class 4
85	69	90	63
82	73	88	64
80	87	85	80
79	81	84	77
77	85	80	81
69	87	76	83
67	88		85
65	90		
	84		

Summarize your results in a summary table.

5. A study was conducted to compare three different methods for training typists. Students were randomly assigned to the three methods and given a standardized typing test at the conclusion of the instruction. The results (in words typed per minute) are shown in the table.

Typing Method

A	B	C
74	71	85
94	81	76
81	84	97
72	72	93
78		

Test the null hypothesis $H_0: \mu_A = \mu_B = \mu_C$ at the 0.05 level of significance. Assume the assumptions for ANOVA hold.

6. Three different army companies participated in a marksmanship competition. Soldiers were randomly selected from each company for the competition. For each company, the mean number of points per ten shots per soldier was obtained. The results were recorded in the table.

Company

A	B	C
$\bar{x} = 87.7$	$\bar{x} = 92.2$	$\bar{x} = 88.4$
$s = 14.3$	$s = 17.4$	$s = 16.5$
$n = 20$	$n = 25$	$n = 30$

At $\alpha = 0.01$, do the data indicate that the companies differ in marksmanship?

7. Three contestants were in a talent contest involving five judges. Each contestant was judged on a scale of 1 to 10. The results are contained in the accompanying table.

	Candidates		
Judge	A	B	C
1	9	8	9
2	7	8	8
3	8	7	6
4	5	3	4
5	9	7	5

a. Use ANOVA to test for differences in each factor.
b. Follow up significant tests on main factors with tests for pair-wise differences in means. Use $\alpha = 0.05$ for all tests.

8. A study was made by a television picture tube manufacturer to determine the effects of two glass types and three phosphor types on the light intensity of its tubes. Light intensity is measured by the amount of current (in microampheres) flowing in series with the tube to produce light output; the higher the current, the poorer the tube is in output. Three observations were made under each of the six experimental conditions in a 2×3 factorial design. The following data were recorded:

		Phosphor Type		
		A	B	C
Glass type	I	285	305	275
		295	315	290
		290	300	295
	II	235	265	225
		240	245	230
		245	240	235

a. Use $\alpha = 0.05$ and two-way ANOVA to test the effect of glass type, phosphor types, and interaction on the current flow from cathode to anode.
b. Where appropriate, use $\alpha = 0.05$ and Bonferroni's procedure to test for pairwise differences.

Computer Applications

1. Four different four-cylinder engines were used in a certain automobile used by a taxi service. The following data represent the mileage (in thousands of miles) that each engine was run before it needed to be rebuilt.

Model A		Model B			Model C			Model D		
68	80	68	89	82	61	67	80	76	74	63
74	90	74	78	73	64	71	93	60	74	87
74	83	73	88	90	80	87	81	68	100	79

Model A		Model B			Model C			Model D		
70	91	83	68	79	74	79	65	84	71	74
62	69	95	76	97	62	71	71	80	85	67
78	57	66	76	75	84	84	62	71	82	82
67	66	86	75	60	60	85	76	80	95	79
79	74	71	79	91	74	62	75	80	75	77
70	71	72	74	73	61	64	67	91	70	79
77										

Use a computer program and $\alpha = 0.05$ to determine if there is a difference between the mean lifetimes of the four engines before rebuilding.

2. Four different methods of teaching American history were compared. Students were randomly assigned to the four classes and a standardized final examination was administered to each student at the end of the course. Scores on the final examination for the four classes follow:

Method A: 68 80 68 89 82 61 67 80 76 74 63
74 90 74 78 73 64 71 93 60 74 87

Method B: 74 83 73 88 90 80 87 81 68 100 79
70 91 83 68 79 74 79 65 84 71 74
62 69 95 76 97 62 71 71 80 85 67

Method D: 78 57 66 76 75 84 84 62 71 82 82
67 66 86 75 60 60 85 76 80 95 79

Method E: 79 74 71 79 91 74 62 75 80 75 77
70 71 72 74 73 61 64 67 91 70 79

Use a computer program and $\alpha = 0.01$ to determine if there is a difference between the four methods of teaching history.

3. Use a computer program to do parts a and b of exercise 14 of exercise set 13.5.

EXPERIMENTS WITH REAL DATA

Treat the 720 subjects listed in the database in appendix C as if they constituted the population of all U.S. subjects.

1. Do the mean systolic blood pressures for subjects aged in their 20s, 30s, and 40s differ? Select random samples of size 25 for each of the three age groups.
   a. Conduct an ANOVA to determine whether differences

exist among the three groups' mean systolic blood pressures. Use $\alpha = 0.05$.
   b. Use Bonferroni's procedure to compare the means. Use an overall significance level of 0.05.
   c. What assumptions are necessary in order to use the procedures of parts a and b?

■ CHAPTER ACHIEVEMENT TESTS ■

1. Consider the following data set:

Sample 1	Sample 2	Sample 3
6	5	7
3	4	3
1		5
2		

Find each of the following:
a. SST
b. SSB
c. SSW
d. df_b
e. df_w
f. MSB
g. MSW
h. s_b^2
i. s_w^2
j. F

2. For the data in exercise 1, complete the following ANOVA summary table and test $H_0: \mu_1 = \mu_2 = \mu_3$ at $\alpha = 0.05$.

ANOVA Summary Table

Source of variance	SS	df	MS	F
Between samples				
Within samples				
Total				

3. For the data in question 1, calculate the value of $\hat{\omega}^2$, Hays' omega-square statistic.

4. Consider the following data:

Block	1	2	3	4
I	11	13	12	14
II	27	26	27	28
III	14	14	14	16

Treatments

Calculate the value of the F statistic to determine if there is a difference in the treatment means.

5. A two-way ANOVA is used to analyze the data for a factorial experiment. The MINITAB printout for the data is shown here.

SOURCE	DF	SS	MS
A	2	32.69	16.34
B	3	39.28	13.09
INTERACTION	6	13.06	2.18
ERROR	12	23.12	1.93
TOTAL	23	108.16	

a. How many treatments are involved in the experiment?
b. How many replications are there for each treatment?
c. How many levels of factor A are there?
d. How many levels of factor B are there?
e. Calculate the value of the F statistic to test for interaction between factors A and B. How many degrees of freedom are associated with this statistic?
f. Calculate the values of the F statistics to test for main effects. How many degrees of freedom are associated with each statistic?

6. What do we know about the null hypothesis if the value of the F statistic is less than or equal to 1?

14

Linear Regression Analysis

CHAPTER OBJECTIVES

In this chapter we will investigate

▷ Two different regression models.

▷ The assumptions for the linear regression model.

▷ How to determine if linear regression is appropriate.

▷ How to test the parameters in the linear regression model.

▷ How to summarize our results using an ANOVA summary table.

▷ Correlation analysis.

▷ The coefficient of determination and how we interpret it.

▷ Multiple linear regression analysis and how it is used.

▷ The multiple R^2 coefficient and how to interpret it.

|||| MOTIVATOR 14 ▶

*T*he United States is facing a new energy crisis. Oil imports have risen from 37% of the oil consumed in the United States in the early 1980s to over 46% in 1989 and are projected to surpass 50% within the next few years. Global warming, caused in large part by energy consumption, threatens worldwide environmental disruption; already, six of the hottest years on record have occurred during this past decade. Other energy-related environmental problems, such as urban smog, continue to worsen. A key contributor to these problems is the automobile.

In 1989, cars, trucks, and other U.S. transportation activities consumed 3.6 billion barrels of crude oil—36% more oil than the total of all domestic petroleum production. This oil use accounted for more than 27% of all U.S. energy consumption and is equal to over five times the maximum annual capacity of the Alaskan oil pipeline.

Automobiles consume the greatest share of this energy, accounting for almost 41% of all the fuel used for transportation activities in 1989. Further, each gallon of gasoline burned to power an automobile generates approximately 19.7 pounds of carbon dioxide (CO_2), the primary "greenhouse" gas. In 1989, automobiles released approximately 677 million tons of CO_2 into the atmosphere—about 11% of total U.S. CO_2 output of almost 170 typical coal-fired electrical generating plants. Thus, the combustion of petroleum by automobiles is a major source of the emissions causing global warming. The following table lists the twenty 1990 cars that produce the most CO_2.[46]

The Twenty 1990 Cars that Produce the Most CO_2

Class	Model	Engine size/cylinders (size in liters)	Composite mpg	Lifetime CO_2 (tons)	Lifetime gasoline (gallons)
2-Seaters	Lamborghini Countach	5.2/12	7.80	124.77	12,666.67
Subcompact	Rolls-Royce Bentley Continental	6.8/8	11.35	85.74	8,704.85
Mid-Size	Rolls-Royce Bentley Eight/Bentley Mulsan	6.8/8	11.35	85.74	8,704.85
Subcompact	Rolls-Royce Corniche III	6.8/8	11.35	85.74	8,704.85
Mid-Size	Rolls-Royce Silver Spirit II/Silver Speer	6.8/8	11.35	85.74	8,704.85
2-Seaters	Ferrari Testarossa	4.9/12	12.25	79.44	8,065.31
Subcompact	Ferrari Mondial T/Cabriolet	3.4/8	14.25	68.29	6,933.33
Mid-Size	BMW 750 IL	5.0/12	14.70	66.20	6,721.09
2-Seaters	Jaguar XJ-S Convertible	5.3/12	14.80	65.76	6,675.68
Mid-Size	Mercedes-Benz 560 SEL	5.6/8	14.80	65.76	6,675.68
Subcompact	Jaguar XJ-S Coupe	5.3/12	15.25	63.82	6,478.69
Compact	Mercedes-Benz 560 SEC	5.6/8	15.35	63.40	6,436.48
2-Seaters	Ferrari 348 TB/TS	3.4/8	15.70	61.99	6,292.99
Large Cars	Ford LTD Crown Victoria	5.8/8	15.70	61.99	6,292.99
Minicompact	Porsche 928 S4	5.0/8	15.70	61.99	6,292.99
Mid-Size	Audi V8	3.6/8	15.80	61.59	6,253.16
2-Seaters	Mercedes-Benz 500 SL	5.0/8	15.80	61.59	6,253.16
Mid-Size	Maserati 228	2.8/6	16.25	59.89	6,080.00
Minicompact	Maserati 222E	2.8/6	16.35	59.52	6,042.81
Mid-Size	Maserati 228	2.8/6	16.35	59.52	6,042.81

You might expect that there is a relationship between lifetime CO_2 (in tons) (the dependent variable) and engine size, number of cylinders, composite miles per gallon (mpg), and lifetime gasoline (in gallons) (the independent variables). Using the given data, can we conclude that there is a linear relationship between lifetime CO_2 (in tons) and engine size, number of cylinders, composite mpg, and lifetime gasoline (in gallons)? If so, what is the relationship? You should be able to answer these questions after studying this chapter.

Chapter Overview

In chapter 4 we studied **correlation analysis** and viewed correlation primarily from a descriptive point of view; scattergrams and the magnitude of the linear correlation coefficient *r* were used to determine if a linear relationship existed between two variables, and if it did we used **regression analysis** to describe it. Correlation coefficients and regression equations were computed from samples of bivariate data. The regression equation was used to describe the manner in which two variables were related. No attempt was made in chapter 4 to draw inferences concerning linear relationships for paired data in populations of bivariate data from which the samples were obtained. The purpose of this chapter is to study regression and correlation from an inference-making point of view using the descriptive statistics developed in chapter 4. We begin by discussing the linear regression model and related concepts.

SECTION 14.1   *Linear Regression Model*

Linear **regression** involves predicting the value of one variable from one or more other variables. The dependent variable is sometimes called the **response variable** and the independent variable is called the **predictor variable.** We will begin our study with the simplest case, the prediction of one variable, y, from another variable, x. In section 14.4 we shall examine the case of predicting the response variable by using more than one independent variable.

The techniques of linear regression and correlation that we will examine in this chapter require that the dependent variable, y, must be measured on at least an interval scale. The independent variable or variables can be measured on any type of measurement scale, although in all of our examples, the independent variable will be measured by at least an interval scale of measurement. The context of a problem will dictate which variable is the dependent variable and which variable is the independent variable.

APPLICATION 14.1   For the following situations, state which is the dependent variable (y), and which is the independent variable (x).

a. An actuary wants to predict the amount of life insurance carried by teachers using their annual salaries.

b. A biologist wants to determine if there is any relationship between the number of pine trees per acre and the height of 15-year-old trees.

c. A manager of a restaurant wants to estimate the number of customers to be expected on a given evening from the number of dinner reservations received by 5 P.M.

d. A weather forecaster wants to determine the relationship between average monthly temperature and average monthly rainfall for a certain region of the country.

Solution:

a. The dependent variable is the response variable, the amount of life insurance carried by a teacher, and the independent variable or predictor variable, is the annual salary of a teacher.

b. In this case the number of trees per acre can be represented by either variable. It makes most sense to determine the height of 15-year-old trees given the number of trees per acre. If this is the case, then the number of trees per acre should be the independent variable, x.

c. The number of customers is the response or dependent variable, y.

d. The assignment of variables is arbitrary in this case. All one is interested in is a relationship; either variable can represent the average monthly rainfall. ■

The Environmental Protection Agency released the information in table 14.1 comparing engine size (in cubic inches of displacement [cid]) and miles-per-gallon (mpg) estimates for eight representative models of 1984 compact automobiles.

TABLE 14.1

EPA Mileage Ratings

Compact Cars	Engine Size (cid)	mpg
Chevrolet Cavalier	121	30
Datson Nissan Stanza	120	31
Dodge Omni	97	34
Ford Escort	98	27
Mazda 626	122	29
Plymouth Horzion	97	34
Renault Alliance/Encore	85	38
Toyota Corolla	122	32

MINITAB can be used to construct a scattergram for the EPA mileage ratings data by using the PLOT command. Computer display 14.1 contains the commands used and the corresponding output.

Computer Display 14.1

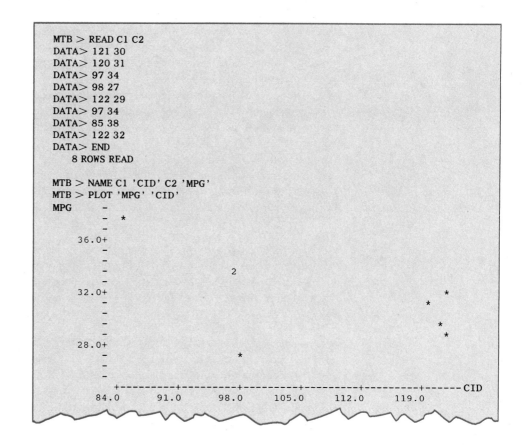

```
MTB > READ C1 C2
DATA> 121 30
DATA> 120 31
DATA> 97 34
DATA> 98 27
DATA> 122 29
DATA> 97 34
DATA> 85 38
DATA> 122 32
DATA> END
    8 ROWS READ

MTB > NAME C1 'CID' C2 'MPG'
MTB > PLOT 'MPG' 'CID'
MPG    -
       -   *
       -
  36.0+
       -
       -
       -                        2
       -
  32.0+                                              *
       -                                          *
       -
       -                                       *
       -                                       *
  28.0+
       -                    *
       -
       -
       -
       +--------+--------+--------+--------+--------+------ CID
      84.0     91.0     98.0    105.0    112.0    119.0
```

Suppose we view the collection of eight data pairs (cid, mpg) as constituting a sample from a population of pairs where the cid measurements can be any value within the range of values extending from 85 to 122. For each possible cid measurement, there are many mpg ratings associated with it. For example, for an engine size of 97, there are a large number of associated mpg ratings, one for each car with engine size 97 cid. Let's assume a linear relationship exists for the population of data pairs of cid and mpg ratings. The following **probabilistic model** can then be used to explain the behavior of the mpg ratings for the eight (six distinct) cid measurements. It is called a **linear regression model** and expresses the linear relationship between cid (x) and mpg (y).

Linear Regression Model

$$y = \beta_0 + \beta_1 x + \epsilon \qquad\qquad (14.1)$$

The right-hand side of equation (14.1) involves a random variable ϵ (lowercase Greek letter epsilon) that expresses the random error involved in measuring y at a fixed value of x. If all the points in the population scattergram fell on a straight line, there would be no need to include the error term in the model, nor to involve probability in our study of the relationship between x and y. In this case for each value of x, y could be determined exactly and the resulting linear model $y = \beta_0 + \beta_1 x$ would be called a **deterministic model.** The two parameters β_0 and β_1 would represent the y-intercept and slope, respectively, of the line. The expression $\beta_0 + \beta_1 x$ is sometimes referred to as the **deterministic component** of the linear regression model. Our sample data consisting of eight data pairs will be used to estimate the parameters β_0 and β_1 of the deterministic component.

The main difference between a probabilistic model and a deterministic model is the inclusion of a random error term in the probabilistic model. For our sample data, a probabilistic model is appropriate since the Mazda 626 and Toyota Corolla have different mpg ratings (29 and 32) for the same size engine (122 cid). The different mpg ratings for the same engine size are accounted for by the error term ϵ in the regression model. For each engine size, the deterministic component of the regression model is equal to a constant, say C. That is, for a fixed engine size x, the miles-per-gallon rating is equal to the constant C plus a random error. That is,

$$y = C + \epsilon$$

The values of y (mpg ratings) vary directly with the values of the random error term. The terms y and ϵ that occur in (14.1) are random variables.

We shall make the following assumptions concerning the error term ϵ in the linear regression model $y = \beta_0 + \beta_1 x + \epsilon$:

Assumptions for Linear Regression Model

1. For each value of x, the random variable ϵ is normally distributed.
2. For each value of x, the mean or expected value of ϵ is 0; that is, $E(\epsilon) = \mu_\epsilon = 0$.
3. For each value of x, the variance of ϵ is the constant σ^2 (called **error variance**).
4. The values of the **error term** ϵ are independent.

As a result of these four assumptions, we can make the following observations concerning the linear regression model:

1. The values of x are fixed. For our application, we have six distinct values: 85, 97, 98, 120, 121, and 122.

2. The values of the parameters β_0 and β_1 are constant, but unknown. Based on the four assumptions and the sample statistics, their values can be estimated.

3. Since the value of y for a fixed value of x is determined by adding the constant $\beta_0 + \beta_1 x$ to the random variable ϵ, the values for y will depend on the values for ϵ. Therefore, y is a random variable.

4. For a fixed value of x, the sampling distribution of y is normal, since the values of y depend on the values of ϵ and the values of ϵ are normally distributed (see fig. 14.1).

FIGURE 14.1

Sampling distribution of y for different values of x

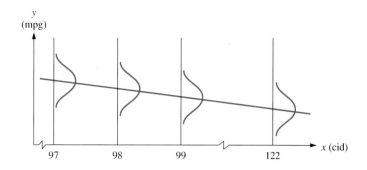

5. The sampling distribution of y for a fixed value of x has a mean, denoted by $\mu_{y\,|\,x}$. The symbol $y\,|\,x$ is read "y given x." That is, $E(y\,|\,x) = \beta_0 + \beta_1 x$, since $E(\epsilon) = 0$. The equation $E(y\,|\,x) = \beta_0 + \beta_1 x$ is called the **population regression equation** (see fig. 14.2).

FIGURE 14.2

Graph of population regression equation

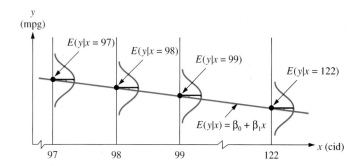

6. The variance of the sampling distribution of y for any value of x is σ^2. This means that the y values for any two distinct x values have the same variance. As a result, all the normal distributions of the y values corresponding to the x values have the same shape as shown in figure 14.2.

7. For a fixed x value, the value of y can be predicted.

8. For a fixed x value, the average value of y can be estimated.

9. The unknown error variance σ^2 can be estimated.

EXAMPLE 14.1

Let's return to the application involving engine size and mpg ratings for the eight 1984 compact cars and calculate the estimated **regression equation** (or least squares prediction equation). For convenience, let's represent the estimated regression equation by $\hat{y} = b_0 + b_1 x$. Remember that we call $\hat{y}$ "y hat," the predicted value of y for a particular value of x. To indicate the dependence of $\hat{y}$ on x, the predicted value of y for a given value of x is sometimes written as $\hat{y}(x)$. The constant b_0 will serve as a point estimator for β_0 and the constant b_1 will serve as a point estimator for β_1. Note that in chapter 4 we represented the estimated regression equation by $\hat{y} = b + mx$. Toward computing the values for b_0 and b_1, recall the following formulas from chapter 4.

$$SS_x = \Sigma x^2 - \frac{(\Sigma x)^2}{n}$$

$$SS_y = \Sigma y^2 - \frac{(\Sigma y)^2}{n}$$

$$SS_{xy} = \Sigma xy - \frac{(\Sigma x)(\Sigma y)}{n}$$

$$b_1 = \frac{SS_{xy}}{SS_x}$$

$$b_0 = \bar{y} - b_1 \bar{x}$$

where n represents the number of data pairs. We will use these formulas to calculate the least squares estimates of β_0 and β_1. The following table organizes the computations for the gas mileage data given in table 14.1:

x	y	x^2	y^2	xy
121	30	14,641	900	3,630
120	31	14,400	961	3,720
97	34	9,409	1,156	3,298
98	27	9,604	729	2,646
122	29	14,884	841	3,538
97	34	9,409	1,156	3,298
85	38	7,225	1,444	3,230
122	32	14,884	1,024	3,904
862	255	94,456	8,211	27,264

Determine SS_x:

$$SS_x = \Sigma x^2 - \frac{(\Sigma x)^2}{n} = 94,456 - \frac{862^2}{8} = 1575.5$$

Determine SS_y:

$$SS_y = \Sigma y^2 - \frac{(\Sigma y)^2}{n} = 8211 - \frac{255^2}{8} = 82.875$$

Determine SS_{xy}:

$$SS_{xy} = \Sigma xy - \frac{(\Sigma x)(\Sigma y)}{n} = 27,264 - \frac{(862)(255)}{8} = -212.25$$

Determine b_1:

$$b_1 = \frac{SS_{xy}}{SS_x} = -\frac{212.25}{1575.5} = -0.1347$$

Determine $\bar{x}$, $\bar{y}$, and b_0:

$$\bar{x} = \frac{\Sigma x}{n} = \frac{862}{8} = 107.75$$

$$\bar{y} = \frac{\Sigma y}{n} = \frac{255}{8} = 31.875$$

$$b_0 = \bar{y} - b_1\bar{x}$$
$$= 31.875 - (-0.1347)(107.75)$$
$$= 46.3889$$

The least squares prediction equation thus becomes

$$\hat{y} = b_0 + b_1x = 46.3889 - 0.1347x$$

The regression equation for our sample of eight pairs of data is $\hat{y} = 46.3889 - 0.1347x$. It is always a good idea when computing the regression constants b_0 and b_1 to carry as many significant digits as possible in order to minimize the effects of rounding errors. For the remainder of this chapter we will use at least four-decimal-place accuracy in all computations involving b_0 and b_1. Figure 14.3 contains the graph of our regression equation.

FIGURE 14.3

Scattergram and regression
equation for data
of table 14.1

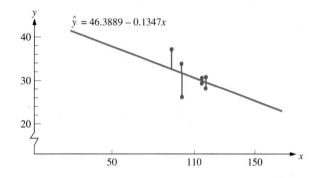

The errors $e = y - \hat{y}$ in using the equation $\hat{y} = 46.3889 - 0.1347x$ to predict the mpg ratings y for each engine size value are indicated on the graph in figure 14.3 by vertical-line segments. The errors e are frequently called **residuals**. Recall from section 4.3 that by using the least-squares criterion to obtain the line that best fits our data, we can calculate the minimum value for the **sum of squares for error,** SSE $= \Sigma (y - \hat{y})^2$.

The variance of the errors e is called **residual variance** and is denoted by s_e^2. The residual variance s_e^2, is found by dividing SSE $= \Sigma (y - \hat{y})^2$ by $n - 2$.

Residual Variance
$$s_e^2 = \frac{SSE}{n-2}$$

(14.2)

where $n - 2$ is the number of degrees of freedom associated with the variance of the residuals. Recall from section 3.2 that we divided SS by $n - 1$ to obtain s^2, an estimator for σ^2. We used $n - 1$ in the denominator because one degree of freedom was lost by using $\bar{x}$ to estimate μ in computing SS $= \Sigma (x - \mu)^2$. So to estimate σ^2 by using SSE $= \Sigma (y - \hat{y})^2 = \Sigma (y - b_0 - b_1x)^2$, we lose two degrees of freedom because β_0 and

β_1 are estimated by b_0 and b_1, respectively. To obtain the estimator s_e^2 for σ^2 we thus divide SSE by $n - 2$ degrees of freedom. The estimator s_e^2 provides a measure of how the points are scattered about the regression line. The dispersion varies directly with the magnitude of s_e^2. The positive square root of the residual variance is called the **standard error of estimate** and is denoted by s_e. We will use it in section 14.2 to draw inferences concerning the linear regression model.

Recall from section 4.3 that the sum of squares for error, SSE, can be found by using the computational formula

$$\text{SSE} = \text{SS}_y - b_1\text{SS}_{xy} \tag{14.3}$$

where b_1 has been substituted for m.

EXAMPLE 14.2

The sum of squares for error for our application is found to be

$$\text{SSE} = \text{SS}_y - b_1\text{SS}_{xy} = 82.875 - (-0.1347)(-212.25) = 54.2849$$

Hence, the residual variance is

$$s_e^2 - \frac{\text{SSE}}{n-2} = \frac{54.2849}{8-2} = 9.0475$$

If we assume that our data satisfy the probabilistic model $y = \beta_0 + \beta_1 x + \epsilon$, then we can use the following point estimators to estimate values associated with the model:

Model Parameter	Estimator
β_0	b_0
β_1	b_1
σ^2	s_e^2
$E(y\mid x)$	$\hat{y}$

EXAMPLE 14.3

For our application, we have the estimates shown in the table.

Parameter	Estimate
β_0	$46.3889\ (= b_0)$
β_1	$-0.1347\ (= b_1)$
σ^2	$9.0475\ (= s_e^2)$
$E(y\mid x = 98)$	$33.1883\ (= \hat{y}(98))$

The estimate for $E(y\mid x = 98)$, $\hat{y}(98)$, was found by substituting $x = 98$ into the regression equation $\hat{y} = 46.3889 - 0.1347x$ and solving for $\hat{y}$. The value $\hat{y}$ is used to predict the value of y and to estimate the value of $E(y\mid x)$. For an engine size of 98 cid, we can use $\hat{y}(98) = 33.2$ mpg to estimate the average mpg rating for all Ford Escort cars, as well as to predict the mpg rating for a particular Ford Escort car.

Predict or Estimate

In statistics we estimate only parameters, not random variables. The mean value of y when $x = 98$ is greatly different from some value of y chosen at random from the collection of all y values for which x equals 98. To make this distinction clear, we shall say that we predict the value of a random variable and estimate the value of a parameter.

Relevant Range of Prediction

To use the sample regression equation for prediction purposes, we must consider only the relevant range of the independent variable x. For our example, the relevant range is the set of six specifically selected engine sizes. Any interpolations between or beyond these six values must be done with extreme caution, because the estimated values are valid only for our six distinct values of x. Even in situations where the linear trend can be expected to continue, extrapolation should be used only with extreme caution.

EXAMPLE 14.4

Suppose a high linear correlation was found between the yields of a certain variety of tomato plant and the amount of fertilizer used per plot, and the amount of fertilizer that was used varied from 10 to 60 pounds per plot in 10-pound increments. At 50 pounds of fertilizer per plot, suppose a bumper tomato crop is predicted. The residual corresponding to $x = 50$ will be small. If the fertilizer level is raised to 100 pounds per plot, the *predicted* tomato yield might triple. But if this much fertilizer is used, the harvest would yield no tomatoes, since all the plants would die as a result of being over fertilized. Too much fertilizer is as bad for crop yields as is too little fertilizer. The point to be made is that the regression equation is to be used for prediction purposes only for those values close to the values used to arrive at the point estimates b_0 and b_1.

Effects of Outliers on Regression

EXAMPLE 14.5

Recall that an outlier is a piece of data that is not typical of the remaining data. Its effect in a regression analysis can be considerable. For example, the first four points in the accompanying table were generated from the equation $\hat{y} = x + 2$. The fifth point (4, 1) was recorded by mistake and lies below the line. Let's see how the point (4, 1) distorts the relationship.

x	y
0	2
1	3
2	4
3	5
4	1

The following table organizes the calculations for determining b_0 and b_1.

	x	y	x^2	y^2	xy
	0	2	0	4	0
	1	3	1	9	3
	2	4	4	16	8
	3	5	9	25	15
	4	1	16	1	4
Sum	10	15	30	55	30

The sum of squares for x is

$$SS_x = 30 - \frac{10^2}{5} = 10$$

The sum of cross products is

$$SS_{xy} = 30 - \frac{(10)(15)}{5} = 0$$

The slope of the regression equation is

$$b_1 = \frac{SS_{xy}}{SS_x} = \frac{0}{10} = 0$$

The mean of the x values is $\bar{x} = 2$ and the mean of the y values is $\bar{y} = 3$. The y-intercept is

$$b_0 = \bar{y} - b_1\bar{x} = 3 - 0(2) = 3$$

The resulting regression equation is $\hat{y} = 3$, which is quite different from the relationship that actually exists.

Recall from section 3.4 that a piece of data is an outlier if it lies more than three standard deviations (s_e) from the mean. One of our regression assumptions stipulated that at each x value, the values of y are normally distributed, and more than 99% of the distribution lie within three standard deviations of the mean. Thus, if the absolute value of a residual $y - \hat{y}$ is more than three times the standard error of estimate, it is typically a candidate for an outlier.

Fixed and Random x Values
If the x values are fixed, it makes sense to predict the y values; the y values are random variables. But it makes no sense to predict the values of x from the y values if the values of x are fixed. That is, given fixed values of x, it makes no sense to find the prediction equation for x given the values for y.

EXAMPLE 14.6
For the application involving engine size and mpg rating, it would make no sense to find the line relating engine size to mpg rating. In particular, solving the prediction equation $\hat{y} = 46.3889 - 0.1347x$ for x in terms of $\hat{y}$ and then using the resulting equation to predict engine sizes given mpg ratings makes no sense and should not be attempted, since the engine sizes are fixed values.

Summary
In summary, if there is a linear relationship between x and y, the population regression model can be expressed as

$$y = \beta_0 + \beta_1 x + \epsilon$$

This functional relationship can be reexpressed as

$$y = E(y|x) + \epsilon$$

where $E(y|x) = \beta_0 + \beta_1 x$. Since we do not have access to the entire population to determine the values of the constants β_0 and β_1, we can estimate their values from a bivariate sample using the method of least squares. The sample regression equation is $\hat{y} = b_0 + b_1 x$.

EXERCISE SET 14.1

Basic Skills

For exercises 1–5, assume that the linear model $y = \beta_0 + \beta_1 x + \epsilon$ holds.

1. a. Write the regression equation if $\beta_0 = 3$ and $\beta_1 = 4$.
    b. For the equation in part a, find the value of y for $x = 2$ if $\epsilon = 0.05$.
    c. For the equation in part a, find the error ϵ if $x = 2$ and $y = 6$.

2. a. Write the regression equation if $\beta_0 = -5$ and $\beta_1 = 7$.

    b. For the equation in part a, find the value of y for $x = 4$ if $\epsilon = 0.03$.
    c. For the equation in part a, find the error ϵ if $x = -2$ and $y = 5$.

3. If the regression equation for a sample of pairs is $\hat{y} = 4x - 8$, find
    a. the predicted value $\hat{y}(3)$.
    b. a point estimate of β_0.
    c. a point estimate of β_1.

d. a point estimate of the average value of y if $x = 3$, $E(y|x = 3)$.

4. If the regression equation for a sample of pairs is $\hat{y} = -3x + 7$, find
   a. the predicted value $\hat{y}(2)$.
   b. a point estimate of β_0.
   c. a point estimate of β_1.
   d. a point estimate of the average value of y if $x = 2$, $E(y|x = 4)$.

5. Jones Realty collected the accompanying data comparing the selling price y of new homes with the size x of the living space (in hundreds of square feet).

Living space x (in hundreds of square feet)	Selling price y (in thousands of dollars)
20	116
22	118
18	91
30	145
23	105
25	121

   a. Find a point estimate for the error variance σ^2.
   b. Find a point estimate for β_0.
   c. Find a point estimate for β_1.
   d. Find a point estimate for $E(y|x = 18)$.
   e. Predict a value for $y(24)$.
   f. If it has been determined that $\beta_1 = 0$, what is the best point estimate for y? for $E(y)$?

6. A major discount department store chain keeps extensive records on its salespeople. A sample of six salespeople produced the data given in the table concerning months on the job x and monthly sales in thousands of dollars y.

x	2	3	6	11	12	7
y	2.4	3.7	7.6	14.2	15	8

   a. Find a point estimate for β_0.
   b. Find a point estimate for β_1.
   c. Find a point estimate for the error variance.
   d. Find a point estimate for the average monthly sales for a salesperson who has been on the job 11 months.
   e. Predict a value for the monthly sales for a salesperson who has been on the job 6 months.
   f. If it has been determined that $\beta_1 = 0$, what is the best point estimate for y? for $E(y)$?

7. A car rental firm compiled the accompanying data concerning car maintenance costs y and miles driven x for seven of its automobiles.

Car	Miles driven x (in thousands of miles)	Maintenance Costs y (in dollars)
A	55	299
B	27	160
C	36	215
D	42	255
E	65	350
F	48	275
G	29	207

   a. Find a point estimate for β_0.
   b. Find a point estimate for β_1.
   c. Find a point estimate for the error variance σ^2.
   d. Find a point estimate for the average cost of driving a car 36,000 miles.
   e. Predict the cost of driving a car 29,000 miles.
   f. If it has been determined that $\beta_1 = 0$, what is the best point estimate for y? for $E(y)$?

SECTION 14.2   ***Inferences Concerning the Linear Regression Model***

In order to use the regression equation $\hat{y} = b_0 + b_1x$ for predictive purposes, we want to be reasonably sure that the slope β_1 of the regression equation $E(y|x) = \beta_0 + \beta_1x$ is not 0. For if $\beta_1 = 0$, then for every value of x, $E(y|x)$ would be identically equal to β_0, as shown in figure 14.4. Toward the goal of determining whether the slope of the population regression equation is different from 0, we shall separate SS_y into two components, SSE and SSR.

FIGURE 14.4
_____

$E(y|x) = \beta_0$ for all x when $\beta_1 = 0$

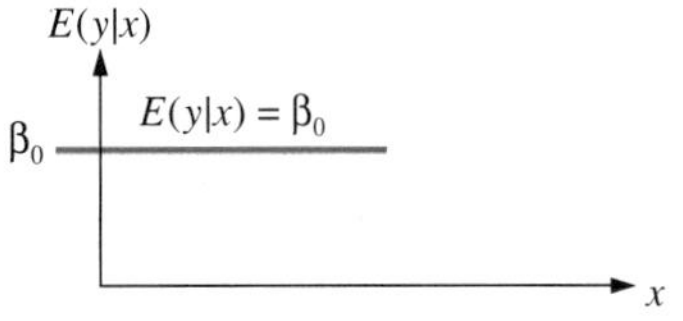

Partitioning SS_y

The deviation score $y - \bar{y}$ can be partitioned into the following two deviation scores:

$$y - \bar{y} = (y - \hat{y}) + (\hat{y} - \bar{y})$$

Consider the following diagram:

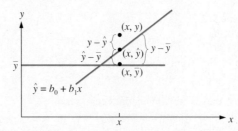

The component measurement $y - \hat{y}$ is due to error and the component measurement $\hat{y} - \bar{y}$ is due to linear regression. That is,

$$\underset{\substack{\uparrow \\ \text{Deviation} \\ \text{score}}}{y - \bar{y}} = \underset{\substack{\uparrow \\ \text{Due to} \\ \text{error}}}{(y - \hat{y})} + \underset{\substack{\uparrow \\ \text{Due to} \\ \text{regression}}}{(\hat{y} - \bar{y})}$$

It can be shown algebraically that SS_y is equal to the sum of squares for error (SSE) plus the sum of squares for regression (SSR), or

$$\underset{SS_y}{\underset{\uparrow}{\Sigma (y - \bar{y})^2}} = \underset{SSE}{\underset{\uparrow}{\Sigma (y - \hat{y})^2}} + \underset{SSR}{\underset{\uparrow}{\Sigma (\hat{y} - \bar{y})^2}}$$

where $\Sigma (\hat{y} - \bar{y})^2$ is called the **sum of squares for regression** and is denoted by SSR. Thus, we have the following relationship

$$SS_y = SSR + SSE \qquad (14.4)$$

Recall from section 14.1 that $SSE = SS_y - b_1 SS_{xy}$. It thus follows that SSR must be identically equal to $b_1 SS_{xy}$. To calculate SSE, this relationship is easier to use than the formula $SSR = \Sigma (\hat{y} - \bar{y})^2$ and is called the *computational formula for computing SSR:*

Computational Formula for SSR

$$SSR = b_1 SS_{xy} \qquad (14.5)$$

It can be shown that

$$df_y = df_E + df_R$$

where df_y is the number of degrees of freedom for SS_y, df_E is the number of degrees of freedom for SSE, and df_R is the number of degrees of freedom for SSR. Since $df_y = n - 1$ and $df_E = n - 2$, it follows that the degrees of freedom for SSR is

$$df_R = df_y - df_E = (n - 1) - (n - 2) = 1$$

Hence, we see that SSR has one degree of freedom associated with it.

Testing the Appropriateness of the Linear Model

We saw earlier that a value of s_e^2 provides an estimate for the error variance σ^2. If we assume that $\beta_1 = 0$ is true, then SSR provides another independent estimate for σ^2. Hence, if the slope of the population regression equation is 0, then we have two independent estimates for σ^2. The sampling distribution of the statistic SSR/s_e^2 is an F distribution with df $= (1, n - 2)$. That is, if $H_0: \beta_1 = 0$ is assumed to be true, then the F statistic serves as our test statistic, where F is defined as

$$F = \frac{SSR}{s_e^2}$$

with df $= (1, n - 2)$, or df $= 1$ for SSR and df $= n - 2$ for s_e^2. The F statistic can be used to determine if β_1 is different from 0. If the slope of the population regression equation is different from 0, then the equation can be used for prediction purposes.

APPLICATION 14.2

For the EPA gas mileage data in section 14.1, test using $\alpha = 0.05$ to determine if $\beta_1 \neq 0$.

Solution: The following values were computed in section 14.1:

$$SS_{xy} = -212.25$$
$$b_1 = -0.1347$$

Compute the sum of squares for regression SSR using (14.5):

$$SSR = b_1 SS_{xy} = (-212.25)(-0.1347) = 28.5901$$

Compute the estimated error variance s_e^2:

$$s_e^2 = \frac{SSE}{n - 2} = \frac{54.2849}{8 - 2} = 9.0475$$

Compute the value of the F statistic:

$$F = \frac{SSR}{s_e^2} = \frac{28.5901}{9.0475} = 3.16$$

Find the critical value: $F_{0.05}(1, 6) = 5.99$. Since $F = 3.16 < F_{0.05}(1, 6) = 5.99$, we fail to reject $H_0: \beta_1 = 0$. Thus, we conclude that the equation $\hat{y} = 46.3889 - 0.1347x$ should not be used for predictive purposes, and we have no evidence to support that a linear model is correct for our data. A nonlinear model, such as $y = \beta_0 + \beta_1 x + \beta_2 x^2 + \epsilon$, may be appropriate. ▪

Mean Squares

Recall that any sum of squares divided by its associated degrees of freedom provides a variance estimate, called a **mean square,** denoted by MS. That is,

$$MS = \frac{SS}{df}$$

As a result, we can define the **mean square for regression** MSR and **the mean square for error** MSE:

$$MSR = \frac{SSR}{1} = SSR$$

$$MSE = \frac{SSE}{n - 2} = s_e^2$$

Notice that SSR is a mean square and thus a variance estimate.

By using the concept of mean squares, we can write the F statistic as SSR/s_e^2, since SSR and s_e^2 are variance estimates. The computations involved in testing the null hypothesis $H_0: \beta_1 = 0$ against the alternative hypothesis $H_1: \beta_1 \neq 0$ can then be summarized by using an **ANOVA summary table,** the general form of which is shown in table 14.2.

TABLE 14.2

General Form of ANOVA Summary Table

Source of Variation	SS	df	MS	F
Regression	SSR	1	SSR	SSR/s_e^2
Error	SSE	$n-2$	$s_e^2 = SSE/(n-2)$	
Total(y)	SS_y	$n-1$		

EXAMPLE 14.7

For the gasoline mileage example, the results are summarized in table 14.3.

TABLE 14.3

ANOVA Summary Table for Data in Table 14.1

Source of Variance	SS	df	MS	F
Regression	28.5901	1	28.5901	3.16
Error	54.2849	6	9.0475	
Total	82.875	7		

To summarize the discussion thus far, note that the following computational steps are typically followed when testing $H_0: \beta_1 = 0$:

1. Compute SS_y using the formula $SS_y = \Sigma y^2 - (\Sigma y)^2/n$.
2. Compute SSR using the formula $SSR = b_1 SS_{xy}$.
3. Compute SSE using the relation $SSE = SS_y - SSR$.
4. Compute s_e^2 using the formula $s_e^2 = SSE/(n-2)$.
5. Compute F using $F = SSR/s_e^2$.
6. Find the critical value $F_\alpha(1, n-2)$ in the F tables.
7. Compare the F test statistic with the critical value $F_\alpha(1, n-2)$; if $F \geq F_\alpha(1, n-2)$, then we reject H_0.

Computer Display 14.2

MINITAB can be used to carry out the analyses for the EPA gas mileage data. Computer display 14.2 contains the commands used and corresponding output.

```
MTB > READ C1 C2
DATA> 121 30
DATA> 120 31
DATA> 97 34
DATA> 98 27
DATA> 122 29
DATA> 97 34
DATA> 85 38
DATA> 122 32
DATA> END
   8 ROWS READ
MTB > NAME C1 'ENG_SIZE'
MTB > NAME C2 'MPG'
```

```
MTB > BRIEF LEVEL 1
MTB > REGRESSION C2 1 C1

THE REGRESSION EQUATION IS MPG = 46.4 - 0.135 ENG_SIZE

PREDICTOR      COEF      STDEV    T-RATIO       P
CONSTANT      46.391     8.234       5.63   0.001
ENG_SIZE     -0.13472   0.07578     -1.78   0.126

S = 3.008   R-SQ = 34.5%   R-SQ(ADJ) = 23.6%

ANALYSIS OF VARIANCE

SOURCE        DF       SS       MS      F       P
REGRESSION     1    28.594   28.594   3.16   0.126
ERROR          6    54.281    9.047
TOTAL          7    82.875
```

Notice that the value of the F statistic agrees with the value we computed and that the F test is significant at the 12.6% level. The values in the summary table differ only slightly from the values we computed; the differences are due to the greater precision in the calculations done by MINITAB.

Testing H_0: $\beta_1 = 0$ Using the t Distributions

We mentioned in section 13.1 that $t^2 = F$ for an F statistic with df $= (1, n - 2)$. Since $t = \sqrt{F}$, the t statistic can be used to test the null hypothesis H_0: $\beta_1 = 0$. The t statistic is defined as

$$t = \frac{\sqrt{\text{MSR}}}{s_e}$$

Since MSR $= b_1 \text{SS}_{xy}$ and $b_1 = \text{SS}_{xy}/\text{SS}_x$, we have

$$\text{MSR} = b_1^2 \text{SS}_x$$

By using these facts, we can express the t statistic as

$$t = \frac{\sqrt{b_1^2 \text{SS}_x}}{s_e} = \frac{b_1}{s_e/\sqrt{\text{SS}_x}}$$

Hence, if the null hypothesis H_0: $\beta_1 = 0$ is assumed to be true, then the test statistic can be expressed as

> **t Statistic for Testing H_0: $\beta_1 = 0$**
>
> $$t = \frac{b_1}{s_e/\sqrt{\text{SS}_x}}$$
> where df $= n - 2$

(14.6)

The two critical values, $t_{\alpha/2}$ and $F_\alpha(1, n - 2)$ are related as follows:

$$t_{\alpha/2}^2 = F_\alpha(1, n - 2)$$

where df $= n - 2$.

APPLICATION 14.3

For the EPA gas mileage data, test to determine if $\beta_1 \neq 0$ by using the t test and $\alpha = 0.05$.

Solution: By using formula (14.6), we have

$$t = \frac{b_1}{s_e/\sqrt{\text{SS}_x}} = \frac{-0.1347}{\sqrt{9.0475}/\sqrt{1575.5}} = -1.7775$$

The critical values $\pm t_{0.025}$ for df $= 6$ are ± 2.447. Since $-t_{0.025} < t < t_{0.025}$, we fail to reject H_0: $\beta_1 = 0$. Hence, we have no evidence to suggest that the linear model is appropriate for our data. Note that

$$t^2 = (-1.7775)^2 \approx 3.16 = F$$

and for df $= 6$, we have

$$(t_{0.025})^2 = (2.447)^2 \approx 5.99 = F_{0.05}(1, 6) \quad \blacksquare$$

Confidence Intervals for β_1

The limits for a $(1 - \alpha)100\%$ confidence interval for β_1 can be found by using

> **$(1 - \alpha)100\%$ Confidence Interval Limits For β_1**
>
> $$b_1 \pm t_{\alpha/2}\frac{s_e}{\sqrt{SS_x}}$$ (14.7)
>
> where df $= n - 2$

By using formula (14.7), we can determine a 95% confidence interval for β_1, the slope of the population regression equation for the EPA gas mileage data. The confidence interval limits are given by

$$b_1 \pm t_{\alpha/2}(s_e/\sqrt{SS_x})$$
$$-0.1347 \pm (2.447)(0.076)$$
$$-0.1347 \pm 0.1854$$

Thus, a 95% confidence interval for β_1 is $(-0.3201, 0.0507)$. Since the interval contains 0, we cannot conclude that $\beta_1 \neq 0$. We can be 95% confident that the value of β_1, the population regression slope, is between -0.3201 and 0.0507.

Confidence Intervals for $E(y|x_0)$

Once the sample regression equation has been found and it has been determined that the model $E(y|x) = \beta_0 + \beta_1 x$ is appropriate for the data, then we can use the regression equation $\hat{y} = b_0 + b_1 x$ to make predictions. In so doing, we will want to estimate the average value for y given x, $E(y|x)$, as well as to predict the random y value for a given value x.

The limits for a $(1 - \alpha)100\%$ confidence interval for $E(y|x_0)$ are given by the following expression:

> **$(1 - \alpha)100\%$ Confidence Interval Limits for $E(y|x_0)$**
>
> $$\hat{y} \pm t_{\alpha/2}s_e\sqrt{\frac{1}{n} + \frac{(x_0 - \bar{x})^2}{SS_x}}$$ (14.8)
>
> where df $= n - 2$

The confidence interval is shortest when $x_0 = \bar{x}$, and the length of the interval increases as x_0 moves away from the mean $\bar{x}$. Consider application 14.4, which shows the construction of a $(1 - \alpha)100\%$ confidence interval for $E(y|x_0)$.

A P P L I C A T I O N 14.4

In a study of how corn yield depends on the amount of fertilizer, the following data were obtained for six different plots of land:

x (fertilizer in 100 lb/acre)	1	2	3	4	5	6
y (yield in bu/acre)	40	50	50	70	65	80

Determine if the linear model is appropriate using $\alpha = 0.05$ and construct a 95% confidence interval for $E(y|x = 4)$.

Solution: With the aid of a hand held calculator we determined the following:

Regression equation: $\hat{y} = 32.6667 + 7.5714x$

$$SS_x = 17.5$$
$$SS_y = 1120.8333$$
$$SS_{xy} = 132.5$$
$$s_e^2 = 29.4123$$

The value of $\hat{y}(4)$ is found by using the regression equation:

$$\hat{y}(4) = 32.6667 + 7.5714(4) = 62.9523$$

To determine if β_1 is different from zero, we use formula (14.6) to determine the value of the t statistic corresponding to b_1.

$$t = \frac{b_1}{s_e/\sqrt{SS_x}}$$

$$= \frac{7.5714}{5.4233/\sqrt{17.5}} = 5.84$$

The positive critical value is found to be $t_{0.025} = 2.776$ for df $= 4$. Since $t = 5.84 > t_{0.025} = 2.776$, we reject H_0: $\beta_1 = 0$ and conclude that the linear model is appropriate. Therefore, we can use the equation $\hat{y} = 32.6667 + 7.5714x$ for prediction purposes.

By using formula (14.8), the limits of a 95% confidence interval for $E(y|x = 4)$ are found as follows:

$$\hat{y} \pm t_{0.025}s_e \sqrt{\frac{1}{n} + \frac{(x_0 - \bar{x})^2}{SS_x}}$$

where df $= 6 - 2 = 4$.

$$62.9523 \pm (2.447)(5.4233) \sqrt{\frac{1}{6} + \frac{(4 - 3.5)^2}{17.5}}$$

$$62.9523 \pm 5.6452$$

Thus, a 95% confidence interval for the average yield of corn (in bushels per acre) given $x = 400$ pounds of fertilizer per acre is (57.31, 68.60). ■

Figure 14.5 illustrates the *95% confidence interval band* for the line representing the regression equation $\hat{y} = 32.6667 + 7.5714x$ from application 14.4. One side of the band was formed by joining the lower limits of the 95% confidence intervals for

FIGURE 14.5

Confidence interval band for application 14.4

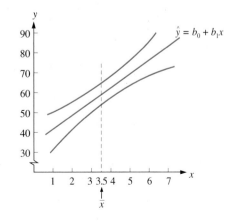

the average values of y at each value of x, whereas the other side of the band was formed by joining the upper limits of the 95% confidence intervals for the average values of y at each value of x. Note that the band is narrowest at $x = \bar{x}$ and gets wider as x gets farther away from $\bar{x}$ (see also table 14.4).

TABLE 14.4

95% Confidence Interval Limits for $E(y|x)$

x	L_1	L_2	Width
1	30.63	49.84	19.21
2	40.60	55.02	14.42
3	49.74	61.03	11.29
3.5	53.75	64.58	10.83
4	57.31	68.60	11.29
5	63.31	77.73	14.42
6	68.49	87.70	19.21

Prediction Intervals for y

EXAMPLE 14.8

There are occasions when we shall want to predict an individual value of y for a given value of x, instead of estimating the average value for all the y values for a particular x value. With the corn fertilizer data of application 14.4, for example, we may want to predict the corn yield in bushels per acre for a plot planted with $x = 400$ pounds of fertilizer per acre instead of estimating the average yield for all plots that are fertilized at a level of 400 pounds per acre.

The limits for a $(1 - \alpha)100\%$ prediction interval for the value of a single random y value for a given value of $x = x_0$ are given by the following expression.

$$(1 - \alpha)100\% \text{ Confidence Interval Limits for } y \text{ at } x = x_0$$

$$\hat{y} \pm t_{\alpha/2}s_e\sqrt{1 + \frac{1}{n} + \frac{(x_0 - \bar{x})^2}{SS_x}} \qquad (14.9)$$

$$\text{where df} = n - 2$$

Note once again that the prediction band widens as the distance between x_0 and $\bar{x}$ increases. The basic difference between the formulas for the limits of a prediction interval for y given x_0 and the limits for a confidence interval for the average value of y given x_0 is the inclusion of the numeral 1 in the expression under the radical sign of formula (14.9).

APPLICATION 14.5

For the corn-fertilizer data in application 14.4, construct a 95% prediction interval for the corn yield of a particular field if 400 pounds per acre of fertilizer are used.

Solution: We use formula (14.9) to find the prediction interval limits:

$$\hat{y} \pm t_{\alpha/2}s_e\sqrt{1 + \frac{1}{n} + \frac{(x_0 - \bar{x})^2}{SS_x}}$$

where df $= n - 2$.

$$62.9523 \pm (2.447)(5.4233)\sqrt{1 + \frac{1}{6} + \frac{(4 - 3.5)^2}{17.5}}$$

$$62.9523 \pm 14.4216$$

$$(48.53, 77.37)$$

Thus, the 95% prediction interval for the yield of corn when 400 pounds of fertilizer per acre are used is (48.53, 77.37). We can be 95% confident that the corn yield is at most 77.37 bushels per acre and at least 48.53 bushels per acre. Note that this prediction interval has a width $w = 77.37 - 48.53 = 28.84$, while the width for the 95% confidence interval for $E(y|4)$ found in application 14.4 is $w = 68.60 - 57.31 = 11.29$. In general, it can be shown that for fixed values of x_0, and α, the width of the prediction interval for y given x_0 will exceed the width of the confidence interval for $E(y|x_0)$. ▪

In section 14.1 we saw that the population regression model can be expressed as $y = E(y|x) + \epsilon$, where, for a fixed value of x, y is the observed value, $E(y|x)$ is the average value of y, and ϵ is random error. This model can be approximated by the least-squares equation, $\hat{y} = b_0 + b_1x$. For a particular value of x, for example, x_0, the vertical distance between a particular value of y, y_0, and the predicted value of y, $\hat{y}$, represents the error of prediction. This error is shown in figure 14.6. We can see that the vertical distance between the two lines represents the error of estimate when $\hat{y}$ is used to estimate the average value of y, $E(y|x)$. When $\hat{y}$ is used to predict y_0, the prediction error is the sum of two errors, the error of estimate and random error. As a consequence of this relationship, for a fixed value of x, the error of predicting a particular value of y is usually larger than the error of estimating the average value of y. And for a particular value of x, the width of the prediction interval for y is usually greater than the width of the confidence interval for $E(y|x)$. We can also see that both the error of estimate and the error of prediction decrease as x_0 approaches $\bar{x}$ (the two lines cross at the point where $x = \bar{x}$).

FIGURE 14.6

Errors of estimating the average value of y and predicting a future value of y for a given value of x

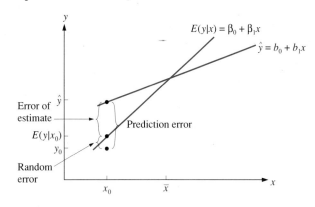

EXERCISE SET 14.2

Further Applications

1. The population (in thousands) of seven countries (x) and the corresponding number of practicing physicians (y) are given in the table.

x	38	52	75	32	60	43	57
y	20	23	35	20	26	25	22

a. Use the F test and $\alpha = 0.05$ to determine if the data satisfy the linear regression model.

b. Construct a 95% confidence interval for β_1.
c. Use the t test and $\alpha = 0.05$ to determine if the linear model is appropriate for the data.

2. The number of murders (y) committed annually in the State of Maryland from 1978 to 1982 (x) using a blunt object are listed in the accompanying table:

Year x	1978	1979	1980	1981	1982
Number of Murders y	10	12	17	22	20

a. By coding the years using the relation $x =$ year $- 1978$, determine if the pairs (x, y) satisfy the linear model. Use the F test and $\alpha = 0.05$.

b. Construct a 95% confidence interval for β_1.

c. Use the t test and $\alpha = 0.05$ to determine if the linear model is appropriate for the data.

3. The data in the table indicate the length of confinement in days (x) and the cost in dollars (y) at Memorial Hospital.

x	3	1	11	4	8
y	685	270	2310	885	1700

a. By using the F test and $\alpha = 0.05$, test to determine if $\beta_1 \neq 0$.

b. Construct a 95% confidence interval for the average cost for four days of hospital confinement.

c. Construct a 95% prediction interval for the cost of a four-day stay in the hospital.

d. Construct a 95% prediction interval for the cost of an eight-day stay in the hospital.

e. What is the additional cost for one more day in the hospital?

f. Use the t test and $\alpha = 0.05$ to determine if the linear model is appropriate for the data.

4. The following data resulted from an experiment to study the relationship between age (x) in months and size of vocabulary (y) for young children aged 18 to 48 months.

x	18	24	30	36	42	48
y	100	250	460	890	1210	1530

a. Construct a 99% confidence interval for β_1.

b. Construct a 99% confidence interval for $E(y\,|\,x = 36)$.

c. Construct a 99% prediction interval for $y(36)$.

d. Use the t test and $\alpha = 0.01$ to determine if the linear model is appropriate for the data.

5. The data in the table lists the highway mpg ratings (x) of the 1990 minicompact automobiles and their corresponding lifetime amounts (in tons) of carbon dioxide (CO_2) emitted (y).[47]

Model	mpg	CO_2 Emitted
Porsche 928	19	61.99
Maserati 222E (L4)	18	59.52
Porsche 928	19	57.93
Maserati 222E (M5)	20	56.42
Porsche 911 (M5)	22	53.62
Porsche 911 (S4)	22	52.04
Porsche 911 (M5)	24	49.65
Porsche 944	26	46.23
Nissan 240SX (L4)	25	43.74

Model	mpg	CO_2 Emitted
Nissan 240SX (M5)	27	42.04
Volkswagon Cabriolet (A3)	28	38.54
Volkswagon Cabriolet (M5)	32	34.57

a. By using the F test and $\alpha = 0.05$, test to determine if $\beta_1 \neq 0$.

b. Construct a 95% confidence interval for β_1.

c. Construct a 95% confidence interval for $E(y\,|\,x = 24)$.

d. Construct a 95% prediction interval for $y(25)$.

6. The data in the accompanying table lists the highway mpg ratings (x) of five 1990 two-seated automobiles and their corresponding lifetime amounts (in tons) of carbon dioxide (CO_2) emitted (y).[48]

Model	mpg	CO_2 Emitted
Lamborghini Countach	10	124.77
Ferrari Testarossa	15	79.44
Jaguar XJ-S	17	65.76
Ferrari 348 TB/TS	19	61.99
Mercedes-Benz 500SL	18	61.59

a. By using the F test and $\alpha = 0.05$, test to determine if $\beta_1 \neq 0$.

b. Construct a 95% confidence interval for β_1.

c. Construct a 95% confidence interval for $E(y\,|\,x = 17)$.

d. Construct a 95% prediction interval for $y(15)$.

7. The accompanying table lists the weights (in pounds) and the heights (in inches) of a sample of ten five-year-old boys.

Height (x)	38	39	40	41	42	43	44	45	46	47
Weight (y)	34	35	36	38	39	41	44	46	47	49

a. By using the F test and $\alpha = 0.05$, test to determine if $\beta_1 \neq 0$.

b. Construct a 95% confidence interval for β_1.

c. Construct a 95% confidence interval for $E(y\,|\,x = 40)$.

d. Construct a 95% prediction interval for $y(45)$.

8. The accompanying table lists the number of calories and the amount of carbohydrates (in grams) of a sample of six 100-gram cans of fruit.

Canned Fruit	Calories (y)	Carbohydrates (x)
Applesauce	72	18.6
Bananas	84	21.6
Peaches	81	20.7
Pears	66	17.1
Plums with tapioca	94	24.3
Prunes with tapioca	86	22.4

a. By using the F test and $\alpha = 0.01$, test to determine if $\beta_1 \neq 0$.

b. Construct a 99% confidence interval for β_1.

c. Construct a 99% confidence interval for $E(y \mid x = 80)$.

d. Construct a 99% prediction interval for $y(85)$.

Going Beyond

9. For the data in exercise 4, show that if each x value is multiplied by $1/12$ and each y value is multiplied by 0.1, then the value of the t statistic for testing H_0: $\beta_1 = 0$ versus H_1: $\beta_1 \neq 0$ for the new data pairs is equal to the value of the t statistic for the original data pairs.

10. Suppose the value of the t statistic for testing H_0: $\beta_1 = 0$ versus H_1: $\beta_1 \neq 0$ is t_0 and c and d are positive numbers. If each value of x is multiplied by c and each value of y is multiplied by d, show that the value of the t statistic for testing H_0: $\beta_1 = 0$ versus H_1: $\beta_1 \neq 0$ for the transformed data is t_0.

11. The regression slope b_1 can be viewed as a random variable. After examining formula (14.6) and recalling the definition of the t statistic, speculate as to the mean and standard error of b_1.

<h2>SECTION 14.3   Correlation Analysis</h2>

In this section we want to reexamine the linear correlation coefficient r that was introduced in section 4.2 and extend its use to forming inferences about the population correlation coefficient. The population correlation coefficient is commonly denoted by the Greek letter ρ (rho). Let's recall several formulas from section 4.2:

Pearson's Correlation Coefficient

$$r = \frac{SS_{xy}}{\sqrt{SS_x SS_y}}$$

(14.10)

Relationship Between b_1 and r

$$b_1 = r \frac{s_y}{s_x}$$

The regression slope b_1 and the correlation coefficient r for a sample of bivariate data are related by the second formula above; the same relationship holds for the population of bivariate data:

$$\beta_1 = \rho \frac{\sigma_y}{\sigma_x}$$

(14.11)

A statistical test for the significance of a linear relationship between x and y can be performed by testing the null hypothesis H_0: $\rho = 0$ against the alternative hypothesis H_1: $\rho \neq 0$. By examining formula (14.11) we see that β_1 and ρ agree in sign (either both positive or both negative) and β_1 is 0 if and only if $\rho = 0$. As a result, testing the null hypothesis H_0: $\rho = 0$ against the alternative hypothesis H_1: $\rho \neq 0$ is equivalent to testing the null hypothesis H_0: $\beta_1 = 0$ against the alternative hypothesis H_1: $\beta_1 \neq 0$.

In some situations, we are not interested in determining the linear regression equation for predictive purposes. In other situations, if both x and y are random variables (instead of just y), then the linear regression model we developed is not appropriate. Therefore, it may be desirable to test for a significant linear relationship between the two variables without performing a regression analysis. If we assume that samples of n data pairs (x, y) are drawn from two normal populations, the x values from one and the y values from the other, then we can perform a hypothesis test to determine if $\rho \neq 0$. If the null hypothesis H_0: $\rho = 0$ is assumed to be true, it can be shown that the sampling distribution of the statistic

$$r \frac{\sqrt{n - 2}}{\sqrt{1 - r^2}}$$

is a t distribution with $(n - 2)$ degrees of freedom. Thus, r has the following t value:

t Value for r
$$t = r\frac{\sqrt{n - 2}}{\sqrt{1 - r^2}}$$
with df $= n - 2$

(14.12)

APPLICATION 14.6

For the EPA gas mileage data of section 14.1, test using $\alpha = 0.05$ to determine if $\rho < 0$.

Solution: The statistical hypotheses are

$$H_0\text{: } \rho \geq 0 \qquad H_1\text{: } \rho < 0$$

The sample correlation coefficient is

$$r = \frac{\text{SS}_{xy}}{\sqrt{\text{SS}_x\text{SS}_y}} = \frac{-212.25}{\sqrt{(1575.5)(82.875)}} = -0.5874$$

From formula (14.12), the t value for r is

$$t = r\frac{\sqrt{n - 2}}{\sqrt{1 - r^2}} = -0.5874\frac{\sqrt{8 - 2}}{\sqrt{1 - (-0.5874)^2}} = -1.78$$

By using the t table (see the back endpaper) we find $-t_{0.05} = -1.943$ for df $= 6$. Since $-t_{0.05} = -1.943 < t = -1.78$, we fail to reject the null hypothesis H_0: $\rho \geq 0$. ∎

Coefficient of Determination

Recall from section 14.2 that SS_y can be partitioned into two sums of squares, SSE and SSR:

$$\text{SS}_y = \text{SSR} + \text{SSE}$$

Dividing both sides of this equation by SS_y, we have

$$1 = \frac{\text{SSR}}{\text{SS}_y} + \frac{\text{SSE}}{\text{SS}_y}$$

Notice that both SSR/SS_y and SSE/SS_y are nonnegative numbers that sum to 1. The expression SSR/SS_y represents the percentage of the total sum of squares that is explained by the regression equation and is called the **coefficient of determination.** It is denoted by r^2. Thus, we have

Coefficient of Determination
$$r^2 = \frac{\text{Sum of squares explained by regression}}{\text{Total sum of squares (before regression)}} = \frac{\text{SSR}}{\text{SS}_y}$$

Since

$$1 = \frac{SSR}{SS_y} + \frac{SSE}{SS_y} = r^2 + \frac{SSE}{SS_y}$$

and

$$1 - r^2 = \frac{SSE}{SS_y}$$

the expression $1 - r^2$ represents the percentage of the total sum of squares that cannot be attributed to regression, but can be attributed to error. The coefficient of determination is quite useful in estimating the strength of the linear relationship between two variables. It can be shown that the coefficient of determination r^2 is also the square of the correlation coefficient. This is the reason the coefficient of determination is denoted by r^2.

APPLICATION 14.7   For the EPA gas mileage data in table 14.1, determine the coefficient of determination and interpret your result.

Solution: From application 14.6 we have $r = -0.5874$.
     The value of the coefficient of determination is

$$r^2 = (-0.5874)^2 = 0.345$$

Approximately 35% of the total sum of squares (before regression) can be attributed to the linear relationship between mileage rating and engine size, and approximately 65% of SS_y is not explained by the linear model $y = \beta_0 + \beta_1 x + \epsilon$. A nonlinear model, such as the quadratic model $y = \beta_0 + \beta_1 x + \beta_2 x^2 + \epsilon$, may account for a larger percentage of SS_y. ▪

     In regression analyses involving more than one independent variable, there are many kinds of correlation coefficients. One important coefficient is the multiple coefficient of determination. Like r^2, the **multiple coefficient of determination** R^2 gives the proportion of the total sum of squares SS_y that can be attributed to (or explained by) all the independent variables. We shall examine the multiple coefficient of determination R^2 in section 14.4.

EXERCISE SET 14.3

Further Applications

1. A small retail business has determined that the correlation coefficient between monthly expenses and profits for the past year is $r = 0.56$. Assuming that both expenses and profits are approximately normal, test at the 0.05 level of significance the null hypothesis H_0: $\rho = 0$ against the alternative hypothesis H_1: $\rho \neq 0$.

2. A medical pathologist has found from a sample of 12 human skeletons that body size and bone size have a correlation $r = 0.46$. Based on this one sample, should bone size from skeletal remains be used to predict body sizes? Use $\alpha = 0.01$ and assume that bone sizes and body weights are normally distributed.

3. A small business determined that the correlation between weekly sales (in dollars) and weekly costs (in dollars) of advertising was $r = 0.61$ for a ten-week period. Determine if there is a positive correlation between weekly sales and weekly advertising costs. Assume that weekly sales and weekly advertising cost follow normal distributions. Use $\alpha = 0.01$.

4. A bookstore owner, using a sample of 30 books, found the linear correlation between book cost and book thickness to be $r = 0.81$. Assuming that book cost and thickness are normally distributed, test the null hypothesis H_0: $\rho \leq 0$ against the alternative hypothesis H_1: $\rho > 0$. Use $\alpha = 0.05$.

5. In a study to determine the relationship between weights of 1-year-old babies and their weights 24 years later, a sample of 20 medical files for females showed that $r = 0.57$. Test to determine if the population correlation coefficient is positive, assuming that the population of weights of 1-year-olds and the population of weights of 25-year-old weights are normally distributed. Use $\alpha = 0.05$.

6. The level of serum cholesterol and fat intake for a sample of 25 men was shown to be 0.78. Determine if there is a positive correlation between level of serum cholesterol and level of fat intake. Assume that the cholesterol and fat levels follow normal distributions. Use $\alpha = 0.05$.

7. In an effort to determine the relationship between annual wages for employees and the number of days absent from work due to sickness, a large corporation studied the personnel records for a random sample of twelve employees. The paired data are provided in the table.

Employee	Annual Wages (in thousands of dollars)	Days Missed
1	15.7	4
2	17.2	3
3	13.8	6
4	24.2	5
5	15.0	3
6	12.7	12
7	13.8	5
8	18.7	1
9	10.8	12
10	11.8	11
11	25.4	2
12	17.2	4

a. Determine Pearson's correlation coefficient r.

b. Test at $\alpha = 0.05$ to determine if the number of days missed is related to annual wages.

c. Test the null hypothesis H_0: $\rho \leq 0$ against the alternative hypothesis H_1: $\rho > 0$ by using formula (14.12) and $\alpha = 0.05$.

d. Calculate the coefficient of determination and interpret the result.

8. The heights x (in inches) and weights y (in pounds) for eight athletes are given in the accompanying table.

Heights (x)	70	67	69	74	72	75	73	70
Weights (y)	173	163	195	196	167	220	191	175

a. Determine Pearson's correlation coefficient r.

b. Test at $\alpha = 0.05$ to determine if height is related to weight.

c. Test the null hypothesis H_0: $\rho \leq 0$ against the alternative hypothesis H_1: $\rho > 0$ by using formula (14.12) and $\alpha = 0.05$.

d. Calculate the coefficient of determination and interpret the result.

Going Beyond

9. For exercise 1, construct a 95% confidence interval for ρ.

10. For exercise 2, construct a 99% confidence interval for ρ.

11. Prove that $b_1 = r s_y / s_x$.

12. Explain how to define the linear correlation coefficient r in terms of the coefficient of determination.

13. One interpretation of r^2 is that it indicates the percentage of the variation in y accounted for by the variation in x. Explain.

S E C T I O N 1 4 . 4   *Multiple Linear Regression*

In sections 4.3 and 14.2 we used the least squares procedure to construct a linear regression equation where one variable, x, can be used to predict the value of another variable, y. The regression equation has the form: $\hat{y} = b_0 + b_1 x$. For the gas-mileage data of section 14.1, the regression equation is $\hat{y} = 46.3889 - 0.1347x$, where x represents the size of the engine (in cid) and y represents the rate of gasoline consumption (in mpg). We should be able to make a better prediction of gasoline consumption if we knew more information about the cars involved in the study. For example, in addition to the size-of-engine variable, we would expect other variables, such as the weight of a car, the type of transmission, the type of driving (highway or city), and the size and type of tires, to also influence the rate of fuel consumption.

When more than one independent variable is used to predict the values of a dependent variable, the process is called **multiple regression analysis.** Multiple regression analysis includes the use of linear and nonlinear equations. Linear regression equations have the form

$$\hat{y} = b_0 + b_1x_1 + b_2x_2 + b_3x_3 + \cdots + b_kx_k$$

where $\hat{y}$ is the variable to be predicted; $x_1, x_2, x_3, \ldots,$ and x_k are k variables with known values; and $b_0, b_1, b_2, \ldots,$ and b_k are numerical coefficients to be determined from the known information.

Examples of **nonlinear regression equations** are

$$\hat{y} = b_0 + b_1x_1 + b_2x_2^2 + b_3x_3^3$$
$$\hat{y} = b_0 + b_1x_1 + b_2x_1x_2 + b_3x_3^2$$
$$\hat{y} = b_0 + b_1\ln x_1 + b_2x_2 + b_3x_3$$

In this text, we shall only be concerned with linear regression equations.

EXAMPLE 14.9

Many premedical programs use average MCAT scores of students exiting their programs as an indication of the quality of their programs. Variables known to influence average MCAT scores (y) are combined SAT math and verbal scores (x_1) and GPAs (x_2) of premed students. Table 14.5 shows measurements on x_1, x_2, and y for six students who have gone through a premed program and have taken the MCAT.

TABLE 14.5

SAT Scores, GPAs, and
MCAT Scores for Six
Premed Students

Student	Combined SAT Score (x_1)	GPA (x_2)	Average MCAT Score (y)
A	1200	3.8	12.4
B	1350	3.4	13.3
C	1000	2.9	9.2
D	1250	3.3	10.6
E	1425	3.9	13.2
F	1340	3.1	11.2

With this information we can find a linear equation that will enable us to predict the average MCAT score of a student in terms of the student's GPA and combined SAT score.

The linear equation for the data of example 14.9 has the form $\hat{y} = b_0 + b_1x_1 + b_2x_2$. The values of b_0, b_1, and b_2 can found by using the least squares method. As was the case in chapter 4, this method minimizes the sum of squares, $\Sigma (y - \hat{y})^2$, where the values of y are the observed values and the values of $\hat{y}$ come from a linear equation in two independent variables. In principle, the method is very similar to the two-variable case studied in chapter 4, but the calculations are somewhat more involved. The method of least squares in this case involves solving three linear equations in three unknowns. These equations, known as **normal equations,** are as follows:

$$\Sigma y = nb_0 + b_1(\Sigma x_1) + b_2(\Sigma x_2)$$
$$\Sigma x_1y = b_0(\Sigma x_1) + b_1(\Sigma x_1^2) + b_2(\Sigma x_1x_2)$$
$$\Sigma x_2y = b_0(\Sigma x_2) + b_1(\Sigma x_1x_2) + b_2(\Sigma x_2^2)$$

EXAMPLE 14.10

Let's determine the normal equations for the data in example 14.9. Table 14.6 organizes the computations.

TABLE 14.6

Data for Six Premed Students

x_1	x_2	x_1^2	x_2^2	x_1x_2	y	x_1y	x_2y
1,200	3.8	1,440,000	14.44	4,560	12.4	14,880	47.12
1,350	3.4	1,822,500	11.56	4,590	13.3	17,955	45.22
1,000	2.9	1,000,000	8.41	2,900	9.2	9,200	26.68
1,250	3.3	1,562,500	10.89	4,125	10.6	13,250	34.98
1,425	3.9	2,030,625	15.21	5,557.5	13.2	18,810	51.48
1,340	3.1	1,795,600	9.61	4,154	11.2	15,008	34.72
7,565	20.4	9,651,225	70.12	25,886.5	69.9	89,103	240.2

The normal equations for this example are

$$69.9 = 6b_0 + 7,565b_1 + 20.4b_2$$
$$89,103 = 7,565b_0 + 9,651,225b_1 + 25,886.5b_2$$
$$240.2 = 20.4b_0 + 25,886.5b_1 + 70.12b_2$$

To solve these three linear equations for b_0, b_1, and b_2 is tedious, to say the least. Matrices are typically used to simplify the process. Nowadays, these kinds of computations are relegated to the computer.

The values of b_0, b_1, and b_2 for the linear regression equation $\hat{y} = b_0 + b_1x_1 + b_2x_2$ can be found by using MINITAB. Computer display 14.3 contains the commands and printout.

Computer Display 14.3

```
MTB > SET C1
DATA> 1200 1350 1000 1250 1425 1340
DATA> END
MTB > SET C2
DATA> 3.8 3.4 2.9 3.3 3.9 3.1
DATA> END
MTB > SET C3
DATA> 12.4 13.3 9.2 10.6 13.2 11.2
DATA> END
MTB > NAME C1 'SAT' C2 'GPA' C3 'MCAT'
MTB > REGRESSION C3 2 C1 C2

THE REGRESSION EQUATION IS
MCAT = - 2.54 + 0.00543 SAT + 2.16 GAP


PREDICTOR        COEF      STDEV    T-RATIO      P
CONSTANT       -2.537      3.756     -0.68    0.548
SAT          0.005425   0.003115      1.74    0.180
GPA             2.161      1.201      1.80    0.170


S = 0.8642      R-SQ = 82.8%      R-SQ(ADJ) = 71.3%

ANALYSIS OF VARIANCE

SOURCE         DF        SS        MS       F       P
REGRESSION      2    10.7547    5.3773    7.20   0.072
ERROR           3     2.2403    0.7468
TOTAL           5    12.9950
```

From the printout, we can see that $b_0 = -2.537$, $b_1 = 0.005425$, $b_2 = 2.161$, and that the regression equation is

$$\hat{y} = -2.537 + 0.005425x_1 + 2.161x_2$$

By using this equation, a student that has a combined SAT score of 1300 and a GPA of 2.00 can be expected to obtain an average MCAT score of

$$\hat{y} = -2.537 + 0.005425(1300) + 2.161(2.00) = 8.8375 = 8.8$$

Let's examine the ANOVA summary table in the printout of computer display 14.3. The sum of squares of y, called the **total sum of squares** and written as SST, is partitioned into two components: sum of squares for regression, written as SSR, and sum of squares for error, written as SSE. That is,

$$\text{SST} = \text{SSR} + \text{SSE}$$

The sum of squares for regression is that part of the total sum of squares that is accounted for by the independent variables. The sum of squares for error is that portion of the total sum of squares of y that is not accounted for by the independent variables, thus the name sum of squares for error. These components are listed under the source column in the summary table. From the table we can determine the following information:

$$\text{SST} = \Sigma\,(y - \bar{y})^2 = 12.9950$$
$$\text{SSE} = \Sigma\,(y - \hat{y})^2 = 2.2403$$
$$\text{SSR} = \text{SST} - \text{SSE} = 10.7547$$

The DF column indicates the degrees of freedom associated with elements in the source column. Since there are six observations, the total degrees of freedom, written as df_T, is $n - 1 = 6 - 1 = 5$. The degrees of freedom for regression, written as df_R, is equal to the number of independent variables. Since we used k to represent the number of independent variables, $\text{df}_R = k$. Thus, $k = 2$ for our example. The degrees of freedom for error, written as df_E, is usually found by subtracting the degrees of freedom for regression from the total degrees of freedom. Thus, $\text{df}_E = (n - 1) - k = n - (k + 1)$. For our example, we have $\text{df}_E = 5 - 2 = 3$. Notice that $(k + 1)$ represents the number of constants in the regression equation. If we lose one degree of freedom for each constant, then the degrees of freedom for error is simply the difference $n - (k + 1)$. In summary, we have the following formulas for degrees of freedom:

Degrees of Freedom for Regression

$$\text{df}_T = \text{df}_R + \text{df}_E$$
$$\text{df}_T = n - 1$$
$$\text{df}_R = k$$
$$\text{df}_E = n - 1 - k = n - (k + 1)$$

The entries in the MS (mean square) column of the ANOVA summary table contained in computer display 14.3 are found by dividing the SS entry by the df entry.

That is,

$$\text{MSR} = \frac{\text{SSR}}{\text{df}_R} = \frac{10.7547}{2} = 5.3773$$

$$\text{MSE} = \frac{\text{SSE}}{\text{df}_E} = \frac{2.2403}{3} = 0.7468$$

As was the case for one independent variable, in order to determine whether the linear model adequately describes the data, the F test is used. For the data of example 14.9, the hypotheses are:

$$H_0: \beta_1 = \beta_2 = 0$$
$$H_1: \beta_1 \neq 0 \quad \text{or} \quad \beta_2 \neq 0$$

If the null hypothesis is true, then the regression coefficients in the model are zero and are of no use in predicting the values of the dependent variable, average MCAT scores in our example. The value of the F statistic is found by dividing MSR by MSE. That is,

$$F = \frac{\text{MSR}}{\text{MSE}}$$

For the data of example 14.9, we can verify the value of the F statistic. From the printout in computer display 14.3, we have

$$F = \frac{5.3773}{0.7468} = 7.20$$

By examining the printout we can also determine that the p-value for the F test is 0.072. Since $0.05 < 0.072$, we cannot reject H_0 at the 0.05 level of significance. In this case, we have no statistical evidence to suggest that the population regression coefficients are different from zero. It would thus be risky to use the regression equation for prediction purposes.

The multiple coefficient of determination R^2 (denoted by R-SQ in the printout of computer display 14.3) is similar to the coefficient of determination r^2 of section 14.3 and indicates the percentage of the total variation in y that is explained by the variation in the independent variables or regression. It is found by dividing SSR by SST. That is,

> **Multiple Coefficient of Determination**
>
> $$R^2 = \frac{\text{SSR}}{\text{SST}}$$

For the data of example 14.9, we have from the printout

$$R^2 = \frac{10.7547}{12.995} = 0.8276 \approx 82.8\%$$

This means that approximately 83% of the variation in average MCAT scores is accounted for by the variation in the SAT and GPA variables. Only 17% of the variation in the dependent variable remains unaccounted for.

APPLICATION 14.8

Table 14.7 lists the fuel consumption (in miles per gallon) under normal driving, curb weights (in pounds), and engine displacements (in cc) for six 1990 sports cars.[49]

TABLE 14.7

Information for 1990 Sports Cars

Sports Car	Displacement	Weight	Fuel Consumption
Chevrolet Corvette	5735	3330	17.9
Jaguar XJ-S	5344	4015	18.7
Mercedes-Benz 500 SL	2174	2865	16.5
Porsche 911	3600	3320	17.0
Maserati 228	2790	3020	15.5
BMW 325i	2494	3100	22.0

a. Determine a regression equation for predicting average fuel consumption by using engine displacement, and calculate r^2, the coefficient of determination.

b. Determine a regression equation for predicting average fuel consumption by using engine displacement and weight, and calculate R^2, the multiple coefficient of determination.

Solution: We shall use MINITAB to determine the information for parts a and b.

a. We are seeking an equation of the form $\hat{y} = b_0 + b_1 x$. Computer display 14.4 contains the MINITAB commands and output to determine the values of b_0, b_1, and r^2:

Computer Display 14.4

```
MTB > SET C1
DATA> 5735 5344 2174 3600 2790 2494
DATA> END
MTB > SET C2
DATA> 3330 4015 2865 3320 3020 3100
DATA> END
MTB > SET 3
DATA> 17.9 18.7 16.5 17.0 15.5 22.0
DATA> END
MTB > NAME C1 'ENG_SIZE' C2 'WEIGHT' C3 'MPG'
MTB > REGRESSION C3 1 C1

THE REGRESSION EQUATION IS
MPG = 17.7 + 0.000068 ENG_SIZE

PREDICTOR        COEF        STDEV      T-RATIO       P
CONSTANT        17.681       2.962        5.97      0.004
ENG_SIZE      0.0000683    0.0007518      0.09      0.932

S = 2.546      R-SQ = 0.2%      R-SQ(ADJ) = 0.0%

ANALYSIS OF VARIANCE

SOURCE        DF        SS         MS       F        P
REGRESSION     1      0.053      0.053    0.01     0.932
ERROR          4     25.920      6.480
TOTAL          5     25.973
```

From the printout, we can determine that the regression equation is $\hat{y} = 17.681 + 0.0000683 x_1$, where $\hat{y}$ represents fuel consumption and x_1 represents the engine displacement. The coefficient of determination, $r^2 = 0.2\%$, indicates that engine displacement has very little effect on fuel consumption. The linear model is not appropriate using $\alpha = 0.05$.

b. The regression equation has the form $\hat{y} = b_0 + b_1 x_1 + b_2 x_2$. The command for obtaining the values of b_0, b_1, b_2, R^2, and corresponding output is REGRESSION C3 2 C1 C2. The entry "2" indicates that 2 independent variables are to be used. Computer display 14.5 contains the commands and output.

Computer Display 14.5

```
MTB > REGRESSION C3 2 C1 C2

THE REGRESSION EQUATION IS
MPG = 10.9 - 0.00050 ENG_SIZE + 0.00270 WEIGHT

PREDICTOR          COEF       STDEV     T-RATIO         P
CONSTANT          10.91       12.90        0.85     0.460
ENG_SIZE      -0.000496    0.001329       -0.37     0.734
WEIGHT         0.002702    0.004982        0.54     0.625

S = 2.805      R-SQ = 9.1%      R-SQ(ADJ) = 0.0%

ANALYSIS OF VARIANCE

SOURCE        DF        SS        MS       F        P
REGRESSION     2     2.368     1.184    0.15    0.866
ERROR          3    23.605     7.868
TOTAL          5    25.973
```

From the printout, we can determine the regression equation to be $\hat{y} = 10.91 - 0.000496 x_1 + 0.002702 x_2$, where x_1 represents the size of the engine, x_2 represents the weight of the car, and $\hat{y}$ represents the predicted fuel-consumption rating. The multiple coefficient of determination, $R^2 = 9.1\%$, indicates that weight has more of an effect on fuel consumption than engine size. But approximately 91% of the variation in fuel ratings still remains unaccounted for. Perhaps other variables, such as gear-ratio ratings and horsepower ratings, might help to explain more of the variation in fuel consumption. By examining the value of the F statistic ($F = 0.15$), which is significant at the 0.866 level, we can conclude that the linear model would not be appropriate using $\alpha = 0.05$ for the data in this case. ■

EXAMPLE 14.11

Let's return to the MCAT data of example 14.9. Suppose we want to determine, in addition to SAT and GPA scores, if IQ has any effect on the MCAT variable. The additional information for the six students is contained in table 14.8.

TABLE 14.8

Information for Six
Premed Students

Student	Combined SAT Score (x_1)	GPA (x_2)	IQ(x_3)	Average MCAT Score (y)
A	1200	3.8	131	12.4
B	1350	3.4	140	13.3
C	1000	2.9	115	9.2
D	1250	3.3	116	10.6
E	1425	3.9	137	13.2
F	1340	3.1	122	11.2

MINITAB is used to obtain the regression constants in the new linear equation:

$$\hat{y} = b_0 + b_1 x_1 + b_2 x_2 + b_3 x_3$$

Computer display 14.6 contains the commands and corresponding output.

Computer Display 14.6

```
MTB > READ C4
DATA> 131 140 115 116 137 122
DATA> END
MTB > NAME C1 'SAT' C2 'GPA' C3 'MCAT' C4 'IQ'
MTB > REGR C3 3 C1 C2 C4

THE REGRESSION EQUATION IS
MCAT = -6.92 + 0.00275 SAT + 0.781 GPA + 0.0981 IQ

PREDICTOR        COEF      STDEV     T-RATIO       P
CONSTANT       -6.917      1.612      -4.29    0.050
SAT          0.002747   0.001232       2.23    0.155
GPA            0.7812     0.5128       1.52    0.267
IQ            0.09814    0.02076       4.73    0.042

S = 0.3034    R-SQ = 98.6%    R-SQ(ADJ) = 96.5%

ANALYSIS OF VARIANCE

SOURCE          DF        SS        MS       F        P
REGRESSION       3   12.8109    4.2703   46.40    0.021
ERROR            2    0.1841    0.0920
TOTAL            5   12.9950
```

From the printout, we can determine the regression equation to be

$$\hat{y} = -6.917 + 0.002747x_1 + 0.7812x_2 + 0.09814x_3$$

The F value of 46.40 is significant at the 0.021 level. This means that the null hypothesis H_0: $\beta_1 = \beta_2 = \beta_3 = 0$ should be rejected at the 0.05 level of significance. The linear model is adequate for describing the relationship between the independent variables and the dependent variable. As a result, at least one population regression coefficient is different from zero.

The next step is to test the coefficients individually. There are three hypothesis tests that we can perform.

1. For combined SAT scores:

$$H_0: \beta_1 = 0$$

2. For GPA:

$$H_0: \beta_2 = 0$$

3. For IQ:

$$H_0: \beta_3 = 0$$

We will test each null hypothesis at the 0.05 level. For each test, the t distribution with $df_E = n - (k + 1)$ is used. Since we have six students, $n = 6$. The number of independent variables is $k = 3$. Hence, the t distribution with df = 2 is used.

$$df_E = n - (k + 1) = 6 - (3 + 1) = 2$$

The column labeled STDEV in the Minitab output contained in computer display 14.6 estimates the standard deviation of the sampling distribution of sample regression coefficients. That is,

$$\sigma_{b_i} \approx \text{stdev} = s_{b_i}$$

Under the assumption that $\beta_i = 0$, the sampling distribution of the statistic

$$\frac{b_i}{s_{b_i}}$$

is a t distribution with df $= n - (k + 1)$. Thus, we have

$$t = \frac{b_i}{s_{b_i}}, \qquad \text{df} = n - (k + 1)$$

Consider the following partial printout from computer display 14.6:

PREDICTOR	COEF	STDEV	T-RATIO	P
CONSTANT	−6.917	1.612	−4.29	0.050
SAT	0.002747	0.001232	2.23	0.155
GPA	0.7812	0.5128	1.52	0.267
IQ	0.09814	0.02076	4.73	0.042

For example, in order to test whether β_1 is different from zero, we determine the t ratio for b_1, the SAT coefficient from the printout.

$$t = \frac{b_1}{s_{b_1}} = \frac{0.002747}{0.001232} = 2.23$$

which agrees with the t ratio in the printout. Since the p-value for the test is 0.155, we fail to reject H_0 at the 0.05 level of significance. By examining the p values in the partial printout, we can determine that only the IQ coefficient is significantly different from zero, using $\alpha = 0.05$.

EXAMPLE 14.12

The constant term, β_0, of the MCAT regression model can also be tested to determine if it is different from 0. For this purpose, the t ratio b_0/s_{b_0} is used. For the MCAT data contained in table 14.7, we can see from the printout in computer display 14.6 that the t value for $b_0 = -6.917$ is

$$t = \frac{-6.917}{1.612} = -4.29$$

which is significant at the 0.05 level. Note also that this t value agrees with the t ratio in the printout.

From the above tests, we can conclude that the SAT and GPA variables are not significant predictors of MCAT success. They should be removed from the study. Only the IQ variable appears to have a significant influence on the MCAT variable. The three variables taken together account for 98.6% of the variance in the MCAT measurements. Let's use MINITAB to construct the regression equation used to predict MCAT scores from the IQ scores. Computer display 14.7 contains the commands and output.

Computer Display 14.7

```
MTB > SET C1
DATA> 131 140 115 116 137 122
DATA> END
MTB > SET C2
DATA> 12.4 13.3 9.2 10.6 13.2 11.2
DATA> END
MTB > NAME C1 'IQ' C2 'MCAT'
MTB > REGR C2 1 C1
```

THE REGRESSION EQUATION IS MCAT $= -6.70 + 0.145$ IQ

PREDICTOR	COEF	STDEV	T-RATIO	P
CONSTANT	−0.697	2.609	−2.57	0.062
IQ	0.14465	0.02051	7.05	0.002

$S = 0.4918$ R-SQ $= 92.6\%$ R-SQ(ADJ) $= 90.7\%$

ANALYSIS OF VARIANCE

SOURCE	DF	SS	MS	F	P
REGRESSION	1	12.028	12.028	49.74	0.002
ERROR	4	0.967	0.242		
TOTAL	5	12.995			

From the printout, we can determine that $r^2 = 92.6\%$, which means that 92.6% of the variance in MCAT scores is esplained by the variance in the IQ variable. By adding the two variables, SAT and GPA, to the equation, we only increase the value of R^2 by $98.6\% - 92.6\% = 6\%$.

It is interesting to note that the value $s = 0.4918$ in the printout of computer display 14.7 represents the standard error of estimate, s_e, studied in section 14.2. Recall that it represents an estimate of how the points in the population of pairs are scattered about the population linear model.

In the case of linear regression involving two variables, the population model $E(y|x_1, x_2) = \beta_0 + \beta_1 x_1 + \beta_2 x_2$ represents a plane. In this case the standard error of estimate, S, estimates the variation in points (x_1, x_2, y) about the multiple regression plane. The formula for the **multiple standard error of estimate** is similar to the formula studied in section 14.2.

Multiple Standard Error of Estimate

$$S_e = \sqrt{\text{MSE}} = \sqrt{\frac{\text{SSE}}{n - (k + 1)}}$$

Note from computer display 14.5 for application 14.8 part b, the multiple standard error of estimate is $S_e = 2.805$. This value can be verified by noting that $n = 6$, $k = 2$, and

$$S_e = \sqrt{\text{MSE}} = \sqrt{\frac{\text{SSE}}{n - (k + 1)}} = \sqrt{\frac{23.605}{6 - 3}} = \sqrt{7.868} \approx 2.805$$

EXERCISE SET 14.4

Basic Skills

1. A pain reliever in capsule form contains three different amounts (measured in milligrams) of medications A, B, and C. Each capsule contained one of three amounts of medication A: 15, 30, or 45 mg (variable x_1); one of two amounts of medication B: 20 or 30 mg (variable x_2); and one of two amounts of medication C: 10 or 20 mg (variable x_3). Each capsule varied in theurapeutic effectiveness for reducing pain from 45% effective to 98% effective (variable y). By using capsules with varying amounts of the three ingredients in a clinical study, the following regression equation was obtained:

$$\hat{y} = -2.33 + 0.9x_1 + 1.27x_2 + 0.9x_3$$

Find the predicted percentage of effectiveness if a capsule contained

a. 30 mg of A, 20 mg of B, and 20 mg of C.
b. 15 mg of A, 30 mg of B, and 30 mg of C.
c. 45 mg of A, 10 mg of B, and 10 mg of C.
d. 25 mg of A, 25 mg of B, and 15 mg of C.

2. A study was done in a rural community to investigate the relationship between the selling price of a home, the living area, the lot size, and the real estate taxes per year. The following regression equation was found for predicting selling price from living area, lot size, and amount of taxes:

$$\hat{y} = 54.2 + 1.36x_1 + 42.36x_2 + 4.32x_3$$

where $\hat{y}$ is the predicted selling price (in dollars), x_1 is the living area (in square feet), x_2 is the taxes (in dollars), and x_3 is the lot size (in acres). Find the predicted selling price if

a. $x_1 = 14{,}500$ square feet, $x_2 = \$1800$, and $x_3 = 3.4$ acres.
b. $x_1 = 2200$ square feet, $x_2 = \$2400$, and $x_3 = 2.1$ acres.
c. $x_1 = 1350$ square feet, $x_2 = \$1700$, and $x_3 = 0.9$ acres.
d. $x_1 = 1200$ square feet, $x_2 = \$1650$, and $x_3 = 1.3$ acres.

Further Applications

3. Refer to the accompanying MINITAB printout to answer the given questions.

PREDICTOR	COEF	STDEV	T-RATIO	P
CONSTANT	−60.25	56.30	−1.07	0.320
IQ	1.0330	0.5184	1.99	0.087
HOURS	2.438	1.355	1.80	0.115

S = 13.66   R-SQ = 60.5%   R-SQ(ADJ) = 49.3%

ANALYSIS OF VARIANCE

SOURCE	DF	SS	MS	F	P
REGRESSION	2	2005.6	1002.8	5.37	0.039
ERROR	7	1306.8	186.7		
TOTAL	9	3312.4			

a. What is the regression equation for the regression analysis?
b. What is the value of R^2?
c. At the 0.05 level of significance, does the linear regression model adequately describe the data?
d. What regression constants are significantly different from 0 using the 0.05 level of significance?
e. What is the predicted score given IQ = 125 and HOURS = 10?

4. Consider the accompanying MINITAB printout.

PREDICTOR	COEF	STDEV	T-RATIO	P
CONSTANT	−0.081	1.924	−0.04	0.968
SCORE	1.6359	0.4601	3.56	0.024
RATING	−0.5541	0.6342	−0.87	0.432

S = 2.043   R-SQ = 82.9%   R-SQ(ADJ) = 74.4%

ANALYSIS OF VARIANCE

SOURCE	DF	SS	MS	F	P
REGRESSION	2	81.027	40.513	9.71	0.029
ERROR	4	16.688	4.172		
TOTAL	6	97.714			

a. What is the regression equation for the regression analysis?
b. What is the value of R^2?
c. At the 0.05 level of significance, does the linear regression model adequately describe the data?
d. What regression constants are significantly different from 0 using the 0.05 level of significance?
e. What is the prediced value of y given SCORE = 5 and RATING = 3?

5. Consider the accompanying partial MINITAB printout.

PREDICTOR	COEF	STDEV	T-RATIO	P
CONSTANT	41.94	18.77	2.23	0.089
ADVER	2.7598	0.5776	4.78	0.009
SALES	3.630	1.311	2.77	0.050

ANALYSIS OF VARIANCE

SOURCE	DF	SS	MS	F	P
REGRESSION	2	1030.32	515.16	17.60	0.010
ERROR	4	117.11	29.28		
TOTAL	6	1147.43			

a. Determine the value of R^2.
b. Determine S_e, the multiple standard error of estimate.
c. Determine if the linear model is appropriate to describe the data. Use $\alpha = 0.05$.
d. Conduct hypothesis tests to determine which of the independent variables have regression coefficients that are significantly different from zero (use $\alpha = 0.05$).
e. Determine the estimated value of the standard error of the sales coefficient s_{sales}.

6. Consider the accompanying partial MINITAB output.

PREDICTOR	COEF	STDEV	T-RATIO	P
CONSTANT	44898	19430	2.31	0.104
C1	661.9	427.2	1.55	0.219
C2	2290	1372	1.67	0.194

S = 5878   R-SQ = 58.5%   R-SQ(ADJ) = 30.9%

ANALYSIS OF VARIANCE

SOURCE	DF	SS	MS	F	P
REGRESSION	2	146331824	73165912	2.12	0.267
ERROR	3	103668168	34556056		
TOTAL	5	250000000			

a. Determine the value of R^2.
b. Determine S_e, the multiple standard error of estimate.
c. Determine if the linear model is appropriate to describe the data. Use $\alpha = 0.05$.
d. Conduct hypothesis tests to determine which of the independent variables have regression coefficients that are significantly different from zero. Use $\alpha = 0.05$.
e. Determine the estimated standard error of C1, s_{c_1}.

7. Ornithologists have noted that more types of birds are present breeding in woodlands than in fields of similar sizes.

The data in the accompanying table were gathered to investigate the relationship that foliage diversity (FHD) and plant species diversity (PSD) have on bird species diversity (BSD).[50]

Site	BSD	FHD	PSD
A	0.639	0.043	0.972
B	1.266	0.448	1.911
C	2.265	0.745	2.344
D	2.403	0.943	1.768
E	1.721	0.731	1.372
F	2.739	1.009	2.503
G	1.332	0.577	1.367
H	2.285	0.859	1.776
I	2.277	1.021	2.464
J	2.127	0.825	2.176
K	2.567	1.093	2.816

a. By using a statistical software program, determine the linear regression equation to predict BSD using FHD and PSD.

b. Determine the value of R^2.

Going Beyond

8. The adjusted multiple coefficient of determination (written as R-SQ(ADJ) in a MINITAB regression printout) is defined as $R_{adj}^2 = 1 - (1 - R^2)\dfrac{n-1}{n-k-1}$. Verify this for the accompanying partial MINITAB output.

PREDICTOR	COEF	STDEV	T-RATIO	P
CONSTANT	−0.081	1.924	−0.04	0.968
SCORE	1.6359	0.4601	3.56	0.024
RATING	−0.5541	0.6342	−0.87	0.432

S = 2.043   R-SQ = 82.9%   R-SQ(ADJ) = 74.4%

ANALYSIS OF VARIANCE

SOURCE	DF	SS	MS	F	P
REGRESSION	2	81.027	40.513	9.71	0.029
ERROR	4	16.688	4.172		
TOTAL	6	97.714			

9. The adjusted multiple coefficient of determination (written as R-SQ(ADJ) in a Minitab regression printout) is defined as $R_{adj}^2 = 1 - (1 - R^2)\dfrac{n-1}{n-k-1}$. Consider the accompanying partial MINITAB output.

PREDICTOR	COEF	STDEV	T-RATIO	P
CONSTANT	−6.697	2.609	−2.57	0.062
IQ	0.14465	0.02051	7.05	0.002

ANALYSIS OF VARIANCE

SOURCE	DF	SS	MS	F	P
REGRESSION	1	12.028	12.028	49.74	0.002
ERROR	4	0.967	0.242		
TOTAL	5	12.995			

Determine
   a. R^2.   b. R^2(adj).   c. S_e^2.

10. Multiple linear regression techniques can be used to determine nonlinear regression equations. The equation $\hat{y} = b_0 + b_1 x_1 + b_2 x_2^2$ is called a quadratic regression equation. Use the methods of multiple linear regression with $x_1 = x$ and $x_2 = x^2$ to determine the quadratic regression equation for the accompanying data.

x	1	2	3	4	5	6	7
y	12	7	4	3	4	7	12

11. The accompanying normal equations can be used to provide the constants in the regression equation $\hat{y} = b_0 + b_1 x_1 + b_2 x_2$.

$$\Sigma y = nb_0 + b_1(\Sigma x_1) + b_2(\Sigma x_2)$$
$$\Sigma x_1 y = b_0(\Sigma x_1) + b_1(\Sigma x_1^2) + b_2(\Sigma x_1 x_2)$$
$$\Sigma x_2 y = b_0(\Sigma x_2) + b_1(\Sigma x_1 x_2) + b_2(\Sigma x_2^2)$$

By using the normal equations and assuming that $b_2 = 0$, derive the normal equations for a linear regression equation having x_1 as the only independent variable. Solve this system by elimination and derive the formulas for b_0 and b_1 given in section 14.1.

12. Use the normal equations of exercise 11 to determine the linear regression equation for the following data.

x_1	1	3	2
x_2	1	1	2
y	4	2	6

13. Use a computer software program to determine the linear relationship for predicting the lifetime amount of carbon dioxide emitted (Y) by using engine size (X_1), number of cylinders (X_2), composite mph (X_3), and the lifetime amount (in gallons) of gasoline consumed (X_4) for automobiles listed in motivator 14.

CHAPTER SUMMARY

In chapter 4 we learned that correlation analysis is used to determine whether a linear relationship exists for a population of bivariate data. If a linear relationship exists, regression analysis can be used to describe the relationship. In this chapter we have examined correlation and regression from an inference-making point of view. We saw how inferences about the population regression equation can be formed by using a sample regression equation. To judge the accuracy of prediction, we learned how to draw inferences concerning the parameters in the linear regression model by using hypothesis testing and interval estimation. We also learned how to construct prediction intervals for a single value of y for a given value of x and confidence intervals for the average value of y for a given value of x. The coefficient of determination is useful for interpreting the correlation coefficient. It is used to determine the strength of the linear relationship between the two variables. And it suggests the percentage of the variance in the dependent variable that is accounted for by the variance in the independent variable. We also learned the methods of multiple linear regression, the process of finding a linear equation that relates two or more independent variables with a dependent variable. The multiple R^2 coefficient measured the strength of this linear relationship.

CHAPTER REVIEW

■ *IMPORTANT TERMS* ■

The following chapter terms have been mixed in ordering to provide you better review practice. For each of the following terms, provide a definition in your own words. Then check your responses against the definitions given in the chapter.

ANOVA summary table	predictor variable	regression
coefficient of determination	total sum of squares	normal equations
correlation analysis	mean square	residuals
deterministic model	deterministic component	residual variance
response variable	error term	slope of regression line
population regression equation	error variance	standard error of estimate
multiple coefficient of determination	linear regression model	sum of squares for regression
probabilistic model	mean square for error	multiple regression analysis
regression analysis	mean square for regression	nonlinear regression equations
regression equation	sum of squares for error	multiple standard error of estimate

■ *IMPORTANT SYMBOLS* ■

ϵ, error term

β_0, y-intercept of linear model

β_1, slope of linear model

σ^2, error variance

$E(y \mid x)$, average value of y given x for the linear model

$\hat{y}$, predicted value for y

$\hat{y}(x)$, predicted value for y at x

b_0, y-intercept of least square regression line (one independent variable) and the constant term in a multiple linear regression equation

b_1, slope of sample regression line (one independent variable)

SSE, sum of squares for error

s_e^2, residual variance

s_e, the standard error of estimate

$y - \bar{y}$, deviation due to the mean

$y - \hat{y}$, component of deviation due to error

$\hat{y} - \bar{y}$, component of deviation due to linear regression

SSR, sum of squares for regression

MSR, mean square for regression

MSE, mean square for error

r, linear correlation coefficient

ρ, population correlation coefficient

r^2, coefficient of determination

R^2, the multiple coefficient of determination

k, the number of variables in multiple linear regression

b_i, the coefficient of x_i in a multiple linear regression equation

s_{b_i}, an estimate of the standard deviation of the sampling distribution of b_i

S_e, multiple standard error of estimate

■ *IMPORTANT FACTS AND FORMULAS* ■

Linear regression model: $y = \beta_0 + \beta_1 x + \epsilon$.   (14.1)

For each value of x, the expected value of the error term ϵ in the linear regression model is zero: $E(\epsilon) = 0$.

For each value of x, the variance of the error term ϵ in the linear regression model is: $\sigma_\epsilon^2 = \sigma^2$.

Sample regression equation (one independent variable): $\hat{y} = b_0 + b_1 x$.

Sum of squares for error: $\text{SSE} = \Sigma (y - \hat{y})^2$.

Average value of y for a given x value: $E(y|x) = \beta_0 + \beta_1 x$.

Prediction error or residual: $e = y - \hat{y}$.

Residual variance or estimate of σ^2 (for one independent variable): $s_e^2 = \dfrac{\text{SSE}}{n-2}$.

Computational formula for SSE: $\text{SSE} = \text{SS}_y - b_1 \text{SS}_{xy}$. (14.3)

$\text{SS}_y = \text{SSR} + \text{SSE}$.   (14.4)

Sum of squares for regression (one independent variable): $\text{SSR} = \Sigma (\hat{y} - \bar{y})^2$.

Computational formula for sum of squares for regression (one independent variable): $\text{SSR} = b_1 \text{SS}_{xy}$ (14.5)

F statistic used to determine whether the linear model is appropriate (one independent variable): $F = \dfrac{\text{SSR}}{s_e^2}$.

Mean square for regression (one independent variable): $\text{MSR} = \text{SSR}$.

Mean square for error (one independent variable): $\text{MSE} = \dfrac{\text{SSE}}{n-2}$.

t statistic used to determine whether the linear model is appropriate (one independent variable):

$t = \dfrac{b_1}{s_e/\sqrt{\text{SS}_x}}$.   (14.6)

Limits for interval estimate of β_1: $b_1 \pm t_{\alpha/2}\dfrac{s_e}{\sqrt{\text{SS}_x}}$ where df $= n - 2$.   (14.7)

Limits for confidence interval estimate of $E(y|x_0)$:

$\hat{y} \pm t_{\alpha/2} s_e \sqrt{\dfrac{1}{n} + \dfrac{(x_0 - \bar{x})^2}{\text{SS}_x}}$, where df $= n - 2$. (14.8)

Limits for confidence interval estimate of y:

$\hat{y} \pm t_{\alpha/2} s_e \sqrt{1 + \dfrac{1}{n} + \dfrac{(x_0 - \bar{x})^2}{\text{SS}_x}}$, where df $= n - 2$. (14.9)

Computational formula for Pearson's correlation coefficient: $r = \dfrac{\text{SS}_{xy}}{\sqrt{\text{SS}_x \text{SS}_y}}$.   (14.10)

Relationship between b_1 and r: $b_1 = r\dfrac{s_y}{s_x}$.

t value for r: $t = \dfrac{r\sqrt{n-2}}{\sqrt{1-r^2}}$.   (14.12)

Coefficient of determination (one independent variable): $r^2 = \dfrac{\text{SSR}}{\text{SS}_y}$.

Normal equations for a linear regression equation having two independent variables:

$$\Sigma y = n b_0 + b_1(\Sigma x_1) + b_2(\Sigma x_2)$$
$$\Sigma x_1 y = b_0(\Sigma x_1) + b_1(\Sigma x_1^2) + b_2(\Sigma x_1 x_2)$$
$$\Sigma x_2 y = b_0(\Sigma x_2) + b_1(\Sigma x_1 x_2) + b_2(\Sigma x_2^2)$$

Degrees of freedom for multiple linear regression:
  a. $\text{df}_T = \text{df}_R + \text{df}_E$
  b. $\text{df}_T = n - 1$
  c. $\text{df}_R = k$
  d. $\text{df}_E = n - (k + 1)$

Mean square for regression: $\text{MSR} = \dfrac{\text{SSR}}{\text{df}_R} = \dfrac{\text{SSR}}{k}$

(k independent variables).

Mean square for error: $\text{MSE} = \dfrac{\text{SSE}}{\text{df}_E} = \dfrac{\text{SSE}}{n - (k + 1)}$

(k independent variables).

F statistic for multiple regression: $F = \dfrac{\text{MSR}}{\text{MSE}}$.

Multiple coefficient of determination: $R^2 = \dfrac{\text{SSR}}{\text{SS}_y} = \dfrac{\text{SSR}}{\text{SST}}$.

t statistic for b_i: $t = \dfrac{b_i}{s_{b_i}}$ for df $= n - (k + 1)$.

Multiple standard error of estimate:

$S_e = \sqrt{\text{MSE}} = \sqrt{\dfrac{\text{SSE}}{n - (k + 1)}}$.

REVIEW EXERCISES

1. Over a six-month period, a large department store compared the number of salespeople available with gross sales receipts. A sample of six pairs of data yielded $r = 0.47$. Assuming that gross sales and number of salespeople are approximately normal, test at $\alpha = 0.05$ to determine if $\rho \neq 0$.

2. A study of 15 cars revealed a correlation of 0.48 between engine size and average gasoline mileage. Assuming engine sizes and average gasoline mileages are normally distributed, use the 0.05 level of significance to test whether r is different from 0.

3. In an attempt to determine the relationship between attendance and concession sales at football games, a high school principal obtained the data given here for five randomly chosen football games during the previous three seasons.

Attendance, x (in hundreds of people)	Sales, y (in hundreds of dollars)
5	15
13	36
10	20
21	64
11	32

a. Using $\alpha = 0.01$, determine if the linear model is appropriate.
b. Construct a 99% confidence interval for β_1.

4. The accompanying data indicate the disposable income and the amount of money spent on food per week for a family of four.

Disposable income, x (in thousands of dollars)	Food costs per week, y (in dollars)
30	106
36	109
27	81
20	77
25	83
24	50

a. By using $\alpha = 0.05$ determine if $\beta_1 > 0$.
b. Construct a 95% confidence interval for the average food cost per week for a family of four with a disposable income of $25,500.
c. For each additional $1000 of disposable income, approximate the increase in food costs per week.

5. A new medication has been developed to lower diastolic blood pressure. It was administered to six patients over a two-week period with the resulting data on diastolic blood pressure shown in the table.

Before treatment x	107 120 92 127 114 105
After treatment y	96 120 70 117 109 90

a. If a patient has a diastolic blood pressure reading of 100 before drug treatment, what is her predicted diastolic reading after treatment?
b. Construct a 95% prediction interval for the patient in part a.
c. Construct a 95% confidence interval for the average diastolic reading after treatment for patients who had a reading of 100 before treatment.

6. To determine the effectiveness of a particular diet the accompanying data were recorded for a random sample of eight dieters.

Number of Weeks x on Diet	Weight Loss y (in pounds)
1	8
4	22
3	17
2	15
3	19
1	10
2	11
4	20

a. Calculate the value of r.
b. By using $\alpha = 0.05$, determine if $\rho > 0$.
c. By using $\alpha = 0.05$, determine if the linear model is appropriate.
d. Suppose a person who initially weighs 290 pounds and is 70 inches tall is on the diet for a year. What do you predict his weight to be at the end of one year of dieting? Explain.

7. A credit bureau compiled the tabled data relating annual family income to savings for six families.

Annual Income, x (in thousands of dollars)	Annual savings, y (in hundreds of dollars)
12	6
15	11
9	2
22	24
16	12
36	36

a. Using $\alpha = 0.01$ determine if the linear model is appropriate.
b. For each increase of $1000 of income, what is the expected increase in savings?
c. How much in savings would you predict for a family whose annual income is $20,000?
d. Construct a 95% confidence interval for the average savings for a family having an income of $17,000.
e. Answer part c by constructing a 95% prediction interval for $y(20)$.

8. Thirteen randomly selected arthritis patients were involved in a study to test a new pain-killing drug. The time elapsed (in minutes) from taking the drug until a noticeable relief in pain is detected is to be predicted from the dosage (in grams) and the age of the patient (in years).

MINITAB was used to carry out the regression analysis. Refer to the accompanying MINITAB printout to answer the given questions.

```
PREDICTOR     COEF     STDEV    T-RATIO       P
CONSTANT    34.663     4.853      7.14   0.000
DOSAGE       1.3002    0.6711     1.94   0.081
AGE         -0.5396    0.1164    -4.64   0.000

S = 6.793   R-SQ = 68.3%   R-SQ(ADJ) = 62.0%

ANALYSIS OF VARIANCE

SOURCE       DF      SS      MS      F       P
REGRESSION    2   994.25  497.12  10.77  0.003
ERROR        10   461.44   46.14
TOTAL        12  1455.69
```

a. What is the regression equation for the regression analysis?
b. What is the value of R^2?
c. At the 0.05 level of significance, does the linear regression model adequately describe the data?
d. What regression constants are significantly different from 0 using the 0.05 level of significance?
e. What is the predicted elapsed time given DOSAGE = 2 grams and AGE = 60 years?

9. A study was conducted to determine the effect of age, height, and weight of elementary school students on a physical fitness examination. MINITAB was used to determine the multiple linear regression equation relating scores to the ages, weights, and heights. Refer to the accompanying printout to answer the following questions.

```
PREDICTOR      COEF     STDEV    T-RATIO      P
CONSTANT   -207.15     78.28     -2.65   0.029
AGE          18.39     15.61      1.18   0.272
HEIGHT        4.934     2.087     2.36   0.046
WEIGHT       -1.824     1.397    -1.31   0.228

S = 13.77   R-SQ = 63.8%   R-SQ(ADJ) = 50.3%

ANALYSIS OF VARIANCE

SOURCE       DF      SS      MS      F       P
REGRESSION    3   2676.6   892.2   4.71  0.035
ERROR         8   1516.4   189.5
TOTAL        11   4192.9
```

a. What is the regression equation for the regression analysis?
b. What is the value of R^2?

c. At the 0.05 level of significance, does the linear regression model adequately describe the data?

d. What regression constants are significantly different from 0 using the 0.05 level of significance?

e. What is the predicted score given AGE = 9, HEIGHT = 52 inches, and WEIGHT = 75 pounds?

Computer Applications

1. A regression experiment was performed. Nine observations were made at each of the five levels of x as follows:

x	3	5	7	9	11
y	5.20	5.46	3.98	4.23	6.27
	4.66	3.58	3.64	4.65	4.71
	4.70	5.23	2.05	4.75	4.90
	5.43	4.86	5.49	5.26	4.61
	3.48	4.13	4.18	5.32	4.12
	5.35	4.85	4.73	5.09	3.42
	5.77	3.86	3.73	5.72	6.79
	4.95	6.10	5.82	6.51	3.55
	5.34	5.17	5.87	4.86	4.59

a. Draw a scatter diagram.
b. Calculate the value of the correlation coefficient.
c. Determine the equation of the regression line.
d. Construct a 95% confidence interval estimate for the value of β_1.

2. Homeowners tended to move more often in the 1980s. The following table contains the average number of years homeowners stayed in their homes for each state for the years 1980 and 1988.[51]

State	Average Number of Years	
	1988	1980
Louisiana	26.3	14.8
Wyoming	20.8	11.4
Utah	19.2	10.7
Idaho	18.5	13.9
Mississippi	18.5	18.3
Montana	18.2	13.5
Oklahoma	17.2	9.5
Washington	16.1	12.0
Alabama	15.9	17.4
Nebraska	15.2	14.5
Colorado	14.7	8.2
North Dakota	14.5	15.0
Texas	14.1	9.8

State	Average Number of Years	
	1988	1980
Wisconsin	14.1	15.8
Indiana	13.9	15.7
Michigan	13.7	13.8
Iowa	13.5	14.0
Missouri	13.3	15.2
Ohio	13.3	16.1
New York	13.2	21.8
Illinois	12.3	17.3
Kentucky	12.3	18.1
Arkansas	12.2	11.5
New Hampshire	12.2	14.2
South Dakota	12.2	12.9
Rhode Island	12.0	24.6
West Virginia	11.9	15.5
Hawaii	11.6	13.9
Georgia	11.5	13.0
Kansas	11.5	10.5
Florida	11.4	9.2
Minnesota	11.4	12.7
Oregon	11.4	10.4
Massachusetts	11.1	19.7
South Carolina	11.0	13.8
Alaska	10.9	8.4
Connecticut	10.9	19.8
Tennessee	10.3	16.0
Arizona	10.2	8.8
New Jersey	10.2	16.0
New Mexico	9.9	10.7
Pennsylvania	9.8	19.4
Nevada	9.4	9.3
North Carolina	9.1	13.4
Washington, D.C.	8.9	11.3
Virginia	8.8	9.8
Maryland	8.5	12.6
Vermont	8.5	12.6
Delaware	8.4	13.5
California	8.0	8.6
Maine	7.9	17.3
U.S.	11.5	12.9

a. Calculate the correlation coefficient between the average number of years homeowners stayed in their homes in 1980 and in 1988.

b. Plot a scattergram for the data.

c. Determine the regression equation that relates average length of stay in 1980 to average length of stay in 1988 for each state using average number of years homeowners stayed in their homes in 1980 as the independent variable.

3. Determine the linear regression equation for the data in motivator 14.

4. Consider the following data.

x	1	2	3	4	5	6
y	4	5	6	7	8	6

Determine the value of b (to two decimal places) in the one-parameter family of straight lines $(\hat{y} - \bar{y}) = b(x - \bar{x})$ that minimizes SSE. (*Hint:* Let $b = 0$ and compute SSE. By using a trial-and-error approach, change b in increments so that SSE continues to decrease.)

5. Refer to the data in computer application 4. Instead of using SSE as the criterion to select the best line, use $SAE = \Sigma \, |\hat{y} - y|$, the sum of the absolute values of the errors. That is, by using a trial-and-error approach, identify the value of b (to two decimal places) in the linear equation $(\hat{y} - \bar{y}) = b(x - \bar{x})$ that minimizes SAE. Compare this equation to the equation identified in computer application 4.

6. For the EPA gas mileage data in table 14.1, determine the value of b (to two decimal places), using a trial-and-error approach, in the one-parameter family of straight lines $(\hat{y} - \bar{y}) = b(x - \bar{x})$ that minimizes SSE.

7. Refer to computer application 6. Instead of using SSE as the criterion to select the best line, use $SAE = \Sigma|\hat{y} - y|$, the sum of the absolute values of the errors. That is, by using a trial-and-error approach, identify the value of b (to two decimal places) in the linear equation $(\hat{y} - \bar{y}) = b(x - \bar{x})$ that minimizes SAE. Compare this equation to the equation identified in computer application 6.

EXPERIMENTS WITH REAL DATA

Treat the 720 subjects listed in the database in appendix C as if they constituted the population of all U.S. subjects.

1. Can a person's age be used to predict diastolic blood pressure?

a. Draw a random sample of size 25 and determine the least squares fit of the linear model for predicting systolic blood pressure from age.

b. Construct a scattergram for the 25 pairs along with the graph of the fitted straight line.

c. Interpret the estimated standard deviation of the error term.

d. Evaluate the usefulness of the linear model.

e. Estimate the diastolic blood pressure of a 30-year-old person.

f. Estimate the mean diastolic blood pressure of a 30-year-old person.

g. Determine the sample correlation between age and diastolic blood pressure. Use this value to determine if the population correlation coefficient is significantly different from zero. Use $\alpha = 0.05$.

2. Can a subject's systolic blood pressure be predicted from his/her diastolic blood pressure?

a. Use a random sample of size 30 to determine the least squares fit of a linear model. Use $\alpha = 0.05$ to determine if the model is appropriate for the data.

b. Use the fitted model to determine the mean systolic blood pressure for a person possessing a diastolic blood pressure of 83.

c. Use the fitted model to determine the systolic blood pressure for a person possessing a diastolic blood pressure of 83.

d. Use the sample of 30 pairs to determine the correlation coefficient between systolic and diastolic blood pressure readings. Is the population correlation coefficient significantly different from zero? Use $\alpha = 0.05$.

3. Can we accurately predict diastolic blood pressure using age and weight? Draw a random sample of 25 subjects. Use a computer program for multiple linear regression to determine the least squares fit of the model $E(y) = \beta_0 + \beta_1 x_1 + \beta_2 x_2$.

a. Test to determine if the model is appropriate. Use $\alpha = 0.05$.

b. Test to determine if the β_i are significantly different from zero. Use $\alpha = 0.05$.

c. Predict the diastolic blood pressure of a 40-year-old subject having a weight of 150 pounds.

d. Determine the coefficient of multiple determination and interpret the result.

▪ CHAPTER ACHIEVEMENT TEST ▪

1. A random sample of 20 salespeople took a test to measure assertiveness. The score of each salesperson was paired with their last six-month gross sales figure. The correlation coefficient was found to be $r = 0.72$. Assuming test scores and gross sales are normally distributed, test at $\alpha = 0.05$ to determine if $\rho \neq 0$. In addition provide the
   a. value of the test statistic.
   b. critical value(s).
   c. decision.

2. For the accompanying data, determine if the linear model is appropriate. Use $\alpha = 0.05$ and provide the
   a. value of the t statistic.
   b. critical value(s).
   c. null hypothesis.
   d. decision.
   e. regression equation.

x	1	2	3	2
y	3	5	6	4

3. For the data in test question 2, determine
   a. a point estimate for b_0.
   b. a point estimate for $E(y\,|\,3)$.
   c. a point estimate for σ^2.
   d. r^2 and interpret your result.

4. For the accompanying data, construct a 95% confidence interval for $E(y\,|\,2)$.

x	1	5	4	2
y	1	2	5	4

5. Consider the accompanying data and the corresponding regression analysis supplied by MINITAB.

x_1	1	2	3	1
x_2	2	1	3	1
y	8	7	3	5

PREDICTOR	COEF	STDEV	T-RATIO	P
CONSTANT	8.667	3.590	2.41	0.250
X1	-1.333	2.211	-0.60	0.655
X2	-0.333	2.211	-0.15	0.905

   $S = 2.828 \quad R\text{-}SQ = 45.8\% \quad R\text{-}SQ(ADJ) = 0.0\%$

   ANALYSIS OF VARIANCE

SOURCE	DF	SS	MS	F	P
REGRESSION	2	6.750	3.375	0.42	0.736
ERROR	1	8.000	8.000		
TOTAL	3	14.750			

   a. If $x_1 = 1$ and $x_2 = 8$, what is the value of $\hat{y}$?
   b. If $x_1 = 1$, $x_2 = 8$, $y = 4$, and the regression equation is used to predict y, what is the error of prediction?
   c. Is the linear regression model appropriate for the data? Explain.
   d. Provide an interpretation of R-SQ = 45.8%.
   e. Interpret the entries in the T-RATIO column in the printout.
   f. What is the significance of S = 2.828 in the printout? How was this value obtained?

15 *Nonparametric Tests*

CHAPTER OBJECTIVES

In this chapter we will investigate

▷ *What nonparametric methods are and under what circumstances they are usually used.*

▷ *The advantages of using nonparametric tests.*

▷ *The disadvantages of using nonparametric tests.*

▷ *How to use and interpret the sign test.*

▷ *How to use and interpret the signed-rank test.*

▷ *How to use and interpret the rank-sum test.*

▷ *How to use and interpret the Kruskal-Wallis test.*

▷ *How to use and interpret Friedman's test.*

▷ *How to use and interpret the runs test.*

▷ *How to use and interpret Spearman's rank correlation coefficient.*

|||| **MOTIVATOR 15** ▶

*E*very year since 1903 (with the exception of 1904), the winner of the American League baseball pennant has played the winner of the National League pennant in the World Series. Since 1905, the winner of a World Series is the first team to win four games out of seven. The results of the World Series from 1960 to 1989 are shown in the table.[52]

Year	Winner	League	Loser	Games Won/Lost
1960	Pittsburgh Pirates	NL	New York Yankees	4–3
1961	New York Yankees	AL	Cincinnati Reds	4–1
1962	New York Yankees	AL	San Francisco Giants	4–3
1963	Los Angeles Dodgers	NL	New York Yankees	4–0
1964	St. Louis Cardinals	NL	New York Yankees	4–3
1965	Los Angeles Dodgers	NL	Minnesota Twins	4–3
1966	Baltimore Orioles	AL	Los Angeles Dodgers	4–0
1967	St. Louis Cardinals	NL	Boston Red Sox	4–3
1968	Detroit Tigers	AL	St. Louis Cardinals	4–3
1969	New York Mets	NL	Baltimore Orioles	4–1

Year	Winner	League	Loser	Games Won/Lost
1970	Baltimore Orioles	AL	Cincinnati Reds	4–1
1971	Pittsburgh Pirates	NL	Baltimore Orioles	4–3
1972	Oakland Athletics	AL	Cincinnati Reds	4–3
1973	Oakland Athletes	AL	New York Mets	4–3
1974	Oakland Athletics	AL	Los Angeles Dodgers	4–1
1975	Cincinnati Reds	NL	Boston Red Sox	4–3
1976	Cincinnati Reds	NL	New York Yankees	4–0
1977	New York Yankees	AL	Los Angeles Dodgers	4–2
1978	New York Yankees	AL	Los Angeles Dodgers	4–2
1979	Pittsburgh Pirates	NL	Baltimore Orioles	4–3
1980	Philadelphia Phillies	NL	New York Yankees	4–2
1981	Los Angeles Dodgers	NL	New York Yankees	4–2
1982	St. Louis Cardinals	NL	Milwaukee Brewers	4–3
1983	Baltimore Orioles	AL	Philadelphia Phillies	4–1
1984	Detroit Tigers	AL	San Diego Padres	4–1
1985	New York Yankees	AL	St. Louis Cardinals	4–3
1986	New York Mets	NL	Boston Red Sox	4–3
1987	Detroit Tigers	AL	St. Louis Cardinals	4–3
1988	Los Angeles Dodgers	NL	Oakland Athletics	4–1
1989	Baltimore Orioles	AL	San Francisco Giants	4–0

Does the list contain evidence that factors other than chance were operating in determining whether a National League team or an American League team won the World Series? After studying this chapter you will know several methods to help you determine the answer to this question.

Chapter Overview

Nonparametric procedures are used, for the most part, when assumptions underlying other tests cannot be satisfied and a data analysis is desired. Usually nonparametric methods assume no knowledge whatsover about the shape of the sampled populations. For this reason, they are sometimes referred to as **distribution-free methods.** Because the *t* and *F* tests require sampling from normal distributions, they are not distribution-free methods, but have traditionally been referred to as **parametric tests.**

As with parametric methods, nonparametric methods have many advantages and disadvantages. The chief disadvantages are that they are usually less efficient and not very sensitive for detecting real differences when they exist, particularly when sampling is from normal populations. They are less efficient because they waste information contained in a sample. For example, several methods (such as Spearman's rank correlation coefficient), replace measurements with their corresponding ranks. Not being sensitive to real differences usually results in a higher probability for Type II errors and a reduction in statistical power. To offset some of this loss of power, larger samples frequently need to be collected.

In addition to being distribution-free, **nonparametric tests** enjoy certain advantages not enjoyed by their parametric counterparts. They are, on the whole, very quick and easy to carry out. Second, perhaps the most important advantage is that they can be used with qualitative data, such as ranks or categorical data. Third, nonparametric tests require fewer restrictive assumptions than their parametric counterparts. Usually nonparametric tests only require sampling from continuous populations that are symmetrical. Generally speaking, nonparametric tests perform nearly as well as their parametric counterparts on normal distributions, and often perform better than parametric methods on nonnormal distributions.

In this chapter we will study seven nonparametric tests: the sign test, the signed-rank test, the rank-sum test, the Kruskal-Wallis test, Friedman's test, the runs test, and Spearman's test. All are founded on the basis of probability theory. We have previously studied one nonparametric test, the chi-square test for analyzing categorical data.

SECTION 15.1 Sign Test (Large Samples)

The **sign test** is a one-sample test that can be used to test a statistical hypothesis regarding the median of a continuous population. If $\tilde{\mu}$ denotes the population median and $\tilde{\mu} = 0$ is assumed to be true, then we would expect any finite random sample of values to have as many positive values as negative values on the average. If a random sample has more positive values than negative values, there is some evidence to suggest that $\tilde{\mu} > 0$. For testing the null hypothesis $H_0 : \tilde{\mu} \leq 0$ against the alternative hypothesis $H_1 : \tilde{\mu} > 0$, the test statistic becomes the number of positive values contained in a sample. The null hypothesis is then rejected if a large number of positive values occur. The level of significance α determines how many positive values a sample must have in order for the null hypothesis to be rejected. If each positive value in the sample is identified as a success, each negative value as a failure, and zero values are discarded, the number of positive values in a sample constitutes a binomial experiment with $p = 0.5$. Testing a null hypothesis of the form $H_0 : \tilde{\mu} = 0$ using the sign test is equivalent to testing the null hypothesis $H_0 : p = 0.5$, where p is the proportion of positive values (successes) in a binomial population. For sample sizes $n \geq 10$, the normal approximation to the binomial can be used to determine the critical values, whereas for $n < 10$, the binomial distribution should be used.

The sign test is also appropriate whenever paired data need to be analyzed to determine if two population means are different. In this case, the null hypothesis $H_0 : \mu_1 = \mu_2$ is equivalent to the null hypothesis $H_1 : \mu_d = 0$, where d represents the distribution of difference scores. Each data pair in the sample is replaced by a $+$ or $-$ sign; zero differences are ignored. The resulting sample of signs is used in the analysis. The test statistic becomes the number of $+$ signs, and the statistic has a sampling distribution that is binomial. The normal approximation to the binomial is used to determine critical values whenever $n \geq 10$.

Recall that if X is the number of successes resulting from a binomial experiment, then,

$$\mu_x = np$$

and

$$\sigma_x = \sqrt{np(1 - p)}$$

In addition, if $np \geq 5$ and $n(1 - p) \geq 5$, then a normal distribution can be used to approximate the binomial. The z value for the number of $+$ signs x is

$$z = \frac{x - np}{\sqrt{np(1 - p)}} \tag{15.1}$$

If we assume that H_0: $p = 0.5$ is true, then equation (15.1) becomes

$$z = \frac{x - (n/2)}{\sqrt{n/4}} = \frac{x - (n/2)}{\sqrt{n}/2} = \frac{2x - n}{\sqrt{n}}$$

Thus, the test statistic for the sign test becomes

Test Statistic for Sign Test

$$z = \frac{2x - n}{\sqrt{n}} \tag{15.2}$$

and formula (15.2) can be used whenever $np \geq 5$ and $n(1 - p) \geq 5$, or whenever $n(1/2) \geq 5$, since $p = 0.5$. This means that formula (15.2) can be used whenever $n \geq 10$.

APPLICATION 15.1

The data in table 15.1 represent the weights of 15 men who have been on a weight reduction diet for one month.

TABLE 15.1

Diet Information Data for Application 15.1

Male	Weight Before	Weight After
1	210	195
2	175	162
3	187	179
4	189	185
5	198	199
6	205	200
7	198	193
8	178	178
9	164	162
10	176	169
11	192	188
12	187	184
13	210	198
14	178	172
15	205	206

Solution: Let's use the sign test and $\alpha = 0.01$ to determine if the diet is effecive in weight reduction. Suppose μ_B and μ_A represent the means of the population of before-diet weights and after-diet weights, respectively.

1. The null hypothesis is H_0: $\mu_B - \mu_A \leq 0$.
2. The alternative hypothesis is H_1: $\mu_B - \mu_A > 0$ (one-tailed test).
3. The sampling distribution is binomial. Note that the alternative hypothesis specifies the order in which we subtract. We let the random variable X denote the number of plus signs, since we are interested in positive "before-minus-after" differences (weight loss). The signs of the differences are recorded in the last column of table 15.2.

Male	Weight Before	Weight After	Sign
1	210	195	+
2	175	162	+
3	187	179	+
4	189	185	+
5	198	199	−
6	205	200	+
7	198	193	+
8	178	178	0
9	164	162	+
10	176	169	+
11	192	188	+
12	187	184	+
13	210	198	+
14	178	172	+
15	205	206	−

TABLE 15.2

Signs for Data Used in Application 15.1

Since we are interested only in plus or minus signs (binomial experiment), we exclude the data for male number 8 from the analysis.

The number of signs is $n = 14$, and the number of plus signs is $x = 12$. By using formula (15.2), the z value for $x = 12$ is

$$z = \frac{(2)(12) - 14}{\sqrt{14}} = 2.67$$

4. The level of significance is $\alpha = 0.01$.

5. The critical value is $z_{0.01} = 2.33$.

6. Decision: Since $z = 2.67 > z_{0.01} = 2.33$, we reject H_0: $\mu_B - \mu_A = 0$ and conclude that the diet is effective in weight reduction.

7. Type of error possible: Type I, $\alpha = 0.01$.

8. p-value: p-value $= 0.5 - 0.4962 = 0.0038$. ▪

APPLICATION 15.2

The sign test also has the advantage of being applicable to data that are dichotomous, such as yes-no data. In a study to determine beer drinkers' preference for two new brands of beer, A and B, a random sample of 20 beer drinkers indicated 4 preferred brand A and 16 preferred brand B. Use the sign test at $\alpha = 0.05$ to determine if brand B is preferred over brand A.

Solution: Let p represent the proportion of all beer drinkers who prefer brand B. We proceed as follows:

1. The null hypothesis is H_0: $p \le 0.5$.

2. The alternative hypothesis is H_1: $p > 0.5$.

3. The sampling distribution is binomial. If x represents the number of beer drinkers who prefer brand B, then $x = 16$ and $n = 20$. By using formula (15.2), we obtain the value of the test statistic:

$$z = \frac{2x - n}{\sqrt{n}} = \frac{32 - 20}{\sqrt{20}} = 2.68$$

4. The level of significance is $\alpha = 0.05$.

5. The critical value is $z_{0.05} = 1.65$.

6. Decision: Since $z = 2.68 > z_{0.05} = 165$, we reject H_0 and conclude that brand B is preferred over brand A.

7. Type of error possible: Type I, $\alpha = 0.05$.

8. *p*-value: *p*-value $= 0.5 - 0.4963 = 0.0037$. ■

APPLICATION 15.3

The sign test can also be used to test null hypotheses of the form H_0: $\tilde{\mu} = \tilde{\mu}_0$, where $\tilde{\mu}$ is the population median and $\tilde{\mu}_0$ is a fixed value, as this application illustrates. The following data represent the number of minutes patients had to wait to see Doctor John on previous office visits: 22, 30, 31, 40, 37, 25, 29, 14, 30, 17, 23, 32, 20, 40, 28, 26, 33, 25, 34, and 21. Use the sign test with $\alpha = 0.05$ to test the doctor's claim that a typical patient will have to wait no more than 25 minutes before being seen by him.

Solution Let $\tilde{\mu}$ represent the median amount of time that a patient waits to see the doctor.

1. H_0: $\tilde{\mu} \leq 25$.

2. H: $\tilde{\mu} > 25$ (one-tailed test).

3. Sampling distribution: Binomial. A plus sign corresponds to a time greater than 25 minutes. If x represents the number of patients who have to wait longer than 25 minutes to see the doctor, then $x = 12$ and $n = 18$ (we discard the two measurements of 25). By using formula (15.2), we obtain the value of the test statistic:

$$z = \frac{2x - n}{\sqrt{n}} = \frac{24 - 18}{\sqrt{18}} = 1.41$$

4. Level of significance: $\alpha = 0.05$.

5. Critical value: $z_{0.05} = 1.65$.

6. Decision: We fail to reject H_0, since $z = 1.41 < z_{0.05} = 1.65$. Hence, there is no statistical evidence to suggest that a typical patient has to wait longer than 25 minutes to see the doctor.

7. Type of error possible: Type II, β is unknown.

8. *p*-value: The *p*-value is $0.5 - 0.4207 = 0.0793$. ■

The STEST command in MINITAB can be used to test the doctor's claim in application 15.3. Computer display 15.1 contains the commands used and corresponding output.

Computer Display 15.1

```
MTB > SET C11
DATA > 22 30 31 40 37 25 29 14 30 17
DATA > 23 32 20 40 28 26 33 25 34 21
DATA > END
MTB > STEST MEDIAN = 25 C1;
SUBC > ALTERNATIVE = +1.

SIGN TEST OF MEDIAN = 25.00 VERSUS G.T. 25.00

          N     BELOW     EQUAL     ABOVE    P-VALUE     MEDIAN
C1        20        6         2        12     0.1189      28.50
```

The *p*-value of 0.1189 differs from the value 0.0793 obtained in application 15.3 because MINITAB does not approximate the sampling distribution of the number of + signs with a normal distribution; instead, a binomial distribution is used. When $n > 50$, MINITAB employs a normal approximation.

Exercise Set 15.1

Further Applications

1. The accompanying data were randomly collected on 16 engineering students at the end of the semester at a certain college to determine if engineering students do better in mathematics than physics.

Student	Math	Physics	Student	Math	Physics
1	97	92	9	80	80
2	78	80	10	82	92
3	85	82	11	96	89
4	84	83	12	85	80
5	92	96	13	50	74
6	33	32	14	100	99
7	62	65	15	82	72
8	80	72	16	80	64

Use the sign test at $\alpha = 0.05$ to determine if the medium math grade of engineering students is higher than the median physics grade.

2. The following nicotine contents (in milligrams) of 15 randomly selected brand A cigarettes were obtained: 2.1, 4.0, 6.3, 5.4, 4.8, 3.7, 6.1, 2.3, 3.3, 6.4, 5.4, 2.5, 3.1, 4.1, and 4.7. At $\alpha = 0.05$, use the sign test to determine if the data indicate that $\tilde{\mu} < 4.5$.

3. Of 50 people randomly interviewed concerning a particular issue, 33 were in favor and 17 were against the issue. Use the sign test at the 5% level of significance to determine if the proportion of people in the population in favor of the issue is different from the proportion against the issue.

4. A random sample of teachers at a certain university yielded the following ages: 26, 49, 55, 55, 60, 61, 68, 57, 74, 75, 68, 68, 68, 67, 64, 64, and 63. Use the sign test at $\alpha = 0.05$ to determine if the median age is different from 64 years.

5. Two different brands of fertilizer were used to test the effects on peach tree yields. One season 16 randomly selected peach trees were treated with fertilizer A, and the following season they were treated with fertilizer B. The number of good peaches (in bushels) for each tree is recorded in the table.

Tree	First Season	Second Season
1	2.0	2.5
2	2.5	3.0
3	2.0	2.0
4	3.5	3.0
5	3.0	2.5
6	2.5	2.0
7	2.5	3.0
8	3.0	2.5
9	2.0	3.5
10	1.5	1.5
11	2.5	2.0
12	3.0	2.0
13	3.5	2.5
14	2.5	3.0
15	1.5	2.0
16	3.0	3.5

Use the sign test at $\alpha = 0.05$ to determine if there is a difference in median peach tree yields for the two types of fertilizer.

6. The data in the accomanying table indicate the systolic blood pressure measurements of 15 subjects before and one hour following an exercise program. At $\alpha = 0.01$ does the exercise program help lower systolic blood pressure?

Person	Before	After	Person	Before	After
1	164	162	9	138	139
2	146	144	10	124	124
3	148	146	11	175	170
4	154	156	12	146	147
5	143	145	13	160	157
6	160	159	14	160	140
7	150	145	15	153	148
8	148	150			

Going Beyond

7. Answer exercise 6 if only the first six people are involved in the data analysis.

8. Two skin lotions were formulated to treat a certain type of skin rash. The lotions were tested on 177 patients who

had this rash on both sides of their bodies, so that each patient was able to compare the effectiveness of the two lotions by applying lotion A to one side of their body and lotion B to the other side. The results are as follows.

Results	Number of Patients
No difference between A and B	96
Much better with A than B	19
Moderately better with A than B	36

Results	Number of Patients
Much better with B than A	5
Moderately better with B than A	21

The statistician's report indicated that the results were not significant at the 5% level. Do you agree with this conclusion? Explain.

9. Solve the problem posed in motivator 15.

SECTION 15.2   Signed-Rank Test (Large Samples)

The sign test of section 15.1 is a test that is easy to apply when paired data are being analyzed and it is known that the samples were not from normal distributions having the same standard deviation. The sign test only uses the signs of the differences between the observations and the hypothesized median for the one sample case and the paired differences for the two sample case, ignoring the magnitude of each difference. The **signed-rank test** (also called **Wilcoxon's signed-rank test**) is a test that takes into account the magnitude of each difference or paired difference; as such, it is usually more sensitive in detecting real differences when they exist. The signed-rank test (for the two-sample case) only requires that the two samples come from populations having a common distribution.

The signed-rank test uses the ranks of the absolute values of the paired differences, assigning rank 1 to the smallest difference in absolute value, rank 2 to the next smallest difference in absolute value, and so on. Zero differences are discarded, and if there are ties involved, each observation in a tie is assigned to mean of the ranks that the tied observations occupy. Each positive rank is then assigned the sign of its corresponding difference. The sum of the positive ranks is denoted by T^+, the sum of the negative ranks is denoted by T^-, and the larger of T^+ and T^- is denoted by T. It is common practice to use T as the test statistic. If the number of nonzero differences (denoted by n) is 15 or more, then the sampling distribution of T is approximately normal. The mean and standard deviation of the sampling distribution of T are

Mean and Standard Deviation of T Statistic

$$\mu_T = \frac{n(n + 1)}{4} \qquad (15.3)$$

and

$$\sigma_T = \sqrt{\frac{n(n + 1)(2n + 1)}{24}} \qquad (15.4)$$

Hence, for cases where $n \geq 15$, the statistic $(T - \mu_T)/\sigma_T$ has a sampling distribution that is approximately equal to the standard normal distribution.

Statistic for Signed-Rank Test When $n \geq 15$

$$z = \frac{T - \mu_T}{\sigma_T}$$

If $n < 15$, the sampling distribution of T is poorly approximated by a normal distribution. In these cases, special tables are available that provide critical values for the signed-rank test.

EXAMPLE 15.1

To show how ranks are assigned when ties are present, consider the following measurements x:

x	Rank if No Ties Present	Rank	
10.4	1	1	
11.2	2	3	
11.2	3	3	The average of the ranks is 3.
11.2	4	3	
12.3	5	5.5	
			The average of the ranks is 5.5
12.3	6	5.5	
13.7	7	7	
14.8	8	8.5	
			The average of the ranks is 8.5
14.8	9	8.5	

APPLICATION 15.4

Refer to application 15.3 of section 15.1. The data representing the number of minutes patients had to wait to see Doctor John on previous office visits is 22, 30, 31, 40, 37, 25, 29, 14, 30, 17, 23, 32, 20, 40, 28, 26, 33, 25, 34, and 21. Use the signed-rank test to test the claim made by the doctor that the typical patient doesn't have to wait any longer than 25 minutes before receiving his attention.

Solution: We first determine the differences $(x - 25)$, where x denotes the time in minutes. The absolute values of the differences are then ranked, as shown in table 15.3.

TABLE 15.3

Signed Ranks of Data from Application 15.4

Time x	Difference $(x - 25)$	Rank of $|x - 25|$	Signed Ranks Positive ranks	Signed Ranks Negative ranks
22	-3	3.5		3.5
30	5	8	8	
31	6	10	10	
40	15	17.5	17.5	
37	12	16	16	
25	0	—	—	—
29	4	5.5	5.5	
14	-11	15		15
30	5	8	8	
17	-8	12.5		12.5
23	-2	2		2
32	7	11	11	
20	-5	8		8
40	15	17.5	17.5	
28	3	3.5	3.5	
26	1	1	1	
33	8	12.5	12.5	
25	0	—	—	—
34	9	14	14	
21	-4	5.5		5.5
			124.5	46.5

Note:

1. The value 3.5 in the first line and third column, for example, is considered as being negatively ranked (or having a negative sign), since the difference $(22 - 25)$ is negative.

2. The differences -3 and 3 in column two, for example, each receive ranks of $(3 + 4)/2 = 3.5$.

3. The two values of 15 in column two have positive ranks. Since they occupy ranks 17 and 18, we assign each the rank 17.5. But since they both belong to the same group, we could have assigned one the rank 17 and the other the rank 18 without affecting the value of the test statistic.

The sum of the positive ranks is $T^+ = 124.5$ and the sum of the negative ranks is $T^- = 46.5$ Hence $T = 124.5$, the larger of the two. Notice also that there are $n = 18$ signed ranks, since the two values of zero are discarded from any further analysis.

The sum of ranks from 1 to $n = 18$ is

$$1 + 2 + 3 + \cdots + 18 = 171$$

It is no coincidence that $T^+ + T^- = 124.5 + 46.5 = 171$, the sum of the ranks from 1 to 18. This will always be the case. The sum of the ranks from 1 to n can be simply found by using the formula

$$\text{Sum of ranks} = \frac{n(n + 1)}{2}$$

Hence, we can write the following relationship between T^+, T^-, and n:

$$T^+ + T^- = \frac{n(n + 1)}{2} \tag{15.5}$$

We could have obtained the value of T^- by subtracting 46.5 from 171 instead of adding the negative signed ranks. Relationship (15.5) is especially helpful for verifying that no computational errors were made in determining the value of T.

The mean of the sampling distribution can be determined by using formula (15.3).

$$\mu_T = \frac{n(n + 1)}{4} = \frac{(18)(18 + 1)}{4} = 85.5$$

And the standard deviation can be determined by using formula (15.4).

$$\sigma_T = \sqrt{\frac{n(n + 1)(2n + 1)}{24}} = \sqrt{\frac{(18)(18 + 1)[(2)(18) + 1]}{24}} = 22.96$$

If the null hypothesis $H_0 : \mu \leq 25$ is true, we would expect most of the ranks to be in the vicinity of $\mu_T = 85.5$. Stated differently, we would expect the values of T^+ and T^- to be close to 85.5. Is $T = T^+ = 124.5$ far enough into the right-tail region of the sampling distribution of T to warrant the rejection of the null hypothesis $H_0 : \tilde{\mu} \geq 25$? We can answer this question by determining the z value for $T = 124.5$ and comparing it to $z_{0.05} = 1.65$. The z value for $T = 124.5$ is

$$z = \frac{T - \mu_T}{\sigma_T} = \frac{124.5 - 85.5}{22.96} = 1.70$$

Since $z = 1.70 > 1.65$, we can reject the null hypothesis and conclude that the doctor's patients had to wait longer than 25 minutes to get his attention. Notice that this conclusion is opposite to the conclusion we arrived at by using the sign test. We would expect this since the signed-rank test used more of the information contained in the data. ■

APPLICATION 15.5

The signed-rank test can also be used to test for differences when paired data are involved. For example, we can use the signed-rank test to determine if the diet described in application 15.1 of section 15.1 is effective in weight reduction. The data representing the weights of 15 men who have been on a weight-reduction diet for one month are repeated here for convenience.

Male	Weight Before	Weight After
1	210	195
2	175	162
3	187	179
4	189	185
5	198	199
6	205	200
7	198	193
8	178	178
9	164	162
10	176	169
11	192	188
12	187	184
13	210	198
14	178	172
15	205	206

Solution: Suppose μ_B and μ_A represent the means of the population of before-diet weights and after-diet weights, respectively.

1. The null hypothesis is H_0: $\mu_B - \mu_A \leq 0$.
2. The alternative hypothesis is H_1: $\mu_B - \mu_A > 0$ (one-tailed test).
3. Note that if the alternative hypothesis is true, we would expect more positive ranks than negative ranks if we subtract after-diet weights from before-diet weights. Therefore, the value of T^+ will tend to be large, so we will let the test statistic be $T = T^+$, the sum on the positive ranks. Table 15.4 contains the difference scores, the ranks of the differences, and the signed ranks.

TABLE 15.4

Signed-Rank Information for Application 15.5.

Male	Weight Before	Weight After	Difference	Ranks of Difference	Negative Ranks	Positive Ranks
1	210	195	15	14		14
2	175	162	13	13		13
3	187	179	8	11		11
4	189	185	4	5.5		5.5
5	198	199	−1	1.5	1.5	
6	205	200	5	7.5		7.5
7	198	193	5	7.5		7.5
8	178	178	0	—		—
9	164	162	2	3		3
10	176	169	7	10		10
11	192	188	4	5.5		5.5
12	187	184	3	4		4
13	210	198	12	12		12
14	178	172	6	9		9
15	205	206	−1	1.5	1.5	

From table 15.4 we can observe that the number of nonzero differences is $n = 14$, the sum of the negative ranks is $T^- = 3$, and the sum of the positive ranks is $T^+ = 102$. Hence, the value of the test statistic is $T = 102$. As a check, notice that $T^- + T^+ = 3 + 102 = 105 = [(14)(14 + 1)]/2$, since $n = 14$.

Using formula (15.3), the mean of the sampling distribution of T is

$$\mu_T = \frac{n(n + 1)}{4} = \frac{(14)(14 + 1)}{4} = 52.5$$

And using formula (15.4), the standard deviation of the sampling distribution of T is

$$\sigma_T = \sqrt{\frac{n(n + 1)(2n + 1)}{24}}$$

$$= \sqrt{\frac{(14)(14 + 1)[(2)(14) + 1)]}{24}} = 15.93$$

The z value for $T = 102$ is

$$z = \frac{T - \mu_T}{\sigma_T} = \frac{102 - 52.5}{15.93} = 3.11$$

4. The level of significance is $\alpha = 0.01$.
5. The critical value is $z_{0.01} = 2.33$.
6. Decision: Since $3.11 > 2.33$, we reject H_0: $\mu_B - \mu_A \leq 0$ and conclude that the diet is effective in weight reduction.
7. Type of error possible: Type I; $\alpha = 0.01$.
8. *p*-value: *p*-value < 0.001.

It is interesting to note that we could have used the sum of the negative ranks as the value of the test statistic. In this case, $T = T^- = 3$. The z value for $T = 3$ is

$$z = \frac{T - \mu_T}{\sigma_T} = \frac{3 - 52.5}{15.93} = -3.11$$

Notice that this value is the negative of the z value for $T = T^+$. In this case, the critical value would become $-z_{0.01} = -2.33$, and we would arrive at the same conclusion as above regarding the test. ■

MINITAB can be used to conduct the analysis of a signed-rank test. The WTEST command is used to carry out the analysis on a column of difference scores. Paired data are entered using the READ command. The column of differences is created by using the LET command. Computer display 15.2 contains the commands used and corresponding output for the data of application 15.5.

Computer Display 15.2

```
MTB > READ C1 C2
DATA > 210 195
DATA > 175 162
DATA > 187 179
DATA > 189 185
DATA > 198 199
DATA > 205 200
DATA > 198 193
DATA > 178 178
DATA > 164 162
DATA > 176 169
DATA > 192 188
DATA > 187 184
DATA > 210 198
DATA > 178 172
DATA > 205 206
DATA > END
   15 ROWS READ
MTB > LET C3 = C1 − C2
MTB > WTEST C3;
SUBC> ALTERNATIVE = 1.

TEST OF MEDIAN = 0.000000 VERSUS MEDIAN G.T. 0.000000

         N FOR WILCOXON ESTIMATED
         N    TEST    STATISTIC   P-VALUE    MEDIAN
C3      15     14       102.0      0.001      5.000
```

The estimated median of 5 represents the median of the 14 nonzero difference scores. If there is no difference between the two groups, one would expect an estimated median value of 0. The larger the estimated median, the greater the evidence for rejecting the null hypothesis.

APPLICATION 15.6   A drug company wants to determine whether or not there are any changes in body temperature for colon cancer patients taking their chemical treatment. Sixteen colon cancer patients are randomly selected for study. Their temperatures are measured before and after taking the treatment. The results are listed in the accompanying table.

Patient	Temperatures (in degrees Fahrenheit) Before	After	Patient	Temperatures (in degrees Fahrenheit) Before	After
1	97.1	98.2	9	97.8	97.4
2	98.3	98.6	10	98.8	99.7
3	98.8	99.1	11	98.4	98.0
4	99.2	98.9	12	98.2	98.2
5	99.7	101.3	13	99.8	102.4
6	99.5	99.0	14	99.6	102.0
7	98.2	99.4	15	98.4	98.6
8	98.0	100.7	16	99.4	99.0

By using a significance level of $\alpha = 0.05$, determine if patients subjected to the chemical treatment have a change in body temperature.

Solution: Suppose μ_B and μ_A represent the means of the population of before-treatment temperatures and after-treatment temperatures, respectively.

1. The null hypothesis is H_0: $\mu_B - \mu_A = 0$.
2. The alternative hypothesis is H_1: $\mu_B - \mu_A \neq 0$ (two-tailed test).
3. Table 15.5 contains the difference scores, the ranks of the differences, and the signed ranks.

TABLE 15.5

Signed-Rank Information for Application 15.6.

Patient	Temp. Before	Temp. After	Difference	Ranks of Difference	Negative Ranks	Positive Ranks
1	97.1	98.2	−1.1	10	10	
2	98.3	98.6	−0.3	3	3	
3	98.8	99.1	−0.3	3	3	
4	99.2	98.9	0.3	3		3
5	99.7	101.3	−1.6	12	12	
6	99.5	99.0	0.5	8		8
7	98.2	99.4	−1.2	11	11	
8	98.0	100.7	−2.7	15	15	
9	97.8	97.4	0.4	6		6
10	98.8	99.7	−0.9	9	9	
11	98.4	98.0	0.4	6		6
12	98.2	98.2	0	—		—
13	99.8	102.4	−2.6	14	14	
14	99.6	102.0	−2.4	13	13	
15	98.4	98.6	−0.2	1	1	
16	99.4	99.0	0.4	6		6
					91	29

Note that $n = 15$ and $91 + 29 = 120 = (15)(15 + 1)/2$. The value of the test statistic is $T = T^- = 91$.

Using formula (15.3) the mean of the sampling distribution of T is

$$\mu_T = \frac{n(n + 1)}{4} = \frac{(15)(15 + 1)}{4} = 60$$

Using formula (15.4) the standard deviation of the sampling distribution of T is

$$\sigma_T = \sqrt{\frac{n(n + 1)(2n + 1)}{24}} = \sqrt{\frac{(15)(16)(31)}{24}} = 17.61$$

The z value for $T = 91$ is

$$z = \frac{T - \mu_T}{\sigma_T} = \frac{91 - 60}{17.61} = 1.76$$

4. The significance level is $\alpha = 0.05$.
5. The positive critical value is $z_{0.025} = 1.96$.
6. Decision: Since $1.76 < 1.96$, we fail to reject H_0: $\mu_B - \mu_A = 0$. We have no evidence that the chemical treatment alters body temperature.
7. Type of error possible: Type II; β is unknown.
8. p-value: p-value $= 2(0.5 - 0.4608) = 0.0784$ ▪

EXERCISE SET 15.2

Basic Skills

In exercises 1–4, use the given paired scores (x, y) and the Wilcoxon signed-ranks procedures to

a. find the values of T^+, T^-, and T.
b. check to determine if $T^+ + T^- = n(n + 1)/2$.
c. find the z value for T.

1.
x	83	78	92	74	98	79	70	81
y	80	85	94	78	99	79	80	96

2.
x	56	48	49	47	52	43	45	50
y	48	42	47	42	43	42	35	54

3.
x	78	71	86	54	62	70	75	61	68	90
y	77	72	84	57	63	65	85	81	68	93

4.
x	4.2	3.9	2.7	3.5	4.0	3.8	2.9	3.9	4.5	3.2	3.0	6.1
y	3.4	3.6	3.0	2.5	3.1	3.3	3.1	2.4	2.5	2.5	3.0	6.2

For exercises 5–10, assume $\alpha = 0.05$. Use the given information to find the z value for T, the critical value(s), and the decision regarding the null hypothesis.

5. One-tailed test, $n = 18$, and $T = 15$.

6. One-tailed test, $n = 18$, and $T = 156$.

7. Two-tailed test, $n = 20$, and $T = 105$.

8. Two-tailed test, $n = 25$, and $T^+ = 100$.

9. One-tailed test, $n = 25$, and $T^- = 225$.

10. One-tailed test, $n = 28$, and $T^+ = 300$.

Further Applications

11. The accompanying data (final examination grades in mathematics and physics) were collected on 16 randomly selected engineering students at a certain college to determine if engineering students do better in mathematics than physics.

Student	Math	Physics	Student	Math	Physics
1	97	92	9	80	80
2	78	80	10	82	92
3	85	82	11	96	89
4	84	83	12	85	80
5	92	96	13	50	74
6	33	32	14	100	99
7	62	65	15	82	72
8	80	72	16	80	64

Use the signed-rank test and $\alpha = 0.05$ to determine if the mean math grade of engineering students is higher than the mean physics grade.

12. Two different brands of fertilizer were used to test the effects on peach tree yields. One season 16 randomly selected peach trees were treated with fertilizer A, and the following season they were treated with fertilizer B. The number of good peaches (in bushels) for each tree is recorded in the table.

Tree	First Season	Second Season
1	2.0	2.5
2	2.5	3.0
3	2.0	2.0
4	3.5	3.0
5	3.0	2.5
6	2.5	2.0
7	2.5	3.0
8	3.0	2.5
9	2.0	3.5
10	1.5	1.5
11	2.5	2.0
12	3.0	2.0
13	3.5	2.5
14	2.5	3.0
15	1.5	2.0
16	3.0	3.5

Use the signed-rank test at $\alpha = 0.05$ to determine if there is a difference in mean peach tree yields for the two types of fertilizer.

13. The following nicotine contents (in milligrams) of a random sample of 15 brand A cigarettes were obtained: 2.1, 4.0, 6.3, 5.4, 4.8, 3.7, 6.1, 2.3, 3.3, 6.4, 5.4, 2.5, 3.1, 4.1, and 4.7. At $\alpha = 0.05$, use the signed-rank test to determine if the data indicate that $\mu < 4.5$.

14. A random sample of teachers at a certain university yielded the following ages: 26, 49, 55, 55, 60, 61, 68, 57, 74, 75, 68, 68, 68, 67, 64, 64, and 63. Use the signed-rank test at $\alpha = 0.05$ to determine if the mean age is different from 64 years.

15. The data in the accompanying table indicate the systolic blood pressure measurements of fifteen randomly selected subjects before and one hour following an exercise program. Use the signed-rank test and $\alpha = 0.01$ to determine if the exercise program helps to lower systolic blood pressure.

Person	Before	After	Person	Before	After
1	164	162	9	138	139
2	146	144	10	130	124
3	148	146	11	175	170
4	154	156	12	146	147
5	143	145	13	160	157
6	160	159	14	160	140
7	150	145	15	153	148
8	148	150			

16. A study involving 16 randomly chosen coffee drinkers was conducted to determine the effects of drinking coffee on pulse rates. The pulse rates for the sixteen subjects before and after drinking two cups of coffee are given below. By using $\alpha = 0.01$ and the signed-rank test, determine if drinking two cups of coffe increases pulse rates.

Pulse Rates

Person	Before coffee	After coffee
A	73	74
B	77	81
C	74	73
D	75	78
E	75	77
F	79	81
G	71	71
H	76	76
I	65	70
J	66	67
K	72	71

Pulse Rates

Person	Before coffee	After coffee
L	80	90
M	78	83
N	66	68
O	72	77
P	70	70

Going Beyond

17. Under the assumptions that the null hypothesis is true and there are no ties or zeros in the n difference scores, show that $\mu_T = n(n + 1)/4$. (*Hint:* Use the fact that $1 + 2 + 3 + \cdots + n = n(n + 1)/2$.)

18. Under the assumptions that the null hypothesis is true and there are no ties in the n difference scores, show that

$$\sigma_T = \sqrt{\frac{n(n + 1)(2n + 1)}{24}}$$

(*Hint:* Use the fact that $1^2 + 2^2 + 3^2 + \cdots + n^2 = n(n + 1)(2n + 1)/6$.)

19. With $n = 10$ pairs of data, find the smallest and largest values of T.

20. With $n = 20$ pairs of data, find the smallest and largest values of T.

21. Assume that the absolute values of the differences of four pairs of data are ranked from smallest to largest with no ties and no differences of zero. Determine the probability distribution of T^+.

SECTION 15.3   *Wilcoxon Rank-Sum Test (Large Samples)*

The **Wilcoxon rank-sum test** (also commonly known as the **Mann-Whitney U Test**), a nonparametric test, is used to compare data from two continuous populations having the same shape and spread. It is the nonparametric alternative to the two-sample t test for independent samples and can be used when the normality assumption cannot be satisfied.

The rank-sum test involves the assignment of ranks to the data after the two samples have been combined and arranged in increasing order. To apply the rank-sum test, the data from the two samples are pooled together and ranked from smallest to largest. In case of ties, the tied ranks are assigned the average of the ranks had there been no ties. For example, if the fourth, fifth, and sixth observations are equal, a rank of 5 is assigned to each of the three identical observations. After we assign ranks to the data, we choose the smaller sample and find the sum of the ranks, denoted by W, for that sample. Either sample can be used to determine the sum of the ranks W if both samples are of the same size. If the sampled populations have unequal means, we would expect most of the lower ranks to be in the sample from the population with the smaller mean and we would expect most of the higher ranks to be in the sample from the population with the higher mean.

The rank sum test is based on the test statistic U defined by

Test Statistic for Rank Sum Test

$$U = W - \frac{n_1(n_1 + 1)}{2}$$

(15.6)

where n_1 is the size of the smaller sample. The number $n_1(n_1 + 1)/2$ is the smallest value that W can assume, and the U statistic measures the distance between W and its smallest value. The U statistic is closely related to W; the values of U vary directly with the values of W. If the U test statistic is large, then W is large and the sample used to generate W should belong to the population with the larger mean. If we choose notation so that μ_1 is the mean of the population whose sample size is n_1, then for the value of U to be large means that H_0: $\mu_1 \leq \mu_2$ is rejected in favor of H_1: $\mu_1 > \mu_2$. And if the value for U is small, H_0: $\mu_1 \geq \mu_2$ is rejected in favor of H_1: $\mu_1 < \mu_2$.

If the samples come from continuous and identical populations and there are no ties in the ranks, then the sampling distribution of U has a mean and standard deviation given by

Mean and Standard Deviation for U Statistic

$$\mu_U = \frac{n_1 n_2}{2}$$

(15.7)

$$\sigma_U = \sqrt{\frac{n_1 n_2 (n_1 + n_2 + 1)}{12}}$$

(15.8)

where n_1 is the size of the smaller sample and n_2 is the size of the larger sample. In addition, if n_1 is greater than 8, the sampling distribution of U is approximately normal. The z value for U is determined by

$$z = \frac{U - \mu_U}{\sigma_U}$$

If $n_1 \leq 8$, then special tables must be used to determine critical values for the U statistic. These tables are contained in most statistical handbooks. Applications 15.7 and 15.8 illustrate the use of the rank-sum test.

APPLICATION 15.7

Two groups of students were taught statistics using different methods, A and B. A final examination was given to both groups at the end of the course. The results are given in table 15.6.

TABLE 15.6

Data for Application 15.7

Method A	73	67	72	46	83	75	62	90	95	
Method B	71	47	68	87	77	92	65	86	79	57

Use the rank-sum test at $\alpha = 0.05$ to determine if there is a difference in the average final examination scores for students taught by methods A and B.

Solution: The null hypothesis is H_0: $\mu_A - \mu_B = 0$ and the alternative hypothesis is H_1: $\mu_A - \mu_B \neq 0$. The following table will be used to organize the ranking.

Data	Rank	Sample	Data	Rank	Sample
46	(1)	A	73	(10)	A
47	2	B	75	(11)	A
57	3	B	77	12	B
62	(4)	A	79	13	B
65	5	B	83	(14)	A
67	(6)	A	86	15	B
68	7	B	87	16	B
71	8	B	90	(17)	A
72	(9)	A	92	18	B
			95	(19)	A

Since sample A is the smaller sample, $n_1 = 9$ and $n_2 = 10$. The sum of the ranks for sample A (shown in parentheses in the table) is determined to be

$$W = 1 + 4 + 6 + 9 + 10 + 11 + 14 + 17 + 19 = 91$$

The value of the U test statistic is found by using formula (15.6):

$$U = W - \frac{n_1(n_1 + 1)}{2} = 91 - \frac{(9)(10)}{2} = 91 - 45 = 46$$

The mean of the sampling distribution of U is found by using formula (15.7):

$$\mu_U = \frac{n_1 n_2}{2} = \frac{(9)(10)}{2} = 45$$

And the standard deviation of the sampling distribution of U is found by using (15.8):

$$
\begin{aligned}
\sigma_U &= \sqrt{\frac{n_1 n_2 (n_1 + n_2 + 1)}{12}} \\
&= \sqrt{\frac{(9)(10)(9 + 10 + 1)}{12}} \\
&= \sqrt{150} = 12.25
\end{aligned}
$$

Since $n_1 > 8$ the sampling distribution of U is approximately normal and the z statistic for U is

$$z = \frac{U - \mu_U}{\sigma_U} = \frac{46 - 45}{12.25} = 0.08$$

The critical values are $\pm z_{0.05} = \pm 1.96$. Since $-1.96 < z = 0.08 < 1.96$, we cannot reject the null hypothesis. Therefore, there is no significant difference between the average final examination scores for students taught by the two methods. ■

The MANN-WHITNEY command in MINITAB can be used to analyze the data in application 15.7. Computer display 15.3 contains the commands used and the corresponding output.

Computer Display 15.3

```
MTB > SET C1
DATA > 73 67 72 46 83 75 62 90 95
DATA > SET C2
DATA > 71 47 68 87 77 92 65 86 79 57
DATA > END
MTB > MANN-WHITNEY C1 C2

MANN-WHITNEY CONFIDENCE INTERVAL AND TEST

C1    N =  9      MEDIAN =    73.00
C2    N = 10      MEDIAN =    74.00
POINT ESTIMATE FOR ETA1-ETA2 IS      1.50
95.5 PCT C.I. FOR ETA1-ETA2 IS (   -14.00,    16.00)
W =    91.0
TEST OF ETA1 = ETA2 VS. ETA1 N.E. ETA2 IS SIGNIFICANT AT 0.9674

CANNOT REJECT AT ALPHA = 0.05

MTB >
```

MINITAB uses a variation of the Mann-Whitney test that involves the sampling distribution of W and medians (see exercise 11 in exercise set 15.3). Note that ETA1 and ETA2 represent the medians of the two populations. A small value of W indicates that ETA1 < ETA2; a large value indicates that ETA2 < ETA1.

It is interesting to note that the value $n_1(n_1 + 1)/2 = 45$ in application 15.7 represents the smallest value that W could possibly be for sample A. This would occur if all the measurements of sample A were less than every measurement in sample B. In this case, W would become

$$W = 1 + 2 + 3 + \cdots + 9 = 45$$

the same value found by evaluating $n_1(n_1 + 1)/2$ for $n_1 = 9$. The value for U in this case would then be

$$U = W - \frac{n_1(n_1 + 1)}{2} = 45 - 45 = 0$$

APPLICATION 15.8

Patients with a certain disease are treated with two different drugs, and patients are evaluated in terms of complete recovery in days. A random sample of nine patients was treated with drug A and another random sample of nine patients was treated with drug B. The time in days to complete recovery is recorded for each patient in table 15.7.

TABLE 15.7

Data for Application 15.8

Drug A	13	10	12	14	14	15	16	16	17
Drug B	8	17	9	11	15	11	14	12	18

Use the rank-sum test at the 0.05 level of significance to determine whether the mean recovery time for patients treated with drug B is less than the mean recovery time for patients treated with drug A.

Solution:

1. H_0: $\mu_B - \mu_A \geq 0$.

2. H_1: $\mu_B - \mu_A < 0$.

3. The following table is used for ranking the data:

Data	Rank	Sample	Data	Rank	Sample
8	(1)	B	14	10	A
9	(2)	B	14	(10)	B
10	3	A	15	12.5	A
11	(4.5)	B	15	(12.5)	B
11	(4.5)	B	16	14.5	A
12	6.5	A	16	14.5	A
12	(6.5)	B	17	16.5	A
13	8	A	17	(16.5)	B
14	10	A	18	(18)	B

Since both samples are of the same size (9), we choose sample B for finding the sum of the ranks W, since the alternative hypothesis is H_1: $\mu_B < \mu_A$ and the sum of the ranks corresponding to sample B should be less than the sum of the ranks corresponding to sample A. In this case, the left tail of the distribution should be chosen as the rejection region. (Had the size of sample A been smaller than the size of sample B, then sample A would have been chosen for finding the sum of the ranks, and the right tail of the U distribution would have been chosen as the rejection region.)

The sum of the ranks corresponding to sample B (shown in parentheses in the table) is

$$W = 1 + 2 + 4.5 + 4.5 + 6.5 + 10 + 12.5 + 16.5 + 18 = 75.5$$

The value of the U statistic is found by using formula (15.6):

$$U = W - \frac{n_1(n_1 + 1)}{2} = 75.5 - \frac{(9)(10)}{2} = 30.5$$

The mean of the sampling distribution of U is given by formula (15.7):

$$\mu_U = \frac{n_1 n_2}{2} = \frac{(9)(9)}{2} = 40.5$$

The standard error of U is given by formula (15.8):

$$\sigma_U = \sqrt{\frac{n_1 n_2 (n_1 + n_2 + 1)}{12}} = \sqrt{\frac{(9)(9)(19)}{12}} = \sqrt{128.25} = 11.32$$

Since $n_1 > 8$, the sampling distribution of U is approximately normal and the value of the z statistic is found to be

$$z = \frac{U - \mu_U}{\sigma_U} = \frac{30.5 - 40.5}{11.32} = -0.88$$

4. $\alpha = 0.05$.

5. Critical value: $-z_{0.05} = -1.65$.

6. Decision: Since $z = -0.88 > -1.65$, we fail to reject H_0. Thus, there is no statistical evidence to suggest that the mean recovery time for patients treated with drug B is less than the mean recovery time for patients treated with drug A.

7. Type of error possible: Type II, β is unknown.

8. p-value $= 0.5 - 0.3106 = 0.1894$. ▪

It is interesting to note that had sample A been chosen in application 15.8 to determine the sum of the ranks W, we would have had the following results:

$$W = 3 + 6.5 + 8 + 10 + 10 + 12.5 + 14.5 + 14.5 + 16.5 = 95.5$$

$$U = W - \frac{n_2(n_2 + 1)}{2} = 95.5 - \frac{(9)(10)}{2} = 50.5$$

$$z = \frac{U - \mu_U}{\sigma_U} = \frac{50.5 - 40.5}{11.32} = 0.88$$

The z values obtained here differs only in sign from the z value obtained in application 15.8. In general, this relationship holds for the z values computed for the two U statistics corresponding to the two samples of data.

If W_1 denotes the sum of the ranks for sample A and W_2 denotes the sum of the ranks for sample B, then $W_1 + W_2$ can be found by using the formula

$$W_1 + W_2 = \frac{(n_1 + n_2)(n_1 + n_2 + 1)}{2} \qquad (15.9)$$

EXAMPLE 15.2

As a consequence of formula (15.9), for application 15.8 we have

$$W_1 + W_2 = \frac{(18)(10)}{2} = 171$$

Since the value of W_2 was found to be 75.5, the value of W_1 could have been found easily by subtraction:

$$W_1 = 171 - W_2 = 95.5$$

which agrees with the result found above. The relationship (15.9) can be used as a check to ensure that the value of W has been calculated correctly.

EXERCISE SET 15.3

Further Applications

1. The useful lives (in months) before failure of color cathode ray tubes made by two manufacturers, A and B, are as shown in the table.

A	42	35	50	41	45	39	47	49	40
B	52	51	46	57	44	58	54	53	43	48

Use the rank-sum test and $\alpha = 0.05$ to determine if there is a difference in the average useful lifetimes of service for the tubes made by the two manufacturers.

2. The accompanying data represent the number of violent crimes committed in a certain city during a nine-week spring period and a ten-week fall period:

Spring	51	42	57	53	43	37	45	49	46
Fall	40	35	30	44	33	50	41	39	36	38

Use the rank-sum test and $\alpha = 0.01$ to determine if there is, on the average, a difference in the weekly number of violent crimes committed for the two periods.

3. Suppose the scores of trainees in a current training program are to be compared with the scores obtained from an independent sample of trainees from a previous program. The following results were obtained:

Previous: 7.5 6.9 6.7 6.4 6.2 6.0 5.9 5.8 5.6 5.4
5.3 5.0 4.9 4.9 4.3 3.8 3.0

Current: 6.7 6.6 6.3 5.4 5.2 5.5 4.9 4.6 4.5 4.3
3.9 3.7 3.2 2.9 2.3

a. By using the rank-sum test, determine whether the previous program produced higher average scores than the current program. Use $\alpha = 0.05$.

b. Use the *t* test to determine whether the previous program produced higher average scores than the current program. Use $\alpha = 0.05$.

c. Compare your results from parts a and b. Which test should you use? Does it make a difference? Explain.

4. A study was made to investigate the effects of vitamin C on the common cold. A group of 20 students who had developed a cold were randomly assigned to two groups. Group A acted as the control group and received only a sugar tablet, whereas students in group B received 1 gram of vitamin C. The data listed in the table indicate the duration (in days) of cold symptoms for each subject.

Group A	13 18 11 20 24 15 19 27 9
Group B	8 14 23 25 17 16 30 7 21 10 22

a. At $\alpha = 0.01$, test by using the rank-sum test to determine if the students in group B have a shorter average duration of cold symptoms than those in group A.

b. Use the *t* test at $\alpha = 0.01$ to determine if the students in group B have a shorter average duration of cold symptoms than students in group A. Compare your results with those obtained in part a.

c. Compare your results from parts a and b. Which test should you use? Does it make a difference? Explain.

5. An algebra teacher teaches factoring by two different methods. To determine if method B is better than method A, two groups of algebra students were taught using the two different methods, one method for each group. A factoring examination was given to all students at the end of the factoring unit and the student scores for both methods are provided in the accompanying table.

Method A	58 65 81 62 96 60 55 70 63 84
Method B	65 68 59 86 95 70 100 98 80 69

a. At the 0.05 level of significance, determine if the average score for students taught factoring by method B is greater than the average score for students taught factoring by method A.

b. At $\alpha = 0.05$ use the two-sample *t* test to determine if the average score for students taught factoring by method B is greater than the average score for students taught factoring by method A.

c. Compare your results from parts a and b. Which test should you use? Does it make a difference? Explain.

6. The data provided in the table represent the number of absences from a Math 101 class for two random samples of males and females.

Male	8 2 7 6 0 2 13 5 14 9 18
Female	3 20 1 11 5 4 2 11 10 3 1 13 21

a. At $\alpha = 0.05$ use the two-sample *t* test to determine if there is a difference in the average number of days missed for males and females.

b. At $\alpha = 0.05$ use the rank sum test to determine if there is a difference in the average number of days missed for males and females.

c. Compare your results from parts a and b. Which test should you use? Does it make a difference? Explain.

7. Two groups of adults were tested for critical-thinking ability. Group A consisted of 11 women and group B consisted of 9 men. Both groups were administered the Watson Glaser (WG) test to measure critical thinking. Their scores are given in the accompanying table.

Group A	72 66 39 101 90 86 48 109 118 64 73
Group B	148 97 83 75 33 67 98 133 70

a. By using the rank-sum test determine if there is a difference in the average critical thinking scores for the two groups. Use $\alpha = 0.01$ and determine the *p*-value.

b. Test to determine if the population variances are different. Use $\alpha = 0.10$.

c. By using the *t* test and the raw data, determine if there is a difference in the average scores for the two groups. Use $\alpha = 0.01$.

8. Consider the following data, which were obtained by taking random samples of size 15 from two normal populations with means 15 and 17 and standard deviations 2 and 2, respectively.

Sample A		Sample B	
14.4	18.3	15.5	11.9
12.1	13.3	16.9	15.5
15.8	20.5	17.0	15.1
12.0	14.2	19.5	14.1
11.0	16.7	16.8	14.6
16.6	11.0	18.8	12.0
13.1	10.5	16.1	18.3
13.8			15.2

a. Use the *t* test at $\alpha = 0.05$ to determine if $\mu_A < \mu_B$.

b. Use the rank-sum test at $\alpha = 0.05$ to determine if $\mu_A < \mu_B$.

c. Compare your results in parts a and b with respect to sensitivity for detecting real differences. Would you expect the rank-sum test to be more sensitive than the *t* test? Why?

Going Beyond

9. Suppose sample A of size $n_1 = 2$ is taken from population I and sample B of size $n_2 = 3$ is taken from population II. The accompanying table contains the ten possible arrangements for the ranks of the five pooled measurements, where a is a member of sample A and b is a member of sample B.

Rank					Statistic	
1	2	3	4	5	W	U
a	b	b	b	a	6	3
a	b	b	a	b		
a	a	b	b	b		
a	b	a	b	b		
b	a	b	a	b		
b	a	a	b	b		
b	b	b	a	a		
b	b	a	a	b		
b	a	b	b	a		
b	b	a	b	a		

a. For each of the ten possible arrangements, compute the values of W and U corresponding to sample A.
b. Construct a probability distribution table for W, the sum of the ranks for A.
c. Construct a probability distribution table for U.
d. Find μ_W and σ_W.

e. Find μ_U and σ_U. Compare these values with those obtained from the formulas contained in the text.
f. Can $\alpha = 0.05$ be used as a significance level for a rank-sum test with $n_1 = 2$ and $n_2 = 3$? Why? Specify the smallest level α that is possible for such a test. (Your decision rule cannot be "never reject H_0.")

10. Refer to the 10 arrangements listed in exercise 9.
a. Compute the values of W and U corresponding to sample B.
b. If W_1 and U_1 correspond to sample A and W_2 and U_2 correspond to sample B, how are W_1 and W_2 related? How are U_1 and U_2 related?

11. A variation of the rank-sum test involves the sampling distribution of W, instead of U. It can be shown that for $n_1 > 8$ and $n_2 > 8$, the sampling distribution of W is approximately normal, with $\mu_W = n_1(n_1 + n_2)/2$ and $\sigma_W = \sqrt{n_1 n_2(n_1 + n_2 + 1)/12}$. Use this result to do exercise 5.

12. By using the relationship $U = W - n_1(n_1 + 1)/2$, show that $\mu_W = n_1(n_1 + n_2 + 1)/2$ and $\sigma_W = \sigma_U$.

13. If $U_1 = W_1 - n_1(n_1 + 1)/2$ and $U_2 = W_2 - n_2(n_2 + 1)/2$, show that the z value for U_1 is the negative of the z value for U_2.

SECTION 15.4   Kruskal-Wallis Test

The **Kruskal-Wallis (KW) test** is a useful alternative when the normality assumption underlying a one-factor analysis of variance ANOVA cannot be satisfied. It is a nonparametric method useful for testing a null hypothesis H_0 that k independent samples are from identical continuous populations. It is a generalization ($k > 2$) of the rank-sum test studied in section 15.3.

If the null hypothesis H_0: $\mu_1 = \mu_2 = \mu_3 = \cdots = \mu_k$ is assumed to be true and the N measurements from all the samples are ranked in ascending order, then each sample can be expected to have random ranks. If the data contain ties, ranks are assigned as in the rank-sum test. The tied values are assigned the average of the ranks occupying the tied positions. If H_0 is false, then some samples will have mostly small ranks and other samples will have mostly large ranks. If R_{ij} denotes the rank of x_{ij} and R_i and $\bar{R}_i$ denote the sum and average, respectively, of the ranks in the ith sample, then when H_0 is true the average value of R_{ij} is $(N + 1)/2$ and the average value of R_i is also $(N + 1)/2$. The KW test measures the extent to which the average ranks of the samples deviate from their average value of $(N + 1)/2$ much in the same fashion that MSB in ANOVA measures the extent to which each sample mean deviates about the grand mean. The test statistic H for the KW test is given by

$$H = \frac{12}{N(N + 1)} \Sigma n_i \left(\bar{R}_i - \frac{N + 1}{2} \right)^2 \tag{15.10}$$

where $N = \Sigma\, n_i = n_1 + n_2 + n_3 + \cdots + n_k$. A computational formula analogous to SSB for the test statistic H is given by

> **Computational Formula for KW H Statistic**
>
> $$H = \frac{12}{N(N+1)} \Sigma \frac{R_i^2}{n_i} - 3(N+1)$$ (15.11)

If each value of n_i is greater than or equal to 5, then the sampling distribution of H is approximately a χ^2 distribution with df $= k - 1$. Consider applications 15.9 and 15.10.

APPLICATION 15.9

The data in table 15.8 represent the suicide rates (per 1000 population) for samples of cities in three sections of the United States.

TABLE 15.8

Data for Application 15.9

Eastern	Central	Western
2.8	2.1	2.1
5.0	2.4	4.2
7.2	3.5	4.3
8.3	7.0	4.8
10.0	12.1	6.4
13.2	13.6	6.6
13.6	14.9	8.4
	15.6	8.9

By using the KW test and the 0.05 level of significance, determine if the population mean suicide rates differ for the three sections of the country.

Solution: Let μ_E, μ_C, and μ_W represent the mean suicide rates for the east, central, and western sections of the United States, respectively.

1. H_0: $\mu_E = \mu_C = \mu_W$
2. H_1: At least two means are different.
3. Toward determining the value of the test statistic, we first rank the combined data from smallest to largest. The following table is used to organize the ranking:

Data	Rank	Group	Data	Rank	Group
2.1	1.5	C	8.3	14	E
2.1	1.5	W	8.4	15	W
2.4	3	C	8.9	16	W
2.8	4	E	10.0	17	E
3.5	5	C	12.1	18	C
4.2	6	W	13.2	19	E
4.3	7	W	13.6	20.5	E
4.8	8	W	13.6	20.5	C
5.0	9	E	14.9	22	C
6.4	10	W	15.6	23	C
6.6	11	W			
7.0	12	C			
7.2	13	E			

Next, we determine n_i and R_i for each area.

East: $n_1 = 7$, $R_1 = 4 + 9 + 13 + 14 + 17 + 19 + 20.5 = 96.5$

Central: $n_2 = 8$, $R_2 = 1.5 + 3 + 5 + 12 + 18 + 20.5 + 22 + 23 = 105$

Western: $n_3 = 8$, $R_3 = 1.5 + 6 + 7 + 8 + 10 + 11 + 15 + 16 = 74.5$

The total number of observations is $N = n_1 + n_2 + n_3 = 23$.

Note: As a check on our calculations, the following relation always holds:

$$\Sigma R_i = \frac{N(N + 1)}{2}$$

The sum of ranks R_i is

$$\Sigma R_i = 96.5 + 105 + 74.5 = 276$$

which is the same value as

$$\frac{N(N + 1)}{2} = \frac{(23)(24)}{2} = 276$$

The value of the H test statistic is found by using formula (15.11).

$$H = \frac{12}{N(N + 1)} \Sigma \frac{R_i^2}{n_i} - 3(N + 1)$$

$$= \frac{12}{(23)(24)}\left(\frac{96.5^2}{7} + \frac{105^2}{8} + \frac{74.5^2}{8}\right) - 3(23 + 1) = 1.96$$

The sampling distribution of H is approximately a χ^2 distribution with df $= k - 1 = 3 - 1 = 2$. Note that n_1, n_2, and n_3 are all greater then or equal to five.

4. $\alpha = 0.05$.

5. The critical value is $\chi_{0.05}^2(2) = 5.991$.

6. Decision: We fail to reject H_0: $\mu_1 = \mu_2 = \mu_3$. Hence, there is no statistical evidence that mean suicide rates for the three sections of the country differ.

7. Type of error possible: Type II, β is unknown.

8. p-value: The p-value is greater than 0.10. ■

The KRUSKAL-WALLIS command in MINITAB can be used to analyze the data of application 15.9. Computer display 15.4 contains the commands and corresponding output.

Computer Display 15.4

```
MTB > SET C1
DATA> 2.8 5.0 7.2 8.3 10.0 13.2 13.6
DATA> SET C2
DATA> 2.1 2.4 3.5 7.0 12.1 13.6 14.9 15.6
DATA> SET C3
DATA> 2.1 4.2 4.3 4.8 6.4 6.6 8.4 8.9
DATA> END
MTB > STACK C1 C2 C3 C4;
SUBC > SUBSCRIPTS C5.
MTB > KRUSKAL-WALLIS C4 C5
```

LEVEL	NOBS	MEDIAN	AVE. RANK	Z VALUE
1	7	8.300	13.8	0.84
2	8	9.550	13.1	0.58
3	8	5.600	9.3	-1.39
OVERALL	23		12.0	

```
H = 1.961
H(ADJ. FOR TIES) = 1.963

MTB > INVCDF 0.95;
SUBC> CHISQUARE DF=2.
  0.9500  5.9915
MTB >
```

| APPLICATION 15.10 | A study was done to compare the effectiveness of four different fertilizers on the growth of twenty plants of approximately the same size. Each fertilizer was used on five randomly chosen plants. The growth in inches was recorded for each plant after three weeks. The results shown in table 15.9 were obtained. |

TABLE 15.9

Data for Application 15.10

A	B	C	D
8.1	5.0	7.6	6.2
6.8	6.6	7.8	5.8
7.3	5.8	8.1	6.0
7.4	6.1	8.5	5.7
7.7	5.6	7.4	5.9

Determine using $\alpha = 0.05$ whether there is a significant difference between the fertilizers in terms of average growth.

Solution:

1. H_0: $\mu_A = \mu_B = \mu_C = \mu_D$.

2. H_1: At least two means are unequal.

3. In order to calculate the value of the H test statistic we shall proceed as follows. First rank the combined data, as indicated in table 15.10.

TABLE 15.10

Ranked Data for Application 15.10

Data	Rank	Group	Data	Rank	Group
5.0	1	B	6.8	11	A
5.6	2	B	7.3	12	A
5.7	3	D	7.4	13.5	A
5.8	4.5	D	7.4	13.5	C
5.8	4.5	B	7.6	15	C
5.9	6	D	7.7	16	A
6.0	7	D	7.8	17	C
6.1	8	B	8.1	18.5	C
6.2	9	D	8.1	18.5	A
6.6	10	B	8.5	20	C

The sample sizes are $n_1 = n_2 = n_3 = n_4 = 5$ and $N = 20$. The sum of the ranks for each group are $R_A = 71$, $R_B = 25.5$, $R_C = 84$, and $R_D = 29.5$. Using formula (15.11) the value of the H test statistic is

$$H = \frac{12}{N(N+1)} \Sigma \frac{R_i^2}{n_i} - 3(N+1)$$

$$= \frac{12}{(20)(21)} \left(\frac{71^2}{5} + \frac{25.5^2}{5} + \frac{84^2}{5} + \frac{29.5^2}{5} \right) - 3(21) = 14.81$$

4. $\alpha = 0.05$ (one-tailed test).

5. Critical value: df $= k - 1 = 4 - 1 = 3$ and $\chi_{0.05}^2(3) = 7.815$.

6. Decision: Since $H = 14.81 > 7.815$, we reject H_0 and conclude that there is a significant difference among the four fertilizers in terms of average growth. The population means are judged unequal, but we do not know which means are different.

7. Type of error posisble: Type I, $\alpha = 0.05$. ■

EXERCISE SET 15.4

Further Applications

In exercises 1–5 use the KW test to test the null hypothesis of equal population means. For each test use $\alpha = 0.05$.

1. Readability scores from exercise 1 of the exercise set 13.1.

Test A	50	49	54	57	58
Test B	48	57	61	55	60

2. Laboratory analysis results from exercise 4 of exercise set 13.1.

Lab A:	58.6	60.7	59.4	59.6	60.5
Lab B:	61.6	64.8	62.8	59.2	60.4
Lab C:	60.7	55.6	57.3	55.2	60.2

3. Typing scores from exercise 5 of exercise set 13.1.

Person

A	B	C	D
82	55	69	87
79	67	72	61
75	84	78	82
68	77	83	61
65	71	74	72

4. Test scores from exercise 5 of exercise set 13.2.

Teacher

A	B	C
75	90	17
91	80	81
83	50	55
45	93	70
82	53	61
75	87	43
68	76	89
62	58	73
47	82	73
95	98	93
38	78	58
79	64	81
	80	70
	33	
	79	

5. A test of basic reading skills was given to a random sample of five students in each of four schools. The results are as follows:

School

A	B	C	D
20	24	16	19
21	21	21	20
22	22	18	21
24	25	13	20
22	23	17	20

Going Beyond

6. Consider the following three samples of data. The data were obtained by taking random samples of size 10 from three normal populations with means 10, 12, and 14 and standard deviations 2, 2, and 2, respectively.

A: 9.6 7.5 9.5 10.9 7.8 9.9 6.7 9.7 11   8.7
B: 12.9 10.8 10.5 7.8 12.2 11.4 10.4 10.4 10.5 9.0 12.7
C: 14.1 14.9 10.5 13.6 16.0 16.9 17.2 13.8 14.0 17.3

a. By using ANOVA and $\alpha = 0.05$, determine if the population means are unequal.
b. By using the Kruskal-Wallis test and $\alpha = 0.05$, determine if the population means are different.
c. Compare your results to parts a and b. Would you expect the Kruskal-Wallis test to be more sensitive than ANOVA? Explain.

7. Prove that computational formula (15.11) is equivalent to formula (15.10).

8. For three samples of size four, what are the smallest and largest values of the H statistic?

9. For $k = 2$, show that the KW test is equivalent to the rank-sum test by showing that $H^2 = z$ where z is the test statistic for the rank-sum test.

10. A correction factor can be used for compensating for many tied observations when the KW test is used by adjusting the value of the H statistic. The correction factor C is defined by

$$C = 1 - \frac{\Sigma (t^3 - t)}{N^3 - N}$$

where t is the number of tied observations in a tie situation and N is the total number of observations in the k samples. The adjusted H value is found by dividing H by C.

a. Show that C is a value less than one if ties are present and that the adjusted value of H is always larger than H.
b. For application 15.9, determine the adjusted value of H. Compare this to the value found in computer display 15.4.

SECTION 15.5   *Friedman's Test*

The standard nonparametric test for analyzing data having a randomized block design is **Friedman's test.** In chapter 13 we used the *F* test and two-way ANOVA to analyze data in a randomized block design. In order to use the *F* test for this purpose the following assumptions had to be satisfied:

1. The *k* sampled populations are normal.
2. The *k* populations have equal variances.

Friedman's test offers a useful alternative when the normality assumption can not be met.

    Friedman's test is very similar to the Kruskal-Wallis test; the basic difference is the manner in which ranks are assigned. For the Kruskal-Wallis test the *k* treatment groups are combined into one large group and ranks are assigned in ascending order. With Friedman's test ranks are assigned in ascending order to the *k* measurements within each block. Tied ranks are assigned the average of the ranks occupying the tied positions. Each block (row) of ranks will necessarily be a permutation of the integers from 1 to *k*. The sum of the ranks within each block will equal $1 + 2 + 3 + \cdots + k = k(k + 1)/2$ and the mean of the ranks within each block will be equal to $(k + 1)/2$. If the null hypothesis $H_0: \mu_1 = \mu_2 = \cdots = \mu_k$ is assumed to be true, then each treatment sample can be expected to have random ranks, while if the null hypothesis is false, some treatment samples will have mostly large ranks and some will have mostly small ranks. That is, when H_0 is true, the means of the ranks within each sample $\overline{R}_j$ will tend to be close to their mean $(k + 1)/2$; and when H_0 is false, the $\overline{R}_j$'s will tend to be quite different from $(k + 1)/2$. The Friedman test statistic *S* measures the discrepancy between the $\overline{R}_j$'s and their mean $(k + 1)/2$. The test statistic for Friedman's test is given by

$$S = \frac{12n}{k(k + 1)} \Sigma \left(\overline{R}_j - \frac{k + 1}{2} \right)^2 \tag{15.12}$$

where *n* is the number of blocks, *k* is the number of treatments, and $\overline{R}_j$ is the mean of the ranks of the *j*th treatment group. If $n > 5$, then the sampling distribution of *S* is approximately a χ^2 distribution with df $= k - 1$. A computational formula for the test statistic *S* is given by

Computational Formula for Friedman's *S*

$$S = \frac{12}{nk(k + 1)} \Sigma R_j^2 - 3n(k + 1) \tag{15.13}$$

Notice that formula (15.13) is very similar to the KW *H* statistic.

APPLICATION 15.11   Graduates within five different departments from four universities are compared using starting salaries for the first job following graduation. The results (in $100) are contained in the following table:

	University			
Department	A	B	C	D
Mathematics	286	299	281	275
Statistics	269	284	291	285
English	220	185	159	231
History	180	156	163	193
Recreation	213	185	161	210

Test for a difference in the mean starting salaries of graduates from the four universities. Use $\alpha = 0.01$.

Solution: Let μ_A, μ_B, μ_C, and μ_D represent the mean starting salaries of graduates from universities A, B, C, and D, respectively.

1. H_0: $\mu_A = \mu_B = \mu_C = \mu_D$.
2. H_1: At least two populations means are different.
3. We calculate the value of the test statistic, S.

Step 1 Rank the data within each department (block). This information is contained in table 15.11.

TABLE 15.11

Ranks for the Blocks

	University							
Department	A	Rank	B	Rank	C	Rank	D	Rank
Mathematics	286	3	299	4	281	2	275	1
Statistics	269	1	284	2	291	4	285	3
English	220	3	185	2	159	1	231	4
History	180	3	156	1	163	2	193	4
Recreation	213	4	185	2	161	1	210	3
	$R_1 = 14$		$R_2 = 11$		$R_3 = 10$		$R_4 = 15$	

Step 2 Compute the sum of the ranks for each university (column). The results are contained in the bottom row of table 15.11.

Step 3 Compute the value of the test statistic S using formula (15.13). Note that $n = 5$ and $k = 4$.

$$S = \frac{12}{nk(k+1)} \Sigma R_j^2 - 3n(k+1)$$

$$= \frac{12}{(5)(4)(5)}(14^2 + 11^2 + 10^2 + 15^2) - 3(5)(5)$$

$$= \frac{3}{25}(642) - 75 = 2.04$$

The sampling distribution of S is approximately a χ^2 distribution with df $= k - 1 = 4 - 1 = 3$.

4. $\alpha = 0.05$.
5. The critical value is $\chi_{0.05}^2(3) = 7.815$.
6. Decision: Since $2.04 < 7.815$, we fail to conclude that there is a difference between the mean starting salaries for the four universities.
7. Type error possible: Type II; β is unknown. ▪

MINITAB can be used to perform the Friedman test for the data in application 15.10. Computer display 15.5 contains the READ command and data.

Computer Display 15.5

```
MTB > READ C1 C2 C3
DATA> 1 1 286
DATA> 1 2 299
DATA> 1 3 281
DATA> 1 4 275
DATA> 2 1 269
DATA> 2 2 284
DATA> 2 3 291
DATA> 2 4 285
DATA> 3 1 220
DATA> 3 2 185
DATA> 3 3 159
DATA> 3 4 231
DATA> 4 1 180
DATA> 4 2 156
DATA> 4 3 163
DATA> 4 4 193
DATA> 5 1 213
DATA> 5 2 185
DATA> 5 3 161
DATA> 5 4 210
DATA> END
   20 ROWS READ
```

Computer display 15.6 contains the commands used to perform the analysis and corresponding output.

Computer Display 15.6

```
MTB > NAME C1 'DEPT' C2 'UNIV' C3 'SALARY'

MTB > FRIEDMAN C3 C2 C1

FRIEDMAN TEST OF SALARY BY UNIV BLOCKED BY DEPT

S = 2.04 D.F. = 3 P = 0.564

             EST.  SUM OF
UNIV    N    MEDIAN     RANKS
   1    5    210.13     14.0
   2    5    194.87     11.0
   3    5    186.37     10.0
   4    5    219.13     15.0

GRAND MEDIAN = 202.62
```

Notice that the value of the S statistic agrees with the value we computed and that the p-value of the test is 0.564.

EXERCISE SET 15.5

Further Applications

1. Before buying a laser printer, the purchasing department of a large company decides to compare three laser printers having similar prices. Each is used in six departments for a period of a month. The average number of problems with each machine is recorded. The results are contained in the following table.

| | Printer | | |
Department	I	II	III
A	9.0	12.1	8.7
B	8.5	9.9	8.6
C	16.4	20.3	15.6
D	5.3	5.2	3.9
E	10.2	12.1	11.3
F	9.4	11.0	9.8

Use $\alpha = 0.05$ and Friedman's test to determine whether there are any differences in average number of problems among the laser printers.

2. Eight infants born at a particular hospital are involved in a study to compare weight gains for three particular feeding regimens. Four-week-old infants are matched in blocks of size three by sex, size, and bone structure. One child from each block is fed using one of three regimens. The weight gains (in ounces) over a period of two weeks is contained in the following table.

| | Regimen | | |
Block	A	B	C
1	12.5	11.9	13.8
2	14.2	15.2	17.0
3	10.1	9.4	10.2
4	12.1	11.1	12.2
5	11.9	13.3	13.0
6	13.5	12.3	12.6
7	13.1	12.3	12.5
8	11.3	11.0	11.1

Use $\alpha = 0.05$ and Friedman's test to determine if there is a difference in weight gains produced by the three feeding regimens.

3. Three real estate appraisers are employed by a large banking firm to determine market prices of property for which financing is being arranged. The bank decides to conduct a study to compare the three appraisers. Their appraisals (in $1000) on six randomly selected properties are provided in the accompanying table. Use $\alpha = 0.05$ and

Friedman's test to determine whether the three appraisers obtain significantly different results.

| | Appraiser | | |
Property	I	II	III
A	58.9	63.7	65.3
B	61.3	62.9	63.5
C	76.4	81.2	77.1
D	81.0	83.4	81.5
E	90.5	91.4	86.2
F	225.0	276.5	245.5

4. In an effort to expand its services, a regional transit authority conducted an experiment to determine which of four routes to take from an airport to the center of the business district of the city. The following data indicate the travel times (in minutes) along each of the four routes.

Day	Route 1	Route 2	Route 3	Route 4
Monday	20	22	22	24
Tuesday	23	24	26	26
Wednesday	22	25	27	25
Thursday	27	23	30	27
Friday	28	26	30	27

By using Friedman's test and $\alpha = 0.05$ to determine if there is a difference in average travel times for the four routes.

5. An electronics manufacturing firm operates 24 hours a day, five days a week. Three eight-hour shifts are used, and the workers rotate shifts each week. A management team conducted a study to determine whether there is a difference in the mean number of 14-inch video monitors produced when employees work on the various shifts. A random sample of five workers is selected and the number of 14-inch video monitors they produce for each shift is recorded here.

| | Monitors Produced | | |
Employee	Morning	Afternoon	Night
Jones	10	4	14
Miller	12	5	12
Phillips	7	3	9
Ross	9	8	7
Stevens	7	5	6

By using the 0.05 level of significance and Friedman's test, can we conclude there is a difference in the mean production by shift?

6. Listed below are the average sentences (in years) given felons by three judicial processes and by crime.[53]

Crime	Trial by Jury	Trial by Judge	Guilty Plea
Murder	28	21	14
Robbery	24	15	10
Rape	18	14	11
Aggravated assault	14	9	7
Burglary	10	5	6
Drug trafficking	8	10	5
Larceny	4	4	4

By using the 0.05 level of significance and Friedman's test, can we conclude there is a difference in the mean sentence for the three judicial processes?

7. Four laboratories, A, B, C, and D, are used by food manufacturing companies for making nutrition analyses of its products. The following data are the fat contents (in grams) of two tablespoons of three similar types of peanut butter:

Peanut Butter	Laboratory			
	A	B	C	D
Brand 1	16.6	17.7	16	16.3
Brand 2	16	15.5	15.6	15.9
Brand 3	16.4	16.3	15.9	16.2

By using the 0.05 level of significance and Friedman's test, can we conclude there is a difference in the mean laboratory analyses?

8. The manager of a large department store conducted an experiment to determine if there is a difference in the average weekly sales of three salespersons in the store's appliance department. The following data indicate sales (in $100) for seven consecutive weeks for the three employees.

Week	Salesperson		
	A	B	C
1	27.6	28.7	26.4
2	31.2	29.3	30.3
3	28.8	28.4	28.0
4	30.6	29.8	28.7
5	30.0	31.0	32.3
6	28.4	29.9	29.6
7	30.9	29.5	31.1

By using Friedman's test and the 0.05 level of significance, can we conclude there is a difference in the mean sales for the three salespersons and in the mean sales by week?

9. An experiment was conducted to compare automobile gasoline mileages for four brands of gasoline, A, B, C, and D. Three automobiles of the same make and model used each of four brands of gasoline for the experiment. The data (in miles per gallon) are recorded in the accompanying table.

Auto Type	Gas Brand			
	A	B	C	D
I	18.8	20.1	20.4	19.2
II	20.3	21.2	21.0	20.8
III	19.2	20.6	19.9	20.9

Use $\alpha = 0.05$ and Friedman's test to determine if there is a difference in mean mileage per gallon for the four gasoline brands and a difference in mean mileages for the three automobiles.

10. Four different kinds of fertilizer, A, B, C, and D, are used to study the yield of soy beans. The soil is divided into twelve plots of the same size. The soil is divided into three blocks, each containing four homogeneous plots. Fertilizers were randomly assigned to the plots within each block. The yields (in pounds per acre) are as follows.

Block	Fertilizer			
	A	B	C	D
1	52.8	49.4	58.6	42.9
2	60.1	48.1	61.0	50.3
3	62.0	56.4	63.6	61.2

Use $\alpha = 0.05$ and Friedman's test to determine if the fertilizers and blocks are effective.

11. Four different machines are under consideration for the assembly of a product. A study is conducted and six different operators are selected to operate the machines and are given initial instruction and practice on their use. The machines are to be compared in time (in seconds) to assemble a product. The results are as follows:

Machine	Operator					
	I	II	III	IV	V	VI
A	32.4	29.2	29.5	29.8	32.8	33.5
B	29.7	30.0	30.4	32.2	32.4	33.0
C	30.1	30.4	31.2	33.3	34.8	35.0
D	31.2	32.1	33.4	34.1	35.8	32.2

Use $\alpha = 0.05$ and Friedman's test to determine if the machines perform at the same mean rate of speed and the operators perform at the same mean rate of speed.

Going Beyond

12. Prove that formula (15.12) is equivalent to formula (15.13).

SECTION 15.6   *Test for Nonrandomness (Large Samples)*

A nonparametric test for determining the nonrandomness of data is the **runs test.**
Suppose the following sequence consisting of Ms and Fs represents the successive
occurrences of births of males (M) and females (F) in a particular hospital:
MMMMMMMMMM FFFFFFFFFF MMMMMMMMMM FFFFFFFFFF. Would
you question the randomness of occurrences? Probably so, but why? There are twenty
of each symbol F and M occurring, so we have the same proportions of male births
and female births. Undoubtedly, you question the order of occurrences rather than the
frequency of occurrences. A random order should possess no pattern of occurrences.
Thus, AABBAAABBBAAAABBBB is *probably* not a random sequence. A **run** is defined to be a sequence of identical symbols that are followed or preceded by different
symbols or none at all.

EXAMPLE 15.3

Runs are underlined in the following sequences of two symbols:

1. A BBB A BBB AA B A B AA, nine runs

2. $++++$ $----$ $+++$ $---$ $+$, five runs

For a sequence of five symbols of one kind and three symbols of another kind, the
minimum number of runs is 2 and the maximum number of runs is 7. Why? We would
not expect a sequence of two symbols to be random if it had too few or too many runs.
For any random sequence of two symbols, we denote the number of runs by R, the
number of one kind of symbol by n_1, and the number of the other kind of symbol by
n_2. The sampling distribution of the number of runs R has a mean defined by

Mean of R, the Runs Statistic

$$\mu_R = \frac{2n_1n_2}{n_1 + n_2} + 1 \qquad (15.14)$$

and a standard deviation defined by

Standard Deviation of R, the Runs Statistic

$$\sigma_R = \sqrt{\frac{2n_1n_2(2n_1n_2 - n_1 - n_2)}{(n_1 + n_2)^2(n_1 + n_2 - 1)}} \qquad (15.15)$$

If both n_1 and n_2 are greater than 10 or if either n_1 or n_2 is greater than 20, the
sampling distribution of R is approximately normal. For situations where n_1 and n_2 are
both less than or equal to 10, the sampling distribution of R is poorly approximated
by a normal distribution and special tables must be used that contain the critical values.
Such tables are provided in most handbooks of statistical tables.

EXAMPLE 15.4

Practical applications of the runs test include the situation where a quality control study was
conducted at a plant producing coil springs used in a certain type of camera. According to specifications, for each gram of pull, the spring should lengthen 0.01 mm. Since the springs are made
to such exacting requirements, they are relatively expensive. Table 15.12 displays the order in
which 50 springs were produced, as well as their elongations for 1 gram of force.

TABLE 15.12

Quality Control Chart

Order	Elongation (in 0.01 mm)	Order	Elongation (in 0.01 mm)	Order	Elongation (in 0.01 mm)	Order	Elongation (in 0.01 mm)
1	1.3	14	1.0	27	1.0	39	1.0
2	1.3	15	1.0	28	0.9	40	0.9
3	1.2	16	1.1	29	1.0	41	0.8
4	1.2	17	1.1	30	0.9	42	1.0
5	1.1	18	1.0	31	1.0	43	0.8
6	1.2	19	1.1	32	0.9	44	0.8
7	1.1	20	1.1	33	1.0	45	0.7
8	1.3	21	1.1	34	1.0	46	0.9
9	1.1	22	1.0	35	0.9	47	0.8
10	1.1	23	1.1	36	0.9	48	0.9
11	1.2	24	1.2	37	0.9	49	0.8
12	1.0	25	1.0	38	0.9	50	0.6
13	1.1	26	1.2				

After examining the table, we suspect that something is wrong, since the spring elongation measurements appear to be decreasing over time. The runs test for randomness can help us verify our suspicion that the production process is in trouble or the instrument used for testing is defective. If changes in elongation are recorded $+$ or $-$ ($+$ for increase and $-$ for decrease) as the springs are produced on the assembly line, the following sequence of changes results:

$$- - + - + - + - + - + - + - + + - + - - - + - +$$
$$- + - + - - + - - + - + - -$$

There are $R = 31$ runs. The number of $+$ symbols is $n_1 = 16$ and the number of $-$ symbols is $n_2 = 21$.

We can test to determine if this sequence is nonrandom. The statistical hypotheses are

H_0: The sequence is random.

H_1: The sequence is not random.

Using formula (15.14) the mean of the sampling distribution of R is

$$\mu_R = \frac{2n_1 n_2}{n_1 + n_2} + 1 = \frac{(2)(16)(21)}{16 + 21} + 1 = 19.16$$

Using formula (15.15) the standard error of R is

$$\sigma_R = \sqrt{\frac{2n_1 n_2 (2n_1 n_2 - n_1 - n_2)}{(n_1 + n_2)^2 (n_1 + n_2 - 1)}}$$

$$= \sqrt{\frac{(2)(16)(21)(635)}{(37)^2(36)}} = 2.94$$

The z value for $R = 31$ is

$$z = \frac{R - \mu_R}{\sigma_R} = \frac{31 - 19.16}{2.94} = 4.03$$

For the 5% level of significance, the critical values are $\pm z_{0.05} = \pm 1.96$. We can reject H_0 and conclude that the sequence is not random, since $4.03 > 1.96$. The sign test in section 15.1 can be used to verify the downward trend in elongation as the springs come off the assembly line.

To aid in analyzing the data, a frequency table (table 15.13) and line chart (fig. 15.1) were constructed.

TABLE 15.13

Frequency Table for
Elongations

Elongation (in 0.01 mm)	Frequency
0.6	1
0.7	1
0.8	5
0.9	10
1.0	12
1.1	12
1.2	6
1.3	3

By examining the frequency graph, we see that it appears to be approximately symmetric and centered close to the specification point of 1.0 mm. However, by examining the original data (fig. 15.2), we can detect a general drift downward as the springs are taken off the assembly line.

FIGURE 15.1

Frequency graph for
elongations

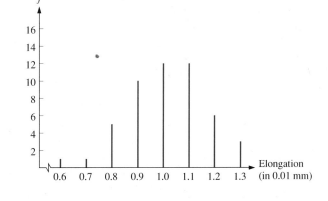

FIGURE 15.2

Original data for elongations

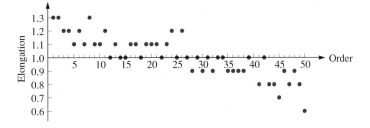

APPLICATION 15.12

The runs test is sometimes used to test a sequence of stock price changes for evidence of nonrandomness. The symbol $+$ denotes a price increase and the symbol $-$ denotes a decrease. Test the following sequence of stock market changes for nonrandomness by using the 0.05 level of significance:

$$- - - + + + - - + - - + - + - - - + + - - + - + - - + - + +$$

Solution: Here $R = 18$, $n_1 = 17$, and $n_2 = 13$. Thus, the mean of the sampling distribution of R is determined by using formula (15.14):

$$\mu_R = \frac{2n_1n_2}{n_1 + n_2} + 1 = \frac{(2)(17)(13)}{30} + 1 = 15.73$$

The standard error of R is found by using formula (15.15):

$$\sigma_R = \sqrt{\frac{2n_1n_2(2n_1n_2 - n_1 - n_2)}{(n_1 + n_2)^2(n_1 + n_2 - 1)}}$$

$$= \sqrt{\frac{(2)(17)(13)[(2)(17)(13) - 17 - 13)]}{(17 + 13)^2(17 + 13 - 1)}} = 2.64$$

The z value for $R = 18$ is

$$z = \frac{R - \mu_R}{\sigma_R} = \frac{18 - 15.73}{2.64} = 0.86$$

To test for nonrandomness using $\alpha = 0.05$, we perform the following steps:

1. H_0: The sequence is random.
2. H_1: The sequence is not random.
3. Sampling distribution: Distribution of the number of runs, R. The test statistic is the z value for $R = 18$.
4. $\alpha = 0.05$ (two-tailed test, too many runs or too few runs).
5. Critical values: $\pm z_{0.025} = \pm 1.96$.
6. Decision: Since $z = 0.86 < 1.96 = z_{0.025}$, we fail to reject H_0. Thus, the number of runs is neither too small nor too large for us to conclude that the sequence of stock prices is not random. ■

APPLICATION 15.13

Last semester a 2:00 P.M. statistics class met every Monday, Wednesday, and Friday at a certain college for 22 weeks. The number of absences was recorded each Friday. The number of absences, in the order that they occurred, are 12, 1, 8, 7, 3, 10, 5, 7, 15, 12, 9, 18, 12, 17, 1, 7, 18, 6, 14, 5, 2, and 11. By using $\alpha = 0.05$, do the data indicate that Friday absences are not random?

Solution: We will test for nonrandomness by using the median value. By ranking the data, we find the median value is 8.5. We will compare each number in the original sequence to $\tilde{\mu} = 8.5$ and indicate that a number is below the median by using a "b" and that a number is above the median by using an "a." The resulting sequence of a's and b's is given by

$$a\ b\ b\ b\ b\ a\ b\ b\ a\ a\ a\ a\ a\ a\ b\ b\ a\ b\ a\ b\ b\ a$$

Let n_1 denote the number of a's and n_2 denote the number of b's. Then we have $n_1 = 11$ and $n_2 = 11$. The number of runs is $R = 11$. The sampling distribution of R is approximately normal, since both n_1 and n_2 are greater than 10. Using formula (15.14) the mean of the sampling distribution of R is

$$\mu_R = \frac{2n_1n_2}{n_1 + n_2} + 1 = \frac{(2)(11)(11)}{11 + 11} + 1 = 12$$

Using formula (15.15) the standard error of R is

$$\sigma_R = \sqrt{\frac{2n_1n_2(2n_1n_2 - n_1 - n_2)}{(n_1 + n_2)^2(n_1 + n_2 - 1)}}$$

$$= \sqrt{\frac{(2)(11)(11)[(2)(11)(11) - 11 - 11]}{(11 + 11)^2(11 + 11 - 1)}} = 2.29$$

The z value for $R = 11$ runs is

$$z = \frac{R - \mu_R}{\sigma_R} = \frac{11 - 12}{2.29} = -0.44$$

To test for nonrandomness, we have the following information:

1. H_0: The sequence of a's and b's is random.
2. H_1: The sequence of a's and b's is not random.
3. Sampling distribution: Distribution of the number of runs, R. The test statistic is $z = -0.44$.
4. $\alpha = 0.05$ (two-tailed test).
5. Critical values: $\pm z_{0.025} = \pm 1.96$.
6. Decision: We fail to reject H_0, since $-1.96 < -0.44 < 1.96$. Hence, we have no statistical evidence to suggest the sequence is nonrandom.
7. Type of error possible: Type II, β is unknown.
8. p-value: $2P(z < -0.44) = (2)(0.33) = 0.66$. ∎

MINITAB can be used to analyze the data of application 15.13. The median must first be found. Computer display 15.7 contains the input data and the command used to obtain the median.

Computer Display 15.7

```
MTB > SET C1
DATA> 12 1 8 7 3 10 5 7 15 12 9
DATA> 18 12 17 1 7 18 6 14 5 2 11
DATA> END
MTB > MEDIAN C1
  MEDIAN = 8.5000
MTB >
```

The RUNS command is used to test for nonrandomness. Computer display 15.8 contains the RUNS command and the corresponding output.

Computer Display 15.8

```
MTB > RUNS 8.5 C1

  C1

  K =   8.5000

THE OBSERVED NO. OF RUNS = 11
THE EXPECTED NO. OF RUNS = 12.0000
11 OBSERVATIONS ABOVE K  11 BELOW
      THE TEST IS SIGNIFICANT AT 0.6623
      CANNOT REJECT AT ALPHA = 0.05
```

Note that the runs test for nonrandomness is always a nondirectional test. There can either be too many runs or too few runs to indicate nonrandom behavior of a sequence of symbols. In addition, the runs test can be applied to numerical data as well as nominal data (such as symbols). Number sequences can be tested for nonrandomness by noting whether each number is above or below the median. Thus, number sequences generated by hand-held calculators or computers can be tested for randomness by noting the number of runs above and below 4.5, the median of the integers from 0 to 9.

EXAMPLE 15.5

The sequence of digits

$$\underline{1}\ \underline{9765}\ \underline{43}\ \underline{7}\ \underline{41}\ \underline{7865}\ \underline{321}$$

has $R = 7$ runs. We could also test this sequence for nonrandomness by noting the number of runs of odd-numbered digits and even-numbered digits. For the given example, we would have $R = 11$ runs if runs were determined by using odd and even digits:

$$\underline{197}\ \underline{6}\ \underline{5}\ \underline{4}\ \underline{37}\ \underline{4}\ \underline{17}\ \underline{86}\ \underline{53}\ \underline{2}\ \underline{1}$$

EXERCISE SET 15.6

Basic Skills

1. Find R, n_1, n_2, μ_R, σ_R, and z for each of the following sequences:
   a. M M F M F F M F M F F M M M F M M F F F M F M
   b. T F F T F T F F T F T T T F F F F T F T T F T T

2. Find R, n_1, n_2, μ_R, σ_R, and z for each of the following sequences:
   a. + − + + − − − + + − + − + − − + − − + − + + −
   b. a a b b b a a a b a b b b b a a b a b b a a b a

Further Applications

3. In order to determine whether speeders and nonspeeders occur nonrandomly on a certain stretch of highway, a state policeman monitors speeds using radar. Each time a vehicle exceeds the 55 mph speed limit, he writes F and each time a vehicle goes the speed limit or less, he writes S. He recorded the following results:

   S S F S F F F F S F S F S F F S S F F F
   F S S F S S S F F S F F F F S F S F

   Test for nonrandomness at $\alpha = 0.05$.

4. The daily high temperature in a Maryland town was recorded one winter for a 25-day period. Each day was classified as above normal (A) or below normal (B). The following sequence was obtained:

   A A B B B A B A A A B B B B A A B A B B B B A A A

   Determine using $\alpha = 0.01$ if the daily high temperatures follow a pattern of nonrandomness.

5. At $\alpha = 0.05$ determine if the first 50 digits in the random number table (table 8.1 in section 8.1) are in a nonrandom

order by:
   a. Using the median.
   b. Letting E represent an even digit and O represent an odd digit.

6. The following is a calculation of π that has been carried out to 200 decimal places:

$\pi = 3.14159$	16535	89793	23846	26433	83279
50288	41971	69399	37510	58209	74944
59230	78164	06286	20899	86280	34825
34211	70679	82148	08651	32823	06647
09384	46095	50582	23172	53594	08128
48111	74502	84102	70193	85211	05559
64462	29489	54930	38196		

Test for nonrandomness of the digits with $\alpha = 0.05$ by
   a. Using the median
   b. Letting E represent an even digit and O represent an odd digit.

7. The following are down times (in minutes) during which a college's computer system was not functional during a one-year period: 50, 47, 40, 36, 33, 45, 49, 34, 37, 45, 39, 30, 36, 44, 32, 29, 34, 25, 30, 37, 40, 37, 33, 22, 30, 41, 34, 40, 38, 39, 43, 46, 22, 29, 32, 25, 33, 34, 38, 49, 55, 52, 46, 41, 32, 43, 24, 42, 53, and 39. Use the method of runs below and above the median and the 5% level of significance to test for nonrandomness.

8. The total number of televisions (in thousands) produced by a certain firm during the years 1965–1985 were 95, 115, 140, 146, 114, 145, 137, 103, 155, 138, 117, 133, 120, 107, 105, 117, 138, 150, 120, 112, and 163. By using the 5% level of significance determine if there is a significant trend in the data.

Going Beyond

9. Would it make any sense to construct a confidence interval for R, the number of runs? Explain.

10. This problem develops the logic of the runs test.
    a. How many different arrangements (sequences) are possible for $n_1 = 3$ symbols of one kind and $n_2 = 3$ symbols of another kind? List all the possibilities.
    b. By using the list compiled in part a, calculate the number of runs R for each sequence and construct a frequency table for the number of runs.
    c. Construct a vertical line graph for the sampling distribution of R.
    d. For the sampling distribution of R, find the mean and standard deviation using the frequency table constructed for part b.
    e. Verify your answers to part d by using the formulas listed in the text for finding the mean and standard deviation of the sampling distribution of R.
    f. If the level of significance is 40%, find the critical values for the runs test, that is, find the values of a and b such that $P(R < a \text{ or } R > b) = 0.4$.

11. Solve the problem posed in motivator 15 by using the runs test. Use $\alpha = 0.05$.

SECTION 15.7   *Spearman's Correlation Coefficient*

There are many occasions where it is desirable to measure the association between two variables measured on an ordinal scale. Spearman's rank correlation coefficient r_s can be used for this purpose. Ranks are assigned to the x measurements and to the y measurements. The correlation coefficient of the ranked data is then found by using the formula $r = \text{SS}_{xy}/\sqrt{\text{SS}_x \text{SS}_y}$. When ranks are used instead of the original measurements, the correlation coefficient is called **Spearman's correlation coefficient** and is denoted by r_s. If there are no values for either set of data that occur with a frequency greater than 1 (that is, there are no ties), then the formula for r_s can be simplified to

Spearman's Correlation Coefficient

$$r_s = 1 - \frac{6\Sigma\, d^2}{n(n^2 - 1)} \qquad (15.16)$$

where d is the difference between ranks for each pair and n is the number of pairs. Application 15.14 illustrates the procedures.

APPLICATION 15.14

A sample of ten students from a statistics class had the SAT math and verbal scores shown in table 15.14. Determine Spearman's correlation coefficient.

TABLE 15.14

Data for Application 15.14

Math	Verbal
425	535
358	375
515	500
672	550
378	414
397	435
715	750
638	515
478	482
350	410

Solution: We first rank both sets of data in increasing order:

Math	Rank	Verbal	Rank
425	5	535	8
358	2	375	1
515	7	500	6
672	9	550	9
378	3	414	3
397	4	435	4
715	10	750	10
638	8	515	7
478	6	482	5
350	1	410	2

For illustrative purposes, we will find r_s using formulas (14.10) and (15.16).

1. The following table organizes the computations for r_s:

	x	y	x^2	y^2	xy
	5	8	25	64	40
	2	1	4	1	2
	7	6	49	36	42
	9	9	81	81	81
	3	3	9	9	9
	4	4	16	16	16
	10	10	100	100	100
	8	7	64	49	56
	6	5	36	25	30
	1	2	1	4	2
Sums	55	55	385	385	378

The values of SS_x, SS_y, and SS_{xy} are found as follows:

$$SS_x = \Sigma x^2 - \frac{(\Sigma x)^2}{n} \qquad\qquad SS_y = \Sigma y^2 - \frac{(\Sigma y)^2}{n}$$

$$= 385 - \frac{55^2}{10} = 82.5 \qquad\qquad = 385 - \frac{55^2}{10} = 82.5$$

$$SS_{xy} = \Sigma xy - \frac{(\Sigma x)(\Sigma y)}{10}$$

$$= 378 - \frac{(55)(55)}{10} = 75.5$$

Pearson's correlation coefficient is computed using formula (14.10):

$$r = \frac{SS_{xy}}{\sqrt{SS_x SS_y}}$$

$$= \frac{75.5}{\sqrt{(82.5)(82.5)}} = 0.91515$$

2. The following table organizes the computations involving the paired ranks:

x	y	d	d^2
5	8	−3	9
2	1	1	1
7	6	1	1
9	9	0	0
3	3	0	0
4	4	0	0
10	10	0	0
8	7	1	1
6	5	1	1
1	2	−1	1
			$14 = \Sigma d^2$

The value of Spearman's correlation coefficient is obtained from formula (15.16):

$$r_s = 1 - \frac{6\Sigma d^2}{n(n^2 - 1)} = 1 - \frac{(6)(14)}{10(100 - 1)} = 1 - \frac{84}{990} = 0.91515$$

We can see by comparing the two methods that both formulas (14.10) and (15.16) provide identical results, but formula (15.16) involves less work; hence, it is the preferred formula to use when paired data are involved with no ties. ▪

MINITAB can be used to determine Spearman's correlation coefficient for the SAT data of application 15.14. The data are first ranked using the RANK command. This is followed by using the CORRELATION command with the ranking data. The commands used and the output are contained in computer display 15.9.

Computer Display 15.9

```
MTB > READ C1 C2
DATA> 425 535
DATA> 358 375
DATA> 515 500
DATA> 672 550
DATA> 378 414
DATA> 397 435
DATA> 715 750
DATA> 638 515
DATA> 478 482
DATA> 350 410
DATA> END
   10 ROWS READ
MTB > RANK C1 C3
MTB > RANK C2 C4
MTB > CORRELATION C3 C4

CORRELATION OF C3 AND C4 = 0.915

MTB >
```

FIGURE 15.3

Monotonic relationships

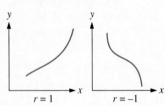

Spearman's correlation coefficient is used for a variety of reasons. It provides an alternative when the normality assumption is tenuous or when we need a more general coefficient that is not restricted by a linear relationship, but is monotonic. A relationship is **monotone increasing** if increases in one variable correspond to increases in the other; a relationship is **monotone decreasing** if increases in one variable correspond to decreases in the other variable. Values of r_s near 1 indicate a monotone increasing relationship and values or r_s near -1 indicate a monotone decreasing relationship (see fig. 15.3). In addition, r_s can be used to determine the strength of relationship or agreement when the original data consist of paired ranks. Such would be the case, for example, when two judges rank the floats in a homecoming parade and a measure of consistency for the two judges is desired.

EXAMPLE 15.6

Note that Σd^2 provides a measurement of the agreement between two sets of ranks. For example, for the data in application 15.14, $\Sigma d^2 = 14$. If Σd^2 is small there is high agreement between the ranks, and if Σd^2 is large, there is little agreement between the ranks. Total agreement occurs when the ranks for each pair are equal. Such would be the case for the following pairing of data sets:

Student	A	B	C	D	E	F	G	H	I	J
Math SAT	1	2	3	4	5	6	7	8	9	10
Verbal SAT	1	2	3	4	5	6	7	8	9	10

In this case, $\Sigma d^2 = 0$. Total disagreement occurs when the data are ranked in reverse order as follows:

Student	A	B	C	D	E	F	G	H	I	J
Math rank	1	2	3	4	5	6	7	8	9	10
Verbal rank	10	9	8	7	6	5	4	3	2	1

For this case, $\Sigma d^2 = 9^2 + 7^2 + 5^2 + 3^2 + 1^2 + 1^2 + 3^2 + 5^2 + 7^2 + 9^2 = 330$. Thus, for ten pairs of ranks, the value of Σd^2 must fall somewhere between 0 and 330, that is

$$0 < \Sigma d^2 < 330$$

Since $\Sigma d^2 = 14$ for application 15.14, we have $0 < 14 < 330$. If this interval is standardized (transformed) so that the interval [0, 330] becomes the interval [-1, 1], then $\Sigma d^2 = 14$ is transformed to the value for r_s. The interval [0, 330] can be transformed to [-1, 1] by carrying out the following steps on the inequalities $0 < 14 < 330$:

Steps	Interval ($0 < 14 < 330$)
1. Multiply by -1	$0 > -14 > -330$
2. Rewrite	$-330 < -14 < 0$
3. Divide by 165 ($= 330/2$)	$-2 < \dfrac{-14}{165} < 0$
4. Add 1	$-1 < 1 - \dfrac{14}{165} < 1$

Note that the value $1 - (14/165)$ is equal to $r_s = 0.91515$.

In practice, Spearman's formula is also used when ties are present in the data sets. In such a case, each of the tied measurements is assigned the average of the ranks that would have resulted had there been no ties.

APPLICATION 15.15

For ten automobiles made recently, weight (x, in thousands of pounds) and average miles per gallon of gasoline (y) were recorded in table 15.15. Determine Spearman's correlation coefficient r_s.

TABLE 15.15

Data for Application 15.15

x	2.7	4.1	3.5	2.7	2.2	3.9	2.2	2.2	3.5	2.2
y	28	19	22	30	30	26	19	38	26	45

Solution: We first rank the x values. Since there are four measurements of magnitude 2.2, we assign each of the four values the rank of the average of the first four rank values:

$$\frac{1 + 2 + 3 + 4}{4} = 2.5$$

Next, since the fifth and sixth measurements are equal they are each assigned the average rank of 5.5. The seventh and eighth values are equal and are assigned the average rank of 7.5. Similarly, ranks are provided for the y measurements. After ranks have been assigned, we compute the square of the difference d for each pair. The sum of the squared differences is then used in formula (15.16) to find r_s. The computations are summarized in the following table:

x	Rank x	y	Rank y	d	d^2
2.7	5.5	28	6	−0.5	0.25
4.1	10	19	1.5	8.5	72.25
3.5	7.5	22	3	4.5	20.25
2.7	5.5	30	7.5	−2	4
2.2	2.5	30	7.5	−5	25
3.9	9	26	4.5	4.5	20.25
2.2	2.5	19	1.5	1	1
2.2	2.5	38	9	−6.5	42.25
3.5	7.5	26	4.5	3	9
2.2	2.5	45	10	−7.5	56.25
					$\Sigma d^2 = 250.5$

The value of Spearman's correlation coefficient is

$$r_s = 1 - \frac{6\Sigma d^2}{n(n^2 - 1)} = 1 - \frac{(6)(250.5)}{(10)(100 - 1)} = -0.52$$

The value $r_s = -0.52$ indicates a monotone decreasing relationship for automobile weights and gasoline mileages; the heavier the car, the lower the gasoline mileage. ▪

EXAMPLE 15.7

If the number of ties is small compared to the number of data pairs, little error results when using Spearman's formula instead of formula (14.10). For example, by using formula (14.10) on the ranked pairs in application 15.15, we find that $r = -0.59$. For this data set, the total number of tied observations is 14, which is 70% of the total number of x and y observations. These ties reflect an error of 0.07 when using Spearman's formula to compute the correlation coefficient r for the ranked pairs.

Testing H_0: $\rho_s = 0$

If $n > 10$ and the population correlation coefficient ρ_s of ranked data is 0, the sampling distribution of r_s is approximately normal with a mean of 0 and a standard deviation given by

$$\sigma_{r_s} = \frac{1}{\sqrt{n-1}}$$

As a consequence, for a population of ranked pairs we can determine if $\rho_s \neq 0$ by first finding the z value for r_s under the assumption that $\rho_s = 0$:

$$z = \frac{r_s - 0}{1/\sqrt{n-1}} = r_s\sqrt{n-1}$$

The value of the test statistic for testing the null hypothesis H_0: $\rho_s = 0$ is given by

> **z Value for Spearman's r_s**
>
> $$z = r_s\sqrt{n-1}$$

(15.17)

APPLICATION 15.16

For the paired ranked data in application 15.14, test the statistical hypothesis that there is no correlation between SAT math scores and SAT verbal scores. Use the 0.05 level of significance.

Solution:

1. H_0: $\rho_s = 0$.
2. H_1: $\rho_s \neq 0$.
3. Sampling distribution: Distribution of r_s. The test statistic is the z value for r_s. Using formula (15.17) we have

$$z = r_s\sqrt{n-1} = (0.92)\sqrt{10-1} = 2.76$$

4. $\alpha = 0.05$.
5. Critical values: $\pm z_{0.025} = \pm 1.96$.
6. Decision: Reject H_0, since $z = 2.76 > z_{0.025} = 1.96$. Hence, we have evidence to suggest that SAT math scores and SAT verbal scores are correlated.
7. Type of error possible: Type I, $\alpha = 0.05$.
8. p-value: $2P(z > 2.76) = 2(0.5 - 0.4971) = 0.0058$. ▪

APPLICATION 15.17

For the gasoline mileage data in application 15.15 test the null hypothesis H_0: $\rho_s \geq 0$ against the alternative hypothesis H_1: $\rho_s < 0$ using the 0.01 level of significance.

Solution:

1. H_0: $\rho_s \geq 0$.
2. H_1: $\rho_s < 0$ (left-tailed test).
3. Sampling distribution: Distribution of r_s. The test statistic is the z value for r_s. Using formula (15.17) we have

$$z = r_s\sqrt{n-1} = (-0.52)\sqrt{10-1} = -1.56$$

4. $\alpha = 0.01$.
5. Critical value: $-z_{0.01} = -2.33$.

6. Decision: Since $z = -1.56 > -z_{0.01}$, we fail to reject H_0, and we cannot conclude that the data suggest that weights of cars and gasoline mileages have a monotone relationship.

7. Type of error possible: Type II, β is unknown.

8. *p*-value: $p(z < -1.56) = 0.5 - 0.4406 = 0.0594$. ▪

Further Applications

1. Ten baseball players were ranked by a scout on running speed and hitting ability as they demonstrated their baseball-playing talents at a major league try-out. The results are shown in the accompanying table.

Player	Speed	Hitting
A	1	7
B	2	9
C	3	3
D	4	4
E	5	1
F	6	6
G	7	8
H	8	2
I	9	10
J	10	5

a. Calculate Spearman's correlation coefficient r_s to measure the association between speed and hitting.

b. At the 0.05 level of significance, test to determine if r_s is different from zero.

2. Two teachers rated ten mathematics students according to ability. The ratings are as given in the table.

Student	A	B	C	D	E	F	G	H	I	J
Teacher X	1	9	6	2	5	8	7	3	10	4
Teacher Y	3	10	8	1	7	5	6	2	9	4

a. Calculate Spearman's correlation coefficient to measure the consistency of the ratings.

b. Test the null hypothesis H_0: $\rho_s \le 0$ against the alternative hypothesis H_1: $\rho_s > 0$ at the 0.01 level of significance.

3. In an effort to determine the relationship between annual wages for employees and the number of days absent from work due to sickness, a large corporation studied the personnel records for a random sample of twelve employees. The paired data are provided in the table.

Employee	Annual Wages (in thousands of dollars)	Days Missed
1	15.7	4
2	17.2	3
3	13.8	6
4	24.2	5
5	15.0	3
6	12.7	12
7	13.8	5
8	18.7	1
9	10.8	12
10	11.8	11
11	25.4	2
12	17.2	4

a. Determine Spearman's correlation coefficient r_s.

b. Test at $\alpha = 0.05$ to determine if the number of days missed is related to annual wages.

4. The accompanying table indicates the midterm and final exam grades made by twelve students.

Student	Midterm Score	Final Exam Score
A	77	89
B	76	88
C	63	71
D	73	85
E	87	91
F	71	62
G	54	53
H	86	89
I	85	88
J	62	45
K	52	61
L	98	87

a. Determine Spearman's correlation coefficient r_s.

b. Test at $\alpha = 0.05$ to determine if the midterm exam score is related to the final exam score.

5. The following table gives the IQs of mother and daughter for ten families:

Family	Mother	Daughter
A	110	125
B	135	115
C	133	117
D	161	145
E	115	135
F	123	115
G	138	149
H	118	105
I	108	105
J	130	107

a. Determine Spearman's correlation coefficient r_s.
b. Test at $\alpha = 0.05$ to determine if the mother's IQ is related to the daughter's IQ.

6. The EPA rated eleven comparable automobiles on gasoline mileage and versatility. The results are as follows:

Automobile	Gasoline Mileage	Versatility
A	23.1	8
B	26.1	6
C	31.1	6
D	30.0	4
E	27.2	10
F	28.6	9
G	33.1	5
H	25.6	7
I	28.1	2
J	34.5	1
K	32.6	3

a. Determine Spearman's correlation coefficient r_s.
b. Test at $\alpha = 0.05$ to determine if the gasoline mileage is related to versatility.

7. Refer to exercise 3.
a. Find Pearson's correlation coefficient using formula (14.10).
b. Test the null hypothesis $H_0: \rho = 0$ against the alternative hypothesis $H_1: \rho \neq 0$ by using formula (14.11) and $\alpha = 0.05$.
c. Compare the results with those obtained from exercise 3.

8. Refer to exercise 4.
a. Find Pearson's correlation coefficient using formula (14.10).

b. Test the null hypothesis $H_0: \rho = 0$ against the alternative hypothesis $H_1: \rho \neq 0$ by using formula (14.11) and $\alpha = 0.05$.
c. Compare the results with those obtained from exercise 4.

9. The accompanying data show the highway mpg ratings and lifetime amounts (in tons) of CO_2 emitted by the 16 best small 1990 station wagons rated by *Public Citizen*.[54]

Model	Highway mph	Lifetime CO_2 (tons)
Honda Civic Wagon	34	30.08
Ford Escort Wagon	36	31.34
Nissan Sentra Wagon	36	31.34
Dodge Colt Wagon	34	31.70
Mitsubishi Mirage Wagon	34	31.70
Plymouth Colt Wagon	34	31.70

a. Determine Spearman's correlation coefficient r_s.
b. Test at $\alpha = 0.05$ to determine if the gasoline mileage is related to the lifetime amount of carbon dioxide emitted.

10. The accompanying table lists the ten states with the most number of automobiles and their corresponding populations for 1989.

State	Registrations	Population
California	16,496,522	23,667,565
Florida	8,713,198	9,746,324
New York	8,558,985	17,558,072
Texas	8,455,744	14,229,288
Ohio	7,003,826	10,797,624
Illinois	6,403,462	11,426,518
Pennsylvania	6,253,550	11,863,895
Michigan	5,556,109	9,262,078
New Jersey	5,222,761	7,364,823
Georgia	3,690,981	5,463,105

a. Determine Spearman's correlation coefficient r_s.
b. Test at $\alpha = 0.05$ to determine if the number of automobile registrations is related to population.

Going Beyond

11. Show that the maximum value of Σd^2 is $(n^3 - n)/3$ for any set of n paired ranks.

12. By using the result of exercise 11, show that $r_s = 1 - \dfrac{6\Sigma d^2}{n(n^2 - 1)}$.

13. Suppose two science fair judges rank three entries without using ties. If the first judge ranks entries A, B, and C as 1, 2, and 3, respectively,
    a. list all the possible ways for the second judge to rank the entries A, B, and C.
    b. by using part a, determine the sampling distribution of r_s.
    c. find the mean and standard deviation of the sampling distribution of r_s given in part b.
    d. determine $P(r_s > 0.9)$.

14. When $n < 10$, the critical values for r_s can be approximated by using

$$\pm \sqrt{\frac{t^2}{t^2 + n - 2}}$$

where t is the t value corresponding to df $= n - 2$ and the level of significance. Use this approximation to find approximate critical values for the test statistic r_s if
    a. $n = 8$ and $\alpha = 0.05$.   b. $n = 7$ and $\alpha = 0.01$.
    c. $n = 42$ and $\alpha = 0.05$.

CHAPTER SUMMARY

In this chapter we presented seven popular nonparametric techniques, the sign test, the Wilcoxon signed-rank test, the Wilcoxon rank-sum test, the Kruskal-Wallis test, Friedman's test, the runs test, and Spearman's correlation coefficient. These methods do not require any special assumptions about the sampled populations and commonly use ranks instead of the raw data. The sign test and the signed-ranks test are the nonparametric counterparts of the two-sample t test for paired data. The signed-ranks test is more sensitive for detecting real differences when they exist, and is usually preferred over the sign test. The rank-sum test is the nonparametric counterpart of the two-sample t test for independent samples. The Kruskal-Wallis test is used instead of

ANOVA when the normality assumptions for ANOVA cannot be met. Friedman's test is used for the randomized block design instead of two-way ANOVA when the normality assumption can not be satisfied. We also learned that the runs test is a test for randomness. Finally, we investigated Spearman's correlation coefficient, which is the nonparametric counterpart to Pearson's correlation coefficient. As is the case with many nonparametric tests, Spearman's correlation coefficient is based on ranks. We also learned that Spearman's correlation coefficient can be used to determine if ordinal data form a monotone relationship. The nonparametric tests discussed in this text are summarized in the table.

Test	Assumptions	Sampling Distribution	Test Statistic	Parametric Counterpart
Spearman's correlation	$n \geq 10$	z	$z = r_s \sqrt{n - 1}$	t test for ρ
Chi-square a. independence b. goodness-of-fit c. multinomial parameters d. population proportions e. homogeneity	a. at least 80% of E's ≥ 5 b. no E's < 1	χ^2	$\chi^2 = \Sigma \dfrac{(O_{ij} - E_{ij})^2}{E_{ij}}$	None
Runs test	$n_1 > 10, n_2 > 10$	z	$z = \dfrac{R - \mu_R}{\sigma_R}$ $\mu_R = \dfrac{2n_1 n_2}{n_1 + n_2} + 1$ $\sigma_R = \sqrt{\dfrac{2n_1 n_2 (2n_1 n_2 - n_1 - n_2)}{(n_1 + n_2)^2 (n_1 + n_2 - 1)}}$	None

(continued)

Test	Assumptions	Sampling Distribution	Test Statistic	Parametric Counterpart
Rank-sum	a. independent samples b. populations have same continuous distributions c. $n_1 > 8$, $n_2 > 8$	z	$z = \dfrac{U - \mu_U}{\sigma_U}$ $U = W - \dfrac{n_1(n_1 + 1)}{2}$ $\mu_U = \dfrac{n_1 n_2}{2}$ $\sigma_U = \sqrt{\dfrac{n_1 n_2 (n_1 + n_2 + 1)}{12}}$	t test (independent samples)
Sign test	$n \geq 10$	z	$z = \dfrac{2x - n}{\sqrt{n}}$	t-test (dependent samples)
Kruskal-Wallis	a. independent samples b. populations have same continuous distributions c. $n_t \geq 5$	χ^2	$H = \dfrac{12}{N(N + 1)} \Sigma \dfrac{R_i^2}{n_i} - 3(N + 1)$	One-Way ANOVA
Friedman's Test	a. independent sample b. populations have same continuous distributions c. $n > 5$	χ^2	$S = \dfrac{12}{nk(k + 1)} \Sigma R_j^2 - 3n(k + 1)$	Randomized Block (ANOVA)
Signed-ranks	$n \geq 15$	z	$z = \dfrac{T - \mu_T}{\sigma_T}$	t test (dependent samples)

CHAPTER REVIEW

■ IMPORTANT TERMS ■

The following chapter terms have been mixed in ordering to provide better review practice. For each of the following terms, provide a definition in your own words. Then check your responses against the definitions given in the chapter.

Mann-Whitney U test	nonparametric tests	runs test
Friedman's test	Wilcoxon's signed-rank test	monotone decreasing
run	parametric tests	sign test
Kruskal-Wallis (KW) test	Wilcoxon rank-sum test	signed-ranks test
monotone increasing	distribution-free methods	Spearman's correlation coefficient

■ *IMPORTANT SYMBOLS* ■

x, number of plus signs

T^+, the sum of the positive ranks

T^-, the sum of the negative ranks

T, the larger of T^+ and T^-

W, sum of ranks

U, Wilcoxon test statistic

R_{ij}, rank associated with x_{ij}

R_i, sum of the ranks in the ith sample

$\overline{R}_i$, average of the ranks in the ith sample

H, Kruskal-Wallis test statistic

S, Friedman's test statistic

R, number of runs

r_s, Spearman's correlation coefficient

ρ_s, Spearman's population correlation coefficient

■ *IMPORTANT FACTS AND FORMULAS* ■

Test statistic for sign test:

$$z = \frac{2x - n}{\sqrt{n}} \quad (15.2)$$

Mean of the sampling distribution of T:

$$\mu_T = \frac{n(n + 1)}{4} \quad (15.3)$$

Standard deviation of the sampling distribution of T:

$$\sigma_T = \sqrt{\frac{n(n + 1)(2n + 1)}{24}} \quad (15.4)$$

$$T^+ + T^- = \frac{n(n + 1)}{2} \quad (15.5)$$

Wilcoxon U statistic: $U = W - \frac{n_1(n_1 + 1)}{2}$ (15.6)

Mean of Wilcoxon U statistic: $\mu_U = \frac{n_1 n_2}{2}$ (15.7)

Standard deviation of Wilcoxon U statistic:

$$\sigma_U = \sqrt{\frac{n_1 n_2(n_1 + n_2 + 1)}{12}} \quad (15.8)$$

Kruskal-Wallis H statistic:

$$H = \frac{12}{N(N + 1)} \Sigma\, n_1 \left(\overline{R}_i - \frac{N + 1}{2} \right)^2 \quad (15.10)$$

Computational formula for Kruskal-Wallis H statistic:

$$H = \frac{12}{N(N + 1)} \Sigma\, \frac{R_i^2}{n_i} - 3(N + 1) \quad (15.11)$$

Friedman's S statistic: $S = \frac{12n}{k(k + 1)} \Sigma \left(\overline{R}_j - \frac{k + 1}{2} \right)^2$

(15.12)

Computational formula for Friedman's S statistic:

$$S = \frac{12}{nk(k + 1)} \Sigma\, R_j^2 - 3n(k + 1) \quad (15.13)$$

Mean of runs statistic, R: $\mu_R = \frac{2n_1 n_2}{n_1 + n_2} + 1$ (15.14)

Standard deviation of runs statistic, R:

$$\sigma_R = \sqrt{\frac{2n_1 n_2(2n_1 n_2 - n_1 - n_2)}{(n_1 + n_2)^2(n_1 + n_2 - 1)}} \quad (15.15)$$

Spearman's correlation coefficient: $r_s = 1 - \frac{6\Sigma\, d^2}{n(n^2 - 1)}$

(15.16)

z-value for Spearman's r_s: $z = r_s\sqrt{n - 1}$ (15.17)

REVIEW EXERCISES

1. Two different types of brake lining were tested for difference in wear. Two independent samples were tested with the results (in thousands of miles) listed in the table.

Brand A	42 58 64 40 47 50 62 54 42 38 66 52
Brand B	48 40 30 44 54 38 32 42 40 52 50 34

Use the rank-sum test to test for a difference in average brake lining wear at the 0.05 level of significance.

2. In order to evaluate two methods of teaching German, students were randomly assigned to two groups. Students in one group were taught by method A and students in the other group were taught by method B. At the end of the

instructional units, both groups were given the same achievement test. The scores for the two groups are given in the accompanying table.

Group A	84	86	91	93	84	88	69	74	81	82
Group B	90	88	92	94	84	85	92	89	80	

At the 0.05 level of significance determine if method B produced higher average results than method A using the rank-sum test.

3. Eleven workers performed a task using two different methods. The completion times (in minutes) for each task are given in the table.

Worker	Method A	Method B
1	15.2	14.5
2	14.6	14.8
3	14.2	13.8
4	15.7	15.6
5	14.9	15.3
6	15.2	14.3
7	15.6	15.5
8	15.0	15.0
9	16.2	15.6
10	15.7	15.2
11	15.6	14.8

Use the sign test to test for a difference in average completion times for the two methods. Use $\alpha = 0.01$.

4. An auto manufacturing plant hopes to institute a new employee incentive plan. To evaluate the new plan, five employees will be under the incentive plan for an experimental period. Their work outputs before and after implementing the new plan are recorded in the table.

Employee	Output Before	Output After
A	20	23
B	17	19
C	23	24
D	20	23
E	21	23

At $\alpha = 0.05$, use the sign test to determine if the new incentive plan results in greater average output.

5. The military conducted a study to determine the accuracy of two types of rifles. A random sample of equally proficient soldiers was chosen to test the accuracy of the rifles. The soldiers were divided into two groups; each group shot only one type of rifle. A high score indicates a more accurate rifle. The results are provided in the table.

Rifle A	88	84	88	90	86	92	88	89	87
Rifle B	90	94	92	90	88	86	91	87	90

Use the rank-sum test at the 0.01 level of significance to determine if, on the average rifle B is more accurate than rifle A.

6. The following sequence represents the order in which the last 30 babies were born at a local hospital (the symbol M represents a male and the symbol F represents a female).

F F M F M M F F M F M M M F M
F F M M F M F M M F M M F F M

Test at $\alpha = 0.05$ to determine if the births occur in non-random order.

7. For review exercise 2, use the Kruskal-Wallis test to determine if methods A and B produce different average results. Use $\alpha = 0.05$.

8. The following sequence of M's and P's shows the order in which cars passed the Mason-Dixon Line on U.S. Route 40. A car with a Maryland license plate is denoted by M and a car with a Pennsylvania license plate is denoted by P.

M P P M P M M P M M M P
P M P M P P M P M M P M

Test for nonrandomness using $\alpha = 0.05$.

9. The accompanying table shows the miles per gallon of gasoline a driver obtained with eighteen tanks of three different kinds of gasoline.

A	B	C
27	22	16
15	15	27
29	19	19
24	29	29
21	20	21
26	18	
28		

Use the Kruskal-Wallis test to determine if there is a difference in the true average mileage-per-gallon ratings for the three kinds of gasoline. Use $\alpha = 0.01$.

10. Two judges ranked each of ten students who had entered a project in a mathematics fair. The rankings are given in the accompanying table.

Student	1	2	3	4	5	6	7	8	9	10
Judge A	6	3	8	1	4	7	2	10	5	9
Judge B	4	1	10	8	3	5	2	9	6	7

a. Compute the value of r_s.

b. At $\alpha = 0.05$ test the null hypothesis H_0: $\rho_s = 0$ against the alternative hypothesis H_1: $\rho_s \neq 0$.

11. The data listed in the table represent SAT math scores and scores made in an introductory statistics course for a sample of ten college students.

SAT math score	476	525	619	515	475	379	517	415	405	616
Statistics score	77	90	69	88	72	62	92	68	73	95

a. Compute the value of the correlation coefficient r_s.

b. Test to determine if $\rho_s > 0$ using $\alpha = 0.05$.

12. In an attempt to determine the relationship between attendance and concession sales at football games, a high school principal obtained the data given here for five randomly chosen football games during the previous three seasons.

Attendance x (in hundreds of people)	Sales y (in hundreds of dollars)
5	15
13	36
10	20
21	64
11	32
10	18
15	44
12	35
11	30
14	40

a. Calculate the value of r_s for the data.

b. Test using $\alpha = 0.01$ to determine if $\rho_s > 0$.

13. To determine the effectiveness of a particular diet the accompanying data were recorded for a random sample of eight dieters.

Number of Weeks on Diet x	Weight Loss y (in pounds)
1	8
4	22
3	17
2	15
3	19
1	10
2	11
4	20

a. Calculate the value of r_s.

b. By using $\alpha = 0.05$, determine if $\rho_s > 0$.

14. The manager of a large department store conducted an experiment to determine if there is a difference in the average weekly sales of three salespersons in the store's appliance department. The following data indicate sales (in $100) for seven consecutive weeks for the three employees.

	Salesperson		
Week	A	B	C
1	27.6	28.7	26.4
2	31.2	29.3	30.3
3	28.8	28.4	28.0
4	30.6	29.8	28.7
5	30.0	31.0	32.3
6	28.4	29.9	29.6
7	30.9	29.5	31.1

By using the 0.05 level of significance and Friedman's test, can we conclude there is a difference in the mean sales for the three salespersons?

15. The accompanying data represent the time (in seconds) required for 19 operators to complete an assembly task using two methods, A and B.

Operator	Method A	Method B
A	56	50
B	99	82
C	51	53
D	43	38
E	40	39
F	45	41
G	70	61
H	66	71
I	73	62
J	61	59
K	50	40
L	47	47
M	63	61
N	54	45
O	84	62
P	59	53
Q	72	72
R	59	56
S	55	43

Use $\alpha = 0.01$ and the signed-rank test to determine if one method takes less time than the other method.

Computer Applications

1. The runs test can be used to evaluate a computer's random number generator. Use a computer program to generate 100 random integers from a uniform distribution of integers between 0 and 9, inclusive. Test the resulting sample for randomness of even and odd numbers using $\alpha = 0.05$.

2. Repeat computer application 1 by comparing each value to the median value, 4.5, of the integers from 0 to 9, inclusive.

3. Use a computer program to test the effects of outliers for the seven nonparametric tests studied in this chapter. For some tests, they have dramatic effects; in others, the effects are minimal or nonexistent. Summarize your findings.

4. Randomly generate 10 pairs of data from the integers 1 to 5. Determine Spearman's correlation coefficient for the paired data. Repeat the experiment 100 times. What is the mean of the 100 values of r_s? Construct a histogram for the values.

EXPERIMENTS WITH REAL DATA

Treat the 720 subjects listed in the database in appendix C as if they constituted the population of all U.S. subjects.

1. Use Spearman's correlation coefficient to determine if there is a monotonic relationship between height and weight. Use $\alpha = 0.05$ and a random sample of 25 subjects.

2. Use Spearman's correlation coefficient to determine if there is a monotonic relationship between diastolic and systolic blood pressures. Use $\alpha = 0.05$ and a random sample of 25 subjects.

3. Do the mean systolic blood pressures for subjects aged in their 20s, 30s, and 40s differ? Select random samples of size 25 for each of the three age groups and use the Kruskal-Wallis test and $\alpha = 0.05$.

4. Use Wilcoxon's signed-ranks test to determine if the mean systolic blood pressure for smokers aged in their 20s is different from the mean systolic blood pressure for non-smokers aged in their 20s. Use random samples of size 30 and $\alpha = 0.05$.

■ CHAPTER ACHIEVEMENT TEST ■

1. For the following sequence of symbols, let n_1 denote the number of A's and n_2 denote the number of B's; test at the 0.05 level of significance for nonrandomness. In addition, provide the following information.

$$A\ B\ A\ B\ B\ A\ B\ B\ A\ B\ B\ A\ A$$
$$A\ B\ A\ B\ A\ B\ B\ A\ A\ B\ A\ A$$

   a. value of R **b.** value of μ_R
   c. value of σ_R **d.** z value for R
   e. critical value   **f.** decision

2. Consider the following paired data:

Pair	1	2	3	4	5	6
Sample 1	11.2	10.6	10.2	11.6	10.9	11.2
Sample 2	10.5	10.8	9.8	11.1	11.3	10.3

Pair	7	8	9	10	11
Sample 1	11.6	11.0	12.2	11.7	11.6
Sample 2	11.5	11.0	11.6	11.2	10.8

Use the sign test to test $H_0: \mu_1 - \mu_2 = 0$ against $H_1:$ $\mu_1 - \mu_2 \neq 0$ at the 0.01 level of significance and provide

the following information:
   a. value of the test statistic
   b. critical values
   c. p-value for the test

3. Refer to achievement test question 2. Construct a 95% confidence interval for the true proportion of $+$ signs.

4. Refer to achievement test question 2. Determine the value of T if the signed-rank test is used to analyze the data.

5. In order to evaluate two methods of teaching statistics, students were randomly assigned to two groups. Students in one group were taught by method A and students in the other group were taught by method B. At the end of instruction, both groups were given the same achievement test. The scores for the two groups follow:

Group A	89	91	96	98	89	93	87	90	94	
Group B	95	93	97	99	89	90	97	88	92	98

At the 0.01 level of significance use the rank-sum test to determine if method B produces higher average results than method A.

6. For achievement test question 5, use the Kruskal-Wallis test to determine if methods A and B produce different average results. Use $\alpha = 0.05$.

7. Consider the following two sets of ratings for five brands of beer.

Brand	Rating 1	Rating 2
A	2	1
B	3	2
C	1	3
D	5	4
E	4	5

a. Find the value of r_s.
b. Find the value of μ_{r_s}, assuming $\rho_s = 0$.
c. Find the value of σ_{r_s}.
d. Find the z value for r_s.
e. At $\alpha = 0.05$, test H_0: $\rho_s = 0$ against H_1: $\rho_s \neq 0$.

8. Refer to achievement test question 2. Use the KW test and $\alpha = 0.01$ to determine if $\mu_1 - \mu_2 \neq 0$.

References

1. David L. Strum and Ronald P. Church, "Hitting the Long Ball for the Customer," *Training & Development Journal,* 44(3), March 1990, p. 45.

2. Sandra Chofon, *The Big Book of Kid's Lists,* World Almanac Publications, 1985.

3. "The Latest on Diet and Your Heart," *Consumer Reports,* July 1985, pp. 423–27.

4. "How To Tell the Real Thing in Coke," *Consumer Reports,* August 1985, p. 447.

5. "The Murky Hazards of Second Hand Smoke," *Consumer Reports,* February 1985, pp. 81–84.

6. "Fords in Reverse," *Consumer Reports,* September 1985, pp. 520–23.

7. *Cumberland Times/News,* p. 7, September 18, 1985.

8. "Customer-Service Perceptions and Reality," Wendy S. Becker and Richard S. Wellins, *Training & Development Journal,* 44(3) March 1990, pp. 49–51.

9. *U.S. News & World Report,* February 5, 1990, p. 74.

10. *U.S. News & World Report,* March 26, 1990, p. 73.

11. *U.S. News & World Report,* April 23, 1984, p. 12.

12. *U.S. News & World Report,* September 10, 1984, p. 16.

13. *U.S. News & World Report,* April 23, 1984, p. 14.

14. *U.S. News & World Report,* April 30, 1984, p. 65.

15. *Shape,* July 1985, p. 23.

16. *U.S. News & World Report,* December 18, 1989, p. 82.

17. Data from American Society of Plastic and Reconstructive Surgeons.

18. *U.S. News & World Report,* June 25, 1990, p. 66.

19. *Readers Digest,* December 1988, pp. 107–108.

20. *U.S. News & World Report,* September 5, 1983.

21. *Journal of Sports Medicine,* 1982, pp. 17–22.

22. *U.S. News & World Report,* June 25, 1984.

23. *U.S. News & World Report,* September 5, 1983.

24. *U.S. News & World Report,* September 26, 1983.

25. *U.S. News & World Report,* June 25, 1984.

26. *The Washington Post,* April 8, 1990, p. C2.

27. *U.S. News & World Report,* September 5, 1983.

28. *The Videodisc Monitor,* February 1990, p. 30.

29. *U.S. News & World Report,* January 22, 1990, p. 28.

30. *The Washington Post,* April 8, 1990, p. C2.

31. *NEA Today,* April 1990, p. 8.

32. Margaret Cozzens, Editor, *Consortium,* COMAP, Inc., Spring 1990, p. 2.

33. *U.S. News & World Report,* September 24, 1984.

34. G. R. Linsey, "An Investigation of Strategies in Baseball," *Operations Research,* II: 477–501, 1963.

35. *The American Statistician,* 39(1): 80, 1985.

36. Kenneth M. Nowack, "Getting Them Out and Getting Them Back," *Training & Development Journal,* 44(4): 82–85, April 1990.

37. *The Sun,* November 11, 1984.

38. *The Cumberland News,* August 15, 1985.

39. *The Cumberland News,* September 13, 1985.

40. Albert E. Cordes, "Using Computers in the Physics Laboratory," *Journal of Computers in Mathematics and Science Teaching,* 9(3): 53–63, Spring 1990.

41. Paul E. Jones and Robert E. Wall, "Components of Computer Anxiety," *Journal of Educational Technology System,* 18(2): 161–168, 1989–90.

42. H. N. Lilliefors, "On the Kolmogorov-Smirnov Test for Normality with Mean and Variance Unknown," *Journal of the American Statistical Association,* 62:399–402, 1967.

43. Jean McEnery and Gaston DesHarnais, "Cultural Shock," *Training & Development Journal,* 44(4): 43–47, April 1990.

44. Gary L. Reglin, "Effects of Computer-Assisted Instruction on Mathematics and Locus of Control," *Journal of Educational Technology Systems,* 18(2): 143–49, 1989–90.

45. *U.S. News & World Report,* April 23, 1990, p. 78.

46. Ken Bossong, et al., *Driving Up the Heat: A Buyer's Guide to Automobiles, Fuel Efficiency, and Global Warming,* Public Citizen Critical Mass Energy Project, 215 Pennsylvania Ave., S.E., Washington, DC 20003, 42 pages, $5.00.

47. Ibid.

48. Ibid.

49. *Road & Track Sports & GT Cars,* Special Series, 1990.

50. Steven Kolmes and Kevin Mitchell, "Information Theory and Biological Diversity," *UMAP Journal,* Spring 1990, p. 46.

51. *U.S. News & World Report,* February 5, 1990, p. 74.

52. *The New York Public Library Desk Reference, Webster's New World,* p. 574–75, 1989.

53. *U.S. News & World Report,* April 23, 1990, p. 78.

54. *A Buyer's Guide to Automobiles, Fuel Efficiency, and Global Warming,* Public Citizen: Critical Mass Energy Project, 215 Pennsylvania Ave., S.E., Washington, DC, 20003, p. 13, July 1990.

Appendix A Summation Notation and Rules

Throughout the text the abbreviated notation ΣX is used to mean the sum of the X measurements. This abbreviated notation does not make it explicitly clear what or how many measurements are being added. To take care of this problem, we use the more formal summation notation:

$$\sum_{i=1}^{n} X_i = X_1 + X_2 + X_3 + \cdots + X_n$$

where i is called the **index of summation** and 1 and n are called the **limits of summation.**

EXAMPLE A1

Suppose X_i and Y_i have the values shown in the table. By inserting these values in the examples, we can see how the process of summation works.

i	1	2	3	4	5
X_i	2	6	8	3	4
Y_i	1	3	4	0	2

a. $\displaystyle\sum_{i=1}^{4} X_i = X_1 + X_2 + X_3 + X_4$

$= 2 + 6 + 8 + 3 = 19$

c. $\displaystyle\sum_{i=1}^{3} 3X_i = 3X_1 + 3X_2 + 3X_3$

$= 6 + 18 + 24 = 48$

b. $\displaystyle\sum_{i=2}^{5} X_i = X_2 + X_3 + X_4 + X_5$

$= 6 + 8 + 3 + 4 = 21$

d. $\displaystyle\sum_{i=2}^{3} X_i^2 = X_2^2 + X_3^2$

$= 36 + 64 = 100$

e. $\displaystyle\sum_{i=1}^{3} (X_i^2 + Y_i^2) = (X_1^2 + Y_1^2) + (X_2^2 + Y_2^2) + (X_3^2 + Y_3^2)$

$= (2^2 + 1^2) + (6^2 + 3^2) + (8^2 + 4^2)$
$= (4 + 1) + (36 + 9) + (64 + 16)$
$= 5 + 45 + 80 = 130$

f. $\displaystyle\sum_{i=2}^{4} X_i Y_i^2 = X_2 Y_2^2 + X_3 Y_3^2 + X_4 Y_4^2$

$= (6)(3^2) + (8)(4^2) + (3)(0^2)$
$= (6)(9) + (8)(16) + (3)(0)$
$= 54 + 128 + 0 = 182$

g. $\displaystyle\sum_{i=2}^{3} (2X_i + 3Y_i) = (2X_2 + 3Y_2) + (2X_3 + 3Y_3)$

$= [(2)(6) + (3)(3)] + [(2)(8) + (3)(4)]$
$= (12 + 9) + (16 + 12)$
$= 21 + 28 = 49$

h. $\displaystyle\sum_{i=1}^{3} X_{i+1} = X_{1+1} + X_{2+1} + X_{3+1}$

$$= X_2 + X_3 + X_4$$
$$= 6 + 8 + 3 = 17$$

There are two basic rules for summations, which when developed lead to a third. They are as follows:

Rule 1: $\displaystyle\sum_{i=1}^{n} (X_i + Y_i) = \sum_{i=1}^{n} X_i + \sum_{i=1}^{n} Y_i$

Rule 2: $\displaystyle\sum_{i=1}^{n} KX_i = K \sum_{i=1}^{n} X_i$, where K is a constant.

Rule 1 can be verified by noting that

$$\sum_{i=1}^{n} (X_i + Y_i) = (X_1 + Y_1) + (X_2 + Y_2) + (X_3 + Y_3) + \cdots + (X_n + Y_n)$$

$$= (X_1 + X_2 + X_3 + \cdots + X_n) + (Y_1 + Y_2 + Y_3 + \cdots + Y_n)$$

$$= \sum_{i=1}^{n} X_i + \sum_{i=1}^{n} Y_i$$

Similarly, rule 2 can be verified by noting that

$$\sum_{i=1}^{n} KX_i = KX_1 + KX_2 + KX_3 + \cdots + KX_n$$

$$= K(X_1 + X_2 + X_3 + \cdots + X_n)$$

$$= K \sum_{i=1}^{n} X_i$$

Suppose for each value of the index i, $X_i = 1$. Then as a consequence of rule 2, we have

$$\sum_{i=1}^{n} K = \sum_{i=1}^{n} (K)(1) = \sum_{i=1}^{n} (K)(X_i) = K \sum_{i=1}^{n} X_i$$

$$= K(1 + 1 + 1 + \cdots + 1) = (K)(n)$$
$$\uparrow$$
$$n \text{ 1s}$$

As a result, we have the following rule:

Rule 3: $\displaystyle\sum_{i=1}^{n} K = nK$

These basic rules can be used to prove statistical properties involving discrete probability distributions.

Frequently, we have occasion to use double sigma notation. In such expressions we could first expand the expression on the index j. For example, to expand the expression $\displaystyle\sum_{i=1}^{3} \sum_{j=1}^{2} X_{ij}$, we first evaluate the expression $\displaystyle\sum_{j=1}^{2} X_{ij}$.

$$\sum_{i=1}^{3} \sum_{j=1}^{2} X_{ij} = \sum_{i=1}^{3} (X_{i1} + X_{i2}) = (X_{11} + X_{12}) + (X_{21} + X_{22}) + (X_{31} + X_{32})$$

Since addition is a commutative operation, we could have expanded the above expression on the index i first, as shown here.

$$\sum_{i=1}^{3} \left(\sum_{j=1}^{2} X_{ij} \right) = \sum_{j=1}^{2} X_{1j} + \sum_{j=1}^{2} X_{2j} + \sum_{j=1}^{2} X_{3j}$$
$$= (X_{11} + X_{12}) + (X_{21} + X_{22}) + (X_{31} + X_{32})$$

This is the same result as we found by expanding first on the index j.

EXERCISE SET APPENDIX A

Basic Skills

1. Write each of the following without the summation symbol:

a. $\sum_{i=1}^{4} X_i$

b. $\sum_{i=1}^{4} (X_i^2 + 2)$

c. $\sum_{i=1}^{4} (X_i^2 + Y_i)$

d. $\sum_{i=1}^{3} X_i^2 Y_i$

e. $\sum_{i=1}^{5} f_i X_i^2$

f. $\sum_{i=1}^{2} \sum_{j=1}^{3} X_i Y_j^2$

i. $\sum_{i=1}^{5} (Y_i - 2)$

j. $\sum_{i=1}^{5} (X_i - 3)(Y_i - 2)$

4. Suppose X_i and Y_j take on the following values:

i	1	2	3	4	5
X_i	2	3	1	2	3

j	1	2	3	4
Y_j	1	3	1	2

Find the values of the following expressions:

a. $\sum_{i=1}^{3} \sum_{j=1}^{2} (X_i + Y_j)$

b. $\sum_{i=1}^{3} \sum_{j=1}^{3} X_i Y_j$

c. $\sum_{i=2}^{4} \sum_{j=3}^{4} X_i Y_j^2$

d. $\sum_{j=1}^{4} \sum_{i=1}^{5} 3X_i$

e. $\sum_{i=1}^{2} X_i^2 + \sum_{j=1}^{3} Y_j^2$

f. $\sum_{i=1}^{5} \sum_{j=1}^{4} 2$

2. Write the following expressions using summation notation:
   a. $X_1 Y_1 + X_2 Y_2 + X_3 Y_3$
   b. $X_1 Y_2 + X_2 Y_3 + X_3 Y_4 + X_4 Y_5$
   c. $X_2^2 + X_3^2 + X_4^2$
   d. $X_1^2 + 2X_2^2 + 3X_3^2 + 4X_4^2$
   e. $(2X_3 + 3) + (2X_4 + 3) + (2X_5 + 3) + (2X_6 + 3)$

3. Suppose X_i and Y_i take on the following values:

i	1	2	3	4	5
X_i	3	4	5	2	1
Y_i	1	3	0	2	4

Find the value of the following expressions:

a. $\sum_{i=1}^{5} (X_i + Y_i)$

b. $\Sigma (3X_i - Y_i)$

c. $\left(\sum_{i=1}^{3} X_i \right)^2$

d. $\sum_{i=1}^{3} X_i^2$

e. $\sum_{i=1}^{3} X_i Y_i$

f. $\left(\sum_{i=1}^{2} X_i \right) \left(\sum_{i=1}^{2} Y_i \right)$

g. $\sum_{i=1}^{5} (X_i^2 - Y_i^2)$

h. $\sum_{i=1}^{5} (X_i - 3)$

Going Beyond

5. Prove that $\sum_{i=1}^{n} i = n(n + 1)/2$.

6. Prove that $\sum_{i=1}^{n} i^2 = n(n + 1)(2n + 1)/6$.

7. Suppose f is a function defined on the integers. Prove that
$$\sum_{i=a}^{b} f(i) = \sum_{i=a-c}^{b-c} f(i + c), \text{ where } a \le b, a > c, b > c,$$
and f is a function of i.

8. Suppose f is a function defined on the integers. Prove that

$$\sum_{i=a}^{c} f(i) = \sum_{i=a}^{b} f(i) + \sum_{i=b+1}^{c} f(i), \text{ where } a \leq b < c.$$

9. Express each of the following using summation notation:
   a. $3 + 5 + 7 + 9 + \cdots + 23$
   b. $9 + 16 + 25 + 36 + \cdots + 400$
   c. $1 + 6 + 15 + 28 + \cdots + 190$
   d. $12 + 20 + 30 + 42 + \cdots + 90$
   e. $5 + 6 + 7 + \cdots + 74$
   f. $2 + 4 + 6 + 8 + \cdots + 40$

10. a. Prove that $1 + 3 + 5 + \cdots + (2n - 1)$
    $$= \sum_{i=1}^{n} (2i - 1).$$

    b. Prove that $\sum_{i=1}^{n} (2i - 1) = n^2.$

Appendix B TABLES

TABLE 1
Binomial Probabilities

n	x	.01	.05	.1	.2	.3	.4	.5	.6	.7	.8	.9	.95	.99	x
1	0	.990	.950	.900	.800	.700	.600	.500	.400	.300	.200	.100	.050	.010	0
	1	.010	.050	.100	.200	.300	.400	.500	.600	.700	.800	.900	.950	.990	1
2	0	.980	.903	.810	.640	.490	.360	.250	.160	.090	.040	.010	.003	.000	0
	1	.020	.095	.180	.320	.420	.480	.500	.480	.420	.320	.180	.095	.020	1
	2	.000	.003	.010	.040	.090	.160	.250	.360	.490	.640	.810	.903	.980	2
3	0	.970	.857	.729	.512	.343	.216	.125	.064	.027	.008	.001	.000	.000	0
	1	.029	.135	.243	.384	.441	.432	.375	.288	.189	.096	.027	.007	.000	1
	2	.000	.007	.027	.096	.189	.288	.375	.432	.441	.384	.243	.135	.029	2
	3	.000	.000	.001	.008	.027	.064	.125	.216	.343	.512	.729	.857	.970	3
4	0	.961	.815	.656	.410	.240	.130	.063	.026	.008	.002	.000	.000	.000	0
	1	.039	.171	.292	.410	.412	.346	.250	.154	.076	.026	.004	.000	.000	1
	2	.001	.014	.049	.154	.265	.346	.375	.346	.265	.154	.049	.014	.001	2
	3	.000	.000	.004	.026	.076	.154	.250	.346	.412	.410	.292	.171	.039	3
	4	.000	.000	.000	.002	.008	.026	.063	.130	.240	.410	.656	.815	.961	4
5	0	.951	.774	.590	.328	.168	.078	.031	.010	.002	.000	.000	.000	.000	0
	1	.048	.204	.328	.410	.360	.259	.156	.077	.028	.006	.000	.000	.000	1
	2	.001	.021	.073	.205	.309	.346	.313	.230	.132	.051	.008	.001	.000	2
	3	.000	.001	.008	.051	.132	.230	.313	.346	.309	.205	.073	.021	.001	3
	4	.000	.000	.000	.006	.028	.077	.156	.259	.360	.410	.328	.204	.048	4
	5	.000	.000	.000	.000	.002	.010	.031	.078	.168	.328	.590	.774	.951	5
6	0	.941	.735	.531	.262	.118	.047	.016	.004	.001	.000	.000	.000	.000	0
	1	.057	.232	.354	.393	.303	.187	.094	.037	.010	.002	.000	.000	.000	1
	2	.001	.031	.098	.246	.324	.311	.234	.138	.060	.015	.001	.000	.000	2
	3	.000	.002	.015	.082	.185	.276	.313	.276	.185	.082	.015	.002	.000	3
	4	.000	.000	.001	.015	.060	.138	.234	.311	.324	.246	.098	.031	.001	4
	5	.000	.000	.000	.002	.010	.037	.094	.187	.303	.393	.354	.232	.057	5
	6	.000	.000	.000	.000	.001	.004	.016	.047	.118	.262	.531	.735	.941	6
7	0	.932	.698	.478	.210	.082	.028	.008	.002	.000	.000	.000	.000	.000	0
	1	.066	.257	.372	.367	.247	.131	.055	.017	.004	.000	.000	.000	.000	1
	2	.002	.041	.124	.275	.318	.261	.164	.077	.025	.004	.000	.000	.000	2
	3	.000	.004	.023	.115	.227	.290	.273	.194	.097	.029	.003	.000	.000	3
	4	.000	.000	.003	.029	.097	.194	.273	.290	.227	.115	.023	.004	.000	4
	5	.000	.000	.000	.004	.025	.077	.164	.261	.318	.275	.124	.041	.002	5
	6	.000	.000	.000	.000	.004	.017	.055	.131	.247	.367	.372	.257	.066	6
	7	.000	.000	.000	.000	.000	.002	.008	.028	.082	.210	.478	.698	.932	7
8	0	.923	.663	.430	.168	.058	.017	.004	.001	.000	.000	.000	.000	.000	0
	1	.075	.279	.383	.336	.198	.090	.031	.008	.001	.000	.000	.000	.000	1
	2	.003	.051	.149	.294	.296	.209	.109	.041	.010	.001	.000	.000	.000	2
	3	.000	.005	.033	.147	.254	.279	.219	.124	.047	.009	.000	.000	.000	3

Table 1 Binomial Probabilities ■ 723

TABLE 1 (continued)

								p							
n	x	.01	.05	.1	.2	.3	.4	.5	.6	.7	.8	.9	.95	.99	x
	4	.000	.000	.005	.046	.136	.232	.273	.232	.136	.046	.005	.000	.000	4
	5	.000	.000	.000	.009	.047	.124	.219	.279	.254	.147	.033	.005	.000	5
	6	.000	.000	.000	.001	.010	.041	.109	.209	.296	.294	.149	.051	.003	6
	7	.000	.000	.000	.000	.001	.008	.031	.090	.198	.336	.383	.279	.075	7
	8	.000	.000	.000	.000	.000	.001	.004	.017	.058	.168	.430	.663	.923	8
9	0	.914	.630	.387	.134	.040	.010	.002	.000	.000	.000	.000	.000	.000	0
	1	.083	.299	.387	.302	.156	.060	.018	.004	.000	.000	.000	.000	.000	1
	2	.003	.063	.172	.302	.267	.161	.070	.021	.004	.000	.000	.000	.000	2
	3	.000	.008	.045	.176	.267	.251	.164	.074	.021	.003	.000	.000	.000	3
	4	.000	.001	.007	.066	.172	.251	.246	.167	.074	.017	.001	.000	.000	4
	5	.000	.000	.001	.017	.074	.167	.246	.251	.172	.066	.007	.001	.000	5
	6	.000	.000	.000	.003	.021	.074	.164	.251	.267	.176	.045	.008	.000	6
	7	.000	.000	.000	.000	.004	.021	.070	.161	.267	.302	.172	.063	.003	7
	8	.000	.000	.000	.000	.000	.004	.018	.060	.156	.302	.387	.299	.083	8
	9	.000	.000	.000	.000	.000	.000	.002	.010	.040	.134	.387	.630	.914	9
10	0	.904	.599	.349	.107	.028	.006	.001	.000	.000	.000	.000	.000	.000	0
	1	.091	.315	.387	.268	.121	.040	.010	.002	.000	.000	.000	.000	.000	1
	2	.004	.075	.194	.302	.233	.121	.044	.011	.001	.000	.000	.000	.000	2
	3	.000	.010	.057	.201	.267	.215	.117	.042	.009	.001	.000	.000	.000	3
	4	.000	.001	.011	.088	.200	.251	.205	.111	.037	.006	.000	.000	.000	4
	5	.000	.000	.001	.026	.103	.201	.246	.201	.103	.026	.001	.000	.000	5
	6	.000	.000	.000	.006	.037	.111	.205	.251	.200	.088	.011	.001	.000	6
	7	.000	.000	.000	.001	.009	.042	.117	.215	.267	.201	.057	.010	.000	7
	8	.000	.000	.000	.000	.001	.011	.044	.121	.233	.302	.194	.075	.004	8
	9	.000	.000	.000	.000	.000	.002	.010	.040	.121	.268	.387	.315	.091	9
	10	.000	.000	.000	.000	.000	.000	.001	.006	.028	.107	.349	.599	.904	10
11	0	.895	.569	.314	.086	.020	.004	.000	.000	.000	.000	.000	.000	.000	0
	1	.099	.329	.384	.236	.093	.027	.005	.001	.000	.000	.000	.000	.000	1
	2	.005	.087	.213	.295	.200	.089	.027	.005	.001	.000	.000	.000	.000	2
	3	.000	.014	.071	.221	.257	.177	.081	.023	.004	.000	.000	.000	.000	3
	4	.000	.001	.016	.111	.220	.236	.161	.070	.017	.002	.000	.000	.000	4
	5	.000	.000	.002	.039	.132	.221	.226	.147	.057	.010	.000	.000	.000	5
	6	.000	.000	.000	.010	.057	.147	.226	.221	.132	.039	.002	.000	.000	6
	7	.000	.000	.000	.002	.017	.070	.161	.236	.220	.111	.016	.001	.000	7
	8	.000	.000	.000	.000	.004	.023	.081	.177	.257	.221	.071	.014	.000	8
	9	.000	.000	.000	.000	.001	.005	.027	.089	.200	.295	.213	.087	.005	9
	10	.000	.000	.000	.000	.000	.001	.005	.027	.093	.236	.384	.329	.099	10
	11	.000	.000	.000	.000	.000	.000	.000	.004	.020	.086	.314	.569	.895	11
12	0	.886	.540	.282	.069	.014	.002	.000	.000	.000	.000	.000	.000	.000	0
	1	.107	.341	.377	.206	.071	.017	.003	.000	.000	.000	.000	.000	.000	1
	2	.006	.099	.230	.283	.168	.064	.016	.002	.000	.000	.000	.000	.000	2
	3	.000	.017	.085	.236	.240	.142	.054	.012	.001	.000	.000	.000	.000	3

T A B L E 1 (continued)

n	x	.01	.05	.1	.2	.3	.4	.5	.6	.7	.8	.9	.95	.99	x
														p	
	4	.000	.002	.021	.133	.231	.213	.121	.042	.008	.001	.000	.000	.000	4
	5	.000	.000	.004	.053	.158	.227	.193	.101	.029	.003	.000	.000	.000	5
	6	.000	.000	.000	.016	.079	.177	.226	.177	.079	.016	.000	.000	.000	6
	7	.000	.000	.000	.003	.029	.101	.193	.227	.158	.053	.004	.000	.000	7
	8	.000	.000	.000	.001	.008	.042	.121	.213	.231	.133	.021	.002	.000	8
	9	.000	.000	.000	.000	.001	.012	.054	.142	.240	.236	.085	.017	.000	9
	10	.000	.000	.000	.000	.000	.002	.016	.064	.168	.283	.230	.099	.006	10
	11	.000	.000	.000	.000	.000	.000	.003	.017	.071	.206	.377	.341	.107	11
	12	.000	.000	.000	.000	.000	.000	.000	.002	.014	.069	.282	.540	.886	12
13	0	.878	.513	.254	.055	.010	.001	.000	.000	.000	.000	.000	.000	.000	0
	1	.115	.351	.367	.179	.054	.011	.002	.000	.000	.000	.000	.000	.000	1
	2	.007	.111	.245	.268	.139	.045	.010	.001	.000	.000	.000	.000	.000	2
	3	.000	.021	.100	.246	.218	.111	.035	.006	.001	.000	.000	.000	.000	3
	4	.000	.003	.028	.154	.234	.184	.087	.024	.003	.000	.000	.000	.000	4
	5	.000	.000	.006	.069	.180	.221	.157	.066	.014	.001	.000	.000	.000	5
	6	.000	.000	.001	.023	.103	.197	.209	.131	.044	.006	.000	.000	.000	6
	7	.000	.000	.000	.006	.044	.131	.209	.197	.103	.023	.001	.000	.000	7
	8	.000	.000	.000	.001	.014	.066	.157	.221	.180	.069	.006	.000	.000	8
	9	.000	.000	.000	.000	.003	.024	.087	.184	.234	.154	.028	.003	.000	9
	10	.000	.000	.000	.000	.001	.006	.035	.111	.218	.246	.100	.021	.000	10
	11	.000	.000	.000	.000	.000	.001	.010	.045	.139	.268	.245	.111	.007	11
	12	.000	.000	.000	.000	.000	.000	.002	.011	.054	.179	.367	.351	.115	12
	13	.000	.000	.000	.000	.000	.000	.000	.001	.010	.055	.254	.513	.878	13
14	0	.869	.488	.229	.044	.007	.001	.000	.000	.000	.000	.000	.000	.000	0
	1	.123	.359	.356	.154	.041	.007	.001	.000	.000	.000	.000	.000	.000	1
	2	.008	.123	.257	.250	.113	.032	.006	.001	.000	.000	.000	.000	.000	2
	3	.000	.026	.114	.250	.194	.085	.022	.003	.000	.000	.000	.000	.000	3
	4	.000	.004	.035	.172	.229	.155	.061	.014	.001	.000	.000	.000	.000	4
	5	.000	.000	.008	.086	.196	.207	.122	.041	.007	.000	.000	.000	.000	5
	6	.000	.000	.001	.032	.126	.207	.183	.092	.023	.002	.000	.000	.000	6
	7	.000	.000	.000	.009	.062	.157	.209	.157	.062	.009	.000	.000	.000	7
	8	.000	.000	.000	.002	.023	.092	.183	.207	.126	.032	.001	.000	.000	8
	9	.000	.000	.000	.000	.007	.041	.122	.207	.196	.086	.008	.000	.000	9
	10	.000	.000	.000	.000	.001	.014	.061	.155	.229	.172	.035	.004	.000	10
	11	.000	.000	.000	.000	.000	.003	.022	.085	.194	.250	.114	.026	.000	11
	12	.000	.000	.000	.000	.000	.001	.006	.032	.113	.250	.257	.123	.008	12
	13	.000	.000	.000	.000	.000	.000	.001	.007	.041	.154	.356	.359	.123	13
	14	.000	.000	.000	.000	.000	.000	.000	.001	.007	.044	.229	.448	.869	14
15	0	.860	.463	.206	.035	.005	.000	.000	.000	.000	.000	.000	.000	.000	0
	1	.130	.366	.343	.132	.031	.005	.000	.000	.000	.000	.000	.000	.000	1
	2	.009	.135	.267	.231	.092	.022	.003	.000	.000	.000	.000	.000	.000	2
	3	.000	.031	.129	.250	.170	.063	.014	.002	.000	.000	.000	.000	.000	3
	4	.000	.005	.043	.188	.219	.127	.042	.007	.001	.000	.000	.000	.000	4

Table 1 Binomial Probabilities ▪ 725

TABLE 1 (continued)

n	x	.01	.05	.1	.2	.3	.4	.5	.6	.7	.8	.9	.95	.99	x
								p							
	5	.000	.001	.010	.103	.206	.186	.092	.024	.003	.000	.000	.000	.000	5
	6	.000	.000	.002	.043	.147	.207	.153	.061	.012	.001	.000	.000	.000	6
	7	.000	.000	.000	.014	.081	.177	.196	.118	.035	.003	.000	.000	.000	7
	8	.000	.000	.000	.003	.035	.118	.196	.177	.081	.014	.000	.000	.000	8
	9	.000	.000	.000	.001	.012	.061	.153	.207	.147	.043	.002	.000	.000	9
	10	.000	.000	.000	.000	.003	.024	.092	.186	.206	.103	.010	.001	.000	10
	11	.000	.000	.000	.000	.001	.007	.042	.127	.219	.188	.043	.005	.000	11
	12	.000	.000	.000	.000	.000	.002	.014	.063	.170	.250	.129	.031	.000	12
	13	.000	.000	.000	.000	.000	.000	.003	.022	.092	.231	.267	.135	.009	13
	14	.000	.000	.000	.000	.000	.000	.000	.005	.031	.132	.343	.366	.130	14
	15	.000	.000	.000	.000	.000	.000	.000	.000	.005	.035	.206	.463	.860	15
20	0	.818	.358	.122	.012	.001	.000	.000	.000	.000	.000	.000	.000	.000	0
	1	.165	.377	.270	.058	.007	.000	.000	.000	.000	.000	.000	.000	.000	1
	2	.016	.189	.285	.137	.028	.003	.000	.000	.000	.000	.000	.000	.000	2
	3	.001	.060	.190	.205	.072	.012	.001	.000	.000	.000	.000	.000	.000	3
	4	.000	.013	.090	.218	.130	.035	.005	.000	.000	.000	.000	.000	.000	4
	5	.000	.002	.032	.175	.179	.075	.015	.001	.000	.000	.000	.000	.000	5
	6	.000	.000	.009	.109	.192	.124	.037	.005	.000	.000	.000	.000	.000	6
	7	.000	.000	.002	.055	.164	.166	.074	.015	.001	.000	.000	.000	.000	7
	8	.000	.000	.000	.022	.114	.180	.120	.035	.004	.000	.000	.000	.000	8
	9	.000	.000	.000	.007	.065	.160	.160	.071	.012	.000	.000	.000	.000	9
	10	.000	.000	.000	.002	.031	.117	.176	.117	.031	.002	.000	.000	.000	10
	11	.000	.000	.000	.000	.012	.071	.160	.160	.065	.007	.000	.000	.000	11
	12	.000	.000	.000	.000	.004	.035	.120	.180	.114	.022	.000	.000	.000	12
	13	.000	.000	.000	.000	.001	.015	.074	.166	.164	.055	.002	.000	.000	13
	14	.000	.000	.000	.000	.000	.005	.037	.124	.192	.109	.009	.000	.000	14
	15	.000	.000	.000	.000	.000	.001	.015	.075	.179	.175	.032	.002	.000	15
	16	.000	.000	.000	.000	.000	.000	.005	.035	.130	.218	.090	.013	.000	16
	17	.000	.000	.000	.000	.000	.000	.001	.012	.072	.205	.190	.060	.001	17
	18	.000	.000	.000	.000	.000	.000	.000	.003	.028	.137	.285	.189	.016	18
	19	.000	.000	.000	.000	.000	.000	.000	.000	.007	.058	.270	.377	.165	19
	20	.000	.000	.000	.000	.000	.000	.000	.000	.001	.012	.122	.358	.818	20
25	0	.778	.277	.072	.004	.000	.000	.000	.000	.000	.000	.000	.000	.000	0
	1	.196	.365	.199	.024	.001	.000	.000	.000	.000	.000	.000	.000	.000	1
	2	.024	.231	.266	.071	.007	.000	.000	.000	.000	.000	.000	.000	.000	2
	3	.002	.093	.226	.136	.024	.002	.000	.000	.000	.000	.000	.000	.000	3
	4	.000	.027	.138	.187	.057	.007	.000	.000	.000	.000	.000	.000	.000	4
	5	.000	.006	.065	.196	.103	.020	.002	.000	.000	.000	.000	.000	.000	5
	6	.000	.001	.024	.163	.147	.044	.005	.000	.000	.000	.000	.000	.000	6
	7	.000	.000	.007	.111	.171	.080	.014	.001	.000	.000	.000	.000	.000	7
	8	.000	.000	.002	.062	.165	.120	.032	.003	.000	.000	.000	.000	.000	8
	9	.000	.000	.000	.029	.134	.151	.061	.009	.000	.000	.000	.000	.000	9
	10	.000	.000	.000	.012	.092	.161	.097	.021	.001	.000	.000	.000	.000	10
	11	.000	.000	.000	.004	.054	.147	.133	.043	.004	.000	.000	.000	.000	11

T A B L E 1 (continued)

n	x	.01	.05	.1	.2	.3	.4	.5	.6	.7	.8	.9	.95	.99	x
													p		
	12	.000	.000	.000	.001	.027	.114	.155	.076	.011	.000	.000	.000	.000	12
	13	.000	.000	.000	.000	.011	.076	.155	.114	.027	.001	.000	.000	.000	13
	14	.000	.000	.000	.000	.004	.043	.133	.147	.054	.004	.000	.000	.000	14
	15	.000	.000	.000	.000	.001	.021	.097	.161	.092	.012	.000	.000	.000	15
	16	.000	.000	.000	.000	.000	.009	.061	.151	.134	.029	.000	.000	.000	16
	17	.000	.000	.000	.000	.000	.003	.032	.120	.165	.062	.002	.000	.000	17
	18	.000	.000	.000	.000	.000	.001	.014	.080	.171	.111	.007	.000	.000	18
	19	.000	.000	.000	.000	.000	.000	.005	.044	.147	.163	.024	.001	.000	19
	20	.000	.000	.000	.000	.000	.000	.002	.020	.103	.196	.065	.006	.000	20
	21	.000	.000	.000	.000	.000	.000	.000	.007	.057	.187	.138	.027	.000	21
	22	.000	.000	.000	.000	.000	.000	.000	.002	.024	.136	.226	.093	.002	22
	23	.000	.000	.000	.000	.000	.000	.000	.000	.007	.071	.266	.231	.024	23
	24	.000	.000	.000	.000	.000	.000	.000	.000	.001	.024	.199	.365	.196	24
	25	.000	.000	.000	.000	.000	.000	.000	.000	.000	.004	.072	.277	.778	25

T A B L E 2

Poisson Probabilities

x	0.1	0.2	0.3	0.4	0.5	0.6	0.7	0.8	0.9	1.0
						λ					
0	0.9048	0.8187	0.7408	0.6703	0.6065	0.5488	0.4966	0.4493	0.4066	0.3679
1	0.0905	0.1637	0.2222	0.2681	0.3033	0.3293	0.3476	0.3595	0.3659	0.3679
2	0.0045	0.0164	0.0333	0.0536	0.0758	0.0988	0.1217	0.1438	0.1647	0.1839
3	0.0002	0.0011	0.0033	0.0072	0.0126	0.0198	0.0284	0.0383	0.0494	0.0613
4		0.0001	0.0002	0.0007	0.0016	0.0030	0.0050	0.0077	0.0111	0.0153
5				0.0001	0.0002	0.0004	0.0007	0.0012	0.0020	0.0031
6							0.0001	0.0002	0.0003	0.0005
7										0.0001

x	1.5	2.0	2.5	3.0	3.5	4.0	4.5	5.0	6.0	7.0
						λ					
0	0.2231	0.1353	0.0821	0.0498	0.0302	0.0183	0.0111	0.0067	0.0025	0.0009
1	0.3347	0.2707	0.2052	0.1494	0.1057	0.0733	0.0500	0.0337	0.0149	0.0064
2	0.2510	0.2707	0.2565	0.2240	0.1850	0.1465	0.1125	0.0842	0.0446	0.0223
3	0.1255	0.1804	0.2138	0.2240	0.2158	0.1954	0.1687	0.1404	0.0892	0.0521
4	0.0471	0.0902	0.1336	0.1680	0.1888	0.1954	0.1898	0.1755	0.1339	0.0912
5	0.0141	0.0361	0.0668	0.1008	0.1322	0.1563	0.1708	0.1755	0.1606	0.1277
6	0.0035	0.0120	0.0278	0.0504	0.0771	0.1042	0.1281	0.1462	0.1606	0.1490

Table 2 Poisson Probabilities ▪ 727

T A B L E 2 (continued)

	λ									
x	1.5	2.0	2.5	3.0	3.5	4.0	4.5	5.0	6.0	7.0
7	0.0008	0.0034	0.0099	0.0216	0.0385	0.0595	0.0824	0.1044	0.1337	0.1490
8	0.0001	0.0009	0.0031	0.0081	0.0169	0.0298	0.0463	0.0653	0.1033	0.1304
9		0.0002	0.0009	0.0027	0.0066	0.0132	0.0232	0.0363	0.0688	0.1014
10			0.0002	0.0008	0.0023	0.0053	0.0104	0.0181	0.0413	0.0710
11				0.0002	0.0007	0.0019	0.0043	0.0082	0.0225	0.0452
12				0.0001	0.0002	0.0006	0.0016	0.0034	0.0113	0.0264
13					0.0001	0.0002	0.0006	0.0013	0.0052	0.0142
14						0.0001	0.0002	0.0005	0.0022	0.0071
15							0.0001	0.0002	0.0009	0.0033
16									0.0003	0.0014
17									0.0001	0.0006
18										0.0002
19										0.0001

	λ				
x	8.0	9.0	10.0	15.0	20.0
0	0.0003	0.0001	0.0000	0.0000	0.0000
1	0.0027	0.0011	0.0005	0.0000	0.0000
2	0.0107	0.0050	0.0023	0.0000	0.0000
3	0.0286	0.0150	0.0076	0.0002	0.0000
4	0.0573	0.0337	0.0189	0.0006	0.0000
5	0.0916	0.0607	0.0378	0.0019	0.0001
6	0.1221	0.0901	0.0631	0.0048	0.0002
7	0.1396	0.1171	0.0901	0.0104	0.0005
8	0.1396	0.1318	0.1126	0.0194	0.0013
9	0.1241	0.1318	0.1251	0.0324	0.0029
10	0.0993	0.1186	0.1251	0.0486	0.0058
11	0.0772	0.0970	0.1137	0.0663	0.0106
12	0.0481	0.0728	0.0948	0.0829	0.0176
13	0.0296	0.0504	0.0729	0.0956	0.0271
14	0.0169	0.0324	0.0521	0.1024	0.0387
15	0.0090	0.0194	0.0347	0.1024	0.0516
16	0.0045	0.0109	0.0217	0.0960	0.0646
17	0.0021	0.0058	0.0128	0.0847	0.0760
18	0.0009	0.0029	0.0071	0.0706	0.0844
19	0.0004	0.0014	0.0037	0.0557	0.0888
20	0.0002	0.0006	0.0019	0.0418	0.0888
21	0.0001	0.0003	0.0009	0.0299	0.0846

T A B L E 2 (continued)

			λ		
x	8.0	9.0	10.0	15.0	20.0
22		0.0001	0.0004	0.0204	0.0769
23			0.0002	0.0133	0.0669
24			0.0001	0.0833	0.0557
25				0.0050	0.0446
26				0.0029	0.0343
27				0.0016	0.0254
28				0.0009	0.0181
29				0.0004	0.0125
30				0.0002	0.0083
31				0.0001	0.0054
32				0.0001	0.0034
33					0.0020
34					0.0012
35					0.0007
36					0.0004
37					0.0002
38					0.0001
39					0.0001

Table entries are $P(X = x \mid \lambda)$.

T A B L E 3
Values of e^{-x}

x	e^{-x}	x	e^{-x}	x	e^{-x}	x	e^{-x}
0.00	1.0000	2.50	0.0821	5.00	0.0067	7.50	0.0006
0.05	0.9512	2.55	0.0781	5.05	0.0064	7.55	0.0005
0.10	0.9048	2.60	0.0743	5.10	0.0061	7.60	0.0005
0.15	0.8607	2.65	0.0707	5.15	0.0058	7.65	0.0005
0.20	0.8187	2.70	0.0672	5.20	0.0055	7.70	0.0005
0.25	0.7788	2.75	0.0639	5.25	0.0052	7.75	0.0004
0.30	0.7408	2.80	0.0608	5.30	0.0050	7.80	0.0004
0.35	0.7047	2.85	0.0578	5.35	0.0047	7.85	0.0004

Table 3 Values of e^{-x} ▪ 729

T A B L E 3 (continued)

x	e^{-x}	x	e^{-x}	x	e^{-x}	x	e^{-x}
0.40	0.6703	2.90	0.0550	5.40	0.0045	7.90	0.0004
0.45	0.6376	2.95	0.0523	5.45	0.0043	7.95	0.0004
0.50	0.6065	3.00	0.0498	5.50	0.0041	8.00	0.0003
0.55	0.5769	3.05	0.0474	5.55	0.0039	8.05	0.0003
0.60	0.5488	3.10	0.0450	5.60	0.0037	8.10	0.0003
0.65	0.5220	3.15	0.0429	5.65	0.0035	8.15	0.0003
0.70	0.4966	3.20	0.0408	5.70	0.0033	8.20	0.0003
0.75	0.4724	3.25	0.0388	5.75	0.0032	8.25	0.0003
0.80	0.4493	3.30	0.0369	5.80	0.0030	8.30	0.0002
0.85	0.4274	3.35	0.0351	5.85	0.0029	8.35	0.0002
0.90	0.4066	3.40	0.0334	5.90	0.0027	8.40	0.0002
0.95	0.3867	3.45	0.0317	5.95	0.0026	8.45	0.0002
1.00	0.3679	3.50	0.0302	6.00	0.0025	8.50	0.0002
1.05	0.3499	3.55	0.0287	6.05	0.0024	8.55	0.0002
1.10	0.3329	3.60	0.0273	6.10	0.0022	8.60	0.0002
1.15	0.3166	3.65	0.0260	6.15	0.0021	8.65	0.0002
1.20	0.3012	3.70	0.0247	6.20	0.0020	8.70	0.0002
1.25	0.2865	3.75	0.0235	6.25	0.0019	8.75	0.0002
1.30	0.2725	3.80	0.0224	6.30	0.0018	8.80	0.0002
1.35	0.2592	3.85	0.0213	6.35	0.0017	8.85	0.0001
1.40	0.2466	3.90	0.0202	6.40	0.0017	8.90	0.0001
1.45	0.2346	3.95	0.0193	6.45	0.0016	8.95	0.0001
1.50	0.2231	4.00	0.0183	6.50	0.0015	9.00	0.0001
1.55	0.2122	4.05	0.0174	6.55	0.0014	9.05	0.0001
1.60	0.2019	4.10	0.0166	6.60	0.0014	9.10	0.0001
1.65	0.1920	4.15	0.0158	6.65	0.0013	9.15	0.0001
1.70	0.1827	4.20	0.0150	6.70	0.0012	9.20	0.0001
1.75	0.1738	4.25	0.0143	6.75	0.0012	9.25	0.0001
1.80	0.1653	4.30	0.0136	6.80	0.0011	9.30	0.0001
1.85	0.1572	4.35	0.0129	6.85	0.0011	9.35	0.0001
1.90	0.1496	4.40	0.0123	6.90	0.0010	9.40	0.0001
1.95	0.1423	4.45	0.0117	6.95	0.0010	9.45	0.0001
2.00	0.1353	4.50	0.0111	7.00	0.0009	9.50	0.0001
2.05	0.1287	4.55	0.0106	7.05	0.0009	9.55	0.0001
2.10	0.1225	4.60	0.0101	7.10	0.0008	9.60	0.0001
2.15	0.1165	4.65	0.0096	7.15	0.0008	9.65	0.0001
2.20	0.1108	4.70	0.0091	7.20	0.0007	9.70	0.0001
2.25	0.1054	4.75	0.0087	7.25	0.0007	9.75	0.0001
2.30	0.1003	4.80	0.0082	7.30	0.0007	9.80	0.0001
2.35	0.0954	4.85	0.0078	7.35	0.0006	9.85	0.0001
2.40	0.0907	4.90	0.0074	7.40	0.0006	9.90	0.0001
2.45	0.0863	4.95	0.0071	7.45	0.0006	9.95	0.0000

TABLE 4
Bonferroni *t* Values for $\alpha = 0.05$

df					Number of Comparisons, *m*					
	2	3	4	5	6	7	10	15	20
2	6.21	7.65	8.86	9.92	10.89	11.77	14.09	17.28	19.96
3	4.18	4.86	5.39	5.84	6.23	6.58	7.45	8.58	9.46
4	3.50	3.96	4.31	4.60	4.85	5.07	5.60	6.25	6.76
5	3.16	3.53	3.81	4.03	4.22	4.38	4.77	5.25	5.60
6	2.97	3.29	3.52	3.71	3.86	4.00	4.32	4.70	4.98
7	2.84	3.13	3.34	3.50	3.64	3.75	4.03	4.36	4.59
8	2.75	3.02	3.21	3.36	3.48	3.58	3.83	4.12	4.33
9	2.69	2.93	3.11	3.25	3.36	3.46	3.69	3.95	4.15
10	2.63	2.87	3.04	3.17	3.28	3.37	3.58	3.83	4.00
11	2.59	2.82	2.98	3.11	3.21	3.29	3.50	3.73	3.89
12	2.56	2.78	2.93	3.05	3.15	3.24	3.43	3.65	3.81
13	2.53	2.75	2.90	3.01	3.11	3.19	3.37	3.58	3.73
14	2.51	2.72	2.86	2.98	3.07	3.15	3.33	3.53	3.67
15	2.49	2.69	2.84	2.95	3.04	3.11	3.29	3.48	3.62
16	2.47	2.67	2.81	2.92	3.01	3.08	3.25	3.44	3.58
17	2.46	2.66	2.79	2.90	2.98	3.06	3.22	3.41	3.54
18	2.45	2.64	2.77	2.88	2.96	3.03	3.20	3.38	3.51
19	2.43	2.63	2.76	2.86	2.94	3.01	3.17	3.35	3.48
20	2.42	2.61	2.74	2.85	2.93	3.00	3.15	3.33	3.46
21	2.41	2.60	2.73	2.83	2.91	2.98	3.14	3.31	3.43
22	2.41	2.59	2.72	2.82	2.90	2.97	3.12	3.29	3.41
23	2.40	2.58	2.71	2.81	2.89	2.95	3.10	3.27	3.39
24	2.39	2.57	2.70	2.80	2.88	2.94	3.09	3.26	3.38
25	2.38	2.57	2.69	2.79	2.86	2.93	3.08	3.24	3.36
26	2.38	2.56	2.68	2.78	2.86	2.92	3.07	3.23	3.35
27	2.37	2.55	2.68	2.77	2.85	2.91	3.06	3.22	3.33
28	2.37	2.55	2.67	2.76	2.84	2.90	3.05	3.21	3.32
29	2.36	2.54	2.66	2.76	2.83	2.89	3.04	3.20	3.31
30	2.36	2.54	2.66	2.75	2.82	2.89	3.03	3.19	3.30
35	2.34	2.51	2.63	2.72	2.80	2.86	3.00	3.15	3.26
40	2.33	2.50	2.62	2.70	2.78	2.84	2.97	3.12	3.23
45	2.32	2.49	2.60	2.69	2.76	2.82	2.95	3.10	3.20
50	2.31	2.48	2.59	2.68	2.75	2.81	2.94	3.08	3.18
100	2.28	2.43	2.54	2.63	2.69	2.75	2.87	3.01	3.10
200	2.26	2.41	2.52	2.60	2.66	2.72	2.84	2.97	3.06
1000	2.24	2.40	2.50	2.58	2.64	2.70	2.81	2.94	3.03

Table 5 Critical Values of the χ^2 Distributions ■ **731**

TABLE 5
Critical Values of the χ^2 Distributions

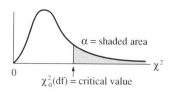

α = shaded area

$\chi^2_\alpha(df)$ = critical value

| | Values of α | | | | | | | | | |
df	0.995	0.99	0.975	0.95	0.90	0.10	0.05	0.025	0.01	0.005
1	0.000	0.000	0.001	0.004	0.016	2.706	3.841	5.024	6.635	7.879
2	0.010	0.020	0.051	0.103	0.211	4.605	5.991	7.378	9.210	10.597
3	0.072	0.115	0.216	0.352	0.584	6.251	7.815	9.348	11.345	12.838
4	0.207	0.297	0.484	0.711	1.064	7.779	9.488	11.143	13.277	14.860
5	0.412	0.554	0.831	1.145	1.610	9.236	11.070	12.832	15.086	16.750
6	0.676	0.872	1.237	1.635	2.204	10.645	12.592	14.449	16.812	18.548
7	0.989	1.239	1.690	2.167	2.833	12.017	14.067	16.013	18.475	20.278
8	1.344	1.646	2.180	2.733	3.490	13.362	15.507	17.535	20.090	21.955
9	1.735	2.088	2.700	3.325	4.168	14.684	16.919	19.023	21.666	23.589
10	2.156	2.558	3.247	3.940	4.865	15.987	18.307	20.483	23.209	25.188
11	2.603	3.053	3.816	4.575	5.578	17.275	19.675	21.920	24.725	26.757
12	3.074	3.571	4.404	5.226	6.304	18.549	21.026	23.337	26.217	28.300
13	3.565	4.107	5.009	5.892	7.042	19.812	22.362	24.736	27.688	29.819
14	4.075	4.660	5.629	6.571	7.790	21.064	23.685	26.119	29.141	31.319
15	4.601	5.229	6.262	7.261	8.547	22.307	24.996	27.488	30.578	32.801
16	5.142	5.812	6.908	7.962	9.312	23.542	26.296	28.845	32.000	34.267
17	5.697	6.408	7.564	8.672	10.085	24.769	27.587	30.191	33.409	35.718
18	6.265	7.015	8.231	9.390	10.865	25.989	28.869	31.526	34.805	37.156
19	6.844	7.633	8.907	10.117	11.651	27.204	30.144	32.852	36.191	38.582
20	7.434	8.260	9.591	10.851	12.443	28.412	31.410	34.170	37.566	39.997
21	8.034	8.897	10.283	11.591	13.240	29.615	32.670	35.479	38.932	41.401
22	8.643	9.542	10.982	12.338	14.042	30.813	33.924	36.781	40.289	42.796
23	9.260	10.196	11.688	13.090	14.848	32.007	35.172	38.076	41.638	44.181
24	9.886	10.856	12.401	13.848	15.659	33.196	36.415	39.364	42.980	45.558
25	10.520	11.524	13.120	14.611	16.473	34.382	37.652	40.646	44.314	46.928
26	11.160	12.198	13.844	15.379	17.292	35.563	38.885	41.923	45.642	48.290
27	11.808	12.879	14.573	16.151	18.114	36.741	40.113	43.194	46.963	49.645
28	12.461	13.565	15.308	16.928	18.939	37.916	41.337	44.461	48.278	50.993
29	13.121	14.256	16.047	17.708	19.768	39.088	42.557	45.772	49.588	52.336
30	13.787	14.954	16.791	18.493	20.599	40.256	43.773	46.979	50.892	53.672
31	14.458	15.655	17.539	19.281	21.434	41.422	44.985	48.232	52.190	55.003
32	15.134	16.362	18.291	20.072	22.271	42.585	46.194	49.480	53.486	56.328
33	15.815	17.074	19.047	20.867	23.110	43.745	47.400	50.725	54.776	57.649
34	16.501	17.789	19.806	21.664	23.952	44.903	48.602	51.966	56.061	58.964
35	17.192	18.509	20.569	22.465	24.797	46.059	49.802	53.203	57.340	60.275
36	17.887	19.233	21.336	23.269	25.643	47.212	50.998	54.437	58.619	61.581
37	18.586	19.960	22.106	24.075	26.492	48.363	52.192	55.668	59.892	62.883
38	19.289	20.691	22.878	24.884	27.343	49.513	53.384	56.896	61.612	64.181
39	19.996	21.426	23.654	25.695	28.196	50.660	54.572	58.120	62.428	65.476
40	20.706	22.164	24.433	26.509	29.050	51.805	55.758	59.342	63.691	66.766

TABLE 6a
Critical Values of the
F Distributions ($\alpha = 0.01$)*

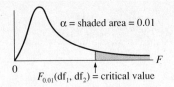

α = shaded area = 0.01

$F_{0.01}(df_1, df_2)$ = critical value

df$_2$	1	2	3	4	5	6	7	8	9
					df$_1$				
1	4052	5000	5403	5625	5764	5859	5928	5982	6022
2	98.5	99.0	99.2	99.3	99.3	99.4	99.4	99.4	99.4
3	34.1	30.3	29.5	28.7	28.2	27.9	27.7	27.5	27.3
4	21.2	18.0	16.7	16.0	15.5	15.2	15.0	14.8	14.7
5	16.3	13.3	12.1	11.4	11.0	10.7	10.5	10.3	10.2
6	13.7	10.9	9.78	9.15	8.75	8.47	8.26	8.10	7.98
7	12.2	9.55	8.45	7.85	7.46	7.19	6.99	6.84	6.72
8	11.3	8.65	7.59	7.01	6.63	6.37	6.18	6.03	5.91
9	10.6	8.02	6.99	6.42	6.06	5.80	5.61	5.47	5.35
10	10.0	7.56	6.55	5.99	5.64	5.39	5.20	5.06	4.94
11	9.65	7.21	6.22	5.67	5.32	5.07	4.89	4.74	4.63
12	9.33	6.93	5.95	5.41	5.06	4.82	4.64	4.50	4.39
13	9.07	6.70	5.74	5.21	4.86	4.62	4.44	4.30	4.19
14	8.86	6.51	5.56	5.04	4.70	4.46	4.28	4.14	4.03
15	8.68	6.36	5.42	4.89	4.56	4.32	4.14	4.00	3.89
16	8.53	6.23	5.29	4.77	4.44	4.20	4.03	3.89	3.78
17	8.40	6.11	5.18	4.67	4.34	4.10	3.93	3.79	3.68
18	8.29	6.01	5.09	4.58	4.25	4.01	3.84	3.71	3.60
19	8.18	5.93	5.01	4.50	4.17	3.94	3.77	3.63	3.52
20	8.10	5.85	4.94	4.43	4.10	3.87	3.70	3.56	3.46
21	8.02	5.78	4.87	4.37	4.04	3.81	3.64	3.51	3.40
22	7.95	5.72	4.82	4.31	3.99	3.76	3.59	3.45	3.35
23	7.88	5.66	4.76	4.26	3.94	3.71	3.54	3.41	3.30
24	7.82	5.61	4.72	4.22	3.90	3.67	3.50	3.36	3.26
25	7.77	5.57	4.68	4.18	3.86	3.63	3.46	3.32	3.22
26	7.72	5.53	4.64	4.14	3.82	3.59	3.42	3.29	3.18
27	7.68	5.49	4.60	4.11	3.78	3.56	3.39	3.26	3.15
28	7.64	5.45	4.57	4.07	3.75	3.53	3.36	3.23	3.12
29	7.60	5.42	4.54	4.04	3.73	3.50	3.33	3.20	3.09
30	7.56	5.39	4.51	4.02	3.70	3.47	3.30	3.17	3.07
40	7.31	5.18	4.31	3.83	3.51	3.29	3.12	2.99	2.89
60	7.08	4.98	4.13	3.65	3.34	3.12	2.95	2.82	2.72
125	6.84	4.78	3.94	3.47	3.17	2.95	2.79	2.65	2.56

*df$_1$ (associated with the numerator)

TABLE 6a (continued)

df$_2$					df$_1$				
	10	12	14	16	20	30	40	50	100
1	6056	6106	6142	6169	6208	6258	6286	6302	6334
2	99.4	99.4	99.4	99.4	99.5	99.5	99.5	99.5	99.5
3	27.2	27.1	26.9	26.8	26.7	26.5	26.4	26.3	26.2
4	14.5	14.4	14.2	14.2	14.0	13.8	13.7	13.7	13.6
5	10.1	9.89	9.77	9.68	9.55	9.38	9.29	9.24	9.13
6	7.87	7.72	7.60	7.52	7.39	7.23	7.14	7.09	6.99
7	6.62	6.47	6.35	6.27	6.15	5.98	5.90	5.85	5.75
8	5.82	5.67	5.56	5.48	5.36	5.20	5.11	5.06	4.96
9	5.26	5.11	5.00	4.92	4.80	4.64	4.56	4.51	4.41
10	4.85	4.71	4.60	4.52	4.41	4.25	4.17	4.12	4.01
11	4.54	4.40	4.29	4.21	4.10	3.94	3.86	3.80	3.70
12	4.30	4.16	4.05	3.98	3.86	3.70	3.61	3.56	3.46
13	4.10	3.96	3.85	3.78	3.67	3.51	3.42	3.37	3.27
14	3.94	3.80	3.70	3.62	3.51	3.34	3.26	3.21	3.11
15	3.80	3.67	3.56	3.48	3.36	3.20	3.12	3.07	2.97
16	3.69	3.55	3.45	3.37	3.25	3.10	3.01	2.96	2.86
17	3.59	3.45	3.35	3.27	3.16	3.00	2.92	2.86	2.76
18	3.51	3.37	3.27	3.19	3.07	2.91	2.83	2.78	2.68
19	3.43	3.30	3.19	3.12	3.00	2.84	2.76	2.70	2.60
20	3.37	3.23	3.13	3.05	2.94	2.77	2.69	2.63	2.53
21	3.31	3.17	3.07	2.99	2.88	2.72	2.63	2.58	2.47
22	3.26	3.12	3.02	2.94	2.83	2.67	2.58	2.53	2.42
23	3.21	3.07	2.97	2.89	2.78	2.62	2.53	2.48	2.37
24	3.17	3.03	2.93	2.85	2.74	2.58	2.49	2.44	2.33
25	3.13	2.99	2.89	2.81	2.70	2.54	2.45	2.40	2.29
26	3.09	2.96	2.86	2.77	2.66	2.50	2.41	2.36	2.25
27	3.06	2.93	2.83	2.74	2.63	2.47	2.38	2.33	2.21
28	3.03	2.90	2.80	2.71	2.60	2.44	2.35	2.30	2.18
29	3.00	2.87	2.77	2.68	2.57	2.41	2.32	2.27	2.15
30	2.98	2.84	2.74	2.66	2.55	2.38	2.29	2.24	2.13
40	2.80	2.66	2.56	2.49	2.37	2.20	2.11	2.05	1.94
60	2.63	2.50	2.40	2.32	2.20	2.03	1.93	1.87	1.74
125	2.47	2.33	2.23	2.15	2.03	1.85	1.75	1.68	1.54

TABLE 6b

Critical Values of the
F Distributions ($\alpha = 0.05$)*

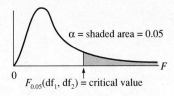

α = shaded area = 0.05

$F_{0.05}(df_1, df_2)$ = critical value

					df_1				
df_2	1	2	3	4	5	6	7	8	9
1	161	200	216	225	230	234	237	239	241
2	18.5	19.0	19.2	19.3	19.3	19.4	19.4	19.4	19.4
3	10.1	9.55	9.28	9.12	9.01	8.94	8.89	8.85	8.81
4	7.71	6.94	6.59	6.39	6.29	6.16	6.09	6.04	6.00
5	6.61	5.79	5.41	5.19	5.05	4.95	4.88	4.82	4.77
6	5.99	5.14	4.76	4.53	4.39	4.28	4.21	4.15	4.10
7	5.59	4.74	4.35	4.12	3.97	3.87	3.79	3.73	3.68
8	5.32	4.46	4.07	3.84	3.69	3.58	3.50	3.44	3.39
9	5.12	4.26	3.86	3.63	3.48	3.37	3.29	3.23	3.18
10	4.96	4.10	3.71	3.48	3.33	3.22	3.14	3.07	3.02
11	4.84	3.98	3.59	3.36	3.20	3.09	3.01	2.95	2.90
12	4.75	3.89	3.49	3.26	3.11	3.00	2.91	2.85	2.80
13	4.67	3.81	3.41	3.18	3.03	2.92	2.83	2.77	2.71
14	4.60	3.74	3.34	3.11	2.96	2.85	2.76	2.70	2.65
15	4.54	3.68	3.29	3.06	2.90	2.79	2.71	2.64	2.59
16	4.49	3.63	3.24	3.01	2.85	2.74	2.66	2.59	2.54
17	4.45	3.59	3.20	2.96	2.81	2.70	2.61	2.55	2.49
18	4.41	3.55	3.16	2.93	2.77	2.66	2.58	2.51	2.46
19	4.38	3.52	3.13	2.90	2.74	2.63	2.54	2.48	2.42
20	4.35	3.49	3.10	2.87	2.71	2.60	2.51	2.45	2.39
21	4.32	3.47	3.07	2.84	2.68	2.57	2.49	2.42	2.37
22	4.30	3.44	3.05	2.82	2.66	2.55	2.46	2.40	2.34
23	4.28	3.42	3.03	2.80	2.64	2.53	2.44	2.37	2.32
24	4.26	3.40	3.01	2.78	2.62	2.51	2.42	2.36	2.30
25	4.24	3.39	2.99	2.76	2.60	2.49	2.40	2.34	2.28
26	4.23	3.37	2.98	2.74	2.59	2.47	2.39	2.32	2.27
27	4.21	3.35	2.96	2.73	2.57	2.46	2.37	2.31	2.25
28	4.20	3.34	2.95	2.71	2.56	2.45	2.36	2.29	2.24
29	4.18	3.33	2.93	2.70	2.55	2.43	2.35	2.28	2.22
30	4.17	3.32	2.92	2.69	2.53	2.42	2.33	2.27	2.21
40	4.08	3.23	2.84	2.61	2.45	2.34	2.25	2.18	2.12
60	4.00	3.15	2.76	2.53	2.37	2.25	2.17	2.10	2.04
125	3.92	3.07	2.68	2.44	2.29	2.17	2.08	2.01	1.95

*df_1 (associated with the numerator)

TABLE 6b (continued)

df$_2$					df$_1$				
	10	12	14	16	20	30	40	50	100
1	242	244	245	246	248	250	251	252	253
2	19.4	19.4	19.4	19.4	19.4	19.5	19.5	19.5	19.5
3	8.78	8.74	8.71	8.69	8.66	8.62	8.60	8.58	8.56
4	5.96	5.91	5.87	5.84	5.80	5.74	5.71	5.70	5.66
5	4.74	4.68	4.64	4.60	4.56	4.50	4.46	4.44	4.40
6	4.06	4.00	3.96	3.92	3.87	3.81	3.77	3.75	3.71
7	3.63	3.57	3.52	3.49	3.44	3.38	3.34	3.32	3.28
8	3.34	3.28	3.23	3.20	3.15	3.08	3.05	3.03	2.98
9	3.13	3.07	3.02	2.98	2.93	2.86	2.82	2.80	2.76
10	2.97	2.91	2.86	2.82	2.77	2.70	2.67	2.64	2.59
11	2.86	2.79	2.74	2.70	2.65	2.57	2.53	2.50	2.45
12	2.76	2.69	2.64	2.60	2.54	2.46	2.42	2.40	2.35
13	2.67	2.60	2.55	2.51	2.46	2.38	2.34	2.32	2.26
14	2.60	2.53	2.48	2.44	2.39	2.31	2.27	2.24	2.19
15	2.55	2.48	2.43	2.39	2.33	2.25	2.21	2.18	2.12
16	2.49	2.42	2.37	2.33	2.28	2.20	2.16	2.13	2.07
17	2.45	2.38	2.33	2.29	2.23	2.15	2.11	2.08	2.02
18	2.41	2.34	2.29	2.25	2.19	2.11	2.07	2.04	1.98
19	2.38	2.31	2.26	2.21	2.15	2.07	2.02	2.00	1.94
20	2.35	2.28	2.23	2.18	2.12	2.04	1.99	1.96	1.90
21	2.32	2.25	2.20	2.15	2.09	2.00	1.96	1.93	1.87
22	2.30	2.23	2.18	2.13	2.07	1.98	1.93	1.91	1.84
23	2.28	2.20	2.14	2.10	2.04	1.96	1.91	1.88	1.82
24	2.26	2.18	2.13	2.09	2.02	1.94	1.89	1.86	1.80
25	2.24	2.16	2.11	2.06	2.00	1.92	1.87	1.84	1.77
26	2.22	2.15	2.10	2.05	1.99	1.90	1.85	1.82	1.76
27	2.20	2.13	2.08	2.03	1.97	1.88	1.84	1.80	1.74
28	2.19	2.12	2.06	2.02	1.96	1.87	1.81	1.78	1.72
29	2.18	2.10	2.05	2.00	1.94	1.85	1.80	1.77	1.71
30	2.16	2.09	2.04	1.99	1.93	1.84	1.79	1.76	1.69
40	2.07	2.00	1.95	1.90	1.84	1.74	1.69	1.66	1.59
60	1.99	1.92	1.86	1.81	1.75	1.65	1.59	1.56	1.48
125	1.90	1.83	1.77	1.72	1.65	1.55	1.49	1.45	1.36

Appendix C Database

The following data represent the gender, age, systolic blood pressure (SBP), diastolic blood pressure (DBP), height, weight, and smoking status for a random sample of 720 adults. Ages are given in years, blood pressure readings are given in millimeters of mercury (mm Hg), heights are given in centimeters (cm), and weights are given in kilograms (kg). These data were collected as part of the Baltimore Longitudinal Study of Aging (BLSA). The data represent only a fraction of the information that has been collected by the study, and it is not intended for scientific use. The BLSA is part of the research activities of the Gerontology Research Center, National Institute of Aging. (Note: This data is available on a computer disk and is supplied with the *Student Study Guide*. For the data file, gender is coded as 0 for female and 1 for male.)

Subject Number	Gender	Age	Current Smoker	SBP	DBP	Height	Weight	Subject Number	Gender	Age	Current Smoker	SBP	DBP	Height	Weight
001	F	22	N	90	60	175.0	58.1	033	F	27	Y	120	82	167.0	55.6
002	F	23	N	106	50	165.5	63.2	034	F	27	Y	100	65	168.8	74.4
003	F	23	Y	90	60	164.7	49.1	035	F	27	N	100	75	170.6	60.0
004	F	23	N	100	65	170.8	64.3	036	F	27	N	114	58	161.2	61.6
005	F	23	N	105	70	156.0	54.9	037	F	27	N	115	80	169.4	55.0
006	F	23	N	128	86	165.7	67.3	038	F	27	N	120	80	157.8	48.8
007	F	24	N	130	80	162.6	53.3	039	F	27	N	110	70	169.6	75.6
008	F	24	N	108	72	170.1	75.0	040	F	27	N	100	60	164.7	52.5
009	F	24	Y	106	66	166.2	56.2	041	F	28	N	105	65	157.4	71.3
010	F	24	N	98	65	177.6	78.9	042	F	28	N	110	65	162.7	47.7
011	F	24	N	120	70	161.0	60.8	043	F	28	N	104	78	174.6	66.3
012	F	24	Y	102	68	168.0	69.3	044	F	28	Y	115	70	175.5	59.4
013	F	24	N	118	65	163.4	58.2	045	F	28	Y	115	80	164.5	57.9
014	F	24	N	115	80	160.5	61.2	046	F	28	N	125	80	160.5	64.7
015	F	24	Y	110	72	160.0	49.3	047	F	28	N	110	80	170.2	70.1
016	F	25	N	115	75	167.4	57.8	048	F	28	N	110	80	175.7	56.1
017	F	25	Y	84	50	160.2	50.9	049	F	28	N	110	70	153.1	53.1
018	F	25	Y	110	80	170.4	60.9	050	F	28	N	90	60	176.0	59.2
019	F	25	N	104	66	156.6	55.2	051	F	29	N	105	65	164.0	61.4
020	F	25	N	126	80	173.1	78.0	052	F	29	N	120	64	161.8	59.6
021	F	25	Y	115	75	171.1	90.7	053	F	29	N	105	65	167.7	57.4
022	F	25	Y	105	65	162.8	65.5	054	F	29	N	90	70	155.6	52.9
023	F	26	N	120	80	175.5	67.0	055	F	29	Y	110	75	167.9	66.7
024	F	26	Y	110	75	162.6	61.3	056	F	29	N	100	60	168.5	59.1
025	F	26	N	138	80	170.0	66.4	057	F	29	N	110	60	155.9	51.7
026	F	26	Y	100	75	169.1	71.1	058	F	29	Y	106	72	170.1	52.3
027	F	26	N	120	80	173.5	61.3	059	F	29	N	105	70	164.8	53.9
028	F	26	N	120	80	162.2	57.0	060	F	29	N	120	80	168.3	55.9
029	F	26	N	125	70	172.9	61.3	061	F	30	N	100	62	161.4	59.5
030	F	26	N	105	70	159.3	50.0	062	F	30	Y	115	80	167.0	57.6
031	F	26	N	95	60	169.1	62.5	063	F	30	N	96	70	166.1	61.5
032	F	27	N	90	58	164.1	65.5	064	F	31	N	108	78	167.4	54.2

Subject Number	Gender	Age	Current Smoker	SBP	DBP	Height	Weight	Subject Number	Gender	Age	Current Smoker	SBP	DBP	Height	Weight
065	F	31	N	116	72	161.5	53.7	114	F	38	Y	118	72	163.1	53.1
066	F	31	Y	124	78	163.7	63.6	115	F	38	Y	108	80	169.2	58.3
067	F	31	N	110	66	167.2	57.0	116	F	38	N	94	65	156.5	60.0
068	F	31	N	118	72	164.0	58.3	117	F	38	N	140	76	162.1	59.5
069	F	31	Y	110	70	169.0	61.2	118	F	39	Y	100	65	158.9	53.5
070	F	31	N	107	70	160.5	53.8	119	F	39	N	102	76	164.4	64.5
071	F	31	N	112	70	166.7	59.5	120	F	39	Y	112	70	161.7	58.1
072	F	31	Y	106	78	176.6	62.6	121	F	40	N	106	74	168.6	56.5
073	F	31	N	122	68	167.5	58.3	122	F	40	N	114	72	172.2	64.6
074	F	31	N	128	66	169.6	55.3	123	F	40	N	140	80	169.7	52.2
075	F	32	Y	116	80	165.8	63.5	124	F	40	N	100	70	172.9	63.0
076	F	32	Y	115	70	176.1	60.6	125	F	40	N	98	60	166.3	56.9
077	F	32	N	110	80	161.7	77.0	126	F	40	Y	110	80	173.9	56.6
078	F	32	N	96	70	172.5	64.3	127	F	40	N	110	60	169.7	77.0
079	F	32	N	110	70	160.8	62.7	128	F	40	Y	108	70	168.5	54.7
080	F	32	N	100	70	168.3	80.2	129	F	40	N	96	70	169.0	75.5
081	F	32	N	110	70	168.7	60.4	130	F	41	Y	110	75	163.4	55.7
082	F	32	Y	120	70	166.5	52.1	131	F	41	N	100	60	164.7	58.4
083	F	32	N	112	75	157.3	49.0	132	F	41	N	105	70	165.3	74.3
084	F	32	N	120	76	157.2	60.2	133	F	41	N	110	65	168.5	66.4
085	F	32	N	90	70	160.0	50.6	134	F	41	N	90	65	160.1	48.7
086	F	33	Y	110	65	166.2	69.0	135	F	41	N	98	40	166.6	57.5
087	F	33	N	110	70	155.4	52.4	136	F	41	N	102	70	169.5	73.7
088	F	33	Y	110	70	166.5	59.6	137	F	42	N	120	88	164.0	56.6
089	F	34	N	120	72	155.5	47.5	138	F	42	Y	104	80	167.3	66.2
090	F	34	N	94	52	172.0	72.1	139	F	42	Y	115	72	163.2	64.4
091	F	34	N	108	65	173.0	79.1	140	F	42	N	120	70	160.0	53.5
092	F	34	N	106	72	164.1	54.6	141	F	42	N	120	90	159.4	49.3
093	F	34	N	122	78	163.5	63.5	142	F	42	N	106	76	169.2	56.4
094	F	34	N	108	74	157.1	49.8	143	F	42	N	120	75	170.5	95.2
095	F	34	Y	130	80	167.8	53.3	144	F	42	N	102	72	168.4	59.0
096	F	34	N	120	70	162.2	49.8	145	F	42	N	110	60	170.9	74.1
097	F	35	N	115	75	158.0	45.0	146	F	42	N	115	80	165.6	49.5
098	F	35	N	100	75	173.8	59.7	147	F	43	Y	110	60	163.3	57.9
099	F	35	N	108	70	159.9	65.4	148	F	43	N	130	78	166.5	60.4
100	F	35	Y	106	76	171.5	53.6	149	F	43	N	130	90	168.0	63.1
101	F	35	N	110	80	176.5	61.9	150	F	43	N	130	90	169.1	89.5
102	F	35	Y	96	60	163.5	47.8	151	F	43	Y	114	80	171.8	71.2
103	F	36	N	110	65	177.0	68.7	152	F	43	Y	128	80	159.3	72.3
104	F	36	Y	100	74	157.4	55.2	153	F	43	N	140	80	164.7	84.8
105	F	36	N	120	70	168.5	70.1	154	F	43	Y	120	60	166.4	53.6
106	F	36	Y	130	84	159.4	45.6	155	F	44	Y	118	90	165.5	67.6
107	F	36	N	110	68	176.3	65.5	156	F	44	N	120	80	169.5	55.5
108	F	37	Y	118	80	153.0	46.2	157	F	44	Y	110	74	173.2	84.2
109	F	37	N	125	82	157.9	52.0	158	F	45	N	124	90	167.4	58.0
110	F	37	N	120	90	157.6	52.9	159	F	45	N	110	64	166.6	76.0
111	F	37	N	125	85	171.0	82.1	160	F	45	N	126	84	166.8	54.7
112	F	37	N	110	78	160.2	68.6	161	F	46	N	115	70	175.6	64.5
113	F	37	N	130	76	160.8	59.2	162	F	46	Y	140	95	156.9	57.8

Subject Number	Gender	Age	Current Smoker	SBP	DBP	Height	Weight	Subject Number	Gender	Age	Current Smoker	SBP	DBP	Height	Weight
163	F	46	N	110	70	154.1	53.2	212	F	55	N	120	88	159.5	53.2
164	F	46	N	100	68	157.2	54.7	213	F	56	Y	138	80	159.0	63.2
165	F	46	N	110	78	163.4	90.5	214	F	56	N	130	80	164.2	69.6
166	F	46	N	114	76	168.1	74.4	215	F	56	Y	96	64	160.0	55.7
167	F	47	N	142	85	152.4	44.8	216	F	56	N	120	70	162.0	60.7
168	F	47	Y	128	76	174.5	70.1	217	F	56	N	150	85	168.6	70.5
169	F	47	N	120	82	165.7	88.7	218	F	57	N	128	78	155.5	60.6
170	F	47	N	105	70	162.0	56.6	219	F	57	N	120	78	153.2	58.1
171	F	47	N	112	80	157.9	54.0	220	F	57	N	120	70	170.7	67.9
172	F	47	N	140	90	171.3	62.0	221	F	57	N	110	70	167.3	69.3
173	F	48	N	120	70	159.5	53.8	222	F	58	N	110	64	146.5	54.6
174	F	48	N	130	90	161.2	61.1	223	F	58	N	120	70	152.3	59.6
175	F	48	N	104	65	168.5	55.3	224	F	58	N	120	82	178.2	72.4
176	F	48	N	138	100	169.3	99.1	225	F	58	N	138	82	158.8	60.6
177	F	48	N	176	118	164.7	63.4	226	F	58	Y	135	85	171.3	80.5
178	F	49	N	118	70	168.0	70.2	227	F	58	N	140	85	168.0	56.7
179	F	49	N	150	90	161.7	66.0	228	F	58	N	140	90	170.4	67.0
180	F	49	Y	104	64	159.6	44.4	229	F	58	N	110	80	160.5	52.0
181	F	50	N	102	70	161.0	52.7	230	F	58	N	132	80	163.4	56.3
182	F	50	N	138	88	166.0	64.2	231	F	58	N	120	78	164.1	77.8
183	F	50	N	125	80	164.1	52.5	232	F	58	N	120	70	161.3	59.7
184	F	50	N	106	70	157.0	50.2	233	F	59	Y	120	80	161.8	57.5
185	F	50	N	100	60	163.4	56.5	234	F	59	N	140	80	157.1	73.4
186	F	50	N	160	92	154.6	48.5	235	F	59	N	120	75	166.2	60.0
187	F	51	N	122	78	156.0	87.9	236	F	59	Y	130	88	166.2	54.2
188	F	51	N	115	75	150.7	52.4	237	F	59	Y	130	80	158.3	44.2
189	F	51	N	105	70	167.6	66.9	238	F	59	N	120	74	163.5	76.3
190	F	51	Y	118	85	171.9	63.9	239	F	59	Y	100	60	163.7	63.3
191	F	52	N	125	80	161.1	67.9	240	F	59	N	158	92	163.3	77.8
192	F	52	Y	130	90	170.2	83.8	241	F	60	N	138	78	175.7	61.2
193	F	52	Y	110	90	164.1	63.1	242	F	60	N	120	70	159.0	57.8
194	F	52	Y	100	70	160.4	53.6	243	F	60	N	130	80	165.1	76.4
195	F	53	N	118	72	172.6	63.3	244	F	61	N	135	85	167.5	58.2
196	F	53	N	125	85	166.5	61.3	245	F	61	N	152	100	150.7	54.8
197	F	53	Y	125	75	163.9	54.2	246	F	61	N	130	86	160.1	56.2
198	F	53	N	132	95	158.7	111.7	247	F	61	Y	110	70	166.8	66.8
199	F	53	N	140	90	167.1	70.1	248	F	61	N	132	70	155.1	47.3
200	F	53	N	120	70	167.0	73.8	249	F	61	N	138	86	157.7	71.2
201	F	53	N	115	80	155.4	81.9	250	F	61	N	118	78	168.7	60.4
202	F	54	Y	120	74	156.5	61.4	251	F	62	N	110	70	164.8	64.5
203	F	54	N	110	70	170.2	63.8	252	F	62	Y	120	85	160.0	52.2
204	F	54	N	120	84	152.1	53.8	253	F	62	N	112	72	166.4	67.6
205	F	54	N	95	60	167.6	72.0	254	F	62	N	128	74	168.3	60.2
206	F	55	N	115	72	156.5	53.8	255	F	62	Y	160	90	159.7	49.1
207	F	55	N	112	74	152.7	51.8	256	F	62	N	130	80	156.0	58.9
208	F	55	N	100	75	158.5	57.4	257	F	62	N	116	84	168.5	86.0
209	F	55	N	140	82	162.5	70.4	258	F	63	N	165	96	172.0	76.3
210	F	55	N	106	68	164.8	61.9	259	F	63	N	120	80	177.1	69.6
211	F	55	N	115	75	156.6	62.4	260	F	63	Y	130	80	171.7	67.4

Subject Number	Gender	Age	Current Smoker	SBP	DBP	Height	Weight	Subject Number	Gender	Age	Current Smoker	SBP	DBP	Height	Weight
261	F	63	N	140	86	169.9	65.4	310	F	71	N	130	72	156.0	54.1
262	F	63	N	125	80	167.7	61.0	311	F	71	Y	126	72	158.2	53.4
263	F	63	N	104	70	159.8	74.2	312	F	71	N	182	80	155.1	65.1
264	F	64	N	170	90	159.0	60.4	313	F	72	N	185	85	164.5	74.6
265	F	64	N	140	70	163.3	64.4	314	F	72	N	152	88	169.3	59.4
266	F	64	N	120	80	155.2	57.8	315	F	72	N	130	80	153.3	59.7
267	F	64	Y	115	84	162.0	58.1	316	F	72	N	130	88	161.0	75.6
268	F	64	Y	150	70	158.6	54.5	317	F	72	N	148	78	160.5	74.1
269	F	64	N	145	75	164.5	71.8	318	F	73	N	110	70	155.5	45.3
270	F	64	Y	120	70	152.3	54.4	319	F	73	N	180	50	153.2	46.6
271	F	65	N	146	90	162.3	57.5	320	F	73	N	154	92	165.4	59.9
272	F	65	N	90	60	156.8	46.5	321	F	73	Y	115	78	157.5	53.1
273	F	65	N	140	90	159.7	66.5	322	F	73	N	140	76	172.4	71.8
274	F	65	N	125	65	159.1	49.7	323	F	73	N	148	65	160.0	56.5
275	F	65	N	140	80	150.4	54.0	324	F	74	N	125	90	161.0	61.5
276	F	65	N	130	82	159.4	62.0	325	F	74	N	168	88	154.8	52.1
277	F	66	Y	122	80	158.4	63.0	326	F	74	N	135	80	155.0	52.9
278	F	66	Y	165	90	160.0	62.5	327	F	74	N	175	95	155.8	41.7
279	F	66	N	108	64	160.0	49.5	328	F	74	N	160	90	161.4	69.3
280	F	66	N	160	80	175.0	85.7	329	F	74	N	140	70	161.0	66.7
281	F	66	Y	120	70	154.8	66.3	330	F	74	N	120	78	153.6	63.8
282	F	66	N	145	95	156.1	57.6	331	F	74	N	162	86	152.4	64.1
283	F	66	Y	190	90	165.0	60.9	332	F	74	N	150	80	153.6	79.1
284	F	66	N	160	94	158.0	57.2	333	F	74	N	146	86	150.2	59.6
285	F	66	N	126	76	160.4	76.4	334	F	74	N	158	94	167.0	74.8
286	F	67	N	110	70	166.4	61.5	335	F	74	N	134	70	161.0	71.0
287	F	67	N	100	72	177.0	64.7	336	F	75	N	158	90	159.0	80.5
288	F	67	N	118	70	169.3	72.7	337	F	75	N	135	78	142.6	55.9
289	F	67	N	140	80	159.4	62.7	338	F	75	N	140	70	163.5	56.1
290	F	68	Y	138	102	166.2	88.9	339	F	75	N	135	70	150.3	42.5
291	F	68	Y	210	110	156.0	67.7	340	F	75	N	112	70	159.3	56.4
292	F	68	N	142	60	149.4	45.7	341	F	76	N	136	82	159.3	62.0
293	F	68	Y	150	60	160.0	51.3	342	F	77	N	126	76	155.0	57.6
294	F	68	N	136	82	152.8	64.0	343	F	77	N	124	72	148.3	46.8
295	F	68	N	142	84	164.9	68.3	344	F	78	N	114	60	157.8	60.2
296	F	68	N	165	95	163.5	76.0	345	F	78	N	160	90	155.8	55.6
297	F	68	N	150	110	158.4	84.9	346	F	79	N	180	110	164.4	73.9
298	F	69	N	135	75	160.0	57.4	347	F	79	N	135	80	156.3	72.8
299	F	69	N	130	80	158.2	63.1	348	F	79	Y	140	70	153.3	53.8
300	F	69	N	156	100	163.6	75.3	349	F	80	N	110	72	147.0	51.9
301	F	70	Y	110	70	158.8	65.4	350	F	80	N	110	70	155.8	40.8
302	F	70	N	155	98	155.7	69.2	351	F	80	N	124	76	146.0	62.6
303	F	70	N	145	85	155.1	61.5	352	F	80	N	200	80	157.6	57.2
304	F	70	N	150	90	165.4	104.9	353	F	81	N	130	75	150.2	63.7
305	F	70	N	128	76	157.1	56.7	354	F	82	N	130	70	160.4	61.1
306	F	70	Y	148	92	164.8	107.2	355	F	84	N	158	70	168.5	68.0
307	F	70	N	132	82	173.2	62.3	356	F	85	N	130	82	152.3	70.2
308	F	70	N	200	100	167.4	59.9	357	F	88	N	145	55	159.0	55.7
309	F	71	N	135	85	161.0	62.4	358	F	91	N	122	68	159.4	58.7

Subject Number	Gender	Age	Current Smoker	SBP	DBP	Height	Weight	Subject Number	Gender	Age	Current Smoker	SBP	DBP	Height	Weight
359	F	94	N	165	85	148.8	53.5	408	M	28	N	115	70	169.5	94.3
360	F	95	N	140	92	158.0	56.5	409	M	28	N	110	80	173.1	77.2
361	M	20	Y	150	86	182.4	85.2	410	M	29	Y	120	78	168.7	58.3
362	M	20	N	110	80	187.0	71.1	411	M	29	N	124	76	196.2	88.6
363	M	21	N	115	65	184.7	81.4	412	M	29	Y	120	80	172.5	68.1
364	M	21	N	128	88	190.3	124.7	413	M	29	N	125	75	189.4	86.8
365	M	21	N	120	80	170.7	78.4	414	M	29	N	120	70	180.0	83.1
366	M	22	N	114	82	180.0	65.9	415	M	29	Y	122	60	187.2	98.5
367	M	22	N	104	70	173.8	73.9	416	M	29	N	110	78	182.5	98.9
368	M	22	Y	110	70	192.5	78.5	417	M	29	N	120	80	176.5	90.0
369	M	23	Y	110	74	174.3	60.2	418	M	29	Y	125	80	173.6	91.2
370	M	23	Y	110	82	175.8	58.8	419	M	29	N	110	78	170.6	74.7
371	M	23	N	136	90	173.5	67.6	420	M	29	N	110	75	180.8	76.8
372	M	24	N	114	64	180.4	78.9	421	M	30	N	126	88	186.0	92.6
373	M	24	Y	140	80	193.4	67.9	422	M	30	Y	110	70	181.6	81.7
374	M	24	Y	116	68	168.6	61.9	423	M	30	N	140	80	173.7	57.5
375	M	24	Y	120	76	178.3	75.3	424	M	30	N	90	60	167.8	67.1
376	M	24	N	105	80	170.3	57.9	425	M	30	Y	120	80	182.0	94.8
377	M	24	N	118	78	175.7	75.5	426	M	31	N	110	70	195.4	68.0
378	M	24	N	150	95	188.8	141.0	427	M	32	Y	120	70	174.0	55.9
379	M	24	N	114	80	184.8	89.6	428	M	32	Y	130	76	191.5	94.4
380	M	25	N	130	70	187.2	80.2	429	M	32	N	120	80	190.0	84.9
381	M	25	N	120	80	172.5	69.3	430	M	32	Y	120	70	179.0	86.4
382	M	25	N	110	70	176.3	77.2	431	M	32	N	124	76	166.2	69.4
383	M	25	N	126	78	169.4	83.1	432	M	32	Y	114	78	183.0	83.1
384	M	25	Y	120	80	180.6	76.5	433	M	32	N	136	74	175.2	72.9
385	M	26	Y	122	68	186.0	86.6	434	M	33	Y	132	70	172.9	67.1
386	M	26	Y	130	72	184.6	85.0	435	M	33	Y	120	70	190.5	96.0
387	M	26	N	136	94	185.0	81.7	436	M	33	Y	155	100	185.0	86.0
388	M	26	N	144	92	179.4	93.0	437	M	33	N	128	90	181.0	84.5
389	M	26	N	110	75	185.8	80.6	438	M	33	Y	130	86	177.5	73.9
390	M	26	N	116	70	184.9	79.5	439	M	33	Y	130	88	164.6	81.5
391	M	26	N	120	90	177.0	76.4	440	M	33	Y	135	80	184.0	83.3
392	M	27	N	122	58	169.2	74.4	441	M	34	N	130	76	177.0	83.2
393	M	27	N	120	80	192.0	93.7	442	M	34	Y	106	74	175.2	78.2
394	M	27	Y	130	80	180.7	74.5	443	M	34	N	110	70	187.4	63.9
395	M	27	N	130	80	171.3	67.7	444	M	34	N	116	66	179.0	75.2
396	M	27	Y	120	80	180.4	80.3	445	M	34	Y	110	70	168.5	63.0
397	M	27	N	122	60	183.5	89.2	446	M	34	Y	110	70	186.8	92.9
398	M	28	Y	120	60	176.9	70.2	447	M	34	N	108	72	190.5	83.8
399	M	28	Y	118	80	185.5	88.8	448	M	34	Y	140	90	183.8	74.7
400	M	28	N	110	70	182.6	77.4	449	M	34	Y	124	82	179.4	86.6
401	M	28	Y	130	70	172.6	61.5	450	M	35	Y	110	64	178.3	70.8
402	M	28	Y	115	80	179.6	95.8	451	M	35	Y	122	80	170.0	69.2
403	M	28	N	140	80	178.5	89.6	452	M	35	Y	120	70	187.4	64.2
404	M	28	N	108	80	174.3	78.6	453	M	35	Y	140	90	175.8	94.5
405	M	28	N	140	90	180.5	78.0	454	M	35	Y	130	84	172.8	78.8
406	M	28	N	105	70	179.9	77.7	455	M	35	N	120	70	176.8	83.4
407	M	28	N	120	80	175.7	68.0	456	M	35	N	110	74	184.1	88.1

Subject Number	Gender	Age	Current Smoker	SBP	DBP	Height	Weight	Subject Number	Gender	Age	Current Smoker	SBP	DBP	Height	Weight
457	M	35	Y	140	90	174.5	77.5	506	M	45	Y	110	70	182.9	87.2
458	M	36	Y	110	70	180.5	83.3	507	M	45	N	110	74	176.2	76.2
459	M	36	N	117	72	170.0	65.1	508	M	46	Y	146	86	164.6	67.7
460	M	36	N	108	76	188.4	96.6	509	M	46	Y	120	70	162.5	57.9
461	M	36	N	110	70	168.0	73.6	510	M	46	N	130	80	171.2	70.1
462	M	36	Y	126	72	174.4	76.0	511	M	46	Y	130	72	178.4	87.6
463	M	36	N	135	70	174.2	68.3	512	M	47	N	140	90	180.5	78.1
464	M	36	Y	120	85	177.0	90.1	513	M	47	Y	110	80	181.6	69.6
465	M	37	Y	114	88	163.5	74.5	514	M	47	Y	140	84	180.3	92.3
466	M	37	N	126	86	187.2	80.8	515	M	47	Y	110	80	181.0	88.4
467	M	37	Y	110	80	175.4	98.9	516	M	47	N	138	76	178.6	74.0
468	M	37	N	110	78	168.0	69.8	517	M	47	N	120	78	176.8	78.4
469	M	37	Y	118	76	177.8	70.5	518	M	48	N	114	74	176.0	67.2
470	M	38	N	98	62	174.0	82.1	519	M	48	Y	120	70	187.0	91.8
471	M	38	N	130	80	178.5	70.1	520	M	48	Y	140	80	178.2	81.3
472	M	38	N	130	80	186.3	83.3	521	M	48	N	130	90	178.8	69.6
473	M	38	N	116	80	167.2	70.0	522	M	48	N	130	80	174.0	61.3
474	M	38	N	128	72	170.2	68.6	523	M	48	Y	130	84	171.0	64.6
475	M	38	N	150	82	178.6	92.0	524	M	48	Y	154	86	164.4	70.1
476	M	38	Y	130	86	167.0	63.4	525	M	48	N	130	94	165.3	71.6
477	M	38	Y	110	80	176.0	57.8	526	M	48	N	124	80	177.0	76.2
478	M	39	Y	110	70	175.8	97.0	527	M	48	N	114	70	172.8	86.2
479	M	39	Y	112	78	172.0	77.8	528	M	48	N	120	78	180.0	85.2
480	M	39	N	130	90	177.0	62.8	529	M	48	Y	130	80	178.6	90.1
481	M	40	N	110	72	165.7	67.8	530	M	48	Y	120	70	176.5	68.9
482	M	40	N	104	67	175.0	59.5	531	M	48	Y	134	84	176.4	74.5
483	M	40	N	120	80	173.6	75.1	532	M	49	N	120	80	172.7	82.7
484	M	41	N	110	70	187.0	82.9	533	M	49	Y	130	88	175.5	75.0
485	M	41	Y	115	70	177.8	79.3	534	M	49	Y	130	85	174.4	74.4
486	M	41	N	124	80	194.4	104.6	535	M	49	Y	170	110	177.8	91.1
487	M	42	N	128	80	184.3	85.3	536	M	49	N	130	80	175.2	80.9
488	M	42	N	120	70	180.4	86.6	537	M	49	Y	150	100	175.5	82.2
489	M	42	N	125	90	179.2	74.9	538	M	49	N	130	90	181.7	77.1
490	M	43	N	130	78	173.0	78.2	539	M	49	N	108	70	178.2	73.0
491	M	43	Y	112	80	173.5	81.6	540	M	49	N	120	84	167.4	102.8
492	M	43	Y	122	78	179.5	75.0	541	M	50	N	120	60	180.3	85.2
493	M	43	N	140	80	171.2	61.3	542	M	50	N	128	70	187.0	89.7
494	M	43	Y	124	86	164.4	66.2	543	M	50	N	114	80	169.2	92.0
495	M	43	Y	120	74	176.0	101.3	544	M	50	N	140	70	181.5	77.9
496	M	43	N	124	80	180.5	76.4	545	M	50	N	140	80	181.2	91.6
497	M	44	Y	130	70	186.2	83.4	546	M	50	N	120	74	172.4	76.5
498	M	44	Y	110	72	179.0	74.2	547	M	50	N	120	70	182.4	75.7
499	M	44	N	128	84	187.3	83.3	548	M	50	N	112	76	171.2	81.2
500	M	45	N	120	70	168.7	70.0	549	M	51	Y	110	70	173.3	69.0
501	M	45	Y	140	70	176.5	69.4	550	M	51	N	120	88	184.7	83.1
502	M	45	N	118	60	174.0	73.5	551	M	51	N	114	78	168.2	76.3
503	M	45	N	120	80	176.2	71.3	552	M	52	N	150	90	177.0	78.0
504	M	45	Y	118	86	177.2	84.9	553	M	52	Y	112	80	175.4	73.3
505	M	45	Y	136	92	176.0	75.2	554	M	52	N	130	82	181.5	73.6

Subject Number	Gender	Age	Current Smoker	SBP	DBP	Height	Weight	Subject Number	Gender	Age	Current Smoker	SBP	DBP	Height	Weight
555	M	52	N	130	80	180.8	72.3	604	M	60	N	150	80	163.7	60.7
556	M	52	Y	120	80	171.5	72.3	605	M	60	Y	140	90	178.9	89.1
557	M	52	N	125	90	180.6	79.7	606	M	61	N	186	120	174.7	71.7
558	M	53	N	155	100	174.7	71.5	607	M	61	Y	118	78	171.5	71.9
559	M	53	Y	130	70	174.5	81.3	608	M	61	N	130	70	171.0	63.7
560	M	53	N	130	80	166.2	84.1	609	M	61	Y	110	70	176.0	75.7
561	M	53	Y	140	80	183.3	64.0	610	M	61	N	200	100	177.0	112.4
562	M	54	N	220	110	174.2	83.4	611	M	61	N	110	60	168.4	74.8
563	M	54	Y	128	76	172.5	76.7	612	M	61	N	130	80	169.2	66.7
564	M	54	N	200	130	184.0	95.6	613	M	61	N	160	100	182.6	78.3
565	M	54	N	120	80	155.0	58.0	614	M	61	N	150	90	182.5	79.4
566	M	54	N	120	80	181.3	82.2	615	M	61	N	170	100	179.0	99.2
567	M	54	N	138	86	180.2	96.1	616	M	61	N	140	82	171.4	76.9
568	M	54	Y	120	80	159.2	58.4	617	M	62	N	128	80	175.5	78.9
569	M	54	Y	130	82	174.8	81.8	618	M	62	N	145	85	175.8	80.4
570	M	54	N	144	82	182.8	68.9	619	M	62	N	117	70	178.0	68.5
571	M	54	Y	118	70	183.2	86.0	620	M	62	N	130	80	175.8	74.5
572	M	54	Y	155	86	168.0	76.6	621	M	62	Y	140	90	164.0	80.8
573	M	55	N	130	82	180.3	81.0	622	M	62	N	132	80	182.2	88.6
574	M	55	N	130	80	178.2	79.8	623	M	62	Y	136	84	175.4	76.0
575	M	55	N	144	78	181.1	82.9	624	M	63	Y	140	80	183.0	79.2
576	M	55	N	140	90	176.2	72.7	625	M	63	N	142	80	172.7	85.0
577	M	55	Y	116	78	172.0	84.1	626	M	63	Y	150	100	160.0	54.0
578	M	56	Y	170	84	182.5	79.2	627	M	63	N	124	80	179.7	79.2
579	M	56	Y	140	90	170.5	67.7	628	M	63	N	140	80	178.5	82.9
580	M	56	N	124	90	184.2	95.5	629	M	63	Y	188	98	181.5	81.9
581	M	57	N	125	70	173.0	70.1	630	M	63	N	118	76	177.5	76.3
582	M	57	Y	140	84	178.8	71.0	631	M	63	Y	104	64	163.4	68.3
583	M	57	N	140	100	180.6	91.3	632	M	64	Y	176	82	178.5	69.8
584	M	57	N	130	80	175.6	69.6	633	M	64	N	142	78	180.2	80.2
585	M	57	N	130	70	177.0	55.9	634	M	64	N	112	78	178.0	78.0
586	M	57	N	120	80	176.0	81.1	635	M	64	N	144	76	174.3	101.2
587	M	57	N	120	80	177.0	83.4	636	M	64	Y	170	110	167.4	81.4
588	M	57	N	130	88	181.6	77.2	637	M	64	N	100	64	171.7	74.9
589	M	57	N	108	80	172.4	75.2	638	M	64	N	120	86	186.4	81.9
590	M	58	N	142	80	168.1	73.3	639	M	64	N	148	84	180.8	91.7
591	M	58	N	110	76	176.8	80.9	640	M	65	N	162	92	173.2	80.7
592	M	58	N	110	70	175.3	68.3	641	M	66	N	110	70	187.2	76.1
593	M	58	N	130	80	168.0	63.8	642	M	66	Y	130	80	172.3	70.4
594	M	58	Y	145	82	187.0	77.4	643	M	66	Y	140	87	195.4	99.0
595	M	58	N	150	86	178.2	74.6	644	M	66	N	112	62	177.2	77.9
596	M	59	N	104	74	181.6	85.1	645	M	67	N	118	76	178.0	67.3
597	M	59	Y	110	70	171.8	64.4	646	M	67	N	132	82	172.0	76.8
598	M	59	Y	170	90	182.7	83.5	647	M	67	N	104	70	176.1	75.6
599	M	59	Y	150	90	181.0	89.1	648	M	68	N	122	72	176.5	84.3
600	M	59	Y	110	70	176.0	73.8	649	M	68	Y	180	95	174.3	79.5
601	M	60	N	138	80	171.2	79.6	650	M	68	Y	168	102	182.2	86.1
602	M	60	N	138	80	169.6	69.0	651	M	68	N	140	84	178.0	84.7
603	M	60	Y	160	94	172.0	89.7	652	M	68	Y	120	72	160.0	70.8

Subject Number	Gender	Age	Current Smoker	SBP	DBP	Height	Weight	Subject Number	Gender	Age	Current Smoker	SBP	DBP	Height	Weight
653	M	68	Y	170	100	179.0	86.5	687	M	72	N	180	100	175.9	76.9
654	M	69	N	120	80	180.4	90.2	688	M	73	N	160	100	175.2	79.0
655	M	69	N	130	74	192.0	100.9	689	M	73	N	150	80	170.9	70.3
656	M	69	N	120	70	159.4	70.5	690	M	73	Y	170	80	174.0	87.3
657	M	69	Y	170	90	174.5	76.5	691	M	73	N	158	90	169.0	57.1
658	M	69	N	130	82	164.7	78.8	692	M	73	N	130	80	175.4	87.4
659	M	69	N	92	60	168.4	74.2	693	M	73	N	180	120	167.4	64.5
660	M	69	N	120	72	176.5	74.6	694	M	73	N	130	70	177.5	80.0
661	M	70	N	132	70	176.0	82.0	695	M	74	N	124	82	167.8	65.0
662	M	70	Y	151	70	178.5	77.0	696	M	74	N	170	70	175.1	76.8
663	M	70	N	140	82	182.9	87.3	697	M	75	N	180	95	175.0	84.0
664	M	70	N	152	70	170.9	71.7	698	M	75	N	134	80	181.6	67.4
665	M	70	N	162	120	161.0	66.7	699	M	75	N	180	84	170.0	75.9
666	M	70	N	130	70	172.3	67.9	700	M	75	N	130	70	169.0	58.1
667	M	70	N	150	88	176.8	76.5	701	M	75	N	140	100	176.5	84.9
668	M	70	N	160	72	165.0	67.5	702	M	75	Y	110	70	157.5	55.7
669	M	70	N	150	70	160.2	64.5	703	M	76	Y	180	80	172.6	61.2
670	M	70	Y	140	75	174.0	68.8	704	M	76	N	140	76	170.6	62.3
671	M	71	Y	140	60	176.0	58.8	705	M	76	N	154	84	172.0	73.9
672	M	71	Y	110	70	166.5	56.8	706	M	77	N	168	94	172.4	86.7
673	M	71	Y	160	90	168.0	54.0	707	M	77	N	170	94	175.4	76.6
674	M	71	N	138	88	175.5	84.9	708	M	77	N	180	120	185.2	90.1
675	M	71	N	150	90	182.8	79.9	709	M	79	Y	140	80	166.0	61.3
676	M	71	N	108	68	173.0	70.5	710	M	79	N	206	110	183.0	72.2
677	M	71	N	95	60	171.0	57.0	711	M	79	N	130	80	184.0	76.6
678	M	71	N	160	90	178.7	75.0	712	M	79	N	140	86	172.3	61.7
679	M	72	N	150	72	174.5	81.5	713	M	79	N	120	80	183.8	86.2
680	M	72	N	140	80	181.0	84.3	714	M	80	N	180	92	166.3	67.1
681	M	72	N	140	80	168.0	72.7	715	M	80	N	160	90	174.8	70.5
682	M	72	N	130	80	170.4	57.5	716	M	83	N	170	104	175.5	65.2
683	M	72	N	132	80	177.8	77.9	717	M	84	N	130	80	175.6	76.4
684	M	72	Y	130	80	177.8	60.0	718	M	86	N	180	120	177.5	74.5
685	M	72	N	96	56	173.0	67.9	719	M	92	N	135	65	178.0	70.8
686	M	72	N	128	76	173.5	76.9	720	M	98	N	170	70	158.4	62.7

Answers to Odd-Numbered Exercises

This section contains answers for odd-numbered basic skills, further applications, and chapter review exercises. In addition, answers for all chapter achievement test questions can be found here.

Note: Your answers may sometimes vary slightly from those listed here depending on the number of decimal places carried in subcomputations. Do not be overly concerned with minor differences between your answers and those given here. Many of these numerical answers were obtained with a hand-held calculator or a computer.

CHAPTER 1

Exercise Set 1.1

1. **a.** The statement creates the impression that the typical 35-year-old person needs disability insurance as much as a typical 65-year-old person.
   b. Are the disabilities job related? One would expect most age-related disabilities to occur with people aged 55 or more.

3. **a.** This statement creates the impression that all the service personnel are experienced.
   b. How many personnel are there? Are they all young?

5. **a.** This statement creates the impression that brand X pain reliever is the pain reliever of choice.
   b. How many doctors? What is the pain reliever? Is it better than aspirin?

7. **a.** This statement creates the impression that all low-fat foods are good for every heart, even the heart of a heavy smoker who is overweight with high blood pressure and diabetes.
   b. In what people? How much of a reduction? Is the reduction significant?

Exercise Set 1.2

1. **a.** The cause of each accident in the state of Maryland for the month of June.
   b. The cause of each accident in the five selected subdivisions of the state for the month of June.
   c. The percentage of accidents where alcohol is a contributing factor for the five subdivisions can serve as an estimate for the percentage of accidents in which alcohol is a factor for the state of Maryland.

3. **a.** The outcomes (cured or not) of all arthritis patients that take cod liver oil and the outcomes (cured or not) of all arthritis patients not taking cod liver oil.

   b. The outcomes (cured or not) of the group of 50 arthritis patients taking cod liver oil and the outcomes (cured or not) of the group of 50 arthritis patients not taking cod liver oil.

5. **a.** The changes in the number of helper and T cells for all individuals subjected to different levels of stress.
   b. The changes in the number of helper and T cells for the 36 people subjected to different levels of stress.

7. A statistic. The poll is most likely for a sample of the households in the community.

Exercise Set 1.3

1. **a.** all responses from the 1500 registered voters
   b. 500 responses from the sampled voters
   c. number of votes for Mr. Jackson in sample
   d. number of votes for Mr. Jackson in population
   e. he will not intensify his campaign efforts.

3. **a.** inferential **b.** descriptive
   c. inferential **d.** inferential

5. **a.** the collection of textbook costs for the 1200 students
   b. the textbook costs for the 25 polled students
   c. 1200, the size of the population, and $135, the average textbook cost for the population
   d. 25, the size of the sample, and $152.25, the average textbook cost for the sample
   e. A mistake had been made, $135 is too low, or the sample was not representative of the population.

CHAPTER 2

Exercise Set 2.1

1. **a.** quantitative **b.** quantitative **c.** qualitative **d.** qualitative

3. **a.** discrete **b.** discrete **c.** continuous **d.** continuous

5. a. column 1: quantitative **b.** discrete
column 2: qualitative
column 3: quantitative
c. column 1: ordinal
column 2: nominal
column 3: ratio

7. a. ordinal **b.** Yes. The results would be easier to analyze and tabulate.

9. apartment numbers, etc.

11. Yes, it would be categorical in nature.

Exercise Set 2.2

1. a. 9 **b.** 13 **c.** 17 **d.** 5 **e.** 6

3. a. 25 **b.** 27.4 **c.** 39 **d.** 28.64 **e.** 19.5

5. a. 16.5–25.5 **b.** 14.45–27.45 **c.** 22.5–39.5 **d.** 23.645–28.645 **e.** 13.55–19.55

7. a.

Class	f	Rel f
1–4	14	0.175
5–8	18	0.225
9–12	12	0.150
13–16	16	0.200
17–20	20	0.250
		1.0

b.

Class	f	Cum f
1–4	14	14
5–8	18	32
9–12	12	44
13–16	16	60
17–20	20	80
		80

c.

Class	f	Cum Rel f
1–4	14	0.175
5–8	18	0.400
9–12	12	0.550
13–16	16	0.750
17–20	20	1.000
	80	

9. a.

x	f	Rel f
4	1	0.0625
7	3	0.1875
8	6	0.3750
9	4	0.2500
10	2	0.1250
		1.000

b.

x	f	Cum f
4	1	1
7	3	4
8	6	10
9	4	14
10	2	16
	16	

c.

x	f	Cum Rel f
4	1	0.0625
7	3	0.2500
8	6	0.6250
9	4	0.8750
10	2	1.0000
	16	

11. a. 2.5, 6.5, 10.5, 14.5, and 18.5 **b.** 0.5–4.5, 4.5–8.5, 8.5–12.5, 12.5–16.5, and 16.5–20.5

13. a. 6 **b.** 7 **c.** 8 **d.** 8 **e.** 10

15.

Class	f
28–38	4
39–49	6
50–60	3
61–71	4
72–82	4
83–93	4

17.

Class	Rel f
28–38	0.16
39–49	0.24
50–60	0.12
61–71	0.16
72–82	0.16
83–93	0.16

19.

Class	f	Cum f
66–70	2	2
71–75	4	6
76–80	10	16
81–85	7	23
86–90	4	27
91–95	1	28

21.

Class	f	Rel f	Cum Rel f
115.9–120.8	10	0.4	0.4
120.9–125.8	6	0.24	0.64
125.9–130.8	5	0.2	0.84
130.9–135.8	3	0.12	0.96
135.9–140.8	1	0.04	1.00

23. a.

X	f	Rel f
2	1	0.033
3	4	0.133
4	8	0.267
5	6	0.200
6	4	0.133
7	2	0.067
8	2	0.067
9	1	0.033
10	1	0.033
11	1	0.033

b.

X	f	Rel f	Cum Rel f
2	1	0.033	0.033
3	4	0.133	0.167
4	8	0.267	0.433
5	6	0.200	0.633
6	4	0.133	0.767
7	2	0.067	0.833
8	2	0.067	0.900
9	1	0.033	0.933
10	1	0.033	0.967
11	1	0.033	1.000

25. a. 3/55. The percentage is 5.45. **b.** 11/40. The percentage is 27.5. **c.** 30/130. The percentage is 23.08. **d.** 20/130. The percentage is 15.38. **e.** 40/130. The percentage is 30.77. **f.** 40/130. The percentage is 30.77.

27. a. 23/35. The percentage is 92. **b.** 7/25. The percentage is 28.

29. a. 5 **b.** 6.95, 11.95, 16.95, 21.95, 26.95 **c.** 4.45– 9.45 **d.** 9/18. The percentage is 50. **e.** 10/18. The percentage is 55.56. **f.** 4/18. The percentage is 22.22.

Exercise Set 2.3

1. a. Grade distribution

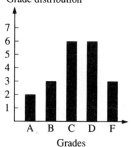

b. Grade distribution

3.

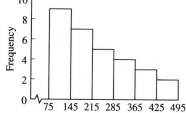

5.

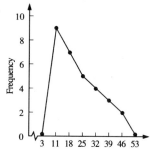

7.

4	5 5 8 9 9 9 9
5	0 0 1 1 1 1 2 2 3 3 3 4 4 9 9
6	0 1 1 1 1 1 2 2 2 2 2 2 3 5 6 7
7	2 5

9.

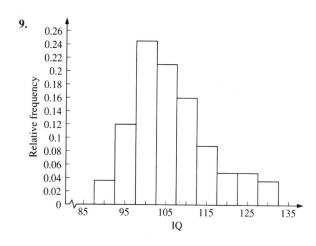

11. a. bar, circle
b. histogram, bar, circle, stem and leaf, line, ogive

13.

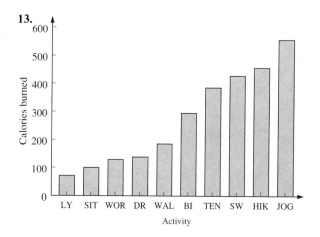

15. a.

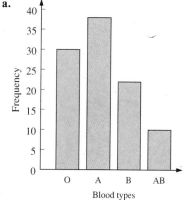

b. Blood types

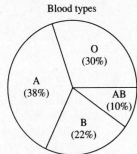

17.

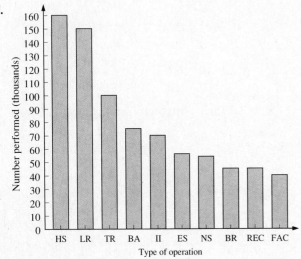

23.

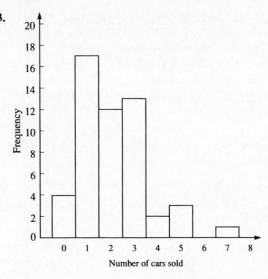

19.

20	2
30	4
40	6 6 6
50	1 4 5 7
60	1 3 5 6 7 9
70	11 4
80	4
90	0

21.

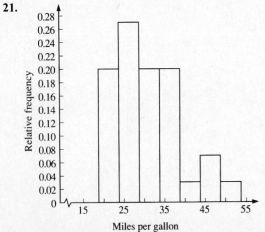

Review Exercises

1. a.

Class	*f*
148–150	6
151–153	5
154–156	6
157–159	6
160–162	5
163–165	7
166–168	3
169–171	7
172–174	3
175–177	2

b.

14b	8 9
15a	0 0 0 0 1 1 2 3 3 4 4
15b	5 5 6 6 7 7 7 8 8 8
16a	0 1 1 2 2 3 3 3 3 4 4
16b	5 7 7 7 9 9 9
17a	0 0 0 1 2 4 4
17b	5 6

c.

d.

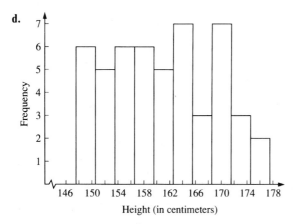

Height (in centimeters)

3. a. ratio **b.** nominal **c.** ratio **d.** nominal **e.** ordinal
f. ratio **g.** nominal **h.** nominal **i.** ordinal **j.** nominal
k. ratio

5. a.

Religion	Frequency (f)
C	5
P	7
J	6
A	2
	20

b.

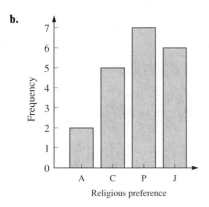

Religious preference

c.

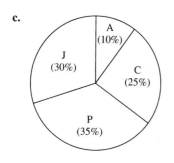

7. 11

9. a.

12	0
13	
14	0 5
15	
16	0 5
17	0 5
18	5 5
19	0 5 5 5
20	5
21	0 0 5
22	0 5 5 5
23	0 5
24	0 5 5
25	
26	0 5
27	
28	5

b.

Class	f
120–136	1
137–153	2
154–170	3
171–187	3
188–204	4
205–221	5
222–238	5
239–255	3
256–272	2
273–289	1

c.

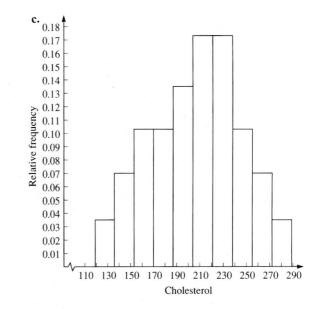

Cholesterol

11.

5b	9
6a	3 3 4 4 4 4
6b	5 5 6 6 6 6 7 7 7 7 7 8 8 8 9 9 9 9
7a	0 0 0 0 1 1 2
7b	
8a	0

13. a.

11	4 5 6		18	0 4 4 4 6 7 9
12	0 3 6 8 9		19	4 5 5
13	1 2 2 3 4 5 5 7 8 9		20	0 1 2 6 7 9
14	2 2 3 6 7			
15	2 7 8			
16	1 4 5 7 8 8			
17	0 0 2 4 4 4 5 5 6 7 7 8			

b.

11a	4
11b	5 6
12a	0 3
12b	6 8 9
13a	1 2 2 3 4
13b	5 5 7 8 9
14a	2 2 3
14b	6 7
15a	2
15b	7 8
16a	1 4
16b	5 7 8 8
17a	0 0 2 4 4 4
17b	5 5 6 7 7 8
18a	0 4 4 4
18b	6 7 8
19a	4
19b	5 5
20a	0 1 2
20b	6 7 9

The diagram shows how many measurements are in the lower half of a stem and how many measurements are in the upper half of a stem. The stems are split, so there not as many leaves on each.

c.

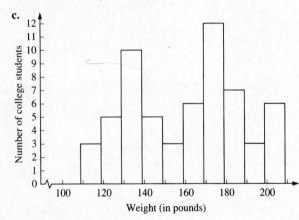

Achievement Test

1.

0a	0 1 1 1 2 2 3 3 3 3 3 4 4 4
0b	5 5 5 5 5 6 7 7 7 8 8 8 8 8 9 9 9 9
1a	0 0 1 1 1 1 1 2 2 2 2 3 4
1b	5 5 7 7 8

2.

Class	f
0–3	11
4–7	12
8–11	16
12–15	8
16–19	3

3. Cum. rel. f

4.

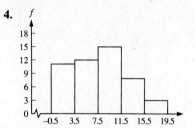

5. a. 11 **b.** 44

6. a. interval **b.** nominal **c.** interval **d.** ordinal **e.** ratio **f.** nominal

7. a. discrete **c.** continuous **e.** discrete   **8.** 8

9. a. 41.67 **b.** 25 **c.** 36.36 **d.** 54.55 **e.** 27.27

CHAPTER 3

Exercise Set 3.1

1. a. $\bar{x} = 12.88$, $\tilde{x} = 10.5$, mode $= 3$, midrange $= 18$
b. $\bar{x} = 11.86$, $\tilde{x} = 7$, modes $= 5, 7$, midrange $= 13.5$
c. $\bar{x} = 7.86$, $\tilde{x} = 7$, mode $=$ none, midrange $= 8.5$
d. $\bar{x} = 8.88$, $\tilde{x} = 11$, mode $=$ none, midrange $= 8$

3. a. 3.08 **b.** 0.37 **c.** -0.36

5. a. $\bar{x} = 0.38$, $\tilde{x} = 0$, mode $= 0$
b. $\bar{x} = 3$, $\tilde{x} = 3$, mode $= 3$
c. $\bar{x} = 2.22$, $\tilde{x} = 2$, mode $= 1, 2, 3, 4$
d. $\bar{x} = 0.13$, $\tilde{x} = 0$, mode $= 0$

7. a. positive
b. symmetric **c.** negative **d.** symmetric

9. 62   **11.** $800,000   **13.** mean, $\bar{x} = 4.1$

15. a. 198.33 **b.** 199.5 **c.** none **d.** 198 **e.** $Q_1 = 187$, $Q_2 = 199.5$, $Q_3 = 208$ **f.** 198

17. a. $5,500 **b.** $9,300 **c.** $7,000 **d.** $17,500 **e.** positive **f.** median **g.** $5,500 **h.** $7,000

19. mode

Exercise Set 3.2

1. $R = 4$ $s^2 = 2$ $s = 1.41$ **3.** $s^2 = 64$ $s = 8$

5. a. $\bar{x} = 1.67$, $s^2 = 13.60$, $s = 3.69$ **b.** $\bar{x} = 5.05$, $s^2 = 0.26$, $s = 0.51$

7. No. SS < 0.   **9.** 10.71   **11.** 0

13. No. For example, $s = 1$ and $s^2 = 1$ or $s = 0.5$ and $s^2 = 0.25$.

15. yes; for example, $\{1, 1\}$

17. The data consist of just one number or repetitions of one number.

19. a. (14.88, 35.12) **b.** 84

21. 96.67. Yes. For Chebyshev's theorem, the percentage is 92.28.

23. The proportion is at most 25%.

25. a. A: 0.94, B: 1.15 **b.** A: 0.0055, B: 0.04
c. A **d.** A **e.** A: 0.1 B: 0.35

27. a. \$24,500 **b.** \$9,300 **c.** \$7,851.49 **d.** \$2,500

29. $\bar{x} = 2.00$, $s = 0.15$, yes

Exercise Set 3.3

1. a. 24.5 **b.** method I: 22.5, method II: 23.65
c. $P_{40} = 21.99$, $P_{65} = 27.18$
d. $Q_3 = 31.5$, $D_3 = 20.33$
e. positive **f.** 53.44 **g.** 7.31

3. a. 40.92 **b.** method I: 40, method II: 40.99
c. 33.39 **d.** 5.78 **e.** $P_{40} = 39.68$, $P_{69} = 44.02$
f. $Q_1 = 37.73$, $D_7 = 44.22$

5. a. 332.58 **b.** 351.29 **c.** 13,288.65 **d.** 115.28
e. $P_{20} = 225.10$, $P_{35} = 291.30$
f. $Q_3 = 433.30$, $D_9 = 473.02$

Exercise Set 3.4

1.

x	80	65	60	11.45	47	92
z	2.2	1.2	0.87	-2.37	0	3

3. a. 12 **b.** 5.66 **c.** $z_4 = -1.41$, $z_8 = -0.707$, $z_{12} = 0$,
$z_{16} = 0.707$, $z_{20} = 1.41$ **d.** $\mu = 0$ and $\sigma = 1$

5.

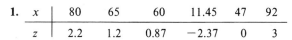

| 0.90 | 1.00 | | 1.10 | 1.20 | | 1.30 | 1.40 |
| | 0.95 | 1.025 | | 1.18 | | 1.29 | 1.41 |

7. Sue's z score $= 0.36$; Mary's z score $= 1$; Mary.

9. e

11. Dick

13. a.

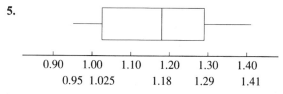

Sales tax
4.0 ↑ 6.0   8.0 ↑ 10.0   12.0   14.0
$L = 5$ $Q_1 = 9$
                $Q_3 = 12$ $U = 14$
            $Q_2 = 10.25$

b. No.

Review Exercises

1. a. $\mu = 5$, $\tilde{\mu} = 5$, no mode, midrange $= 5$, $R = 5$,
$\sigma^2 = 4.67$, $\sigma = 2.16$
b. $\mu = 5$, $\tilde{\mu} = 5$, no mode, midrange $= 5$, $R = 6$,
$\sigma^2 = 5.2$, $\sigma = 2.28$
c. $\mu = 4$, $\tilde{\mu} = 4$, no mode, midrange $= 4.5$, $R = 6$,
$\sigma^2 = 10.8$, $\sigma = 3.29$
d. $\mu = 3$, $\tilde{\mu} = 3$, mode $= 3$, midrange $= 0$, $R = 9$,
$\sigma^2 = 0$, $\sigma = 0$

3. a. 3.5 **b.** -1.88 **c.** -2 **d.** -23.75 **e.** 1

5. $\bar{x} = 5.95$, $\tilde{x} = 6$, mode $= 6$, $s^2 = 2.37$, $s = 1.54$

7. a. sample **b.** $\bar{x} = 85$, $s = 8.94$ **c.** $z_{71} = -1.57$,
$z_{95} = 1.12$

9. a. 7.26 **b.** 3.20 **c.** 0.86
d.

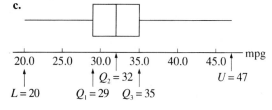

Telebill
2.5   5.0   7.5   10.0   12.5   15.0
↑   ↑
$L = 2.3$ $Q_2 = 7.04$ $U = 15.28$
        $Q_1 = 5.12$ $Q_3 = 9.39$

11. a. $\bar{x} = 32.73$, $s = 6.37$
b. $Q_1 = 29$, $Q_2 = 32$, $Q_3 = 35$, $D_4 = 31$
c.

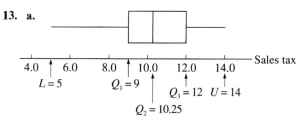

mpg
20.0   25.0   30.0 ↑   35.0   40.0   45.0 ↑
↑   ↑ $Q_2 = 32$ ↑ $U = 47$
$L = 20$ $Q_1 = 29$ $Q_3 = 35$

Achievement Test

1. a. mode, since the data are qualitative **b.** C **c.** 29 **d.** 34.38 **e.** 72.41

2. a. 9 **b.** 8 **c.** 8 **d.** 7.5 **e.** 11.5 **f.** 3.39 **g.** 0.59 **h.** 6

3. a. 4.7 **b.** 2.19

4. a. x is the mean. **b.** x is two standard deviations above the mean. **c.** x is one standard deviation below the mean.

5. c **6.** 4

7.

Values
2.0 ↑ 4.0   6.0 ↑ 8.0   10.0 ↑ 12.0
$L = 3$ $Q_1 = 5$ ↑ $Q_3 = 11$ ↑
                $Q_2 = 8$ $U = 12$

8. 50

9. Yes. There are unfortunate people who have fewer than two feet (or legs). Therefore the average number of feet is less than two.

CHAPTER 4

Exercise Set 4.1

1. a.

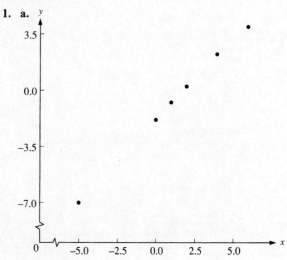

  b. positive
  c. 14.27

3. 24

5. a.

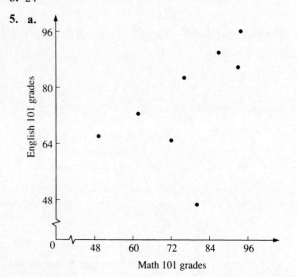

  b. positive
  c. 129.5

7. a.

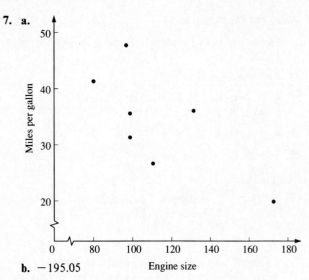

  b. -195.05

9. a.

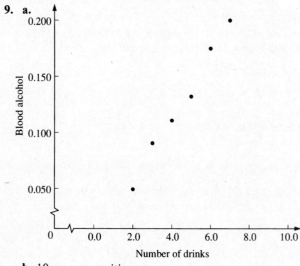

  b. 10   **c.** positive

11. a.

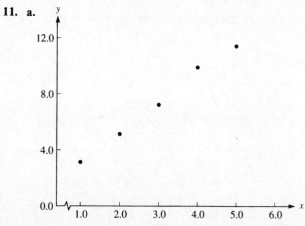

b. 5

Exercise Set 4.2

1. a. 50.83 **b.** 22.83 **c.** 17.83

3. a. positive **b.** none **c.** none **d.** negative

5. 0.94

7. a.

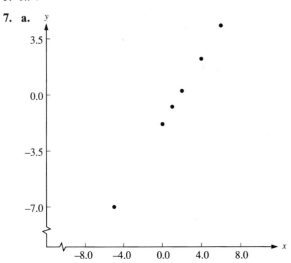

b. positive
c. 1

9. a.

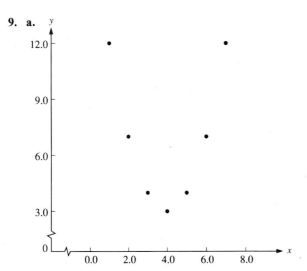

b. 0
c. No linear relationship exists between *x* and *y*.

11. a.

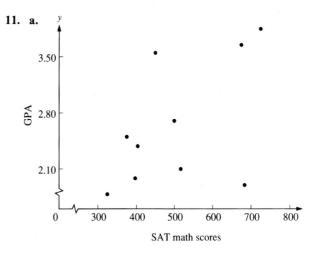

b. positive
c. 0.53

13. 0.49

15. a. −0.91 **b.** 0.94

17. −0.99

Exercise Set 4.3

1. a. $m = 2, b = -3$

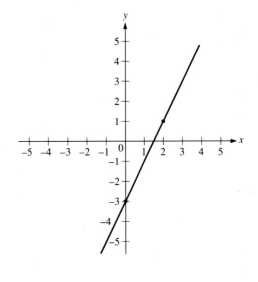

b. $m = 1$, $b = 2$

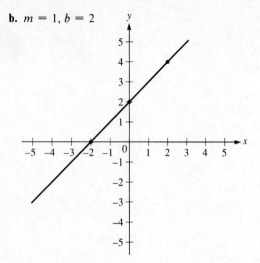

c. $m = -2/3$, $b = 2$

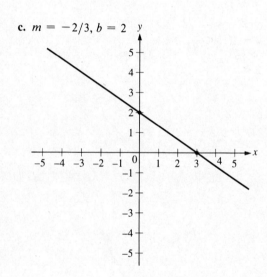

d. $m = -2$, $b = 0$

3. $\hat{y} = 3.138 + 0.3508x$, SSE = 16.581

5. $\hat{y} = -2 + x$, SSE = 0

7. $\hat{y} = 33.75 + 0.5339x$, SSE = 1235.535

9. $\hat{y} = 1.2516 + 2.1647x$, SSE = 72.84

11. **a.** $\hat{y}' = -0.25x'$ **b.** -2.75

13. $\hat{y} = 0.9850 + 0.0100x$

15. $\hat{y} = 305.027 - 1.2553x$, $x = 69$, $\hat{y} = 218.41$ seconds

Review Exercises

1. **a.**

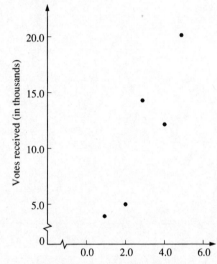

b. 9.75   **c.** 0.93
d. $\hat{y} = -0.7 + 3.9x$, SSE = 23.9 **e.** 12,950 **f.** 3900
g.

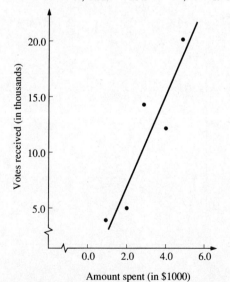

3. a.

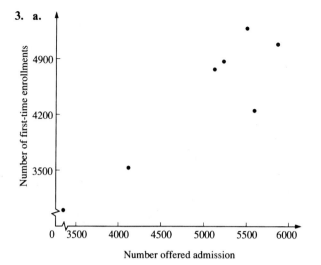

b. 675,249.05
c. 0.91
d. $\hat{y} = 437.008 + 0.785x$, SSE = 696,157.1422
e. 4362.008
f. 785
g.

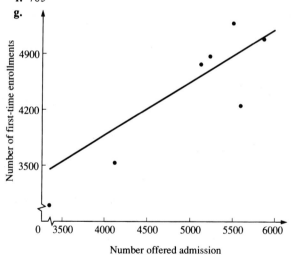

5. a. 0.85 **b.** 194,147.94 **c.** 254.58 **d.** 25.19

7. a.

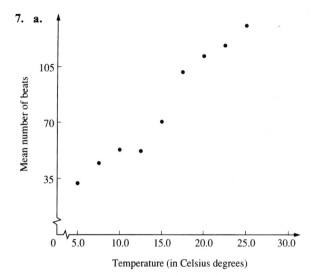

b. 0.98
c. $\hat{y} = 2.8589 + 5.0753x$
d. 409.60
e. 2.86

9. a.

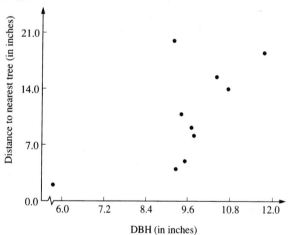

b. 0.67
c. $\hat{y} = 7.7994 + 0.1708x$
d. 12.08

Achievement Test

1. a. 13.1 **b.** increase **c.** 1.4 **d.** No; the number of push-ups cannot be negative.

2. a. 61.33 **b.** 2151.5 **c.** 153 **d.** 2.50 **e.** −22.97
f. $\hat{y} = -22.97 + 2.495x$ **g.** 30.49 **h.** 0.42
i. 1769.77 **j.** 139.21 **k.** 1.39

3. **a.** No. This is an example of spurious correlation.
b. inflation

4. Since $\hat{y} = b + mx$ and $b = \bar{y} - m\bar{x}$, $\hat{y} = (\bar{y} - m\bar{x})$ $+ mx$. Therefore, $\hat{y} = \bar{y} + m(x - \bar{x})$. Substituting $\bar{y}$ for $\hat{y}$ and $\bar{x}$ for x, we get $\bar{y} = \bar{y}$. Hence, the point $(\bar{x}, \bar{y})$ is on the regression line.

5. 6   6. 46.01

CHAPTER 5

Exercise Set 5.1

1. **a.** {HT, TH, TT, HH} **b.** {1T, 2T, 3T, 4T, 5T, 6T, 1H, 2H, 3H, 4H, 5H, 6H} **c.** {yes, no}

3. Two of the many possibilities are: $S_1 = $ {girl, boy}, $S_2 = $ {student 1, not student 1}

5. **a.** It will be hot or it will rain.
b. It will be hot and it will rain.
c. It will not be hot and it will rain.
d. It will not be hot or it will not rain.

7. **a.** 9 **b.** 18 **c.** 27

9. **a.** $S = $ {TTT, HHH, HTH, THH, HHT, THT, TTH, HTT}
b. {HHH, HTH, THH, TTH}   **c.** 4
d. {TTH, THH, THT, HTH, HHT, HHH}
e. 6   **f.** 4   **g.** 4   **h.** 2

11. There are many possibilities, such as $E = $ {(1, 2)} and $F = $ {(3, 4)}.

13. **a.** {1, 2} **b.** {1} **c.** {1, 2, 3, 4, 5, 6} **d.** {1, 4} **e.** {4, 6}
f. {4, 6, 7, 8}

15. **a.** not necessarily **b.** not necessarily **c.** yes

17. {YYY, YYN, YNY, YNN, NYY, NYN, NNY, NNN}

Exercise Set 5.2

1. **a.** 0 **b.** 1 **c.** 0.01 **d.** 0.99   3. **a.** 0.3 **b.** 0.99

5. **a.** 7:3 **b.** 3:7   7. **a.** 3/10 **b.** 1/5 **c.** 1/2

9. For one sample space, S_1, assume that we can tell the marbles apart. Then $S_1 = $ {R1, R2, R3, W1, W2, B1, B2, B3, B4, B5}. The probability of each outcome is 0.1. $S_2 = $ {R, W, B}; $P(R) = 3/10$, $P(W) = 1/5$, $P(B) = 1/2$.

11. 1/50   13. **a.** 9/19 **b.** 10:9 **c.** 9/19 **d.** 9:10 **e.** 9/19

15. **a.** 0.05, 0.48, 0.21, 0.12, 0.05, 0.03, 0.02 **b.** 1
c. 0.05 **d.** 0.30

17. **a.** 17/30 **b.** 13/30 **c.** 1/6 **d.** 1/10 **e.** 3/20

19.

Average annual salary

Exercise Set 5.3

1. **a.** 12 **b.** 10 **c.** 120 **d.** 120 **e.** 35 **f.** 126

3. 30   5. 720   7. 720
9. 70   11. 120   13. 362,880

15. 311,875,200   17. 15

Exercise Set 5.4

1. **a.** 0.236 **b.** 0.26 **c.** 0.47

3. **a.** 0.21 **b.** 0.29 **c.** 0.57 **d.** 0.43

5. 1/24   7. 0.2   9. 2/5, 3/5

Exercise Set 5.5

1. **a.** 0.12; 0.3; 0.58 **b.** no

3. **a.** 0.5 **b.** 0.3 **c.** 0.7 **d.** 0.1 **e.** 1/3
f. 0.5 **g.** 0.4 **h.** 0.3 **i.** 0.4

5. **a.** 0
b. 1/3 **c.** 1

7. **a.** 1/3
b. 2/3 **c.** 1/6 **d.** 1/6 **e.** 1/3

9. **a.** 1/2 **b.** 1/2 **c.** 1/4 **d.** 1/4 **e.** 1/2

11. **a.** 1/3 **b.** 2/15 **c.** 2/3 **d.** 8/15

13. **a.** 0.66 **b.** 22/53 **c.** 22/35 **d.** 0.34

15. **a.** 0.527 **b.** 6/7 **c.** 1/7 **d.** 0.049 **e.** 466/951
f. 0.042 **g.** 42/527

17. **a.** 5/8 **b.** 1/4 **c.** 4/21 **d.** 3/32 **e.** 11/32

Exercise Set 5.6

1. **a.** no; $(0.5)(0.6) \neq 0.1$ **b.** no; $P(E \cap F) \neq 0$

3. a. 0.5 **b.** 0

5. yes; $P(E|F) = P(E)$

7. a. 1/4 **b.** 1/16 **c.** 1/2 **d.** 3/13 **e.** 17/52 = 0.327

9. no; $P(E \cap F) = 0.042 \neq (0.257)(0.049)$

11. no; $P(E \cap F) = 1/10 \neq P(E)P(F) = (1/3)(13/30)$

Exercise Set 5.7

1. a.

x	$P(x)$
0	0.063
1	0.250
2	0.375
3	0.250
4	0.063

b. discrete

c.

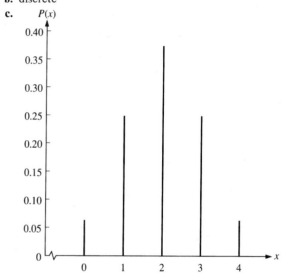

3. $\mu = 2$, $\sigma = 1$ **5.** $\mu_X = 1.5$, $\sigma_X = 0.87$

7. b. 4/55 **c.** 6/55 **d.** 9/11 **e.** 9/55 **f.** 3/55
g. $\mu_X = 7$ **h.** $\sigma_X = 2.45$

9. a. 0.23 **b.** 0.26 **c.** 0.60 **d.** 1.28 **e.** 1.84

11. a.

x	$P(x)$
0	0.044
1	0.087
2	0.128
3	0.234
4	0.297
5	0.155
6	0.030
7	0.025

b. 3.36 **c.** 1.53 **d.** 0.945 **e.** 3.36

13. a. $18,000 **b.** $14,000   **15.** 2.5

Review Exercises

1. a. 0.43 **b.** 0.82 **c.** 0.13

3. a. 0.2 **b.** 0.91

5. a.

x	1	2	3	4	5	6
$P(x)$	1/6	1/6	1/6	1/6	1/6	1/6

b.

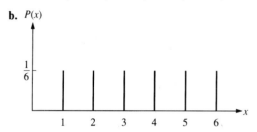

c. 3.5 **d.** 1.71

7. $\mu = \$800$, $\sigma = \$3429.29$

9. a.

x	0	1	2
$P(x)$	1/9	4/9	4/9

b. $\mu_X = 4/3$, $\sigma_X = 2/3$

c.

y	0	1	2
$P(y)$	4/9	4/9	1/9

$\mu_Y = 2/3$, $\sigma_Y = 2/3$

Achievement Test

1. 720

2. a. 6/11 **b.** 14/33 **c.** 8/33 **d.** 8/11 **e.** 4/7
f. 3/10 **g.** no **h.** no

3. a. $S = \{0M, 1M, 2M\}$ **b.** 1 **c.** 3(not counting order, 6(counting order) **d.** 0.3 **e.** 0.6

4. a. 0.33 **b.** 0.62 **c.** 0.29 **d.** 0.91

5. $E(w) = \$0.37$ **6.** $\mu_X = 0.5$, $\sigma_X = 0.61$

CHAPTER 6

Exercise Set 6.1

1. a. The number of ways that eight trials can result in four successes.
   b. The number of ways that six trials can result in no successes.

3. $n = 8$, $p = 0.4$, $0 \leq x \leq 8$; trial: send a student to camp, success: contracting poison ivy.

5. not binomial

7. $n = 30$, $p = 0.4$, $0 \leq x \leq 30$; trial: a student entering college, success: a student graduating

9. a. 120 **b.** 31,824 **c.** 1 **d.** 792 **e.** 1

Exercise Set 6.2

1. a. 0.001 **b.** 0.209 **c.** 0.595 **d.** 0.009

3. a. 0.235 **b.** 0.154 **c.** 0.179

5. a. 0.313 **b.** 0.657 **c.** 0.234 **d.** 0.016

7. a. 0.107 **b.** 0.296 **c.** 0.263 **d.** 0.000 **e.** 0.666

9. a. 0.000 **b.** 0.878 **c.** 0.026 **d.** 0.201

Exercise Set 6.3

1. a. $\mu = 6$, $\sigma^2 = 2.4$, $\sigma = 1.55$ **b.** $\mu = 17$, $\sigma^2 = 2.55$, $\sigma = 1.60$ **c.** $\mu = 12$, $\sigma^2 = 8.4$, $\sigma = 2.90$ **d.** $\mu = 1$, $\sigma^2 = 0.8$, $\sigma = 0.89$

3. $\mu = 1$, $\sigma = 0.99$

5. $\mu = 500$, $\sigma = 15.81$

7. $\mu = 2.7$, $\sigma = 1.64$

9. a.

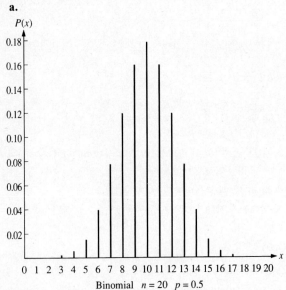

Binomial $n = 20$ $p = 0.5$

b.

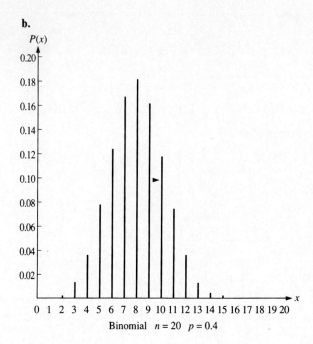

Binomial $n = 20$ $p = 0.4$

c.

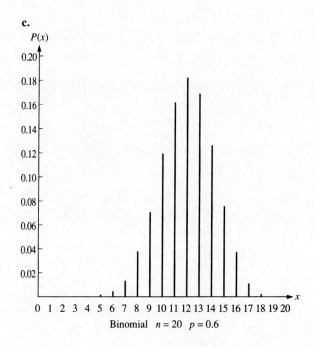

Binomial $n = 20$ $p = 0.6$

d.

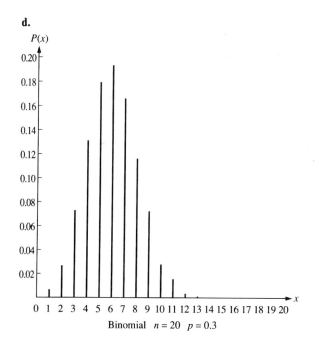

Binomial $n = 20$ $p = 0.3$

e.

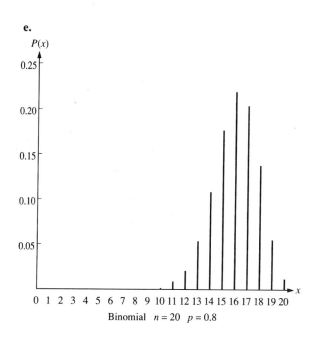

Binomial $n = 20$ $p = 0.8$

f.

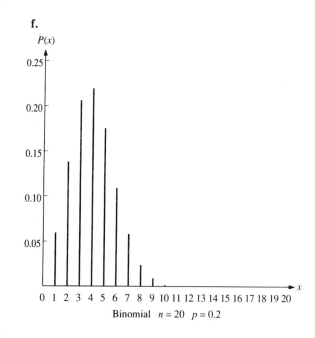

Binomial $n = 20$ $p = 0.2$

Exercise Set 6.4

1. a. 210 **b.** 3,360   **3.** no, $1 + 2 + 0 + 7 \neq 9$

5. a. 0.032 **b.** 0.113 **c.** 0.016 **d.** 0.002

7. a. 0.003 **b.** 0.001 **c.** 0.000

9. a. 0.012 **b.** 0.023 **c.** 0.019

11. a. 0.031 **b.** 0.1 **c.** 0.200

13. a. 0.039 **b.** 0.184

15. 0.043   **17.** 6,930   **19.** 0.375

Exercise Set 6.5

1.

x	$P(x)$
0	$\frac{7}{44}$
1	$\frac{21}{44}$
2	$\frac{7}{22}$
3	$\frac{1}{22}$

3.

x	$P(x)$
0	$\frac{1}{14}$
1	$\frac{8}{21}$
2	$\frac{3}{7}$
3	$\frac{4}{35}$
4	$\frac{1}{210}$

5. a. 0.008 **b.** 0.885 **c.** 0.000

7. a. 0.365 **b.** 0.161 **c.** 0.635

9. a. 0.433 **b.** 0.002

11. 0.063

Exercise Set 6.6

1. a. 0.168 **b.** 0.000 **c.** 0.125 **d.** 0.003

3. a. 0.175 **b.** 0.007 **c.** 0.018 **d.** 0.960

5. a. 0.137 **b.** 0.980 **c.** 0.020 **d.** 0.995

7. a. 0.168 **b.** 0.269 **c.** 0.647 **d.** 0.577

9. a. 0.195 **b.** 0.351 **c.** 0.982 **d.** 0.238

Review Exercises

1. a.

x	$P(x)$
0	0.116
1	0.312
2	0.336
3	0.181
4	0.049
5	0.005

b.

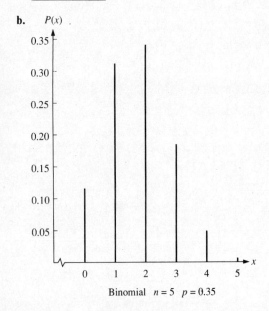

Binomial $n = 5$ $p = 0.35$

c. 1.75 **d.** 1.07 **e.** 0.995

3. a. 0.250 **b.** 0.994 **c.** 0.526

5. 1.53 **7.** $\mu = 8.5, \sigma = 1.13$ **9.** 0.111

11. a. 0.073 **b.** 0.195 **c.** 0.434 **d.** 0.567

13. 0.196 **15. a.** 0.003 **b.** 0.243 **c.** 0.243

Achievement Test

1. a. 0.358 **b.** 0.735 **c.** 0.000 **d.** 0.003

2. a. 0.026 **b.** 0.677 **c.** 1.000 **d.** 0.503

3. a. 2. If a large number of unprepared students guessed at the answers to the 10 questions, the average number of questions answered correctly would be close to 2.

b. 1.6. If a larger number of unprepared students guessed the 10 answers, the variance of the number of questions answered correctly would be close to 1.6.

4. a.

x	$P(x)$
0	0.237
1	0.396
2	0.264
3	0.088
4	0.015
5	0.001

b.

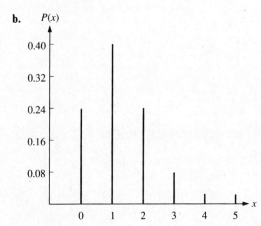

c. $\mu = 1.25, \sigma = 0.97$

5. 0.180 **6.** 0.081 **7.** 0.271 **8.** 4/7

CHAPTER 7

Exercise Set 7.1

1. a. 0 **b.** 1/3 **c.** 5/6 **d.** 7/12 **e.** 1/6

3. a. 1/7 **b.** 6/7 **c.** 6/7 **d.** 0

5. a. 1/4 **b.** 1 **c.** 1/2

7. a. 3/4 **b.** 1/2 **c.** 1 **d.** 1/2

Exercise Set 7.2

1. a. 0.16 **b.** 0.68 **c.** 0.815 **d.** 0.025

3. a. 0.5 **b.** 0.9270 **c.** 0.0933 **d.** 0.0215

5. a. 0.3907 **b.** 0.4904 **c.** 0.9538 **d.** 0.0248

7. a. 0.0277 **b.** 0.9625 **c.** 0.9938

9. a. 1.65 **b.** 1.96 **c.** 1.28 **d.** 2.58

Exercise Set 7.3

1. a. 0.9544 **b.** 0.1587 **c.** 0.9861 **d.** 0.6915 **e.** 0.3551

3. a. 0.0062 **b.** 0.9876 **c.** 0.0013 **d.** 1^-

5. a. 448 **b.** 500 **c.** 552 **d.** 628

7. a. 620 **b.** 98,760 **c.** 130 **d.** 100,000

9. 0.1587

11. 73.33 inches or 6.11 feet

13. a. 69.15 **b.** 30.85 **c.** 30.85 **d.** 81

Exercise Set 7.4

1. a. 0.6563 **b.** 0.6578 **c.** 0.6560

3. a. 0.5438 **b.** 0.2148 **c.** 0.4404 **d.** 0^+

5. a. 0.9732 **b.** 0.7397 **c.** 0.0069 **d.** 0.7257

7. a. 0.0618 **b.** 0.9382

9. 0.9901

Exercise Set 7.5

1. a. 0.0025 **b.** 0.0317 **c.** 0.0462

3. $\mu = 0.1563$, $\sigma^2 = 0.0244$

5. a. 0.3694 **b.** 0.6724 **c.** 0.113

7. 2.56 hours   **9.** 63.92 hours

11. a. 10 minutes **b.** 10 minutes **c.** 0.6501 **d.** 0.3829

13. 0.6357

15. a. 7.33 **b.** 11.09 **c.** 5.55 **d.** 4.09

17. 0.6308   **19.** 4 months

Review Exercises

1. a. 3/8 **b.** 3/4 **c.** 3/8 **d.** 3/8 **e.** 51/80

3. a. 1.04 **b.** −0.25 **c.** 0.25 **d.** 0.84 **e.** −0.52

5. a. 44.8 **b.** 55.2 **c.** 44.8 **d.** 52.5

7. 80.68   **9.** 161.65

11. a. 0.2231 **b.** 0.3834

13. 0.0325, 0.5

15. 0.999^+ or 1^-

Achievement Test

1. a. 5/6 **b.** 5/6 **c.** 2/3

2. a. 0.6950 **b.** 0.8742 **c.** 0.2095 **d.** 0.8729 **e.** 0.4052

3. a. 0.52 **b.** −0.67 **c.** −0.25

4. a. 0.0099 **b.** 0.9082 **c.** 0.0376 **d.** 0.0437

5. a. 16.16 **b.** 20.75 **c.** 22.31   **6. a.** 44.02 **b.** 38.5

7. 0.62   **8.** 0.8925

9. a. 1.000, 1^-, or 0.9999 **b.** 0.0025 **c.** 0.002 **d.** 0.23

CHAPTER 8

Exercise Set 8.1

1. 13: Robert Moon, 04: Ed Doe, 01: Mike Able, 20: Maud Tuck, 12: James Lum, 17: Rick Quest, 07: Pete Gum, 02: Mary Baker, 10: Helen Jewel, 18: Bart Rat

3. Let an even number denote a head and an odd number a tail: TTHHH TTHHH.

5. Assign a two-digit number from 00 to 96 to each of the 97 customers. Then at some random starting point in the table, move horizontally to the right until 15 customers have been chosen.

7.

Digit	f
0	18
1	27
2	34
3	35
4	24
5	21
6	27
7	15
8	27
9	22

We expect each digit to occur 25 times.

9.

Digit	Frequency	Expected Frequency	Sampling Error
0	18	25	−7
1	27	25	2
2	34	25	9
3	35	25	10
4	24	25	−1
5	21	25	−4
6	27	25	2
7	15	25	−10
8	27	25	2
9	22	25	−3
			0

11.

Digit	Frequency	Expected Frequency	Sampling Error
1	1	2	−1
2	3	2	1
3	2	2	0
4	1	2	−1
5	0	2	−2
6	5	2	3
			0

13. a. teachers 057, 013, 044, 193, 112.
  b. 057: teacher 057, 629: teacher 29, 013: teacher 13, 843: teacher 43, 840: teacher 40

15.

Ordered Samples	$\bar{x}$	Sampling Error
{0, 0}	0	$0 - 3 = -3$
{0, 2}	1	$1 - 3 = -2$
{0, 4}	2	$2 - 3 = -1$
{0, 6}	3	$3 - 3 = 0$
{2, 0}	1	$1 - 3 = -2$
{2, 2}	2	$2 - 3 = -1$
{2, 4}	3	$3 - 3 = 0$
{2, 6}	4	$4 - 3 = 1$
{4, 0}	2	$2 - 3 = -1$
{4, 2}	3	$3 - 3 = 0$
{4, 4}	4	$4 - 3 = 1$
{4, 6}	5	$5 - 3 = 2$
{6, 0}	3	$3 - 3 = 0$
{6, 2}	4	$4 - 3 = 1$
{6, 4}	5	$5 - 3 = 2$
{6, 6}	6	$6 - 3 = 3$
		0

Since the average of the sampling errors is 0, $\bar{x}$ is an unbiased estimator of μ.

17. a. no; not random if we select only those people who wore glasses
  b. No; this procedure only selects gym students or students having business in the gym.
  c. yes

19. not unless the sample was arrived at by using a randomization procedure

Exercise Set 8.2

1. a. The sampling distribution of the median is the distribution of all possible values of the median computed from samples of the same size.
  b. The sampling distribution of the variance is the distribution of all possible values of the samples variance computed from samples of the same size.

3. a. formula with correction factor
  b. formula without correction factor
  c. formula without correction factor

5. $\mu_{\bar{x}}$ is the mean of the sampling distribution of the sample mean; μ is the mean of the sampled population.

7. The standard error of the mean changes from 0.75 to 0.5. $\sigma_{\bar{x}}$ gets smaller and approaches 0 as the sample size becomes large.

9. a. 6.45 **b.** 3.62   **11.** $\mu_{\bar{x}} = 3.5$, $\sigma_{\bar{x}} = 2.09$

13. $\mu_{\bar{x}} = 3.5$, $\sigma_{\bar{x}} = 2.09$

15. The sampling distribution of the mode only makes sense when a unique mode exists for every sample drawn from a population; in practice, this is never the case.

17. a.

$P(\bar{x})$	0.16	0.16	0.36	0.16	0.16
$\bar{x}$	12	13.5	15	16.5	18

  b. $15   **c.** $1.90

Exercise Set 8.3

1. a. 0.0212 **b.** 0.7912 **c.** 0.2709 **d.** 0.8438 **e.** 0.9925

3. 0.005   **5.** 0.95   **7.** 0.025   **9.** 12   **11.** −2.14

13. a. 2.718 **b.** 1.796 **15.** 0.975

17. a. 0.8531 **b.** 0.6448 **c.** 0.9826

Exercise Set 8.4

1. a. 6 **b.** 0.77 **c.** normal

3. a. $\mu = 4$, $\sigma = 2.24$ **b.** $\mu_{\bar{x}} = 4$, $\sigma_{\bar{x}} = 1.58$

  c.

$\bar{x}$	1	2	3	4	5	6	7
f	1	2	3	4	3	2	1

  d. $\mu_{\bar{x}} = \mu = 4$; $\sigma_{\bar{x}} \approx \sigma/\sqrt{2} = 1.58$; approximately normal

5. a. $\mu = 4.5$, $\sigma = 3.35$ **b.** $\mu_{\bar{x}} = 4.5$, $\sigma_{\bar{x}} = 1.94$

7. a. 0.7794 **b.** 0.8030 **c.** 0.7422

9. The population is normal. For samples of size 1, the sampling distribution of the mean is identically equal to the sampled population.

11. a.

Sample	Sum	Sum	f	$(Sum - 14)^2$	Sum · f
{1, 1}	2	2	1	144	2
{1, 7}	8	8	2	72	16
{1, 13}	14	14	3	0	42
{7, 7}	14	20	2	72	40

Sample	Sum	Sum	f	$(\text{Sum} - 14)^2$	$\text{Sum} \cdot f$
{7, 1}	8	26	1	144	26
{7, 13}	20			432	126
{13, 13}	26				
{13, 1}	14				
{13, 7}	20				

b. $\mu_{\text{sum}} = 126/9 = 14$, $n\mu = (2)(7) = 14$

c. $\sigma_{\text{sum}} = \sqrt{432/9} = 4\sqrt{3} = 6.93$,
$\sqrt{n}\sigma = (\sqrt{2})(4.898) = 6.93$

13. a. 0.5089 **b.** 0.8078 **c.** 0.9120 **d.** 0.3300

15. a. 0.1190 **b.** 0.7620 **c.** 0.1190

17. a. $\mu_{\bar{x}} = 75$, $\sigma_{\bar{x}} = 1.58$ **b.** $238.8 \approx 239$
c. $291.39 \approx 291$

Exercise Set 8.5

1. a. approximately normal, $\mu_{\hat{p}} = 0.05$, $\sigma_{\hat{p}} = 0.03$
b. 0.8340 **c.** 0.8506 **d.** 0.6255

3. a. 0.2877 **b.** 0.7123 **c.** 0.9875

5. a. 95.44 **b.** 6.68

7. a. 0.0245 **b.** 0.3409 **c.** 0.6827

9. a. 0.0194 **b.** (0.73, 0.77)

Review Exercises

1. a. yes **b.** no; all possible samples of size 5 are not available using this method

3. We cannot tell; we must know the procedure used.

5. a. 4 **b.** 1.29

c.

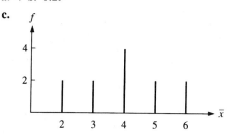

7. a.

Median	f	Sampling Error	$f \cdot$ Sampling Error
1	10	-2	-20
2	22	-1	-22
3	22	0	0
6	10	3	30
	64		-12

b.

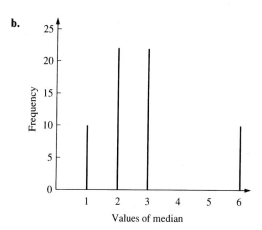

c. -0.1875 **d.** 2.8125 **e.** 1.5297 **f.** 1.5297
g. The results are equal.

9. a. 0.9554 **b.** 0.8187 **c.** 0.9977 **11.** 0.9412

13. a. 0.3108 **b.** 0.0703 **c.** 0.5793 **15.** 1^-

Achievement Test

1. -5

2. Assign each letter a two-digit number from 01 to 26. Begin at an arbitrary point in the table and draw pairs until the three letters are chosen. If a two-digit number greater than 26 is chosen, discard it, choose the next pair, and so on.

3. 1.5 **4. a.** 2.53 **b.** 9 **c.** 2.11

5. a. 0.7492 **b.** 0.5 **c.** 40.3

6. a. 0.1056 **b.** 0.4649

7. -3.16 **8.** 0.65 **9. a.** 4 **b.** 0.7071

CHAPTER 9

Exercise Set 9.1

1. a. 79.94% **b.** 85.02% **c.** 99.02% **d.** 99.42%

3. a. 1.88 **b.** 1.65 **c.** 1.56 **5.** 2.145 **7.** 2.462

9. 0.55

11. 6.20. If $\bar{x}$ is used to estimate μ, we can be 99% confident that μ is within 6.20 of $\bar{x} = 98.2$.

13. 77.38%

15. 1239.7

Exercise Set 9.2

1. a. (27.56, 29.04), $w = 1.48$ **b.** (27.42, 29.18), $w = 1.76$ **c.** (27.14, 29.46), $w = 2.32$
d. (26.96, 29.64), $w = 2.68$

3. a. 1.796 **b.** 0.30 **c.** 0.85 **5.** (70.75, 84.85)

7. ($26.91, $30.01) **9.** (63.20, 68.20)

11. (155.41, 171.59) **13.** (0.98, 1.04)

Exercise Set 9.3

1. a. (0.21, 0.29) **b.** (0.22, 0.28)
c. (0.58, 0.62) **d.** (0.34, 0.46) **e.** (0.06, 0.38)

3. a. 0.058 **b.** (0.45, 0.57)

5. (0.63, 0.70); $E = 0.036$

7. (0.024, 0.056) **9.** (0.47, 0.79)

11. (0.46, 0.60) **13.** (0.41, 0.51)

15. (0.83, 0.91) **17.** (0.58, 0.82)

19. (0.35, 0.61)

Exercise Set 9.4

1. a. 2090 **b.** 2401 **3.** 4096 **5.** 24

7. 7657 **9.** 606

Exercise Set 9.5

1. a. 5.14, 34.27 **b.** 12.40, 39.36

3. a. σ^2: (0.015, 0.039); σ: (0.12, 0.20)
b. σ^2: (478.85, 1817.37); σ: (21.88, 42.63)
c. σ^2: (0.72, 1.57); σ: (0.85, 1.25)
d. σ^2: (480,223.34, 1,172,672); σ: (692.98, 1,082.90)

5. a. 1.61 **b.** (1.37, 1.98) **7.** 6.09

9. σ^2: (0.020, 0.11); σ: (0.14, 0.34)

11. 0.99 **13.** (1.07, 5.14) **15.** (0.0018, 0.0055)

Review Exercises

1. a. (0.55, 0.57) **b.** $n = 9,466$ **3.** 666

5. (0.35, 0.44), $E = 0.049$ **7.** (0.089, 0.50)

9. (57.76, 66.24) **11.** (3.49, 6.29)

13. a. 0.026 **b.** (7.99, 8.05)

15. a. (0.36, 0.46) **b.** 1,068

Achievement Test

1. a. 1.740 **b.** 42.98

2. a. 4.26 **b.** −9.62 **c.** 540

3. (16.47, 17.93) **4.** (0.12, 0.22)

5. (0.97, 2.42) **6.** 547 **7.** 80

8. a. 0.60 **b.** 0.85 **c.** 0.80 **d.** 0.95 **e.** 0.98

CHAPTER 10

Exercise Set 10.1

1. a. B **b.** A **c.** B **d.** B **3.** a, d

5. a. H_0: Effectiveness of new drug is less than or equal to the effectiveness of the old drug.
H_1: Effectiveness of new drug is greater than the effectiveness of the old drug.
Decision: reject H_0, Type I error.
b. H_0: School 1 is as effective as school 2.
H_1: School 1 is not as effective as school 2.
Decision: do not reject H_0, good decision.

7. a. two-tailed **b.** two-tailed **c.** two-tailed

9. a. The drug contains 5 mg of codeine.
b. The drug may be ineffective for lessor amounts or dangerous for greater amounts.
c. Type I error: concluding the drug does not contain 5 mg of codeine when it does.
Type II error: concluding that the drug contains 5 mg of codeine when it doesn't.
A Type II error is more serious; leading to a possible overdose.

11. $1 - \alpha$ is the probability that the null hypothesis is not rejected when true, whereas $1 - \beta$ is the probability that the null hypothesis will be rejected when false.

13. a. Deciding that smoking would be harmful when it really is not harmful.
b. Deciding that smoking is not harmful when it is harmful.
c. Deciding that smoking is harmful when it is harmful or deciding that smoking is not harmful when it is not harmful are good decisions.
A type II error is the most serious.

15. a. Type I: deciding that the discovery method is not better than the expository method when in reality the opposite is true
Type II: deciding that the discovery method is better than the expository method when in reality the opposite is true
b. Type I: deciding that more than 2% of machines are defective when in reality at most 2% are defective
Type II: deciding that at most 2% of machines are defective when in reality more than 2% are defective
c. Type I: deciding that less than 95% of Americans are against war when in reality at least 95% of Americans are against war.

Type II: deciding that at least 95% of Americans are against war when in reality less than 95% are against war.

17. a. false **b.** false **c.** true

19. Because there is a chance that H_0 may be false.

21. a. incorrect, Type I **b.** incorrect, Type II **c.** correct **d.** correct

Exercise Set 10.2

1. a. 479.38 **b.** all $\bar{x} < 479.38$ **c.** $H_0: \mu \geq 500$ **d.** $H_1: \mu < 500$ **e.** $\bar{x}$, 480 **f.** 0.05 **g.** fail to reject H_0 **h.** Type II

3. a. 30,140.25 **b.** All $\bar{x} > 30,140.25$. **c.** $H_0: \mu \leq 30,000$. **d.** $H_1: \mu > 30,000$. **e.** $\bar{x}$, 30,200 **f.** 0.05 **g.** reject H_0 **h.** Type I

Exercise Set 10.3

1. a. $H_0: \mu \geq 95$. **b.** $H_1: \mu < 95$. **c.** 0.05, one-tail test **d.** distribution of the mean **e.** -1.94 **f.** -1.65 **g.** Reject H_0, conclude that 92.25 is significantly less than 95. **h.** Type I error possible, 0.05 is the probability of a type I error. **i.** 0.0262

3. a. $H_0: \mu \geq 10$. **b.** $H_1: \mu < 10$. **c.** 0.01, one-tail test **d.** distribution of the mean **e.** -2.15 **f.** -2.264 **g.** fail to reject H_0 **h.** Type II error possible, probability is unknown. **i.** $0.01 < p\text{-value} < 0.025$

5. a. $H_0: \mu = 70$. **b.** $H_1: \mu \neq 70$. **c.** 0.05, two-tail test **d.** distribution of the mean **e.** 0.67 **f.** 2.262 **g.** fail to reject H_0 **h.** Type II, probability is unknown. **i.** $p\text{-value} > 0.10$

7. a. $H_0: \mu \geq 84$. **b.** $H_1: \mu < 84$. **c.** 0.05, one-tailed test **d.** distribution of the mean **e.** -1.86 **f.** -1.65 **g.** reject H_0 **h.** 0.05 **i.** 0.0314

9. a. $H_0: \mu = 14$. **b.** $H_1: \mu \neq 14$. **c.** 0.05, two-tail test **d.** distribution of the mean **e.** 0.70 **f.** 2.365 **g.** fail to reject H_0 **h.** Type II error, probability unknown. **i.** $p\text{-value} > 0.20$

11. $H_0: \mu \leq 7$.
$H_1: \mu > 7$.
$z = 0.37, z_{0.05} = 1.65$
Since $z = 0.37 < 1.65$, we fail to reject H_0. Hence, we have no evidence that the average stay for surgery patients exceeds seven days.

13. (13.29, 15.31); the interval contains 14. This is consistent with the result of exercise 9, since we failed to reject $H_0: \mu = 14$.

Exercise Set 10.4

1. a. $z = -2.24, \pm z_{0.005} = \pm 2.58$; fail to reject H_0
b. $z = 2, \pm z_{0.005} = \pm 2.58$; fail to reject H_0
c. $z = 2.89, z_{0.01} = 2.33$; reject H_0
d. $z = 0.77, z_{0.01} = 2.33$; fail to reject H_0

3. a. $\chi^2 = 13.5, \chi^2_{0.95}(27) = 16.151$; reject H_0
b. $\chi^2 = 30, \chi^2_{0.01}(24) = 42.980$; fail to reject H_0
c. $\chi^2 = 28.8, \chi^2_{0.025}(36) = 54.437, \chi^2_{0.975}(36) = 21.336$; fail to reject H_0

5. $\chi^2 = 3.14, \chi^2_{0.95}(10) = 3.940$; reject H_0

7. $H_0: p \leq 0.20$; $H_1: p > 0.20$; $z = 2.795$; $z_{0.01} = 2.33$; $p\text{-value} = 0.0026$; reject H_0, therefore more than 20% watch football

9. $z = -1.86, -z_{0.05} = -1.65, p\text{-value} = 0.0314$; reject H_0

11. $H_0: p = 0.60, H_1: p \neq 0.60, z = -1.63, \pm z_{0.025} = \pm 1.96, p\text{-value} = 0.1032$; fail to reject H_0

13. $H_0: \sigma^2 \leq 5, H_1: \sigma^2 > 5, \chi^2 = 12.54, \chi^2_{0.05}(11) = 19.675$; fail to reject H_0

15. (0.42, 0.62) We can not establish that $\sigma^2 \neq 0.60$.

17. a. (0.013, 0.047) **b.** $H_0: \sigma^2 \leq 0.01, H_1: \sigma^2 > 0.01, \chi^2 = 42.12, \chi^2_{0.05}(19) = 30.144$; reject H_0

Review Exercises

1. $H_0: \mu \leq 24, H_1: \mu > 24, z = 3.94, z_{0.05} = 1.65$; decision: reject H_0, $p\text{-value} = 0^+$

3. $H_0: p \geq 0.62, H_1: p < 0.62, z = -1.84, z_{0.05} = -1.65$, decision: reject H_0, $p\text{-value} = 0.0329$

5. $H_0: \mu = 140, H_1: \mu \neq 140, t = 1.38, \pm t_{0.05} = \pm 2.447$; fail to reject H_0

7. $H_0: \sigma^2 \leq 0.05, H_1: \sigma^2 > 0.05, \chi^2 = 33.6, \chi^2_{0.05}(24) = 36.415$; fail to reject H_0

9. $H_0: p \leq 0.6, H_1: p > 0.6, z = 2.91, z_{0.05} = 1.65$; decision: reject H_0

11. $H_0: \mu = 16, H_1: \mu \neq 16, t = -1.21, \pm t_{0.005} = \pm 4.032$; fail to reject H_0

Achievement Test

1. $H_0: \mu = 71, H_1: \mu \neq 71$
a. $t = 1.15$
b. $\pm t_{0.025} = \pm 2.201$
c. fail to reject H_0
d. $p\text{-value} > 0.2$

2. $H_0: p \geq 0.85, H_1: p < 0.85$
a. $z = -1.08$

b. $\pm z_{0.01} = \pm 2.33$
c. fail to reject H_0
d. p-value $= 0.1401$

3. a. 23.75 **b.** 30.144 **c.** fail to reject H_0

4. $H_0: \mu \geq 60{,}000$, $H_1: \mu < 60{,}000$, $z = -3.81$, $z_{0.05} = 1.65$; decision: reject H_0. Thus, the data indicate that $\mu < 60{,}000$ miles.

5. $H_0: \sigma \geq 1{,}200$, $H_1: \sigma < 1{,}200$, $\chi^2 = 22.12$, $\chi^2_{0.01}(34) = 17.789$; fail to reject H_0

CHAPTER 11

Exercise Set 11.1

1. dependent, since the half-acre plots used similar soils

3. dependent, since baits were related to sites and placed in pairs

5. Do not place the bait in pairs.

7. a. Waterproof one of each pair.
b. Randomly select five pairs to waterproof, and compare with the other five pairs not waterproofed.

9. dependent, since pairs of rats competed

Exercise Set 11.2

1. 0.0793

3. a. $H_0: \mu_1 - \mu_2 = 0$, $H_1: \mu_1 - \mu_2 \neq 0$, $z = -1.44$, $\pm z_{0.005} = \pm 2.58$; fail to reject H_0 **b.** $(-0.16, 0.56)$

5. a. $H_0: \mu_1 - \mu_2 = 0$, $H_1: \mu_1 - \mu_2 \neq 0$, $z = 1.98$, $\pm z_{0.005} = \pm 2.58$; fail to reject H_0 **b.** $(-0.42, 3.22)$

7. $z_{0.01} = 2.50$, critical value $= 2.33$, p-value $= 0.0062$; reject H_0

9. $H_0: \mu_1 - \mu_2 \leq 0$, $H_1: \mu_1 - \mu_2 > 0$, $z = 2.67$, $z_{0.05} = 1.65$, p-value $= 0.0038$; reject H_0

Exercise Set 11.3

1. a. 0.2 **b.** 0.061 **c.** -0.66

3. a. $H_0: p_1 - p_2 = 0$, $H_1: p_1 - p_2 \neq 0$, $z = 0.53$, $\pm z_{0.025} = \pm 1.96$, p-value $= 0.5962$; fail to reject H_0
b. $(-0.054, 0.094)$ The interval contains zero; fail to reject H_0.

5. a. $(-0.145, 0.185)$
b. $H_0: p_1 - p_2 = 0$, $H_1: p_1 - p_2 \neq 0$, $z = 0.24$, $\pm z_{0.025} = \pm 1.96$; fail to reject H_0, p-value $= 0.8104$

7. a. $(-0.1082, 0.1442)$

b. $z = 0.37$, critical values $= \pm 2.58$; fail to reject H_0, p-value $= 0.7114$

9. $z = 2.09$, $\pm z_{0.025} = \pm 1.96$; reject H_0, confidence interval is $(0.0053, 0.142)$

Exercise Set 11.4

1. $F = 1.20$ df $= (19, 14)$

3. a. 0.500 **b.** 0.392 **c.** 0.206

5. a. $(1.09, 5.13)$ **b.** $(0.1666, 1.0845)$

7. a. $F = 1.70$, df $= (25, 15)$, $F_{0.05}(25, 15) \approx 2.33$; fail to reject H_0
b. $F = 2.10$, df $= (10, 20)$, $F_{0.05}(10, 20) = 2.35$; fail to reject H_0

9. a. $F = 1.65$, df $= (30, 24)$, $F_{0.01}(30, 24) = 2.58$; fail to reject H_0
b. $F = 2.64$, df $= (20, 23)$, $F_{0.01}(20, 23) = 2.78$; fail to reject H_0

11. a. $F = 1.78$, df $= (20, 30)$, $F_{0.05}(20, 30) = 1.93$; fail to reject H_0
b. $(0.93, 1.83)$

13. a. $F = 1.78$, df $= (20, 19)$, $F_{0.05}(20, 19) = 2.15$; fail to reject H_0
b. $(0.91, 1.94)$
c. We cannot recommend that the new machine is superior to the old machine.

Exercise Set 11.5

1. a. $s_p = 6.9602$, $t = 0.96$, $\pm t_{0.025} = \pm 2.101$, df $= 18$; fail to reject H_0
b. $(-3.57, 9.57)$

3. a. $(-6.73, 0.73)$
b. $s_p = 4.1806$, $t = -1.68$, $t_{0.025} = 2.086$, df $= 20$; fail to reject H_0

5. a. $s_p = 7.908$, $t = 1.50$, $t_{0.005} = 2.831$, df $= 21$; fail to reject H_0
b. $(-14.417, 4.417)$

7. a. $s_p = 4.61$, $t = 3.71$, $t_{0.05} = 1.734$, df $= 18$
b. $(3.38, 12.22)$

9. a. $(-0.53, 2.93)$
b. $s_p = 2.075$, $t = 1.44$, $t_{0.025} = 2.069$, df $= 25$; fail to reject H_0

11. $(1.03, 6.37)$

Exercise Set 11.6

1. a. 1.22 **b.** 3   **3.** $t = 1.20$, df $= 9$

5. a. $t = 3.04$, $t_{0.05} = 2.015$, df $= 5$; reject H_0; the program is effective.
b. $(0.62, 7.38)$

7. a. $\bar{d} = 2.5$, $s_d = 2.9807$, $t = 3.14$, $t_{0.05} = 1.771$, df $= 13$; reject H_0
b. $(0.78, 4.22)$

Review Exercises

1. a. $z = 2.58$, $z_{0.025} = 1.96$; reject H_0, p-value $= 0.0098$
b. $(2.02, 14.82)$

3. $(-0.13, 0.20)$; no difference between population proportions since the interval contains 0

5. $z = 4.82$, $z_{0.025} = 1.96$; reject H_0

7. $z = 1.29$, $z_{0.01} = 2.33$; fail to reject H_0, p-value $= 0.0985$

9. $(0.82, 2.48)$

11. $z = 3.7$, $z_{0.01} = 2.58$; reject H_0. This does not mean that $\mu_b < \mu_a$, only that $\mu_b \neq \mu_a$.

13. a. $t = 1.74$, $t_{0.025} = 2.201$, df $= 11$; fail to reject H_0; $0.10 <$ p-value < 0.20
b. $(-2.21, 19.05)$

15. a. $t = -0.68$, $t_{0.005} = 3.169$, df $= 10$; fail to reject H_0
b. $(-3.38, 2.18)$

17. $(-3.24, -1.16)$ The hypothesis test in exercise 16 was one tailed at $\alpha = 0.05$. It is not comparable to constructing a two sided 95% confidence interval.

19. $z = 3.87$, $\pm z_{0.025} = \pm 1.96$; reject H_0

21. $F = 4.39$, $F_{0.05}(5, 6) = 4.39$, $F = 1.20$; fail to reject H_0. The population variances may be considered equal, so the t test is appropriate for exercise 18.

Achievement Test

1. a. $F = 1.33$, $F_{0.05}(30, 30) = 1.84$; fail to reject H_0
b. $z = 2.83$, $z_{0.05} = 1.65$; reject H_0 **c.** $(1.29, 4.91)$

2. a. $\hat{p} = 0.56$ **b.** $z = 0.32$ **c.** $z_{0.05} = 1.65$; fail to reject H_0 **d.** $(-0.16, 0.22)$

3. a. $(0.72, 2.44)$ **b.** $(0.64, 1.18)$

4. a. 2.83 **b.** 2.12 **c.** $z = 2.47$; reject H_0 **d.** p-value $= 0.0068$ **e.** $(2.33, 11.67)$

5. a. 4.31 **b.** $F = 1.33$, $F_{0.05}(14, 14) = 2.48$; fail to reject the hypothesis of equal variances. **c.** $t = 1.97$, $t_{0.05} = 1.701$, df $= 28$; reject H_0 **d.** $(0.42, 5.78)$
e. $(-5.78, -0.42)$

CHAPTER 12

Exercise Set 12.2

1. a. df $= 5$, $\chi^2_{0.05}(5) = 11.070$ **b.** df $= 20$, $\chi^2_{0.01}(20) = 37.566$

3. a. $z = 1.12$, $\pm z_{0.025} = \pm 1.96$; fail to reject H_0
b. $\chi^2 = 1.25$, $\chi^2_{0.05}(1) = 3.841$; fail to reject H_0
c. $1.12^2 \approx 1.25$

5. $\chi^2 = 17.449$, $\chi^2_{0.05}(2) = 5.991$; reject H_0; we can conclude that the proportions differ.

7. $\chi^2 = 64.7927$, $\chi^2_{0.01}(3) = 11.345$; reject H_0; we can conclude that the proportions differ.

9. $\chi^2 = 2.12193$, $\chi^2_{0.05}(3) = 7.815$; fail to reject H_0

11. $\chi^2 = 0.80324$, $\chi^2_{0.05}(3) = 7.815$; fail to reject H_0

Exercise Set 12.3

1. $\chi^2 = 1.22$, $\chi^2_{0.01}(4) = 13.277$; fail to reject H_0

3. $\chi^2 = 26.6$, $\chi^2_{0.05}(9) = 16.919$; we can conclude that the frequencies differ.

5. $\chi^2 = 32.6$, $\chi^2_{0.05}(4) = 9.488$; we can conclude that the color preferences differ.

7. $\chi^2 = 11.9833$, $\chi^2_{0.05}(3) = 7.815$; we can conclude that they do not follow the specified binomial distribution.

9. $\chi^2 = 6.8$, $\chi^2_{0.05}(9) = 16.919$; fail to reject H_0; the digits may follow with equal frequencies.

11. $\chi^2 = 19.2$, $\chi^2_{0.05}(9) = 16.919$; reject H_0; we can conclude that they do not follow with equal frequencies.

Exercise Set 12.4

1. $\chi^2 = 0.1532$, $\chi^2_{0.05}(2) = 5.991$; fail to reject H_0

3. $\chi^2 = 16.8107$, $\chi^2_{0.05}(4) = 9.488$; reject H_0; we can conclude that interest and ability are related.

5. $\chi^2 = 19.3322$, $\chi^2_{0.01}(6) = 16.812$; reject H_0; we can conclude that social status and high school program are dependent.

7. $\chi^2 = 30.1484$, $\chi^2_{0.05}(10) = 18.307$; reject H_0; we can conclude that cigarette smoking and systolic blood pressure are related.

9. $\chi^2 = 12.3565$, $\chi^2_{0.05}(2) = 5.991$; reject H_0; we can conclude that student reponses are related to grade level.

Review Exercises

1. $\chi^2 = 9.5$, $\chi^2_{0.05}(4) = 9.488$; reject H_0; we can conclude that there are differences in the preferences.

3. $\chi^2 = 8.82$, $\chi^2_{0.05}(1) = 3.841$; reject H_0; we can conclude that the efficacy of the drug differs from 85%.

5. $\chi^2 = 7.6$, $\chi^2_{0.05}(5) = 11.07$; fail to reject H_0; there is no evidence that the die is biased.

7. $\chi^2 = 0.92$, $\chi^2_{0.05}(1) = 3.841$; fail to reject H_0; we cannot conclude that the percentages differ.

9. $\chi^2 = 5.33$, $\chi^2_{0.05}(2) = 5.991$; fail to reject H_0; we cannot conclude that the percentages have changed.

11. $\chi^2 = 1.38$, $\chi^2_{0.01}(3) = 11.345$; fail to reject H_0; we can not conclude that the proportions differ.

13. $\chi^2 = 35.25$, $\chi^2_{0.05}(2) = 5.991$; reject H_0; we can conclude that social class and newspaper subscriptions are related.

15. $\chi^2 = 157.39$, $\chi^2_{0.05}(2) = 5.991$; reject H_0; the data indicate that the categories are not homogeneous.

Achievement Test

1. $z = 3.87$

2. **a.** $P_A \neq 0.9$ **b.** 129.60 **c.** 6.635 **d.** reject H_0

3. **a.** 8.63 **b.** 13.277 **c.** fail to reject H_0

4. **a.** 4.74 **b.** 5.991 **c.** fail to reject H_0; the variables may be independent.

CHAPTER 13

Exercise Set 13.1

1. **a.** $t = 0.8866$, $t_{0.025} = 2.306$, df = 8; fail to reject H_0
   b. $s^2_b = 21.5$, $s^2_w = 16.9$, $F = 0.786$, $F_{0.05}(1, 8) = 5.32$; fail to reject H_0
   c. $t^2 = (0.8866)^2 = 0.7860$

3. $s^2_b = 80.6$, $s^2_w = 56.68$, $F = 1.42$, $F_{0.05}(4, 60) = 2.53$; fail to reject H_0

5. $s^2_w = 85.48$, $s^2_b = 17.4$, $F = 0.20$, $F_{0.05}(3, 16) = 3.24$; fail to reject H_0; there is no evidence that the typing speeds of the four secretaries differ.

7. $s^2_w = 7.71$, $s^2_b = 29.19$, $F = 3.78$, $F_{0.01}(2, 18) = 6.01$; fail to reject H_0; there aren't any differences in the mean completion times for the three disciplines.

9. $s^2_w = 0.0000141$, $s^2_b = 0.000701$, $F = 49.72$, $F_{0.05}(3, 16) = 3.24$; reject H_0; there are differences among the average contaminant levels for the four utility plants.

Exercise Set 13.2

1.

Source	SS	df	MS	F
Between samples	16.9	1	16.9	0.79
Within samples	172	8	21.5	
Total	188.9	9		

3.

Source	SS	df	MS	F
Between samples	52.2	3	17.4	0.20
Within samples	1367.6	16	85.5	
Total	1419.8	19		

5. SST = 13,658.38, SSB = 335.54, SSW = 13,322.84, MSB = 167.77, MSW = 360.08, $F = 0.47$, $F_{0.05}(2, 37) = 4.08$; fail to reject H_0

7. SST = 3722.65, SSB = 321.85, SSW = 3400.8, MSB = 80.46, MSW = 56.68, $F = 1.42$, $F_{0.05}(4, 60) = 2.53$, fail to reject H_0

Exercise Set 13.3

1. **a.** SST = 55,275, SSB = 39,875, SSW = 15,400, MSW = 481.25, MSB = 13,291.67, $F = 27.62$, $F_{0.05}(3, 32) \approx 2.92$; reject H_0; the four diets produce different effects on blood glucose readings.
   b. CD = 29.16, A $\neq$ D, A $\neq$ C, B $\neq$ D, B $\neq$ C
   c. B–A: $(-22.49, 35.83)$; C–A: $(50.84, 109.16)$; D–A: $(25.28, 83.6)$; C–D: $(-3.6, 54.72)$; D–B: $(18.61, 76.93)$; C–B: $(44.17, 102.49)$
   d. $\hat{\omega}^2 = 0.69$

3. **a.** SST = 540.21, SSB = 366.55, SSW = 173.66, MSB = 183.28, MSW = 19.30, $F = 9.50$, $F_{0.05}(2, 9) = 4.26$; reject H_0; the mean rates of penetration differ for the three drill bits.
   b. CD = 9.10, DB3 $\neq$ DB2
   c. DB3–DB1: $(-2.85, 15.35)$; DB3–DB2: $(4.42, 22.62)$; DB1–DB2: $(-1.81, 16.37)$
   d. $\hat{\omega}^2 = 0.59$

5. **a.** SST = 400.37, SSB = 287.15, SSW = 113.22, MSB = 143.57, MSW = 6.29, $F = 22.83$, $F_{0.05}(2, 18) = 3.55$; reject H_0; there are differences in yields for the three fertilizers.
   b. C–A: CD = 3.68, C $\neq$ A; C–B: CD = 3.58; B–A: CD = 3.43, B $\neq$ A **c.** $\hat{\omega}^2 = 0.68$

Exercise Set 13.4

1.

Source	DF	SS	MS	F
Block	3	60.33	20.11	8.24
Treat	2	0.67	0.33	0.14
Error	6	14.67	2.44	
Total	11	75.67		

3.

Source	DF	SS	MS	F
Brand	2	1.620	0.810	3.99
Lab	3	0.740	0.247	1.22
Error	6	1.220	0.203	
Total	11	3.580		

5.

Source	DF	SS	MS	F
Employ	4	33.73	8.43	1.55
Hour	2	62.53	31.27	5.76*
Error	8	43.47	5.43	
Total	14	139.73		

$*p < 0.05$

$F_{0.05}(4, 8) = 3.84$, $F_{0.05}(2, 8) = 4.46$
$CR = 4.45$, $A \neq N$

7.

Source	DF	SS	MS	F
Crime	6	645.24	107.54	12.49
Process	2	172.67	86.33	10.33*
Error	12	103.33	8.61	
Total	20	921.24		

$*p < 0.05$

$F_{0.05}(6, 12) = 3.00$, $F_{0.05}(2, 12) = 3.89$
$CD = 4.36$, $TJury \neq GP$

9. a.

Source	DF	SS	MS	F
Block	2	197.6	98.8	8.30*
Fert.	3	218.2	72.7	6.11*
Error	6	71.4	11.9	
Total	11	487.2		

$*p < 0.05$

$F_{0.05}(2, 6) = 5.14$, $F_{0.05}(3, 6) = 4.76$

b. $CD = 8.03$, $B1 \neq B3$, $\hat{\omega}^2 = 0.35$
c. $CD = 10.87$; no differences in fertilizers are detected; $\hat{\omega}^2 = 0.39$

Exercise Set 13.5

1. a.

Source	DF	SS	MS	F
Factor A	1	533	533	2.17
Factor B	2	1138	569	2.31
Interaction	2	4	2	0.008
Error	6	1475	246	
Total	11	3150		

b. Interaction and main effects are not significant at $\alpha = 0.05$.

3. a. no interaction **b.** no interaction

5. a. 6 **b.** 2 **c.** 2 **d.** 3 **e.** $F = 0.008$, df $= (2, 6)$
f. $A: F = 2.17$, df $= (1, 6)$; $B: F = 2.31$, df $= (2, 6)$

7.

Source	SS	DF	MS	F
A	13.8	2	6.9	69
B	5	4	1.25	12.5
AB	1.2	8	0.15	1.5
Error	3	30	0.1	
Total	23	44		

9. a.

Source	DF	SS	MS	F
Fert.	3	11,789.0	3,929.7	172.36
Variety	2	232.7	116.4	5.11
Interact.	6	563.3	93.9	4.12
Error	12	273.0	22.8	
Total	23	12,858.0		

b. $F_{0.05}(6, 12) = 3.00$; the interaction of fertilizer and variety of corn is significant. Since this interaction is present, it is not wise to test for any main effects of fertilizer or corn variety. One could test, for example, differences in the three corn varieties for each level of fertilizer. This would result in three t tests for each of the four fertilizer levels, or a total of 12 t tests.

11. a.

Source	DF	SS	MS	F
Gear ratio	2	27.63	13.81	2.89
Mixture	2	96.52	48.26	10.10
Interact.	4	31.93	7.98	1.67
Error	18	86.00	4.78	
Total	26	242.07		

b. Since $F_{0.05}(4, 18) = 2.93$, the interaction of mixtures and gear ratios is not significant. We can therefore test for the main effects of mixture and gear ratio. Main effect (air/gasoline mixture): $F_{0.05}(2, 18) = 3.55$. This main effect is significant. $CD = 4.71$. Main effect (gear ratio): not significant.

13.

Source	DF	SS	MS	F
Pressure	1	0.0242	0.0242	1.12
Catal.	2	0.2881	0.1441	6.64
Interact.	2	0.2676	0.1338	6.17
Error	12	0.2601	0.0217	
Total	17	0.8400		

Since $F_{0.05}(2, 12) = 3.89$, interaction between catalyst and pressure is present. It is not wise to test for the main effects of catalysts or pressure. One could test, for example, the effects of the three catalysts at each of the two pressure levels. This would involve a total of six tests. One might also be interested in testing the two pressure levels for each of the three catalysts. This effort would involve three t tests.

15. a.

Source	DF	SS	MS	F
Method	1	24.0	24.0	1.31
Period	2	112.0	56.0	3.06
Interact.	2	144.0	72.0	3.93
Error	18	330.0	18.3	
Total	23	610.0		

b. Since $F_{0.05}(2, 18) = 3.55$, there is interaction between time period and teaching method. As a result, tests for main effects of time period and method of teaching should not be done. Instead, one might want to test for differences in the two methods of teaching for each period of time. This would involve three t tests. Or, one might want to test for a difference in achievement for the three time periods for each method of teaching. This would involve six t tests, three for each teaching method.

Review Exercises

1. SST = 323.87, SSB = 80.27, SSW = 243.6, MSB = 40.14, MSW = 9.02, $F = 4.45$, $F_{0.05}(2, 27) = 3.35$; reject H_0

3. a. SST = 782.97, SSB = 554.81, SSW = 228.16, MSB = 184.94, MSW = 28.52, $F = 6.48$, $F_{0.05}(3, 8) = 4.07$; reject H_0
b. CD = 15.17, Group 1 $\neq$ Group 3
c. $\hat{\omega}^2 = 0.58$

5. SST = 937.08, SSB = 251.53, SSW = 685.55, MSB = 125.76, MSW = 68.56, $F = 1.83$, $F_{0.05}(2, 10) = 4.10$; fail to reject H_0

7. a.

Source	DF	SS	MS	F
Judge	4	36.40	9.10	7.91*
Person	2	4.13	2.07	1.8
Error	8	9.20	1.15	
Total	14	49.73		

*$F_{0.05}(4, 8) = 3.84$, $F_{0.05}(2, 8) = 4.46$

b. There are no significant differences between the three candidates. But since $F_{0.05}(4, 8) = 3.84$, there are significant differences in the average ratings of the judges. CD = 3.35: J2 $\neq$ J4 and J1 $\neq$ J4.

Achievement Test

1. a. 30 **b.** 7.5 **c.** 22.5 **d.** 2 **e.** 6 **f.** 3.75 **g.** 3.75 **h.** 3.75 **i.** 3.75 **j.** 1

2.

	SS	df	MS	F
Between samples	7.5	2	3.75	1
Within samples	22.5	6	3.75	
Total	30.0	8		

3. 0

4.

Source	DF	SS	MS	F
Block	2	494.000	247.000	
Treat.	3	7.333	2.444	5.50
Error	6	2.667	0.444	
Total	11	504.000		

5. a. 2 **b.** 2 **c.** 3 **d.** 4 **e.** 1.13 **f.** *A:* $F = 8.47$, df = (2, 12); *B:* $F = 6.78$, df = (3, 12)

6. We fail to reject H_0.

CHAPTER 14

Exercise Set 14.1

1. a. $y = 3 + 4x + \epsilon$ **b.** 11.05 **c.** -5

3. a. 4 **b.** -8 **c.** 4 **d.** 4

5. a. 83.2796 **b.** 28.1818 **c.** 3.8181 **d.** 96.9076 **e.** 119.8162 **f.** 116

7. a. 57.5582 **b.** 4.4970 **c.** 170.5713 **d.** 219.4502 **e.** 187.9712 **f.** 251.5714

Exercise Set 14.2

1. a. $F = 13.280$, $F_{0.05}(1, 5) = 6.61$; the linear model is appropriate; $\hat{y} = 9.0256 + 0.3020x$.
b. (0.0889, 0.515) **c.** $t = 3.64$, $t_{0.025} = 2.571$ (df = 5); the linear model is appropriate.

3. a. $F = 270,692.1$, $F_{0.05}(1, 3) = 10.1$; the linear model is appropriate; $\hat{y} = 69.7085 + 203.758x$.
b. (879.98, 889.50) **c.** (873.73, 895.75) **d.** (1,688.43, 1,711.11) **e.** \$203.76 **f.** $t = 525.82$, $t_{0.025} = 3.182$ (df = 3); the linear model is appropriate

5. a. $F = 231.649$, $F_{0.05}(1, 10) = 4.96$; the linear model is appropriate; $\hat{y} = 96.753 - 2.0027x$
b. $(-2.296, -1.71)$ **c.** (47.48, 49.90) **d.** (42.34, 51.03)

7. a. $F = 527.2708$, $F_{0.05}(1, 8) = 5.32$; the linear model is appropriate; $\hat{y} = -33.539 + 1.7515x$
b. (1.58, 1.93) **c.** (35.85, 37.19) **d.** (43.55, 47.01)

Exercise Set 14.3

1. $t = 2.137$, $\pm t_{0.025} = \pm 2.228$ (df = 10); there is no evidence that ρ is different from 0.

3. $t = 2.177$, $t_{0.01} = 2.896$ (df = 8); there is no evidence that $\rho > 0$.

5. $t = 2.943$, $t_{0.05} = 1.734$ (df = 18); there is evidence that $\rho > 0$.

7. a. -0.6678 **b.** $F = 8.0372$, $F_{0.05}(1, 10) = 4.96$; reject H_0; there is a relationship. **c.** $t = 2.83$, $\pm t_{0.025} = \pm 2.228$, df = 10; reject H_0; $\rho \neq 0$ **d.** 0.446

Exercise Set 14.4

1. a. 68.07 **b.** 76.27 **c.** 59.87 **d.** 65.42

3. a. $\hat{y} = -60.25 + 1.0330$ IQ $+ 2.438$ hours
b. 0.605 **c.** Yes; $F = 5.37$ with p-value $= 0.039$ < 0.05. **d.** None **e.** 93.255

5. a. 0.8979 **b.** 5.41 **c.** Yes; $F = 17.06$, p-value $= 0.01 < 0.05$ **d.** ADVER and SALES regression coefficients are both significantly different from zero at $\alpha = 0.05$. **e.** 1.31

7. a. BSD $= 0.288 + 1.75$ FHD $+ 0.183$ PSD **b.** 0.92

Review Exercises

1. $t = 1.06$, $\pm t_{0.025} = \pm 2.776$ (df = 4); there is no evidence to indicate that $\rho \neq 0$.

3. a. $F = 58.2175$, $F_{0.01}(1, 3) = 34.1$; the linear model is appropriate; $\hat{y} = -4.9824 + 3.1985x$ **b.** (0.75, 5.65)

5. a. $\hat{y} = -64.0305 + 1.4830x$, 84 **b.** (67, 101)
c. (76, 93)

7. a. $F = 134.852$, $F_{0.01}(1, 4) = 21.2$; the linear model is appropriate. **b.** \$128.41 **c.** \$1,730.70 **d.** (1,070.90, 1,620.04) **e.** (1,010.50, 2,450.90)

9. a. Score $= -207.15 + 18.39$ Age $+ 4.934$ Height -1.824 Weight **b.** 0.638 **c.** Yes; $F = 4.71$, p-value $= 0.035 < 0.05$ **d.** b_0 and height **e.** 78.13

Achievement Test

1. a. 4.4 **b.** ± 2.101 **c.** Reject H_0; there is evidence to support that ρ is different from 0.

2. a. 4.24 **b.** ± 4.3 **c.** $\beta_1 = 0$ **d.** The linear model is not appropriate. **e.** $\hat{y} = 1.5 + 1.5x$

3. a. 1.5 **b.** 6 **c.** 0.25 **d.** 0.90

4. $(-2.73, 8.13)$

5. a. 4.7 **b.** -0.7 **c.** No; $F = 0.42$, p-value $= 0.736$ > 0.05 **d.** 45.8% of the variance in y is accounted

for by the variance in x_1 and x_2. **e.** For example, $2.41 = 8.667/3.59$. In general, $t = b_i/\text{Stdev}$. These statistics are used to determine if β_i are different from 0. **f.** s is the standard error of estimate. It is defined by $s_e = \sqrt{\text{SSE}/(n - 3)}$. Its value is 2.828.

CHAPTER 15

Exercise Set 15.1

1. $z = 1.29$, $z_{0.05} = 1.65$; fail to reject H_0

3. $z = 2.26$, $\pm z_{0.025} = \pm 1.96$; reject H_0

5. $z = 0$, $\pm z_{0.025} = \pm 1.96$; fail to reject H_0

Exercise Set 15.2

1. a. $n = 7$, $T^+ = 25$, $T^- = 3$, $T = 3$ **b.** $3 + 25 = (7)(7 + 1)/2$ **c.** $\mu = 14$, $\sigma = 5.92$, $z = -1.86$

3. a. $n = 9$, $T^+ = 32$, $T^- = 13$, $T = 13$ **b.** $32 + 13 = (9)(9 + 10)/2$ **c.** $\mu = 22.5$, $\sigma = 8.44$, $z = -1.13$

5. $\mu = 85.5$, $\sigma = 22.96$, $z = -3.07$, $-z_{0.05} = -1.65$; reject H_0

7. $\mu = 105$, $\sigma = 26.79$, $z = 0$, $\pm z_{0.025} = \pm 1.96$; fail to reject H_0

9. $\mu = 162.5$, $\sigma = 37.17$, $z = -1.68$, $-z_{0.05} = -1.65$; reject H_0

11. $n = 15$, $T = 44$, $\mu = 60$, $\sigma = 17.61$, $z = -0.91$, $-z_{0.05} = -1.65$; fail to reject H_0

13. $n = 15$, $T = 49$, $\mu = 60$, $\sigma = 17.61$, $z = -0.624$, $-z_{0.05} = -1.65$; fail to reject H_0

15. $n = 15$, $T = 96.5$, $\mu = 60$, $\sigma = 17.61$, $z = 2.07$, $-z_{0.01} = -2.33$; fail to reject H_0

Exercise Set 15.3

1. $U = 13$, $z = -2.61$, $\pm z_{0.025} = \pm 1.96$; there is a difference in the useful lives.

3. a. $U = 79.5$, $z = -1.81$, $-z_{0.05} = -1.65$; there is a difference. **b.** $t = 1.924$, $t_{0.01} = 2.457$, df = 30; fail to reject $H_{:0}$, there is no difference. **c.** Different results; the result of part a is best, since we do not know anything about the assumptions for the t test.

5. a. $U = 29$, $z = -1.59$, $-z_{0.05} = -1.65$; we have no evidence that method B is better.
b. $t = 1.52$, df = 18, $t_{0.05} = 1.734$; we have no evidence that method B is better.
c. The rank sum test is the preferred test to use if the assumptions of the t test can not be satisfied. Neither test produces significant results in this case.

7. a. $U = 58$, $z = 0.65$, $\pm z_{0.005} = \pm 2.58$; fail to reject H_0, p-value $= 0.5156$

b. $F = 2.00$, $F_{0.05}(8, 10) = 3.07$; we have no evidence to suggest that the population variances are different.

c. $t = -0.79$, $\pm t_{0.005} = \pm 2.878$; fail to reject H_0

Exercise Set 15.4

1. $H = 0.70$, $\chi^2_{0.05}(1) = 3.841$; fail to reject H_0

3. $H = 0.54$, $\chi^2_{0.05}(3) = 7.815$; fail to reject H_0

5. $H = 12.79$, $\chi^2_{0.05}(3) = 7.815$; reject H_0

Exercise Set 15.5

1. $S = 6.33$, $\chi^2_{0.05}(2) = 5.991$; reject H_0; there is a difference in the average number of problems among the laser printers.

3. $S = 7.0$, $\chi^2_{0.05}(2) = 5.991$; reject H_0; the three appraisers obtain significantly different results.

5. $S = 4.90$, $\chi^2_{0.05}(2) = 5.991$; fail to reject H_0; we cannot conclude that there is a difference in the mean production by shift.

7. $S = 5.00$, $\chi^2_{0.05}(3) = 7.815$; fail to reject H_0; we cannot conclude that there is a difference in the mean laboratory analyses.

9. Gasoline brands: $S = 5.80$, $\chi^2_{0.05}(3) = 7.815$; fail to reject H_0; we cannot conclude that there is a difference in mean mileage per gallon for the four brands of gasoline. Automobile types: $S = 4.50$, $\chi^2_{0.05}(2) = 5.991$; fail to reject H_0; we cannot conclude that there is a difference in the mean mileages for the three automobiles.

11. Operators: $S = 14.57$, $\chi^2_{0.05}(5) = 11.07$; reject H_0; the operators perform at different rates of speed. Machines: $S = 6.80$, $\chi^2_{0.05}(3) = 7.815$; fail to reject H_0; there is no evidence to suggest that the machines perform at different rates of speed.

Exercise Set 15.6

1. a. $R = 15$, $n_1 = 12$, $n_2 = 11$, $\mu_R = 12.4783$, $\sigma_R = 2.3381$, $z = 1.079$ **b.** $R = 15$, $n_1 = 12$, $n_2 = 12$, $\mu_R = 13$, $\sigma_R = 2.3956$, $z = 0.8349$

3. $R = 20$, $n_1 = 15$, $n_2 = 22$, $\mu_R = 18.838$, $\sigma_R = 2.888$, $z = 0.402$, $\pm z_{0.025} = \pm 1.96$; fail to reject H_0

5. a. Based on the median score, 4.5, $R = 26$, $n_1 = 25$, $n_2 = 25$, $\mu_R = 26$, $\sigma_R = 3.499$, $z = 0$, $\pm z_{0.025} = \pm 1.96$; fail to reject H_0.

b. Based on evens and odds, $R = 26$, $n_1 = 29$, $n_2 = 21$, $\mu_R = 25.36$, $\sigma_R = 3.41$, $z = 0.19$, $\pm z_{0.025} = \pm 1.96$; fail to reject H_0.

7. $\tilde{x} = 37.5$, $R = 19$, $n_1 = 25$, $n_2 = 25$, $\mu_R = 26$, $\sigma_R = 3.499$, $z = -2.00$, $\pm z_{0.025} = \pm 1.96$; reject H_0

Exercise Set 15.7

1. a. 0.0061 **b.** $z = 0.0183$, $\pm z_{0.025} = \pm 1.96$; fail to reject H_0

3. a. -0.78 **b.** $z = -2.62$, $\pm z_{0.025} = \pm 1.96$; reject H_0

5. a. 0.51 **b.** $z = 1.53$, $\pm z_{0.025} = \pm 1.96$; fail to reject H_0

7. a. -0.81 **b.** $t = -4.327$, $\pm t_{0.025} = \pm 2.228$, df $= 10$; reject H_0 **c.** The results are consistent.

9. a. -0.114 **b.** $z = -0.26$, $\pm z_{0.025} = \pm 1.96$; fail to reject H_0

Review Exercises

1. $U = 109$, $z = 2.14$, $\pm z_{0.025} = \pm 1.96$; reject H_0

3. $z = 1.90$, $z_{0.005} = 2.58$; fail to reject H_0

5. $U = 24.5$, $z = -1.41$, $-z_{0.01} = -2.33$; fail to reject H_0

7. $H = 2.54$, $\chi^2_{0.05}(1) = 3.841$; fail to reject H_0

9. $H = 1.56$, $\chi^2_{0.05}(2) = 9.210$; fail to reject H_0

11. a. 0.60 **b.** $z = 1.80$, $z_{0.05} = 1.65$; reject H_0

13. a. -0.95 **b.** $z = 2.85$, $\pm t_{0.025} = \pm 2.228$; reject H_0

15. $n = 17$, $T = 10.5$, $\mu = 76.5$, $\sigma = 21.12$, $z = -3.125$, $-z_{0.01} = -2.33$; reject H_0

Achievement Test

1. a. 17 **b.** 13.48 **c.** 2.44 **d.** 1.44 **e.** ± 1.96 **f.** fail to reject H_0

2. a. 1.90 **b.** ± 2.58 **c.** 0.0574 **3.** $(0.55, 1)$ **4.** 5.5

5. $U = 32.5$, $z = -1.02$, $-z_{0.01} = -2.33$; fail to reject H_0

6. $H = 1.04$, $\chi^2_{0.05}(1) = 3.841$; fail to reject H_0

7. a. 0.60 **b.** 0 **c.** 0.5 **d.** 1.2 **e.** fail to reject H_0

8. $H = 2.18$, $\chi^2_{0.01}(1) = 6.625$; fail to reject H_0

APPENDIX A

1. a. $X_1 + X_2 + X_3 + X_4$ **b.** $(X_1^2 + 2) + (X_2^2 + 2) + (X_4^2 + 2)$ **c.** $(X_1^2 + Y_1) + (X_2^2 + Y_2) + (X_3^2 + Y_3) + (X_4^2 + Y_4)$ **d.** $X_1^2 Y_1 + X_2^2 Y_2 + X_3^2 Y_3$ **e.** $f_1 X_1^2 + f_2 X_2^2 + f_3 X_3^2 + f_4 X_4^2 + f_5 X_5^2$ **f.** $X_1 Y_1^2 + X_1 Y_2^2 + X_1 Y_3^2 + X_2 Y_1^2 + X_2 Y_2^2 + X_2 Y_3^2$

3. a. 25 **b.** 35 **c.** 144 **d.** 50 **e.** 15 **f.** 28 **g.** 25 **h.** 0 **i.** 0 **j.** -7

Index